W9-AGU-568

A Century of Chemistry

The Role of Chemists and the American Chemical Society

A Century of Chemistry

The Role of Chemists and the American Chemical Society

Herman Skolnik, CHAIRMAN
Board of Editors

Kenneth M. Reese, EDITOR

AMERICAN CHEMICAL SOCIETY
WASHINGTON, D. C. 1976

Library of Congress CIP Data

A Century of chemistry.

Includes index.

1. American Chemical Society. 2. Chemistry—History —United States.

I. Skolnik, Herman, 1914– II. Reese, Kenneth M., 1923–

QD1.A583C46 540'.6'273 76-6126
ISBN 0-8412-0307-5 1-468

PRINTED IN THE UNITED STATES OF AMERICA

The Objects of the American Chemical Society

according to its National Charter

To encourage in the broadest and most liberal manner the advancement of chemistry in all its branches;

The promotion of research in chemical science and industry;

The improvement of the qualifications and usefulness of chemists through high standards of professional ethics, education, and attainments;

The increase and diffusion of chemical knowledge; and

By its meetings, professional contacts, reports, papers, discussions, and publications, to promote scientific interests and inquiry;

Thereby

Fostering public welfare and education,

Aiding the development of our country's industries, and

Adding to the material prosperity and happiness of our people.

CONTENTS

Objects of the American Chemical Society v

A Message from the President ix

Preface xi

Introduction xiii

PART ONE

I. Historical Perspectives 1

II. Chemical Education 59

III. Professionalism 78

IV. Publications 94

V. Impact of Government 144

VI. Public Affairs 159

VII. Intersociety Relations 171

VIII. Governance 179

IX. Headquarters Staff and Operations 230

X. ACS Divisions and Their Disciplines 236

PART TWO

XI. The Record 383
- The Presidents 383
- Chairmen of the Board 393
- Elected Members of the Board of Directors 395
- Administrative Officers 395
- Committees of the Board of Directors 396
- Committees of the Council 396
- Council Committees 397
- Joint Board–Council Committees 398
- Joint Board–Council Policy Committees 398
- Society Publications 399
- Divisional Publications 400
- Local Section Publications 400
- Chairmen of Divisions 401
- Local Section Chairmen 406
- ACS Awards 441
- Divisional Awards 445
- Local Section Awards 447
- National Meetings 454
- Membership Statistics 456

Index 457

A MESSAGE FROM THE PRESIDENT

WHEN an organization such as the American Chemical Society reaches the 100-year mark, this fact in itself indicates that the organization has made important contributions to the nation, the science, and the profession it serves. When, moreover, the organization has become the largest of its kind in the world, there is still further reason to regard it as a useful and successful institution.

Observance of the ACS Centennial provides an occasion for looking back over the years and trying to determine what accounts for the Society's truly remarkable success. In a sense, this history represents an effort to discover how the Society has managed to fulfill so well the predictions of Professor Charles F. Chandler and associates in their March 22, 1876, letter to fellow chemists which urged formation of the ACS and continued: "It is believed that the existence of such a Society in this country would prove a powerful and healthy stimulus to original research among us, and that it would awaken and develop much talent now wasting in isolation, besides bringing the members of the association into closer union, and ensuring a better appreciation of our science and its students on the part of the general public."

The real purpose of looking back is not, of course, merely to obtain satisfaction from reflecting on past triumphs; rather, it is to discover as many clues as possible to the likely developments of the future. Our Centennial history, prepared under the general guidance of Dr. Herman Skolnik and an Editorial Board, definitely accomplishes this purpose, and I commend it to all who are interested in the outlook for chemistry, chemists, and the American Chemical Society in the years ahead.

GLENN T. SEABORG

PREFACE

IT has been a most gratifying experience to be involved in planning for the American Chemical Society Centennial. In the mid-1960s, I was a member of an informal ACS staff group concerned with those Centennial arrangements which had to be made early. It was this group that suggested holding the Society's 1976 Spring Meeting in New York—the city where the ACS was organized. This was a departure from the Society's custom of scheduling New York meetings for the fall since meeting attendance then is usually larger than in the spring and New York is one of the cities best equipped to handle large meetings. The decision to meet in New York in the spring of 1976 has made possible several ceremonial events appropriately related to the Society's founding in that city. For example, the Centennial Banquet on April 6 will occur precisely a century—to the hour—after the ACS was officially established there.

Preparation of a centennial history of the Society was proposed at the first meeting of the staff group. The same topic was high on the agenda for the first meeting of the Board–Council Centennial Coordinating Committee on August 30, 1972. An ACS Centennial History Committee was formed soon thereafter under the chairmanship of Dr. Herman Skolnik.

The importance of maintaining historical documentation of Society progress is evident from the histories prepared in 1901, 1926, and 1951. Each of these has a distinctive character, though all three provide valuable information about the development of the Society into the world's largest association serving a single science. The ACS Centennial History Committee reviewed these earlier publications and decided on still another approach. I believe readers will agree with the Committee's decision to tell the story of American chemistry and the American Chemical Society, and their interaction, in easily readable style free from extensive details. This is the "story" you will find in Part One of this book. Part Two contains much detailed information, a very necessary part of any historical document.

BRADFORD R. STANERSON, CHAIRMAN
ACS Centennial Coordinating Committee

INTRODUCTION

AN awareness and appreciation of the historical importance of Priestley's isolation and characterization of oxygen in 1774 led in 1874 to the first American national meeting of chemists only, at Northumberland, Pa., where Priestley had last lived. From today's perspective, the significance of that meeting was the feeling of fraternity it engendered among the 77 American chemists who attended. These chemists, especially those from the New York City area, returned home with fond memories of the commemoration and with the realization that they had found a new way of relating to each other—not by geography, tradition, or belief—but by common interest and experience, by their thinking as chemists, and by how they learned about chemistry. This new way was formalized on April 6, 1876, in New York City and called the American Chemical Society.

Conceived with a deep sense of history, the ACS has made each 25-year period an occasion for commemoration, dedication, and celebration during a national meeting. In addition, the Society has made each 25-year period an opportunity for the preparation and publication of an historical record.

The history of the Society and surveys of progress in chemistry for the ACS silver anniversary (1901) were published (168 pp.) in the "Twenty-fifth Anniversay Number" of the *Proceedings of the American Chemical Society (JACS)* in 1902. A. C. Hale's paper (pp. 36–85) covered the history extensively and tabulated meeting places, officers, membership statistics, and other details. Much of the writing for this special publication was by those who played a part in the formation of the Society.

For the golden anniversary, the semicentennial of the founding of the Society, "A Half Century of Chemistry in America, 1876–1926," edited by C. A. Browne, was published as the Golden Jubilee Number, *J. Am. Chem Soc.*, **48**, 254 pages, Aug. 20, 1926.

A series of articles on the progress of chemistry and chemical technology, published in *Industrial & Engineering Chemistry*, February through June of 1951, was reprinted as a diamond jubilee volume with the title "Chemistry . . . Key to Better Living" and issued in late 1951. C. A. Browne, the ACS historian, began in 1944 to write a comprehensive history of the Society for the diamond anniversary. By the time of his death in 1947 at the age of 77, nine of the contemplated 20 chapters had been completed. Fortunately, he had recruited Dr. Mary Elvira Weeks of the Kresge–Hooker Scientific Library, Wayne State University, to assist him, and she, on Dr. Browne's death, completed the book from his working plans and notes. The book, entitled "A History of the American Chemical Society—Seventy-five Eventful Years," could not be finished in time for the Diamond Jubilee meeting in September 1951. Issued in 1952, it soon became recognized as the definitive history of the Society and as the basic source for all future histories.

When the Board–Council Committee to Plan for the ACS Centennial Celebration was organized in mid-1972, one of its first decisions was to have a new ACS history prepared in time to be issued by or on April 6, 1976. Toward this objective a subcommittee on the ACS Centennial History was formed with the following members: Herman Skolnik, chairman; Jack J. Bulloff; Robert F. Gould; Wyndham D. Miles; and W. A. Noyes, Jr. Our commission was to determine what kind of ACS Centennial History should be written. None had any doubt that one should be written, and if one were written, each of us would want to read it. But not just any history.

Histories, especially those of scientific societies, tend to be a mixture of what can be described as the trees and the woods, that is, the details and the story. To avoid this ambivalence, we decided to relegate the details to a separate section, the second part of the book. This section, a kind of monument of memory, includes a list of past presidents, chairmen and members of the board of directors, editors, award winners, division chairmen, local section chairmen, etc. With this decision, the major part of the book, the first section, could be allocated to the story, from which the reader can gain a sense of what chemistry, chemical technology, chemists, and the ACS looked and felt like at each stage of the evolving history.

Another tendency of most histories that we wanted to avoid as much as possible was the chronological and linear unfolding of events. Except for the early part of its history, there has been no real unity or homogeneity of content in the ACS, unless unity and homogeneity are defined in the most narrow and personal sense. The ACS is characterized more appropriately as a highly integrated whole within which are discernible: a heterogeneous membership of B.S., M.S., Ph.D. chemists, biochemists, chemical engineers, etc., employed in a great variety of areas in industry, academia, government, etc., but yet a membership of over 110,000 whose agora comprises ACS local, regional, and national meetings; an extensive body of subject matter, amounting to more than 450,000 documents per year (the number cited by *Chemical Abstracts* in 1975), in all areas of chemistry and chemical technology; a nonpareil ACS scientific and technological publication program; and a dedicated ACS leadership and staff. We thus proposed the Table of Contents and outlined the subject areas that we considered to be most pertinent to the understanding of ourselves, the nature of our heritage, the interplay of internal and external stresses, the interaction of parts with each other and the whole with society and government, and the evolving professionalism of chemists over the years.

With the contents of the book planned, the ACS Centennial History subcommittee was replaced by a new committee, the Board of Editors for the ACS Centennial History. In the interim, however, I began to recruit writers to provide input for each of the subject areas. Recruitment began in mid-1973. By mid-1974, I was enmeshed in continuing correspondence and telephone communication with the large number of people I had involved in the book. This continued into 1976 even as the book went to press.

Inasmuch as over 50 people were contributing input for the book, it was highly desirable to assign one person, Mr. K. M. Reese, to the rewriting of the input so as to have a uniform format and style and to avoid the inevitable duplication of information. This was particularly essential for the write-ups on the history of each of the 28 ACS divisions and on the evolving nature of the science or technology of each of the divisions. Although the divisions were informed of our objectives relative to their contributions, and communicated with for a period of over two years by letter, telephone, and personal conversation at ACS national meetings and at two Program Coordination meetings, I was not able to get a 100% response. Nevertheless, the response was good, many of the write-ups were excellent, and, in my opinion, this section of the book constitutes a major contribution to the understanding of the evolving nature of chemistry and chemical technology over the past century.

Members of the Board of Editors for the ACS Centennial History were: Herman Skolnik, chairman; Jack J. Bulloff; Robert F. Gould; Robert M. Hawthorne, Jr.; Robert E. Henze; Richard L. Kenyon; Wyndham D. Miles; James H. Stack; B. R. Stanerson; and Ernest H. Volwiler. Their responsibilities were to assist in any necessary rewriting, check the integrity of the contents, and edit and proofread as required.

The following people provided input for the section on the history of the divisions or on the evolving nature of the disciplines of chemistry or chemical technology:

Allen A. Alexander
Herbert R. Appell
John C. Bailar, Jr.
W. W. Binkley
G. L. Bridger
Alfred Burger
Edward M. Burgess

C. K. Cain
D. F. Durso
Philip J. Elving
Henry F. Enos
Galen W. Ewing
Thomas Fitzsimmons
J. Gindler
John T. Hays
Hal G. Johnson
Llewellyn H. Jones
George B. Kauffman
W. G. Kessel
H. B. Klevens
Carl G. Krespan
William A. Lester, Jr.
Peter Lykos
Richard J. Magee
G. Alex Mills
F. M. O'Connor
R. A. Osteryoung
Henry J. Peppler
K. M. Reese
Arthur R. Rescorla
Thomas H. Rogers
Julius Schultz
William A. Sheppard
Herman Skolnik
Bruce Smart
Ellis P. Steinberg
H. Gladys Swope
Ann Tracy Tarbell
D. Stanley Tarbell
R. D. Ulrich
Aaron Wold
J. J. Zuckerman

The following people provided input to the other sections in the book:

Robert W. Cairns
Edward P. Donnell
Robert B. Fox
Robert F. Gould
Henry A. Hill
Barbara R. Hodsdon
Richard L. Kenyon
David M. Kiefer
Marshall W. Mead

Stephen T. Quigley
K. M. Reese
Herman Skolnik
B. R. Stanerson
Cheves Walling
David M. Wetstone

Of the many people involved in the planning, writing, and editing of this book, none gave more of himself than Dr. B. R. Stanerson. His leadership, assistance, and encouragement were invaluable. Special thanks are also due to Mr. James H. Stack for his considerable assistance and to Mr. K. M. Reese whose writing and editing skills contributed materially to the condensation of a century of ACS history to fit the confines of this book.

HERMAN SKOLNIK, Chairman
Board of Editors
ACS Centennial History

CHAPTER I

At the Priestley Centennial meeting, July 31, 1874, in Northumberland, Pa., a motion was made to form a chemical society, but it met strong opposition and was not put to a vote.

HISTORICAL PERSPECTIVES

THE founding of the American Chemical Society, in April 1876, came in a year that saw a series of memorable events. In February, teams from eight cities formed the National League of Professional Base Ball Clubs, now the National League. In March, Alexander Graham Bell received a U.S. patent on the telephone—less than a month after he had filed the application. In June, George Armstrong Custer and 267 officers and men of the 7th U.S. Cavalry were wiped out by the Sioux at the Little Bighorn River in Montana. In August, Colorado became the 38th state to be admitted to the Union. And during the year, Philadelphia entertained some eight million visitors to the Centennial Exposition, the focal point of the nation's celebration of its first 100 years.

Chemists, and indeed scientists in general, in the modern sense were a relatively rare breed in the United States in 1876, but the educational opportunities, though limited, were beginning to grow. Harvard, Yale, and Princeton had established chairs of mathematics and natural philosophy as early as 1800. Rensselaer Polytechnic Institute had been founded in Troy, N.Y. in 1824 to provide instruction in the "application of experimental chemistry, philosophy, and natural history, to agriculture, domestic economy, the arts and manufactures." Harvard's Lawrence School of Science and Yale's Sheffield Scientific School were established in the 1840s. Two decades later, Rutgers instituted laboratory instruction, with student participation, on the Rensselaer pattern. Johns Hopkins, the nation's first distinctively graduate university, opened its doors in 1876; the original staff included Ira Remsen, professor and director of the Chemical Laboratory. Still, in that year the country had fewer than 600 institutions of higher learning, many of them small and inferior, with a total enrollment of about 13,000. Curricula were geared largely to the recognized professions: medicine, law, the military, and the church. Chemistry as a rule was taught as part of medical education. As late as 1890, less than 7% of the 14–17 age group attended high school; very few high-school graduates went on to college, where admission in most cases required knowledge of Latin, Greek, elementary mathematics, history, and geography. Not until the turn of the century would colleges generally begin to accept subjects like modern languages, chemistry, and physics as qualifications for entrance.

The U.S. was industrializing rapidly by 1876, and technology was on the move. In May 1869, the Union Pacific, building westward from Omaha, and the Central Pacific, building eastward from Sacramento, had met at Promontory, Utah to create the nation's first transcontinental rail link. In 1887, a record 13,000 miles of railroad track was laid in this country, which a decade later would become the world's leading producer of pig iron. In 1855, Benjamin Silliman, Jr., professor of chemistry at Yale College, had completed the first scientific analysis of petroleum. Petroleum refining was booming by the early 1880s, when Standard Oil Co. hired George Saybolt, probably the first full-time chemist in the industry. In 1873, Andrew Carnegie began to hire trained scientists to do full-time research on problems in steelmaking. Of such problems he would say later, "Nine-tenths of the uncertainties were dispelled under the burning sun of chemical knowledge." The nation's organic chemicals industry was rudimentary in the 1870s, but the manufacture of inorganic chemicals was well established. Production of sulfuric acid (50° Baumé—about 63%) exceeded 154,000 tons in 1880 and 692,000 tons in 1890; output of what now would be called "chemicals and allied products" was valued at roughly $140 million in 1880. The Manufacturing Chemists Association was formed in 1872 by 14 companies; six more were added in 1873–74.

Chemical science, too, was moving rapidly in the latter part of the 19th century. In 1828, Friedrich Wöhler had made urea by heating ammonium cyanate. With this feat Wöhler started the decline of "vitalism"—the notion that the lack of some "vital force" barred the synthesis of "natural products" in the laboratory—although he did not claim to have done so, and vitalism lost credence only gradually as knowledge of organic compounds increased. (In 1876, Wöhler would be elected an honorary member of the American Chemical Society, along with seven other foreign chemists.) In 1858, Friedrich A. Kekulé and Archibald S. Couper independently published their ideas on the tetravalency and chain-forming properties of carbon atoms, thereby laying the basis for the structural concepts of organic chemistry. In 1869, Dmitri Mendeleev proposed his formulation of the Periodic Law of the elements. In 1876 and 1878, Josiah Willard Gibbs, Jr., professor of mathematical physics at Yale, published the two parts of his landmark paper, "On the Equilibrium of Heterogeneous Substances," in the *Transactions of the Connecticut Academy of Science*. Gibbs, perhaps the greatest American scientist of the 19th century, defined in that paper the entirely new science of chemical thermodynamics. The paper was ignored in this country, and for almost a decade attracted the attention of no one in Europe, excepting James Clerk Maxwell, who died in 1879.

Chemists by 1876 had discovered only 63 of the 106 natural and man-made elements detected by 1976, but the practice of chemistry was emerging speedily in this country as a recognized science and profession. The inevitable result in hindsight was the formation of a national society composed exclusively of chemists. The precipitating event was the Priestley Centennial, in July 1874, but the underlying drive was the fact that it is extremely difficult to be a chemist in isolation. The scientist's need for like-minded companions is evident in a plaint by John Winthrop the Younger (1606–76), Governor of the Connecticut Colony of New England, member of The Royal Society (London), and the leading, if not the only, chemical manufacturer of his time in America. In a letter to Sir Robert Morey, written Sept. 20, 1664, Winthrop remarked, "I had sad and serious thoughts about the unhappinesse of the condition of a Wilderness life so remote from the fountains of learning and noble sciences . . . When I was greatly revived with the special favour of your honor's letter. . . ."

John Winthrop died April 6, 1676, and by the time the American Chemical Society was formed, 200 years later to the day, conditions had much improved. Scientists were still sparse, but small enclaves of chemists had grown up around New York City, Philadelphia, and Washington, D.C. The American Association for the Advancement of Science, formed in 1848, had established a subsection on chemistry in 1874 (the present Section C, Chemistry, would displace the subsection in 1882). Most American chemists, moreover, subscribed to or read Chandler's *American Chemist* and perhaps Silliman's *American Journal of Science*.

The Founding of the Society

In the *American Chemist* for April 1874, Dr. H. Carrington Bolton of the Columbia College School of Mines suggested a meeting of chemists to mark the centennial of Joseph Priestley's isolation and identification of oxygen, Aug. 1, 1774. Prof. Rachel L. Bodley of the Women's Medical College of Pennsylvania then proposed as the site of the celebration the British chemist's last home, at Northumberland, Pa., where he had died in 1804. Upon these suggestions, a committee of the New York Lyceum of Natural History issued a circular appealing to American chemists to attend such a meeting.

The circular was signed by 37 prominent American chemists, and it had the desired effect. On July 31, 1874, 75 chemists from 15 states and the District of Columbia and one each from Canada and Britain met in the lecture room of the Northumberland public school. Dr. Charles F. Chandler of the Columbia College School of Mines was in the chair; the papers presented were published in the *American Chemist,* **5,** 35–114, 195–209, 327–328 (1875).

The Priestley Centennial bred a strong "feeling of fraternity" among the chemists on the scene, so much so that after the concluding paper Prof. Persifor Frazer of the University of Pennsylvania proposed "the formation of a chemical society which should date its origin from this centennial celebration." Frazer's motion was opposed by J. Lawrence Smith, F. W. Clarke, E. N. Horsford, Benjamin Silliman, Jr., and others. They argued that the country was too large (presumably for ease of travel) and chemists too few to warrant a national organization and that chemists' needs could be served adequately by the chemical subsection of the American Association for the Advancement of Science (Prof. Smith had served only recently as AAAS president). Dr. Bolton then modified Frazer's motion to read: "That a committee of five be appointed from this meeting to cooperate with the American Association for the Advancement of Science at their next meeting, to the end of establishing a chemical section on a firmer basis."

Many chemists retained fond memories of the meeting with their colleagues in Northumberland. Some 18 months later a group of them in New York City met at the home of Prof. Chandler and decided to form a local chemical society. Accordingly, they organized themselves as a committee and, on Jan. 22, 1876, wrote to 100 chemists in New York and vicinity inviting their views and cooperation. When this circular letter drew 40 favorable replies, the committee decided to attempt to form a national society rather than a local one. Consequently, on March 22, 1876, a second circular letter was mailed to about 220 chemists throughout the United States (Figure 1). This letter drew 60 favorable responses

New York, March 22d, 1876

Dear Sir: Several weeks ago the undersigned issued an invitation to the chemists of this vicinity, requesting their coöperation toward the organization of a *local* Chemical Society. The response was so unexpectedly satisfactory, that on further consultation it was deemed opportune to attempt the formation of a *national* Society, somewhat on the plan of those in such successful operation in France, Germany, and England. It is believed that the existence of such a Society in this country would prove a powerful and healthy stimulus to original research among us, and that it would awaken and develop much talent now wasting in isolation, besides bringing the members of the association into closer union, and ensuring a better appreciation of our science and its students on the part of the general public. Guided by these considerations, we have, in consultation with other members of the profession, drafted the accompanying Constitution and By-Laws, which we submit respectfully to your kind consideration.

Among the objects contemplated are the fitting up of permanent rooms as the headquarters of the Society, and in connection therewith the establishment of a library of reference, and the gradual collection of a chemical museum. It is also proposed to hold at least one meeting in each year outside of this city, at such a time and in such a place as to make attendance on the part of non-resident members more convenient and representative.

The proceedings of the Society will be published and sent to the members.

If these views and objects meet with your approval, please sign and return the postal card enclosed.

Chas. F. Chandler	Henry Morton
W. M. Habirshaw	Isidor Walz
H. Endemann	Fred. Hofmann
M. Alsberg	P. Casamajor

Figure 1. Circular Mailed to Some 220 Chemists in the United States

from nonresident chemists; these, with the 40 favorable responses from the New York City area, guaranteed 100 members to begin with. Thus the organizing meeting of the American Chemical Society was scheduled for Thursday evening, April 6, 1876, at 8 P.M., in the Lecture Room of the College of Pharmacy of the City of New York, University Building, corner Waverly Place and University Place. Thirty-five chemists attended.

Prof. Chandler, elected president for the meeting, summarized the need for a new society: to bring chemists together in scientific and social intercourse, to establish a reference library and a chemical museum, and to secure rooms for these purposes. After a certain amount of discussion, Dr. Isidor Walz, elected secretary of the meeting, moved that those present "proceed to organize a national chemical society which shall be called the American Chemical Society." The motion was carried with only three dissenting votes. The organizing committee had drawn up a Constitution and Bylaws, which were adopted by vote at this first meeting. In addition, the group appointed a nominating committee, which presented a slate of officers at the second organizational meeting, April 20, 1876. The 27 chemists who attended this second meeting accepted the slate unanimously (Table I). Also at this second meeting, William H.

TABLE I

The First Officers of the American Chemical Society

President: John W. Draper

Vice-Presidents: J. Lawrence Smith
Frederick A. Genth
E. Hilgard
J. W. Mallet
Charles F. Chandler
Henry Morton

Corresponding Secretary: George F. Barker

Recording Secretary: Isidor Walz

Treasurer: W. M. Habirshaw

Librarian: P. Casamajor

Curators: Edward Sherer
W. H. Nichols
Fred'k Hofmann

Committee on Papers and Publications: Albert R. Leeds
Hermann Endemann
Elwyn Waller

Committee on Nominations: E. P. Eastwick
M. Alsberg
S. St. John
Charles Fröbel
Chas. M. Stillwell

Nichols suggested that the New York City members get together every third Thursday "to discuss matters of relatively minor importance not suited to the dignity of the regular sessions" (which, under the bylaws, were to occur the first Thursday of every month). Many of these informal gatherings or "conversaziones" took place; some years later Nichols would recall them as the forerunner of The Chemists' Club of New York (founded in 1898 with 154 charter members).

The first regular meeting of the ACS took place May 4, 1876. By then the plans for organizing the Society had been approved by 133 chemists, 53 of them resident and 80 nonresident (defined by the bylaws as residing more than 30 miles from New York City). The secretary was authorized to print the names of these chemists as organizing members. Attendees at this first regular meeting learned also of plans for the *Proceedings of the American Chemical Society*. The first issue of the *Proceedings* appeared as part of the *American Chemist* for June 1876; it included the minutes of the May meeting and a paper read at that meeting by Hermann Endemann on "The Determination of the Relative Effectiveness of Disinfectants." The *American Chemist* carried the *Proceedings* through the first three meetings of 1877, but then ceased publication. Thereafter, the *Proceedings* carried on alone, becoming the *Journal of the American Chemical Society* with the issue of April 1879.

On June 16, 1876, the Society met in Philadelphia in connection with the Centennial Exposition; the attendance was 70 and included several foreign chemists. Excepting the meeting in Philadelphia, all the Society's regular sessions during its first year took place in the lecture room of the New York College of Pharmacy. Desiring a more permanent home, the Society found suitable quarters on 14th Street at a rental of $700 up to May 1, 1878, and $800 from then until May 1, 1879. The ACS could not lease its new home unless it were incorporated, so a charter of incorporation was obtained in the State of New York on Nov. 10, 1877.

Dissatisfaction and Reform

Membership in the ACS fluctuated from 192 at the end of 1876 to a high of 243 in 1881 and back to 167 in 1889. Even the resident membership fluctuated markedly, from 71 in 1876 to a high of 119 in 1881 and back to 91 in 1889. Several factors contributed to these early difficulties. Although chemists in the New York City area were enthusiastic about the Society, it meant little to those in other parts of the country. Nonresident members understandably resented the dominance of the New Yorkers, but the dominance was necessary if officers were to attend the meetings. Communications was a problem. Of the 75 million people in the United States in 1890, only the one-fourth or so who lived in the larger cities enjoyed mail delivery. Transportation by railroad, horse and buggy, horseback, and steamboat was anything but fast and convenient. It is not surprising, then, that attendance at regular ACS meetings fell sharply after the first flush of enthusiasm. During 1880, six regular meetings lacked even the quorum of 15 required to transact Society business. The decrease in membership was reflected in the declining number of papers submitted to the *Journal,* whose publication as a result was suspended for various months.

Dissatisfaction with the ACS was expressed most tellingly Jan. 31, 1884, by the formation of the Chemical Society of Washington. The organizing members of this group, led by F. W. Clarke and Harvey W. Wiley, comprised 37 chemists associated with the Government and local organizations, and six of private status. By 1887, no resident of the Washington, D.C. area was a member of the ACS.

In 1888, under the umbrella of Section C of the American Association for the Advancement of Science (AAAS), F. W. Clarke initiated steps toward forming a national chemical society based on the concept of local sections. C. F. Munroe of Newport, R.I., with the support of C. F. Chandler, proposed such a move to the ACS at its meeting of Nov. 1, 1889. The establishment of local sections was provided for in a new Constitution adopted by the Society at its meeting of June 6, 1890; the new document also gave absent members the privilege of voting by mail, and provided for meetings outside the New York City area. The Society held its first general meeting at Newport, R.I., Aug. 6–7, 1890; 43 chemists from seven states attended. The second general meeting, in Philadelphia, Dec. 30–31, 1890, was attended by 75 chemists. On Jan. 21, 1891, the formation of the first ACS local section, the Rhode Island Section, was approved. The third ACS general meeting was held in Washington, D.C., Aug. 17–18, 1891. At that meeting, representatives of the ACS and nine other societies (Table II) met to discuss a plan for federation. The committee of representatives, whose organizations had a combined membership of about 1000, passed the following resolution: "That the American Chemical Society be so extended as to include the members of all local societies in its membership; that the New York members . . . be requested to organize a local section of the American Chemical Society in New York."

Pursuit of the spirit of this meeting required a new Constitution, which the Society adopted Nov. 4, 1892. Among other provisions it established the Council as the governing body of the Society. The new American Chemical Society took form at the sixth general meeting, in Pittsburgh, Dec. 28–29, 1892; Harvey W. Wiley, chief of the Division of Chemistry, U.S. Department of Agriculture, was elected Persident.

In 1893, the publishing office of the *Journal of the American Chemical Society (JACS)* was transferred from New York to Easton, Pa., to accom-

TABLE II

Societies that Met in Washington, D.C., August 1891, to Discuss Plan for Federation (membership in parentheses)

American Chemical Society (290)
Section C, American Association for the Advancement of Science (200)
Association of Official Agricultural Chemists (75)
Chemical Section, Brooklyn Institute (75)
Chemical Society of Washington (70)
Chemical Society, Franklin Institute (70)
Chemical Society, University of Michigan (60)
Louisiana Sugar Chemists' Association (52)
Cincinnati Chemical Society (29)
Manufacturing Chemists' Association

modate the newly appointed Editor, Prof. Edward Hart, who was head of the Chemical Publishing Co. of Easton and owner of the *Journal of Analytical and Applied Chemistry* (originally the *Journal of Analytical Chemistry*). A fortunate side effect of the move was its elimination of the New York board of editors of *JACS,* whose intrastate composition had vexed chemists in other parts of the country. Prof. Hart merged his journal into *JACS,* effective with the issue of June 1893, and actively encouraged his former subscribers to become ACS members.

Another positive factor was the World's Congress of Chemists, which was held in Chicago, Aug. 21–26, 1893, during the Columbian Exposition. The Congress was arranged jointly by two committees—one from the ACS and one from the World's Congress Auxiliary—and was the first large international meeting of chemists in the United States. It drew 182 chemists, 83 of them ACS members, and 76 papers. The meeting also was the first at which ACS members presented papers in divisional classifications: Analytical, Bibliography, Inorganic, Organic, Agricultural, Technological, Didactic, Physical, and Sanitary.

The Society still had the problem that its certificate of incorporation in the state of New York required ACS directors to be residents of the state. General Secretary Albert C. Hale took steps to correct the situation, and on April 16, 1895, the Senate and Assembly of New York passed, and the Governor signed, "An Act for the Relief of the American Chemical Society." This done, the Society again revised its Constitution, effective Dec. 2, 1897.

The ACS celebrated its 25th anniversary at the general meeting in New York, April 12–13, 1901. The meeting was held in the quarters of the two-year-old Chemists' Club at 108 West 55th Street. Registration was 147, of whom 100 were from the New York City area. ACS membership at the time totaled 1858, including 1740 "members" and 118 "associates" (those interested in the promotion of chemistry but not trained sufficiently to be recognized as chemists). Membership had passed 1000 in 1896, and at the meeting in April 1901, Harvey Wiley predicted that "the membership will be nearly 10,000" in 1976. In fact, the total passed 2000 in the spring of 1902 and reached 10,000 in 1917, almost 60 years ahead of schedule. (Wiley predicted also that in 1976 the nation's population would be 225 million and the federal budget $4 billion; he was only about 10 million high on the population, but some $345 billion low on the budget.)

On Dec. 31 of the 25th anniversary year, Dr. Edward Hart resigned as Editor of *JACS.* He was succeeded by W. A. Noyes, who took office in 1902. In the same year, on the initiative of the New York Section, the Society did away with the distinction between "member" and "associate." The necessary amendments to the Constitution, adopted Oct. 23,

Some founding members of the Chemists' Club, established in New York City in 1898. Top left, Marston T. Bogert. Bottom row, from left: Herman Frasch, Charles F. Chandler, C. F. McKenna, W. H. Nichols. Bogert, Chandler, and Nichols were ACS Presidents; McKenna was a Director.

1902, provided in part that "any person interested in the promotion of chemistry may be nominated for election as a member."

Chemical Science and Industry ca. 1900

The evolution of the American Chemical Society has always mirrored the evolution of science and technology in general. Thus Harvey Wiley's 1901 forecast of ACS membership in 1976 misfired partly because of six immense breakthroughs that came within a single decade around the turn of the century. These discoveries were: x-rays, the electron, radioactivity, radium, quantum theory, and relativity. Biochemist and Nobel laurate Albert Szent-Gyorgyi wrote in 1975 that these discoveries created modern science; they exposed a world composed of energy and electromagnetic radiation, a world of which man's senses told him nothing. The practical effect of these findings on science and technology was hardly predictable in 1901. Their purely scientific import, however, was recognized by six Nobel prizes: Wilhelm Roentgen, Physics, 1901, x-rays; A. H. Becquerel, Pierre and Marie Curie, Physics, 1903, radioactivity; J. J. Thomson, Physics, 1906, the electron; Marie Curie, Chemistry, 1911, radium and its properties; Max Planck, Physics, 1918, quantum theory; Albert Einstein, Physics, 1921, relativity.

As these developments were emerging in Europe, chemical science in the United States, with some exceptions, was advancing relatively slowly. During the last quarter of the 19th century and into the 20th century, students who wished to do graduate work in chemistry went as a rule to England, France, and, particularly, Germany. In the two decades 1881–1900, schools in this country produced only 418 Ph.D.'s in chemistry, although they produced 591 in the single decade 1901–1910.

Most of the professors in the U.S. in the 19th century studied mineralogy or analytical chemistry. Very few studied inorganic chemistry. Organic chemistry was being studied intensively in Europe by 1876, and many schools there granted a Ph.D. in the field. At the same time, in contrast, few colleges in the U.S. offered even a separate course in organic chemistry. Physical chemistry was barely known in this country until American students began to study with Ostwald at Heidelberg in the late 1880s; the first chair in the subject was established in the U.S. only in 1895, at Cornell. The American Willard Gibbs by 1880 had introduced in his thermodynamics a new language for the underlying processes in chemistry, but few of his countrymen knew his work until well into the 20th century. Gibbs' publications were translated into German by Ostwald and into French by Le Chatelier in the 19th century; they were reprinted in English in 1906 in London and not reprinted in the U.S. until 1928.

Until World War I, Europe was well ahead of the U.S. in chemical technology as well as in chemical science. Perhaps the first serious attempt to use scientists to work out the commercial production of chemical substances came in France during the Napoleonic Wars. A notable result was Nicholas Leblanc's process, developed in 1790, for making sodium carbonate (soda ash) from salt. Much of the chemistry that evolved during the 19th century dealt with the synthesis of chemicals and studies of chemical properties and analytical methods. This work took place predominantly in the universities and, on the whole, had little connection with events in the chemical industry. An important exception was Germany. In most countries during the 19th century, men became scientists by inclination, not through formal education. But Germany had an excellent system of universities and technical colleges that in part was designed specifically to produce scientists. The German government, moreover, encouraged a symbiosis between university research and industry, and this was particularly true of chemical research and industry. By 1913, in consequence, Germany accounted for one quarter of the world's chemical industry and produced 85% of its dyes.

An intimate relationship between chemical science and technology began to develop in the United States only around the turn of the century. In 1860, the country had about 84 chemical companies with 1500 employees; by 1914, it had 395 chemical companies with 32,000 employees. The Schoellkopf Aniline & Chemical Co., started in 1879 in Buffalo, had an active research program; in 1901, the company sold the patent rights to a dye it had developed to German interests, a real mark of distinction, in view of Germany's dominance of the dye industry. General Electric established its research laboratory in 1900 with Willis R. Whitney in charge. Whitney served as ACS President in 1909; in the same year GE hired Irving Langmuir, a future ACS President (1929) and Nobel laureate in chemistry (1932). Du Pont opened its laboratories in 1911, Eastman Kodak in 1912, U.S. Rubber in 1913, and Hercules Powder Co. in 1916. Government laboratories were pioneered in the 19th century by the Department of Agriculture. Harvey Wiley, ACS President in 1893–94, was chief of USDA's Bureau of Chemistry from 1883 to 1912 and was a major force in the passage of the Pure Food and Drugs Act of 1906. Other early government laboratories were the National Bureau of Standards, established in 1901, and the U.S. Bureau of Mines laboratories, established in 1910.

A stronger scientific input could not cure some of the American chemical industry's major ailments of the early 20th century. An important problem was competition from overseas. Chemicals like acids, fertilizers, and sodium carbonate, which were in great demand in this country, could be bought more cheaply in England and Germany, where they were industrial by-products. The U.S. chemical industry nevertheless was making progress. By 1909, for example, the Belgian-born chemist Leo Baekeland (ACS President in 1924) had patented the phenolic resins trade-named Bakelite. These materials found ready markets, especially in the youthful automobile and electrical industries; indeed, the markets for phenolics were still expanding half a century later. But despite such advances, the chemical industry would not begin to accelerate until its lifeline to Europe was broken by the war that erupted in 1914. A second noteworthy, though less cataclysmic, event of that year was the award of the Nobel Prize for Chemistry to Theodore W. Richards, the first American to win it. Richards, who also was ACS President in 1914, determined the atomic weights of 28 elements.

Specialization and the ACS

As chemical knowledge and activity expanded during the 19th century, chemists inevitably began to specialize, presaging the divisional structure that the American Chemical Society would adopt in the early 20th century. In this country the trend became evident with the founding of the *American Chemical Journal* in 1879 by Ira Remsen, of the *Journal of Analytical Chemistry* in 1887 by Edward Hart, and of the *Journal of Physical Chemistry* in 1896 by Wilder D. Bancroft and J. E. Trevor. (Bancroft was ACS President in 1910; Remsen, ACS President in 1902, published in the first volume of his journal his discovery, with a student, of the compound later named saccharin.) By 1900, American chemists were tending to identify themselves by field of interest—inorganic, organic, analytical, physical. Attendance at ACS national meetings during the first 25 years or so was relatively small, however (it would not reach 1000 until 1916); the need hardly existed to segregate papers by area of specialization. Indeed, the presentation of 76 papers in nine categories at the World's Congress of Chemists in 1893 in Chicago was sufficiently novel to be undertaken with some trepidation.

That chemists in this country wanted to be differentiated by specialty was clear in the formation of groups outside the American Chemical Society. The Society was apprehensive over the advent of these organizations. The New York Section of the Society of Chemical Industry, for example, was formed under the chairmanship of Charles F. Chandler, a founder of ACS, in 1906; the section's 1503 members included about one-fourth of the Society's 3079 members. When the American Electrochemical Society (now the Electrochemical Society) was formed, in 1902, its 350 charter members included 103 members of ACS. About half of the charter members of the American Society of Biological Chemists, formed in 1906, were also members of ACS.

As early as 1903, acting on a proposal by A. A. Noyes, the ACS Council appointed a committee of five representatives of different specialties of chemistry to consider the advisability of forming divisions within the Society. The committee recommended to the next Council that five divisions be formed, but the motion did not carry. Nevertheless, at the next general meeting, Dec. 28–31, 1904, the papers submitted were such that it proved useful to organize a general session followed by simultaneous sessions for papers in five specialized areas: physical chemistry; agricultural, sanitary, and physiological chemistry; industrial chemistry; inorganic chemistry; and organic chemistry.

Both philosophically and practically during the early 1900s the ACS had difficulty resolving the problem of unity vs. specialization for its meetings and for its publications. The number of industrial chemists in the Society was approaching the number from academe, and the two groups did not cooperate well. New organizations, with their own publications, were springing up. The need for the divisional structure at last became obvious. The industrial chemists were not alone in desiring a degree of autonomy; the issue was forced also by several academic groups who were working in new and evolving fields.

In the face of these multiple pressures, the ACS Council, on Jan. 2, 1908, established the *Journal of Industrial and Engineering Chemistry,* to be issued beginning in 1909. At the same time the Council established the ACS Division of Industrial Chemists and Chemical Engineers. The remainder of 1908 saw the formation of four additional divisions: Agricultural and Food Chemistry; Fertilizer Chemistry; Organic Chemistry; and Physical and Inorganic Chemistry. With these moves came an appreciable increase in ACS membership, which reached about 5000 in 1910.

For a year or so before the Division of Industrial Chemists and Chemical Engineers was formed, the momentum had been building that led to the formation, also in 1908, of the independent American Institute of Chemical Engineers. Many charter members of AIChE and all of its first officers were ACS members. They were aware of the plans for the analogous ACS division and its companion journal, but they persisted in forming the institute for a single reason—to have rigid requirements for membership. At the time, nomination for membership in ACS was still open to "any person interested in the promotion of chemistry"; membership in AIChE, on the other hand, was restricted to those "not less than 30 years of age . . . proficient in chemistry and in some branch of engineering as applied to chemical problems, and . . . engaged actively in work involving the application of chemical principles."

Despite the upsurge in sister societies in the early 1900s, new ACS divisions formed slowly after the initial five of 1908. The total had reached nine by 1919; eight more were formed in the 1920s and one each in the 1930s and 1940s. By 1951, the year of the ACS Diamond Jubilee, there were 20 divisions; in 1975 there were 28, one of them probationary.

Growth and diversification in chemistry also has affected the ACS presidency. In the early years, the membership was predominantly academic. The eighth President, J. C. Booth, who served in 1883–1885, was the first who was not from academe, although he did teach analytical chemistry. The 18th President, C. B. Dudley, who held office in 1896–97, was the second nonacademic. Of the first 75 presidential terms, 22 were held by 17 nonacademics. In 1943, it became the practice to alternate the office of President between academics and nonacademics; thereafter, with very few exceptions, this practice was maintained.

The early 20th century brought two other milestones in the growth of the American Chemical Society. In 1907, the Society began publication of *Chemical Abstracts* as a separate entity. For the preceding decade, abstracts had been a part of the *Journal of the American Chemical Society*. Also in 1907, Charles L. Parsons of New Hampshire College began his eventful four decades as Secretary of the Society. In 1911, Dr. Parsons became chief chemist of the Bureau of Mines in Washington, D.C.; the office of Secretary, still a part-time function, moved with him in 1912. The event would prove fortuitous in view of the role that chemistry would play in the national interest during the war that was soon to strike.

The First World War

The outbreak of World War I, in July–August 1914, propelled chemists and the chemical industry in this country into public view as never before. The British blockade barred German shipments of chemicals to the U.S., and programs of several ACS local sections in 1914 dealt with the need for American chemical companies to expand into dyestuffs, pharmaceuticals, and other organic chemicals. ACS President Charles H. Herty advised President Wilson of the need by letter of Sept. 25, 1915. The war-stimulated expansion of the steel and petroleum industries, meanwhile, provided growing sources of coal-tar chemicals and petrochemicals that the chemical industry could convert to dyes, drugs, and other products once imported from Europe and especially from Germany. Rising demand for explosives called for sharply increased supplies of toluene and phenol. Critical also to explosives manufacture was nitric acid, made from ammonia. The need for ammonia was met by the Frank–Caro process, which had been developed in Europe and was based on the production of calcium cyanamide. Germany, ironically, was using the much superior Haber–Bosch ammonia process, based in part on Willard Gibbs' phase rule of the late 1870s, which Fritz Haber had learned of from Van't Hoff in about 1903. The Haber–Bosch process did not reach the United States to any degree until the 1920s.

Germany introduced gas warfare in April 1915; in 1917 the Secretary of the Interior assigned the gas problem to the Bureau of Mines, which engaged the Chemistry Committee of the National Research Council to help initiate the work. This group and others in universities and the chemical industry, many of them recruited by ACS Secretary Parsons, were blended into what became the Chemical Warfare Service of the U.S. Army. Among chemists involved in the gas warfare work were a dozen past and future ACS Presidents, one of them (T. W. Richards) a Nobel laureate; altogether, more than 200 officers and 500 enlisted men took part in the program. Much of the research was done at American University. The gases were manufactured at Edgewood Arsenal, Md., and protective equipment in New York City.

In February 1917, Secretary Parsons initiated a census of the nation's chemists, sponsored jointly by the Society and the Bureau of Mines.

The ACS also advised the National Academy of Sciences in organizing the National Research Council, established by President Wilson Sept. 21, 1916. An early activity of the NRC Chemistry Committee was the preparation of an accurate, classified listing of research chemists in the United States. This was done in mid-1917 under the direction of Marston T. Bogert, ACS President in 1907–08.

Wartime activities markedly affected ACS membership, which in 1915–20 more than doubled, to 15,582. One result was that Secretary Parsons resigned from the Bureau of Mines, in 1919, to devote more time to his ACS duties. The period saw other events in chemistry, not all of them related directly to the war. In 1915, E. J. Crane became Editor of *Chemical Abstracts*—located by then on the Ohio State campus in Columbus—which in the subsequent four decades he would build into the world's premier scientific abstracting service. In September of the same year, some 63,000 visitors attended the first Exposition of Chemical Industries, held in the New Grand Central Palace in New York City. The attendance was symptomatic, perhaps, of the high interest in chemistry engendered by its importance to the war effort. The press was devoting increasing attention to the subject, but newspaper accounts often were marred by errors and a tendency to stress the sensational. Newspapermen were aware of the difficulties, and in 1916 a group of them asked for help from the Society's New York Section. The section responded by forming a Press and Publicity Committee whose function was to supply accurate chemical information in a form readily adaptable to the interests of the lay reader. At the ACS national meeting in New York City in September 1916 this committee helped to improve press coverage sharply. Earlier in the year the national Society had formed a committee to study press relations, and in December, impressed by the work of the New York Section's committee, the ACS consolidated the two to form the ACS Press and Publicity Committee. In 1919, the committee's work was assigned to the newly formed ACS News Service, a staff unit whose major duty was and remains the promotion of public understanding of chemistry. In 1923, James T. Grady became Managing Editor of the News Service. His work during the subsequent quarter century was recognized by the Society in 1955 when it established the annual James T. Grady Award for Interpreting Chemistry for the Public.

One further Society action, later regretted, occurred at the 56th national meeting, in September 1918. By unanimous vote, the German chemists Walther Nernst, Wilhelm Ostwald, and Emil Fischer were dropped from the rolls as honorary members, effective retroactively to Aug. 1, 1914. Fischer and Ostwald were Nobel laureates in chemistry; Nernst would win the prize in 1920. With "a return of better judgment" the Society rescinded the move in 1926 for Fischer, by then deceased, and in 1927 for Nernst and Ostwald.

With the coming of peace, on Nov. 11, 1918, the U.S. Government abruptly canceled all war contracts and ceased production at government-owned or -operated plants. Many chemical plants constructed for wartime purposes were not needed by a civilian economy. Up to 1914, for example, the nation's requirement for phenol was about one million pounds annually. By 1918, productive capacity for the compound exceeded 100 million pounds annually. The end of the war left stockpiles of many chemicals far larger than the needs of civilian markets. Many plants were closed, and chemists and chemical engineers were among those facing unemployment. The situation was further exacerbated by the resurgence of the German chemical cartel. ACS membership reacted predictably; it declined in 1921 and would not recover fully until 1928 when, at 16,240, it exceeded the 1920 level for the first time, but only by 4%.

Another product of World War I was the Chemical Foundation, Inc., organized in February 1919 by executive order of President Woodrow Wilson. The foundation was conceived by Francis P. Garvan as a means of breaking the German stranglehold on the nation's organic chemicals industry. It was to do this by making available to U.S. manufacturers enemy-owned (mostly German) patents seized by the Alien Property Custodian, which Garvan headed for a time. Stock in the foundation was purchased by individuals and organizations who would be affected by its work, and the Society was among the original shareholders. In return for a total payment of $269,850, the Alien Property Custodian assigned to the Chemical Foundation 4764 patents, 283 patent applications (of which 196 were granted), 874 trademarks, 492 copyrights, and 56 prewar contracts. During 1919–46, the foundation's income—primarily from sale of nonexclusive licenses—totaled some $8.6 million, about $8.2 million of it collected by 1938. Of the total, the foundation contributed 72% to scientific research and education. No shareholder received dividends, and the president and vice-president worked without salary. ACS operations, including the journals, were substantial beneficiaries of the Chemical Foundation, which, for example, subsidized the *Journal of Chemical Education* for its first nine years.

Besides Francis Garvan's work at the Chemical Foundation, he and his wife contributed personal funds to various aspects of chemical education. In 1929, Mr. Garvan became the third recipient of the Society's Priestley Medal. In 1936, he established the Society's Garvan Medal, which is supported by a fund set up at that time with a donation from Mr. Garvan.

Francis P. Garvan (1875–1937), president of the Chemical Foundation, remains in 1976 the only layman to have received the ACS Priestley Medal.

Chemistry and ACS in the 1920s

The postwar slump was succeeded late in 1922 by a boom that accelerated steadily until the crash of 1929. The 1920s were marked also by the 18th Amendment, which brought prohibition to the nation beginning in January 1920. The democratization of the automobile, meanwhile, along with the explosive growth in radio and motion pictures, keyed an unprecedented wave of consumer demand. These and kindred events of the decade interacted inescapably with activities in chemistry and the American Chemical Society.

The 18th Amendment banned the manufacture, sale, or transportation of intoxicating liquors; the Volstead Act (the National Prohibition Act), which was designed to enforce the Amendment, defined said liquors as those containing 0.5% or more alcohol by volume. The two pieces of legislation created a new industry—bootlegging. And they spurred—even before taking effect—a frightening rise in the incidence of blindness and death from drinking wood (methyl) alcohol. At the request of the ACS, Dr. Reid Hunt at Harvard Medical School prepared a statement for release Jan. 1, 1920, warning the public of the toxic effects of wood alcohol. The Hunt statement was distributed to the press by the ACS News Service. Through its work in chemical nomenclature, the Society also was instrumental in manufacturers' revision of labels to read "methanol" instead of "wood alcohol" or "methyl alcohol." The idea was to avoid tempting people who evidently were prepared to drink any liquid whose label bore the word "alcohol."

Another side effect of prohibition was the series of problems it caused for legitimate industrial users of alcohol. The 18th Amendment recognized alcohol solely as a beverage. The Volstead Act did recognize industrial alcohol, thanks to some last-minute scrambling by the alcohol industry, and by 1920 many formulas of denatured alcohol were authorized for industrial use. Illegal regeneration of denatured alcohol spread rapidly, however, and partly for this reason the law was administered so lopsidedly in the direction of its prohibitory provisions as to impose a heavy regulatory burden on legitimate users of alcohol. In mid-1923, as a means of alleviating the situation, the Commissioner of Internal Revenue organized an Alcohol Trades Advisory Committee whose secretary was Harrison E. Howe, Editor of the ACS journal *Industrial & Engineering Chemistry*. The committee, whose members included Charles L. Reese, a future ACS President (1934), worked effectively to ease the problems of industrial users of alcohol while not sapping the enforcement effort.

The millionth "horseless carriage" had been produced in the United States in 1916, and in the period 1922–29 the number of cars and trucks on the road more than doubled to exceed 25 million. The price of the cheapest car, $825 in 1908, was only $260 by 1925. The bottleneck at the end of the automotive assembly lines was broken in 1922 by the quick-drying nitrocellulose lacquers that allowed cars to be finished in hours instead of days. The impact of these new synthetics was outlined in July 1927 by M. J. Callahan of Du Pont. Sales of nitrocellulose lacquers in this country in the last half of 1926, he reported, had exceeded 10 million pounds, compared with less than one million pounds in all of 1922. Chemical science played a vital role also in the phenomenal expansion of the petroleum and rubber industries that attended the rise of the automobile. It was hardly a coincidence that the ACS Division of Rubber Chemistry was formed in 1919 and the Divisions of Petroleum and Cellulose Chemistry in 1922. Activities in cellulose chemistry, moreover, involved not only lacquers, but also rayon, whose output in this country rose 12-fold in the 1920s.

Callahan delivered his remarks on nitrocellulose lacquers during a four-week Institute of Chemistry held at Penn State College (now University). ACS President (in 1927) George D. Rosengarten gained Chemical Foundation support for the institute, which consisted of brush-up courses and survey lectures on recent progress in chemistry. A second four-week institute was held at Northwestern University in the summer of 1928, but the Society was unable to raise funds for further sessions.

The number of radio broadcasting stations in the U.S. rose from none in 1920, when Westinghouse Laboratories made the first commercial broadcast, to almost 1000 in 1929. The consequent demand for radio housings and parts propelled the adolescent plastics industry quickly into adulthood. In the process it made quick work of the Government's war surplus of 35 million pounds of phenol, which was used to make the phenol-formaldehyde plastics pioneered by Leo Baekeland. Similar growth occurred in the motion picture industry, whose films by 1929 were being shown in 20,000 theaters across the nation. The industry's chemical demands included clearer, stronger, nonflammable film and a variety of photographic chemicals.

Science in the United States progressed apace during this active decade. In the years 1920–29, the number of papers and patents abstracted annually in *Chemical Abstracts* rose some 250%, to 46,949. The Nobel prizes were first awarded in 1901, but before 1920 the prizes in physics, in chemistry, and in physiology or medicine had gone to only three Americans, one in each of the three categories. During the decade 1920–29, no Americans were honored in chemistry or in physiology or medicine, but the physics prize went to Robert Millikan in 1923 and to Arthur Compton (with the Englishman Charles Wilson) in 1927. Thereafter, U.S. scientists would receive about one of every three prizes in the three areas combined. Another indicator of the vitality of science in this country at the time was the production of Ph.D.'s in chemistry. Though still not large by 1976 standards, the figure reached 2323 during 1921–30 and had been about doubling in every decade since 1881. And applied chemical research was reoriented during the decade toward the modern practice of synthesizing new compounds and materials deliberately to fill industrial and consumer needs, as opposed to the earlier concentration on improving and lowering the cost of existing chemicals.

In September 1926, in the midst of this scientific and technological surge, the American Chemical Society celebrated its 50th anniversary. As part of the 72nd general meeting, a group of chemists journeyed to Northumberland, Pa., where the home of Joseph Priestley was dedicated as a permanent memorial. The newly opened Priestley Museum was dedicated at the same time. Immediately afterward in Philadelphia came the second part of the meeting, which was attended by four founding members: S. A. Goldschmidt, J. B. F. Herreshoff, Charles E. Munroe, and H. E. Niese. Although ACS membership remained relatively stable during most of the 1920s, it had grown sevenfold, to 14,704, in the 25 years that ended with the Society's Golden Anniversary. The ACS by 1926 had 15 divisions and 67 local sections nationwide. The practice of chemistry as a profession clearly had come of age.

In the 1920s the Society moved in several areas to adapt itself to the changing face of the science it serves. In the years 1919–27, the ACS formed nine new divisions, more than in any similar period before or since. In 1923, the *Journal of Industrial and Engineering Chemistry* became simply *Industrial and Engineering Chemistry (I&EC)*. In the same year the Society launched the *News Edition* of *I&EC*, and in 1929 came the *Analytical Edition*. By 1928, the ACS had become so large and its operations so complex that incoming Presidents, who served for one year, were hard put to master their responsibilities. Accordingly, the Society's Constitution was revised to provide for a President-Elect who would hold that office for one year of familiarization and automatically take office as President in the following year.

The Depression Years and Professionalism

The giddy 1920s ended abruptly with the stock market crash of 1929. During 1929–32, the nation's Gross National Product fell more than 25%, to $113.6 billion (1950 dollars). In the same period, unemployment climbed nearly eightfold, to 12.1 million. The jobless included many chemists and chemical engineers, which prompted the Society to launch programs that remain important parts of its work in 1976. In the main these efforts reflected chemists' sharply heightened interest in professionalism—education and training, employment aids and counseling, employer–employee relationships, and the status of chemists and chemistry in the eyes of the public.

The effect of the depression on ACS membership was minimized by acceptance of dues on the installment plan, in three equal payments, in 1934 and 1935, but more importantly by a revised dues structure. Membership was 17,426 in 1929; it reached 18,963 in 1931, fell to a low of 17,541 in 1935, and by 1940 had rebounded to 25,414. The Society's annual dues in 1933 were $15, but for this the member received all ACS journals. In the same year the ACS lost subsidies from industry and the Chemical Foundation that were equivalent to perhaps another $5.00 per member. One alternative was to raise dues to $19, but this was considered impractical. Instead, dues were reduced to $9.00 for members who wished to receive only the *News Edition* of *I&EC;* members could receive all journals for $20 and various combinations of journals at intermediate rates. The new dues structure was well received, and the Society credited it with reversing the decline in membership.

In 1931 came the first Gordon Research Conference. It was conceived by Neil E. Gordon, founder of the *Journal of Chemical Education,* as an occasion on which scientists could discuss their work in a leisurely manner in pleasant surroundings. The first conference, the only one in 1931, took place at Johns Hopkins, where Dr. Gordon was on the staff. In 1975, more than 80 Gordon Research Conferences, each four and a half days long, took place at eight schools in New Hampshire and a hotel in California.

Although the economic context of ACS activities in the 1930s was not good, it was somewhat better than for the nation as a whole. The chemical industry by then was a major employer of chemists, and the annual dollar value of the industry's output rose some 40% in the years 1931–39. The comparable value for all industry declined about 10%. In its worst depression year the chemical industry reported gross income 39% below its then-record high of 1929; the decline for all industry was 52%.

The chemical industry's performance in the 1930s stemmed in large part from the rising sales of synthetic materials. The new materials were chiefly the results of the industry's rapid adoption of applied research after World War I. A brief, illustrative list of chemical developments that became commercially important in this country in the two decades ending in 1940 would include rayon, alkyd resins, cellulose acetate, synthetic methanol, fluorocarbon refrigerants, polychloroprene (neoprene) synthetic rubber, bromine from seawater, and nylon. The activity was not restricted to the U.S. The disclosure of a German patent on the first of the sulfanilamides, in 1932, for example, would open a new era in the chemotherapy of bacterial diseases. The effectiveness of applied chemical research as an organized activity would allow Du Pont, a leading practitioner, to report that 40% of its sales in 1937 came from a dozen products developed since 1928. Funding of science in this country in 1930–40 rose at an average annual rate of 9%, even with the decline of 1932–34, and the chemical industry was a major contributor. By 1940, the industry was estimated to be the nation's largest industrial employer of research workers, accounting for one-fifth of total research employment.

Despite the relatively good record of the chemical industry in the 1930s, no one could be immune to a wave of unemployment that struck one of every four working Americans. Many chemists and chemical engineers were jobless. The salaries of many others were reduced, and the majority were working at tasks that today would be termed "underemployment." This situation, and the by-then large fraction of ACS members who were—or wished to be—industrially employed, were the major forces in the rise of professionalism in the Society.

A concrete step in the movement came in 1933 when the Society revised its requirements for membership. At the time, almost all ACS members had been trained as chemists, but for three decades the basic qualification for membership had remained simply an interest "in the promotion of chemistry." The requirements adopted in 1933 called in part for "an adequate collegiate training in chemistry or its equivalent" and active engagement "in some form of chemical work for at least two years."

In April 1935, the ACS Committee on Unemployment of Recent Graduates (appointed in 1934) recommended "That a standing committee . . . be established to give continuing consideration to matters affecting the status of the chemical profession." The Council approved, and President Roger Adams appointed the committee, but a year later it was split into a Committee on Professional Status and a Committee on Accrediting Educational Institutions Offering Instruction in Chemistry. The Committee on Professional Status was discharged in 1942, but its functions have been handled continuously since then by one or another committee of the Society and since 1966 by the Committee on Professional Relations. The Committee on Accrediting Educational Institutions has remained in existence continuously; in 1946 it assumed its present name, Committee on Professional Training.

The absence of jobs during the Depression years induced many young people to continue through high school and even college. That instruction in chemistry was not of high quality in many colleges and universities had first become evident during World War I, and the problem persisted into the 1930s. Thus the charge to the Committee on Professional Training (as it is now known) in 1936 was to draw up suitable minimum standards for accrediting chemistry departments. The committee published its first list of approved departments Oct. 10, 1940, in the *News Edition* of *Industrial and Engineering Chemistry*. ACS interest in college-level chemistry students was reflected also by the introduction of the Student Affiliate program, which offered undergraduate chemists and chemical engineers certain benefits of Society membership at nominal cost. The first chapter of student affiliates was chartered at Lafayette College in November 1937; at the end of 1975 there were chapters at more than 600 colleges and universities.

At the national meeting in Chapel Hill, N.C., in April 1937, the Society undertook informally to arrange contact between job-seeking chemists and prospective employers. Several local sections, notably the Philadelphia Section, had set up employment aids for chemists relatively early in the 1930s and had argued for greater participation by the national Society in helping unemployed chemists. The results of the Chapel Hill experiment led to a formal Employment Clearing House at the next national meeting, in Rochester, N.Y., and the practice was continued at subsequent national meetings. In 1944, this twice-yearly service was expanded to include a full-time Employment Clearing House at Society headquarters as well as Regional Employment Clearing Houses operated by selected ACS local sections.

The professional activities that began in the 1930s within the Society also included an active, if short-lived, role in recommending equitable salaries for chemists and chemical engineers. In the 1930s and well into the 1940s, B.S. graduates were offered starting annual salaries of $1500 by some chemical companies, but considerably less by others. In 1941 the ACS suggested a minimum starting salary of $1500 per year for B.S. chemists and $2400 for Ph.D.'s. Effective with the spring meeting of 1942, these minimums were made requirements for employers who wished to use the services of the Employment Clearing House. This excursion into minimum salary recommendations was stopped, however, by the end of 1943. One reason was the inflation brought on by World War II; another was the fact that some employers assumed the minimums to be maximums.

A different approach to remuneration—a study of the economic status of ACS members—began in November 1941 under the aegis of the ACS Committee on Economic Status, which had been formed in September of that year (and would assume the functions of the Committee on Professional Status when it was discharged, in 1942). Questionnaires were mailed to the entire membership, and more than 75% responded before Jan. 25, 1942. The U.S. Bureau of Labor Statistics analyzed the returns statistically. The resulting publication, "Economic Status of the Members of the American Chemical Society," attracted wide attention when it was issued, in 1942. Periodic surveys made by the Society since that time have given members a valuable measure of their economic status.

The rise of professionalism within the ACS during the 1930s paralleled a longer-running trend—the steady growth of the Society—that was more revealing of the general health of chemistry. In the decade 1930–39, ACS membership rose some 35%, to 23,519; annual disbursements rose by about the same percentage, to $646,630. The national meeting in September 1933, at the Century of Progress Exposition in Chicago, attracted a record turnout of more than 3000 members. That record fell in April 1935 in New York City, where more than 5000 members registered for the national meeting that celebrated the American Chemical Industries Tercentenary.

The Society's Federal Charter

The decade of the 1930s also saw federal legislation with specific impact on chemists and on the Society. Prohibition, which had involved the ACS in problems with the ingestion of wood alcohol and restrictions on industrial alcohols, came to an end in December 1933, when Utah became the 36th state to ratify the 21st Amendment and thus to repeal the 18th.

In June 1938 President Franklin Roosevelt signed the Federal Food, Drug, and Cosmetic Act, the first major revision of the Food and Drugs Act of 1906 that Harvey Wiley had fought so hard for. The new act marked a movement—still accelerating in 1975—that would draw chemists and chemistry more and more deeply into the complexities of making and implementing public policy on science and technology. Of much more immediate interest to the ACS, however, was Public Act No. 358, 75th Congress. This "Act to Incorporate the American Chemical Society" granted the Society a federal charter, effective Jan. 1, 1938.

The events leading to the Charter—events in which Secretary Parsons was a moving spirit—had begun with a Board of Directors' decision that an effort should be made to incorporate the Society under an Act of Congress. At the time, the ACS was still a New York corporation; its legal headquarters were in that state, although operating headquarters had been in Washington, D.C. since 1912, when Parsons moved there from New Hampshire. Overtures were made, and the necessary legislation was submitted in the House by Rep. Walter Chandler of Tennessee and in the Senate by Sen. Walter F. George of Georgia. The proposed action was approved by the five federal departments—Agriculture, Commerce, Interior, Navy, and War—interested most directly in scientific

Board of Directors in 1937, when the Society's national Charter was granted by Congress. Seated, from left: R. E. Wilson, R. T. Baldwin (ACS Treasurer), Thomas Midgley, Jr. (Chairman), E. R. Weidlein, C. L. Parsons (ACS Secretary), F. C. Whitmore, Elisha Hanson (ACS Counsel). Standing, from left: E. M. Billings, G. J. Esselen, Edward Bartow, H. H. Willard, E. K. Bolton, A. J. Hill, T. R. Leigh. Missing: Willard H. Dow, R. E. Swain.

work. It was approved also by the Bureau of the Budget and by the Committees on the Judiciary of the House and Senate. The House passed the measure on Aug. 16, 1937, and the Senate on Aug. 20; President Roosevelt signed it into law on Aug. 25, 1937.

In its Aug. 10 issue, meanwhile, the *News Edition* of *Industrial and Engineering Chemistry* had carried an "Important Notice to Members." The notice called a General Assembly of the Society for Sept. 8 at the national meeting in Rochester, N.Y. The purpose was to act on a Board resolution ". . . to take whatever steps are necessary to reincorporate the Society, either under an Act of Congress or the laws of the District of Columbia." The notice also outlined the benefits of federal incorporation. The proposed move, it said, would establish the Society firmly as a national organization, free to operate in any state or group of states. As a scientific and educational body, the ACS had always been exempt from federal and state taxation, but a national charter would free it from local taxation as well. Such a charter also would insulate the Society from ". . . many petty vexations that are likely to occur through loosely drawn or ill-considered legislation in our 48 states or the District of Columbia."

On Sept. 8, 1937, the ACS General Assembly in Rochester duly approved the Board resolution—by then a *fait accompli.* The Board of Directors gathered on Jan. 7, 1938, in Washington's historic Willard Hotel for the Organization Meeting of the American Chemical Society under Public Act 358. The Directors formally accepted the Charter, ordered that a Certificate of Dissolution of the American Chemical Society be filed with the state of New York, and adopted a Constitution and Bylaws under the federal Act of Incorporation. Federal incorporation was a landmark in the development of the ACS. In the four decades since the Charter was granted it has proved most beneficial to the Society's ability to work effectively in the interests of chemists and chemistry. A companion source of strength has been the location of ACS headquarters at the seat of the Federal Government, increasingly the source of decisions with major impact on scientists and engineers. As the Charter has conferred benefits, however, so has it imposed obligations. The first of these is that the Society pursue the "objects of the incorporation," of which the overriding one is "to encourage in the broadest and most liberal manner the advancement of chemistry in all its branches." The ACS also has felt obliged to respond to requests for information from any branch of the Government. In this vein, moreover, the Society over the years has tended more and more to make itself available to the Government—instead of waiting to be asked—as a source of objective expertise in chemically oriented areas.

The evolution of the ACS during the 1930s was accompanied—indeed, stimulated—by steady progress in chemical science worldwide. The growth of research and development is evident in the number of abstracts published annually by *Chemical Abstracts,* which rose almost 40%, to 65,307, in the decade ending with 1939. American chemistry, meanwhile, was moving rapidly into the front rank. In the period 1931–40, the nation's graduate schools produced 6877 Ph.D.'s in chemistry, three times the output of the previous decade. In 1931, Linus Pauling, a future ACS President (1949) and Nobel laureate (1954–Chemistry; 1962–Peace), won the first American Chemical Society Award in Pure Chemistry, for his research on chemical bonding. In 1932, Irving Langmuir became the second American to win the Nobel Prize for Chemistry. He was honored for diverse theoretical and practical work, including his research on vacuum tubes and the gas-filled incandescent light bulb. In 1934, Harold C. Urey became the third American to receive the prize, for his discovery of deuterium, the hydrogen isotope of mass two.

The chemical community of the decade preceding World War II displayed a style that has largely succumbed to the explosive growth that soon would begin. A case in point is the early history of the ACS Award in Pure Chemistry. The award was initiated by A. C. Langmuir, elder brother of Irving. The two supplied the honorarium ($1000) from 1931 through 1937, but then elected to withdraw. When their decision was announced, James Kendall, then in the chemistry department of the University of Edinburgh, wrote to the Society and offered to provide the money in 1938. He had spent 15 rewarding years at Columbia and New York University, Kendall explained, and he wished to discharge his debt to American chemistry. The Society accepted, and the award for 1938 went to Paul D. Bartlett. The award was not given in 1939, but the chemical fraternity Alpha Chi Sigma has funded it continuously since 1940.

The style of the 1930s is evident also in the chemical reportage of the day, and the ACS Award in Pure Chemistry again is illustrative. When Linus Pauling gave his award address, at the national meeting in Buffalo in September 1931, the projector broke down. The mishap was carefully reported by the *News Edition,* which added cheerfully that it had "provided an unexpected opportunity for him to display his thorough knowledge of his subject." Further insight on the American chemical world of the 1930s is provided by the *News Edition's* report of the death of Charles H. Herty on July 27, 1938. Herty had been the first full-time Editor of *Industrial and Engineering Chemistry* and ACS President in 1915–16; the page that carried his obituary carried also a brief account of the death of Thomas Egan, "well and affectionately known waiter in the dining room of the Chemists' Club (N.Y.). . . ."

Times change, but a record of life—and of chemistry—as it was in the late 1930s lies buried in the Westinghouse Time Capsule, some 50 feet beneath the surface on the grounds of the New York World's Fair of 1939. The Westinghouse Electric and Manufacturing Co. designed the 800-pound capsule to remain buried for 5000 years and, accordingly, sent directions for finding it to some 3000 libraries and other safe repositories around the world. The capsule's contents include many articles and materials in common use at the time as well as a considerable microfilm record. Among the microfilmed items are a Periodic Table of the elements, excerpts of the Encyclopaedia Britannica's treatment of chemistry and applied chemistry, and the presidential address—"A World of Change"—of Edward R. Weidlein, ACS President in 1937. (Westinghouse interred a second, 300-pound Time Capsule next to the first one on Oct. 16, 1965, during the New York World's Fair of 1964–65; a granite monument marks the site.)

The Second World War

The Time Capsule was placed in its crypt Sept. 23, 1938; by the time it was sealed, Sept. 23, 1940, the world had changed drastically. Japan, at odds with China since 1931, had launched a full-scale attack in July 1937. The German blitzkrieg struck Poland on Sept. 1, 1939. By June 22, 1940, when France capitulated, Hitler was master of Europe, and Britain stood alone in the west. These violent events provoked a metamorphosis in the United States. Between 1935 and 1939, Congress had passed neutrality acts that forbade the sale of armaments to any belligerant. By 1940 the national mood had passed from pacifism to extending "to the opponents of force the material resources of this nation" and to accelerating "the use of these resources in order that we . . . may have equipment and training equal to the task of any emergency" (President Roosevelt, in a speech at the University of Virginia, June 10, 1940).

The nation's scientific resources were rallied quickly. On June 15, 1940, President Roosevelt formed the National Defense Research Committee (NDRC), a group of eminent civilian scientists with Vannevar Bush as chairman. NDRC, whose task was to channel federal funds into weapons research, was superseded a year later by the Office of Scientific Research and Development (OSRD), with Bush as Director. Most of the scientific research done for the armed forces during the war moved through NDRC or OSRD. In the fall of 1940, at the request of President Roosevelt, the Society set out to compile the names and qualifications of the country's chemists and chemical engineers. The mechanism was a coded questionnaire designed by Erle M. Billings, then Secretary of the ACS Committee on Professional Training. The resulting data, on punched cards, became part of the National Roster of Scientific and Specialized Personnel, whose purpose was to permit prompt allocation of scientists and engineers in accord with the needs of the nation.

The onset of World War II found the United States far stronger in science and technology than it had been a quarter of a century earlier. The utility of science in World War I had been impressed on many Americans through the popular press and the public information programs of scientific societies, including the American Chemical Society. Chemists, for example, had been publicized widely for their work on explosives, smoke screens, and war gases and the associated protective devices. More important in the long run, both government and industry had recognized the value of science in solving problems; both had begun to spend more liberally on research and development. Universities built more courses in the sciences into their curriculums. Between 1920 and 1940, the number of industrial laboratories in this country rose from about 300 to 3500, and their scientific staff increased from 9000 to 70,000.

Combatants on both sides would exploit their science and technology to the limit during the war, both in weaponry and elsewhere. But the landmark achievement of the period, whether for good or for bad, was the discovery and harnessing of nuclear fission, until then the ultimate manifestation of the great breakthroughs of the turn of the century.

Late in 1938, the German physicists Otto Hahn and Fritz Strassmann at the Kaiser Wilhelm Institute, upon bombarding uranium with neutrons, obtained a small amount of barium. This result meant that the uranium had been split into two almost equal fragments; it meant also that the energy produced in the reaction far exceeded that produced in any previous transmutation reaction. Fortunately for the Allies, they and not Germany converted the discovery eventually to the atomic bomb. Fortunately also, many of Europe's outstanding physicists already were refugees from Hitler's Europe. The Austrians Lise Meitner (one of Hahn's collaborators) and Otto Frisch were in Sweden; the Italian Enrico Fermi and the Hungarian Leo Szilard were in the United States, as was the German Albert Einstein, who had come in 1933.

In March 1939, when the Hahn–Strassmann results were known, Fermi tried and failed to interest the Navy Department in atomic fission. Then, a letter describing the potential of the phenomenon was drafted by Alexander Sachs and Leo Szilard, signed by Einstein, and delivered to the President by Sachs on Oct. 11, 1939. In February 1940, $6000 was made available to begin research on a bomb. Shortly after Japan attacked Pearl Harbor, Dec. 7, 1941, the project was given high priority. In January 1942 came a schedule: to determine by July 1, 1942, whether a chain reaction was possible; to achieve a chain reaction by January 1943; to have a bomb by January 1945. In June 1942 the War Department organized the Manhattan District to assume control of the work from OSRD, and in September Brigadier General Leslie R. Groves took command.

When Groves took over, events already were moving swiftly. Research had been under way for some months at the Metallurgical Laboratory at the University of Chicago, and on Dec. 2, 1942, under the grandstands of Stagg Field, Enrico Fermi and his team demonstrated the first chain reaction or sustained nuclear fission. This was the critical experiment, but there were other problems. Uranium, to be used as fuel in a practical atomic pile, would have to be enriched on a large scale in its fissionable isotope, uranium-235, whose concentration in the natural ore is less than 1%. Plutonium for a bomb would have to be isolated from the fission products created in the pile.

In 1941–42, first at Berkeley and later at the Metallurgical Laboratory, a means of isolating plutonium from fission products and uranium had been devised by a team headed by Glenn T. Seaborg (ACS President in 1976); the group isolated the first visible quantity of the element on Aug. 20, 1942. At Columbia University, Harold C. Urey was at work on the gaseous diffusion process for enriching uranium; scientists at the University of Virginia and Standard Oil Co. (N.J.) were investigating the centrifuge method for the same purpose; E. O. Lawrence and his coworkers at Berkeley were studying electromagnetic methods for extracting uranium-235 and plutonium. New cities were under construction at Oak Ridge, Tenn., Hanford, Wash., and Los Alamos, N.M., the manufacturing sites for specific phases of the project. The Du Pont Co. had been assigned the task of designing, building, and operating the Hanford works; Union Carbide had the administrative responsibility for Oak Ridge. The bomb itself would be designed and assembled at the Los Alamos Scientific Laboratory, headed by J. Robert Oppenheimer. All told, the Manhattan District employed 150,000 people.

The time spans between conception and application in this immense project were remarkably short; the scaleups, from miniscule amounts of materials to huge production plants, were remarkably large. Still, everything seemed to work. The first atomic device was exploded at Alamagordo, N.M., on July 16, 1945. The bomb was dropped on Hiroshima on Aug. 6, 1945, and on Nagasaki on Aug. 9, three months after the war in Europe had ended.

An early problem in conventional weaponry during World War II was the lack of incendiary bombs. The British solved it by adding rubber to gasoline to yield a rubbery jelly, but then the Japanese advance in Southeast Asia cut off the supply of rubber. Chemists at Du Pont found that poly(isobutyl methacrylate) thickened gasoline to a jelly, but the process required too much polymer, which was needed also to make transparent noses and "bubbles" for bombers and other aircraft. At the same time, chemists at A. D. Little and a team under Louis F. Fieser at Harvard discovered that an aluminum soap of naphthenic and palmitic acids would thicken gasoline. Fieser's group named the product napalm (from *na*phthenic and *palm*itic). By 1945, nine companies in this country were producing about 12 million pounds of napalm annually.

It was quite apparent in 1939, when the fighting erupted in Europe, that this country if it entered the war would be critically short of rubber. In mid-1940 the Government set up the Rubber Reserve Co. to stockpile rubber and in May 1941 directed that agency to develop a synthetic rubber industry. Companies were already working on synthetic elastomers, but by the end of 1941 the year's output totaled only 8000 long tons. Rubber Reserve was shifted quickly to a crash basis under a Rubber Director. Production tripled, to 22,000 tons, in 1942 and reached 820,000 tons in 1945, the peak year. The synthetics produced were butyl, neoprene, N-type, and the general purpose GR-S, which accounted for roughly 85% of total output during the war. Of these, GR-S and N-type initially were German developments, although they were greatly improved in this country; neoprene was invented by Wallace Carothers (the inventor of nylon) and A. M. Collins, and butyl by William J. Sparks (ACS President in 1966) and R. M. Thomas. The chemical intermediates for GR-S were styrene and butadiene, both required in tremendous quantities. The chemical industry supplied the styrene; petroleum chemists rather quickly developed a process for making butadiene from butane and butene.

The petroleum industry was in a relatively advanced stage when the war began. Between 1937 and 1942, catalytic cracking and two alkylation processes (sulfuric acid and hydrogen fluoride) were developed to commercial status by U.S. companies. Together the three processes were the source of enormous amounts of 100-octane aviation gasoline used by the Allies during the war. In the Battle of Britain, an early turning point of the conflict, this fuel gave the fighter aircraft of the Royal Air Force a distinct edge in performance over the German air armada. Petroleum chemists also devised means of making toluene, the major ingredient of TNT, which in World War II was produced at 10 times the level of World War I. Until 1939, the major source of toluene had been the coal tar industry.

Research on drugs and other health-related compounds was pursued vigorously in this country during World War II. Chemists at various universities and in the pharmaceutical industry developed a series of antimalarial drugs that saved many American lives in the Pacific. Other contributions were new sulfa, antibiotic, anesthetic, analgesic, hypnotic, autonomic, and cardiac drugs. The Office of Scientific Research and Development during the war had some 600 contracts in this area with approximately 150 institutions and involving about 5500 people. The results of these contracts included methods for large-scale production of atabrine, a malaria suppressive; pure albumin (from blood plasma), which was effective against shock; penicillin, which provided unprecedented control of bacterial infections; and DDT, which was used widely against disease vectors such as lice (typhus), mosquitoes (malaria, yellow fever), and houseflies (gastrointestinl diseases).

Wartime and Aftermath

The wartime contributions of American research, development, and production were prodigious, but with them came manpower-utilization problems that occupied the American Chemical Society for the duration. The Federal Government had no direct policy for the utilization of scientists and engineers. The Society, on the other hand, held that chemists and chemical engineers should be used where they might serve best—whether in the laboratory, in the plant, or in the armed forces. The ACS led the scientific community in support of this position, and the work was the Society's single most important effort of the war.

Essentially all of the nation's scientists and engineers submitted data for the National Roster of Scientific and Specialized Personnel. Relatively few were called from the Roster for special duty, although the data were used regularly in considering deferments from the military draft and for other purposes. Scientists were not exempt from the draft, and the indiscriminate drafting of chemists and chemical engineers troubled both the chemical industry and educational institutions. Relatively few students over 18 who were physically qualified for military service studied science in this country during the war. Professors and students alike left the universities, excepting those individuals involved in programs sponsored by NDRC or OSRD or in "speed-up" programs for educating members or prospective members of the armed forces. (In the four years 1941–44, the ACS Committee on Professional Training certified annually between 1400 and 1800 bachelor-level graduates in chemistry; the figure plunged to 876 in 1945 and 830 in 1946, but rebounded to 1605 in 1947 and 2442 in 1948 as discharged servicemen began to complete their educations.) In the face of these difficulties, the ACS throughout the war battled with Selective Service and its many local draft boards over the essential nature of the work of chemists and chemical engineers in the war effort. In addition, the Society worked to keep employers informed of rules and rights under Selective Service.

Among other war-related activities, the ACS cooperated with the Alien Property Custodian at the close of the war in abstracting 7000 German

The first ACS headquarters building at 1155 Sixteenth St., N.W., in the District of Columbia was occupied by the Society in April 1941.

chemical patents. It worked with the U.S. Office of the Publication Board, established by President Truman (President Roosevelt had died April 12, 1945), in declassifying information that had been produced for the war effort but could be useful to the nation's peacetime economy. When the scope of the Office of the Publication Board was enlarged to include "enemy scientific and industrial information," the Society was involved in disseminating the information. The ACS also took a strong interest in European recovery. In 1946 the Board of Directors took public notice of the loss to the scientific world through the cessation of publication of the Beilstein and Gmelin handbooks. Chairman Roger Adams was authorized to encourage such action as he deemed logical and practical to aid and expedite the resumption of these publications. In addition, the Society shipped a considerable volume of ACS and other journals to Germany and other European countries to help them revive their scientific efforts.

The ACS, like everyone else, felt the dislocations of war. Notable among these were the high cost and short supply of paper for journals and the censorship of scientific information that might be useful to the enemy. *Chemical Abstracts,* which had published 65,000 abstracts in 1939, published only 32,000 in 1945; *JACS,* which had published 1000 articles in 1939, published only 700 in 1944.

The war years and the remainder of the 1940s brought change and progress as well as problems. In April 1941, the Society moved into its own building, a five-story apartment house in the District of Columbia, that it had purchased and converted to office use. In January 1942, the *News Edition* of *Industrial & Engineering Chemistry* became *Chemical & Engineering News;* in January 1947, it was changed from a biweekly to a weekly magazine. In February 1943, Walter J. Murphy joined the Society as Editor of *I&EC,* including its *Analytical Edition,* and as Editor of *C&EN* and Director of the ACS News Service. Murphy succeeded Harrison E. Howe, who had died the previous December after serving the Society for 21 years.

In September 1944, the owners of Universal Oil Products Co., confronted by antitrust and other legal difficulties, advised ACS President Thomas Midgley, Jr., that they were thinking of offering their security holdings in UOP to the Society as a trust whose income would be used to promote fundamental research and advanced scientific education in the petroleum field. A special committee of the ACS Board reviewed Universal's business operations and recommended against any arrangement that would give the Society direct ownership of the company. The problem was resolved under an agreement of October 1944 between the donor companies[1] and the Guaranty Trust Co. of New York. The bank accepted the securities as trustee for the newly established Petroleum Research Fund; the ACS, as recipient of the income from the trust, agreed to use the funds solely to support ". . . advanced scientific education and fundamental research in the 'petroleum field' . . ." and assumed no responsibility for the management of UOP.

From 1946 to 1954, The Petroleum Research Fund provided about $150,000 annually, which was accumulated until sufficient funds were

[1] Shell Oil Co., Standard Oil Co. of California, Standard Oil Co. (Indiana), Standard Oil Co. (New Jersey), The Texas Co., N. V. de Bataafsche Petroleum Maatschappij. Phillips Petroleum Co., not a party to the original agreement, subsequently made a substantial contribution to the ACS, which, pursuant to the terms of the agreement providing for additional contributions, was transferred to the Trustee.

available to initiate a program of research grants. Income rose sharply thereafter, and in 1954 the first PRF grants were awarded. In 1956, in a step designed to increase the income from the Trust and alleviate problems experienced by UOP in operating competitively under a trustee, Guaranty Trust applied for permission from the N. Y. Supreme Court to sell UOP to the public. The eventual sale returned about $70 million to the PRF and enabled UOP to secure the necessary capital to enlarge its operations. The PRF money was invested in a diversified portfolio, with Morgan Guaranty Trust Co. (formed by a merger of Guaranty Trust with J. P. Morgan & Co.) as trustee; the position of the Society as recipient and administrator of the income was unchanged. In 1976, the total research support provided by PRF during its lifetime will pass $60 million.

At the end of 1945, Charles L. Parsons retired after more than 38 years as Secretary of the Society and 52 years as a member. During his service as chief full-time executive, ACS membership had grown from 3000 to 43,000 and its annual budget from a few thousand dollars to $1.5 million. When he stepped down, 99% of the members had known no Secretary but Charlie Parsons. Among his many honors were the Society's Priestley Medal, awarded in 1932 for "distinguished services to chemistry"; in 1952 he would be the first recipient of the Society's Charles Lathrop Parsons Award for "outstanding public service by a member of the American Chemical Society." Dr. Parsons was succeeded as Secretary and Business Manager of the Society by Alden H. Emery. Mr. Emery had joined the ACS in 1936 as Assistant Manager and thus had served almost a decade in that post before becoming Secretary. In 1947, the title was changed from Secretary and Business Manager to Executive Secretary.

The Society's steady growth, from more than 25,000 members in 1940 to more than 63,000 in 1950, dictated change in several areas, including arrangements for national meetings. Registration averaged about 3000 at the 10 national meetings held during 1936–40; despite wartime travel restrictions it averaged about 4600 at the eight meetings of 1941–44. Travel problems barred national meetings in 1945, but at the 10 meetings of 1946–50, average registration climbed to some 7700, with a single-meeting peak of 11,649. Meetings of such size could be accommodated by relatively few cities, such as New York, Chicago, and Atlantic City, and even in those cities meeting rooms and housing were scattered more widely than desirable. In an effort to make the twice-yearly events smaller and more accessible to ACS members throughout the country, the Society decided to hold one of them in two or several cities in successive weeks. The first of these divided meetings came in the fall of 1948 in Washington, St. Louis, and Portland, Ore. The concept was tried again in 1950 and 1952 and then dropped. During the 1960s the ACS experimented with a third annual national meeting, in a city that normally

***Charles Lathrop Parsons** (1867–1954) was the principal architect of the American Chemical Society, which he served as Secretary from 1907 through 1945.*

could not hold a large meeting. This concept, too, was dropped after trials in 1963, 1964, and 1966.

A more successful approach to meetings has been the regional concept. Regional meetings of groups of ACS local sections date back to 1908, but the several series in existence by the 1930s were brought to an end by World War II. In 1944, the first of the modern series of regional meetings, the Southwest Regional Meeting, took place in Baton Rouge, La. These events became an important adjunct of the Society's meetings program. By 1976, local sections were sponsoring nine regional meetings annually and one biennially.

The growth of the Society also brought increasing diversity in members' needs and interests. The Constitution of 1897 had not been amended significantly—except for the establishment of divisions and their Councilors in 1908—and it was clear by the mid-1940s that changes were needed to bring the organization into harmony with the times. The Board authorized a study of the Society's governance, and John M. Hancock, an industrial banker and public servant, agreed to undertake it. The Hancock report, submitted early in 1947, led to a completely revised Constitution and Bylaws, which took effect Jan. 1, 1948. By 1976, the Constitution of 1948 had served without major amendment for 28 years.

Big Science and the Cold War

The American Chemical Society in 1948 was not alone in entering a new era. World War II had demonstrated beyond any doubt the power of science—particularly the "big science" of the Manhattan District—as an instrument of national policy. At the same time, the Cold War was accelerating rapidly. The United States and the western bloc were fighting the Soviet Union and the eastern bloc for ascendancy on many fronts, not least the scientific front. This state of affairs produced an impact on science, technology, and their practitioners that is still evident in 1976.

The preeminent fact of the time was nuclear fission. That the bomb had locked mankind in the grip of a new destiny was clear to those close to the Manhattan District. They knew that no degree of secrecy could long prevent foreign scientists from mastering fission technology, especially when they knew that its product existed. And they feared for the future of the world if the military retained control of that technology. These fears led scientists into the political arena. An early manifestation was the Federation of Atomic Scientists, formed in 1945 and renamed the Federation of American Scientists in 1946. The same fears led to the creation of the Atomic Energy Commission, in August 1946, to assume control of both the military and peaceful uses of atomic energy. The race was on, however. In September 1949, President Truman announced that the Soviet Union had exploded a fission device; in January 1950, he directed AEC to proceed with the development of a fusion (hydrogen) bomb. The U.S. exploded its first fusion device in May 1951; the Soviets followed in August 1953.

In February 1950, meanwhile, Sen. Joseph R. McCarthy had fired his first ill-supported broadside against Communist infiltration of the Federal Government. In June came the Korean conflict, which pitted the United Nations and notably the United States, on the one hand, against North Korea and, eventually, Red China on the other. The war ended inconclusively in July 1953. Sen. McCarthy was censured by the Senate in December 1954. The passing of both episodes was greeted with relief in this country, but the tensions of the Cold War ran on unabated.

International tension was no small factor in the formation of the National Science Foundation, in 1950, to channel federal support into basic research in the U.S. The concept had grown out of the success of the wartime Office of Scientific Research and Development, and OSRD Director Vannevar Bush led the fight for NSF. The foundation's appropriations would grow from $225,000 in fiscal 1951 to $14 million in 1955 and to $153 million in 1960. This growth was paralleled by growth in research and development overall and in the federal share of the total. In the period 1953–60, the nation's spending on R&D grew from $5.13 billion to $13.55 billion and the federal share of the total from 54% to 65%. Expanding support brought steady progress across the full spectrum of science and technolgy. In addition, however, such support, and particularly its large federal component, sowed the seeds of future difficulties. Two of these—the loss of jobs caused by the sudden cessation of R&D programs, and the schism between academic and industrial research—would be of particular interest to the American Chemical Society.

The Diamond Jubilee

For the ACS itself, the single most important event of the 1950s was the 75th anniversary meeting, Sept. 3–7, 1951, in New York City. Planning for the Diamond Jubilee had started in June 1948; it culminated in the most extensive meeting program the Society had ever organized. Registration at the meeting totaled a record 13,466; the 20 ACS divisions arranged 275 technical sessions with 1622 papers. The event was followed immediately by the XVIth Conference of the International Union of Pure and Applied Chemistry. Together the two provided two weeks of continuous activity aptly called the World Chemical Conclave. The Cold War was seldom out of mind, however. President Truman, in his congratulatory letter to ACS President N. Howell Furman, noted that, "It is a striking tribute to our democracy that so great a number of scientists can assemble here free from suspicion of one another, and free from fear of outside interference. This kind of personal freedom is our most precious national asset. It needs to be carefully guarded and zealously watched, for freedom is in more serious danger today than at any time in our national history."

The Diamond Jubilee, the Society's 120th national meeting, opened Monday, Sept. 3, at the scene of the first ACS meeting, April 6, 1876, at the College of Pharmacy of the City of New York, a site occupied in 1951 by the main building of New York University. Retired Secretary Charles L. Parsons unveiled a commemorative plaque in a ceremony whose participants included President Furman, Executive Secretary Alden H. Emery, and Past President (1933) Arthur B. Lamb, for 32 years (1918–49) Editor of the *Journal of the American Chemical Society.*

At the general meeting, in Manhattan Center, President Furman among other ceremonial duties commended John H. Nair and his Diamond Jubilee committee for their work. Roger Adams, Past President (1935) and Board Chairman (1945–49), introduced the honorary co-chairmen of the meeting, Dr. Parsons and Marston T. Bogert, ACS President in 1907–08. The major speakers at the General Meeting were Dr. Furman, Priestley medalist E. J. Crane, then in his 36th year as Editor of *Chemical Abstracts,* and Alfred R. Driscoll, Governor of New Jersey. Crane, in still another echo of the Cold War, argued that excessive secrecy was impeding scientific progress. "The power to withhold scientific information," he contended, "should never be in the hands of those who do not understand it."

Postmaster General Jesse Donaldson had authorized a commemorative three-cent stamp to mark the Diamond Jubilee, and the first sale took place in a program on the steps of New York's main Post Office. Deputy

The ACS Diamond Jubilee commemorative stamp was introduced in a ceremony on the steps of New York City's main Post Office.

Mayor Charles Horowitz proclaimed Sept. 2–8 Chemistry Week in New York. The assembly was addressed by Assistant Postmaster General Osborn Pearson, New York City Postmaster Albert Goldman, and Sen. J. Allen Frear of Delaware (home of Atlas Powder Co., the Du Pont Co., and Hercules Powder Co.), who had been helpful in obtaining the issuance of the stamp.

The ceremonial session of the Diamond Jubilee meeting packed the historic 71st Regiment Armory on the afternoon of Sept. 5. Guests included 41 representatives of the technical societies of 38 nations and 77 representatives of domestic societies and associations. The major speaker was James B. Conant, president of Harvard University and a longtime member of the Society (he had been chairman of its Division of Organic Chemistry in 1931). Formal commemoration of the Diamond Jubilee concluded the evening of Sept. 5 at a banquet at the Waldorf Astoria. The ballroom could not accommodate the more than 2700 who attended; those seated in adjacent rooms followed the proceedings by closed-circuit television (the number of television receivers in this country did not pass 10 million until 1951). Guests at the banquet included 16 of the 20 living former Presidents of the Society. The soloist was Risë Stevens of the Metropolitan Opera. The speakers were Charles A. Thomas, Chairman of the ACS Board of Directors; C. E. Kenneth Mees, vice-president for research at Eastman Kodak Co.; and Alben W. Barkley, Vice-President of the United States.

A notable feature of the Diamond Jubilee meeting was the Younger Chemists' International Project, which grew out of fear that the then-prevalent dollar shortage would keep many foreign chemists at home. Appeals to industry secured support for the attendance of a number of chemists of established reputation, but not of younger chemists. Biochemist Erwin Brand of Columbia took the problem to the Economic Cooperation Administration. The upshot was a six-week program for

James B. Conant, president of Harvard University and a long-time ACS member, addressed the ceremonial session of the Society's Diamond Jubilee meeting on Sept. 5, 1951, at the 71st Regiment Armory in New York City.

younger chemists, funded by ECA and administered by ACS. It included attendance at the Diamond Jubilee and postmeeting tours of U.S. industry and institutions. ECA support was limited to Marshall Plan countries and Southeast Asia, but the Ford Foundation agreed to support 60 chemists from other countries outside the Iron Curtain. Participants in the Younger Chemists' International Program were to be not more than 40 years old (excepting those from Southeast Asia) and to be practicing chemists or chemical engineers and proficient in English. In all, 256 took part.

Outside New York City, more than 100 of the 139 ACS local sections arranged Diamond Jubilee events during 1951. Extensive assistance in these affairs was provided by the ACS News Service under Managing Editor James H. Stack, who had succeeded James T. Grady in 1948. The activities ranged from special displays by merchants and libraries to an open house at Mellon Institute in Pittsburgh (It drew 27,000 visitors.) to a demonstration of firefighting chemicals by the fire department in Schenectady, N.Y.

The Diamond Jubilee came early in another decade of rapid growth for the Society. In the years 1950–60, membership rose from 63,349 to 92,193 and annual expenditures from about $2.84 million to about $8.5 million. In 1955, expenditures at *Chemical Abstracts* exceeded $1 million for the first time, and the deficit over subscription income approached $500,000. In that year the ACS Board of Directors determined to put *CA* on a break-even basis and adjusted subscription prices accordingly. Three years later, in October 1958, E. J. Crane retired after 43 years as Editor of *CA*. He was succeeded by Dale B. Baker, who was appointed Director of the renamed Chemical Abstracts Service. In November 1959, Walter J. Murphy, Editorial Director of the ACS Applied Journals, died after 17 years of service. He was succeeded by Richard L. Kenyon, who had joined the Society's editorial staff in 1946.

In 1952, the ACS Corporation Associates was formed specifically to provide financial support for *Chemical Abstracts* and for the Society's fundamental journals. In later years the group began to support a wide range of ACS projects. It does not initiate projects of its own, but rather responds to requests for funds from other units of the Society. By 1975, almost 200 companies were members of the Corporation Associates. Their dues were expected to provide some $71,000 for various ACS activities in 1976. These activities include the ACS Award for Creative Invention; regional meetings; distribution of two publications to students; Project SEED, an educational and employment program for economically deprived students; and Operation Interface, which is designed to promote better relations between industry and academe.

Housing grew critical in the 1950s both at CAS and at ACS headquarters. In mid-decade, a three-story building, financed jointly by the Society and Ohio State University, was erected on the OSU campus to house *CA* operations. In 1962, the ACS purchased 50 acres next to the OSU campus; a four-story CAS building was completed on that site in 1965 and a second of almost equal size in 1973. Early in 1960, meanwhile, the ACS headquarters staff occupied the Society's new eight-story building in the District of Columbia. The building was on the site of the converted apartment house that had been the Society's home since 1941; the staff had used temporary quarters during construction.

Perhaps the most acrimonious debate of the 1950s within the ACS erupted when members learned that the Society's Committee on Admissions in 1953 had rejected the application of the French physicist Irene Joliot-Curie on the grounds that she was an "avowed Communist." Mme. Joliot-Curie was the daughter of Pierre and Marie Curie; in 1935, she and her husband Frederic had shared the Nobel Prize for Chemistry for

In the spring of 1958, John H. Nair (right), chairman of the planning committee for the ACS building fund campaign, accepted a $10,000 check from E. K. Stevens, manager of the Exposition of Chemical Industries. The campaign raised more than two thirds of the cost of the eight-story structure that the Society built at 1155 16th St., N.W.

In February 1960 the Society occupied its present headquarters building, then newly completed at a cost of about $3 million.

their discovery of artificially induced radioactivity. Her rejection triggered a flood of correspondence—pro and con—so great that by May 1954 the Editor of *Chemical & Engineering News* closed the Letters column to the subject. In the March 22 issue of *C&EN* the Board of Directors stated its support of the Committee on Admissions. The Board's statement noted first that the Society had never identified persons rejected for membership—thus placing the leak elsewhere. The statement held in part that "Support of and compliance with the objects of the American Chemical Society as specified in its Charter are in conflict with the tenets of the Communist Party as enforced by the discipline exercised over its members." This position, though many found it arguable, seemed fair enough in the context of the Cold War.

The Space Race

Since World War II, both East and West had been looking beyond the earth's atmosphere, at first in terms of ballistic missiles to deliver their bombs. Yet both sides cooperated in the International Geophysical Year (1953—1957–58), a symbol of peaceful exploration of space by all advanced nations. Both sides planned to launch earth satellites during IGY, and the Soviets said in 1955 that they would do so sometime in 1957. In the end, they launched two. Sputnik 1 went into orbit Oct. 4, 1957, and Sputnik 2, with the dog Laika aboard, on Nov. 11. This country's program, troubled partly by rivalry among Army, Navy, and Air Force, did not go so well. Vanguard, which collapsed in flames on the launch pad Dec. 6, 1957, was dubbed Kaputnik by the American press. Explorer 1 was launched successfully Jan. 31, 1958, but offered small consolation. Its orbital weight was only 30.8 lb, compared with 184 for Sputnik 1 and 1120 for Sputnik 2. The Soviets clearly were ahead and especially so in propulsion, a critical element in the missile race.

Congress reacted strongly to the Soviet triumph. In 1958 it passed the National Defense Education Act, which provided scholarships; loans and grants to improve teaching in science, mathematics, and foreign languages; and other aids to education. Congress also established the National Aeronautics and Space Administration, which absorbed the existing National Advisory Committee for Aeronautics and began functioning Oct. 1, 1958. NASA's budget, $494 million in the first full year, rose to $1.35 billion in 1960 and $5 billion by the mid-1960s. The NASA effort was enormous. For Project Mercury (1958–63), for example, the agency mobilized a dozen prime contractors, 75 major subcontractors, and about 7000 sub-subcontractors, all of which required the employment of two million people. These exertions produced results. John H. Glenn, Jr., orbited the earth in a Mercury spacecraft in February 1962, 10 months after the Russian Yuri Gagarin had broken the ice in Vostok 1. In May 1961, President John F. Kennedy committed the nation to landing a man on the moon before the end of the decade; Neil Armstrong made the trip with two colleagues in Apollo 11 and set foot on the lunar surface July 20, 1969.

The White House, too, reacted strongly to the Sputniks. In November 1957, in a speech on "Science in National Security," President Dwight D. Eisenhower announced the appointment of James R. Killian as his special assistant for science and technology. Dr. Killian's main job, the President said, would be to speed up the missile program, but he was also to spur basic research and scientific education. Killian would be succeeded by four more science advisers, including chemists George Kistiakowsky (1959–60) and Donald R. Hornig (1964–69), before President Richard Nixon did away with the White House science apparatus (by then the Office of Science and Technology) in mid-1973 and transferred most of its functions to the director of the National Science Foundation.

The glamour of the space program and the federal emphasis on education in the sciences and engineering were not lost on young people. The number of bachelor's degrees granted in chemistry climbed steadily, from 5500 in 1954 to some 12,000 annually by 1975. The number of Ph.D.'s granted in chemistry held level at about 1000 annually during the 1950s, but then rose sharply, to a peak of 2200 in 1969–70. New Ph.D.'s in chemistry totaled about 10,000 during 1950–60, but climbed to 15,000 in 1960–70. The story was the same in the other physical sciences and engineering. In 1954, some 237,000 scientists and engineers were employed in research and development in this country; the number rose to 494,000 by 1965 and to 550,000 by 1970. The flow of research results expanded accordingly. The number of abstracts published annually by *Chemical Abstracts* reached 100,000 in 1957 and passed 200,000 in 1966 and 300,000 in 1971.

Many new chemists and chemical engineers in the two decades after World War II were absorbed by the chemical industry, whose long-term growth was spurred further by the high level of scientific and technological activity in the country. During 1925–50, the industry's physical output grew at an average annual rate of 10%, compared with 3% for all industry. Despite recessions in 1953–54 and 1957–58, the industry's sales rose from about $13 billion in 1947 to $27 billion in 1960. Synthetic organic chemicals set the pace. The industry's annual output of organic chemical intermediates in 1947–60 climbed 370% to 9.6 billion pounds; medicinal chemicals, 230% to 114 million pounds; plastics materials, 490% to 6.1 billion pounds; surface active agents, led by household synthetic detergents, 530% to 1.5 billion pounds; and pesticides and other organic agricultural chemicals, 520% to 648 million pounds.

The rapid growth of the chemical and other industries after about 1950 figured largely in the emergence of a strong, new trend, the environmental movement. "Chemicals," typified by synthetic detergents and pesticides, were early targets; chemistry, especially analytical chemistry, supplied much of the knowledge needed to put environmental research and control on a sound basis. The movement gained impetus in 1962 with the publication of Rachel Carson's "Silent Spring," whose target was pesticides; it fattened in the atmosphere of mistrust and activism fueled during the 1960s by the war in Vietnam.

In 1955 Congress passed a water pollution control act, the first of a series of measures that by 1975 had put the Federal Government, as opposed to the states or smaller political units, directly in charge of the nation's environmental effort. Beginning in the same period, Congress amended the Food, Drug, and Cosmetic Act of 1938 to stiffen controls on specific substances: pesticide residues—the Miller amendment of 1954; food additives—the Food Additives Amendment of 1958, which included the Delaney clause on carcinogens; prescription drugs—the Kefauver-Harris amendments of 1962. On April 22, 1970, environmental demonstrations marked Earth Day around the nation. In mid-1970 President Nixon proposed a reorganization of the Government's pollution control activities and the formation of the Environmental Protection Agency, which began operating in December of that year.

The Society After 1960

This country's efforts in space and the environment, and the general surge in technical activity since World War II, have shaped—and been shaped by—chemistry and the work of the American Chemical Society in ways that emerged clearly during the 1960s. The forces at play have been interrelated and marked, not surprisingly, by high economic content. Well before 1960, injections of federal money into the universities were estranging academic and industrial chemistry. During the 1960s, public awareness of the costs and benefits of competing, tax-supported programs, plus the sheer complexity of the nation's problems, made a major issue of federal policy on science and technology. Late in the decade, the incipient mismatch between supply and demand in chemically trained manpower became a fact. And elements of each of these forces combined to break the close, prewar ties between the Society and the chemical industry, although the two inevitably retained many interests in common.

The ACS itself continued to grow during most of the 1960s. Annual expenditures topped $15 million in 1965; the 100,000th member joined in that year, and membership on Dec. 31 was 102,525. In mid-1965, Alden

Alden H. Emery *(1901–1975) served the Society for more than 29 years, more than 19 of them as Secretary and Executive Secretary.*

H. Emery ended his more than 19 years as Secretary and Executive Secretary, but remained as Honorary Secretary until mid-1966, when he retired; he was succeeded as Executive Secretary by B. R. Stanerson, who had been Deputy Secretary since 1960.

In January 1961, the ACS News Service released the first program in its "Men and Molecules" series, a 15-minute science documentary produced weekly and taped for distribution to radio stations at no charge. "Men and Molecules" was designed to feature noted scientists talking about their work in terms that the layman could understand. Its main goal was to convey the value of research to the public at large. The series was actually the third to be produced by the News Service for radio. The first was "Headlines in Chemistry," which started in the fall of 1947. This 15-minute show featured current developments in research and often an interview with a local scientist. It was succeeded by a second 15-minute program, "Objective," in which professional actors dramatized the stories of famous chemists. "Objective" was succeeded in turn by "Men and Molecules," whose name was changed in 1974 to "Man and Molecules." By the end of 1975, 785 programs in the series had been completed. It was being carried by some 525 radio stations in all 50 states and at a few points abroad, and by the Voice of America and Radio Free Europe. The program was available on tape cassettes also and was going to 199 subscribers by the end of 1975.

By the mid-1960s, the Society was taking a particular interest in the divergent paths of academic and industrial chemistry. The "academic–industrial interface" had become an oft-heard expression. Students of chemistry, particularly graduate students, were being trained to do highly specialized basic research. Their goals included careers in research and peer recognition of a kind not readily achieved in industry, where publication is limited by the secrecy induced by competitive pressures. Companies, on the other hand, though the major employers of new Ph.D. chemists, found them often to be overspecialized, inflexible, and ill-prepared for the team research and economic context of industrial R&D. Still, companies could push their views only by persuasion, since support of research in the schools had shifted largely from private to federal funds.

The Society's reaction to the academic–industrial problem included the ACS Award for Creative Invention, established by the Board of Directors and first awarded in 1968 (the Society's Corporation Associates assumed the sponsorship in 1975). The canvassing committee for the award, in a letter to ACS local section officials in late 1966, noted in part, "Here at last is an opportunity to recognize and honor the chemist or chemical engineer who has applied his scientific knowledge and talents to creating and bringing into being practical applications for better living." A second ACS attack on the problem was a series of articles, "The Chemical

Innovators," published in *Chemical & Engineering News* beginning in January 1970. The series focused on the achievements of innovative chemists and chemical engineers. The academic–industrial interface remained a persistent if understandable problem, however. As recently as October 1975 the Society's 9th Biennial Education Conference devoted much of its program to the design of curricula oriented toward industrial employment.

In April 1964, the Society granted "up to $50,000" to support the work of the Committee for the Survey of Chemistry appointed earlier that year by the National Academy of Sciences. The committee, chaired by Frank H. Westheimer of Harvard, received additional support from the National Science Foundation. The survey committee's charge was to assess the contention that, despite the sharp rise in spending on research and development that had begun in the mid-1950s, opportunities in basic research in chemistry were being missed because of lack of funding. The Westheimer report, "Chemistry: Opportunities and Needs," was published in 1965. It pointed out in part that modern instrumentation was "opening vast new areas of research in chemistry." Advances in theory and instrumentation, for example, had created what Nobel laureate (1965) R. B. Woodward had called a "second great revolution" in organic chemistry, a revolution that permitted the planned syntheses of complicated compounds. In short, the committee agreed with the basic contention and recommended increased funding, both public and private, for basic research in chemistry.

The ACS grant for the Westheimer study was a sign of the Society's growing belief that it ought to speak more openly on issues of public policy in science and technology. For many years the ACS had provided objective expertise to the Government, but as a rule only when asked to do so. This passive approach stemmed mainly from the wish to avoid the fact or appearance of lobbying, which would jeopardize the Society's favored tax status as a scientific and educational organization. Increasingly, however, the Government was making the major decisions on science policy, decisions not only on funding but in many other areas. The ACS concluded finally that it would serve itself and the nation better by a more aggressive approach short of lobbying. The result was the Board Committee on Chemistry and Public Affairs, formed in 1965 and made a Board–Council committee in 1968. The Committee on Chemistry and Public Affairs and its staff regularly have developed testimony and statements for Congress and other arms of the Federal Government. Among other activities the committee has guided the development of books on environmental chemistry, on the function of chemistry in the nation's economy, and on chemistry in medicine.

Federal spending on research and development was in decline when David Perlman (right) moderated a discussion on "Science, Politics, and Money" during the ACS national meeting in San Francisco in April 1968. Participants were George B. Kistiakowsky (left), a former Presidential science adviser; ACS President Robert W. Cairns; and Glenn T. Seaborg, then chairman of the Atomic Energy Commission. The program was one of five arranged during the meeting by the ACS News Service and educational television station KQED.

The late 1960s saw the start of a bad period for the Society and for scientists and engineers in general. In 1968, the Federal Government began to cut back its support of a range of research and development programs in government, in industry, and in academe. At the same time the nation's economy began to suffer from a combination of stagnation and inflation and, finally, from the energy crisis imposed on the world by the oil-producing countries of the Middle East. In 1969, scientists and engineers began to lose their jobs; in the period 1969–71, some 1500 industrial chemists and chemical engineers were affected. For several years new graduates found it difficult even to arrange interviews with employers, let alone find jobs. In the first half of the 1970s, the competition for jobs was so stiff that even academic scientists, with or without tenure, feared for their futures.

The nation's spending on research and development continued to climb during 1967–75, at an average annual rate of 5% in current dollars. But in constant 1967 dollars, spending declined about 1% per year during 1967–75; federal spending, which declined about 3% per year, was partly counterbalanced by nonfederal spending, which increased about 1.8% per year.

ACS operations reflected the economic upset quickly. Employment Clearing House records for the two national meetings in 1967 showed a total of 875 registered job-seekers and 1361 employers. But in 1968, job seekers exceeded employers by 1233 to 1068 and in 1969 by 1446 to 844. ACS membership declined in 1969, for the first time since the depression years of the 1930s, under the combined pressures of unemployment and an increase in dues, from $16 to $25, that took effect Jan. 1, 1970. (Dues had gone from $9.00 to $12 in 1950 and to $16 in 1961.) Membership fell from a high of 116,816 in 1969 to 110,285 in 1973. The Society was not alone in this regard—membership in the American Association for the Advancement of Science dropped from 131,000 to 120,000 and in the American Institute of Aeronautics and Astronautics from 39,000 to 22,000.

These unhappy statistics, and the need for stringent economies, made a bleak beginning for Frederick T. Wall, who took office in October 1969 as Executive Director of the Society, a new post created in a reorganization of full-time staff. B. R. Stanerson, Executive Secretary since 1965, remained in that post until October 1970, when he retired (in 1971, Dr. Stanerson was elected to a three-year term as a Director of the ACS and in 1974 won a second term).

Confronted by unemployment and economic uncertainty, some ACS members, particularly the younger ones, turned quickly toward professionalism, as defined by an interest in the individual chemist rather than

***B. R. Stanerson,** Executive Secretary of the Society from 1965 until his retirement, in 1970, won successive terms as a Director in 1971 and 1974.*

the science of chemistry. A clamor arose for political and social action, for the Society to "do something." Resources for such activities typically are meager, even in a society as large as the ACS. The Society's budget for 1975, for example, was about $35.4 million, but this included large items such as journal subscription income and expense, which tend to be in balance; primary liquid funds available to support programs directly related to membership affairs consist essentially of dues income, which in 1975 amounted to only about $2.8 million. (Dues rose to $28 in 1974 and to $29 in 1975, but the average per member was less because of student and other types of discounts.) Forecast distribution of dues revenue per member to dues-supported programs in 1975 appears in Table III.

Despite its limited resources for the purpose, the Society responded to the employment crisis more strongly than its sister organizations, if not as strongly and not as quickly as some members may have wished. Unemployed members were allowed to defer payment of dues and subscriptions; they could place weekly free employment ads in *Chemical & Engineering News,* enroll free for ACS short courses and audio courses, and use the National Employment Clearing House without paying registration fees at national meetings. A full-time hot line was set up in the headquarters building in Washington, D.C., to counsel members on finding employment. In addition, significant programs of member assistance were initiated by the Council Committee on Professional Relations.

The swing toward professionalism also was reflected in electoral politics within the Society. The most notable manifestation was a series of five petition candidacies for President-Elect, in 1970–74. Two of the three candidates involved were nominated twice by petition, and all three ran on professionalism platforms. Two of the three candidates also won office. Dr. Alan C. Nixon lost in 1970, but was elected in 1971. Dr. Bernard S. Friedman won on his first try, in 1972.

By 1975, although economic conditions were improving for the nation and for scientists and engineers, the new professionalism seemed certain to remain a stronger force within the Society than in earlier years. In 1972, for example, the Council approved the formation of a Division of Professional Relations, after narrowly defeating the move in 1971. The division came into being in 1973 with a membership of 400.

The Society's professional activities were recognized in the major recommendation of the study of the organization conducted by Arthur D. Little, Inc., during 1974. The Boston consulting firm recommended that the ACS should split its governance structure into two parts—one devoted to membership affairs and the other to scientific affairs. At the same time, the report emphasized, members' most widely held feeling

TABLE III

Forecast Distribution of Dues Revenue Per Member to ACS Dues-Supported Programs, 1975

Dues Income/member	$29.00
Expenses/member	
C&EN allocation	5.00
Allotment to local sections	4.23
Awards	.62
Membership records servicing	3.02
Public and member relations	2.25
Press releases	
Radio/TV programs	
Reports/communications	
Exhibits	
National meetings abstracts	
Professional relations and manpower	2.10
Salary/employment surveys	
Supply/demand studies	
Professional handbooks	
Employee/employer guidelines	
Member assistance	
Membership, local sections, divisions, and regions	5.14
Promotion/admissions	
Employment aids	
Speakers tour service	
Regional meeting assistance	
Public affairs	2.81
Liaison with government	
Information on legislation	
Special studies, projects, and symposia	
Educational activities	3.83
Conferences	
Student Affiliates	
Committee on Professional Training	
Career literature	
Continuing education	
Total expenses/member	*29.00*

by far was that, whatever changes were made in the Society, its scientific reputation and the quality and integrity of its journals must be protected. The nine-month, $100,000 study was completed early in 1975. Soon after, ACS Executive Director Robert W. Cairns forecast that the Board and Council, with the A. D. Little report as background, would "evolve appropriate changes in the governance of ACS" during 1975–76. Dr. Cairns, a former ACS President (1968) and Board Chairman (1972) had succeeded Dr. Wall as Executive Director in December 1972.

After 100 Years

To the founders of the American Chemical Society in 1876, an accurate forecast of the chemical world of 1976 might well have seemed the purest science fiction. Gregor Mendel, who founded the science of genetics by studying crossbred garden peas, died only in 1884; by 1976, the molecular basis of the genetic code had been worked out. In 1876, guano and other organic wastes were the only nitrogen fertilizers; 100 years later, nitrogen from the air is fixed in chemical plants that synthesize 1500 tons and more of fertilizer ammonia daily. In the decade 1881–90, American schools granted 136 Ph.D.'s in chemistry; in 1974–75, they granted more than 1700. In 1906, Congress passed the first food and drug act, which was aimed mainly at adulteration; in 1976, a skein of federal regulations deals not only with simple deception but with accidental, unintended, and unexpected effects of a wide range of the products of modern science and technology.

Basic chemical science has made immense progress since 1876. By that year, chemists had discovered only 63 elements; by 1975, they had discovered an additional 29 and synthesized 14 more (through element 106). The nature of biological and synthetic polymers, whose existence was hardly recognized only 50 years ago, has been elucidated in extensive detail. The nature of chemical bonding has been clarified greatly. Biochemical research has illuminated the metabolism and function of fats, proteins, carbohydrates, and trace nutrients; such research has benefited from the use of radioactive and, later, nonradioactive isotopes as tracers. Complicated molecules like chlorophyll and vitamin B_{12} have been synthesized. Paralleling and supporting such developments have been great advances in theory and instrumentation and in the electronic digital computers used widely for both theoretical calculations and data processing.

Progress in chemical science has been reflected in the growth and products of the chemical and other industries. Sales of the chemicals and allied products industries in the United States in 1974 totaled $86.8 billion. Chemical technology is critical also in other sectors of the group loosely defined as the "chemical process industries," which include petroleum and coal products; rubber and miscellaneous plastics products; primary metals; stone, clay, and glass; and paper and allied products. Chemical scientists and technologists are employed as well in a variety of other industries, sometimes in surprisingly large numbers; Bell Laboratories employed well over 300 of them in 1973, IBM Corp. some 2400, General Electric Co. at least as many.

Innovation in industry, whether chemical or otherwise, during the past 100 years has become a highly organized enterprise based on mission-

TABLE IV

A Profile of Chemists in the United States[a]

Gender	
Male	89.1%
Female	10.9%
Age Distribution	
Under 25	2.3%
24 to 34	33.7%
35 to 44	27.6%
45 and over	36.1%
Highest Degree	
Ph.D.	33.7%
(35.7% of the men and 17.9% of the women held a Ph.D.)	
M.S.	16.6%
B.S.	46.6%
Less than Bachelor's	2.8%
Other	0.2%
Field of Highest Degree	
Chemistry	76.8%
Engineering and Related Fields	3.2%
Agricultural & Biological Sciences and Related Fields	12.6%
Health, Business, Education	2.8%
Other	3.4%
Not Reported	1.2%

[a] Covers 104,413 chemists and biochemists as defined by the National Science Foundation from the 1972 Postcensal Survey. Source: "The 1972 Scientist and Engineer Population Redefined," Vol. I, NSF 75–313, National Science Foundation, Washington, D. C., 1975.

oriented research and development by interdisciplinary teams of specialists. This enterprise has created thousands of products, some no doubt trivial in the long view, others immensely useful. A brief selection of those marked by high chemical input would include:

- Antibiotics and other prescription pharmaceuticals
- Catalysts for numerous industrially important reactions
- Films for color photography
- Fibers for easy-care clothing and marine line and cable
- Oral contraceptives
- Phosphors for color TV
- Pest-control agents of many kinds
- Plastics and resins for numerous consumer and industrial uses
- Single-cell protein from petroleum

TABLE V

A Profile of Chemical Engineers in the United States[a]

Gender	
Male	99.2%
Female	0.8%
Age Distribution	
Under 25	4.0%
24 to 34	34.0%
35 to 44	22.0%
45 and over	40.0%
Highest Degree	
Ph.D.	12.7%
(12.8% of the men and 0.0% of the women held a Ph.D.)	
M.S.	21.2%
B.S.	61.8%
Less than Bachelor's	3.8%
Other	0.4%
Field of Highest Degree	
Chemical Engineering	86.0%
Chemistry	7.0%
Agricultural & Biological Sciences and Related Fields	1.0%
Health, Business, Education	3.0%
Other	2.0%
Not Reported	1.0%

[a] Covers 44,113 chemical engineers as defined by the National Science Foundation from the 1972 Postcensal Survey. Source: "The 1972 Scientist and Engineer Population Redefined," Vol. I, NSF 75–313, National Science Foundation, Washington, D. C., 1975.

- Synthetic elastomers, including synthetic "natural rubber"
- Synthetic household detergents
- Synthetic industrial diamonds
- Transistor-grade silicon and other semiconductor materials
- Water-base paints, durable auto finishes, and many other coatings

The science that has worked so well for industry and government has worked equally well for the critics of science. Scientists themselves have detected the minute amounts of organochlorine pesticides in the environment, the adverse effects of oral contraceptives, the radiation hazard of television sets. Some such problems, though solvable, have come as a surprise; others have been expected but accepted because the benefits involved far outweighed the costs. Unanticipated pitfalls can be unsettling, however, and the cost-benefit concept can be difficult to defend. Further concerns flow from corporate mishandling and consumer misuse of troublesome products and practices. The repute of science suffers also when noted experts take positive but directly opposed stands on the same issue—a putative carcinogen, nuclear power, or the SST. Compounding it all is a pervasive communications network in a nation whose population grew some 40%, to 215 million, in the period 1950–75. In such an atmosphere, the public view of science is bound to change. It is not clear that many people have become "antiscience," as is sometimes claimed; but there can be little doubt in 1976 that many are far more skeptical of the works of scientists and engineers than ever in the past.

Skepticism of science played a role, if a difficult one to assess, in the troubles that overtook chemical education and employment in the last five years of the American Chemical Society's first century. After Sputnik, in 1957, the number of schools granting the Ph.D. in chemistry more than doubled, spurred mainly by federal support. Then, in about 1968, began the downturn (in constant dollars) in the nation's funding of research and development and the recession-aided weakening of the chemical job market. These events left many graduate programs overextended. By 1975, academic openings in chemistry seemed likely to be relatively scarce for some years; chemical employment in general had improved since 1970, but new graduates at all degree levels were in markedly lower demand than they had been until the late 1960s.

The effect of federal programs on chemical education and employment typifies the growth of the past century in the scientific role of the Government, increasingly the source of decisions—executive, legislative, regulatory—with great immediate impact on science and technology. The growth in the federal role was perhaps inevitable. The accelerating utility of chemistry, for example, has made the science both a vital tool and a major regulatory target for federal agencies. The uses of science in national policy led to the creation of the presidential scientific advisory apparatus. Although the function was relegated to the National Science Foundation by President Nixon, by 1975 it was evidently on its way back to the White House of President Gerald Ford. The pervasiveness of science-intensive issues also led Congress, in 1972, to create its own Office of Technology Assessment.

The rise of the federal structure for science necessarily injected politics into normally scientific deliberations. It has prompted scientists and engineers and their societies to advance their views to government with unaccustomed force and frequency. On the other hand, the nation faces problems—energy, the environment, materials, health—whose solutions call not only for strong scientific input but for the kind of federal impetus that can be generated only with substantial political support for the chosen courses of action.

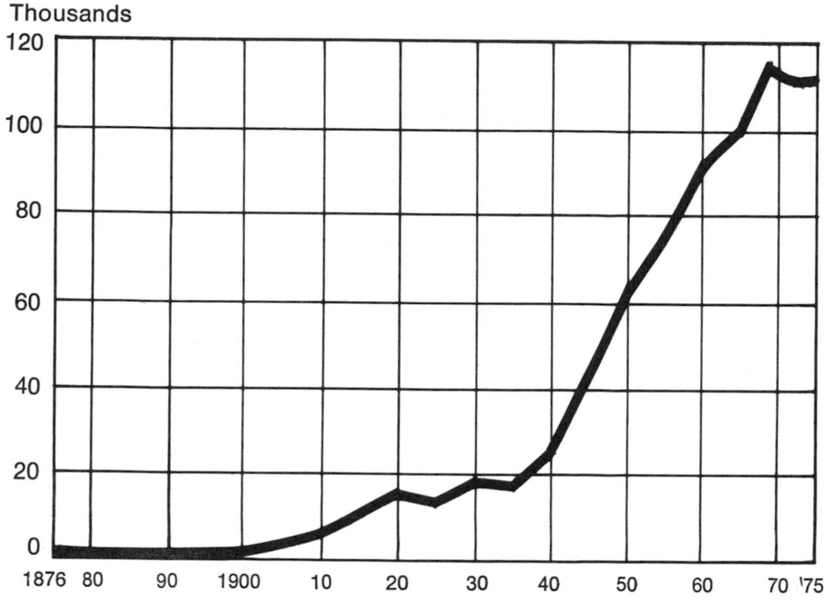

Figure 2. ACS Membership (1876–1975)

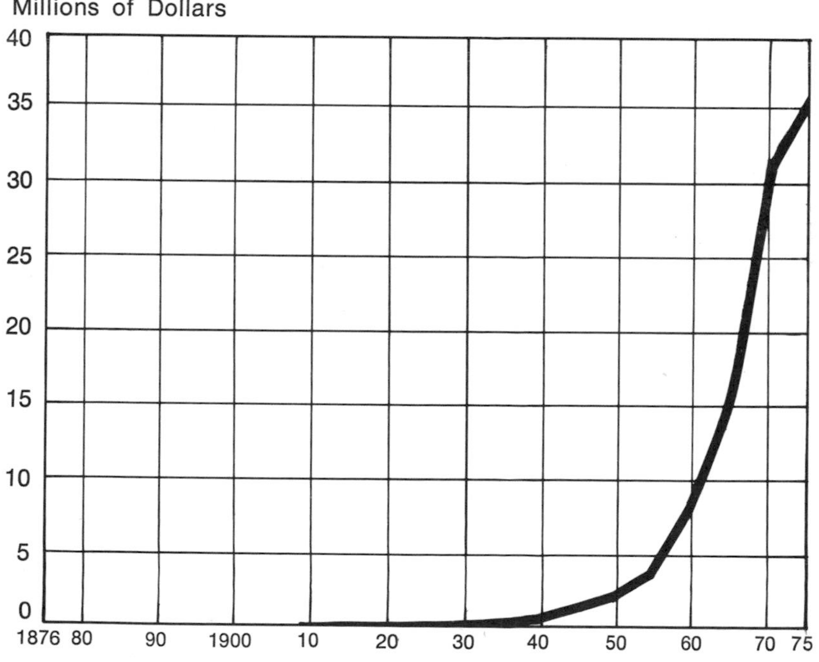

Figure 3. ACS Expenditures (1876–1975)

However the United States elects to cope with events at the beginning of its third century, scientific and technological manpower should be among the least of concerns. In 1974, employed scientists and engineers in this country numbered an estimated 1.6 million; some 528,000, or one third of these, were working in research and development. About 200,000 chemical professionals are employed in the U.S.; census and other data indicate totals in 1970 of about 125,000 chemists, 52,000 chemical engineers, and 15,000 biochemists. About half of these chemical professionals are members of the American Chemical Society. The nation's chemists are distributed broadly: about 60% of them work in industry, but only about 36% in the chemical process industries; about 23% are in colleges and universities, 11% in all levels of government, and 6% in nonprofit institutions and elsewhere. About half of all chemists work in research and development. Futher details on the nation's chemists and chemical engineers appear in Tables IV and V (*see* pp. 54, 51).

During 1975, the ACS was emerging from perhaps the most trying period of its first 100 years. The decade 1966–75 closed with a slight increase in membership, to 110,820, after a steady decline from the record 116,816 of 1969 (Figures 2 and 3). The Society's divisions numbered 28 in 1975, and its local sections numbered 175. Subscriptions to ACS journals and magazines in 1975 totaled 315,837, including 121,458 for the weekly *Chemical & Engineering News.* Stringent economies had kept the Society financially sound. Expenditures in 1974, at $33.8 million, slightly exceeded revenues; a deficit of about $700,000 in 1971 had been offset by comfortable margins in both 1972 and 1973. Expenditures in 1975 were estimated at year's end to be about $35.2 million, slightly under the budgeted amount.

No one can predict the future. But the American Chemical Society approached its Centennial, on April 6, 1976, in a position of strength to serve chemists and chemistry worldwide and thereby to continue to advance the uses of chemical science and technology in filling the needs of mankind.

CHAPTER II

CHEMICAL EDUCATION

AT the time of the founding of the American Chemical Society, a century ago, chemical education and, in fact, higher education as a whole bore little resemblance to the complex structure we know today. In the intervening years the two have grown together along paths involving increasing interaction and mutual benefit.

Training in the physical sciences had grown only slightly in the 100 years leading up to 1876. It is true that the settlers of what was to become the United States had found time to establish schools and colleges as they struggled to clear the wilderness, and that chemically-based factories—glass, naval stores, the smelting of iron and other metals—were early introduced (in fact, their development and British attempts to control them contributed substantially to the friction preceding the Revolution). It is true also that by the end of the 18th century the new nation had a recognizable scientific establishment that included at least one

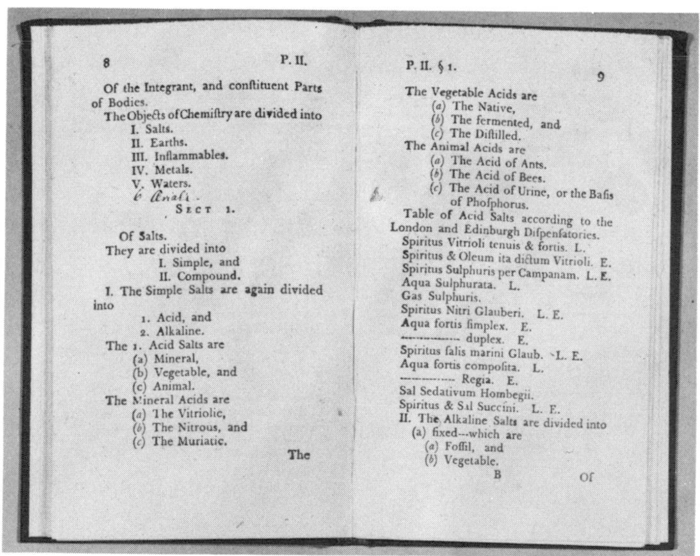

8 P. II.

Of the Integrant, and conſtituent Parts of Bodies.

The Objects of Chemiſtry are divided into
I. Salts.
II. Earths.
III. Inflammables.
IV. Metals.
V. Waters.

SECT 1.

Of Salts.

They are divided into
I. Simple, and
II. Compound.

I. The Simple Salts are again divided into
1. Acid, and
2. Alkaline.

The 1. Acid Salts are
(a) Mineral,
(b) Vegetable, and
(c) Animal.

The Mineral Acids are
(a) The Vitriolic,
(b) The Nitrous, and
(c) The Muriatic.

The

P. II. § 1. 9

The Vegetable Acids are
(a) The Native,
(b) The fermented, and
(c) The Diſtilled.

The Animal Acids are
(a) The Acid of Ants.
(b) The Acid of Bees.
(c) The Acid of Urine, or the Baſis of Phoſphorus.

Table of Acid Salts according to the London and Edinburgh Diſpenſatories.

Spiritus Vitrioli tenuis & fortis. L.
Spiritus & Oleum ita dictum Vitrioli. E.
Spiritus Sulphuris per Campanam. L. E.
Aqua Sulphurata. L.
Gas Sulphuris.
Spiritus Nitri Glauberi. L. E.
Aqua fortis ſimplex. E.
------------ duplex. E.
Spiritus ſalis marini Glaub. L. E.
Aqua fortis compoſita. L.
------------ Regia. E.
Sal Sedativum Hombegii.
Spiritus & Sal Succini. L. E.

II. The Alkaline Salts are divided into
(a) fixed---which are
(a) Foſſil, and
(b) Vegetable.

B Of

The pages are from a syllabus of lectures published as early as 1770 by Benjamin Rush, professor of chemistry at the College of Philadelphia.

figure, Benjamin Franklin, of international repute and that supported Joseph Priestley when he came to this country because of his difficulties in England. Scientific societies had been founded in Philadelphia and Boston, and some instruction in natural philosophy, including chemistry (as it then existed), was given in a few colleges, notably Kings College (renamed Columbia College after independence), the College of William and Mary, and the College of Philadelphia, where Benjamin Rush held the post of professor of chemistry and published a syllabus of his lectures as early as 1770. But despite these beginnings, progress in chemical education during the first century of our independence was slow, and the whole venture was on a very modest scale.

Some instruction in chemistry appeared in the curricula of the established colleges—Dr. John MacLean was appointed professor of chemistry at Princeton in 1795, to be followed by similar appointments at Yale (1803), Bowdoin (1805), Brown (1811), Dartmouth (1820), and at a number of other institutions. Education was erratically supported, however; small colleges waxed and waned, and the emphasis appears to have been largely on providing a smattering of chemistry for medical students (a recurring theme) and training in the analytical methods needed in industry, mining, and mineralogy. Beyond this, the dominant philosophy of higher education, following largely the British model, was not always sympathetic to a laboratory science. Admission requirements commonly were Latin, Greek, mathematics, and English but rarely science. Ira Remsen, when he went to Williams in 1872 as professor of chemistry and physics and requested laboratory facilities (then nonexistent), was told, "the students who come here are not to be trained as chemists. . . . they are to be taught the great fundamental truths of all sciences. The object aimed at is culture, not practical knowledge."

Finally, graduate instruction as we know it was virtually nonexistent in this country in 1876. Yale granted its first earned Ph.D. in 1861 and its first in science to J. Willard Gibbs in 1863; Harvard granted its first Ph.D. in chemistry to Frank Austin Gooch, inventor of the Gooch crucible, in 1877. As a consequence, chemists who desired academic appointments were forced to obtain their advanced training in Europe, a not altogether unfortunate situation, for they brought back with them the most advanced ideas of the time.

Typical of the problems of a young man of the period who wished to enter the profession were those of Charles Frederick Chandler, later an eminent chemist and one of the founders of the American Chemical Society, as well as its President in 1881 and 1889. Chandler's early interest in science apparently was inspired further by an able high school teacher. He entered Harvard in 1853, but in his first year became disappointed with the poor quality of laboratory instruction and the strong prejudice against science students that he found there. Accordingly, he set sail for Europe, where he studied under Wöhler and Rose, met a number of the most distinguished European scientists, and received his Ph.D. in 1856 at the age of 19. On his return to the United States he applied for an assistantship at Union College, but found on his arrival there that the only position available was that of janitor at a salary of $400 per year, although he would be permitted to carry out the duties of assistant as well. Fortunately for Chandler, the professor of chemistry, Charles Joy, resigned shortly thereafter to take another post, and, on reaching the age of 21, Chandler was appointed to fill his position. The action probably reflects both the new professor's ability and the limited knowledge required for such a position at the time.

This, then, was the state of chemical education at the time of the founding of the American Chemical Society. In fact, the entire educational establishment was very small. A. F. Nightingale, in "A Handbook of Requirements for Admission to the Colleges of the United States," published in 1879, reported that the total enrollment of the 20 leading men's colleges, including professional schools, was only 7696. And colleges admitting women were few and far between.

In contrast to these faltering beginnings, chemical education experienced consolidation, reorganization, and very substantial growth during the American Chemical Society's first half century, 1876–1926. The period saw enormous expansion in secondary education, a prerequisite for growth at the college or professional level. The number of public high schools in the country grew from fewer than 100 in 1860 to more than 6000 in 1900. At the same time, high school curricula and college entrance requirements were transformed from their "classical" orientation to something much more like those we know today. Landmarks in this change were the Eliot report of 1894 and the Nightingale report of 1895, both done for the National Education Association. (Charles W. Eliot, president of Harvard 1869–1909, is sometimes described as a chemist. It is true that he taught chemistry at one point early in his career, but he seems to have had little formal training in the subject, a situation further illustrating the minor status of science at the time.) The Nightingale report—after A. F. Nightingale, chairman of the committee that prepared it—is noteworthy for its modern tenor and outlook. In discussing high school chemistry curricula, it stressed that theories and principles should be presented inductively, that classroom work should be supplemented by adequate individual laboratory work, and that chemistry teachers should have had what then must have been a full college major in chemistry, including physics and general, analytical, theoretical, and organic chemistry, an ideal still not always attained. By 1900, as a

Central High School in Washington, D.C. was offering a course in chemistry in 1899.

consequence of these changes, chemistry had become a part of the curriculum of most good high schools.

The graduates of the enlarged high school enterprise fed a growing college and university system. The Morrill Act of 1862, the origin of our land-grant colleges and universities, was further strengthened in 1890; it began to take effect, and the majority of our state university systems either were founded or discovered new vitality during the period. The research-oriented graduate school so familiar today came into being during this period as well and enjoyed rapid growth. In 1881–90, 136 Ph.D.'s in chemistry were granted in the U.S., and the number roughly doubled every decade, reaching 2323 in 1921–30.

The role of the American Chemical Society in these developments, although largely indirect, seems to have been substantial. In the Society's early days the majority of its members belonged to the teaching profession; through its publications and meetings the ACS provided forums for the exchange of ideas and the dissemination of new chemical knowledge. The records of early ACS meetings show that, besides the usual research papers, increasing attention was being given to chemical education and the teaching of chemistry. At the general meeting in 1890, A. A. Breneman discussed "Some Lecture Experiments"; at the national meeting in Chicago in 1895, there was a symposium on Didactic Chemistry, chaired by Dr. W. E. Stone of Purdue, which included topics such as "The Relation of Teaching to Chemistry" and "How Chemistry Is Best Taught."

Problems that still trouble us had already been identified—at the same symposium, George Lunge of Zurich, Switzerland spoke on "The Education of Industrial Chemists" and Henry Pemberton, Jr., on "The Teaching of Industrial Chemistry in Colleges," while R. W. Jones countered with "Laboratory Work Must be Subordinate and Auxiliary to the Presentation of Facts, Laws, and Theories by the Teacher."

By 1909, interest in chemical education within the Society had reached the point that the Council voted to arrange for an Educational Section at ACS meetings, but this section met regularly for only a few years. In 1924, mounting enthusiasm and interest led to its transformation into the Division of Chemical Education, and the following year the division's *Journal of Chemical Education* appeared. Both division and journal would prosper and exert an increasing effect during the subsequent half century and more (*see* Chapter X); the journal by 1975 enjoyed an international reputation as a "living textbook of chemistry."

As the first 50 years of the ACS came to an end in 1926, chemical education had become well established. Its growth at the undergraduate level during the period, in both chemistry and chemical engineering, had mirrored rather closely the expansion of the colleges, but its quality had improved significantly. Graduate departments, while still limited to a relatively small number of leading schools, also were well established; some had built excellent international reputations, so that it was no longer necessary for an academic aspirant to consider foreign study.

Laboratory instruction was part of the chemistry curriculum at Central High School in 1899.

Behind both undergraduate and graduate training, and providing a market for their products, as well as modest support in the form of grants and fellowships, was a developing chemical industry greatly spurred by the need for self-sufficiency and increased production that had been brought on by World War I.

Curricula at the better schools by 1926 had largely standardized on a sequence of general, analytical, organic, and physical chemistry (interestingly, the historical order in which the topics had developed). Compared with the present, much was known about *what* occurred in chemistry, but much less about *why* it occurred. The study of reaction mechanisms was in its infancy; quantum mechanics extended little beyond the Bohr atom and the electron-pair bond. As a result, descriptive chemistry, analytical procedures, and empirical manipulations received relatively heavy emphasis, and thermodynamics was the major "theoretical" topic studied. While this may have made chemistry less intellectually satisfying and mathematically demanding than it is today, much of the descriptive material came from industrially important processes, and the cleavage between academic chemistry and technology was less evident than it would become in later years. As a final contrast, instrumentation in academic laboratories at any level of complexity was almost nonexistent. Separations were made by extraction, crystallization, and distillation; even efficient fractionating columns were rare. Physical measurements usually were restricted to melting points, boiling points, and similar easily measured quantities. Daring investigators in such exotic areas as spectroscopy made do with homemade apparatus and laborious point-by-point measurements.

Viewed from 1976, the first 50 years of the ACS may seem a period of rather steady growth in both the quantity and quality of chemical education. But there is little doubt that in the Society's second 50 years, political and economic changes have superimposed drastic oscillations and sudden changes of direction on the process.

The first of these external forces was the great depression of the 1930s. It produced serious economic problems for the colleges and universities, not to mention their students, as well as a shortage of jobs for graduates. Significantly, although the growth of college enrollments slowed somewhat, the colleges remained a relatively prosperous segment of society. Their situation reflected both the strong U.S. faith in education as a cure for all ills and the fact that, with jobs scarce and poorly paid, many young people chose further education as an alternative. The struggle for the few positions available in chemistry, by a variety of applicants with very different levels of training, appears to have spurred the ACS to involve itself more deeply in the professional training of chemists. This course of action led eventually to the establishment of the ACS Committee on

Professional Training, which has played a major role in chemical education in the United States.

Committee on Professional Training

The first step was taken April 24, 1935, when the Council voted to establish a standing committee to give continuing consideration to matters affecting the status of the chemical profession. Initially this committee consisted of Thomas Midgley, Jr. (Chairman), S. C. Lind, R. E. Swain, E. R. Weidlein, and H. B. Weiser; it was divided into two groups, one to concern itself specifically with the training of chemists and the possible accrediting of schools of chemistry, and the other to consider the more general phases of professional standing.

One year later, the Council voted to separate the two groups. The first group was named the Committee for Accrediting Educational Institutions and was set up on a rotating basis (the second group was a precursor of the Council Committee on Professional Relations). Soon thereafter the name was changed to the Committee on Professional Training of Chemists, which was shortened in 1946 to Committee on Professional Training (CPT). The 1936 action ended with the statement:

> The responsibility for properly accrediting institutions is wholly within the province of the permanent committee, and it is hereby granted the authority commensurate with this responsibility.

Following the meeting, President Edward Bartow appointed F. W. Willard (Chairman), Roger Adams, E. M. Billings, and R. E. Swain to this committee. The committee devoted the next two years to a general review of the problems involved, data collection, consultation with department heads, and the formulation of a plan for the approval of programs, and issued progress reports regularly to the Council.

After considering possible alternatives, the committee recommended first a procedure by which it would formulate a set of minimum standards of eligibility and circulate them to interested institutions with a suitable form of application for approved status (the word "accredited" as first used was replaced by "approved" as the work of the committee progressed, because "accredited" as used by accrediting associations had little in common with ACS objectives). Next, the committee decided that, rather than go to some existing outside agency to secure and evaluate data, it would generate and evaluate its own information. A tentative questionnaire was formulated and sent for comment to the heads of 30 large and well-established departments of chemistry. The 21 replies that were received contained many valuable suggestions and showed a gratifying interest in the project. On the basis of these replies the committee prepared a second questionnaire and sent it to more than 750 colleges and universities. With the questionnaire went a covering letter from Roger Adams, who had become committee chairman, to the president of each institution explaining the purpose of the committee and the reasons for the questionnaire. The results, again, were gratifying; some 450 questionnaires were returned.

At this point, an extensive body of data had been gathered on all aspects of chemistry programs at well over half the U.S. institutions offering a major in chemistry. To organize this mass of information, the committee engaged Prof. Ethel French of the University of Rochester, together with a staff of student assistants. Prof. French was to put the data in a more tractable form and, insofar as possible, to locate current norms on which realistic standards could be based.

The resulting charts and graphs were shown to the Council at the Milwaukee meeting in September 1938. Then the committee studied them in detail as a basis for establishing a set of minimum standards for the approval of programs. The idea was that, wherever such a procedure was applicable, the standard would be set just above the current median of performance for U.S. colleges and universities as a whole.

By the following spring, a set of minimum standards had been formulated. They were reported to the Council at the Baltimore meeting in April, and the report was published in the *News Edition* of *Industrial and Engineering Chemistry* [**17,** 270 (1939)]. Not surprisingly, this report generated many letters from Society members. While generally favorable, the letters raised a number of questions and indicated some need for clarification. Accordingly the committee scheduled a public hearing on Sept. 13, 1939, in advance of the meeting of the Council in Boston. The hearing was well attended and led to frank and helpful exchanges of opinion.

At the subsequent Council meeting the minimum standards were adopted with minor change, and the committee was instructed to proceed with its evaluation of programs for professional degrees in chemistry at interested institutions. In adopting these standards the Council specified that future revisions might raise the standards or criteria, but were not to lower or reduce them.

The procedure developed for approval of professional programs in chemistry was, and has remained, entirely voluntary, and the Society has always borne all costs—some $60,000 annually in the mid–1970s. A department wishing its program to be evaluated would request and complete a questionnaire describing its curricula, staff, and facilities. Next, on invitation from the president of the institution, the committee would send a visiting associate (an experienced academic chemist, chosen carefully to avoid any possible conflict of interest) to obtain a firsthand impression and to answer questions raised by responses to the questionnaire; the visiting associate would report his findings to the committee. Finally, the department head would be invited for a conference with the committee, usually in conjunction with the next national ACS meeting. Using primarily these three sources of information, the committee would then make its decision. Whenever the committee withheld approval, it gave its reasons to the institution, and the matter could be reopened when the department felt that the deficiencies had been met.

Once a department's program was approved, its standards were to be monitored through the submission of annual reports, and it would be revisited when the need arose. Department chairmen were to submit annually a list of students who could be certified as having completed the professional degree (and were thus eligible for full membership in the Society after two years' additional experience). The resulting statistics, together with a list of schools with approved programs, were to be included in the committee's annual report, published in the *News Edition, I&EC* (subsequently *Chemical and Engineering News*). It should be noted at this point that the committee has never implied that all candidates for the major in chemistry at an institution need qualify for the professional degree (and thus for certification) and has recognized the desirability of less intensive programs for students with other objectives.

Although the committee's original minimum standards included masters and doctoral degree programs, approval in practice has always been limited to programs leading to the bachelors degree in chemistry. The committee also has never attempted to evaluate and approve chemical engineering programs. Rather, in its reports, it has listed such programs accredited by the Engineers Council for Professional Development, together with data on degrees conferred.

In October 1940, the committee issued its first list of schools, 65 of them, with approved programs. As of 1976, the procedures that it developed have stood the test of time to a remarkable degree, with only minor changes being necessary. Revised minimum standards have been issued periodically. As the number of schools on the approved list has grown, reports have been put on a biennial and then a triennial basis, and visits to candidate schools have sometimes been omitted when adequate information was available elsewhere. The committee has grown from the original four to 13 members; they are still appointed on a three-year rotating basis and normally meet twice a year in conjunction with ACS national meetings.

The committee's success has been a consequence of the large amount of time and effort that its members and a total of several hundred visiting associates have devoted voluntarily to its operations. The committee's secretaries and the Eastman Kodak Co. also have contributed greatly.

Erle Billings, director of business and technical personnel at Eastman, was a member of the original committee and served as its secretary until the end of 1949 when he retired, and as a consultant to the committee for several years thereafter. He was succeeeded by John H. Howard, the present secretary, who held the same position as Billings at Eastman. Without the continuity of experience and accumulated wisdom of these men, it is difficult to see how the committee could have operated. In addition, Eastman provided office space, equipment, and services for the secretary's office, through which most committee business was conducted until mid-1974, when the last of the office operations were transferred to ACS headquarters in Washington.

The entrance of the U.S. into World War II, in December 1941, placed chemical education essentially in standby status. Enrollments declined, faculty were drawn into wartime research, and the academic community emerged in 1945 into a scene that was greatly changed. The scientific and technical achievements of the war left the country with a strong sentiment in favor of expanded research and scientific education, but it took several years and a boost from the space race to translate the sentiment into concrete action. In the meantime, schools were crammed with returning veterans, and all plans for the future were in a state of flux. Under these circumstances the CPT began an extensive program of revisitation of schools on the approved list and a reexamination of educational policy. These efforts led, in 1949, to a revised set of minimum standards that were printed in pamphlet form and sent to all U.S. institutions offering a degree in chemistry.

A further revision, emphasizing more strongly the need for an adequate mathematical and physical-chemical background for advanced courses, appeared in 1962 and was reissued in slightly altered form in 1965. The most recent version, retitled "Objectives and Guidelines for Undergraduate Programs in Chemistry," appeared in 1972. It reflects the changing viewpoint of the committee and the academic community and contains major changes aimed at achieving greater flexibility without decreasing the total content or final level of the professional degree. Also, it stresses the important commitment of departments of chemistry to the education of students with goals other than becoming professional chemists. The CPT has developed each revision by soliciting detailed information and opinion from department chairmen and other interested persons, by circulating a preliminary draft to departments for comment, and by discussion at either open meetings or special meetings of department chairmen. As a result, the revisions have been well accepted by the academic community.

There seems little doubt that the activities of the CPT have had a real and beneficial effect on chemical education at the undergraduate level.

TABLE I

Institutions Offering Professional Programs in Chemistry (CPT Approved List)

Year	*Number of Institutions*	*Bachelor Degrees*		
		Certified	*Noncertified*	*Total*
1940	65	—	—	—
1944	133	1386	—	—
1949	165	2996	1492	4488
1954	213	1843	1202	3045
1959	270	2305	1554	3859
1964	316	3277	2325	5602
1969	395	3917	3726	7643
1974	514	3844	4868	8712

The number of schools on the approved list and the numbers of their graduates have increased steadily (Table I). In 1975, the approved list included about half of the 1050 schools that offered degrees in chemistry in this country, and those schools were producing more than three-quarters of the degrees granted in chemistry. Further, the desire of institutions to be listed has given chemistry departments significant leverage in procuring the necessary staff and facilities. Chemistry in consequence is very frequently the strongest science department on a campus. The benefits of a strong department, in turn, extend not only to professional majors, but to students who elect less intensive majors in chemistry or simply take chemistry courses as part of another curriculum.

Starting in the immediate postwar period the CPT also undertook an extensive study of graduate programs in chemistry, initially to determine the feasibility of producing a set of minimum standards and an approved list. The first phase, again, involved developing information via questionnaires to graduate departments, individual chemists in industry, and industrial research directors and meetings with panels of university professors active in graduate research and of industrial research directors. The results of these surveys were published in 1947 in a series of articles in *Chemical and Engineering News* and were combined in a reprint, "Graduate Training at the Doctorate Level." Based on the surveys and its other deliberations, the committee also issued a statement, "Philosophy of Graduate Training at the Ph.D. Level," [*C&EN,* **26,** 166 (1948)]. In 1950 the committee supplemented this information by a series of visitations to graduate departments. It continued its studies and in 1957 issued a second report, "Doctoral Training in Chemistry," [*CE&N,* **35,** 56 (1957)], which attempted to define in more detail the criteria common to strong and successful programs.

One outcome of these studies was the CPT's conclusion that flexibility and diversity were such important characteristics of the graduate enterprise that an attempt to produce a set of guidelines and an approved list of programs would be inadvisable. On the other hand, the CPT did recognize a need to provide potential graduate students with better information, both on the requirements of graduate study and the nature of different departmental programs. To meet these needs, a pamphlet, "Planning for Graduate Work in Chemistry," was published and has gone through several editions. More important was the institution, in 1953, of the biennial publication of the "Directory of Graduate Research." The directory lists staff, research publications, and other statistics on all U.S. departments granting the Ph.D. in chemistry or chemical engineering (later extended to biochemistry and medicinal chemistry and to Canadian institutions). This publication, distributed gratis to all U.S. institutions offering a major in chemistry, has transcended its original purpose and become a standard reference for anyone desring data on graduate departments or the location, research interests, and publications of faculty.

After the postwar bulge, the production of undergraduate majors in chemistry grew rather steadily—from a low of 5500/yr in 1954 to roughly 12,000/yr in 1975—but graduate education shows a more erratic pattern (Table II). Ph.D. production was roughly 1000/yr all through the 1950s, shot up to a peak of more than 2100/yr in 1969–70 under the forced draft of the growing college-age population and the post-Sputnik government funding of the 1960s, and declined in 1970–74 in the face of decreasing fellowship support and a tightening job market. The expansion of graduate programs was even more striking. Early in the 1960s the CPT

TABLE II

Ph.D. Production in Chemistry, 1950–1974

Year	*Schools*	*Degrees*	*Year*	*Schools*	*Degrees*
1950–1	84	1013	1962–3	143	1220
1951–2		989	1963–4		1280
1952–3	87	946	1964–5	153	1398
1953–4		993	1965–6		1531
1954–5	98	988	1966–7	164	1690
1955–6		1011	1967–8		1705
1956–7	110	1003	1968–9	176	1912
1957–8		939	1969–70		2145
1958–9	122	1009	1970–1	181	2097
1959–60		1048	1971–2		1929
1960–1	125	1106	1972–3	184	1880
1961–2		1125	1973–4		1733

became seriously concerned that the projections of graduate departments were outrunning any possible supply of qualified students. This was clearly pointed out in the committee's third report on graduate education, "Doctoral Education in Chemistry, a Report of Current Needs and Problems,"[*C&EN,* **42,** 76 (1964)], which discussed in detail the problems that new programs must expect to face.

The statistics the CPT collected made it apparent subsequently that little correlation existed between numbers of undergraduate majors receiving degrees and numbers of students entering graduate school. In fact, the latter had begun to decline by 1966, and the trend continued at least through 1973–74 (whether a low point was reached then was not yet clear as of 1975). Added to this, the chemical job market declined rapidly in the early 1970s, so it became evident that the three factors—supply of students, demand for graduates, and capacity and aspirations of overexpanded graduate schools—were badly out of balance. This problem was a major topic in the CPT's 1972 report, "Doctoral Education in Chemistry: Facing the 1970's" [*C&EN,* **50,** 35 (1972)]. The report concluded that many small and recently started programs could have little prospect for successful survival. The report pointed out also the projected small number of academic openings to be expected in the 1970s and the changing employment prospects for new Ph.D.'s, which appeared to demand greater versatility and flexibility of training and better understanding of the role of chemistry in technology and society as a whole than was emphasized in conventional, academically oriented programs. At the same time, the report returned to the issue of ACS approval of graduate programs, which had excited heated discussion within the Society as the employment situation deteriorated. The committee concluded that an attempt by the ACS to institute a program of approval of graduate programs at that time would arouse opposition from established accrediting agencies and would be difficult to implement. The CPT recommended instead that ACS cooperate with state and regional accrediting agencies that already were attacking the problem.

After the 1972 report, CPT continued its study and data-gathering. Of particular utility were statistics on entering-class enrollments in graduate schools, which permitted a reliable estimate of Ph.D. production some four to five years in the future and indicated a continued decline, at least through 1977.

In September 1974, the U.S. Office of Education dropped the Society—in effect, the Committee on Professional Training—from the list of recognized accrediting agencies that USOE was required by law to maintain. The ACS had been on the list continuously for 22 years. USOE said the Society was deficient in five of its criteria for accrediting agencies; the ACS, though skeptical of parts of the criteria, saw no problem in meeting three of the five and was prepared to meet a fourth—involving on-site visitations—at an additional cost to itself of $50,000 annually. The fifth criterion was that an accrediting agency must meet USOE's definition of "need"—that is, it must establish institutions' (and their students') eligibility for federal funds. This criterion the ACS could not meet: CPT approves only departments; and all departments on its approved list were at institutions already accredited by a recognized agency. USOE agreed that the Society's program upgrades educational programs, but did not include that function in its definition of need. In December 1975, the ACS Board of Directors decided "not to apply at this time for reinstatement" on the USOE list. The CPT program continued to operate without interruption and continued to be recognized by the Council on Postsecondary Accreditation (previously the National Commission on Accrediting).

Growth in Educational Programs

The work of the Committee on Professional Training, and of the Division of Chemical Education and its journal, has been paralleled over the years by a growing variety of other ACS activities in education.

The Society's Sixth Biennial Education Conference took place at Airlie, Va. in October 1968. From left are Herman S. Bloch, O. T. Benfey, Dale B. Baker, Peter E. Yankwich, and Cheves Walling.

As the Society grew, its educational programs came under the aegis of the Council Committee on Chemical Education, formed in 1948, and the Board Committee on Education and Students, formed in 1953. In the late 1950s, the Board committee began to sponsor biennial conferences of national leaders in chemical education, convened by invitation to deal with issues of major concern; the 9th Biennial Education Conference took place in 1975. In 1971, the Board established the Board–Council Chemical Education Planning and Coordinating Committee to coordinate programs and formulate long-range plans. These moves were paralleled by changes in staff organization. In 1958, the educational duties being handled by Assistant Secretary B. R. Stanerson were assigned to Robert E. Henze, who was appointed to the new post of Educational Secretary. Dr. Henze was succeeded in that post by Robert L. Silber in 1960 and by Moses Passer in 1964. In 1970, a Department of Educational Activities was created under Dr. Passer.

In 1937, the Society started its Student Affiliate program for undergraduates. The program allows students of chemistry, chemical engineering, or a related discipline to affiliate with the ACS at nominal cost. Each affiliate receives a subscription to an ACS journal of his choice and enjoys many of the privileges of a member, as well as certain special services provided by the staff Department of Educational Activities. These services include the *Student Affiliate Newsletter,* mailed three times annually to all Student Affiliates and chapter faculty advisers. At the end of 1975, the program included some 7000 Student Affiliates and chapters at more than 600 colleges and universities.

In October 1944, the Society became the administrator of The Petroleum Research Fund, whose sole purpose is to support " . . . advanced scientific education and fundamental research in the 'petroleum field' . . ." The PRF is a trust that was created by a gift of security holdings in Universal Oil Products Co.; in 1956, the holdings were sold to the public and the funds invested in a diversified portfolio with Morgan Guaranty Trust Co. as trustee and the Society as recipient and administrator of the income. The fund awarded its first research grants in 1954; in the period 1960–75 it provided an average of more than $3.4 million annually in research support.

From the outset the Society's program of grants-in-aid supported by income from the PRF has been designed to provide highly flexible research support, mostly for faculty investigators at colleges and universities. Individual programs have been initiated, as required, for fundamental research projects involving faculty members in primarily undergraduate departments, for well-established investigators whose chief need is unrestricted funds, and for starting faculty investigators. Special ad hoc programs were developed to assist projects involving graduate fellows in a period of declining research support (1968–69) and, later, to assist postdoctoral fellows during a period of inadequate job opportunities (1972). Throughout, the Society has adhered to the grant-in-aid philosophy, by which some assistance, though by no means enough to meet all the costs of a project, is extended to institutions which share the objectives of the PRF program. Thus, the benefits of some research support have been spread to a large number of investigators.

The PRF is administered by the ACS staff Department of Research Grants & Awards, under Justin W. Collat since 1970. Policy and programs are set by the Board Committee on Grants and Awards. A PRF Advisory Board was established in 1954 to make specific recommendations on the allocation of funds and to assist in the development of the grants program. It started with 13 members and grew, as the work load expanded, to 21 members in 1975. In order to recommend grants in the various fields of pure science that may afford a basis for subsequent research in the petroleum field, scientists of a variety of technical and professional backgrounds have been appointed. PRF grants in chemistry, chemical engineering, physics, and various earth sciences, among other fields, have been selected for support. The number of grants recommended by the Advisory Board grew from 25 during 1954 to 230 for the year 1974–75. PRF grants in effect at any one time, and grants in effect by virtue of time extension without additional funding, increased steadily through the late 1950s and 1960s to a maximum of 971 in 1970. Subsequent years saw a slight decline in the number of proposals for consideration by PRF and the number of grants in force, but by 1975, PRF still was receiving more scientifically meritorious proposals than could be supported by available funds.

Since 1950, the Society has administered the annual ACS Award in Chemical Education, established in that year by the Scientific Apparatus Makers Association (SAMA). The award, $2000 and a certificate, recognizes outstanding contributions to chemical education. The Society also administers the annual James Bryant Conant Award in High School Chemistry Teaching, which consists of $2000 and a certificate. The award was established in 1965 by E. I. du Pont de Nemours & Co., which sponsored it through 1972. The ACS sponsored the award in 1973–74; CHEM Study (the Chemical Education Material Study) assumed the sponsorship effective in 1975. Candidates for the ACS Regional Award in High School Chemistry Teaching are nominated by ACS local sections, and a winner is picked in each of 10 regions of the U.S. Each regional winner receives a certificate and other recognition determined by the Steering Committee of the region, and becomes automatically a candidate for the James Bryant Conant Award in the three succeeding years. Besides these two national awards, many ACS local sections sponsor awards for science teachers.

In the early 1960s, the Society was instrumental in launching the discussions that led to the development of two new high school curricula: CHEM Study and the Chemical Bond Approach. Most observers were agreed by 1975 that these curricula constituted one of the most significant advances of the time in high school chemistry. In addition to their impact in the United States, they have stimulated experimentation in many parts of the world, to the extent that, in the 1970s, work in curricular innovaion in high school chemistry almost invariably begins with CHEM Study and CBA as the point of departure.

Another activity primarily at the high school level is the monthly magazine *Chemistry,* which the Society has published since September 1962. In that year the ACS acquired the magazine from Science Service; it was redesigned both physically and editorially during the following 15 months and appeared in its new format in January 1964. Although *Chemistry* is edited for the superior high school student, many high school teachers find it stimulating and instructive.

Most Society activities have always embodied elements of post-school or continuing education, but in 1965 the ACS Department of Educational Activities inaugurated programs designed specifically for that purpose. The first of these was the ACS Short Courses: two to five days of instruction by one to three experts on subjects selected expressly to meet the continuing-education needs of chemists. By the end of 1975, some 50 different Short Courses were being presented annually, and some 30,000 individuals had participated since the program began. A second program, ACS Audio Courses, was introduced in 1971; a third and fourth, ACS Film Courses and Correspondence Interaction Courses, were in develop-

Dr. Nicholas J. Turro gave an ACS Short Course in organic photochemistry in Washington, D.C., in February 1974.

ment in 1975. Of these four programs, the ACS Audio Courses appeared by 1975 to be the most promising for the long term; the titles available had reached 30, and about 10 were being added annually. Techniques had evolved in which the courses were being used by groups of students with a discussion leader familiar with the subject matter. One tape cassette of the audio portion of the course serves the entire group, but each student has his own copy of the printed course material. The technique was proving highly effective, both pedagogically and in minimizing the cost per individual. In this vein, an important tenet of the ACS continuing education programs is that they be priced to be self-supporting—though not to make a profit—but so conceived that the cost per individual is reasonable.

In July 1974, the National Science Foundation awarded the Society an $839,000 grant to support the design, production, and distribution of new kinds of continuing education programs. They were to be marked by a much higher level of individualized user control than had been possible in earlier formats. One such project in development in 1975 was a computer-text combination for individual students. The computer would be programmed to ask questions of the student, check his answers, and in case of error refer the student, for example, to a particular section of the written course material. The computer programs would be stored in existing national computer networks, which by 1975 were accessible from most parts of the nation; the student would work at an individual keyboard terminal. A second project in development in 1975 combined an audio cassette and a film cassette that could be used in television-like systems available from at least two manufacturers. The audio cassette drives the film cassette, switching it on and off and running it at variable speeds, as desired. The basic concept is to use film—which is very costly to produce—only for parts of the course where visual presentation is most effective and otherwise to rely on the audio and written material.

In 1967, the Society established a staff office for chemical education in junior colleges, which already were the fastest growing of all academic institutions. Many junior colleges were—and are in 1975—installing two-year, laboratory-oriented programs for training chemical technicians. The ACS became concerned over the unique requirements of these programs, which are not met by the first two years of the typical four-year curriculum, and in July 1969 received a $621,000 grant from the National Science Foundation to develop a two-year curriculum for chemical technicians. The project, called ChemTeC, was carried out by a team of university professors, technology instructors, and industrial chemists headed by Robert L. Pecsok of the University of Hawaii. The work was completed in 1972. By mid-1973, the "Modern Chemical Technology" series was available from the ACS, which had won the distribution rights in a competitive bid procedure stipulated by NSF. The series includes a teacher's manual, a guidebook to general laboratory technology, and eight textbooks. The volumes can be purchased individually or in any desired combination, and some 26,000 had been sold by the end of 1975.

The Society has long prepared and distributed career guidance literature through high school counselors and through the education committees of its local sections, which numbered 175 in 1975. Its college-level services in that year, in addition to the previously-mentioned "Directory of Graduate Research," included: "Academic Openings," a thrice-yearly listing of teaching and postdoctoral research vacancies in colleges, universities, and junior colleges in the U.S. and Canada; "College Chemistry Seniors," a special placement service for soon-to-graduate chemistry majors who are interested in graduate work; and "College Chemistry Faculties," a biennial directory of all college-university teachers of chemistry and related subjects in the U.S. and Canada.

These many efforts indicate the importance that the Society traditionally has attached to chemical education. So, too, does the extensive section devoted to the subject in the ACS publication "Chemistry in the Economy" (1973). The American Chemical Society enters its second century with every intention of continuing this valuable and symbiotic relationship.

The ACS Employment Clearing House saw considerable activity during the national meeting in Detroit in April 1965.

CHAPTER III

PROFESSIONALISM

DURING its early years the American Chemical Society was predominantly academic: its members viewed the ACS primarily as a servant of chemistry; they were confident of their status; they were little concerned with professionalism as the term is understood in the Society's 100th year. This situation changed gradually after World War I and through the 1920s as a strong chemical industry emerged, and it changed decisively during the depression of the 1930s. By then the membership was comparatively large (18,206 in 1930), a significant fraction worked in industry, and many soon found themselves jobless. From these circumstances sprang the Society's first sustained efforts in professionalism. These efforts are aimed today, as they were originally, at the economic well-being of the individual member; in general they encompass education and training, employment aids and counseling, employer–employee relations, and the status of chemists[1] and chemistry in the eyes of the public.

ACS work in professional relations over the years has involved in part the professional status of the chemist in the legal sense. In 1896, 20 years after the Society was founded, the Supreme Court ruled in United States vs. Laws that "The chemist who places his knowledge acquired from a study of the science to the use of others as he may be employed by them, and as a vocation for the purpose of his own maintenance, must certainly be regarded as one engaged in the practice of a profession which is generally recognized in this country." The chemist in the case had been hired in Germany to work on a sugar plantation in Louisiana. The Government had charged the employer (Laws) with illegal importation of foreign labor. In ruling against the Government, the Court held that the law did not apply to professionals and that the chemist involved was a professional, whether self-employed or working for someone else.

The Court did not define "profession" in its opinion of 1896, but Louis D. Brandeis did so in 1912, four years before he joined the Court himself. A profession, Brandeis argued, "is an occupation for which the necessary preliminary training is intellectual in character . . . which is pursued largely for others and not merely for one's self . . . [and] in which the amount of financial return is not the accepted measure of success."

Brendeis' view strikes an idealistic note that is missing from the definition of a "professional" in the Taft-Hartley Act (the Labor Management Relations Act of 1947), the one most widely accepted in the 1970s. The Taft-Hartley definition reflects closely the ideas of the ACS, which testified on the point before the appropriate commitees of the House and Senate as the legislation moved through the Congress. The act defines a professional as one engaged in work:

> predominantly intellectual and varied in character as opposed to routine mental, manual, mechanical, or physical work; involving the consistent exercise of discretion and judgement in its performance; of such a character that the output produced or the result accomplished cannot be standardized in relation to a given period of time; requiring knowledge of an advanced type in a field of science or learning customarily acquired by a prolonged course of specialized intellectual instruction and study in an institution of higher learning or a hospital, as distinguished from a general academic education or from an apprenticeship or from training in the performance of routine, mental, manual, or physical processes.

The Society's contribution to the Taft-Hartley definition of a professional marked a relatively new interest. The ACS had always considered

[1] Taken hereafter to include chemical engineers.

the practice of chemistry a profession, but before the rapid upsurge of the union movement during the 1930s it had displayed no great official interest in the associated legal wordplay. Indeed, whatever interest was present seems for some years to have tended toward Justice Brandeis' idea that, in a profession, financial return is not the accepted mark of success. In 1913, for example, a member of the Board of Editors of the *Journal of Industrial and Engineering Chemistry* could write in an editorial, "The trouble with the American chemists . . . is that they want to draw large salaries before they have really demonstrated to the manufacturer the ability to give ample value in return for the money he pays them. American manufacturers . . . are very shy in buying a 'pig in a poke.' " The author, ironically, was speaking of the dye industry, then monopolized by Germany, which the British blockade of World War I and the skills of the "pigs in a poke" would soon transform into a vigorous domestic enterprise.

The need to maintain a strong chemical industry, a need brought home by the shortages of World War I, was an important factor in the establishment of the ACS News Service, later to be considered an element of the Society's professional programs. The News Service began work Jan. 1, 1919. Its task was to inform the lay public, through the news media, of developments in chemical science and technology; its goal was to gain the public's understanding and good opinion of chemistry and its support for the chemical industry. ACS officials recognized that an effective News Service would act indirectly to improve the professional and economic status of the chemist, but this result was not nearly the issue that it became in the 1930s. Early explanations of the function of the News Service focused on the value of publicizing chemistry and its utility; they said little or nothing of the status of the individual chemist.

In 1923, in its first clearly identifiable professional activity, the ACS entered the long battle to protect the right of chemists to practice their profession in the clinical laboratory. The precipitating event was a bill in the Pennsylvania legislature that proposed in part that "no laboratory procedure for the diagnosis or treatment of human disease shall be performed or reported by persons who have not a license to practice medicine." The Society actively opposed the bill, which was defeated. In March 1924, the ACS, the American Medical Association, and the American Association of Pathologists and Bacteriologists agreed to a statement of principles that fully protected the rights of the clinical chemist. Efforts continued, however, to restrict the operation of clinical laboratories to physicians, and in 1939 the AMA rescinded its approval of the statement of 1924. In December 1939 the ACS Board created a Commitee on Chemical Service to Medicine to pursue the struggle. The work of this group, renamed the Committee on Clinical Chemistry in 1945, resulted in 1950 in the formation of the independent American Board of Clinical Chemistry, whose function was to certify clinical chemists. In 1952 the Society and other organizations formed the National Council on Health Laboratory Services (initially the Intersociety Committee on Laboratory Services Related to Health) to continue the fight to avoid monopolization of clinical laboratories by the medical profession. In recent years the issue has been resolved to a degree by Medicare regulations. The ACS remains active in the field, through both the National Council and the National Registry in Clinical Chemistry, formed in 1967.

Effects of the Depression

The depression of the 1930s made professional needs a major issue within the ACS, which in consequence created the committee structure that in one form or another has dealt with such problems ever since. In 1935, as a result of a study of unemployment among chemists, the Society established a Committee on Professional Standing of the Chemist and directed it "to give continuing consideration to matters affecting the status of the chemical profession." In 1936 the committee was divided into two groups: the Committee on Professional Status and the Committee on Accrediting Educational Institutions Offering Instruction in Chemistry (the present Committee on Professional Training). In September 1941, the Board of Directors created a Committee on Economic Status. In 1942, the Committee on Professional Status was discharged, having "found nothing to do for the past two years," and in 1944 the Committee on Economic Status was renamed the Committee on Professional and Economic Status. In 1948, as a result of the Hancock report of 1947 on the structure of the Society, the latter committee was replaced by the Committee on Professional Relations and Status, a standing committee of the Council; in 1966 the name was shortened to Committee on Professional Relations. In 1959, meanwhile, the Board added professional relations to the duties of its Committee on Member and Public Relations, which was renamed the Committee on Public, Professional, and Member Relations.

Thus since 1935, without a break, at least one ACS committee has been concerned with professional problems. The name, size, and parent body (Board or Council) have changed, but the basic duties have remained the same. The operating pattern of the mid-1970s has been for the Council Committee on Professional Relations to initiate actions for Council decision and for the Board Committee on Public, Professional, and Member Relations to recommend to the Board approval or disapproval of Council-approved actions, particularly in terms of financial support.

The Board committee in addition devises Society policy in professional relations and may initiate actions of its own for Board decision.

The Society's overt move into professional relations in the 1930s raised new issues, not the least of them stemming from the Wagner Act (the National Labor Relations Act) of 1935. The act guaranteed employees the right to organize and bargain collectively. Further, it applied to professionals as well as nonprofessionals and thus compelled the ACS to develop a policy on unionization of chemists. Opinion within the Society was diverse—more so than it might have been in a strong economy—but the Directors decided in the end that their primary duty was to safeguard professional status. The Board advised the members that they were entitled to bargain collectively; also, it expressed the opinion that they as professionals should not be forced to join a union against their wishes "as expressed in a vote conducted by the National Labor Relations Board of all of the employees in a unit composed wholly of professional employees."

The Society's position on unionization was tested in 1941, when the International Federation of Architects, Engineers, Chemists, and Technicians tried to force 200 professionals at Shell Development Co., Emeryville, Calif., to join the union. A group of the professionals requested and received ACS help in resisting the move; Elisha Hanson, the Society's counsel, eventually argued the case before the National Labor Relations Board. In January 1942 the NLRB ruled that the professionals at Shell could not be forced into a heterogeneous union (one with both professional and nonprofessional members), but that they could be represented by the union if they so chose. In an NLRB election in February, the professionals rejected the union.

In its decision in the Shell Development case, the NLRB pointed out that, although it had not forced the professionals into a heterogeneous union, the Wagner Act gave it discretionary power to do so. The Society opposed this provision. In 1947, in Congressional committe hearings on the Taft-Hartley Act, which revised the Wagner Act, ACS counsel Hanson argued that the NLRB should not have the power to decide the appropriateness of a collective bargaining unit for professionals if the unit includes "both professional employees and employees who are not professional employees." This view prevailed and was incorporated in Taft-Hartley along with the definition of "professional" suggested by the Society.

The depression also led to the ACS employment aids program. At the national meeting in Chapel Hill, N.C., in April 1937, the Society informally arranged contacts between job-seeking chemists and prospective employers. Several local sections, notably the Philadelphia Section, had set up employment aids for chemists relatively early in the 1930s and they had argued for greater participation by the national Society in helping unemployed chemists. The results of the Chapel Hill experiment led to a formal Employment Clearing House at the next national meeting, in Rochester, N.Y., and the practice was continued at subsequent national meetings. In 1944, this twice-yearly service was expanded to include a full-time Employment Clearing House at Society headquarters as well as Regional Employment Clearing Houses operated by selected ACS local sections. In 1967, the regional operations in New York, Chicago, and Berkeley were closed, and the Employment Clearing House was centralized in Washington.

For about a decade, beginning in 1949, the licensing of chemists as a means of improving professional status became a major issue within the Society. A Committee on Licensing Inquiry was appointed in 1939 and discharged in 1944. "Should Chemists be Licensed?" was debated in *Chemical & Engineering News* in 1946 by Gustav Egloff (yes) and then-recently-retired Secrtary Charles L. Parsons (no). In 1947, the Committee on Professional and Economic Status polled the membership and found that ". . . the bulk of the respondents are still opposed to both licensure and legal registration in almost any form—permissive, semi-permissive, compulsory, or what not." In September 1948 the Council formally reaffirmed the Society's opposition to licensing or registration of chemists. Responsibility for the status of the chemist in the eyes of the public has remained with the ACS News Service. The task had grown more difficult by the mid-1970s: news of chemistry was not uniformly good, as it had been 25 years earlier, and the news media had grown much more skeptical of the works of chemists.

In the early 1970s, interest revived in certification, licensing, and registration of chemists. In 1973 the ACS Council established a Subcommittee on Regulation to examine the situation and suggest courses of action. The Subcommittee was sponsored jointly by the Council Committees on Professional Relations, Chemical Education, and Membership Affairs and by the Board–Council Committee on Professional Training. In 1975 the subcommittee was at work on a model bill to permit voluntary state registration of chemists. The Society's policy, meanwhile, remained as it had last been stated, in 1963: the ACS endorsed compulsory licensure for clinical chemists, opposed compulsory licensure or registration for other chemists, but did not oppose voluntary registration. At the same time, the Society did not endorse certification as a means of identifying the qualifications of chemists and chemical engineers.

In 1941, the Society for the first and last time took a position on specific salaries for chemists. The Board of Directors voted to deny access to the Employment Clearing House to employers unwilling to meet recommended minimum annual salaries of $1500 for "chemical internes" and $2400 for "professional" grade chemists. This action took effect in the spring of 1942, but was abandoned after the fall of 1943. It foundered in part because it failed to take account of geographical differences in living costs. In addition, the high level of industrial activity that accompanied World War II made the economic status of the chemist considerably less urgent an issue than it had been in the previous decade.

In November 1941 the Committee on Economic Status began a mail survey of ACS members' dollar income (the committee saw no practical way to measure fringe benefits such as bonuses, insurance, or pensions). This was the origin of the Chemists' Salary Survey, a strong contender for the single most important professional relations function of ACS. In the mid-1970s the Society is doing two types of salary surveys annually. For the comprehensive salary survey, questionnaires go to 25% of the domestic, nonstudent members (randomly selected), and the return is about 55%; data are collected on both "incomes" and "salaries." For the starting salary survey, questionnaires go to all chemists and chemical engineers graduating from departments approved by the ACS Committee on Professional Training, and the return is about 40%. Median salaries and incomes determined in the comprehensive salary surveys made in 1974 and 1975 appear in Table I. In addition to its salary surveys, the Society each year conducts an "employment status survey," the first of which was done in 1971. A five-year summary of the results appears in Table II.

Hard times have been a major creator of professional needs among chemists, but changes in the working environment have played a role, too.

TABLE I

1974 Overall Median Salaries and Incomes of Chemists and Chemical Engineers[a]

	Degree Level	*Salary*	*Income*
Chemists:			
	B.S.	$ 17,500	$ 19,000
	M.S.	18,400	19,900
	Ph.D.	21,700	23,000
Chemical Engineers:			
	B.S.	21,300	24,500
	M.S.	22,400	25,000
	Ph.D.	24,800	27,000

Source: American Chemical Society—1974 Report of Chemists' Salaries and Employment Status, 1975 Report of Chemists' Salaries and Employment Status.

[a] Respondents are asked to supply "salary" for current year and "income" for previous year.

TABLE II

Employment Status of Respondents: 1971-1975 (Percentages)

<table>
<tr><th>Employment Category</th><th>1971</th><th>1972</th><th>1973</th><th>1974</th><th>1975</th></tr>
<tr><td>Full-time Employed</td><td>88.2</td><td>88.0</td><td>88.7</td><td>92.5</td><td>91.0</td></tr>
<tr><td>Subprofessionally Employed</td><td>2.5</td><td>2.8</td><td>2.0</td><td>1.0</td><td>2.2</td></tr>
<tr><td>Part-time or Temporarily Employed</td><td>2.3</td><td>1.5</td><td>1.3</td><td>1.0</td><td>1.3</td></tr>
<tr><td>Postdoctoral or Other Fellowship</td><td>1.6</td><td>2.0</td><td>2.9</td><td>2.4</td><td>2.0</td></tr>
<tr><td>Unemployed Seeking Employment</td><td>2.8</td><td>3.1</td><td>1.7</td><td>1.4</td><td>1.6</td></tr>
<tr><td>Retired Seeking Employment</td><td rowspan="2">2.6</td><td>0.4</td><td>0.4</td><td>0.3</td><td>0.4</td></tr>
<tr><td>Not Seeking Employment [a]</td><td>2.2</td><td>3.0</td><td>1.4</td><td>1.4</td></tr>
</table>

Source: 1975 Report of Chemists' Salaries and Employment Status, American Chemical Society.

[a] In 1974 and 1975 members over 64 years of age were not surveyed.

This has been true especially among industrially employed chemists, who in 1974 accounted for 62% of the Society's domestic members. The numbers of chemists have grown steadily in response to demand; ACS membership multiplied from 23,519 in 1939 to almost 116,816 (a peak) in 1969. At the same time, fewer and fewer chemists have been employed in the control functions of the manufacturing plant, and more and more have been employed in activities characterized in varying degree as "research." Increasingly chemists have performed staff rather than line functions; they have done so in "research centers" or "experiment stations" designed specifically for the purpose. The similarity and uniformity of these institutions and the chemist's role in them have attracted comment from several sources:

- The Committee on Professional Status in 1937 sought to develop a model employment contract. Dr. Thomas Midgley, Jr., the chairman, noted in 1939: "Irrespective of the intrinsic merits of such documents for generalized cases, the specific conditions which arise in individual cases and the fact that, under normal conditions, the employer and not the employee dictates the terms of employment, have rendered their applications substantially impractical."
- Past President Roger Adams in 1945: "The term 'professional status' is not easy to define . . . The research man in the laboratory should be provided with the reward his ability and accomplishments justify . . . the proper and formal evaluation of the employment advantages of various companies . . . would then be possible."
- The Hancock report (1947): "and the peculiar detachment he [the chemist] frequently experiences from being allied to neither management

nor labor conspire to develop professional needs which either this Society or some other must supply."

• President Robert W. Cairns, in his State of the Society Editorial in 1968: " 'Professional relations' are deemed by some to be more dignified words than 'pay' and 'prestige,' yet in the proper context, 'pay' and 'prestige' may be more apt labels to describe our human goals. . . . Collectively, we must seek to enhance that [public] image . . . knowing that if we thereby attain broader responsibilities . . . a more satisfactory sharing of pay and prestige will likely result."

The notion that the Society should concern itself with pay—or with members' economic well-being in general—has always been controversial. By the end of World War II, the economic pressures that had spurred the professional activity of the previous decade had vanished; the Society had grown tremendously and saw the need for thorough study and revamping of its entire effort. The resulting Hancock report of 1947 covered professionalism in part in these blunt words:

> . . . a substantial section of . . . membership opinion . . . insists that the Society should adhere strictly to its scientific moorings . . . Others hold with equal firmness that . . . professional yearnings cannot be ignored even though this results somewhat in a gradual weakening of the . . . traditional devotion to science . . . Whether these two convictions are diametrically opposed is perhaps . . . somewhat beside the point. The Society is not confronted with an abstract theory but the solid fact that these two ideas are hopelessly intertwined in the sentiments of its members. It cannot return even if it would to an exclusively impersonal interest in science. Nor can it suddenly completely change its scientific traditions in response to a rising tide of professional awareness.

This assessment would serve as well in the 1970s, perhaps, as it did three decades earlier, notwithstanding the substantial expansion of the Society's professional programs since then.

An unsurprising gamut of reactions attends any discussion of economic compensation in deliberative assemblies of the ACS, whose members include not only many working chemists, but also company research directors, deans, department heads, and other leaders who pass on the performance of subordinates. For a time, opponents of ACS involvement with individual economic welfare argued that it was not covered by the federal Charter and was hazardous to the Society's favored tax status. By the 1970s the majority opinion had swung to the view that the charter and tax status do not bar certain kinds of involvement with individual welfare.

ACS counsel A. B. Hanson has said that, with proper planning, "the Society can accomplish under its present organization every legitimate professional objective which any professional association may accomplish."

The Hancock report recommended the formation of a Council Committee on Professional Relations and Status. It saw the committee as the focal point of professional activities within the ACS, being, as it would be, a unit of the Society's popular deliberative assembly. The report did not recommend an analogous committee of the Board of Directors. The Committee on Professional Relations, as it is now known, in general has filled the role visualized for it in the Hancock report.

Paralleling the fivefold growth of ACS membership since 1940 has been the growth of the American labor movement and of the benefits it has brought to unionized workers. Though scientists are not legally excluded from organizing into unions, few have done so. Estimates vary between 1% and 5% of the total, and chemists probably are toward the low side of that range. Scientists' views of unionization vary from disdainful, to strongly opposed, to neutral, to lukewarm, to a very small percentage who believe that organization is both desirable and inevitable. These attitudes and the interrelationships of issues such as collective bargaining, salaries, Society economic policies, tax status, and the ACS charter have been mirrored in the actions of the Committee on Professional Relations.

Before 1951, the Committee on Professional Relations (CPR) had directed ACS attention toward and suggested policy issues including:

- Qualifications for membership
- Improved standards of training
- Employment aids—Clearing Houses
- Professional recognition
- Protection of the right to practice (formation of American Board of Clinical Chemistry)
- Licensing
- Legal and professional relationships
- Collective bargaining for professional employees
- Compulsory unionization

Herman Bloch, CPR chairman in 1952–56 (and Board chairman 1973–) has said, "This was a period of feeling our way . . . Many years ago, many of our actions were frowned upon but tolerated by the Council." CPR's program grew from what 1957–59 Chairman John K. Taylor called "poorly attended breakfast meetings" to its 1975 format of six subcommittees that meet at national meetings on Friday, Saturday, and Sunday and hold a Monday Open Session, as well as four liaisons to other ACS groups. In addition, the committee was holding day-long symposia at national meetings in alternate years, but by 1975 had relinquished this function to the Division of Professional Relations, which had been formed in 1972. Dr. Taylor also noted, "Instead of acting on matters brought before us, we started to develop a positive program." The committee had begun to initiate actions as well as react to them. This approach is exemplified by the Member Assistance Program and self-policing.

The compensation and other working conditions of the employed chemist traditionally have been resolved between the individual and his employer. When the number of chemists in an establishment exceeds a certain limit, however, jobs must be classified and levels and ranges of salary developed. The chemist begins to lose his individuality. When he is then discharged for reasons beyond his control or understanding he may find himself at best in difficult circumstances and at worst confused and distraught. In such situations, which arise despite the acknowledged skills of management, CPR saw some years ago a need to interpose itself "to assist individual members or groups of members involved in situations which they feel may compromise their professional status or attainment."

The committee's willingness to provide such member assistance was first made known in the late 1950s. During the late 1960s and early 1970s, CPR became involved in as many as 20 individual cases at a time; it handled more than 100 cases during 1968–73. The committee works with both employee and employer on a confidential basis. The chemist first completes a Member Assistance Request. Then members of the CPR or, as the load has grown, consultants to CPR gather the facts, which are presented to a subcommittee and ultimately to the 15-member CPR. Some aggrieved members have approached the committee too late to permit corrective action to be taken. In many instances, however, CPR has been able notably to improve the member's situation. Outcomes of both types appear in the following examples:

> A member employed by a branch of the Government was responsible for a development on which his immediate superior had clamped a secrecy lid. The next phase of the work, which had been done outside the Government by an independent contractor, was not considered so secret, and the contractor was about to publish both phases as if they were his, completely ignoring the prior rights of the member. After investigation by the Committee on Professional Relations, the federal agency was persuaded to acknowledge the member's claim to priority and his right to publish; when the facts were known, the member received a $5,000 award for an individual contribution.

> A member had been essentially solely responsible for a development; he had clung tenaciously to the possibilities of making a particular reaction go when all others had abandoned it. The result of his effort was a chemical compound with unique prop-

erties that had found a commercial market. In a retrenchment, the member had been offered a reduced work load, then termination and retention as a consultant. When approached by the committee, the employer soon saw the inadequacy of the proposal that had been made to the member. For the first time, CPR took the step of suggesting to the employer a specific monetary settlement satisfactory to the committee's view, and the employer agreed to make such a settlement.

A member who worked for a large corporation asked the committee to investigate his dismissal. He had been terminated for poor performance after five years of service and had received termination benefits. The member contended, however, that he had been dismissed because of a personal conflict, and that the employer had used poor performance as a excuse. A consultant investigated the case and learned that the member had been tried on many projects, but was not able to handle the problems assigned to him satisfactorily. The company had planned to let him go for some time, but had waited until the employment outlook improved. In this case, the committee concluded that the chemist had been treated fairly, and the case was closed.

Early in the 1960s, several cases of theft and fraud and the activities of an international ring composed partly of ACS members stirred CPR to action. The committee had available specific procedures in ACS Bylaw 1, Section 7, under which two members may bring charges against a third, leading to expulsion from the Society, for activities that the ACS Constitution, Article IV, Section 3, defines as ". . . conduct which in anywise tends to injure the Society or to affect adversely its reputation or which is contrary to or destructive of its objects."

In one case, CPR did not act until the Supreme Court had refused to review a lower court decision finding the member guilty of theft. Then the chairman and secretary of CPR signed a letter to the Committee on Membership Affairs (CMA) requesting that it proceed to remove the member from the rolls of the Society, which in due course was done. The wisdom of waiting for a legal determination became evident in a subsequent case in which a member's conviction at one level of jurisdiction was reversed on appeal and the lower court did not see fit to retry the case. Another member elected to resign when asked by CMA to do so. Still another member charged under Bylaw 1, Section 7, sought to exhaust his remedies within the Society. He had been asked to resign by CMA. He had been granted a hearing before a Committee of nine Councilors appointed by the ACS President, and he then appealed to the President of the Society requesting permission to present his case before the Council within the specified one-hour limit. A Councilor moved to expel the member, but the Council voted instead to censure him. His plea was that in the activities involved he had not realized all of the implications of his deeds.

The effort on the pension problem was a highlight of CPR's activities in the middle 1960s. For almost 20 years, employment opportunities for trained scientists had been excellent, but then the situation changed. The first rumblings of discontent came to the CPR from the California Coordinating Committee (CCC), which represented the eight ACS local sections in California. The CCC, among other functions, was assisting members harmed by the Government's peremptory contract-award and -termination policies in defense- and aerospace-oriented industries. CCC pointed out that individual scientists could work successively for four or five companies, spend more than half of their professional careers in the process, and acquire no company pension benefits. A committee headed by the late Joseph Stewart made preliminary findings and recognized a rather widespread problem. Stewart then proposed to establish a portable pension plan—Pensions for Professionals (PFP). For the next five years, three as chairman of the Committee on Professional Relations, he spearheaded the drive.

CPR took the position that the pension was a right and not a benefit. To secure this right, CPR sought to establish an early vesting, fully portable pension plan. The committee had to educate itself about pension plans, the pension industry, and the statistical and actuarial bases of the plans. Through detailed surveys, it found that company pension plans varied from good to bad to none. For most companies, the pension was a reward for faithful service, and plans made little allowance for individual mobility. The turning point in CPR's effort came in 1970, when the ACS Board of Directors authorized an appropriation of up to $105,000 to explore the possibility of establishing a pension plan. The realities indicated that short-term employees could not be included—administrative and other costs probably would not permit the plan to vest in fewer than five years. In December 1970 PFP was launched as an independent corporation, and in September 1971 Dr. Arthur Hale became its administrative vice-president.

In November 1972 John Hancock Mutual Life Insurance Co. contracted to act as insurance carrier for PFP. By early 1973, PFP had agreed with the Internal Revenue Service on a form of organization that would not jeopardize the tax status of nonprofit participating societies. By mid-1974 each of eight participating societies had purchased a $5000 debenture to support PFP and had promised an additional $5000; the

ACS had purchased 18 debentures at a total cost of $90,000 against its original authorization of $105,000.

PFP launched a test-marketing program in August, but terminated it in April 1975 for lack of funds. A few articles in society news organs and displays at national meetings elicited more than 700 inquiries on pension problems from members of participating societies. About half of these were considered good prospects for PFP pension plans. Four corporate PFP pension plans were designed and adopted; some 40 additional corporate prospects were identified but could not be followed up for lack of funds. This marketing experience led PFP to conclude that at least three more years would be required to establish a viable pension program, and that the additional cost to the societies involved would be about $300,000. PFP officers felt that if marketing were successful, the initial investment would be recoverable in several years. The results of the marketing effort were reported to the participating societies.

In addition to CPR's work on Pensions for Professionals, the committee has intervened in pension problems of individual ACS members. One example involved two employees who were being terminated by a company with a 15-year vesting policy. On termination, one of these employees would have had service of 14 years, 11 months; at CPR's request, the company extended the termination date by two days, which gave the employee the 15 years required for vesting. The other employee's termination date gave him service of 14 years, four months; the company saw no feasible way to vest the employee without violating its 15-year policy.

The economic slowdown and cutbacks in space and military programs during the late 1960s and early 1970s created an uncertain employment situation for chemists and chemical engineers. New graduates had difficulty finding jobs. Companies in general hired few new scientists, and many reduced the numbers they employed, in some cases by means of mass layoffs. CPR investigated a number of terminations in the light of its experience with the kinds of situations that can arise, the range of treatment that terminated employees receive, and the kinds of problems they face. As a result of these investigations, the committee set out to develop a series of "Guidelines for Employers," an idea first advanced by Dr. Henry A. Hill, CPR chairman in 1968–69. The guidelines were formulated by a Subcommittee on Professional Standards and embodied standards for terms of employment, employment environment, pension plans, and termination conditions. The ACS Council adopted the "Guidelines for Employers" in September 1970; the Board of Directors endorsed them in June 1971.

The Guidelines recommended five-year vesting for pensions, which prompted the ACS as an employer to revise its pension plan to 100%

vesting in five years, effective in 1972; previously, employees had been 50% vested in 10 years and 100% in 20 years.

The Guidelines continued to be revised and refined and in 1975 were renamed "Professional Employment Guidelines" to reflect the inclusion of professional standards for employees as well as employers. Professional standards for chemists had been embodied originally in "The Chemist's Creed," a guide to ethics for the chemical profession, which was developed by CPR in 1965.

In 1970, economic problems and the consequent demands of ACS members for more professional activity led the Board of Directors to consider extending its efforts in that area. In 1971, President Melvin Calvin appointed President-Elect Max Tishler to chair an ad hoc Committee on Professionalism (COP), in addition to the Board's standing Committee on Public, Professional, and Member Relations. Dr. Tishler was succeeded as chairman in 1972 by President-Elect Alan Nixon. COP's basic assignment was to determine what level of professional activity was consistent with the requirements of the Society's federal Charter and favored tax status; when the committee voted to disband, at the end of 1972, it recommended that the ACS not change its legal structure because all proposals that had significant support could be implemented without such change.

One result of COP's work was the Professional Enhancement Program (PEP). To support its activities, PEP raised about $120,000 from individual ACS members. The money thus raised, though considerably less than the hoped-for $1 million, helped to support expanded employment-aid activities by ACS staff as well as permanent, intensified monitoring of employment supply-demand factors, including both the legislative and administrative activities of the Federal Government.

In employment conflicts that cannot be resolved by CPR mediation, the chemists involved may have legal recourse. A lawsuit can be costly, however, to the point of not being a feasible mechanism for the chemist whose employment problems have reduced or eliminated his income. To remove or at least to ease the financial difficulty, the Board Committee on Public, Professional, and Member Relations proposed that the Society establish a Legal Aid Fund; the move was approved by the Council and by the Board in June 1973. A chemist, whether or not an ACS member, may borrow up to $10,000 from the Legal Aid Fund. If his case has precedent-setting legal characteristics, the ACS Board may move to waive repayment of the loan.

The American Chemical Society has steered a difficult course between professional matters and its more traditional scientific and educational functions. In future years that course no doubt will take new directions in response to prevailing needs, and to help keep abreast of those needs the Society in 1974 formed a joint Board–Council Professional Program Planning and Coordinating Committee. Although interest in professionalism appears to wax and wane with the condition of the economy, it does seem clear, as the Hancock report put it, that the ACS ". . . cannot return even if it would to an exclusively impersonal interest in science." In fact, the membership ratified a Constitutional amendment in this vein toward the end of the employment inflection of 1968–72. The amendment, which took effect Aug. 10, 1972, added to the Society's Objects a new section: "To foster the improvement of the qualifications and usefulness of chemists, the Society shall be concerned with both the profession of chemistry and its practitioners."

The 24 journals and magazines published by the American Chemical Society and its divisions in 1975 had a total worldwide circulation of some 345,000.

CHAPTER IV

PUBLICATIONS

THE founders of the American Chemical Society and of most other early scientific societies considered the circulation of new scientific findings among members one of their primary goals. Accordingly, the first *Proceedings of the American Chemical Society* appeared in 1876, the founding year of the ACS. From that time on, an irregular but continuing series of steps has led to new journals, absorption of non-ACS publications, splitting, and reorganization. By the 1960s, the publications program, including *Chemical Abstracts* (whose history is related independently later in this chapter), was so voluminous that the Society was looked on by its admirers as one of the world's greatest publishers of scientific information and by its critics as nothing but a publishing house. By 1975, the ACS and its divisions were publishing 24 primary periodicals in chemistry (Table I), several annuals, and a considerable number of books and other nonperiodical publications. In addition, five ACS divisions were producing preprints or reprints of meeting papers, and many ACS local sections were publishing newsmagazines or newsletters in various forms. All ACS journals and magazines published at headquarters were being produced on microfilm as well as in hard-copy form, and the research journals were available on microfiche.

The Society has derived much income from advertising in its publications, but not without problems. For a time immediately after World War II, *Industrial & Engineering Chemistry* was so popular that its advertising volume was limited only by the amount of paper that could be allotted to it. Yet, a quarter-century later the core unit of that backbone of the applied publications—the group designed originally to serve industrial chemists and chemical engineers—was discontinued after several years of serious deficits. In the late 1950s and early 1960s, *Chemical*

TABLE I.

ACS Journals and Magazines—Total Circulation, 1975

Publication	*Foreign*	*Domestic*	*Total*
ACS Journals			
Chemical & Engineering News	11,377	110,081	121,458
Chemistry	2,542	25,896	28,438
Chemical Technology	3,536	11,719	15,255
Journal of Physical and Chemical Reference Data	417	742	1,159
Analytical Chemistry	9,614	22,826	32,440
Environmental Science & Technology	3,110	20,153	23,263
Accounts of Chemical Research	2,628	8,515	11,143
Chemical Reviews	2,561	3,051	5,612
Industrial & Engineering Chemistry—			
Process Design and Development	3,144	2,914	6,058
Fundamentals	2,946	2,934	5,880
Product R&D	3,072	2,878	5,950
Journal of Agricultural and Food Chemistry	2,473	2,510	4,983
Journal of Chemical and Engineering Data	1,000	1,089	2,089
Biochemistry	2,463	4,600	7,063
Inorganic Chemistry	1,628	3,273	4,901
Journal of the American Chemical Society	5,648	9,886	15,534
Journal of Chemical Information and Computer Sciences	797	1,158	1,955
Journal of Medicinal Chemistry	1,626	2,611	4,237
Journal of Organic Chemistry	3,508	5,967	9,475
Journal of Physical Chemistry	2,332	2,738	5,070
Macromolecules	888	1,537	2,425
Single Article Announcement	106	1,343	1,449
Total	67,416	248,421	315,837
Divisional Journals			
Journal of Chemical Education[a]	7,500	17,000	24,500
Rubber Chemistry and Technology[b]	1,300	4,000	5,300

[a] Owned and published by the Division of Chemical Education, Inc. Figures are for January 1976.

[b] Owned and published by the Rubber Division, Inc. Figures are for December 1975.

& Engineering News averaged net contributions to the Society of a half million dollars a year; a few years later, the magazine was running deficits of comparable size. It passed through the low point in 1970–71, and advertising thereafter improved, albeit slowly. On balance, however, during the Society's second half century, advertising in ACS publications contributed substantially to the support of publications and other activities as well.

In all but dollar-volume, the ACS publishing program, excepting Chemical Abstracts Service, reached a peak in 1967. In that year, 340,484 subscriptions carried a total of 38,052 editorial pages to subscribers; page impressions, including advertising pages, totaled more than 1.4 billion. While in the early years ACS publications were produced primarily with the members in mind, nonmember subscriptions had reached 135,090 by 1967, and ACS journals were going to 125 countries of the world.

In 1974, the Society's journals and magazines produced 896,677,347 page impressions and filled 323,115 subscriptions, of which 132,068 or 41% were from nonmembers. Subscription patterns had changed since the days when every ACS member received "all" publications: *JACS, Chemical Abstracts,* and the Industrial, Analytical, and News editions of *I&EC*. Since the beginning of 1934, the *News Edition, I&EC,* which became *Chemical & Engineering News* in late 1942, has been the only publication sent automatically to all members, except for occasional introductory issues of new publications. By 1975, the number of personal subscriptions by members had declined to the point where fewer than 45% of them were receiving any ACS periodical other than *C&EN*.

Printing

Early ACS relationships with printers seem to have been relatively unstable, since five different printing firms were involved in 17 years. But in 1893, long term stability set in when the Chemical Publishing Co. of Easton, Pa., became the printer. In 1900, Harvey F. Mack joined that company as bookkeeper and secretary. He soon owned part of the company and in 1905 formed the Eschenbach Printing Co., which thereafter handled the ACS printing.

In 1926, the company became Mack Printing Co. In 1956, when Harvey Mack retired as president, he was lauded by ACS leaders as a significant contributor to the success and standing of the American Chemical Society. Mr. Mack remained chairman of the company, but died in 1956. His successors as president, Cyrus S. Fleck, Harold S. Hutchison, and J. Wilbur Mack, his nephew, have not exerted so great an influence on the Society as Harvey Mack, but all have maintained active relationships with the ACS staff. The same was true of Mrs. Harvey Mack, who

Harvey F. Mack *(1878–1956) joined the Chemical Publishing Co., later Mack Printing Co., in 1900.*

succeeded her husband as chairman of Mack Printing and held the post until her retirement in 1972. She died in 1975.

The Mack Printing Co., although a substantial printer of non-ACS technical material, has remained close to the Society and its publications staff. Very importantly, the company has responded to ACS-expressed needs for modernization. Major evolution began in 1957 with the installation of a web-fed rotary press for printing *Chemical & Engineering News.* Later came such steps as photo-offset printing, computer-assisted typesetting, and accompanying technological developments that have kept Mack in step with the most modern systems practically available. At the end of 1975, under the ACS policy of regular review of Mack's competitive position, the company still held the ACS printing contract after 83 years.

Advertising

Advertising appeared in ACS publications from the earliest days, reaching a volume of $4286.12 in 1908. In 1909, however, the *Journal of Industrial & Engineering Chemistry,* a medium attractive to industry, came on the scene, and advertising became a more serious business. It appears that the Society's first advertising sales agent was Harvey F. Mack, head of the printing company. At Mr. Mack's request this arrangement was short-term (1910–11); another agent served for six years, then the ACS handled its own sales for three years.

In 1920, the Chemical Catalog Co., later Reinhold Publishing Co., became advertising sales agent for ACS publications. Through Reinhold Publishing came another figure from outside the chemical world who, like Harvey Mack, took a strong interest in ACS publications. Philip H. Hubbard succeeded Ralph Reinhold as president of that company in 1945 and held the position until 1968. While Mr. Hubbard ran a successful company, he acted from the position that ACS publications were of great value and ought to be supported beyond the normal commercial effort. This philosophy helped to bring *Analytical Chemistry* to strength, but failed to save *Industrial & Engineering Chemistry* in its decline in the 1960s.

In 1965, Mr. Hubbard, looking toward retirement, merged Reinhold with Medical Economics, Inc. Within two years, the combined operation was sold to Litton Industries, whose Reinhold division continued to sell advertising space in ACS publications. But at this time, advertising sales in general were declining steadily, and chemical advertising was no exception. In 1970, the 50-year relationship with Reinhold was ended, and a new organization, Century Communications, Ltd., soon to become Centcom, Ltd., was set up with ACS encouragement to sell advertising space exclusively for the Society. Unfortunately, the decline in sales had not yet ended, and predictions of upturns in advertising proved incorrect. Problems of capital plagued Centcom, and in 1973 the Society took over the organization as a wholly owned but separately operated corporation. The company proceeded under the presidency of Thomas N. J. Koerwer, formerly a vice-president of Reinhold Publishing Co., with a Board of Directors, a majority of whom are ACS officers or staff.

For a time *Industrial & Engineering Chemistry* was the leading source of advertising income to ACS, but *C&EN* assumed that position after World War II and continues to maintain it. *Analytical Chemistry* has developed a strong volume of advertising, and some of the other journals derive helpful levels of income from advertising.

In 1926, the Society's advertising income exceeded $100,000 for the first time. By 1940, it exceeded total ACS subscription income, including *Chemical Abstracts* subscriptions, and by 1943 was more than twice subscription income. Advertising volume reached $1,219,000 in 1950 and achieved a peak in 1967 with a total gross advertising income of $6,943,000 and income to ACS of $4,426,000. In 1974, the Society's income from advertising was $2,414,000.

Copyright

The ACS in its early days had not been especially sensitive to copyright; until 1940, in fact, the Society did not copyright its publications. Scientific journals in those days were not major items of commerce in the eyes of most publishers. In the interest of dissemination of information, the ACS had arranged with the Library of the U.S. Department of Agriculture for a photocopying service. ACS members could buy books of coupons which could be used to purchase photocopies of pages of any publications held in the USDA Library. On recommendation of its legal counsel, the Society ended this practice in 1957. In 1965, the Board of Directors took action with a bylaw stating that, "The Society shall own the copyright for the original and any renewal term thereof in any writing of an author which is published by the Society."

In 1969, a Board–Council Committee on Copyrights was established and began to study the matter as the effort by Congress to revise the copyright law focused attention on the unsatisfactory state of the existing law. The Williams & Wilkins case, in which that publisher sued the National Library of Medicine and the National Institutes of Health over photocopying of scientific journals, loomed as a landmark case. The Society filed briefs of *amicus curiae* in support of the Williams & Wilkins position in the Court of Claims Commissioner's hearing and in the full Court of Claims hearing, when the decision favorable to Williams & Wilkins was reversed. The ACS filed such briefs twice in the Supreme Court, first in urging the Court to accept the case and later for the actual hearing, which in 1974 resulted in a four-to-four deadlock with one Justice abstaining, thus leaving the decision favorable to the government libraries.

In Congressional hearings, Executive Director Robert Cairns twice testified on the ACS position, once before the pertinent Senate subcommittee in 1973 and once before the House subcommittee in 1975. The Society presented written positions and commentaries on several other occasions. When the Register of Copyrights and the Executive Director of the National Commission on Libraries and Information Sciences (NCLIS) established a Conference on the Resolution of Copyright Issues, the ACS was included. The Society's representative was among a small group of representatives of publishers and librarians that met extensively to work toward a compromise on the particularly sticky section of the House and Senate bills. The issue was the definition of "fair use" and the prohibition of the systematic production and distribution of single or multiple copies. The ACS held that copying should not be prohibited, as it could aid dissemination, but that some royalty should be paid. Without such payment, and in the face of dwindling subscription income, the cost of processing research papers to the stage of printing the first copy would have to be loaded on a decreasing number of subscribers to the point of serious financial harm to the journal system.

By late 1975, the subcommittee of publishers and libarians of the Conference had developed with the NCLIS a proposal for a study of copying in libraries and the testing of a mechanism for collecting royalties. Perhaps the most concrete step by Congress and the President was the appointment of a National Commission on New Technological Uses of Copyrighted Works. The commission's final report was to be submitted by Dec. 31, 1977.

The Publications Program

With the first issue of the *Proceedings of the American Chemical Society,* the ACS became the first society in this country publishing a journal specifically for chemists. At the first regular meeting of the ACS, on May 4, 1876, it was reported that arrangements had been made to publish the *Proceedings* in the *American Chemist* with reprints to be issued to ACS members under the title *Proceedings of the American Chemical Society.* The print order was 500 for the first issue of the *Proceedings,* which appeared in the June issue of the *American Chemist.* This practice continued into 1877, when the *American Chemist* ceased publication and the *Proceedings* became a publication on its own. The *Proceedings* of 1876 and 1877 were combined and issued as Volume I. The second volume covered 1878; then, in 1879, the name was changed to the *Journal of the American Chemical Society,* and the journal began to appear monthly.

During the Society's first three years, the *Proceedings* and *Journal* were headed by a Committee on Papers and Publications, of which Dr. Hermann Endemann, the first Editor of the *Journal,* was a leading figure. Endemann, German-born and German-educated, came to the U.S. in 1867 to become assistant to Prof. Chandler at Columbia College School of Mines. Dr. Gideon Moore, born in New York, educated in Germany, and a chemist at the Passaic Zinc Co., became Editor in 1880, but resigned in 1881. Dr. Endemann returned for a while, but in 1882 the reelected Committee on Papers and Publications resigned after a few months; a new committee was elected but no Editor was appointed.

In 1884, a new committee was elected, and a new Editor, A. A. Breneman, was appointed from its ranks. Dr. Breneman was the first American-educated Editor, having been trained at Pennsylvania State College. During his nine years in the post, the Society and its journal went through a difficult period, and a special committee was appointed to seek pledges of money as well as scientific contributions. But by 1892, the Society was reorganized and assured of its position as the national chemical society, and the journal entered on its path of growth and success.

Late in 1892, Dr. Edward Hart, a professor at Lafayette College and owner and editor of the *Journal of Analytical and Applied Chemistry,* was persuaded by his friend Harvey Wiley, then ACS President, to merge his successful and remunerative journal with *JACS* in January 1893 and

PROCEEDINGS

OF THE

AMERICAN CHEMICAL SOCIETY.

VOL. I. NO. 1.

CONTENTS:

I. Proceedings of the First Meeting for Organization.......... 3

II. Proceedings of the Second Meeting for Organization 18

III. Proceedings, May 4th, 1876....... 20

IV. Proceedings, June 1st, 1876...... 22

V. On the Determination of the Relative Effectiveness of Disinfectants. By Hermann Endemann, Ph.D...................... 24

VI. On the Amalgamation of Iron, and of some other Metals. By P. Casamajor.................... 49

VII. On the General Occurrence of Vanadium in American Magnetites. By Isidor Walz, Ph.D.... 58

VIII. Action of Impure Rain Water on Lead Pipes. By Paul Schweitzer, Ph.D.............. 66

IX. On the Cause of Discrepancies in the Estimation of Silver in Pig Lead. By Paul Schweitzer, Ph.D............................ 67

X. Analysis of the Gneiss of Manhattan Island. By Paul Schweitzer, Ph.D....................... 69

XI. Bismuth in Lead in the Manufacture of White Lead. By Hermann Endemann, Ph.D......... 70

XII. Kerosene Oil. By H. B. Cornwall. 71

XIII. Preliminary Note on Litmus. By Howard W. Mitchell........... 79

NEW YORK:

JOHN F. TROW & SON, PRINTERS.

1876.

to accept the nonpaying editorship of the latter. Furthermore, Hart urged his subscribers, contributors, and other supporters to join the Society. It was at this time that the Chemical Publishing Co., owned and operated by Prof. Hart, became the printer for ACS. Dr. Hart, it seems, was a man of parts, as he also led the development of a successful company, Baker and Adamson, and served as its president until it was bought by General Chemical Co., later a part of Allied Chemical & Dye Co., now Allied Chemical Corp.

With Hart's skills as editor, printer, and businessman and Wiley's ability to get papers submitted, *JACS* began to prosper. During 1897 to 1901 it averaged about 1300 pages, compared with 419 pages for 1892. Total expenditures in 1901 were about $4700. Receipts from nonmember subscriptions, sales of back numbers, and advertisements totaled about $2700. The net cost of about $2000 for the 1935 ACS members who received the journal without subscription charge was not considered high. More than two-thirds of the $8750 in dues collected by the Society in 1901 was available for other expenses, and the officers considered the journal successful, both scientifically and financially.

Review of American Chemical Research was a journal established by Dr. Arthur A. Noyes (ACS President in 1904) of the Massachusetts Institute of Technology, to provide abstracts or brief reviews of chemical papers appearing in other journals. Its first two volumes appeared as part of MIT's *Technology Quarterly,* but its third was published as part of the *Journal of the American Chemical Society.* In 1902, Dr. Noyes announced that MIT would no longer be able to provide the review. Just before then, in January 1902, Dr. W. A. Noyes, of Rose Polytechnic Institute, Terre Haute, Ind., and later of the U.S. Bureau of Standards and the University of Illinois, became Editor of *JACS.* He also took over and reorganized *Review of American Chemical Research,* which had been appearing since 1897 as a part of *JACS.* The improvement and enlargement stimulated further the rising demand for an American worldwide abstract journal, which the Society launched in 1907 as *Chemical Abstracts* with Noyes as Editor.

In 1907, the Society appointed a committee to study the establishment of an industrial journal and in January 1909 launched the *Journal of Industrial & Engineering Chemistry.* Dr. W. D. Richardson, a chemist with Swift & Co., was *I&EC*'s first Editor. Richardson was succeeded in 1911 by Dr. M. C. Whittaker, professor of chemical engineering at Columbia University. When Whittaker resigned, in 1916, to become vice-president of the U.S. Industrial Alcohol Co., he recommended that a full-time editor be appointed. Accordingly, in 1917, Charles H. Herty, professor of chemistry and Dean of the School of Applied Science of the University of North Carolina, as well as ACS President in 1915 and 1916,

The Journal of Industrial and Engineering Chemistry

PUBLISHED BY

The American Chemical Society

EASTON, PA.

Vol. 1, No. 1 | JANUARY, 1909 | MONTHLY $6.00 Per Year

Board of Editors

Table of Contents

EDITORIALS:

The Industrial Chemist and His Journal. By T. J. Parker . . 1
The Ethics of Engineering Supplies. By William H. Walker . 2
The Sugar Industry in its Relations to the United States. By W. D. Horne . . . 4
The Fixation of Nitrogen. By F. B. Carpenter . . . 4
Standard Methods of Analysis. By W. D. Richardson . . . 5

ORIGINAL ARTICLES:

Free Lime in Portland Cement. By Alfred H. White . . . 5
Notes on Anthracite Producer Practice. By George C. Stone . . . 11
A Comparison of the Calculated and Determined Viscosity Numbers (Engler) and Flashing and Burning Points in Oil Mixtures. By H. C. Sherman, T. T. Gray, and H. A. Hammerschlag . . . 13
A New Bomb Calorimeter. By Chas. J. Emerson . . . 17
The Manufacture of Oil of Lemon and Citrate of Lime in Sicily. By E. M. Chace . . . 18
Turpentine and Its Adulterants. By Arthur E. Paul . . . 27
The Determination of Total, Fixed and Volatile Acids in Wines. By Julius Hortvet . . . 31
A Comparsion of Methods for the Preparation of Milk Serum. By Hermann C. Lythgoe and Lewis I. Nurenberg . . . 38

NOTES:

On the Use of Incandescent Lamps with Volatile Solvents . 40
The Bureau of Standards' Analyzed Samples . . . 41
Report of the Committee on the Analysis of Phosphate Rock 41
A Test for Skin Pulp in Tomato Catsup . . . 44
The Committee on Analysis of Fats, Soaps, and Glycerine . . 44

QUOTATIONS:

Chilean Nitrate Fields . . . 45
Catalytic Reduction of Fats and Oils . . . 47
Indian Indigenous Dyes . . . 47
Concrete Railway Ties . . . 48

BOOK REVIEWS:

Metallurgy; Electro-Metallurgy; Der Betriebs-Chemiker; The Chemical Analysis of Iron; Laboratory Guide of Industrial Chemistry; Soils and Fertilizers; The Manufacture of Lubricants, Shoe-Polishes, and Leather Dressings; Liquid and Gaseous Fuels and the Part They Play in Modern Power Production; The Power Handbooks; Methods and Devices for Bacterial Treatment of Sewage . . . 49

NEW BOOKS . . . 52

SCIENTIFIC SOCIETIES:

American Chemical Society; American Electrochemical Society; Western Association of Technical Chemists and Metallurgists; Association of Official Agricultural Chemists; American Association for the Advancement of Science; American Institute of Mining Engineers; American Leather Chemists' Association . . . 53

TRADE AND INDUSTRIAL NOTES . . . 57

OFFICIAL REGULATIONS AND RULINGS . . . 59

Application made at the Post-office at Easton, Pa., for entry as Second-class Matter

Papers intended for publication in the Journal of Industrial and Engineering Chemistry should be sent to the editor, W. D. RICHARDSON, 4306 Forrestville Ave., Chicago, Ill.

Papers intended for publication in the Journal of the American Chemical Society should be addressed to the editor, WILLIAM A. NOYES, University of Illinois, Urbana, Ill. Remittances for annual dues or for subscriptions, changes of address, orders for subscriptions or back numbers, claims for missing numbers and all matter intended for the Proceedings should be sent to the Secretary, PROF. CHARLES L. PARSONS, New Hampshire College, Durham, N. H.

Claims for missing numbers will not be allowed if due to failure to give sufficient notice of change of address and in no case if received later than sixty days from date of issue.

became the first full-time Editor of *I&EC*. These developments brought the Society—now with a strong national organization having a fundamental journal, an industrial journal, and an abstract journal—to the U.S. entrance into World War I and the resulting isolation from Europe.

Between World Wars I and II, ACS membership more than doubled, and the journals grew and prospered along with the profession of chemistry. One important change was the accession of Harrison E. Howe as Editor of *Industrial & Engineering Chemistry* at the beginning of 1922. Howe had begun his career as an industrial research chemist. In 1919 he became chairman of the Division of Research Extension of the National Research Council, a position he left to assume the editorship that he held until his untimely death in December 1942. This energetic, dapper man, with his wing collar and elegant Vandyke beard, moved about the Society with stimulating effect and furthered the development of *I&EC* as a publication of high standing. Howe's editorship was characterized by an eminent professor, whose papers were appearing mostly in *JACS,* who once remarked that when he wanted to know where chemistry was really going he read *I&EC*.

Howe took one step that had a lasting effect on the Society's publishing history and perhaps on the shape and character of the ACS. He divided *I&EC* into three editions—Industrial, Analytical, and News. Each of these was to become an individual publication of wide influence, and the latter two were to outlive the parent. The *News Edition,* born in 1923, appeared on the 10th and 20th of each month, and the *Industrial Edition* appeared on the 1st. The *Analytical Edition* first appeared in January 1929 as a quarterly.

W. A. Noyes was succeeded as Editor of the *Journal of the American Chemical Society* in 1918 by Arthur B. Lamb of Harvard University. An earlier history of the ACS notes that the remarkable success of *JACS* under Lamb's editorship "was due largely to the breadth and thoroughness of his education and his great versatility." Lamb's bachelor's and master's degrees were in biology; he received a Ph.D. in organic chemistry at Tufts College and a Ph.D. in physical chemistry from Harvard in the same year.

Lamb was faced with the growing size of *JACS,* a problem that Noyes had struggled with and that remains one of the journal's greatest difficulties 58 years later. Another problem was the practice of publishing a paper in more than one journal, which Lamb ended in *JACS,* noting that his actions "occasioned us some embarrassment." He also pressed against publication in the journal of doctoral theses in full. Lamb found that the burdens of selecting papers for publication had become too heavy for the board of associate editors. He developed a system in which each manuscript was sent to one or two referees who were specialists in that field or in one closely related and who remained anonymous to the author. In cases of continuing doubt the matter was settled by the associate editors. This system, under a certain amount of continuous fire, has persisted ever since. Lamb commented, and most editors probably would agree, that most of the success of *JACS* was due to the "devoted, arduous, largely unrecognized, unrequited, and unselfish services" of referees.

Shortly after World War I, the Society began to edge into book publishing. In 1919, it started the "ACS Monograph Series," with Scientific Monographs edited by W. A. Noyes and Technological Monographs by John Johnston and Harrison Howe. These books were manufactured and marketed by the Chemical Catalog Co. Following the deaths of the original editors, F. W. Willard, president of the Nassau Smelting and Refining Co., a subsidiary of Western Electric Co., assumed the editorship of both series. In 1944, the two were fully merged and became the "ACS Chemical Monographs." Upon Willard's death, in 1947, William Hamor of the Mellon Institute assumed the editorship.

The ACS undertook one other new type of publication in the decade following World War I. In 1923, the Division of Chemistry and Chemical Technology of the National Research Council suggested to the Society the desirability of a special journal for review articles. The ACS decided to issue such a journal, under its own auspices and control, if arrangements could be made for its publication without cost or financial risk to the Society. A suitable contract was negotiated with the Williams & Wilkins Co. of Baltimore; W. A. Noyes took on one more editorship, and in April 1924 the quarterly *Chemical Reviews* appeared. After three years, Gerald K. Wendt, Dean of the School of Chemistry and Physics at Pennsylvania State College, became Editor of *Chemical Reviews.* The publication grew and in 1932 became a bimonthly. When Wendt retired from the editorship, a new Noyes—W. Albert Noyes, Jr.—appeared on the ACS editorial scene and began the long and productive part of his career devoted to that kind of work. Noyes served for 12 years as Editor of *Chemical Reviews,* resigning at the end of 1949 to succeed Arthur Lamb as Editor of the *Journal of the American Chemical Society.* The son succeeded the Editor who had succeeded his father, and all three were ACS Presidents—Noyes, Sr., in 1920, Lamb in 1933, and Noyes, Jr., in 1947.

While the ACS Board of Directors was responsible for the origin and development of the publications that have been described, two journals originated with ACS divisions in the period between the wars. Both have remained divisional property and have become important chemical publications. One is the *Journal of Chemical Education;* the other is *Rubber Chemistry & Technology.*

The *Journal of Chemical Education* was started in 1924 as a result of the efforts of Neil Gordon of the University of Maryland and later of Wayne State University. In proposing the journal to the Society, in 1923, Gordon was backed by the Division (then the Section) of Chemical Education and strengthened by some $2000 worth of advertising he had sold in advance. The Board, convinced, approved the idea with the stipulation that the ACS would be in no way responsible. Gordon did the first year's work on the struggling publication almost single-handedly.

The *Journal of Chemical Education*'s difficulties were solved early by the interest of the Chemical Foundation, Inc., headed by Francis P. Garvan. Through the Foundation, Mr. Garvan supported the journal and even paid the Editor's salary for nine years. The Division of Chemical Education and the Chemical Foundation also undertook responsibility for a publication for high school students, *The Chemistry Leaflet,* edited by Pauline Beery Mack of Pennsylvania State College (no relation to the Macks of Mack Printing).

The great depression caused the Chemical Foundation to withdraw its support in 1932. The Division signed an agreement with Mack Printing Co. for the support of the *Journal of Chemical Education,* which was saved only through the interest of Harvey Mack. The *Journal*'s offices were moved into the Mack plant at Easton, Pa.; the Chemical Education Publishing Co. was organized to handle its business affairs while the division remained responsible for editorial content. *The Chemistry Leaflet* was left to Mrs. Mack (it came back to the ACS in 1962 under the name *Chemistry*).

In 1933, Neil Gordon resigned as Editor, and Otto Reinmuth of the University of Chicago moved into the chair. Norris Rakestraw of Brown University became Editor in 1940, the year in which income exceeded expenses for the first time. William Kiefer of the College of Wooster became Editor in 1955 and Thomas Lippincott of Ohio State University and later the University of Arizona in 1967.

Rubber Chemistry & Technology was started in 1928 by the ACS Division of Rubber Chemistry. It published in its first issue all papers presented at an earlier meeting of the Division, plus some papers from abroad. Under its first Editor, C. C. Davis of the Boston Woven Hose & Rubber Co., the journal worked effectively to reflect the activity in rubber chemistry and technology throughout the world. At first it carried predominantly reprints, including translations of foreign papers, of the most important papers on fundamental research, technical developments, and chemical engineering problems related to rubber. In recent years, the journal has published an increasing volume of original papers, which by 1975 comprised about two thirds of its editorial content. *Rubber Chemistry & Technology* is a quarterly, but since 1957 has published a fifth issue, *Rubber Reviews,* comprised entirely of invited reviews in rubber chemistry and technology.

Davis was succeeded as Editor by David Craig of B. F. Goodrich Co. in 1957. On Craig's death in 1964 the editorship was assumed by Edward M. Bevilacqua of U.S. Rubber who held it until his death in 1968. Earl Gregg of Goodrich then took the position and was succeeded in 1975 by H. K. Frensdorff of Du Pont.

In 1933, Wilder D. Bancroft of Cornell University gave his publication, *Journal of Physical Chemistry,* to the Society. Bancroft had founded the publication, in 1896, and financed it personally during its early struggles. It was aided in the early part of the great depression by the Chemical Foundation, but that support ceased in 1933. Upon receiving the journal from Prof. Bancroft, the ACS negotiated a contract with the Williams & Wilkins Co. for its publication and appointed S. C. Lind of the University of Minnesota as Editor, starting in 1933.

The *Journal of Physical Chemistry* had a special relationship with the ACS Division of Colloid Chemistry from 1933 through 1950, under which it published the papers of the division's annual Colloid Symposium in its June issue. This relationship was recognized in 1947 by a change in name, to the *Journal of Physical and Colloid Chemistry;* in 1951, the journal reverted to its original name. In 1952, W. Albert Noyes, Jr., Editor of *JACS,* also became Editor of the *Journal of Physical Chemistry,* succeeding S. C. Lind, and the ACS took on full operation of the journal, ending its contract with Williams & Wilkins.

In February 1943, Walter J. Murphy succeeded Harrison E. Howe as Editor of *Industrial & Engineering Chemistry.* Murphy, who was trained as a chemist, had been in industrial research, sales, and management; he entered editorial work with *Chemical Marketing,* which became *Chemical Industries,* of which he was Editor and Manager (it was later purchased by McGraw-Hill and converted to the weekly *Chemical Week*). As Murphy told the story, ACS Secretary Charles L. Parsons came to him after Howe's rather sudden death and offered him the editorship. Murphy told Parsons he liked the idea and was sure he would like working with Parsons, whom he admired, but was equally sure that if he had to work under him (as Howe had), they would be fighting futilely from the beginning. Parsons was agreeable to a different arrangement, and the ACS structure was set up so that both the Secretary and the Editor of *Industrial & Engineering Chemistry* answered to the Board of Directors. The pair worked well together during the remainder of Parsons' service.

Murphy also became Editor of *Chemical & Engineering News,* renamed only the previous year after its 19 years as the *News Edition, I&EC.* He had a feeling that the chemical industry needed and would support a newsmagazine that could cover events on the spot, but his staff was small and located entirely in Washington except for David Killeffer in New York. Murphy, therefore, set about convincing the Board of Directors of the value of a geographical network of reporters. The first step was to hire Robert F. Gould, Editor of *The MemphIon*, the publication of the ACS Memphis Section, and one of Murphy's voluble critics, as an associate editor in an office in Chicago in 1945. Within 18 months, Murphy had also established news offices in Houston and San Francisco, and his program was on its way. In 1947, *C&EN* was converted from a biweekly to a weekly.

The field staff worked with the Industrial and Analytical Editions of *Industrial & Engineering Chemistry* as well as with *C&EN*. F. J. Van Antwerpen, who had worked under Howe and was Managing Editor of the Industrial Edition, left early in 1946 for a post with the American Institute of Chemical Engineers, and Murphy replaced him with D. O. Myatt, whom he had just hired for the field staff.

The ACS headquarters, when Murphy arrived, had been almost a family-type operation—so much so that during the humid summer months in Washington, before office air conditioning was widespread, the staff picked up typewriters and files and moved as a unit to the Marine Biological Laboratory at Woods Hole, Mass. In those days the applied publications staff enjoyed the services of a number of ladies who had virtually started their careers with the ACS and, through reviewing and editing manuscripts and assisting Howe, had built a wide acquaintance among members of the Society. The names of Gladys Gordon, Stella Anderson, Helen Newton, and Nellie Parkinson in Washington and Bertha Reynolds and Charlotte Sayre of the editing staff in the printing plant in Easton, Pa., were familiar to chemists all over the country.

A new group now joined the editorial staff under Murphy, who still operated very much in the fatherly mode. These new staff members, all trained as chemists or chemical engineers, were journalistically awkward, but enthusiastic. They charged about the country to endless meetings, wrote masses of recitative copy, and took rolls of line-up meeting photographs. But it was a start. It began the conversion of a good society house organ into an international chemical newsmagazine. By the summer of 1946, Murphy had James M. Crowe as his executive editor, two reporters in New York, and one each in Chicago, San Francisco, and Houston. Joseph H. Kuney, a chemist who had worked his way through school in the printing business, joined the group and later became production manager.

The Analytical Edition, meanwhile, although it had not yet blossomed, had not been neglected. In 1943, Murphy enlisted two eminent analytical chemists, Lawrence T. Hallett of General Aniline & Film as associate editor and Ralph H. Müller of New York University as columnist. By

1948, the Analytical Edition had been made a separate publication called *Analytical Chemistry* with Hallett as Managing Editor.

In 1949, the applied publications group launched the "Advances in Chemistry Series" with James Crowe as managing editor. This series of books was designed to publish collections of papers from symposia that were considered valuable contributions but that could not be published in full in the various journals of the Society. By the spring of 1951, four volumes had appeared, several more were in process, and the concept of the series was being broadened to include collections of highly specialized data, such as "Azeotropes and Nonazeotropes."

In 1950, Murphy took one more major step by sending Richard Kenyon from the Chicago office to open an office in London. Kenyon's task was to report on the postwar recovery of the European chemical community for the applied publications and to lay a base for future relations. The publications have maintained operations abroad ever since.

In 1952, the *Journal of the American Chemical Society,* because of its volume of papers, was converted from a monthly to a semimonthly. *JACS,* a strong center of attention, almost always was immersed in controversy. At this time the ferment concerned the handling of Communications. In 1951, the Council Committee on Publications undertook a study of author and reader attitudes regarding "Communications to the Editor." A large majority favored continuation. Authors showed concern over priority treatment, the follow-up with a definitive paper, and the choice of journal into which that paper should go. These problems remained subjects of debate for many years, but no specific change in policy was adopted.

In 1953, the Society established the biweekly *Journal of Agricultural and Food Chemistry* with Walter Murphy as Editor. The publication differed somewhat from any existing ACS journal in that, in addition to publishing research papers, it drew on the applied publications field staff as well as on a central staff of its own for a considerable part of its content. The journal also published feature articles by invited authors. It was willing to carry advertising, but the response was not strong.

Also in 1952 there developed a body of feeling, particularly among organic chemists, that the Society ought to have an outlet for papers on organic chemistry in addition to *JACS.* The Board of Directors authorized negotiations to acquire the *Journal of Organic Chemistry* from Williams & Wilkins; in 1955, *JOC* was finally brought into the ACS stable. George Coleman of Wayne State University was Editor. The journal had been conceived by Morris Kharasch of the University of Chicago; the first issue had appeared in 1936.

The push to broaden the publications program was obvious in the early 1950s. The Board voted down a proposal to split *JACS* into two editions, one for physical and the other for organic chemistry. The problems of distribution of papers among the prestigious *JACS*, the *Journal of Organic Chemistry,* and the *Journal of Physical Chemistry* were discussed at the summer conference of the Board Committee on Publications in 1954. A plan was advanced to convert *JPC* into a quarterly symposium journal with all other physical chemistry papers turned over to *JACS,* which then would appear monthly in three sections—general interest papers, physical and inorganic papers, and organic and biological papers. Subscribers would receive the general section and a choice of the other two. No change in policy was made, however.

The 1954 summer publications conference did produce one clear mandate: each publication should stand on its own feet financially. At that time, the cost per volume of every ACS journal was greater than the member subscription price, and nonmember subscription rates covered costs for only two journals. The new *Journal of Agricultural and Food Chemistry* was not gaining the projected amount of advertising income and was operating at a considerable deficit. The Board voted to convert it from a biweekly to a monthly to trim losses.

In 1955, in an important policy step, the Board established a journal reserve fund with two purposes: to avoid arbitrary curtailment of ACS journals when the annual income of one or more did not cover annual expenses; and to finance special publications and projects. In that year the Board was still wrestling with the problem of volume of pages and the resulting deficits of the journals. The summer publications confer-

***Walter J. Murphy** (1899–1959) became Editor of Industrial & Engineering Chemistry in 1943 and was Editorial Director, Applied Journals, at his death.*

ence received a suggestion from the Council Committee on Publications that scientific articles in ACS journals be published in condensed form with the full text available only in miniprint.

At about this time the Board reorganized the applied publications. A new position, Director of Publications, Applied Journals, was established with responsibility for the editorial, advertising, circulation, promotion, and production functions. Late in 1955, C. B. Larrabee, who had been successively editor, president, and chairman of *Printers Ink,* a trade publication serving the publishing industry, was appointed to the post. Walter Murphy was made Editorial Director, Applied Journals. A few months later, in 1956, editors were appointed to succeed him on each of the applied publications, as follows—*C&EN,* Richard L. Kenyon; *I&EC,* Will H. Shearon, Jr.; *Analytical Chemistry,* Lawrence T. Hallett; *Journal of Agricultural and Food Chemistry,* Rodney N. Hader.

In 1958, the Society's contract with Williams & Wilkins for publication of *Chemical Reviews* came to an end, and the ACS assumed control of the journal, which was printed thereafter at Mack Printing Co. In 1959, the ACS created the quarterly *Journal of Chemical & Engineering Data* —designed to publish data of interest to a limited audience—from the Chemical Engineering Data Series begun in 1956. Walter Murphy died late in 1959 and was succeeded as Editorial Director by Richard L. Kenyon. In 1960, Robert F. Gould was appointed Editor of the "Advances in Chemistry Series."

The Society's publications program expanded considerably in 1960–62. Leaders in several ACS divisions had been vigorously expressing needs for ACS journals devoted specifically to their disciplines. As one result, three journals authorized in 1958 came into being in 1961–62.

The *Journal of Chemical Documentation,* with Herman Skolnik of Hercules, Incorporated as Editor, began publication as a quarterly in 1961. It published papers from the Division of Chemical Literature as well as papers in related fields. The bimonthly *Biochemistry,* with Hans Neurath of the University of Washington as Editor, appeared in 1962. And the quarterly *Inorganic Chemistry,* with Robert W. Parry of the University of Michigan as Editor, appeared in the same year.

During 1961, the ACS had purchased the *Journal of Medicinal and Pharmaceutical Chemistry* from Inter-Science Publishers, Inc. The Society took over the publication of the journal at the beginning of 1962, with Alfred Burger of the University of Virginia as Editor. A year later the name was changed to *Journal of Medicinal Chemistry.*

Even before the advent of these new publications, the problems of managing the Society's research journals and the associated financial matters had become so complex that the Board felt that it could not properly plan the development of the journals, nor could an already overburdened Executive Secretary. Thus a new position, Director of Planning, Fundamental Journals, reporting to the Executive Secretary, was established, effective in September 1960. Richard H. Belknap, who had been business manager for the Society, was appointed to the post.

The early 1960s also saw changes in the older research journals. Frederick D. Greene of Massachusetts Institute of Technology succeeded George H. Coleman as Editor of the *Journal of Organic Chemistry,* beginning in 1962. W. Albert Noyes, Jr., who had been carrying the editorship of both *JACS* and the *Journal of Physical Chemistry,* asked to be relieved of the *JACS* post. Marshall Gates of the University of Rochester, who had been assisting Noyes for some years on *JACS,* succeeded him as Editor in 1962.

In 1961, the Federal Council for Science and Technology announced the policy that page charges could be paid from federal research grants when results of work supported by those grants were published. The policy was based on the principle that research is not completed until the results are published; in general, the Council felt that page charges should be such as to pay the cost of bringing a paper to the point of readiness for printing the first copy of the journal. The Society previously had taken a conservative position on such matters, believing that it should bear all costs for all of its programs. By this time, however, costs were rising, and the arguments for accepting assistance were compelling. Late in 1961, the Board voted that papers accepted in all ACS journals except *C&EN, Chemical Reviews, I&EC, Analytical Chemistry,* and the *Journal of Agricultural and Food Chemistry* should be subject to page charges.

In 1962, the Board voted that after that year the journals should receive income from subscriptions, page charges (where assessed), advertising revenue, member dues allocation, Corporation Associates dues, investment income from the journals fund, and investment income from the Corporation Associates fund. Also in 1962, the Board established a financial policy that the journals should be managed so that, as a group, they would not incur a long-term net loss or gain. Short-term net losses or gains were acceptable for the group and for individual journals so long as the group broke even in the long run.

Also in 1962, the Society organized a staff corps to handle journal manuscripts after their acceptance by editors; the group occupied quarters at the Mack Printing plant in Easton, Pa. To direct the new group, the ACS secured the services of Allen D. Bliss, professor of analytical chemistry at Simmons College in Boston. In 1927, when Bliss was a graduate student at Harvard, he had begun handling final preparation of manuscripts and proofs for *JACS* under Arthur Lamb and had continued that work even after he had become a professor at Simmons. In

1950, he was made an assistant editor of *JACS*. In late 1962, he showed his devotion to the Society's fundamental journals by resigning his post at Simmons and moving to Easton to direct the new editing group. Within a year of that move, unfortunately, Bliss fell ill and died. He was succeeded by Charles Bertsch, formerly a chemist with Pennsalt Co. and currently head of the Editorial Processing Department in Easton. Dr. Bertsch has led the development of a skilled group at the printing plant that is responsible for perfecting manuscripts as well as for the work of the printer in developing better techniques for moving from manuscript to printed page.

Because of the growing complexity of the ACS publishing program a Special Committee of Editors was formed in 1964 to improve the services of the fundamental journals to authors and subscribers. Annual conferences of editors and editorial staff were begun to deal with the attendant problems, both great and small. In the same period, the manuscript reviewing and editing groups in Washington and Easton began jointly to develop a manual for the guidance of authors. This was no easy task. The number of journals, editors, and editorial and advisory boards by then was considerable. Rules of style and other standards often had been developed autonomously. Even such matters as bibliography and placement of footnotes differed so much that an author whose paper was written for one journal but later offered to another faced a burdensome amount of reorganization. Individual journal editors exercised their autonomy and expressed their convictions in firm fashion. Still, by 1967 the first edition of the "ACS Handbook for Authors" was produced. It was made available to all subscribers to the Society's journals, and a copy was given to each ACS member enrolled after January 1968. Perfect consistency of style had not been achieved among the journals, but progress had been made.

The Applied Journals also began to change at the beginning of the 1960s. *Industrial & Engineering Chemistry* was not maintaining the status it had once enjoyed, and it was not doing well financially, particularly because of lack of advertising. Federal support of research was high; trends toward more fundamental and less "applied" research in the universities were strong. *I&EC* was criticized in a number of quarters as being too much applied and not offering a desirable publishing medium for leaders in research fundamental to chemical engineering and industrial chemistry. In response to these attitudes, the Society decided to publish three quarterly supplements for research papers, with *Industrial & Engineering Chemistry* continuing as the core publication and carrying general papers relating to industrial chemistry. The three research supplements appeared in 1962. *I&EC Fundamentals,* with Robert L. Pigford of the University of Delaware as Editor, began publishing papers on the fundamentals of chemical engineering. It was followed by *I&EC Process Design and Development,* with Hugh Hulburt of American Cyanamid Co. and later Northwestern University as Editor, and *I&EC Product Research and Development,* with Byron Vanderbilt of Esso Research & Engineering as Editor.

These new supplements, along with the *Journal of Chemical & Engineering Data,* formed a group of specialized journals around the original *Industrial & Engineering Chemistry* which provided a clear categorization of papers for the reader. The change also made some authors more comfortable, as a number of the younger academic chemical engineers had been objecting to having their fundamental papers published in the same journal with papers on applied research and development. The subscriber received the general edition of *I&EC* plus the quarterly of his or her choice and could receive the other quarterlies at moderate additional cost.

In 1962, Gordon Bixler, Managing Editor of *Chemical & Engineering News,* was appointed Editor. He succeeded Richard L. Kenyon, who had become Editorial Director of the Applied Journals in 1959 and had continued to edit *C&EN* since that time. Late in 1962, C. B. Larrabee retired as Director of Publications, Applied Journals; he was succeeded by Richard L. Kenyon.

The same period saw the Society's return to a publication for secondary-level students. *The Chemistry Leaflet* had been given up by the ACS in 1932 for lack of funds. Since then it had been taken over by Science Service, renamed *Chemistry,* and published as a pocket-size magazine. In 1962, Watson Davis, head of Science Service, came to the ACS with problems of finanical support for *Chemistry.* After some discussion, the Society agreed to buy the publication and publish it.

Rodney Hader, Editor of the *Journal of Agricultural and Food Chemistry* and by then Executive Assistant to the Editorial Director, Applied Journals, became Acting Editor of *Chemistry* and began the process of developing it into an ACS publication. An advisory panel chaired by L. Carroll King of Northwestern University undertook a study that led to the conclusion that a need existed for a publication to better serve students at the secondary-school level. Dr. O. T. Benfey of Earlham College and later of Guilford (N.C.) College agreed to become Editor and assumed his duties in September 1963. The magazine's physical dimensions were enlarged to 8½ x 11 and its layout redesigned; the first issue in the new format appeared in January 1964.

Also in 1963, Will H. Shearon, Jr., was relieved as Editor of *Industrial & Engineering Chemistry* because of ill health and became Consulting Editor. He was succeeded as Editor by David E. Gushee, who had joined the *C&EN* staff in 1956 and became Managing Editor of *I&EC* in 1962.

Shearon remained Editor of the *Journal of Chemical and Engineering Data* until his death in September 1963, when Rodney N. Hader became Acting Editor.

A number of changes in editorships occurred during this period. Robert Parry asked to be relieved as Editor of *Inorganic Chemistry* and was succeeded in 1964 by Edward L. King of the University of Colorado. In that year both *Inorganic Chemistry* and *Biochemistry* became monthlies. W. Albert Noyes, Jr., retired as Editor of the *Journal of Physical Chemistry,* and for the first time in the 20th century, no ACS publication was being edited by a Noyes. Frederick T. Wall of the University of California at San Diego was appointed Editor of *JPC* beginning in 1965.

The "Monograph Series" had been allowed to fall into a poor state during the ill health of William Hamor, the Editor, who had died in 1961, and the Board of Directors was inclined to discontinue it. On the other hand, there had been indications from members that quite a number of chemists felt that if the Society did not publish books in certain areas not highly attractive to commercial publishers, such books would not get into print. After considerable agonizing the Board decided that the series should be revived and continued, but along more scientific lines than had been pursued in the industrially oriented monographs of then-recent years. Near the end of 1963, Frederick M. Beringer of the Polytechnic Institute of Brooklyn was named Editor of "ACS Monographs." The first volume under the new Editor was "Formaldehyde," issued in 1964. The new scientific orientation began to show with the publication of "Molecular Basis of Virology," by H. Frankel-Conrat, in 1968.

In January 1965, the Society took a major step by merging the Applied Journals and the Fundamental Journals into a single unit. Richard L. Kenyon was appointed Director of Publications and Richard H. Belknap, Assistant Director of Publications and Director of Research Journals. Also at this time, the graphic arts research under way in the Applied Journals group since 1957 was recognized with the creation in the Division of a research department under a Director of Publications Research. This post was filled by Joseph H. Kuney, who at the same time was appointed Director of Business Operations for the ACS publications. These moves placed all of the journal, magazine, and book publishing operations of the Society under a single, unified management, a structure that had been taking shape gradually for a number of years. The editors of journals retained a high degree of autonomy, being appointed by and responsible to the Board of Directors.

Also by 1965, a policy change of 1962 for the Applied Journals, which originally had staff editors, was gradually being implemented: all editors appointed for journals publishing research papers were professionally involved with research in the field concerned. Thus the new Editor appointed in 1965 for the *Journal of Agricultural and Food Chemistry* was Philip K. Bates, Director of Research of the Carnation Co. He succeeded Rodney N. Hader, Editor since 1956. Similarly, Bruce Sage of the California Institute of Technology agreed to become the first Editor of the *Journal of Chemical and Engineering Data*, effective in January 1965. *JC&ED* had operated from its birth as an adjunct of *Industrial & Engineering Chemistry* but had reached a size and level of interest that warranted a separate editor.

The *Industrial & Engineering Chemistry* complex—the general edition and its three quarterlies—were established by the mid-1960s, but subscribers were indicating a preference for completely separate subscriptions. In 1965, in consequence, each unit of the group was made available as a separate publication. Byron Vanderbilt retired as the Editor of *I&EC Product Research & Development* in 1965; Rodney N. Hader became Acting Editor. At the end of the year Dr. Hallett retired from the editorship of *Analytical Chemistry* and was succeeded by Herbert Laitinen of the University of Illinois.

The early 1960s saw growing attention to environmental pollution and increasing interest within the ACS Division of Water, Air, and Waste Chemistry in an ACS publication in its field. In 1965, the division and the publications staff undertook studies that led to a proposal that such a publication be established. In accepting the proposal, the Board debated whether the publication should be limited to environmental chemistry. The argument that chemists historically had been too self-limiting in describing their work won out, and the publication was named *Environmental Science & Technology*. In view of the growing importance of technology, economics, legislation, and social factors in the environmental field, the Board decided also that the publication should contain staff developed magazine-type material as well as authored feature articles in addition to the research papers that would comprise a considerable part of its contents. James J. Morgan of the California Institute of Technology was appointed Editor, and the first issue of *ES&T* appeared in January 1967.

During the late 1950s and early 1960s, as support of scientific research reached record levels, strong and continuing pressure arose for more scientific reportorial content written by research scientists in *Chemical & Engineering News*. Proposals extended from reviews of scientific features to reports of scientific meetings and even to a suggestion that communications to the editor be moved from the journals into *C&EN* to take advantage of the timeliness of the weekly issue. Alternative proposals included a chemical science news magazine that would carry such material. After much study and debate the Society settled on a current chemical science monthly containing concise, up-to-date reviews of fields particularly active at the time of publication. The journal would be a supplement to *C&EN,* and all ACS members would receive it in the experimental stages. The Board requested that the combination of *C&EN* and its supplement be operated with a minimum financial target of break-even. Joseph Bunnett of the University of California at Santa Cruz was appointed Editor, and the publication was named *Accounts of Chemical Research.* Its contents consisted of short reviews of limited fields, written by leading specialists. The journal-supplement was first issued in January 1967; it went to all ACS members for the first six months and to all who asked to receive it for the second six months. Beginning with the second year, 1968, subscribers were asked to pay for the journal.

Polymer chemists also were pressing in the mid-1960s for an ACS outlet for their papers. A study was made of the demand and potential use for a research journal in the polymer field, which already had some strong publications. The response was by no means unanimously favorable, but it indicated clearly that a substantial fraction of the polymer community desired such a publication. Purchase of an existing journal was considered but dropped in favor of starting a new venture. The Board approved, and the journal, *Macromolecules,* was established to publish high quality, full-length research reports, communications, and an occasional review at the discretion of the Editor, all emphasizing fundamental aspects of polymer chemistry. Field H. Winslow of Bell Telephone Laboratories was appointed Editor, and the first issue appeared in January 1968.

The volume of scientific publication by the ACS, by other societies, and by commercial publishers was climbing rapidly by the late 1960s. Libraries were being strained for space, and users were being frustrated by problems of access. To ease such problems, the Society decided in 1967 to issue microfilm editions of its journals on an experimental basis. Volumes covering the five most recent years of publication were to be leased subject to a license-to-photocopy fee of five cents per editorial page. All other volumes were to be made available at established back-issue prices and sold on order. Announcement of a five-cent-per-page license fee for photocopying brought storms of protest from libraries; study and negotiation led to a revised fee of 1.5 cents per printed journal page. Licensees could make as many copies as desired for use at the location of the microfilm.

In this same period, the computer was looming as a potent tool for handling scientific information, and increasing effort was being devoted to the development of information systems. The ACS, with its Chemical Abstracts Service as an advanced information operation and with a leading set of journals, was a natural focus of attention. The National Science Foundation had for several years supported research in both ACS groups and now was pushing the Society to take the lead, both at home and abroad. Accordingly, in the fall of 1967 the Board of Directors established an Office of Planning for Information Systems with Richard L. Kenyon as Director. During the next two years, that office worked particularly to develop a formal structure of relations with the European community. Milton Harris, ACS Board Chairman, Robert W. Cairns, Fred Tate of *Chemical Abstracts,* and Kenyon traveled frequently to Europe for negotiations that led to formal agreements by the ACS with The Chemical Society (London) and the Gesellschaft Deutscher Chemiker. These agreements committed the three groups to active cooperation in the development of an international chemical-information system centered in the Chemical Abstracts Service in Columbus, Ohio.

The continuing fiscal pressure on the journals is particularly evident in the minutes of the Board Committees on Finance and on Publications in 1968. At that time, the chairman of the Committee on Publications was asked to bring to the attention of the Editors of the journals the serious financial problems facing the Society and to urge each to cooperate in holding down expenses. The page charge was increased to $50 and extended to the *Journal of Agricultural and Food Chemistry* and the three I&EC Quarterlies, which had not previously carried it. Also, the practice of providing free reprints for authors paying page charges was discontinued.

In 1968, Howard L. Gerhart, vice-president for Research for Pittsburgh Plate Glass Co., was appointed Editor of *I&EC Product Research & Development;* he replaced the Acting Editor, Rodney N. Hader. In the same year, Marshall Gates expressed his desire to retire as Editor of *JACS;* he was succeeded in 1969 by Martin Stiles, University of Michigan.

Before he retired, Dr. Gates revived a suggestion of his predecessor, W. Albert Noyes, Jr., that the fundamental journal system be reorganized with *JACS* as an umbrella journal and the various disciplinary journals as subunits of it. The revised system would comprise the *Journal of the American Chemical Society, Organic Chemistry Section; JACS, Physical Chemistry Section,* and so on. Another problem would be attacked through this arrangement by establishing a *JACS, Communications.* As this proposal surfaced, so did the feelings of authors, editors, and bystander ACS members. The furor was considerable. A semiformal survey made by the chairman of the Council Committee on Publications among authors of journal papers showed a fairly clear split: those whose papers were published in *JACS* largely opposed the change; those whose papers were not published there supported it. Considerable heat and some interesting views were generated, but no action resulted.

Early in 1969, Gordon Bixler, who had been Editor of *C&EN* since late 1962, asked to be relieved; Patrick P. McCurdy, the Managing Editor, was appointed Editor. Also at the beginning of that year, Edward King resigned as Editor of *Inorganic Chemistry,* and M. Frederick Hawthorne of the University of California at Riverside succeeded him.

In October 1969, Dr. Frederick T. Wall of the University of California at San Diego assumed the post of Executive Director, ACS, with responsibility for the publications program and all other units of the Society's administrative structure. The new position had been created in a major reorganization in which the Board also established a Publications Division under Dr. Richard L. Kenyon, who had been Director of Publications since 1965. In mid-1970, Dr. Wall initiated a further reorganization. A Books and Journals Division was formed under the direction of Dr. John K. Crum, who had recently become Group Manager for Journals in the Publications Division. At the same time, Dr. Kenyon became head of a new Division of Public Affairs and Communication, which included the ACS magazines, the circulation and advertising operations of all publications, the Department of Chemistry and Public Affairs, and the Department of Public and Member Relations.

Effective Jan. 1, 1970, the Board consolidated the Journals Fund into the General Fund. Allocation of dues income to direct support of journals, an ACS practice since 1876, was discontinued.

This same period saw several further developments. In 1969, the contract with Van Nostrand Reinhold for publication of the "ACS Monograph Series" had been ended, and the Society took over all publishing operations for that series. In 1970, Dr. Bryce Crawford was appointed Editor of the *Journal of Physical Chemistry,* succeeding Dr. Wall. In 1971, Dr. Philip S. Portoghese was appointed Editor of the *Journal of Medicinal Chemistry,* succeeding Dr. Alfred Burger, who retired.

By 1970, the ACS publications that relied heavily on advertising were in serious trouble. Advertising for *Chemical & Engineering News* had dropped continuously since 1967, while publication costs had risen. A special committee considered various alternatives to the existing design and operation of *Chemical & Engineering News,* but concluded that the magazine was the one continuous contact between the Society and its members—"the glue that holds the Society together." *C&EN* retained its existing form, but sharply reduced its budget, which required some reduction of staff and considerable reduction in page volume.

In the same period, the ACS decided to replace the monthly edition of *Industrial & Engineering Chemistry* with a new publication, *Chemical Technology.* The new magazine was to be designed to serve readership needs better as determined by studies made during 1969 and 1970; its stated objective was to "serve industrially-oriented, chemically-trained professionals with a current-awareness publication directed toward chemical technology presented in the job-oriented way and designed to help them in their jobs."

Each ACS division that felt it served a field of applied chemistry appointed a representative to an advisory panel for *ChemTech.* The panel took the responsibility of seeing that the content of the publication was in keeping with its stated objectives. Benjamin J. Luberoff of the Lummus Co. was appointed Editor under a type of arrangement similar to that used earlier by the ACS Applied Journals. He did not become a regular staff member, but undertook the work on a contract basis, maintaining his office in Summit, N.J. Production, circulation development, and other aspects of publishing the magazine were to be handled by the ACS headquarters staff in Washington.

The first issue of *ChemTech* appeared in January 1971. In order to expose all possible interested ACS members to the new magazine, it was made available free to members during its first six months and then was converted to a subscription basis.

Another publishing venture undertaken by the Society in the early 1970s was the production and sale of audio-tape cassettes. A weekly news tape, "Chemical Executives Audionews," was developed and offered on a subscription basis beginning in 1970, but after a year was discontinued for lack of subscription income. Also, audio tapes were made of selected symposia presented at ACS meetings. The "Men and Molecules" tapes, which had been offered to radio stations since 1961 and were designed to acquaint the general public with chemical developments that were influencing their lives, were also put on audio-tape cassettes. These were offered for sale through a communication products department of the Division of Public Affairs and Communication. In 1974, "Men and Molecules" was renamed "Man and Molecules."

During that same period, the National Bureau of Standards had become concerned that the output of its National Standard Reference Data Service (NSRDS) was not reaching a large enough audience to provide the service it might. Discussions with the Society and the American Institute of Physics led to a three-party contract for producing a journal, the *Journal of Physical and Chemical Reference Data.* NSRDS was to produce camera-ready copy, the American Institute of Physics was to handle the printing and production of the journal, and the ACS was to manage subscription development and sales. The first issue of the new journal appeared in 1972; it is published four times a year, with occasional special supplements composed of particularly large compilations of data in a given area. Reprints of the individual articles of the journal are made available for separate sale.

In 1972, the ACS Board of Directors approved the development of an experimental publication, *Clinical Laboratory Digest,* designed to provide digests of articles of value to clinical chemists. The Society arranged with the Institute for Scientific Information, in Philadelphia, to fill orders, received through the ACS, for full copies of those articles digested. The first issue of *Clinical Laboratory Digest* appeared in 1973. However, it was unable to meet the requirements of financial break-even and ceased publication after the January 1976 issue.

In 1973 the Society began to publish and distribute the text and teaching materials developed in the Chemical Technician Curriculum Project. The project had been initiated and carried out by the ACS under a grant from the National Science Foundation. Its purpose was to develop training materials in an area that the ACS and others felt had been neglected in the past and that was important in the operation of chemical laboratories.

By the 1970s, although Society policy on the authority and responsibility of ACS editors was generally understood, the volume of the publications had become so large as to suggest that a specific policy be set down and applied to all of the Society's publications. In December 1972, the Board approved an official policy:

> The journals and magazines of the Society shall be edited in a manner consistent with the objects of the Society and with the maintenance of the dignity, the worth, and the professional standing of its members. Toward that end, editors and their staffs are expected to show a sense of responsibility toward the membership and a sensitivity to the conflicting views and interests of the various sets of chemists which constitute the ACS membership; and they are expected to edit the journals so that the articles, headings, and editorials reflect fairly, fully, impartially, and in balance the facts involved.
>
> Within these limitations, journal editors retain final responsibility for editorial decisions relative to their publications, and no attempt should be made by the Board, its members, or its committees to instruct editors in the day-to-day direction of their editorial activities. Should an editor's judgment prove too often questionable or unacceptable, the Board should replace him rather than attempt to supervise him in the performance of his assigned duties.

A number of refinements were developed under this policy; in view of the increasing member-interest in *Chemical & Engineering News* and the efforts to renew its strength following the bad financial period 1969–72, a specific policy on the objectives of that magazine was developed and refined by 1973:

> The *C&EN* contribution to achieving the objectives of the Society shall be threefold: 1) to keep members informed about the policies and

activities of the ACS, 2) to keep ACS members as well as other readers well informed on the activities of the chemical world through news reporting and through calling attention to issues of consequence to chemists and chemical engineers and to their contributions to society, and 3) to present the beneficial contributions of the ACS and of its members to the broad goals of society at large.

To support and guide the Editor of *Chemical & Engineering News,* a *C&EN* Editorial Board was established in 1973. It was composed of the Board Committee on Publications, the President-Elect and immediate Past President of the Society, and a representative of the Council Committee on Publications. The Editorial Board is required to meet at least quarterly and to advise the Board Committee on Publications in setting broad policy guidelines for *C&EN*.

In 1974 Albert F. Plant, who had been Editor of *Industrial Research,* published by the Dun·Donnelley Publishing Corporation, became Editor of *Chemical & Engineering News,* succeeding Patrick McCurdy, who had resigned.

By 1974, the Editor and advisory board of the *Journal of Chemical Documentation* had noted for some time that many papers in the area of computer science that might be of interest to *JCD* readers were being submitted to other journals on the assumption that *JCD* would not accept them. To clarify the role of the publication and to encourage members of the new ACS Division of Computers in Chemistry to submit papers, the name was changed to the *Journal of Chemical Information and Computer Science* effective January 1, 1975.

For some time there had been occasional inquiries into possible outright purchase by members of microfilm versions of ACS journals. The Board of Directors voted in 1974 to establish a member rate of $10.00 per reel of microfilm editions of back volumes of journals at least three years old. The stipulation was attached that the purchaser guarantee not to sell or furnish the reels to a library or reading room. The interest in microfiche was growing. The Editor's Committee on Improved Publishing Formats, established in 1972 and headed by Bryce Crawford, developed a number of proposals, including a recommendation that microfiche versions of journals be made available on a subscription basis. In 1975, all ACS research journals became available on microfiche for direct purchase.

The Books Department, which grew out of "Advances in Chemistry Series" and was set up as a unit in 1971 with Robert F. Gould as head, added a new series in 1974—"ACS Symposium Series." Whereas "Advances" volumes are edited by staff and typeset in metal, volumes in the new series are edited by symposium chairmen or other editors named by sponsors and printed directly from authors' manuscripts to minimize the cost and the lapse of time between the presentation of the papers and their availability in book form. By the end of 1975, 21 volumes of this series had been produced, and "Advances in Chemistry" had published its 147th volume. A new policy was instituted with the beginning of the Symposium Series; royalties were paid to the organized units serving as symposium sponsors and editors, both for the "ACS Symposium Series" and "Advances in Chemistry Series."

The Books Department also had been experimenting with the publication of reprint collections whose editorial content came mainly from the Society's journals. The first of these volumes, "Collected Accounts of Transition Metal Chemistry," was published in 1973. The Board Committee on Publications voted in 1975 to make this program a regular part of the Books Department's activities. Also, the ACS Monograph Series was revived in late 1973 under the management of the Books Department; the series had been dropped in 1972.

Financial problems continued to hover about the journals even though nonmember prices had been raised in 1972 to three times the member prices. Problems became particularly severe in 1974, when paper prices rose sharply. Nonmember prices for journals were increased to four times the member prices, and the nonmember price for *C&EN* went from $9.00 to $15.00 effective in 1975. Also, page charges were increased on *JACS* as a test, to be followed later by increases for other journals. The percentage payment of page charges, which had been as high as 75% in the mid-1960s, stood at about 55% by the end of 1975.

In 1975, Cheves Walling, University of Utah, became Editor of the *Journal of The American Chemical Society*, replacing Martin Stiles, who had asked to be relieved. In the same year, Russell F. Christman of the University of North Carolina became Editor of *Environmental Science & Technology*, succeeding James J. Morgan, who also had asked to be relieved. In mid-1975 D. H. Michael Bowen, head of the Journals Department, was promoted to Director of the Books and Journals Division, succeeding John Crum who was made Treasurer and Chief Financial Officer of the Society.

By the mid-1970s, very rapidly rising costs and the failure of advertising volume to rise were continuing to create difficulties for *C&EN*, but a determined effort was made to rebuild it. A detailed study of the attitudes of members indicated a high level of readership and interest in *C&EN* and a desire to bring back some types of material, especially scientific features, which had been dropped in the severe cutbacks of 1970–71. After detailed committee study, the Board and Council voted a $3.00 increase in member subscription, to $8.00, and a consequent increase in basic ACS dues effective January 1976.

Publications Research and Development

While the ACS occasionally changed the design of its publications over the years, there was no organized research and development effort on the general technology of producing journals and magazines until 1957. In October of that year, Joseph Kuney, Production Manager for the Applied Journals, noted in a memo to C. B. Larrabee, "On the scene after many years of promises is electronic photocomposition. If the cost picture proves competitive, this type of composition holds real promise for the future." Later that year the Board voted a 1958 budget item of $62,500 for research projects in the application to the publication program of new developments in the graphic arts. Of this amount, $50,000 was to be used to purchase a Photon photocomposition machine. That sum was about 10% of the total annual composition costs of the ACS journals and magazines at the time.

For the first year or two progress was slow, and the Photon machine was dismantled much of the time. But the staff learned the capabilities of the machine and began to produce pages experimentally. Computer-assisted photocomposition was the goal, and meeting it required more staff, system development, and programming. In 1962, the Board appropriated an additional $100,000 for the work. As the hoped-for results might be valuable to all scientific publishers, the National Science Foundation responded to an ACS proposal with a grant of $225,800 in 1965. Over the next five years, NSF support reached a total of about $500,000.

By 1966, production-line results were visible. In February 1966, the *Journal of Chemical Documentation* noted in its opening article, "This issue marks the beginning of regular production of this journal using computer-aided typesetting methods and an important first step in evaluating the encoding of scientific manuscripts in machine-readable form at the time of primary publication." This was the first scientific journal to be produced by a computer-assisted photocomposition system. It was followed, on a production basis, by the *Journal of Chemical & Engineering Data* and the *I&EC Product Research and Development* and *Process Design and Development* quarterlies. These journals provided a practical testing system for the continuing research and development.

By 1970, the publications' financial problems had become severe. NSF was reassessing its commitments and pressing for a single graphics research and development activity at the ACS, which decided that the work should be done at the Chemical Abstracts Service. By July 1, 1971, the last member of the publications photocomposition team had been reassigned, the photocomposition machine had been sold, and the operation had been closed down. Something of real significance, however, had been started. By 1976, all ACS publications were being photocomposed; the resulting savings were 15–20% of the cost of using the older methods.

The publications research group also had studied the problems of publishing large journals for limited readership. A study of the *Journal of Organic Chemistry* indicated that the average reader started reading only 17% of the papers in a given issue. In 1970, experimental work was done on the *Journal of Organic Chemistry,* with the cooperation of a few authors, aiming toward a short-paper journal. The principle involved was the publication of a short paper—two to four pages—with the full paper available on request. Eleven such papers were published; the long versions were published in the same, experimental issue of *JOC* for comparison. The reactions of the authors were mixed, with some cautiously favorable.

Although the research department was eliminated in the reorganization of 1971, the ideas were not all discarded. By 1972, a new research effort had been mounted in the Books and Journals Division with Dr. Seldon Terrant as head of Research and Development. During the next year a modest staff was assembled and work was resumed on typewriter composition of journals, miniprint, and, on a longer-range basis, a dual journal system for publication separately of synopses and archival papers. The ACS Single Article Announcement service, in operation since 1971, was continued as a study of reprint demand. Tables of contents of research journals were published twice a month, and from this periodical reprints of the articles could be ordered. Cooperative work was begun with Chemical Abstracts Service on the development of primary-secondary publication interlinks. Typewriter composition of *Inorganic Chemistry* was begun at Chemical Abstracts Service in 1973 as the first phase of a computer-based primary publications system. This was followed by the development of computer-based photocomposition of *Inorganic Chemistry* in 1975 and by preparation for production of all ACS journals by that system.

Perhaps the most intensively watched activity of the Books and Journals Department during 1975 was the evaluation of the dual journal concept. This work is being done with financial support from the National Science Foundation. The aim is to obtain the reactions of the journal-using community to a dual system. In such a system, the general circulation issue would carry only synopses of papers; a separate issue would carry the full archival papers, perhaps accompanied by the synopses.

Conclusion

The Society moved into the year of its Centennial with a publications program that was strong and progressive. The program was not without problems, but there was good evidence that the ACS had the resources to deal with them. The Society could look back with pride and satisfaction on a century of publishing scientific literature in which it had achieved a leadership that was hard to dispute. It had continued to respond to needs without losing the support of the research scientists who, although devoted to the search for new knowledge, are highly sensitive to perturbation of the system within which they work. The ACS could claim with pride and justification that its scientific information system in 1976 was the standard of the world.

CHEMICAL ABSTRACTS SERVICE

The first American venture into chemical abstracting came in 1895 when Arthur A. Noyes of the Massachusetts Institute of Technology, convinced that U.S. chemists were not being recognized adequately for their accomplishments, founded the *Review of American Chemical Research.* The *Review* published abstracts of American chemical papers only. It appeared initially as a supplement to MIT's *Technology Quarterly,* but in 1897 became part of the *Journal of the American Chemical Society.* W. A. Noyes, Sr., a distant cousin of Arthur Noyes, assumed the editorship of *JACS* in 1902. He became a strong advocate of the Society's publishing a more comprehensive abstracting journal. After several abortive efforts to arrange with The Chemical Society (London) for joint sponsorship of such a journal, the ACS Council in 1906 authorized publication of *Chemical Abstracts.* It was born in January 1907 with W. A. Noyes, Sr., as Editor.

CA was created primarily to serve American chemists, but over the years it became the principal abstracting and indexing service for most of the world, with some 60% of its circulation outside the United States. What began as virtually a one-man enterprise grew eventually into the largest division of the Society, with an annual budget in excess of $28 million in 1975. By then the more than 1100 full-time employees of the Chemical Abstracts Service Division were operating a highly automated information-processing system that was producing not only *Chemical Abstracts* but also a growing family of auxiliary publications and computer-based information files.

Noyes edited *CA* through its first two years, working first from the Bureau of Standards in Washington, where he was chief chemist, and later from the University of Illinois, where he became chairman of the chemistry department in September 1907. He was succeeded as Editor in 1909 by Austin M. Patterson, who had served earlier as chemical editor for "Webster's New International Dictionary." At the invitation of William McPherson, then head of the chemistry department at Ohio State University, *CA*'s editorial offices were moved to the Ohio State campus in Columbus to be nearer the Patterson home in Xenia, Ohio. Patterson was forced by impaired health to resign as editor in 1914, but he maintained a lifelong association with the publication as consultant and adviser. He was succeeded briefly as Editor by John J. Miller, who resigned at the end of 1914.

The next four decades in the history of *CA* are the story of Evan J. Crane. He had come to *CA* in 1911, at the age of 22, when Miller and Patterson had asked the Ohio State chemistry department to recommend "someone with industry, intelligence, personality and stamina" to fill an associate editor's position. Crane became Acting Editor when Miller resigned and was named Editor in the spring of 1915. He developed and nurtured *CA* through very difficult times—money was so scarce at one point that he financed business travels out of his own meager salary. In his 43 years as Editor, Crane built the fledgling publication into the model and pacesetter for all scientific abstracting and indexing journals. He became the first Director of Chemical Abstracts Service when the growing *CA* editorial organization was renamed and made a division of the ACS in 1956.

E. J. Crane's contributions to chemical documentation brought him many honors during his lifetime. They included the American Chemical

E. J. Crane *(1889–1966) was Editor of* Chemical Abstracts *1915–58 and Director of Chemical Abstracts Service 1956–58.*

Society's Priestley Medal, the Chemical Industry Medal of the Society of Chemical Industry (American Section), and the Austin M. Patterson Award of the ACS Dayton Section. Until 1938 Crane always said that he held only two degrees: B.A. and *CA*. That year Ohio State conferred on him the honorary degree of Doctor of Science.

Crane's zest for his work and his sense of humor were both reflected in *The Little CA,* a bulletin for *CA* workers that he issued three times yearly from 1930 until his death, in December 1966. Along with pep talks and instructions for abstractors, Crane used "terse verse" to egg *CA* workers on. Some examples:

On style—
Queer style may rile;
Clear style—worthwhile.
If you're obscure,
Please find a cure.

On accuracy—
To authors it's fairer
The rarer an error.

Even after his retirement, in 1958, Crane continued to abstract for *CA* and to write *The Little CA*. His lifelong attitude toward his work is evident in lines he wrote for the final issue (number 110) of *The Little CA* a few weeks before his death: "Abstracting is a wonderful way to spend an evening . . . to go off to one's study, bedroom, kitchen table, or wherever is quiet with chemical reading to do having a double purpose: (1) to learn and savor more chemistry and (2) make useful abstracts."

Coverage

The growth of *CA* reflected the phenomenal growth of chemical science and technology in the 20th century. In 1907, its first year, *CA* published just over 12,000 abstracts—almost half of them based on work done in Germany. The total increased almost every year thereafter.

From the beginning *CA* strove to cover all of the world's chemical literature. The first claim to complete coverage occurred in the Editor's report for 1912, although a claim was made as early as 1910 that *CA* had become "as complete in its coverage of chemistry as any other similar publication." The definition of "completeness" varied somewhat over the years, however. In particular, W. Russell Stemen, who joined the editorial staff in 1922, was influential in expanding *CA's* coverage of the industrial applications of chemical science and technology. Stemen recognized also the importance of patents as sources of new chemical information and pressed for broader coverage of the patent literature.

At times *CA* had to go to considerable lengths to maintain a semblance of complete coverage. During the early years of World War II, when Britain's sea blockade began to take effect, arrangements were made with European publishers to send journals through Italy, whose ships at first were not blockaded. After Italy entered the war, *CA* arranged to get journals via Siberia, but that ended when Germany invaded Russia. With the aid of W. Nowacki of Berne, *CA* then organized a staff of abstractors in Switzerland. Their abstracts—in German or French—were typed on thin paper and sent by clipper-plane mail to Columbus, where they were translated and published. This arrangement held up until the fall of France cut communication between Switzerland and the U.S. After that, papers, often on microfilm, were obtained for abstracting with the aid of the Office of Scientific Research and Development, the Interdepartmental Committee on the Acquisition of Foreign Periodicals, the Alien Property Custodian, and the American Library Association's Joint Committee on Importations. *CA* drew also on *Chemisches Zentralblatt,* obtained on microfilm, and took advantage of the readiness-to-cooperate of some American chemists who were interned in Germany and China. Japanese journals remained inaccessible until many months after the war ended, but, with the aid of the Scientific and Technical Division of the Army of Occupation, *CA* eventually obtained those needed to fill the wartime void.

In 1942 *CA* established an office in the Library of Congress Annex to help channel some of the German documents to Columbus from their recipients in Washington. This office played an even larger role after the war. It was charged then with examining the Soviet scientific literature received by the Library of Congress from various sources and selecting that suitable for abstracting in *CA*.

Before *CA* was started, a list of 396 candidate journals for coverage had been drawn up. By 1912 the number of journals monitored had grown to 600; it passed 1000 in 1922 and 2000 in 1932. The number of journals in which *CA* found papers suitable for abstracting passed 5000 in the early 1950s and 10,000 in the early 1960s.

The pace accelerated after World War II. *CA* needed 32 years—until 1939—to publish its first million abstracts; the second million required only 18 years, the third million eight years, the fourth million less than five years, and the fifth and sixth million just over three years each. By the eve of World War II, *CA* was publishing more than 65,000 abstracts annually. The number rose to 100,000 in 1957 and passed 200,000 in 1966 and 300,000 in 1971.

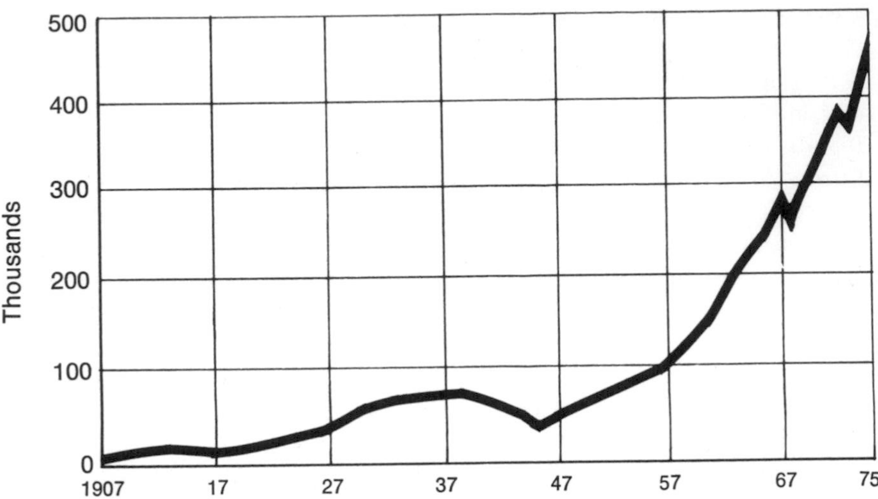

Figure 1. Number of documents cited annually in Chemical Abstracts

In 1975 CAS monitors some 14,000 journals, patents issued by 26 nations, and books, dissertations, and conference proceedings from around the world; CAS abstracts and indexes and cites more than 450,000 papers, patents, and reports annually. About three quarters of the material abstracted originates outside the United States.

Indexes

E. J. Crane was the first *CA* editor to recognize that indexes are a highly important part of an abstract journal as extensive as *CA*. Somewhat meager author and subject indexes were published annually from the beginning, but serious attention was not given to indexing until 1916. Crane personally made the first *CA* Decennial Index possible. Not only did he do much of the work on the index himself; he also solicited enough advance subscriptions to convince the Society's Board and officers that its publication would not be a financial disaster.

In the first nine volumes of *CA,* chemical compounds were indexed mostly under the names the authors used for them in the papers abstracted. When *CA* began to consider a 10-year cumulative index, it became evident that some systematic means of naming and indexing compounds should be used. Otherwise, references to a particular compound would be scattered throughout the index, and related compounds would not be grouped together in the alphabetically ordered index. A means of naming and indexing compounds systematically was devised in 1916, primarily by Carleton E. Curran and Austin Patterson, whom Editor Crane had pressed back into service. It had a profound effect on chemical nomenclature in general, and the *CA* offices subsequently became a world center for nomenclature development. The same basic principles still are used to name and index compounds for *CA,* although the system has been improved and updated continually over the years. In this work, Leonard Capell, Mary A. Magill, Russell Stemen, and Cecil C. Langham made major contributions. Austin Patterson served *CA* as a consultant on nomenclature until early 1956 and personally named all new ring systems for the journal until that time.

As the volume of material abstracted continued to grow, indexing assumed even greater importance. Eventually it became more than half of the task of *CA*'s full-time editorial staff. In 1920, *CA* added an annual formula index, the first to be published by an abstract journal. Numerical patent indexes appeared in *CA* in 1911 through 1913, but were discontinued because of lack of funds; they were reinstituted in 1935 and have appeared continuously since.

CA's annual indexes exceeded 2000 pages for the first time in 1930. In 1934 the format was changed to accommodate twice as many index entries per page. Nevertheless, the indexes exceeded 2000 pages again in 1950 and by 1959 were running more than 5000 pages annually.

In 1956 author indexes were added to the then semimonthly abstract issues. Numerical patent indexes were published in each issue beginning in 1958. A keyword subject index, organized and composed by computer, was added to the issues in 1963, as was a computer-produced patent concordance that correlates the patents issued by various countries on a given invention.

Staff members at Chemical Abstracts *dictated index entries in 1949, two decades before the advent of automated information processing methods at Chemical Abstracts Service beginning in the late 1960s.*

Beginning in 1962, Volume Indexes were published semiannually rather than annually. Still, these indexes were exceeding 20,000 pages per year by the early 1970s. To facilitate both their production and use, the subject indexes at that point were subdivided into separate general subject and chemical substance indexes. Cross references and scope notes formerly scattered through these indexes were brought together in a separate Index Guide.

Collective indexes continued to be produced at 10-year intervals until 1956, when the collective period was changed to five years. The last 10-year index (1947–56) totaled some 22,000 pages. The eighth collective index (1967–71), at 75,000 pages, was more than three times as large. CAS anticipates that the ninth collective index (1972–76) will total 100,000 pages.

Abstractors

From the beginning, *CA* depended heavily on volunteers for the bulk of its abstracting. In addition, each subject section was overseen by one or more volunteer section editors, subject-matter authorities who monitored the coverage of each section and the quality of the abstracts. In 1907, 129 volunteer chemists contributed to *CA*. The total grew steadily, to a peak of more than 3200 in the mid-1960s.

Crane was fond of referring to certain of his volunteers as "the iron men of *CA*," and many indeed deserved the label. Charles A. Rouiller, who contributed to the first volume in 1907 and became a section editor in 1912, continued to work for *CA* until his death, in 1968. C. H. Kerr, whose first abstract appeared in 1910, served as ceramics section editor from 1921 until 1961. William G. Gaessler started abstracting articles on pharmaceutical chemistry in 1913 and was still doing so in early 1976. The names of Louis E. Wise, who wrote his first abstract in 1912, and Joseph Hepburn, who began in 1914, still appeared on the masthead at that time, as did those of 13 others who had been abstractors or section editors for *CA* for 50 years or more.

As *CA* grew more international in scope and reputation, the volunteer abstractor corps took on a distinctly international flavor. At one point in the late 1960s, the corps included chemists in 70 different nations. Formally organized groups of *CA* abstractors grew up in some nations. By far the largest is the Japanese *CA* Abstractors Association, which currently boasts more than 250 members.

Until the early 1960s, virtually all abstracts were prepared by volunteers. The 1960s and early 1970s, however, saw a need for greater currency in abstracting and indexing and the advent of specialized requirements generated by computer processing of abstracts. These developments led to a gradual shifting of the abstracting effort to full-time employees in Columbus. While volunteers still make a major contribution, more than 65% of all abstracts in *CA* now are processed completely in the Columbus offices, either as staff-prepared abstracts or staff-modified author abstracts.

Finances

With E. J. Crane's retirement, in October 1958, Dale B. Baker, another Ohio State alumnus, was named Director of Chemical Abstracts Service. Baker, a chemical engineer, had come to CAS from Du Pont in 1946. When he became Director, Charles L. Bernier was named Editor, and Leonard T. Capell, who had succeeded Austin Patterson as CAS's leading authority on nomenclature, was appointed Executive Consultant.

One of the major tasks confronting this new team was to shepherd CAS through the transition from a subsidized operation to a financially self-supporting one. Financial difficulties were nothing new for *CA;* they had existed from the beginning. The basic problem was always the same. The volume of chemical literature being published in the world grew rapidly, and since *CA* was committed to complete, accurate, and timely coverage of this literature, the cost of performing its mission grew exponentially.

Through 1933, *CA* had been financed solely by ACS member dues, and all members who wished received the publication free. After 1933, a small subscription fee was imposed to supplement the allocation from dues. This arrangement met financial requirements until the end of World War II, but thereafter operating expenses rose rapidly with the expanding growth of scientific publication and the inflationary spiral of the postwar period. Subscription prices were increased moderately, but did not keep pace with costs. In 1952 a Corporation Associates plan was instituted through which industry helped make up the deficit, which nevertheless continued to mount. Growing alarm was evident during the mid-1950s in headlines in *Chemical & Engineering News:* "Chemical Abstracts Service—Good Buy or Good-by?"; "*Chemical Abstracts*—Millstone or Milestone?"

In 1955, operating expenses exceeded $1 million for the first time, and the deficit over subscription revenues was almost $500,000. The ACS Board of Directors that year changed subscription policies and prices drastically, decreeing that *CA* should be priced henceforth to break even. The nonmember subscription rate was increased from $60 to $350 for industrial, commercial, and governmental organizations and to $80 for colleges and universities, although ACS members still were allowed to subscribe for personal use for $20. Board Chairman Ernest Volwiler,

The Chemical Abstracts Service complex in Columbus, Ohio in 1975 comprised two buildings, the first completed in 1965, the second in 1973.

writing in *C&EN,* said that the net result of the pricing change was to make financing of *CA* "a joint responsibility of the profession and of those governmental, industrial, and commercial organizations that have a direct stake in its availability."

By the early 1960s, annual operating costs were exceeding $2 million, and the nonmember subscription price had risen to $1000. At that point, member subscriptions to *CA* were increased from $40 to $500. As an alternative, subscribers for personal use were offered a choice of five Section Groupings, each containing about one fifth of the abstracts published in *CA,* for $25 per section per year; they could buy all five sections for $125 per year. At the same time, a flat $500 grant to colleges and universities against the nonmember *CA* subscription rate was instituted. The special member subscription rate for *CA* was eliminated entirely in 1966.

CAS now is fully self-supporting. Revenues from the sale or lease of its publications and computer-readable information files meet essentially all operating expenses. In the 1960s and early 1970s, the National Science Foundation and other federal agencies provided financial support for CAS's research and development on new information-handling methods and systems. Since 1956, however, no governmental funds or ACS member dues have gone toward subsidizing the cost of producing CAS publications or services. (An exception was a one-time grant of $190,000 from NSF in 1962 that helped CAS catch up on indexing for *Chemical Abstracts.*)

Besides CAS's fiscal problems, Director Baker was confronted shortly by the problem of housing for the growing organization. When *CA* moved to the Ohio State campus, in 1909, a 15-by-30-foot room had been adequate for the four-member staff. In a few years this was augmented by an adjoining room, about twice that size, and when the McPherson Chemistry Laboratory was built, in 1928, *CA* was allocated about 1600 square feet in it.

By the mid-1950s, the staff had grown to about 100, and a three-story building, financed jointly by the ACS and OSU, was erected on the campus especially to house *CA* operations. In 1961 a fourth floor was added to this building, but the organization by then had grown to 300 and was straining both the building and the university's hospitality.

In 1962, the Society purchased 50 acres adjacent to the Ohio State campus. A four-story, multimillion-dollar Chemical Abstracts Service building was completed in 1965. This building, too, was soon outgrown, and a second of almost equal size was completed and dedicated in June 1973.

Modernization

Baker recognized early that CAS would have to modernize its information handling and publishing procedures. The postwar spurt in scientific and technical activity had created an unprecedented outpouring of new published information on chemical science and technology. Conventional methods for producing abstracts and indexes were being stretched to the limit. Costs were rising rapidly, and services were becoming increasingly less timely and effective.

Some groundwork already had been laid. In 1955 CAS had become the first operating organization dealing with the scientific literature to establish a research and development department. In 1959 Baker recruited G. Malcolm Dyson, an Englishman noted for developing one of the first linear notation systems for representing chemical structures, to direct an expanded research and development program. Baker also sought and received financial support from the National Science Foundation for the R&D effort. Dyson worked part-time in Columbus from 1959 through 1963. He was responsible for a number of the basic concepts that underlie the automated information processing system that emerged at CAS in the late 1960s and early 1970s.

The first finished product of Dyson's research and development efforts was *Chemical Titles* (*CT*), introduced in 1961. It was the world's first periodical to be organized, indexed, and composed almost totally by computer. In producing this biweekly listing of new papers appearing in selected chemical journals, the emphasis was on speed. The issue of *CT* that listed a given article often reached the subscriber ahead of the issue of the journal that contained the article. Along with other innovations, *CT* was the first periodical to employ the keyword-in-context indexing technique developed by H. P. Luhn of IBM Corp. In the *CT* index, significant words in each published title were arranged alphabetically down the center of a column, with the context of the title printed on either side.

Dyson also conceived the idea of an alerting service to the literature on the biological activities of chemical substances. After much experimentation, and with some financial assistance from the National Institutes of Health, CAS introduced such a service, *Chemical-Biological Activities* (*CBAC*), in 1965. *CBAC* was issued simultaneously in printed form and on magnetic tape for computer searching. It was the first computer-produced service to include the full text of abstracts; as such, it served essentially as a pilot plant for the *CA* of the future. While *CT* had been composed through a standard 48-character computer output printer, a special 120-character print chain was developed to compose *CBAC*. This made it possible to print both upper and lower case letters and numerical superscripts and subscripts along with additional punctuation and mathematical symbols.

In 1961 Baker recruited Fred A. Tate, then manager of the Scientific Information Section of Wyeth Laboratories, as Assistant Director. Tate, who earlier had been an assistant and associate editor of *CA* (1953–56), devoted his energies to reorganizing work flow and implementing modern information-handling systems and procedures across the entire range of CAS operations. He worked also to obtain government financial support to augment CAS's meager resources for research and development.

When Charles Bernier departed, in 1961, Tate assumed the additional post of Acting Editor, which he held until Russell J. Rowlett, Jr., was appointed Editor in 1967. Rowlett, another former *CA* associate editor (1947–52), had served as director of research and development for Virginia-Carolina Chemical Corp. and as assistant director of the Virginia Institute for Scientific Research before returning to CAS.

CA underwent a number of evolutionary changes during the 1960s. Some were instituted mainly to facilitate handling of the rapidly growing volume of information to be abstracted and indexed; others were designed to improve the ease and range of use of the publication. Production of volume and collective indexes was accelerated during the early 1960s by shifting from monotype composition at Mack Printing Co. to cold type composition in the Columbus offices. Volume indexes for 1962 through 1966 were produced by Varityping index entries on individual cards and filming these entries, a line at a time, with a special camera to produce negatives from which offset printing plates were produced. These same cards then were merged, reordered, and refilmed to produce the Seventh Collective Index.

Abstract issues were changed from semimonthly to biweekly in 1961 and to a weekly schedule in 1967. The number of subject sections into which abstracts were classified for publication was increased from 33 to 74 in 1962 and to 80 in 1967.

To make current abstracts more widely available to individuals, CAS in 1962 and 1963 began publishing *CA* abstracts in the five separate groupings of sections mentioned earlier: applied chemistry and chemical engineering; biochemistry; macromolecular chemistry; organic chemistry; and physical and analytical chemistry. In 1965, to simplify use of the accumulated store of *CA* abstracts for retrospective searching, CAS made all abstracts published since 1907 available on 16mm microfilm.

In the last half of the 1960s, with financial support from the National Science Foundation, CAS set out to build a highly automated processing system that would produce printed abstracts and indexes more efficiently and economically and at the same time create a machine-readable data base that could provide the basis for new forms of secondary information services. In the man–machine partnership that began to emerge at CAS in the late 1960s, the results of human intellectual analysis of documents —citations, abstracts, and index entries—were recorded in computer-readable form and subsequently passed from stage to stage through the necessary processing with a minimum of human intervention. Contents of publications and services were organized according to predetermined instructions, converted to the appropriate character set and format, and photocomposed in a form suitable for conversion to offset printing plates, all by computer programs.

Production of *CA* through a computer-based system called for techniques and technology that were beyond the state of the art when CAS began the effort. Partly because of the complexities of chemical nomenclature, some 1500 different characters and symbols (including various type sizes and faces) are necessary to compose *CA* and its indexes, and the same element of data—an author's name, for example—may be printed in a different type format in an abstract issue and in each of several indexes. One development that was critical to the system was a computer-driven composition method that did not require detailed typographic instructions to be recorded along with the data. No commercially produced computer-operated composing device in the 1960s had the needed capabilities. In 1967 CAS acquired an IBM 2280 film recorder unit—one of only three such devices delivered by IBM in the U.S.—and developed a software-based, computer-driven composition system around it. This unit was used to compose the *CA* indexes until commercial equipment with comparable capabilities became available in the early 1970s.

To assure continuity of its publications and services, CAS introduced new processing technology stepwise. Beginning in 1968, abstract headings were converted to computer processing, and the author indexes were derived automatically from these data. By the end of 1970, all *CA* indexes were being organized and composed by computer, and CAS had begun

Data for Chemical Abstracts *and its indexes in the mid-1970s are recorded in computer-readable form through disk-recording keyboards and passed from stage to stage through the necessary processing with minimum human intervention.*

By 1975, structure diagrams in the Chemical Abstracts *indexes were being composed from computer files of structure images recorded through cathode-ray-tube terminals.*

to produce the eighth collective index (1967–71) by merging computer-readable files generated in producing the corresponding volume indexes. Gradual computerization of abstract processing began in 1972 and was completed in 1975. By the mid-1970s, annual author and keyword subject indexes for most of the ACS primary journals also were being produced by the computer system from data recorded in the system in the course of analyzing papers in the journals for coverage in *CA;* indexes for a number of other U.S. and European journals were being produced in the same way.

Considerable savings in manpower, time, and cost have been realized through computerization of CAS's operations. Each element of data now need be recorded only once. The name of an author of a paper, for example, is recorded only once though it will appear in the heading of an abstract, in the weekly, semiannual, and collective indexes of *CA,* and possibly in *Chemical Titles.* On the average, each character recorded in the computer-readable data base results in about 2.5 printed characters in *CA* and its indexes. Routine manual work such as sorting and alphabetizing index entries has been eliminated, as has most of the proofreading formerly required. Data that can be characterized precisely—structure diagrams, systematic index names of substances, journal titles (recorded in coded form), and standardized index entries—now are checked by computer for recording errors. Thus human reviewers can focus on a relatively small portion of the data being processed, and the staff can devote a far greater fraction of its effort to the intellectual task of analyzing the primary literature.

In 1969 The Chemical Society (London) entered into a joint working agreement with the ACS under which the United Kingdom Chemical Information Service provides CAS with abstracts and index entries for papers and patents published in the UK, and markets publications and information services derived from the CAS data base in the UK and Ireland. Earlier, The Chemical Society's research unit had been one of the first organizations to work with CAS's experimental computer-readable files and developed some of the first search techniques and programs for these files. When publication of *Chemisches Zentralblatt* was discontinued, at the end of 1969, West Germany's chemical society, Gesellschaft Deutscher Chemiker, also began to provide input to the CAS data base and to market CAS publications and services in West Germany. In 1975 the agreement with GDCh was replaced by a new, long-term agreement with Internationale Dokumentationsgesellschaft für Chemie, which operates a centralized computer-based information system for the West German chemical community, and a new, broader agreement was concluded with The Chemical Society. Both the British and West German organizations have assumed a share of the cost of producing the CAS data base equal to the percentage of documents abstracted by CAS that originate in their nations (currently, each nation accounts for about 7% of the total). Both, in turn, have been granted the exclusive right to market CAS publications, microform services, computer-readable files, and services derived from them in their nations. The Chemical Society in addition provides computer-readable abstracts and index entries for British chemical papers and patents for input to the CAS data base; the value of this input is credited as part of the British share of the data-base production costs. These cooperative undertakings with sister organizations in the UK and West Germany might be viewed as the first step toward a formal international network for chemical information.

Chemical Registry System

Perhaps the most far-reaching development to come out of CAS's work on mechanized information-handling was the Chemical Registry System. The Registry, too, derived from a concept of Malmolm Dyson's. He felt that the key to effective mechanized handling of chemical information was some means of identifying chemical substances and relating this fundamental identification to information on the substances.

Chemical nomenclature was rejected at an early stage as a basis for the Registry. The rules of nomenclature change continually and would require constant reorganization of the files, which necessarily would be massive, containing information on several million substances. Early efforts at developing a Registry centered around the use of notation systems—linear algebraic codes—to define a substance's structure in a form that could be stored on punched cards and manipulated mechanically. However, the rules for generating such notations were complex and required so much professional judgment in their application as to make the cost of creating a massive file prohibitive.

In the early 1960s, Harry L. Morgan of CAS, using largely fundamental work by D. J. Gluck and others at Du Pont, developed an algorithm for generating a unique and unambiguous computer-readable record of a substance's two-dimensional structure from a tabular description of its structure diagram. Morgan's scheme was simple enough to be handled by clerical personnel with relatively little special training. The algorithm, along with means of adding stereochemical data to the record, became the foundation of the CAS Chemical Registry System. The manual generation of connection tables was replaced at an early stage by computer generation of the tabular description from structure diagrams recorded through a chemical structure typewriter.

In June 1965 the Office of Science Information Service of the National Science Foundation awarded CAS a two-year, \$2-million contract to demonstrate the technical feasibility of the Registry System. The National Institutes of Health and the Department of Defense joined in funding this initial contract. NSF continued to provide substantial financial support for the development of the Registry into the 1970s.

Beginning in 1965 all chemical substances indexed for *Chemical Abstracts* have been registered—their structures and names were recorded in computer-readable files, and each unique substance was assigned, by computer, a permanent identifying number, the CAS Registry Number. At the same time, CAS used Registry techniques to establish a separate computer-based file of substances of interest to the National Cancer Institute's Division of Cancer Treatment. This special NCI registry was expanded in 1974 into a full-scale Drug Research and Development

Chemical Information System, designed and programmed by the University of Pennsylvania and operated by CAS under contract for NCI. During the late 1960s and early 1970s, CAS also worked with the National Library of Medicine and the Food and Drug Administration on computer-based substance identification techniques; NLM used CAS Registry technology in creating the CHEMLINE on-line chemical dictionary that helps to locate substance-related information in its TOXLINE toxicology information retrieval service.

By the early 1970s, the Registry was an integral part of the indexing operations for *CA*. A more advanced version of the system, with extended capabilities for supporting substance naming and the potential for recreating structure diagrams algorithmically from the computer-readable record for display and publication, was put into operation in 1974. The three-millionth unique structure was recorded in the system's files in January 1975.

The Registry now provides mathematically precise controls for structure diagrams and index names of chemical substances processed by *CA*, and CAS's information analysts no longer need generate index names for the majority of substances selected for indexing. The analyst feeds into the system either a name or a structure from the paper or patent being indexed; for about 75% of the substances selected, the index name is retrieved automatically through name or structure matching. Only if the substance is not among the more than three million registered since 1965 is it necessary to generate the index name, and the system helps by retrieving for the analyst the index names of structurally similar substances named and registered previously. CAS also has developed techniques for automatically converting systematic index names for organic substances into machine-readable structure representations and is developing programs that generate systematic names directly from the Registry structure record.

The CAS Registry Number for each unique substance also provides a simple and invariant identification tag that bridges the inconsistencies and variations of chemical nomenclature. This number now is part of the entry for each chemical substance in *CA* volume and collective indexes. Registry numbers also are published in the abtracts in certain sections of *CA* and are being used in a growing number of journals, handbooks, and governmental and private information files.

New Services

Processing *CA* and its indexes by computer also made possible a variety of new approaches to locating and retrieving chemical and chemical engineering information. As the system progressed, CAS introduced a series of new publications, information services, and computer-readable files derived from it.

Polymer Science & Technology (*POST*), launched in 1967, applied lessons learned from *CBAC* in a somewhat more refined approach to producing an abstracting and indexing service through a computer system. Like *CBAC, POST* (which covered the journal, patent, and report literature of macromolecular chemistry) was issued simultaneously in printed and computer-readable form.

Also in 1967 CAS introduced its first information service oriented toward chemists and chemical engineers engaged primarily in management or marketing. *Plastics Industry Notes* was a weekly, computer-produced digest of and index to business and technical news items selected from key trade journals and newspapers and of importance to the polymer and plastics industries. In 1972 *Plastics Industry Notes* was expanded into *Chemical Industry Notes,* which provides similar coverage for the chemically-based industries as a whole.

CA Condensates, introduced in 1968, was the first computer-readable file to cover the full range of documents abstracted by *CA*. It contains the complete abstract headings (which include the titles of papers, patents, and reports; the names of authors, patent assignees, and patentees; and full bibliographic citations) and keyword index terms for all articles, patents, and reports abstracted in *CA*.

The *CA Source Index,* first published in 1969, was a much expanded, computer-compiled version of the *List of Periodicals* that *CA* had issued at various intervals from 1918 onward to help users identify and locate the primary publications abstracted and indexed. This comprehensive, periodically updated listing includes bibliographic data on some 35,000 scientific and technical periodicals, conference proceedings, and other document collections published since 1830 along with information on which of the world's principal libraries hold copies.

With the conversion of the *CA* subject indexes to computer-based production, in the early 1970s, *CA* introduced a computer-readable file based on them. The *CA Subject Index Alert,* issued biweekly, consists of segments of the computer file of index entries being processed for the semiannual chemical substance and general subject indexes of *CA*.

The printed editions of *CBAC* and *POST* were discontinued at the end of 1971 with the conversion of abstracts in the corresponding sections of *CA* to computer processing, but the computer-readable *CBAC* and *POST* files continue to be issued. In 1975 CAS made available additional computer-readable files that brought together citations, abstracts, and indexing terms in four broad subject areas: energy sources, production, and use; ecology and environment; production, properties, and uses of industrially important materials; and agriculture and soil chemistry and the production, preservation, and consumption of food. Together, these six computer-readable files contain about half the total content of *CA* abstract issues.

To give individuals and smaller organizations access to its computer-readable files, CAS, beginning in the late 1960s, licensed a number of organizations in the U.S. and abroad to act as intermediaries in developing and providing public information services from these files. As of early 1976, 38 organizations in 19 nations were providing information services from CAS computer-readable files; several were offering remote, on-line access to the files through nationwide and overseas telecommunication networks.

Conclusion

For most of its 69 years, CAS provided access to the new and chemically significant information in the world's published scientific and technical literature through a single publication, *Chemical Abstracts*. In the mid-1970s, CAS is in transition from an abstracting and indexing service to a computer-based chemical information-accessing system. While the printed *CA* is likely to remain the most widely used product of the system for some years to come, it is possible also to derive from this system a variety of new information services for chemists and chemical engineers. What forms these services take will depend on the needs and desires of the users of chemical information.

CHAPTER V

IMPACT OF GOVERNMENT

The staff of the division of chemistry, U.S. Department of Agriculture, in 1886 included chief chemist Harvey W. Wiley (third from left).

THE impact of the Federal Government on chemists and their discipline had become massive indeed by the 1970s. The Government itself in 1975 employed some 8000 chemists, the largest single group of physical scientists in its hire. But the number of scientists and engineers on the federal payroll only hints at the pervasive effects of federal activities on chemistry and on science and technology in general.

The ties that bind the Federal Government to science and technology have evolved through several stages. Through most of the nation's history, Congress and the administrative agencies displayed relatively little interest in science, either pure or applied. Few scientists had any direct contact with Government bodies, and the Government's own outlays for research and development were negligible by the standards of 1976. The two World Wars brought dramatic changes in that situation. Wartime needs for new materials and new weapons systems—especially during World War II and the development of the atom bomb—forged a link between science and the Government that was to remain strong with the return of peace. In the 30 years since 1945, international tensions and competition, as in exploring space or developing nuclear or other sources of energy, have steadily expanded the Government's need for ever more sophisticated technology. And as the Government has become increasingly a consumer of scientific information, it has found itself in a much stronger position to call the shots for science and technology.

More recently, the Government has become more than just a major motivator and supporter of research and technological change. In response to widespread fears that unbridled science and technology may be a threat as well as a boon to public well-being, it is assuming the additional role of controller. In policing research and innovation, both public and private, the government establishment will be reaching for much more powerful leverage on the scientific community. It is still not clear just how effectively it can put its imprint on something as diffuse and little understood as the dynamic processes of technological change. But if laissez-faire technology must give way to greater regulation during the final quarter of the 20th century, in much the same way that laissez-faire economics gave way in the preceding 100 years, the resulting impact on how research is done and on national policies and priorities for science could be far-reaching.

The Government's changing role in science and technology, however, is likely to develop—as it has in the past—in an evolutionary rather than a revolutionary manner. Even during the long decades when government treated science generally with benign neglect, its influence, though sporadic, was by no means negligible. This was especially true for chemistry.

The early ties between chemistry and the federal structure stemmed in part from chemistry's importance as a "practical" science with broad applications in industry, agriculture, and health care. Those ties stemmed also in part from the military uses to which chemistry historically has been put; the relationship is evident in the Society's federal Charter, which requires that it ". . . shall, whenever called upon by the War or Navy Department, investigate, examine, experiment, and report upon any subject in pure or applied chemistry connected with the national defense. . . ." The historical mandate to provide for the common defense and promote the general welfare spawned a federal involvement with chemical technology that began with the birth of the nation. Much that has taken place at the government–science interface within just the past decade or so was foreshadowed, if only hazily, much earlier by developments bearing directly on chemistry.

Chemistry in the Nation's Defense

During the Revolution, the Continental Congress and several colonial governments attempted to make up severe shortages of salt by offering bounties for saltworks to be built along the coast. And one of the Congress's first acts following the Declaration of Independence was to proclaim a need for domestic supplies of saltpeter and to issue a manual detailing steps for producing this vital ingredient of gunpowder.

Chemistry, however, played only a minor role in the Government's wartime efforts during the 19th century. Even on the eve of U.S. entry into World War I, the science was accorded scant attention by military officials. Chemist James B. Conant, who served as a major in the Chemical Warfare Service in 1918 (and later as president of Harvard), recalled that when the ACS offered the help of the nation's chemists to the War Department in 1917, Secretary of War Newton Baker replied that, although he appreciated the offer, he believed that it was unnecessary because "he had looked into the matter and found the War Department already had a chemist." (If the story is not apocryphal, Baker changed his mind; in 1919, he commended the Society for its war work.)

World War I did prove nevertheless to be a strong stimulus—for chemical technology if not for chemical research. The U.S. military establishment's long-sleeping interest in technology began to awaken, if only fitfully, as war spread across Europe after 1914. The Navy created a Naval Consulting Board of civilian inventors and engineers which led to a Congressional appropriation of $1 million in 1916 to furnish a laboratory with a staff of "civilian experimentors, chemists, physicists, etc." Little came of this proposal directly, although it did provide a basis for the formation seven years later of the present Naval Research Laboratory. The threat of war likewise led to the creation in 1916 of the National Research Council (which included a chemical committee) out of the long-dormant National Academy of Sciences to serve as a defense research agency. (NAS itself had been established during the Civil War.) NRC became a focal point for wartime scientific work that gained permanency in 1918 and successfully weathered the transition to a peacetime world, although not without considerable shrinkage.

Perhaps the most publicized chemical development of World War I was poison gas. The Army's Chemical Warfare Service, formed in mid-1918, took over earlier work on gases and gas masks conducted by the Bureau of Mines (including a laboratory set up by the bureau in 1917 at American University) and commissioned a number of chemists, including half a dozen past or future Presidents of the Society. More than a quarter of the service's officers were ACS members, even though many chemists at the time were dubious of military domination of any type of research, even that centering on chemical agents for military use. An active recruiter for the Chemical Warfare Service was ACS Secretary Charles L. Parsons, who at the time was also chief chemist of the Bureau of Mines.

But the war's most significant imprint on American chemistry dealt less with research than with industrial chemistry. Like most of the rest of the world, the country was dependent on distant Chilean nitrate deposits as a source of nitrogen for fertilizer and munitions. Even before the U.S. entered the war, Congress appropriated $20 million for a program to develop domestic nitrogen supplies. At the time, the only North American plant capable of fixing atmospheric nitrogen was American Cyanamid's unit based on the cyanamide process and completed in late 1909 at Niagara Falls, Ont. The Government decided to build five nitrogen plants: Haber and cyanamide process units at Muscle Shoals, Ala.; cyanamide plants at Toledo and Broadwell, Ohio; and a modified Haber process plant at Indian Head, Md. Only the Muscle Shoals plant was completed before the Armistice brought a quick shutdown of the nitrogen program, and the Haber process plant there never operated on more than an experimental basis.

The government nitrogen program had cost $107 million. Proposals that the Government either sell the two Muscle Shoals units or use them to make nitrogen fertilizers got tangled in more than a decade of heated political controversy, with the chemical industry fighting federal operation and farmer interests supporting it. In the end, the Muscle Shoals facilities, including associated hydroelectric power installations, were incorporated into the Tennessee Valley Authority when it was created, in 1933.

The wartime nitrogen program had few tangible operating results. But it did pave the way for the giant postwar fertilizer industry and provided practical design and construction experience to aid in building the commercial ammonia plants that sprang up during the 1920s. And the long wrangle over disposal of the Muscle Shoals operations aired, if inconclusively, many still troublesome questions about government competition with industry.

The U.S. potash industry similarly was stimulated when the war cut off German supplies. Lake brines were exploited, especially in California and Nebraska, kelp was harvested along the Pacific coast as a source of potash, and some material was even obtained by the long-obsolete process of leaching wood ashes. The Government set up an experimental kelp products plant at Summerland, Calif., which operated for a half dozen years. But when the war ended, German potash once again captured the U.S. market, and domestic production quickly collapsed. Nevertheless, the seeds of a U.S. industry had been planted. And vestiges of the wartime operations remained, especially at Searles Lake, Calif., until the vast potash deposits of the Southwest were opened up in the late 1920s to make the U.S. independent of overseas supplies.

A third government-backed chemical enterprise to spring from military demands was helium production. After a crash development program, three government-owned plants were built in Texas to extract helium from natural gas, although production was just getting under way as the war ended.

More than any single development, however, the war marked the coming of age of the American chemical industry as a whole, especially the segment of the industry producing organic chemicals. Between 1914 and 1919, the number of plants producing chemicals in the U.S. nearly doubled, to 781, while the industry's employees more than doubled, to 71,000. The value of chemicals made in the U.S. leaped from $192 million to $644 million, with coal-tar products soaring from $13.5 million to $133.5 million and other organic chemicals from $16.4 million to $72.1 million. The war made the nation at last self-sufficient in organic chemicals and dyes.

U.S. producers, fearful of a resurgent German dyestuff and coal-tar chemical industry after the war, pressed for tariff protection. The ensuing debate in Congress aired much high praise for the fledgling industry. One Senator noted: "Chemistry is what is governing the world today, both in war and in peace, and progress in chemistry cannot be maintained unless you preserve an actually going concern." Another added: "It is the man working in the chemical laboratory who is to blaze the way for human progress. . . The two things—research in the laboratory and the successful conduct of a chemical industry—go hand in hand. Neither can proceed without the other." The Society also was pressing such views; one of the functions of the ACS News Service when it was formed, in 1919, was to build public support for the maintenance of a strong domestic chemical industry.

Arguments that a strong coal-tar chemical industry was vital to the U.S., on both economic and defense grounds, carried the day. Congress passed the Tariff Act of 1922, with its unusual provision for American valuation or American selling price as the basis for ad valorem duties on coal-tar intermediates and chemicals. The 1922 tariff provided a wall behind which the U.S. synthetic organic chemicals industry could expand between the two World Wars, relatively well insulated from entrenched foreign competition.

The 1922 act was by no means the first tariff barrier offered to the chemical industry by the Government. Protective tariffs to shield immature American industries had been a political issue ever since the War of 1812. Increased duties on such chemical products as red and white lead, alum, sulfuric acid, and Glauber's and Epsom salts did much to encourage the establishment of a small but nonetheless significant U.S. inorganic chemicals industry during the 1820s. And a committee of chemical producers, in a report issued in 1831 during a General Convention of Friends of Domestic Industry, persuasively made a case for tariff protection for the then $1 million, 30-plant, 900-employee chemical industry of the U.S. After the 1830s, a shift toward lower tariffs tended to slow—although certainly it did not halt—expansion of the industry during the rest of the 19th century. It was the heavy inorganic chemical industry, however, that chiefly benefited from the demand for chemicals generated by growing industrialization after the Civil War. The struggling coal-tar chemicals and dyes business was largely stifled by low-cost imports from Europe after Congress bowed to pressure from textile and paper producers in 1883 to lower tariffs on those products.

World War I's boost for U.S. chemical technology was repeated during World War II. In much the way that the foundations for a large-scale synthetic ammonia industry were laid during the earlier war by the Government's defense program, the stage was set for today's petrochemical industry by government-built plants, many on the Gulf Coast of Texas, for producing synthetic rubber and its styrene and butadiene intermediates, as well as by war-bred demand for DDT and for synthetic materials such as nylon and polyethylene.

World War II also rekindled federal interest in scientific research, but with more long-lasting effect than earlier. The scientific advisory structure established by the Government in World War I was largely demobilized when the war ended—except for a frail National Research Council. With the threat of war again hanging over the U.S., a National Defense Research Committee was set up in June 1940 to serve as a channel for government funds in support of weapons research. NDRC was superseded a year later by the broader, more comprehensive Office of Scientific Research and Development. OSRD itself did not outlast the war by long, but it broke ground for a much greater federal involvement in support of research, especially basic research. The ambitious research programs of the Department of Defense and the Atomic Energy Commission were an outgrowth of seeds planted by OSRD. Perhaps more important, the agency's success during the war helped to open the way for the creation of the National Science Foundation in 1950 as a focus for federal funding of basic science.

NSF has been a major source of government backing for chemical research, although its interests encompass much more than chemistry. Only about 10% of NSF's budget for supporting research projects, for example, is ticketed for chemical research, although additional funds go into closely related areas such as biochemistry and materials science. The agency also has funded a number of ACS projects. By 1975, for example, it had provided some $23 million for the automation research and development work at Chemical Abstracts Service. But in any event, no federal agency has aided chemistry to the degree that medical science has found a home at the National Institutes of Health, physics at the Atomic Energy Commission, aeronautical engineering and astrophysics at the National Aeronautics and Space Administration, and agricultural research at the Department of Agriculture. On the other hand, because of the breadth of its applications, chemistry is a significant part of the research programs of all the federal agencies involved with science and technology.

Chemists have found that this position has both advantages and disadvantages. Dependent on no single agency or limited number of programs for government money, chemistry has been more insulated than some other disciplines from changing fashions in research or shifts in national priorities. At the same time, no single arm of government is responsive specifically to chemistry's particular needs. Consequently, many chemists have believed that funding for their work has been inadequate, compared with that available to other scientists. This concern was spelled out in 1965 in the National Academy of Sciences' report, "Chemistry: Opportunities and Needs" (the Westheimer report).

The Westheimer report, which was funded in part by the Society, was widely applauded as a model of low-key but thorough and persuasive pleading for increased financial aid for scientific research, especially in the universities. There was no serious disagreement with its conclusions. And more federal money did flow into chemical research following its publication, at least until the late 1960s, when the Government began to cut back. But in the intensely competitive struggle for research grants, many chemists continue to feel that federal generosity often falls short of matching their "opportunities and needs." In the light of other, shifting demands impinging on the national budget, this feeling perhaps is inevitable.

Chemistry for the General Welfare

Government programs for promoting chemical technology certainly have not been limited to the needs of war and national defense. To encourage native inventive genius and "promote the progress of science and useful arts," a provision for granting exclusive patents to inventors was written into the Constitution. The U.S. Patent Office long has been a focus for the Government's support of chemistry. In fact, the nation's first patent, issued July 31, 1790, went to Samuel Hopkins of Pittsford, Vt., for "the making of pot ash and pearl ash by a new apparatus and process" (the first patent to be issued in North America was granted by the Massachusetts General Court nearly 150 years earlier to Samuel Winslow for a new method of making salt). The first trademark to be registered by the Patent Office, in 1870, covered a design used by Averill Chemical Paint Co., New York City, showing an eagle holding a can of paint in its beak, with "chemistry" emblazoned across the bottom.

Potash production from wood ashes was the first U.S. chemical business. Export trade reached a peak about 1809, when more than 12,000 tons of potash, worth about $1.5 million, were shipped to England, and then slowly declined. Output of potash for domestic industrial use, however, continued to flourish long into the 19th century until synthetic soda ash displaced it.

Potash also was one of the first U.S. industrial products to be targeted for government investigation and regulation. In the early 1830s the state of New York made an effort, apparently with limited success, to end the widespread practice of adulterating potash with salt, lime, or other foreign materials. The state selected Lewis C. Beck, a professor of chemistry at Rutgers University, to study potash production, to analyze samples of the substance available on the market, and to develop standards for the product. His findings were published in 1836.

But efforts by the Federal Government to either encourage or regulate the development of chemical technology were of relatively little consequence until well into the 20th century. Its earliest involvement stemmed from a push to develop agriculture—not surprising in a country whose economy, even after the Civil War, depended largely on farming and where farmers had considerable political power. The first government-funded publication to deal with technology concerned agricultural chemistry, specifically the cultivation of sugarcane and the refining of sugar. Congress authorized the preparation of the manual in 1830; the 122-page report, published in 1833, was completed under the direction of Benjamin Silliman the Elder—professor of chemistry at Yale and the country's most influential teacher of chemistry at the time—in collaboration with several other chemists.

More important was the interest in agricultural development shown by the Patent Office during the years following 1835 when it was headed by Henry L. Ellsworth. With an appropriation of $1000 from Congress in 1839, Ellsworth began to collect agricultural statistics and to distribute information and seeds to farmers. In 1848, Congress appropriated $1000 so that the Patent Office could set up a system for analyzing grains and flour. The same Professor Beck who had earlier studied potash problems for New York was engaged for the job. Dr. Beck prepared for the Patent Office two reports on breadstuffs in which he discussed methods for determining their value and detecting such foreign materials as alum, copper sulfate, carbonates, chalk, and plaster of Paris. Thus was born federal interest in product purity and safety. (An earlier concern for safety had led Congress in 1830 to authorize an investigation into the causes of boiler explosions on steamboats; this study, by the Franklin Institute in Philadelphia, led to a report in 1836 and to subsequent regulatory legislation.)

The Patent Office's agricultural division was the direct forerunner of the U.S. Department of Agriculture, established by Congress in 1862. Among other tasks, the department was to gather information by "practical and scientific experiments" utilizing the services of "chemists, botanists, entomologists, and other persons skilled in the natural sciences pertaining to agriculture." USDA was created at a time when many people had become convinced that progress in agriculture must be based on a better understanding of the chemical nature of crops, soils, and products used by farmers. Its first scientist was German-trained chemist Charles M. Weatherill, who published a report on the "Chemical Analysis of Grapes" in 1862.

Much, although by no means all, of the department's early chemical work was relatively routine analysis. Nevertheless, it was the first federal agency to emerge with a strong mandate to do scientific work and a staff and structure designed to carry out that mandate. Earlier scientific endeavors by government agencies—in exploration, mapmaking, astronomy, or meteorology—had been for the most part tacked on, often with questionable legal authorization, to functions whose primary goals were nonscientific.

The birth of USDA, therefore, marked a turning point in federal involvement with science. By incorporating divisions specifically assigned to chemistry, botany, entomology, and horticulture, USDA paved the way for other federal agencies to sponsor scientific research. Its small, ill-equipped laboratories in Washington foreshadowed not only USDA's present far-flung regional laboratories, but also such federal research facilities as the National Bureau of Standards, formed in 1901, and the National Institutes of Health. NIH was established in 1930 as an outgrowth of the Public Health Service's Hygienic Laboratory, which in turn had been set up in 1909 when a small Public Health Service laboratory dating from 1887 was moved to Washington. USDA also pioneered in providing federal financial aid for university research in working with the land-grant colleges that were established under the Morrill Act of 1862 and with the state experiment stations authorized by the Hatch Act of 1887.

USDA's chemical arm grew haltingly, however. By 1888, for example, the department's Bureau of Chemistry was staffed by a single chemist, two assistant chemists, and seven laboratory assistants. As the only government laboratory of any significance in Washington, moreover, the Bureau frequently was called upon to do routine testing or to answer scientific questions for most of the other federal agencies.

Still, USDA's scientific efforts were setting precedents that would blaze a path for much of the future development of government science. Its employees were learning how to launch problem-oriented, frequently interdisciplinary attacks on the difficulties faced by the nation's farmers—for example, in soil erosion or plant and animal breeding. It nurtured a stable staff of capable scientists who were gaining experience in working with groups outside their own laboratories—in the universities and Congress or among the public at large—and thereby were developing public support and a political constituency for their work. And perhaps most important, they were finding that hand in hand with success in solving scientific and technological problems came a need to prevent or control those problems by taking on regulatory or law enforcement functions as well.

In 1883, Harvey W. Wiley—ACS President in 1893–94 and one of the most influential chemists ever to work for the Government—was appointed chief chemist of USDA's chemistry division. Dr. Wiley at first continued the work begun by his predecessor, Peter Collier, on the chemistry of sugar and sugar-producing crops. A project to develop sorghum as a source of sugar proved unfeasible, but when Wiley switched his investigation to beets his research led eventually to the development of the U.S. beet sugar industry. More important, though, he became intrigued by questions of food adulteration and the evils of useless or even harmful patent medicines, issues that also had interested Collier. Work in these areas increasingly monopolized Wiley's attention.

Chemical Products and Public Safety

Congressional concern about pure food had already surfaced. The first bill to prohibit the adulteration of food was introduced in 1879. More than 100 similar proposals were considered in subsequent Congresses during the next quarter of a century, several of them receiving favorable committee action. A bill banning adulteration of food and drugs passed the Senate unanimously in 1892, and two similar bills won House approval in 1902 and 1904. But none was enacted.

Congress, for one thing, was still uncertain of its constitutional powers for regulating commerce; it was reluctant to encroach on the traditional concepts of states' rights and police powers. Federal control of the production and distribution of foods and drugs was a new and controversial legislative doctrine during the late 19th century and could evolve only very slowly. The form such legislation might take and the way in which it might be implemented brought out sharp differences between the House and Senate which neither body was willing to compromise. Finally, the enactment of any food and drug legislation was strongly opposed by many politically powerful manufacturers and sellers of these products as an unwarranted extension of government authority over private business.

Within USDA, meanwhile, Harvey Wiley was broadly expanding his investigations of food and drug adulteration. His studies of food additives, analytical methods, and standards led to important progress in the science of nutrition. He also became a strong and publicity-wise critic of the food and drug industries, capturing public attention and launching a national crusade for regulation. His indefatigable efforts culminated with the passage by Congress of the Federal Food and Drugs Act of 1906, which designated USDA's Bureau of Chemistry, still headed by Wiley, as the regulatory agency.

Law enforcement is a legal and political as well as a scientific function, and in his new role Wiley soon found himself attacked by private interests and frequently hamstrung by higher officials, all the way up to President Theodore Roosevelt. An unsuccessful effort was made in 1911 to remove him from office for misuse of government funds. Unwilling to compromise his zeal, he resigned from the department in 1912, blasting his superiors for restricting his powers to administer the law as he saw fit and for putting him in an inhospitable environment. "I saw the fundamental principles of the Food and Drugs Act, as they appeared to me, one by one paralyzed or discredited," he wrote. Wiley continued his crusade in an editorial post with *Good Housekeeping*.

Enforcement of the food and drugs law led to marked changes in the Bureau of Chemistry. Its staff jumped from 20 in 1897 to 110 in 1906 and 425 in 1908 and its appropriations from $131,000 in 1906 to $698,000 in 1908 as Wiley shifted its operations away from experimental research to inspection and regulation. However, the Bureau continued to make significant contributions to food technology, for example in the use of freezing and cold storage for preservation.

The 1906 law was ground-breaking legislation, but political considerations led to its being forged by compromise. Consequently, despite its broad scope, it was less than a perfect instrument for doing the job needed. Its enforcement, too, called for new and untried practices on the part of government officials in an area that likewise was steeped in controversy. Proving adulteration or misbranding in court was often very difficult, and many cases failed to result in conviction. Hence, neither in theory nor practice could the 1906 law eliminate all the abuses it was designed to overcome. Dr. Wiley himself, a year before he died, in 1930, was moved to publish a "History of a Crime against the Food and Drugs Act." In it he detailed the assaults made against the law and decried the splitting, in 1927, of the Bureau of Chemistry into a Bureau of Chemistry and Soils and a Food, Drug, and Insecticide Administration (the latter being the direct predecessor of the present Food and Drug Administration, now part of the Department of Health, Education, and Welfare).

But the 1906 law did set the stage for further federal control over medicinal and food chemicals. It was strengthened by the Sherley Amendment of 1912, which banned false or fraudulent statements about the effects of proprietary drugs. The Food, Drug, and Cosmetic Act of 1938 was passed after several people had been poisoned when a drug manufacturer marketed a solution of sulfanilamide in diethylene glycol. The act required testing for safety of any new pharmaceutical sold in interstate commerce and extended the reach of the law to a wider range of medical products as well as to cosmetics. The 1906 law also was a harbinger of the Insecticide and Fungicide Act of 1910, the first federal legislation to deal specifically with pesticides. This act, however, was passed to protect farmers from inferior products rather than to protect the public from misuse of poisonous substances in agriculture.

The Food and Drug Administration long had been a target of criticism for lax enforcement of the law, for ineffective organization, and for what many observers saw as ties too close to the industries it was intended to regulate. These attacks were not stilled by the 1938 law, but led to still further tightening and reform following tragedies later with thalidomide and a number of other drugs that showed unexpected hazards. The Kefauver-Harris Amendments enacted by Congress in 1962, for example, considerably extended FDA's control over the drug industry. Among other things, they required that new drugs must be proved with "substantial evidence" to be effective as well as safe, provided for tighter FDA control over the testing of drugs and their manufacture, and gave the agency greater authority to order drugs found to be harmful off the market.

The 1938 law had been broadened by the Miller Pesticide Amendment of 1954 to cover residues on crops. It had been broadened also by the Food Additives Amendment of 1958 and the Color Additive Amendment of 1960, which prohibit the use of any chemical in foods, except for additives "generally recognized as safe" through years of widespread acceptance, until tests have shown it to be safe. The Delaney clause of the 1958 amendment bans the use in food of any chemical that has been found to act as a carcinogen in man or animals when ingested in any amount. This provision has opened up questions about threshold levels of biological activity and about zero tolerances that have proved particularly irksome for producers of food additives and pesticides, especially as more sophisticated and sensitive techniques have permitted analysts to pick up trace amounts of materials that previously would have gone undetected. In the 70 years since 1906, federal regulation of food and drugs has become firmly established in our legal structure. But controversy continues over means of implementing and strengthening existing laws, with many producers viewing them as undercutting free enterprise and innovation while many consumerists call for further safeguards.

Increasing federal control over pharmaceuticals, food additives, and pesticides, in fact, has only foreshadowed steadily widening regulation of other chemical products on grounds of worker safety, consumer safety, and environmental protection. The Clean Air and Occupational Safety and Health Acts of 1970, the Consumer Product Safety and Environmental Pesticide Control Acts of 1972, and a much strengthened (in 1972) Water Pollution Control Act have all tightened federal control over the use of chemical materials. New toxic substances control legislation goes a step further by calling for premarketing tests to determine the safety of all newly developed chemical products.

The current or potential impact of all this regulation on chemical research and development, on the processes of innovation and technology transfer, and on the economics of chemical production and marketing has been the subject of much debate. Many industrial executives see the dead hand of bureaucracy stifling development and repressing research. The pharmaceutical industry, for example, tends to blame the procedures and policies of the Food and Drug Administration in enforcing the 1962 drug amendments for the sharp decline during the past decade in the number of new drugs reaching the marketplace. Lower productivity from research, however, probably reflects several other factors as well. Many of the easier problems of medical science already have been adequately worked out, so that research now must often be aimed at developing more sophisticated and more potent drugs, including those that must be used in large doses or over long periods to alleviate chronic ailments. As the more readily accomplished objectives of medical research are successfully siphoned off, progress seems to be achieved more slowly. But there is little question that the need for much more intensive and comprehensive testing of new products before they can be passed on to the public has increased the cost of research and development, in terms of both time and money, perhaps to the point where product innovation is pushed beyond the financial reach of many smaller firms.

Control of Technological Change

The issues involved are a part of a broader overall movement toward greater federal involvement in all types of scientific and technological development, a movement that in recent years has marched under the banner of "technology assessment." Technology assessment, as its proponents see it, is a process that would uncover and attempt to weigh the competing benefits and risks resulting from a proposed technological change. It would, moreover, concern itself not only with the immediate or obvious consequences of new technology—in safety, efficiency, profitability, and the like—but would try to scrutinize the secondary or unintended social and environmental effects as well.

The precursors of technology assessment can be found in the new drug applications required under the food and drug laws or in the provisions of the National Environmental Policy Act calling for environmental impact statements for all new government projects that impinge on the environment. Technology assessment itself, in a formal sense, has barely come onstage within the Government. In 1972, Congress established an Office of Technology Assessment, whose first completed project, in 1974, was a short evaluation of the bioequivalence of drugs. And the National Science Foundation has funded several assessments under its Research Applied to National Needs program. The implementation of the concept in a practical or meaningful fashion, however, is still to be accomplished.

Government attempts to assess and control technological change, on whatever scale they are undertaken, seem likely to make the problems of establishing research priorities and policies even more difficult and to raise new obstacles to innovation. In the evolution of new technology, timing is often a vital element. Opportunities delayed may be opportunities lost. Yet regulators, faced with unresolved questions about offsetting costs and benefits, often may find it easier to block a proposal than to approve it or to take no action at all while calling for more data.

The new regulatory bodies, moreover, tend to have very sweeping mandates. FDA and most of the other older agencies are responsible for overseeing relatively specific industries or economic functions. In contrast, the Environmental Protection Agency or the Occupational Safety and Health Administration have responsibilities cutting across wide segments of activity. As a result, some observers fear, they may be less likely to consider all of the narrow impacts, economic or otherwise, of their decisions on specific industries.

And as the control mechanism encompasses broader and broader aspects of impact on society, the uncertainties become more and more difficult to deal with. The comprehensive testing now demanded for new drugs and food additives and pesticides, for example, can generally be carried out in the controlled and reproducible environment of a laboratory or clinic. But a sweeping assessment of all the social, political, environmental, cultural, and economic impacts of technological change would have to deal with many factors difficult or impossible to quantify. One result is likely to be that decisions on research and development will have to be hammered out in a more politicized, public-interest arena than in the past. Strictly scientific considerations alone will no longer be the determinants.

All this is a far cry from the seeds of federal regulation planted in 1906 within the Department of Agriculture. Then, control focused on adulteration of food and drugs as a result of intentional attempts to deceive. Now, control has evolved into much more complex and far-reaching concerns dealing with accidental, unintended, and unexpected effects. Such effects, to be sure, are no less a threat to man and his environment than those resulting from pure and simple fraud. But they are much more difficult to weigh.

Yet this evolution toward greater federal regulation probably was inevitable—and certainly seems irreversible. Chemistry and its products have become much more pervasive in the past 50 years, touching on nearly everything the nation produces and consumes; it has become more and more apparent, too, that chemical products can have a more critical impact on the national health and environmental integrity than anyone could have envisioned even a decade or two ago. This situation, in fact, was a major consideration in the Society's decision of 1965 to form its Committee on Chemistry and Public Affairs as a mechanism for taking a more active role in the formulation of national policy on chemically related issues. Perhaps if the United States can spell out what kind of life it wants to lead, it can indeed manage to impose a better design and control on its technology. But it is yet to be demonstrated that this can be done by grand design rather than in a step-by-step, even trial-and-error manner.

CHAPTER VI

PUBLIC AFFAIRS

IN March 1965, ACS President Charles C. Price proposed to the Board of Directors and the Council Policy Development Committee that the Society form a committee to work actively on issues of public policy related to chemistry. "Only if the public and its officials are adequately informed," Dr. Price argued, "can they make the best decisions on the future role of chemistry in public policy matters. The Society should play a much more prominent role in producing information and responsible opinion on public policy questions related to chemistry."

Dr. Price's proposal, which led quickly to an active ACS program in public affairs, embodied more a change in emphasis than a fundamentally new approach to public policy and the public interest. The Society had a long history of positive relations with government. Indeed, the ACS on occasion had treated legislative and other issues a good deal more aggressively before and during World War II than later. It was a return to the activist approach, albeit in much changed circumstances, that mainly marked the Price proposal of 1965.

The Society's first sustained effort in public affairs was precipitated by World War I. The cessation of imports of dyes and other chemicals from Germany in 1914 led the New York and other ACS local sections to study means of expanding this country's chemical industry. In September 1915,

September 1954 saw the first meeting of the Society's newly established Committee Advisory to the National Bureau of Standards. From left are N. Howell Furman, Norman A. Shepard, Milton Harris, Wallace R. Brode (assistant director of NBS), C. S. Marvel, C. F. Rassweiler, and J. R. Ruhoff.

with the unanimous approval of the Council, ACS President Charles H. Herty urged President Woodrow Wilson by letter to ask Congress for legislation to protect domestic manufacture of dyes and chemicals from foreign competition. In mid-1915, Secretary of the Navy Josephus Daniels asked President Herty to nominate two ACS members to serve on a Naval Consulting Board organized to advise on methods of strengthening the Navy. The posts went to Willis R. Whitney and Leo Baekeland. In the early months of 1916, at the request of President Wilson, Dr. Herty nominated 50 more ACS members, one from each state and territory, to help the Naval Consulting Board "in the work of collecting data for use in organizing the manufacturing resources of the country for the public service in case of emergency."

The Society's wartime public services were marked sometimes by a peculiar circumstance: ACS Secretary Charles L. Parsons was also chief chemist of the U. S. Bureau of Mines. The Bureau of Mines' research on gas warfare, for example, was subject to the approval of a Committee Advisory to the War Work of the Bureau; the committee was made up of ACS members, including Secretary Parsons, who thus in a sense was advising himself. Similarly, a census of American chemists, the brainchild of Dr. Parsons, was conducted jointly by the Bureau of Mines and the ACS just before the U. S. declared war in April 1917. The purpose of the census was to help insure the most effective use of scientific manpower, an interest that would occupy the Society extensively in subsequent years. In September 1919, 10 months after the Armistice, Secretary of War Newton D. Baker acknowledged the Society's contributions to the war effort in an address at the ACS national meeting in Philadelphia.

Growing Federal Contact

The steady growth of science and of ACS activities in the years between the two World Wars brought an increasing range of contacts with the Government. Partly for this reason it became desirable for the Society to seek the federal Charter that came into being in August 1937 when President Franklin Roosevelt signed Public Act No. 358, 75th Congress. With the Charter the ACS assumed a general obligation to put its expertise at the service of the Government and the nation.

The outbreak of World War II plunged the Society into another sustained round of public-interest activities, although the problems differed markedly from those of World War I. In the fall of 1940, at the request of President Roosevelt, the ACS set out to identify the nation's chemists and chemical engineers and their qualifications. The data were obtained for the National Roster of Scientific and Specialized Personnel "to enable prompt and correct allocation of chemists and chemical engineers among other scientists, if and when needed because of an actual emergency."

Efficient use of scientific manpower remained by far the Society's major concern during World War II. In January 1943, 13 months after the U.S. entered the war, ACS President Per K. Frolich outlined the basic problem: ". . . As a profession we do not want class deferment. But, technically trained men already doing essential war work in industry, or capable of filling such jobs, must not be drafted for nontechnical service." The Society launched an effort to keep draft boards and Selective Service officials informed about the work of chemists and chemical engineers and its value to the war effort. Throughout the war the ACS campaigned to assure adequate technical manpower for industry.

The end of World War II, in May 1945 in Europe and in August in the Pacific, did not bring an end to ACS involvement in manpower policy. The Selective Service legislation that emerged in the late 1940s greatly concerned the technical community because of its impact on science students. In 1948, ACS President Charles A. Thomas testified on the subject before the Senate Armed Services Committee and was asked to help write a new bill on Selective Service. In the fall of 1950—the Korean conflict had erupted in June—Dr. Thomas, by then ACS Board Chairman, reported to the Council that, despite the best efforts of the Society, the problem of blanket drafting of technical manpower had not been solved. He asked that the Council adopt a resolution, which it did along with the Board, calling on other major technical societies to join the ACS in cooperating with the Government to establish "a program for the proper training and use of scientific personnel in time of national emergency." A copy of this resolution was sent to President Harry S. Truman.

In 1955, Board Chairman Ernest H. Volwiler testified before Congress three times on the military and other aspects of manpower policy. In so doing he was armed with an ACS statement of policy which read in part that every step should be taken "to assure maximum and uninterrupted growth of scientific and technological developments by promoting a strong educational system at all levels which will produce an adequate flow of specialized personnel of outstanding qualifications." The effects of these efforts on legislation were termed "discouraging" by Executive Secretary Alden H. Emery. Two years later, in October 1957, Sputnik 1 went into orbit, and the ACS view suddenly became fashionable.

Impact of Public Funding

The Society's concern for the educational system was valid, but science and technology in this country nevertheless had expanded rapidly from the huge base built up during World War II. At the same time, financing of research and development had shifted increasingly from private to public hands. The missile and space races, the National Defense Education Act of 1958, and similar factors extended this trend to the point where federal funding decisions were dominant elements in the health of chemistry and other sciences. In 1964, for example, federal money accounted for some 80% of the explicit support for basic chemical research in the universities.

The preponderance of federal support moved the ACS to a significant, if indirect, effort to influence public policy. Large sums were being spent on research and development by the early 1960s, but influential chemists contended that opportunities in basic research were being missed because of lack of funding, particularly government funding. To assess this contention, early in 1964 the National Academy of Sciences, at the instigation of its Committee on Science and Public Policy, appointed a Committee for the Survey of Chemistry with Frank H. Westheimer of Harvard as chairman. The Society was not directly involved in the study, but in April 1964 it granted "up to $50,000" to support the committee's work. The 222-page Westheimer report, "Chemistry: Opportunities and Needs," was published in 1965. Not surprisingly, it urged greatly expanded support for chemical research from both public and private sources.

The true effect of the Westheimer report is moot, but the ACS grant for the study was clearly a mark of the Society's growing inclination to speak more openly on public policy. A second sign of the trend was the establishment in 1963 of the Council Committee on Patent Matters and Related Legislation, which in 1965 was made a joint Board–Council committee. The first chairman of the committee, Pauline Newman, was succeeded in 1970 by John T. Maynard, who was succeeded in 1975 by Willard Marcy. In 1964, several advisory committees appointed by ACS President M. H. Arveson concurred on the need for the Society to move further into public policy. And in 1965 came President Price's proposal that a committee be formed specifically for that purpose.

In June 1965, the ACS Board of Directors adopted the Price proposal by establishing the Board Committee on Chemistry and Public Affairs. The charge to the committee was:

• To initiate and conduct studies, and prepare and publicize to the membership and to the public, reports on problems involving the role of chemistry in public affairs with which the Committee believes the Society must concern itself in order to fulfill the obligations imposed by its National Charter.

• To recommend to the Board and other Society units appropriate studies and actions which they may undertake.

ACS President Charles C. Price told the Council in April 1965 that the Society needed a more effective mechanism for participating in the formation of public policy.

• To advise ACS staff members on their government relations in other than routine matters.

The first chairman of the 15-member Committee on Chemistry and Public Affairs (CCPA) was Dr. Price, who served during 1965–68. Franklin A. Long assumed the post in 1969 and was succeeded in 1973 by Charles G. Overberger. In 1968, the CCPA was made a joint Board–Council committee. In 1965, shortly after the CCPA itself was established, the Board authorized a staff Office of Chemistry and Public Affairs under Stephen T. Quigley. In 1967, the office was made a department, which in 1975 was moved into the Office of the Executive Director with Dr. Quigley as Director of Chemistry and Public Affairs.

Major ACS Studies

Among the first major projects undertaken by the CCPA was a study of the chemical science and technology of environmental improvement. The work began in 1966 and culminated in September 1969 with publication of the 250-page paperback, "Cleaning Our Environment: The Chemical Basis For Action." In October 1969, members of the CCPA subcommittee and task force that had prepared the report were asked by the President's Science Advisory Committee to refine and highlight the most important of its 73 recommendations. The result was a 20-page supplement to "Cleaning Our Environment," published by the Society in January 1971. Through 1975, the ACS had distributed more than 21,000 copies of "Cleaning Our Environment," free of charge, to key legislators, federal administrators, and national, state, and local policy-makers and public officials. The Society had sold an additional 52,000 copies of the volume at a nominal price. More than 200 colleges and universities had used or were using it as a text or reference book for environmental courses, and it had been translated into Arabic, Italian, and Japanese.

The Society introduced "Cleaning Our Environment: The Chemical Basis for Action" on Capitol Hill in September 1969. From left: Rep. Emilio Q. Daddario (back to camera) (D.—Conn.), Rep. George P. Miller (D.—Calif.), Wallace R. Brode, Milton Harris, W. O. Baker, Lloyd M. Cooke, Franklin A. Long, T. E. Larson (head hidden), Daniel MacDougall, James P. Lodge, Jr., James J. Morgan. Dr. Cooke headed the CCPA subcommittee that guided the development of the book; Dr. Larson headed the task force that assembled it.

A major outgrowth of the environmental study within the ACS was the establishment of the Board–Council Committee on Environmental Improvement (CEI), in 1969. In 1973, under chairman T. E. Larson, CEI undertook a revision of "Cleaning Our Environment," with publication scheduled for 1976. Dr. Larson had headed the task force that prepared the first edition of the book, which included chapters on air, water, solid wastes, and pesticides. The revised edition was to include four additional chapters, on analysis and monitoring, energy, radiation, and toxicology.

A second major study initiated by CCPA was designed to provide a broad and thoughtful assessment of the function of chemistry in the nation's economy. The work was supported partly by the Society and partly by the National Science Foundation. The result was the 600-page paperback, "Chemistry in the Economy," published by the ACS in October 1973. By the end of 1975, the Society had sent more than 3000 complimentary copies of the book to legislators, government administrators, and other policymakers. It had sold an additional 11,500 copies, and the volume was in use at a dozen colleges and universities.

A third study initiated by CCPA was "Chemistry in Medicine," scheduled for publication in 1976 and designed to elucidate the impact of chemistry in medicine and related fields. Two further CCPA studies were under way in 1975, though not yet scheduled for publication. One of these was an examination of the science and technology of energy production, storage, transmission, and conversion, all from the point of view of environmental impact. The second of the two studies involved the myriad chemical aspects of the nation's and the world's material resources.

Statements on Public Policy

In addition to its major studies of chemically or scientifically oriented issues, the Society develops "official public policy statements and communications." These documents take three basic forms: (1) oral presentation before a committee of the Congress or before an administrative law judge during a federal agency rule-making procedure; (2) statements submitted for the official record of hearings where oral presentations are made or occasionally submitted to a court of law, usually as an *amicus curiae* brief; and (3) communications, generally letters, to public officials, which are in the public domain but are not part of any formal record of proceedings of any of the three branches of Government. The number of these public policy documents grew steadily after the CCPA was formed, in 1965. In the period October 1972 through December 1975, the Society submitted 50 official statements and communications to the three branches of the Government.

In May 1975, ACS Executive Director Robert W. Cairns (center, head bowed) testified at House hearings on copyright reform legislation. To Dr. Cairns' right are ACS staff members Richard L. Kenyon and Stephen T. Quigley.

ACS policy statements are prepared under the direction of the Committee on Chemistry and Public Affairs with staff support from the Department of Chemistry and Public Affairs. Important to the process are open meetings held by ACS committees at the Society's national meetings, a procedure that gives members an opportunity to comment or raise questions on topics of concern. In addition, the CCPA and other committees seek opinion both outside their memberships and outside the ACS in order to develop a consensus of the chemical or scientific community on the issue at hand.

The scope of the Society's public affairs interests broadened markedly in the last decade of its first century. The ACS has taken official stands on topics that include environmental pollution, patent protection for inventors, copyright protection for scientific publishers, pension reform, science information systems, and toxic substances control. Forging a consensus on such issues can be difficult for a heterogeneous body like the ACS, because the issues commonly involve social, political, and economic elements as well as scientific fact.

Once adopted, the Society's stand on a given issue normally will remain valid in the face of all but major changes in circumstance. In 1966, for example, the ACS endorsed conversion to the metric system in the United States; in the subsequent decade it held to this view consistently before the appropriate Congressional committees. In 1970, on the other hand, the Society largely reversed its stance of half a century and supported the federal administration's decision to endorse the Geneva Protocol with qualifications; in 1973, the ACS reversed its traditional stand altogether and endorsed the Protocol without qualifications. In 1925, the Government had refused to sign the Geneva Protocol, which bans the use of chemical and biological weapons. The Government's position had been supported by the Society, which was instrumental in the formation of the Chemical Warfare Service (later the Army Chemical Corps) in World War I and for some years maintained a Committee Advisory to the Chemical Corps. The conflict in Vietnam in the 1960s, however, highlighted the drastic changes that had come about in chemical and biological warfare. The use of such weapons was no longer technologically restricted to combatants in a small battle zone, but could readily be extended to noncombatants over large areas. On these grounds the Society reversed its position on the Geneva Protocol.

Member Involvement

The CCPA concluded early in its existence that a sound ACS program in public affairs required that the members be exposed regularly to important questions of public policy. To this end, the committee urged the Board to inaugurate a series of public affairs symposia at ACS national meetings. The goal was to stimulate exchange among chemists from many subdisciplines and members of legislative and other policymaking bodies. The first of these symposia was held in September 1966 at the ACS national meeting in New York City. Its topic, "The Synthesis of Living Systems," was inspired by the CCPA's conviction that research

TABLE I

Selected Public Affairs Symposia Presented at ACS National Meetings

Title	*Date*
The Synthesis of Living Systems [a]	September 1966
Basic Research: Its Function & Its Future [a]	September 1968
Herbicides and Pesticides—Policies and Perspectives [a]	September 1971
Nutrition and Public Policy in the United States [a]	April 1972
Air Pollution and U.S. Public Policy [a]	August 1972
Water Pollution and U.S. Public Policy [a]	August 1973
Public Hearing on Compensation for Employed Inventors [b]	August 1973
Assessing the Problem of Lead-Based Paint Hazards [c]	August 1975

[a] Sponsored by the Committee on Chemistry and Public Affairs
[b] Sponsored by the Committee on Patent Matters and Related Legislation
[c] Sponsored by the Committee on Environmental Improvement

A Symposium on Public Policy Aspects of Education in Chemistry was sponsored jointly by the Division of Chemical Education and the Committee on Chemistry and Public Affairs during the ACS national meeting in Chicago in September 1970. From left are Edward Creutz, Herschel Cudd, George S. Hammond, and Jack Martinelli.

on the structure and control of living systems was destined to raise critical questions of public policy. This symposium and some of those held subsequently (Table I) were reduced to printed form and distributed widely as a public service. For the Society's Centennial meeting in April 1976 in New York City, the CCPA organized a Centennial Symposium, "Energy, Food, Population, and World Interdependence."

The CCPA, with staff support, has also worked to involve the Society's local sections and divisions in public affairs activities. Local sections have planned cooperative programs with local schools and school administrators, local civil defense committees, legislative advisory groups, and local study groups working on science-related issues. Some local sections have established legislative screening committees and appointed public affairs coordinators. The Divisions of Medicinal Chemistry and Nuclear Chemistry and Technology have established committees on public affairs.

In 1973, as part of its effort to bring ACS members' expertise to bear more broadly on public issues, the Committee on Chemistry and Public Affairs established its Congressional Science Counselor Program. The Counselors are ACS member-volunteers who are available to consult with their Congressmen on chemically related issues; the Department of Chemistry and Public Affairs coordinates the assignment of Counselors to the Senators and Representatives in their states and districts. The Counselors remain in their full-time jobs; they are encouraged to serve at least three years, but appointments are made for one year and reviewed annually. A Counselor is expected to establish a personal working relationship with his or her Congressman on a constituent–Congressman basis. The Counselor represents the Society officially only when informing the Senator or Representative of official ACS positions on legislative issues. At the beginning of 1976, Counselors were assigned to 500 of the nation's Senators and Representatives.

In 1974, the Society established its Chemistry and Public Affairs Fellowship program for chemists and chemical engineers. One Public Affairs Fellow is appointed annually and receives a suitable stipend. After a brief orientation period, the Fellow works in the legislative and executive branches of the Government under the supervision of the ACS Department of Chemistry and Public Affairs. The actual research and liaison assignments depend on the individual's background and interests as well as the interests and responsibilities of the Society. The Fellow prepares a final report, which includes recommendations for potential ACS actions or for further research. The first Public Affairs Fellow served in 1975–76 at the National Science Foundation and in a Congressman's office, pursuing policy problems in energy and control of toxic substances. The second Fellow began his orientation period in January 1976.

Aid for the Underprivileged

In September 1968 in San Francisco, the ACS Council adopted a resolution calling on the Society to take appropriate steps to ease the problems of underprivileged segments of the nation's population, particularly in relation to lack of education and unemployment. The responsibility for implementing the resolution was assigned to the CCPA, which organized a Subcommittee on the Education and Employment of the Disadvantaged (SEED). The subcommittee then developed a program to assist in the training and employment of secondary and college-age students. This effort includes summer research assistantships for high school students (the Catalyst Program); tuition aid for ACS Short Courses to upgrade the level of instruction in chemistry in underfinanced colleges; assistance to guidance counselors; and tutorial and motivational work with students from difficult socioeconomic backgrounds. Project SEED's most active effort, the Catalyst Program, by 1975 had placed more than 650 high school students in summer programs in chemistry laboratories at colleges and universities. Through 1975, solicitations for Project SEED raised an average of about $50,000 annually from ACS members, the Society's Corporation Associates, and other organizations and individuals. In 1974, the administration of Project SEED was transferred from the Department of Chemistry and Public Affairs to the ACS Department of Educational Activities, but policy direction remained with the CCPA's subcommittee for the program.

As the ACS enters its second century, its basic role in the formulation of public policy has become perforce a response to two interdependent facts: the health of chemical science and technology depends very largely on how policymakers choose to solve major problems that include energy, health care, nutrition, and population growth; and the ability to solve such problems depends heavily on the further development and use of chemical knowledge. The Society comes to this complex role armed with a reputation for excellence and with the large reservoir of expertise represented by its members. Such weapons, however, are not enough. ACS efforts in public affairs must be realistic, in terms of both the Society's resources and the positions it adopts; they must not be self-serving in fact or in appearance. Given these conditions, the American Chemical Society will continue successfully to stimulate open debate on the meeting ground of public policy and the profession and practice of chemistry.

CHAPTER VII

INTERSOCIETY RELATIONS

The ACS News Service arranged a press conference with Nobel laureates at the First Chemical Congress of the North American Continent, in Mexico City in December 1975. From left: Donald A. Glaser (Physics, 1960), Gerhard Herzberg (Chemistry, 1971), Severo Ochoa (Medicine, 1959, with Arthur Kornberg), Paul J. Flory (Chemistry, 1974), Glenn T. Seaborg (Chemistry 1951, with Edwin M. McMillan).

FROM its beginnings the American Chemical Society has worked to maintain good relations with other scientific societies. In fact, when the formation of an American Chemical Society was first proposed, during the Priestley Centennial celebration in 1874, chemists who were then members of the American Association for the Advancement of Science expressed concern that formation of a separate chemical society might weaken the broader AAAS, which had been organized in the 1840s. The Society's interest in AAAS continues to this day; the ACS remains one of the nearly 300 scientific groups affiliated with AAAS, and many ACS members are active in the association's Section C (Chemistry).

The ACS also has long maintained relations with the several chemically oriented societies that formed during its early years (Table I). The

TABLE I

Chemically Oriented Societies Formed During the Early Years of the American Chemical Society

Society	Year
American Water Works Association	1881
Association of Official Agricultural Chemists[a]	1884
New York Section, Society of Chemical Industry	1896
American Ceramic Society	1899
American Electrochemical Society[b]	1902
American Leather Chemists' Society	1903
American Society of Biological Chemists	1906
American Institute of Chemical Engineers	1908
American Oil Chemists Society	1909

[a] Now the Association of Official Analytical Chemists
[b] Now the Electrochemical Society

establishment of these groups reflected mainly the urge to specialize that was emerging around the turn of the century; their advent played a role in the Society's move toward a divisional structure, in 1908.

International Cooperation

Seventeen years after its founding, the ACS ventured for the first time into the international arena. The occasion was the World's Congress of Chemists, held in Chicago in August 1893 in connection with the Columbian Exposition. One hundred eighty-two chemists attended, many from abroad. ACS President Harvey Wiley, in his opening address to the Congress, called for "a wider and deeper fraternity of chemists." He suggested that a triennial international congress on chemistry be established and that it meet in different countries. In the following year, L'Association Belge des Chimistes convoked such a meeting, an International Congress of Applied Chemistry. It took place in Antwerp and Brussels in August 1894 and was the first of eight such events that were held before World War I intervened. The ACS was a major organizer of the eighth congress, which took place in Washington and New York in September 1912. In 1911, in a second venture abroad, the ACS had joined the International Association of Chemical Societies, but this short-lived (1911–12) body, like the international congresses, succumbed to the war.

In July 1919, at meetings in London and Brussels, the International Union of Pure and Applied Chemistry was organized; Charles L. Parsons, Secretary of the Society and a member of the U.S. delegation, was elected a vice-president. IUPAC has operated continuously ever since, except during World War II. In 1975, chemists in 44 nations were represented in the organization. Its objectives are to promote cooperation among chemists internationally; to study chemical topics of international importance that require standardization, regulation, or codification; to cooperate with other international organizations that deal with chemical topics; and to contribute to the advancement of pure and applied chemistry in all its aspects.

The ACS participates in IUPAC only indirectly because in this country the National Academy of Sciences is the organization that adheres officially to the union. The president of the academy appoints U.S. representatives to the IUPAC council. Academy presidents consistently have asked the Society to recommend prospective appointees, and most such recommendations have been honored. IUPAC presidents have included three American chemists: Marston T. Bogert (1938–47), W. Albert Noyes, Jr. (1959–63), and Robert W. Cairns, who began a two-year term in September 1975 while serving as Executive Director of the Society. All three have been ACS Presidents as well.

IUPAC devotes a good deal of attention to chemical nomenclature, one of many areas in which the steady evolution of chemistry creates a need for continuous standardization. In this work, the union leans heavily on leadership from several divisions of the ACS, from the ACS Council Committee on Nomenclature, and from the staff of the Chemical Abstracts Service. IUPAC's standards-setting efforts are exemplified by its adoption, in 1961, of carbon-12 as the base for atomic weight values. The International Union of Pure and Applied Physics adopted the same base at the same time. Before, the reference point for the atomic weight scale had been natural oxygen, whose atomic weight was set at 16, but chemists and physicists had been using minutely different scales since the discovery, 30 years earlier, of the isotopes ^{17}O and ^{18}O. The differences, though small, had become troublesome by the 1950s; hence the IUPAC–IUPAP change to the carbon-12 scale.

The Society's international efforts also include joint meetings with foreign groups. In May 1970 in Toronto, for example, the ACS held its spring national meeting jointly with the Chemical Institute of Canada in celebration of CIC's 25th anniversary. In December 1975 in Mexico City came the First Chemical Congress of the North American Continent. The participants were the ACS, the Chemical Institute of Canada, and three Mexican groups—Sociedad Quimica de Mexico, Instituto Mexicano de Ingenieros Quimicos, and Asociacion Farmaceutical Mexicana. In 1977 the Society will meet jointly in Montreal with the Chemical Institute of Canada and in 1979 in Honolulu with chemical societies from Australia, Japan, and New Zealand. The Second Chemical Congress of the North American Continent is scheduled for 1980 in San Francisco.

In December 1967, presidents of nine national chemical societies met at ACS headquarters at the instigation of President Charles G. Overberger. At top: Paul Giguere, Canada (left); Edward C. Kooyman, the Netherlands. At center: Hellmut Ley, Federal Republic of Germany (left); Dr. Overberger; Carl O. Gabrielson, Sweden; Sir Harry Melville, Great Britain. At bottom: Harusada Suginome, Japan (left); Guido Sartori, Italy; Jacques Benard, France.

Other, less formal international activities include a series of biennial meetings of elected presidents of national chemical societies. ACS President Charles G. Overberger organized the first such meeting, which took place in 1967 in Washington, D.C. On hand were the presidents of the

Chemical Institute of Canada, the Chemical Society (London), Société Chimique de France, Gesellschaft Deutscher Chemiker (West Germany), Societa Chemica Italiana, Nippon Kagakukai, Koninklijke Nederlandse Chemische Vereniging, and Svenska Kemistaamfundet. The purpose of the gathering was to talk freely about problems common to chemists all over the world. The meeting had its limitations, especially in transmitting the concerns of the group to constituent chemists in home countries, but it was sufficiently productive to warrant continuance on a biennial basis. In the years since, additional national societies have joined in the sessions; 20 nations were represented at the 1973 meeting in Italy and 14 at the 1975 meeting in Spain.

Domestic Cooperation

On the domestic scene, the American Chemical Society works with other scientific and engineering bodies in several ways. These efforts range from participation in special-purpose organizations and meetings, through nomination of candidates for non-ACS awards, to membership in associations of full-time professional employees of societies. Such activities have expanded in recent years as chemistry itself has expanded and as the problems that confront science and technology in general have grown increasingly interdisciplinary.

The Society is active in a number of special-purpose organizations. Representative examples include: the National Council on Health Laboratory Services, founded in 1952; the Intersociety Committee on Methods of Air Sampling and Analysis, founded in 1962; the National Registry in Clinical Chemistry, founded in 1967; and the League for International Food Education, founded in 1968.

The National Council on Health Laboratory Services was formed originally as the Intersociety Committee on Laboratory Services Related to Health. In 1968 the name was changed to Intersociety Committee for Health Laboratory Services and in 1973 to its present form. The National Council is a group of organizations whose original purpose was to serve as an information exchange on laboratory services, such as clinical chemistry, related to health and through meetings and joint efforts to improve such services. In its earlier years, the council devoted considerable effort to supporting the principle that legal permission (usually licensure) to direct a health laboratory should be based on scientific and professional qualifications, not the possession of a specific degree, such as an M.D. The American Medical Association disagreed with this position. The issue was resolved to some extent by regulations established under Medicare. In October 1975, the National Council undertook its first cooperative effort, sponsoring and organizing the second National Conference on Proficiency Testing (of health laboratory personnel), in Washington, D.C. The council has grown steadily, from seven member organizations in 1952 to 18 in 1975.

The Intersociety Committee, which currently includes representatives from 13 scientific and engineering societies, first engaged the interest of the ACS Division of Analytical Chemistry. In 1971, the Society became an active participant. In its early years the committee dealt with air pollution analysis methods, such as the sampling and analysis of ambient air and emissions from stacks. More recently it has turned to analyses of airborne contaminants in the work environment. Analytical methods recommended by the Intersociety Committee were compiled in "Methods of Air Sampling and Analysis," published in 1972 by the American Public Health Association, a member of the committee. In 1975, the committee was at work on a second edition, which will contain recommended methods of analysis for both ambient and workplace atmospheres.

The National Registry in Clinical Chemistry is sponsored by the ACS and five other U.S. organizations. The Registry certifies professionals who are qualified to perform services of a chemical nature in the nation's clinical laboratories. To be certified, applicants must meet standards of education and experience and pass an examination. Certification is at two levels: Clinical Chemistry Technologist for applicants with recent degrees in chemistry or other disciplines who regularly perform clinical chemistry determinations; and Clinical Chemist for more experienced graduates who have majored in chemical science and are active in clinical chemistry. The Registry deals primarily with holders of bachelor's or master's degrees; the American Board of Clinical Chemistry, a sponsor of NRCC, certifies clinical chemists with the doctorate and extensive experience.

The League for International Food Education (L.I.F.E.), a consortium of nine U.S. scientific professional societies, including the ACS, seeks to help developing nations solve problems in nutrition and food technology. L.I.F.E. maintains liaison with universities, food processors, government, and volunteer agencies in this country and abroad; recruits personnel for overseas assignments; keeps special information files on solutions to food problems; publishes a monthly newsletter and other documents; and conducts seminars and workshops in the U.S. and abroad on key problems in nutrition and food technology. The ACS Divisions of Agricultural and Food Chemistry and Microbial Chemistry and Technology are especially active in L.I.F.E.

Through its Board Committee on Grants and Awards, the Society has long aided the American Section of the U.K.'s Society of Chemical Industry by nominating one or two candidates each year for the Perkin

Medal. The medal recognizes outstanding contributions to applied chemistry; it was founded in 1906 to honor the discoverer of mauve dye, Sir William Perkin. Besides the ACS, the American Institute of Chemical Engineers, the American Institute of Chemists, the Electrochemical Society, and the American Section of the Société de Chimie Industrielle nominate candidates for the Perkin Medal.

The ACS and its divisions take part regularly in joint, interdisciplinary scientific meetings. One example is the annual Intersociety Energy Conversion Engineering Conference. The 10th conference was held at the University of Delaware in August 1975; the 7th, in San Diego in 1972, was administered by the ACS, which was one of seven sponsoring societies. A second example is the biennial Joint Conference on Sensing Environmental Pollutants. The third of these meetings, held in September 1975 in Las Vegas, was sponsored by the ACS, five other societies, and four federal agencies (the conference was merged with the International Symposium on Environmental Monitoring under a new name, International Conference on Environmental Sensing and Assessment).

In recent years, elected officers of scientific societies have felt a need for an organization through which they could coordinate their efforts on national issues important to their members. In 1973, under the leadership of ACS President Alan C. Nixon, a group of these officials formed the Committee of Scientific Society Presidents (CSSP). Officers of 12 societies attended the committee's first meeting. The group meets two or three times a year and has expanded to include the presidents of more than 30 organizations.

Although CSSP deals with matters of national concern to scientists, particularly science policy, it does not speak officially for the societies. CSSP does influence those societies, however, through its proposals for action in certain areas and through its statements on pending legislation, on the organization of science within the Federal Government, and the like. CSSP is young and has operated informally so far; it is directing some effort toward defining its duties more precisely and creating a secretariat.

The ACS participates also in the Council of Engineering and Scientific Society Executives (CESSE), which was formed in 1949 (originally as the Council of Engineering Society Secretaries). CESSE provides a forum for information exchange among full-time executives of engineering and scientific societies on subjects of common interest. The organization holds national meetings and publishes a newsletter, *Quill*. CESSE carries out projects of service to members, holds workshops and panel discussions on a variety of society operations, exhibits society projects and procedures, and attracts prominent outside speakers to its meetings. Former ACS staff member R. M. Warren was president of CESSE in 1965–66.

The Society was active in a second organization, very similar in purpose to CESSE but involving only staff of scientific societies. The Managing Officers of Scientific Societies (MOSS) was organized in 1958 and met approximately annually through 1970. MOSS has not convened since then and is unlikely to do so because scientific society executives now may participate in CESSE. The societies represented in MOSS were the American Association for the Advancement of Science, American Chemical Society, American Geological Institute, American Institute of Physics, American Mathematical Society, American Meteorological Society, American Psychological Society, and Federation of American Societies for Experimental Biology.

For a few years, the Society was a member of the American Society of Association Executives. ASAE is much larger than CESSE and MOSS because it includes professional and trade associations as well as scientific and engineering societies. This in fact is why the ACS left ASAE. Although scientific societies and trade associations have management problems in common, their objectives and programs differ appreciably.

Besides the many opportunities for the American Chemical Society to cooperate formally with other groups, good relationships often have developed more spontaneously. When other societies or organizations observe anniversaries or dedicate new facilities, for example, the Society frequently is invited to extend greetings, and the President as a rule appoints an appropriate representative to do so.

Because the ACS has long been among the largest of the scientific societies, it has worked to avoid the fact or appearance of dominating intersociety activities. There are good indications that it has succeeded during its first 100 years. There are good indications also that effective intersociety cooperation will grow increasingly useful during the Society's second century as the problems of the nation and the world impose progressively more complex demands on science and technology and their practitioners.

CHAPTER VIII

GOVERNANCE

THE governance of the American Chemical Society has always been based on a Constitution and Bylaws, and the Society was operating under its fifth such instrument as it entered its centennial year. The first Constitution lasted 14 years, the second two years, the third only five years. But the fourth ACS Constitution endured for half a century, and the fifth was 28 years old on Jan. 1, 1976. Familiar elements of the Society's governance—President, Council, committees—have been present from the beginning, by name if not by function. A Board of Directors has existed for more than 98 years, local sections and their Councilors for 85 years, divisions and their Councilors for 68 years.

The Society's first Constitution, adopted April 6, 1876, provided for a President, six Vice-Presidents, two Secretaries (recording and corresponding), a Treasurer, a Librarian, and three Curators. These 14 officers were

Henry A. Hill (left) and Anna J. Harrison were nominated for ACS President-Elect during the national meeting in Philadelphia in April 1975. Dr. Hill won. Standing is Raymond P. Mariella, an ACS Director in 1975.

to be elected by majority vote at the annual meeting each December. The document specified also that the ACS would be governed by a Council of 16 members: the officers, and the chairmen of the two committees, on Nominations and on Papers and Publications. On Nov. 10, 1877, the Society incorporated in the state of New York in order to lease meeting rooms. This required that the Council be replaced by a Board of Directors, to act as the legal representative of the ACS. The new Board was composed of the officers, excepting the three nonresident Vice-Presidents, and the committee chairmen. Under the charter of incorporation, a majority of the Directors had to be residents of New York, and the annual meeting and election of officers had to take place in New York City. These requirements created substantial friction with members from other states, who felt deprived of an effective voice in governance. The result was a fracturing of the Society and a secession movement that led to a revised Constitution, adopted June 6, 1890.

The Constitution of 1890 did much to heal the breach with the members from states other than New York. Its most significant features were the establishment of local sections and the creation of an Advisory Council separate from the Board of Directors. The Board and Council were both to be elected at the annual meeting in December; absentee and proxy voting were allowed. The Board included the President, two local (New York City) Vice-Presidents, two Secretaries, a Treasurer, and six other Directors. The Advisory Council included the President, all chairmen of local sections, and 12 other Councilors, nine of whom were to be nonresidents (living beyond 50 miles of New York City). The Board was required to "consult" the Advisory Council in framing amendments to the Constitution, and such amendments, to be adopted, had to be approved by two thirds of the members at a regular meeting. The Board also was required to consult the Council on the selection of sites for general meetings and on single expenditures of more than $500 from the General Fund (which was to represent income less expenditures and was to be invested only in Registered U.S. Government Bonds). The Council otherwise was to act as "an advisory body in all matters affecting the general policy of the Society."

Under the third Constitution, adopted Nov. 4, 1892, the Board of Directors remained the legal representative of the ACS and the custodian of its property and funds. This document extended the powers of the Council to include elections to membership in the Society, designation of new local sections, appointment of all standing committees, approval or disapproval of the Editor of the ACS publications, determination of the salaries, "if any," to be paid to the Editor and the General Secretary, and approval of all expenditures of $250 or more. The Council was also empowered to approve or disapprove every proposed amendment to the

Constitution before the Board presented it to the members for a vote. Amendments still had to be approved by two-thirds of the members, but the vote was to be taken by mail, not at a regular meeting of the ACS. The chairmen of local sections continued to be members of the Council and also were designated Vice-Presidents of the Society, ex officiis. The Constitution of 1892 created considerable overlap of responsibilities between Board and Council. During the immediately succeeding years, the ACS considered reorganizing to form only one governing body, but never did so.

As the Society continued to grow, it became necessary that the Council represent the members more equitably than simply by including the chairmen of the local sections. The Constitution of Dec. 2, 1897, the Society's fourth, was the first to give local sections, which by then numbered nine, proportional representation on the Council. The Constitution of 1897 was a tangled skein of governance. The Council was characterized as the "general governing body" of the ACS, responsible for general management and policy. Its members were the President (elected for one year), the Past Presidents, the Secretary and the Editor (one-year terms), one Councilor elected by each local section for each 100 members, or fraction thereof, in good standing and 12 Councilors at large elected by the members for three-year terms. In seeming duplication at the head of the Society was the Board of Directors, "who shall be the legal representatives of the Society," to have, hold, and administer all of the property of the Society in trust under the general direction of the Council and in conformity with the statutes of the state of New York. The Board was composed of the President and Secretary, ex officiis, and four other Directors elected by the Council from among its own members for two-year terms. The presiding officers of all local sections were Vice-Presidents of the Society, ex officiis, but were not on Council or Board unless elected to those bodies. Under a special act of relief that the Society had obtained April 16, 1895, from the state of New York, Directors no longer had to be New York residents and national meetings no longer had to be held in that state.

The Constitution of 1897 specified three standing committees: Finance, Membership, and Papers and Publications. Each had three members, all appointed by the Council to one-year terms. The Finance Committee was accountable to both Council and Board, the Committee on Membership to the Council only. The Committee on Papers and Publications was accountable only to itself as the final authority in the "publication of any matter."

This fourth Constitution of the ACS provided for a Council on a basis that continued through 1947, save only for the amendments of 1908 that allowed for divisions and divisional Councilors. In 1928, the post of

President-Elect was established. By 1930, the Board of Directors had been expanded to 14 members: the President, the President-Elect, the Secretary, the Treasurer, four Directors-at-Large, and six Regional Directors. From 1930 through 1947, the composition of the Board was altered only by the addition of the immediate Past President, in 1936. In 1931, the election of the Secretary and the Treasurer was made a responsibility of the Board, and the same year saw the creation of the position of Business Manager. The ACS as a whole was affected during this period by the granting of the national Charter, effective in 1938, by act of Congress. The Charter solved permanently whatever governance problems may have remained from being incorporated in New York.

By 1945 the ACS was still operating successfully, but the need for change in organization and governance is suggested by a few key figures:

Year	*Members*	*Expenditures*
1898	1,415	$1,409.57
1930	18,206	$626,301.65
1945	43,075	$1,012,075.48

In December 1945, faced by internal friction and the increasing diversity of members' needs and interests, the Board authorized a study of the Society's organization and of member opinion. A special committee of the Board—Directors Charles A. Thomas and Ernest Volwiler and President Bradley Dewey—asked John M. Hancock, a noted investment banker and public servant, to undertake the study. To assist him, Hancock engaged experts from McKinsey, Kearney and Co. and Opinion Research Corp. The Hancock report was received by ACS Secretary Alden H. Emery on Jan. 25, 1947, and appeared as a 55-page insert in *Chemical & Engineering News* for Feb. 17, 1947. It spurred sweeping changes in the Society's organization, including the completely revised Constitution and Bylaws that took effect Jan. 1, 1948. This fifth governing instrument expanded the powers of the Board of Directors and spoke of the Council as deliberative and advisory. Among its other major innovations were these:

- The President-Elect and six Regional Directors were to be elected directly by the membership. The Committee on Nominations and Elections, to be elected by the Council from among its members, was made responsible for soliciting four nominees for each of these offices; from these the Council would select, for each office, two nominees to be voted on by the membership. Nomination by petition was also provided for; petition nominees would be added to those selected by the Council.

• The Council Policy Committee was empowered to coordinate Council business. In addition, a series of standing committees were established with permission to hold open meetings.

• Councilors were to be elected for three-year terms. Councilors from "larger" local sections were to number between 280 and 320, and each division was to have two Councilors.

The Constitution of 1948, without major revision, had served for more than a quarter century by 1974, when the ACS once again undertook a self-assessment in the spirit of the Hancock study. In that year, the Society retained Arthur D. Little, Inc. (founded, appropriately enough, by the ACS President of 1912–13), to study its organization, structure, and business management. The study was completed early in 1975. Thus, on the eve of its centennial, the ACS was considering anew the actions needed to bring its governance, operations, and functions into harmony with the times and with the wishes and needs of the membership.

The Presidency

The presidency has been from the beginning the most prestigious post in the ACS, although the chairmanship of the board has long been the most powerful. This is not to say that the President is a figurehead. Since 1936, the three-year presidential succession–President-Elect, President, immediate Past President—has carried with it three years as a voting Director. Under the Constitution of 1948, the President is president and presiding officer of the Council and chairman of its most influential committee, the Council Policy Committee; in addition, he appoints all members and the chairmen of the Council's nine standing committees. But ACS Presidents for many years tended not to use the undoubted, if short-lived, powers of the office aggressively. They discharged their constitutional and ceremonial duties, but typically they left the direction of the Council Policy Committee to its elected vice-chairman; equally typically they relied heavily on full-time staff in selecting members and chairmen of Council committees.

A more active approach to the ACS presidency began to appear in the early 1960s. In 1962, M. H. Arveson became the first candidate to campaign vigorously and overtly for the office of President-Elect. In that post and as President (1964) he attended meetings of about half of the Society's 165 local sections in an unprecedented effort to improve communication between the Society's leadership and staff and the ordinary member. In 1965, President Charles C. Price pressed for greater ACS effort in advising government on questions involving science and technology. In the same year, in consequence, the Society formed its Committee on Chemistry and Public Affairs with Dr. Price as the first chairman. In 1967, President Charles G. Overberger organized the first meeting of elected presidents of national chemical societies to strengthen international bonds through discussion of mutual interests and problems. This biennial event, held first in Washington, D.C., occurred for the fifth time in October 1975 in Alicante, Spain.

The active approach to the presidency—and to ACS politics in general—drew strength from the unemployment problems that struck the chemical profession in 1968 and did not begin to taper off until about 1972. The crisis markedly spurred members' interest in professionalism, especially in California, where cutbacks in the aerospace industry had thrown many chemists out of work and where local sections were mounting member-assistance programs. A direct result, in 1970, was the nomination by petition of Alan C. Nixon of the ACS California Section for President-Elect. Dr. Nixon was only the fourth petition candidate for the office. The first, Vladimir Kalichevsky, had been defeated (along with Edgar C. Britton) by Ernest H. Volwiler in 1948. The second, N. Howell Furman, had won in 1949. The third, Lawrence H. Flett, had lost (along with Eger V. Murphree) to Albert L. Elder in 1958. Dr. Nixon campaigned in 1970 on a strong professionalism platform. He was supported actively by a nationwide group that originated in the California Section and later was named Grassroots. Despite these efforts, however, victory went to Max Tishler who, with Lloyd M. Cooke, had been nominated by the Council.

In 1971, Dr. Nixon again ran as a petition candidate on a professionalism platform holding that "the first responsibility of the ACS is to its members"; again he had the active support of Grassroots. This time he won, defeating George S. Hammond and William A. Mosher, the Council nominees. In 1972, with the approval of Board and Council, President-Elect Nixon organized a Professional Enhancement Program (PEP) to run through 1973, his year as President. The effort included some new, but mainly, expanded programs in the Society's Employment Aids Office and Departments of Chemistry and Public Affairs, Professional Relations and Manpower Studies, and Public and Member Relations. Important elements of these programs were continued when PEP itself came to an end.

The Nixon petition candidacies for President-Elect in 1970–71 were followed by three more. In 1972, the Grassroots candidate, Bernard S. Friedman, defeated the Council nominees, Milton Harris and Henry A. Hill. In 1973, petition candidate Emerson Venable ran third to winner William J. Bailey, who had "unsolicited" Grassroots support, and Bryce Crawford. In 1974, petition candidate Venable ran third again, trailing winner Glenn T. Seaborg and Anna Harrison; Grassroots did not endorse a candidate. In 1975, in the first race since 1969 without a petition candidate, Henry A. Hill, who was supported by Grassroots, narrowly defeated Dr. Harrison.

The rise of professionalism that began in the ACS in the late 1960s affected not only electoral politics, but also procedures and, for a time, at least, the turnout of voters. The succession of petition candidates for President-Elect led to revised election procedures that took effect in 1973. When more than two candidates ran, the voter had to mark a first and second choice for the ballot to be valid (in 1975 the second choice was made optional). If no candidate received a majority of the first-choice votes, the winner was determined by combining the first-choice votes of the first- and second-running candidates with their second-choice votes from the ballots of the third-running candidate. Under this procedure in 1973, winner Bailey received 48% of the first-choice votes, but almost 63% of the combined first- and second-choice votes. The procedure was not used in 1974, when 54.9% of the first-choice votes went to Dr. Seaborg, the only candidate in five successive three-way races to win a majority of the votes cast. At the same time, however, only about 32% of the membership voted in 1974. This was about the traditional level, but represented a steady decline from the record 40% turnout of 1971, the year of Dr. Nixon's winning petition candidacy. In 1975, the number of votes cast for President-Elect shrank even further, to 29% of the membership. The four-year decline at least suggests that interest in professionalism will not produce a higher voter turnout unless times are bad, as they were in 1971.

The Committees of the ACS

Although the American Chemical Society is governed by its Board of Directors and Council, it has always relied on the collaborative efforts of its committees in pursuing its stated goals and meeting its obligations. In the broadest sense, the Society's Board is a committee; so, too, is the Council with its paralegislative functions. The concern here, however, is with the 2-to-30 member bodies that do much of the work of the ACS in liaison with the permanent staff. This effort is wide-ranging. It may include "housekeeping" duties such as review and approval of local section bylaws by the Council Committee on Constitution and Bylaws. It may relate to a continuing technical issue, such as chemical nomenclature, analytical reagents, or patents. It may involve creating recommendations for official statements of Society policy, which is part of the work of the Board-Council Committee on Chemistry and Public Affairs. Or a committee may deal with matters vital to individual ACS members such as professional relations or economic status. Society committees may provide an operating or administrative function, as does the Board

Committee on the Chemical Abstracts Service. And ad hoc committees frequently are formed to handle specific problems. A roster of committees is an index of ACS activity at any given time; Tables I and II show that this activity has been very large indeed.

During the quarter century after 1950, most of the committee work of the Society was done for the Board of Directors or the Council. The

TABLE I

Elected and Standing Committees of the Board of Directors 1960–1975

Committee
Executive (Elected; Formed 1931)
Chemical Abstracts Service (Formed 1966)
Education and Students (Formed 1953)
Finance (Formed 1893)
Grants and Awards (Formed in 1971 from Awards and Recognitions (Formed 1953) and Grants and Fellowships (Formed 1959)
Public, Professional, and Member Relations (Formed in 1960 from Member and Public Relations (Formed 1949)
Publications (Formed 1948)

TABLE II

Elected and Standing Committees of the Council, 1949–75

Committee[a]
Council Policy Committee (Elected)
Nominations and Elections (Elected)
Committee on Committees (Elected; Formed 1975)
Chemical Education
Constitution and Bylaws
Local Section Activities
Membership Affairs
National Meetings and Divisional Activities[b]
Meetings and Expositions (Formed 1971)
Divisional Activities (Formed 1971)
Program Review (Formed 1970)
Professional Relations[c]
Publications

[a] Except as indicated, each was formed under the Constitution of 1948 as first adopted, although predecessor committees existed in some cases.

[b] Became the two committees indicated in 1971.

[c] Name changed from Professional Relations and Status, effective 1967.

The Committee on Finance of the Board of Directors met with staff members during the national meeting in Atlantic City in September 1968.

latter years of the period saw an increasing number of joint Board–Council committees. They saw also a greater degree of interaction between Board and Council and consequent pressure to reduce the degree of overlap of their respective committees. "Official" functions of the Board and Council have been the province of the elected and standing committees of Tables I and II.

The Board of Directors has only one elected committee, the Executive Committee, which was formed in 1931 and is empowered to act ad interim for the Board. The Chairman of the Board, who is elected by the Directors, is ex officio a member of the Executive Committee and is its chairman. All other committees of the Board are appointed by the Chairman.

The Council has three elected committees: Council Policy, Nominations and Elections, and the Committee on Committees, whose duties include making recommendations on the duties, formation, and discharge of Council committees. The chairman of the Council Policy Committee, as mentioned earlier, is the President of the Society; the other two elected committees elect their own chairmen. All other Council committees and their chairmen are appointed by the President. In addition to the President, the President-Elect, the immediate Past President, and the Executive Director serve on the Council Policy Committee, and the chairmen of the elected and standing committees of the Council are nonvoting members.

The Council Policy Committee has existed since 1923 and with that name since 1931; it plays a major role in the activities of the Council. It serves as the executive committee and, like the Executive Committee of the Board, is empowered to act ad interim for its parent body. The Council Policy Committee has long proposed nominees for the Committee on Nominations and Elections and in 1971 was made responsible for long-range planning. In addition to other duties, such as scheduling

TABLE III

Subcommittees of the Council Policy Committee, 1954–75

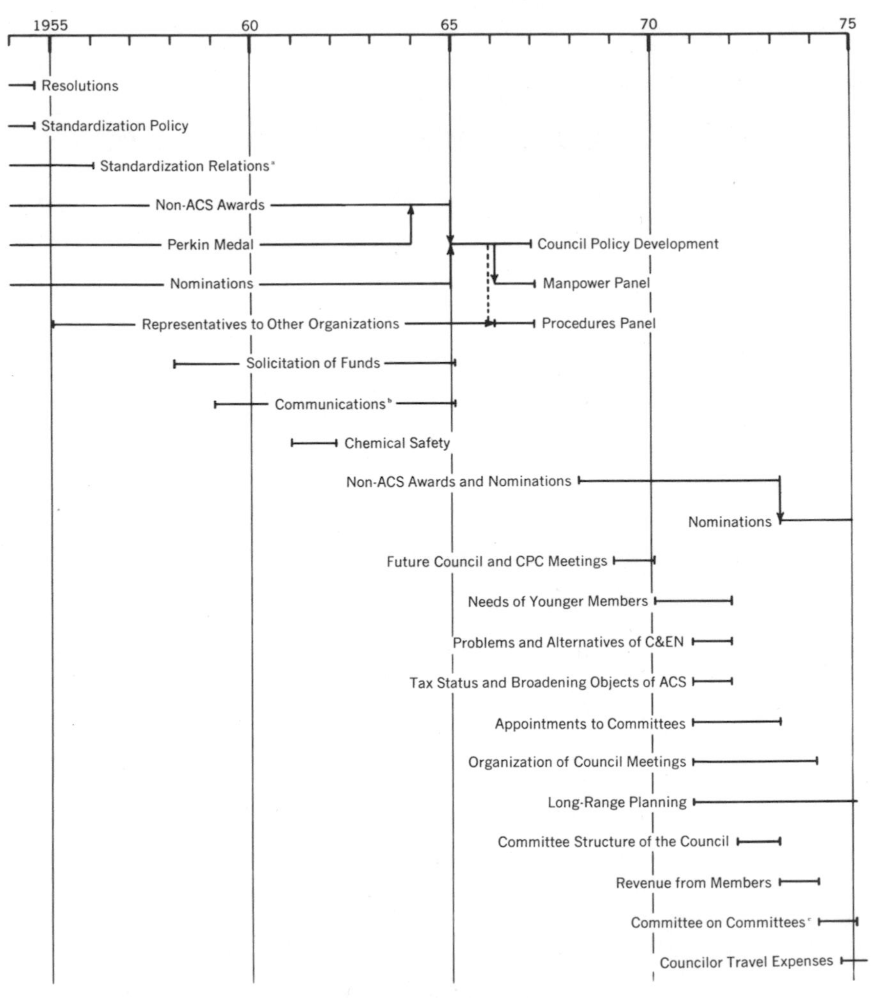

[a] Became a Committee of the Council, 1956
[b] Improved Council-Member Communications, 1959–61
[c] Became an Elected Council Committee, 1975

business sessions, the Committee plans the detailed Council agenda "to facilitate effective action" by the Council. Although the Council Policy Committee is chaired by the President of the Society, it elects a vice-chairman, who usually guided its activities until the advent of the activist Presidents of the 1960s and later, who tended to take a stronger hand in the committee's work. Like most committees, the Council Policy Committee has found it useful to work through subcommittees. Since these subcommittees cover a fairly wide range of activities important to the Council, and since they typify the operations of one ACS committee over a period of time, they are outlined in Table III.

Both Board and Council, in addition to their elected and standing committees, are authorized to form other committees whose members need not be members of the parent body. These committees have been called "task force," "board," "panel," and "subcommittee," as well as "committee," and their names have contained terms such as "special," "continuing," "advisory," "ad hoc," and even "other." Some have been so short-lived that they existed—if they did exist—only in Minutes stating that a committee would be formed. "Other" committees date back to the very early days of the Society: the Committee on Admissions, for example, has existed in one form or another since the ACS was founded.

The Board of Directors has been the strongest generator of new committees in the Society, a reflection of its small size and the myriad problems it faces. Selected Board committees that have been active since 1954 appear in Table IV; many more could be named. In a few instances the Board has performed the function of a given committee for years before the committee was created formally. An example is the Board's continuing attention to *Chemical Abstracts* before the Standing Committee on Publications was formed, in 1948 (a standing committee on Chemical Abstracts Service was formed in 1966). Many Board committees have been ad hoc, at least initially. Most of these are not listed in Table IV; indeed, some do not appear in the Board's minutes, yet a few of them, at least, must have existed as preliminaries to the appearance of committees in full-blown activity.

Committees of the Council other than those that are elected or standing appear in Table V. Most of those formed after about 1965 are devoted either to professional matters or to organizational problems; those formed earlier tend to deal more with the traditional activities of a learned society. This is true also of the Joint Board–Council Committees, which appear in Table VI. Impetus for the formation of joint committees has come from both Board and Council and, on occasion, has come simultaneously from both bodies, as was the case with the Younger Chemists Task Force. With the trend toward a single set of

TABLE IV

A Selection of Other Committees of the Board of Directors, 1954–75

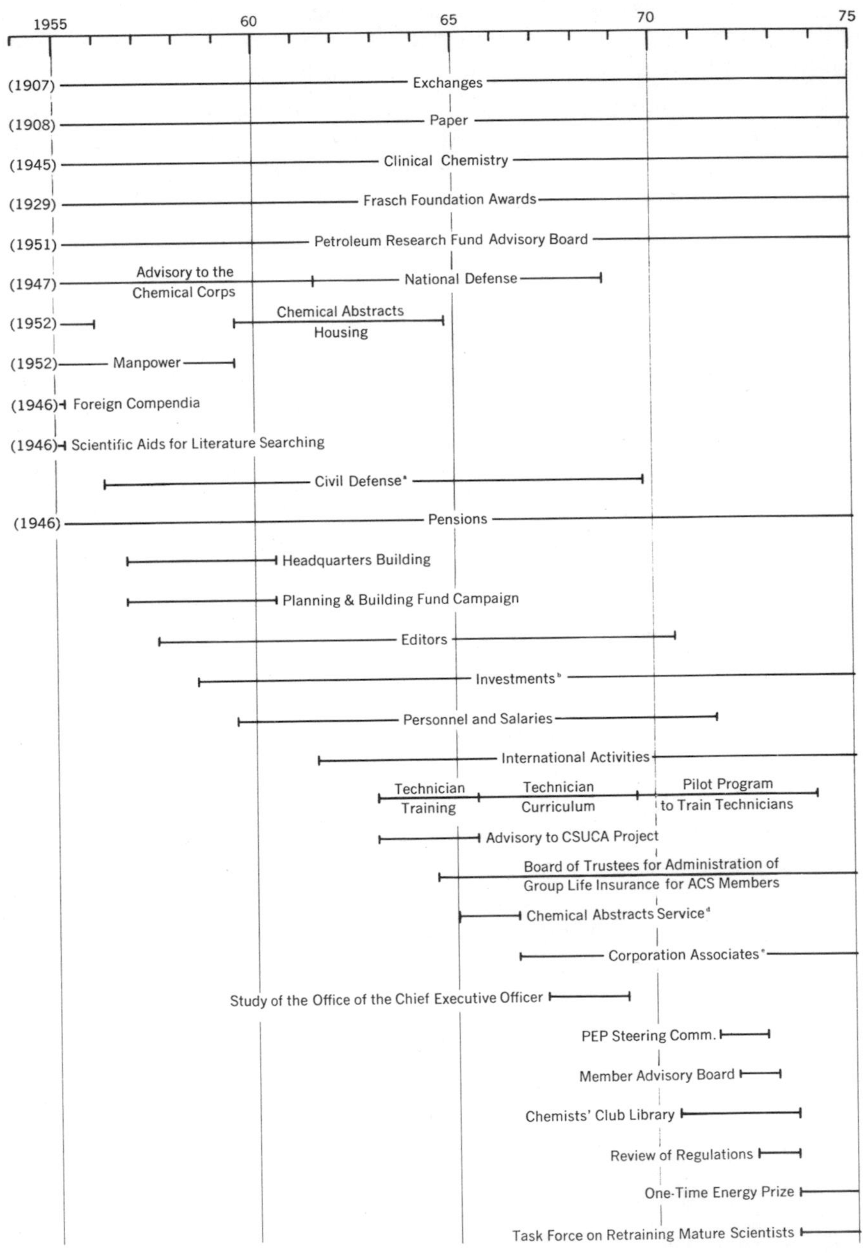

[a] Civil Defense and Disaster, 1968–70
[b] Standing Committee prior to 1959
[c] Salaries only, prior to 1963

TABLE V

Other Committees of the Council, 1954–75

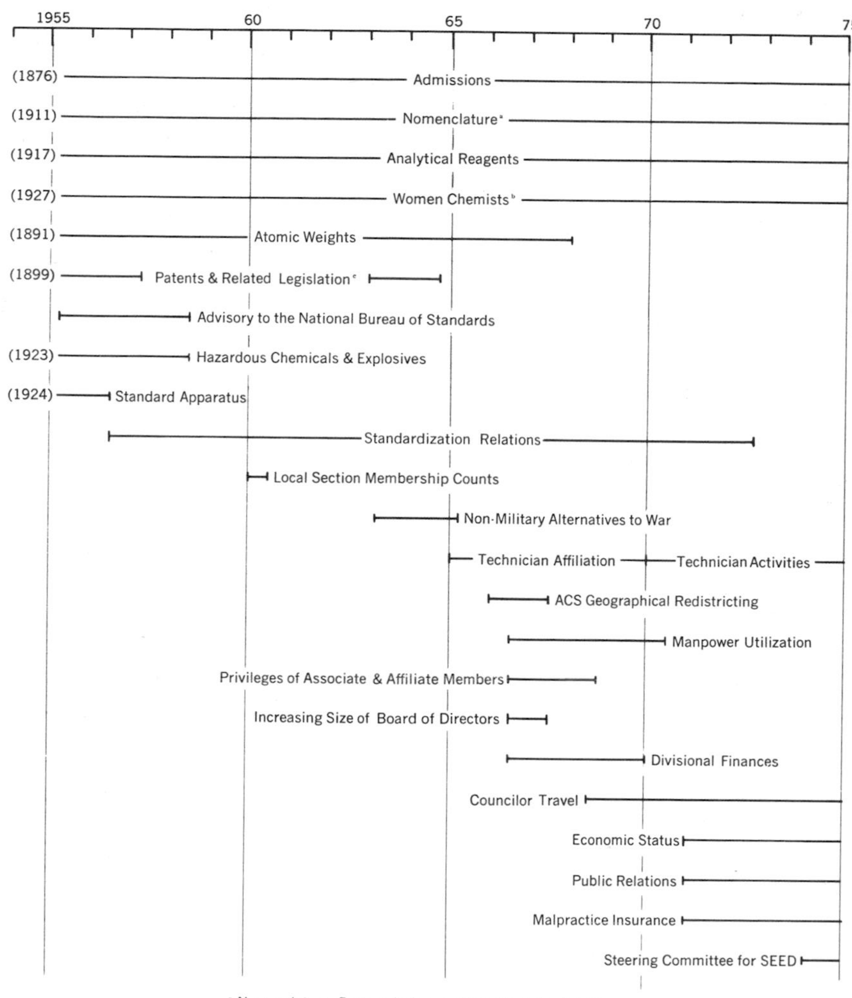

Society committees, an increasing number of joint committees may be expected.

A number of other types of committees exist within the Society. Two presidential ad hoc committees—one on professionalism and the other on manpower policy—were appointed and later discharged in the early 1970s. Many divisional committees, such as those on nomenclature, interact strongly with the corresponding Society committee. The Council

Committee on Nomenclature itself has been authorized to interact directly with foreign nomenclature committees. Editorial advisory boards were created to advise the editors of Society publications and, through the editors, may influence the actions of the publications committees of the Board and Council. Regional Meeting Steering Committees and Regional Councilors' Caucuses exist, and the latter, especially, have exercised some influence on Council affairs. Local section committees abound in both numbers and activities.

Some 55 committees are listed in Tables I–VI as being active at the beginning of 1975, and they were simply the tip of the iceberg. ACS committees represent a tremendous wealth of membership manpower

TABLE VI

Joint Board-Council Committees, 1954–75

1955 60 65 70 75

(1936) Professional Training[a]
(1952) Air Pollution
Proposed Increase in Dues
Plans for Future National Meetings
Divisional Problems
Patents and Related Legislation[b]
Chemistry and Public Affairs[c]
Profession-Wide Pension Plan[d]
Environmental Improvement
Future Financing Requirements and Dues
Copyrights
Younger Chemists[e]
Chemical Education Planning and Coordinating
Centennial Coordinating
Revenues from Members
Coordinating Professional Programs
To Oversee Study of Structure, Governance, and Management[f]
Feasibility of a Member Investment Program
Feasibility of an Endowment Fund

[a] Committee of the Council prior to 1968
[b] Committee of the Council 1963–5 and 1899–1957
[c] Board Committee 1965–8
[d] Board of Trustees, Pensions for Professionals, Inc., after 1972
[e] Younger Chemists Task Force prior to 1974
[f] Joint Board-Council Policy Committee committee

which is, in the final analysis, the Society's greatest resource. The success of the utilization of this manpower through committees is evident in the Society's record of accomplishment.

The Board of Directors

The ACS Board of Directors, the legal representative of the Society since 1877, was subservient to the Council under the first three Constitutions and, initially, under the fourth, the Constitution of 1897. Thereafter, power flowed steadily from Council to Board. In 1897, the Council was the "general governing body" of the ACS; the Board was to "have, hold, and administer" the Society's property under the direction of the Council. In 1947, the last year of the fourth Constitution, the Council was "an advisory body"; the Board was to "have, hold, and administer all the property, funds, and affairs of the Society." This Council–Board relationship was retained in the Constitution of 1948 and remained largely unchanged as the Society entered its 100th year.

The shift of power from Council to Board was perhaps inevitable. As the ACS grew, the Board remained small and efficient. But the Council, designed as a representative body, grew automatically with the membership—by 1947 there were 573 Councilors—and became increasingly unwieldly as a management tool. An able, decisive governing body certainly was needed, if only on financial grounds. The ACS was no longer a simple, dues-paying organization. In 1934, expenses had about equaled dues; in 1944 they were three times dues, in 1964 they were 10 times dues, and in 1974, at $33.8 million, they were 15 times dues. As ACS operations expanded, the Board expanded accordingly, from six members in 1897 to 15 in 1947. The number was reduced to 13 under the Constitution of 1948, but increased again to 15 effective in 1969. In 1971, the Executive Director was made a nonvoting member of the Board, bringing its size to 16: the President, the President-Elect, the immediate Past President, and the Executive Director (all ex officiis), six Regional Directors elected directly by the memberships of six geographical regions, and six Directors-at-Large elected by the Council.

The ACS Board of Directors discharges an enormous amount of business in the course of a year. Much of it is routine, but notable decisions have been required of the Board during its years at the helm. Moves that entailed new responsibilities, if not great cost, were the decisions to seek a federal charter, granted in 1937, and to assume the administration of the trust income from the Petroleum Research Fund, established in 1944. Both steps were taken during the chairmanship of Thomas Midgley, Jr., who in 1944 was also ACS President. Important financial undertakings of the Directors have involved housing for the headquarters staff and housing and certain other aspects of Chemical Abstracts Service.

The ACS was in its 65th year before it bought its own headquarters building. The five-story apartment building in Washington, D.C., on the site of the current headquarters, was purchased for $150,000 late in 1940, again during the tenure of chairman Thomas Midgley, Jr. In September 1957, under chairman Ralph Connor, the Board decided to proceed with plans to replace the old building with the one in use in 1976. The dedication plaque of the new structure was unveiled in October 1960 by chairman Arthur C. Cope. The building cost about $3 million; a fund drive raised almost $2.3 million, about half of it from 34,482 ACS members and half from 436 companies.

The boom in scientific publishing after World War II created both operating deficits and housing problems at *Chemical Abstracts*. In 1955 the Board of Directors, chaired by Ernest H. Volwiler, decided that *CA* should be priced to break even and revised nonmember subscription charges drastically upward, although members could still subscribe to *CA* for their own use for $20. By 1962, the financial problem had grown even more severe, and the Board, under chairman Cope, raised the member subscription charge for the complete *CA* from $40 to $500. The outcry from the members was great, but the Board stood firm and, indeed, had little choice. The policy of pricing CAS so that the operation breaks even remained in effect in the Society's 100th year.

The Board, under chairman Cope and his successor Milton Harris, continued to face heavy responsibilities at Chemical Abstracts Service. In 1962, after careful study by a Board committee, the ACS bought a 50-acre, $750,000 site next to the Ohio State University campus in Columbus, Ohio. CAS dedicated a $6.8 million facility on that site in June 1965 and a second, $7 million facility in June 1973. Also in 1965, at Board initiative, the ACS concluded a contract under which the National Science Foundation supplied $2 million to assist in automating *Chemical Abstracts*. The Directors took the step despite the fears of federal interference in the operation of CAS, fears that proved unwarranted. Some $23 million in federal money had gone into automation research and development at CAS by 1975, when the work was decades beyond the stage that the Society could have reached in that year on its own.

At the ACS spring meeting of 1965, President (and Director) Charles C. Price, in a landmark speech to the Council, urged forcefully that the Society should establish closer relations with the Government and assume a more prominent role in national affairs by providing significant information and responsible opinion on questions of public policy related to chemistry. He proposed that a new committee be formed to undertake the task. Later that year the Board created the previously mentioned

Committee on Chemistry and Public Affairs and a staff office to support its activities. The Board is the governance body charged with speaking externally for the Society, and the committee reported exclusively to it until the fall of 1968. At that time the committee was made a joint Board–Council committee.

The Council

The Council is by name the oldest governance body of the Society. The present Council, however, is not precisely a lineal descendant of the first—which became a Board of Directors in 1877—but appears rather to have evolved from the Advisory Council created in 1890. The proportional representation of local sections provided by the Constitution of 1897 during its 50-year life caused steady growth and a corresponding loss of influence by the Council. By the time of the Hancock study, in 1946, the Council had become too large (more than 500 voting Councilors) and too poorly organized to be either an effective "governing body"—giving "general direction" to the Board—or an effective deliberative body. The Hancock report recommended, and the Society adopted, four major changes in Council organization, affecting represensation, length of term, and mode of accreditation of Councilors, agenda, and committee structure.

The chief change in representation applied to the local sections. Since 1897 they had been entitled to one Councilor per 100 members or principal fraction thereof, and in 1946 this provision was creating almost 450 local section Councilors. The problem was solved by creating the concept of the "divisor"—essentially the number of members to be represented by a local section Councilor. The divisor is set each year by the Council Policy Committee; it was set initially at about 200, but by 1975 exceeded 300. All sections are entitled to at least one Councilor; those with memberships of one half the divisor or more ("larger" sections) are entitled to one Councilor for each divisor of members or principal fraction thereof. Since the Bylaws permit no more than 320 Councilors from "larger" sections, a continuing lid is maintained on the size of local section representation. This was, of course, the original intent of the Hancock report, although the local section representation created at the time was larger than the report had suggested. The Council, moreover, has grown since 1947, in part because of the increase from 18 to 27 divisions (each entitled to two Councilors) and the increase in the number of living Past Presidents. Thus the voting membership of the Council in 1975 stood at 450. This number begins to approach the number that was a cause of concern in the Hancock report and that—during current discussions of reorganization—is once again a cause of concern.

The second of the four major changes in Council organization adopted from the Hancock report was the replacement of one-year terms for Councilors by three-year terms, which were put on a rotating basis to provide continuity of experience. Mechanisms were created also for providing alternates and for orderly election and certification of Councilors and alternates well before they are seated.

A strong source of contention during the decade preceding the Hancock report had been the degree to which operating control of the Society, including the Council, had been dominated by the professional staff. Insofar as the Council was concerned, this probably was attributable at least as much to ineffectiveness of Council operations as to staff overzealousness. The report identified several sources of the Council's debility, among them the failure to control its own agenda, a document then created entirely by staff. The situation was rectified by making the Council Policy Committee responsible for Council agenda. In addition, the membership of the Council Policy Committee was altered from being predominantly the officers, secretariat, and editors, ex officiis, to being predominantly Councilors elected by the Council.

A second shortcoming of Council organization at that time was the lack of provision for a regular, continuing committee structure to bear the burden of the detailed study of issues before the assembly. The Constitution in 1946 provided only for a Board Executive Committee, a Council Policy Committee, and a Society Membership Committee; any other committees were to be appointed, as required, on an annual basis. The Hancock report recommended the creation of seven standing committees: Chemical Education, Divisional Activities, Local Section Activities, Membership Affairs, Nominations and Elections (made an elected committee), Professional Relations and Status ("and Status" was dropped in 1967), and Publications. These standing committees of the Council were established and—together with others created since—were still in existence in 1975. The new Bylaws, moreover, included a provision permitting standing committees to hold open hearings "to gather Councilor and member opinions." By 1960, open hearings were to be a prerequisite for Council action on major legislative matters.

The new Constitution was approved Sept. 13, 1947, by 95% of the Councilors voting, and was ratified shortly thereafter by the membership. It took effect, together with new Bylaws, on Jan. 1, 1948. With these changes, the Council assumed the form that in general it had retained for almost three decades by 1976.

Council History After 1960

From 1948 until about 1960, the Society was content to make few changes in its Constitution and Bylaws and instead was adapting to the new procedures. The Council, particularly, was learning to perform in new patterns of operation with its committees and in relation to staff. With a few exceptions these were uneventful years for the Council.

The first significant change in the provisions of 1948 came in the spring of 1960, when the duties of the Council's standing committees were redefined and enlarged. At the same time, committee hearings, open to the membership, were mandated for all major committee recommendations to be decided by the Council and were required to be held "at a previous national meeting." In another effort to open the governance to the membership, the Council Policy Committee voted to admit ACS members as observers to meetings of the Council. In 1970 this measure was broadened to include all registrants at a national meeting.

During the early 1960s, the workload of the Board of Directors increased substantially as the budget and membership of the Society multiplied. On several occasions Directors inquired of the Council Policy Committee and of the Committee on Constitution and Bylaws as to the willingness of the Council to enlarge and possibly restructure the Board. An ad hoc committee was appointed in the fall of 1965 to study the need for increasing the number of Regional Directors. No increase resulted, but a number of the regional lines were redrawn so as not to split local sections. In 1967 the Board appointed a committee (later to include elected Councilors) again to study the need for increase in the number of Directors. That fall the Council adopted the committee's recommendations, redefining the six existing regions to have essentially equal member populations and creating a system for redrawing the boundaries to maintain these populations within 5% of the mean. Finally, in the spring of 1968, the Council increased the number of Directors-at-Large from four to six and decreased the term from four years to three.

The increase in workload of the Board during the 1960s reflected a similar trend in the business management of the Society. For a number of years staff activities had been divided into four operations: membership (Executive Secretary); finance (Treasurer); publications; and *Chemical Abstracts;* to which had been added, in 1967, a planning and information systems function. Each of these activities reported directly to the Board—in practice, to its Chairman, who was functioning as a chief executive officer. Not surprisingly, the workload of the Chairman, a member-volunteer, became intolerable, and the Board began to seek alternatives.

In the fall of 1967, the Council was informed officially that the Board was planning to consolidate all staff operations under one salaried executive officer who would report to the Board. This official, to be called Executive Director, was also to assume all duties of the post of Executive

Secretary, including the secretaryship of the Council and of the Council Policy Committee. The following fall, upon request of the Board, the Council endorsed the concept; the necessary amendments to the Constitution and Bylaws were adopted by the Council in the fall of 1971.

As the operations of the Society grew increasingly complex, the Board became ever more deeply immersed in its management responsibilities. Conversely, for lack of full information, the Council found it increasingly difficult to fulfill its role "as an advisory body in matters pertaining to the general management of the Society." This was particularly true of Councilors who were not on standing committees. Clearly, a central problem was that the growing quantities of management information were not adequately available to all those who were charged with helping to make decisions.

In the spring of 1968, several steps were taken to remedy this condition. As an experiment, Councilors were admitted to parts of executive sessions of standing committees of which they were not members (an arrangement made permanent the following year). The chairmen of certain Council standing committees were invited to meetings of the comparable Board committees, and the reverse. In addition, regular but informal meetings were established, at national meetings, between Society officers and groups of Councilors.

The need for adequate communication among Councilors became especially acute as proposals to amend the Constitution and the Bylaws—essentially, the statutes of the Society—became increasingly numerous and often controversial. Recognizing that such proposals deserved at least as much deliberation as major committee recommendations on other topics, the Committee on Constitution and Bylaws proposed that, except for urgent matters, each amendment petition be publicized and debated in committee at a national meeting prior to that at which it is voted. The Council adopted these new amendment procedures in the spring of 1968, and the membership subsequently ratified them.

During this period, many of the more controversial issues—particularly those related to the Society's assuming a larger role in "professional" matters—were first raised by local sections on the West Coast. Their Councilors perceived a common bond in these matters and began to "caucus" informally a day or two before each Council meeting. The western regional Councilors' caucus first met formally in the spring of 1968. The years thereafter saw meetings of Councilors from the Mid-Atlantic states, the Northeast, the Southeast, the Central Region, and the Midwest.

In 1969, the Council began to look more closely at the management of financial affairs, which had been exclusively the province of the Board of Directors since the early years of the Society. The precipitating event

***Roger Adams** (1889–1971), a Past President, Chairman, and Priestley Medalist, received birthday congratulations during the Council meeting in Minneapolis in April 1969. He had turned 80 in January.*

was a major increase in dues—from $16 to $25— adopted by the Council in April 1969, mainly as a result of the general rise in the cost of living. The lengthy debate in Council over the move focused particularly on many of the dues-supported programs. The Board Committee on Finance also wished to involve the Council to a greater extent, although some Directors had misgivings. At any rate, the Council formed an ad hoc Committee on Program Review to review the effectiveness of the programs supported by dues income. In the fall of 1969, the Council amended a Bylaw to make Program Review a standing committee.

The Board ratified the new Bylaw, but some of its members still had misgivings about the practicality of this apparent sharing of financial responsibility with the Council. The feeling was further intensified in the fall of 1971 (financial conditions having become more severe) when the Council authorized the committee to review all proposals for new dues-supported activities and estimate their financial impact. At that meeting, the committee made its first major recommendations for curtailing several Society activities or changing their sources of support. These recommendations were endorsed by the Council and, with one exception, largely accepted by the Board. The exception was the recommended elimination of the national radio program "Men and Molecules." This the Board chose to continue because of its great popularity and a reluctance to discontinue it until alternate sources of support could be studied.

The Men and Molecules issue was remembered with some heat when, in the fall of 1974, the Council made the Committee on Program Review responsible for reviewing and evaluating all "member-oriented programs," including those not directly supported by dues. Some Board members actively opposed this proposal on the floor of Council as an improper and impractical invasion of Board responsibility. It was debated strongly also at the Board meeting, which subsequently ratified—by a narrow majority—the necessary Bylaw amendment. With this sequence of events, the Council reestablished for the first time in more than a quarter century its right to share in the financial management of the Society.

Toward the end of the 1960s, the decline of the national economy and of chemical employment exposed the Council to strong pressures for the ACS to recognize more overtly its presumed responsibilities to its members, to undertake more extensive "professional" activities. In many instances, the pressures came down to demands that the Society act to protect members' jobs. For some time, many ACS leaders had argued that the principal reason the Society could not engage heavily in "professional" activities was that its Objects, as stated in the Charter and Constitution, obligated the Society to serve "chemistry," not "chemists"; they held also that such activities might jeopardize the Society's favored tax status.

President Melvin Calvin conducted the Council meeting in Boston in April 1972. From left are Rodney N. Hader, assistant to the President; Dr. Calvin; Pauline Newman, vice-chairman of the Council Policy Committee; and Executive Director Fred Wall.

Others argued that these were not real impediments to expanded professional activities, provided that such activities did not dominate ACS affairs. In the spring of 1972, after a year of debate, the Council adopted and the members ratified a Constitutional amendment that specified that the Society should be concerned with "both the profession of chemistry and its practitioners." Since the Objects in the federal Charter could be changed only by an act of Congress, the device was adopted of adding a second section to the existing Objects in the Constitution.

Concern for the professional status of members naturally came to encompass also concern for minority members, particularly women. At the fall meeting in 1972 the Council adopted a resolution calling for equality of treatment for all chemists and urging chemists in academic institutions to "encourage qualified women students to continue in chemistry." This same Council meeting, ironically, was disrupted for several minutes by female members of Scientists and Engineers for Social and Political Action who were also ACS members. One woman seized the podium and passionately protested governmental and industrial discrimination against women. Some Councilors were outraged, others amused, but the incident, though surely unprecedented, was largely benign. In any event, it was certainly true by then that women chemists were represented—though thinly—on both the Council Policy Committee and the Board of Directors. And the first woman candidate for President-Elect, Anna Harrison, was nominated by the Council in both 1974 and 1975. The Council's first black nominee for the post had been Lloyd Cooke, in 1970; Henry Hill was the second, in 1972 and in 1975, when he won.

The elections of 1971–72, in which petition candidates Nixon and Friedman defeated the Council nominees for President-Elect, were viewed by many Councilors as the clear voice of the members calling for greater concern for their well-being than the Council had demonstrated previously. Councilors, it was said, were unresponsive to the members' wishes mainly because they were elected by undemocratic procedures in the local sections and divisions and, once elected, were substantially unaccountable to the members. The Councilors who held this view mounted a campaign for reform during the early 1970s with a veritable blizzard of amendment petitions seeking to "democratize" the Society. A significant number of these eventually were adopted, though not without prolonged debate; others would supply part of the impetus for the second major study of the Society's governance.

One reform proposal called for a record vote (by the name of each Councilor) on any matter before the Council other than an election, if so ordered by a three-tenths vote of the Councilors voting. The Council adopted this Bylaw amendment in the fall of 1972. But despite repeated attempts by some Councilors to employ the mechanism on controversial issues, it had been invoked only once through 1975. An even more controversial proposal would have required a general membership referendum on any Constitutional amendment defeated or twice deferred by the Council if 2% of the members so petitioned. This proposal had a checkered history, but by 1976 it had been defeated twice in Council and the ACS still had not enacted a provision for membership referendum.

Many of the proposals to democratize the Society took the form of requiring changes in the Councilor election procedures of local sections and divisions. Some elections, for example, were held at poorly attended meetings rather than by mail, or were uncontested, or did not require candidates to make position statements. Other proposals sought to increase or decrease the representation of divisions in Council. Still others sought to have Directors-at-Large elected by the membership rather than the Council, or to require mail ballots of the Council for election to its elected committees. These many proposals would have had interlocking and sometimes conflicting consequences if all had been adopted, but all reflected the common theme of governance reform. In the fall of 1971, fortuitously, the Council had assigned to the Council Policy Committee a long-range planning function; CPC, therefore, was assigned to unravel these governance-reform proposals and similar ones that might be offered in the future. CPC formed a Long-Range Planning Subcommittee, whose members concluded that they faced not just a set of conflicting amendment petitions but the obligation to reexamine and perhaps revise the entire governance structure of the Society. This decision, in addition to its ultimate consequence—a major study of ACS governance—produced two intermediate results.

The first of these dealt with the manner in which Councilors were appointed (by the President) to the standing committees of the Council. As in any large deliberative assembly, the detailed substance of most legislative decisions by Council came from its standing committees. A Councilor who wished to influence such decisions perforce sought membership on a standing committee (multiple membership was prohibited). For the approximately 390 Councilors who were by statute or custom eligible for such membership, however, there existed at most 135 such positions. Because of the number of Councilors who could not be appointed each year there had always been some dissatisfaction with the selection procedures employed by the Presidents, and this and other dissatisfactions became substantial toward the end of the 1960s.

In 1971 the Council Policy Committee suggested a number of reforms. They included ways in which the Council could exercise greater influence over the membership of its standing committees (then at the sole discretion of the President) and ways that more Councilors could be utilized in the committee system. Because of circumstances at the time, these recommendations when first offered were largely ignored, which engendered strong resentment in many Councilors.

The Long-Range Planning Subcommittee eventually recommended the establishment of a third elected committee of the Council—a Committee on Committees—to "advise" the President on committee appointments and to "consent" to them. The new committee also was to make recommendations to Council on the size and duties of Council committees and to coordinate their meetings and agenda. The Council created the Committee on Committees in the spring of 1974, but without the controversial "consent" clause; the President remained free of Council constraint, though not of Council influence, in making committee appointments.

The second intermediate result that emanated from the Long-Range Planning Subcommittee was a proposal for a major restructuring of the Society's governance. In the proposal, the Council would establish Society policy, and the Board of Directors would implement it, an arrangement reminiscent of the early years of the ACS. In addition, the increasing overlap of Council and Board committee functions would be eliminated by having one set of all-Society committees. The subcommittee felt that some significant alterations in governance probably were desirable and offered its proposal more as a focus for discussion than as a plan to be ultimately adopted. It was endorsed by the Council Policy Committee and placed on the Council agenda for vote at the 1974 fall meeting, but—on recommendation of the sponsors—was deferred to the fall of 1975. At that time it was deferred again, to the spring of 1977, in the full expectation of its sponsors that its purposes would by then have been served—that the governance reorganization it was designed to stimulate would be well under way.

Council Functions and Organization, 1975

The prime responsibility of the Council, perhaps, since the inception of the Society, has been to act as keeper of the governing documents, the Constitution and the Bylaws. It has been the duty of the Council to revise and amend these documents in consonance with the needs of the Society as an institution and with the wishes of the membership. But although the Council initiates these changes, the members must ratify Constitutional amendments and the Board of Directors must approve Bylaw amendments. These governing documents deal in considerable detail with the affairs of the Society. In consequence, many of the legislative actions taken by the governance bodies of the ACS must be in the form of Council-adopted amendments to one or both documents. The principal functions of the Council are given in Table VII; the organization of the Council in 1975 appears in Table VIII.

TABLE VII

Principal Functions of the Council, 1975

Amend the Constitution and Bylaws
Establish standards of membership in or affiliation with the Society*
Charter local sections; designate headquarters and boundaries
Establish rules for the assignment of members to local sections*
Establish divisions
Approve bylaws and articles of incorporation for local sections and divisions
Set dues for membership and affiliation*
Set monetary allocations to the local sections*
Approve journal subscription rates and allocations to *C&EN* as recommended by the Board of Directors
Set the election procedures of the Society*
Nominate candidates for President-Elect and all directors
Supervise all elections; set the election regions
Determine the divisor for local section Councilors
Establish all elected and standing committees of the Council and their duties and composition*
Establish all other committees of the Council and their duties and composition
Elect the elected committees of the Council
Oversee the operations of all committees of the Council
Charter chapters of student affiliates
Collaborate with the Board of Directors in setting governance meeting dates
Expel members of the Society
Advise the Board of Directors

* Indicates that the Council must amend the Constitution, the Bylaws, or both to perform the function.

TABLE VIII

Organization of the Council, 1975

The composition of the Council is as follows:

Local Section Councilors:	
54 Sections with membership under 160:	54
121 Sections with membership over 159:	302
Division Councilors (two per Division):	54
Ex officio Councilors:	
President, President-Elect, Executive Director	3
Elected Directors	12
Past Presidents	23

Bylaw Councilors (a one-year continuation of members of elected committees who have failed of re-election as Councilors):	2
Voting Councilors	450
Nonvoting Councilors:	
Non-Councilor Chairmen of Other Committees of the Council	11
Senior Members of Staff	6
Total Council Membership	467

The organization of the Council is as follows:

The President and Executive Director of the Society are President and Secretary of the Council, respectively, ex officiis

There are three Elected Committees:

- Council Policy Committee (executive committee):
  - Voting members: President, President-Elect, Immediate Past President, Executive Director, all *ex officiis;* 12 members, elected by and from the voting Councilors
  - Nonvoting members: Chairmen of the other Elected Committees and Chairmen of the Standing Committees of the Council, all ex officiis
  - Chairman: President, ex officio
  - Vice-Chairman: elected by and from the voting members of the Committee
  - Secretary: Executive Director, ex officio
- Committee on Nominations and Elections:
  - 15 members, elected by and from the voting Councilors
  - Chairmen: elected by and from the Committee
- Committee on Committees:
  - President-Elect, ex officio; 15 members, elected by and from the voting Councilors
  - Chairman: elected by and from the Committee

There are nine Standing Committees:

- Chemical Education
- Constitution and Bylaws
- Divisional Activities
- Local Section Activities
- Meetings and Expositions
- Membership Affairs
- Professional Relations
- Program Review
- Publications

The chairman and members of each are appointed by the President. There are 21 Other Committees, for special purposes, of which eight are continuing, two are ad hoc, and 11 are joint with the Board of Directors.

Council Actions, Post-1947

At each of its meetings, the Council is concerned mainly with the procedures and duties that derive from its functions. Nominations are announced and elections conducted. Reports are heard from each of the elected and standing committees. Reports from other committees are heard by prior arrangement with the Council Policy Committee. Virtually all legislative actions result from recommendations by a committee, usually a standing committee, but most are routine rather than controversial or tending to set precedent. Yet, the Council and its committees have been the source of many new directions for the ACS. Sometimes in support of the Board of Directors, sometimes itself setting the course and pace, the Council has helped, for example, to thrust the Society strongly into such areas as professional concerns. And at many Council meetings, decisions are made that will affect chemistry and chemists in some important way. Following are the innovative or more important directions and decisions of the Council (and, occasionally, refusals to act), arranged by subject area and dating (with one exception) from about the time of the Hancock report.

Local Sections and Divisions

The Constitution of June 6, 1890, authorized the establishment of local sections, and the Council first authorized a division in 1908. By 1975 the Society had 175 local sections and 27 divisions, plus one probationary division. As a result of the Hancock report, in 1947, the minimum size of a local section was set at 50 members, and it was required that the bylaws of local sections and divisions be approved by the Council (in the fall of 1967 the Council delegated this authority to the Committee on Constitution and Bylaws). To help new local sections begin operation, the Council in the fall of 1962 authorized the establishment of a set of charter bylaws that local sections could use until each saw fit to revise or amend them; it adopted the actual text a year later. Similarly, in the spring of 1966 the Council adopted a set of bylaws for divisions in probationary status.

Nominations and Elections

During most of the history of the ACS, its members have agreed generally on its purposes and directions, and the criteria for the election of officers therefore have tended to rest largely on the professional eminence of the candidates. Nevertheless, in the fall of 1959 the Committee on Nominations and Elections announced a two-year experiment in which each nominee for national office would be allowed a voluntary statement of opinion on matters affecting the Society; the statements would be limited to 150 words and would be mailed with the ballot. The Council Policy Committee opposed the scheme, but an informal show of hands at that meeting indicated that half the Councilors favored it. The Committee on Nominations and Elections found very few candidates willing to submit a statement, however. And though the committee did continue the practice after the two-year experiment, it was never willing to make submission of a statement mandatory, though urged by many Councilors to do so.

Toward the end of the 1960s, however, it became plain that serious questions were beginning to arise on the directions of the Society. Many Councilors felt that the members were entitled to choose a President-Elect at least as much for his perception of the Society's future as for his scientific eminence. Thus, in the fall of 1968, the Council voted to ask members to submit "timely and significant questions pertaining to the Society" to the Committee on Nominations and Elections. The Committee was to prepare a list of such questions, to be approved by the Council Policy Committee; candidates' responses were to be included with the biographical material mailed to the members. This procedure, at first an experiment, became permanent policy.

The method of electing Councilors was an active issue during the Council meeting in Atlantic City in September 1974.

The proponents of this policy sought for some time also to make platform statements customary, if not obligatory, for candidates for local section and division Councilor. Indeed, the question of dictating mandatory election procedures to local sections and divisions was discussed often during the 1960s by the Committee on Constitution and Bylaws and was always substantially opposed. Even though it was asserted—correctly—that many local sections and divisions did not use mail ballots, or did not use contested elections with statements, or elected Councilors by executive committee action, many Council members were reluctant to interfere with section and division autonomy. And a proposal to require that all Councilors be selected in popular, mail-ballot elections was defeated by the Council in the spring of 1972. A new proposal to the same end was introduced in the spring of 1974, deferred to the fall meeting, and then further deferred for one year because of the expected new study of the Society. By the fall of 1975—when a provision requiring contested elections was added to the proposal—discussion of governance reorganization suggested that increased Council authority would be one likely result. A substantial body of Council opinion felt that with increased authority the manner by which Councilors were chosen could no longer be left almost entirely to the determination of each local jurisdiction. Thus, the popular mail-ballot portion of the 1975 proposal was adopted for election of Councilors and Alternate Councilors, though the contested election portion was not adopted. The mail-ballot provision was ratified by the membership and took effect Jan. 1, 1976.

The numerical results of ACS elections had never been publicized, but in the spirit of more open governance the Committee on Nominations and Elections voted in the spring of 1972 to publicize the numerical results of final balloting for President-Elect and all Directors. This procedure later was applied also to balloting for seats on all elected committees of the Council.

Membership

Until recent years the Society was said to be devoted primarily to the "advancement of chemistry," but the Council invariably has been reluctant to grant membership to those interested in or associated with chemistry who were not otherwise professionally qualified—this, though it has tinkered constantly with the detailed criteria of membership. Persons not qualified for membership were permitted "affiliation" with a local section or division but not with the ACS nationally. The Division of Chemical Education and some local sections, however, were finding many high school teachers and chemical technicians who—though unqualified for membership—were quite active as affiliates and pleaded for national status. The matter first came to a head in the spring of 1970 when the Council defeated a proposal for a category of "national affiliate." The move reflected in part a heightened feeling by many Councilors of a need to maintain professional solidarity in the face of a then-deepening unemployment crisis.

In December 1971, however, the Board of Directors called for a reevaluation of "the basic structure of the Society with respect to possible means by which the ACS might accommodate such various chemical populations as high school chemistry teachers, chemical technicians, chemical industry executives, and chemists in biomedical fields. . ." The Council Committee on Membership Affairs, after more than two years of study, proposed to Council a category of National Affiliates strikingly similar to that rejected in 1970. Many Councilors' views clearly had changed, for the proposal was adopted in the fall of 1975 with little debate.

Another way in which a membership society manifests solidarity is to expel members who violate its standards of conduct. It is curious, however, that the ACS found cause to expel three members and censure a fourth during the period 1965–67, though it had done so only once previously, in 1924. This abrupt appearance, after so many years, of several membership terminations for unprofessional conduct may not have been entirely by chance. In the fall of 1965 the Coucil adopted the "Chemist's Creed," a statement of principles of professional ethics. It is not unlikely that work on this document had sharpened the sensitivity of the Council to incidents of unprofessional behavior.

The Constitution of 1948 provided not only for individual membership (senior grade and junior grade) but also for corporate membership. Subsequently, however, corporate membership was felt not to be entirely appropriate and was abolished by the Council in the spring of 1951. In the spring of 1962, the "grades" of membership were eliminated in favor of the categories of Member and Associate Member.

Member Finance

In 1948, the membership dues of the Society were $9.00. The Council raised this to $12 in the spring of 1949, to $16 in the spring of 1960, and to $25 in the spring of 1969 (each increase effective in the year following). Comparable increases were made in each case in the local section allotment schedule. Modest as these rates may appear in retrospect, each increase occasioned prolonged Council debate and deep examination of the cost of member activities.

The 1969 increase to $25 was, in percentage, a large one, and there was every expectation that it would give the Society and the local sections an opportunity to build a reserve over several years. The steeply rising cost of living ruled otherwise, however, and the Society headed into deficit budgeting by the early 1970s. Unless the Council wanted to undertake ever more frequent debates on dues increases, it was clear that some procedure would be required to keep the dues abreast of the cost of living. In the fall of 1973 the Council adopted such a scheme: a base rate of $28 and an automatic escalator clause attached to the cost of living for services. In the spring of 1975, the Council modified the procedure slightly, giving itself the right each year to overrule the escalation if it wished. At the same time the Council raised the dues allocation to *Chemical and Engineering News* to $8.00, a $3.00 increase, and raised the dues base accordingly, to $31. The actual dues for 1976, including the escalator, were $35.

Travel to national meetings also became a financial burden beginning in the late 1960s; many employers, particularly academic institutions, were obliged to cut back or eliminate such travel allowances. It was greatly feared, and there was some evidence, that attendance at Council would be severely impaired. In the fall of 1971, therefore, an ad hoc committee was appointed to study the feasibility of paying travel expenses for at least some Councilors. In the spring of 1972, the Council and Board approved a plan of selective support at $20,000 annually, which plan was continued in subsequent years.

National Meetings

In the spring of 1961, the Council approved the experiment of holding a third national meeting in 1963 and 1964. Thereafter the practice was dropped except for one additional third meeting planned later and held in 1966. Other than this, and an earlier experiment on divided meetings, almost the only significant national meeting actions the Council has taken in the current era have been those to avoid efforts to move meeting sites to protect some or all of the attending members from discrimination.

At the spring meeting of 1956, it was moved that "meeting sites be selected which will permit equal opportunity to all members for free exchange of ideas and information and for common access to public facilities." The motion could not be made applicable to the forthcoming Miami meeting (spring 1957) and was "tabled" to the fall of 1956 so as not to offend—it was stated—the Society's current hosts (Dallas). At the 1956 fall meeting, however, the Council toyed with the wording and again tabled the motion. At the spring 1957 meeting, the Board of Directors adopted a resolution stating in part ". . . in seeking to secure for all ACS members equal facilities at national meetings, the Executive Secretary has taken a position in accordance with the Constitution of the American Chemical Society." The statement was presented to Council for action and, at the same time, the spring 1956 motion was again removed from the table. Considerable debate ensued in which various substitute language was suggested, some of it frankly addressing "legal restrictions or social customs" requiring "racial segregation." Finally, both motions were tabled and never again considered.

One further attempt to deal with discrimination against members at meetings was made in the spring of 1958. The Division of Biological Chemistry reported a resolution it had adopted that "the Division [would] refrain from participating in national meetings in such segregated areas. . ." A motion was then made to create a committee to study member sentiment on the problem and to formulate a solution, but it, too, was tabled and never again considered.

Possible political discrimination was a concern raised by Councilors after the 1968 Democratic National Convention in Chicago. An attempt was made at the spring 1969 meeting to eliminate Chicago as a national meeting site on those grounds, but it was defeated. Chicago as a meeting site was again the target at the spring meeting of 1975, this time in connection with the Chicago Section's National Chemical Exposition. The section had been organizing the events since 1940 and had slated the 17th for the ACS fall meeting in 1977 in Chicago. The Committee on Meetings and Expositions reported that, after extensive negotiations with the Chicago Section, the section still did not wish to comply with the uniform policy on exposition income approved by the Council and Board. The Committee recommended, and the Council concurred, that the Board of Directors be asked to withdraw the fall 1977 meeting from Chicago. The Board agreed, but later the Chicago Section agreed to abide by the ACS uniform policy, and the meeting was returned to Chicago.

Technical Affairs

The Council does not as a rule become involved in questions predominantly technical in nature, but there are some exceptions. In the spring of 1951, on the recommendation of the Committee on Nomenclature, the Council adopted, for American usage, names and symbols for beryllium and aluminum and for 13 of the newer elements. Further recommendations of the Committee, bearing on nomenclature rules, were endorsed by the Council in the fall of 1952, spring of 1954, fall of 1955, spring of 1956, spring of 1957, and spring of 1968 when rules for naming boron compounds and polymers were adopted. The one other technically related decision was made in the spring of 1968 when the Council (and the Board) endorsed the use of eye protective devices in the chemistry laboratories of all educational institutions.

Publications

Aside from approving journal subscription rates set by the Board of Directors, the Council generally exercises no policy voice in the publication affairs of the Society, although it plays a strong advisory role through its Committee on Publications. The one exception occurred in the spring of 1964 when the Council voted to relinquish the Society's first right of publication for papers presented at ACS national meetings.

Professional Concerns

The desire of many ACS members that the Society be as concerned with the well-being of chemists as with that of chemistry is by no means new, nor are strenuous differences of opinion among members as to the appropriate degree of Society involvement. Nevertheless, the record of the Council during the almost three decades since the reorganization of 1947 suggests that those who have favored a strictly scientific orientation—free of professional considerations—for the ACS have fought a long but losing battle against an overwhelming combination of forces: changing employment practices; changing values, particularly among younger chemists; and, in the late 1960s, a steadily deteriorating economic situation that especially affected science.

It must be said at once that discussions of professional concerns within the ACS refer overwhelmingly to the more than 60% of the members employed by industry. The Council (and Board) has been strongly disinclined, until recently, to deal with the special problems of the academic community, perhaps because other organizations are thought better equipped to do so. In the spring of 1949, for example, the Council received a report of a member with "academic freedom" problems and was asked to refer the matter to the Committee on Professional Relations and Status. But after learning that the Board had refused to act on the matter, the Council tabled the motion, and it died. By contrast, in the spring of 1973 the committee recommended to the Board that the ACS enter a private suit for academic sex discrimination as *amicus curiae,* and the Board agreed to do so. The main historical thread, however, leads through the development of concepts of professional identity and ethics to the promulgation of employment guidelines—largely related to industrial employment—and the growth of an enforcement mechanism through publicity.

Earlier conceptions of how the status of chemists could be improved are illustrated by the vote of the Council in the fall of 1958 asking the Board to adopt an editorial policy to include the highest earned degree with each author's name in journals published by the ACS, and also in news items and letters to the editor if the writer gave the degree. In the spring of 1959, the Board returned the request to the Council for further study and it was assigned to the Committee on Professional Relations and Status. By the fall of 1959, the Council had second thoughts and voted to rescind its earlier request, but the practice had already been started by *C&EN*.

While some members associated status with titles, others felt that the establishment of identifiable ethical principles would help more. During 1960, the committee wrestled with a proposed code of ethics but, by the fall meeting, tabled the idea in favor of a statement of Principles of Professional Conduct. This proposal was presented to the Council in the spring of 1961, but it was tabled and died. It had become evident meanwhile that behavior codes for chemists are difficult to formulate without defining what a chemist is. A definition of a chemist was therefore adopted by the Council in the spring of 1963. In the fall of 1965 the Committee on Professional Relations and Status finally presented to the Council a statement of ethical principles, the "Chemist's Creed." This document the Council adopted, partly on its merits—which many felt were limited—but partly in recognition of the fact that further years of effort probably would not produce a superior result.

Many Councilors believed that adoption of ethical principles was not the ultimate goal in assisting the member in employer relations, but rather was a necessary prerequisite to asserting what those relations should be. The development of these assertions was to take many years, however. In the fall of 1959, at a meeting of the Committee on Professional Relations and Status, the suggestion was first made that the ACS should form a committee to consider cases of improper treatment of professionals by employers. It was not until 1964, however, that ACS staff—with the guidance of the committee—began to assist members by investigating their complaints of unprofessional treatment by employers. By 1965, the committee had approved formal procedures for handling member-assistance cases. Thereafter, the need to codify the proper professional relationship between employer and employee became increasingly apparent. Finally, in the fall of 1970, the Committee recommended, and the Council adopted, the first Guidelines for Employers—a statement of hiring, employment, and termination practices that an employer should follow in dealing with professional chemists. In the spring of 1975 the Council readopted the Guidelines, by then revised and renamed Professional Employment Guidelines to reflect the inclusion of statements on proper behavior by professional employees.

The issuance by a major technical society of statements on employment practices was unprecedented. It led other organizations, including the various engineering societies and the American Physical Society, to consider and undertake similar steps. But it was clear that issuing statements and making them stick were quite different matters; the question of enforcement inevitably arose.

In adopting the original Guidelines, in 1970, the Council was acutely sensitive to the then-soaring unemployment of members, brought on in part by the exceptional number of mass layoffs—later referred to by industry as "multiple terminations"—in the chemical industry. Indeed, the controversial nature of the Guidelines stemmed mainly from the strong provisions on termination conditions and benefits, provisions that many companies appeared not to meet although some were reluctant to acknowledge the fact. Here, then, was the basis of enforcement: inform the members publicly of the names of companies that did not comply with the Guidelines when laying off chemists.

In the spring of 1971, with the approval of legal counsel, the Committee on Professional Relations publicly reported to Council the names of two employers who were "significantly at variance" with the ACS Guidelines and of one who had responded inadequately to requests for information. This was, again, an unprecedented step for a technical society. But once it was taken, the committee continued to investigate layoffs and report its results on the floor of Council. Companies in substantial compliance with the Guidelines were cited also, nine of them being named before Council in the fall of 1971. All of this information was reported in *C&EN*.

While many Councilors admitted the propriety of the ACS's giving major concern to professional matters, this did not imply necessarily that equal concern for economic matters was appropriate. The Hancock report had identified the issue this way: "It is a foregone conclusion that economic interests should occupy a distinctly subordinate role in relation to the Society's scientific and professional objectives. Distinct occupational classifications place many of the membership in opposite camps. Their private economic interests often have little in common. . ." Against this background the Council, in the fall of 1952, heard several petitions from local sections urging the ACS to use its influence and prestige to improve the economic conditions and professional positions of chemists. In response, the Council adopted a motion reaffirming its "interest in the economic status of its members" and urging that they be compensated "in a manner consistent with their training, ability, contributions, and professional standing."

The Council was not again asked formally to intervene in economic matters for more than a decade. Many Councilors were irked, however, that the Society had no mechanism even for studying economic matters of concern to members. Finally, in the fall of 1966, the Council heard a proposal to create a standing committee on economic status. It was defeated on the grounds that the stated purposes, in the opinion of the

Committee on Constitution and Bylaws, lay largely outside the Objects of the Society. Instead, the Committees on Membership Affairs and on Professional Relations and Status were instructed to determine what steps might be taken, within the scope of the Charter, to meet the intended objectives.

In the spring of 1968, the Council again was asked to create a committee (nonstanding) on economic status with the same goals and again refused for the same reason, after lengthy debate. In the spring of 1969, however, the Council authorized the Committee on Professional Relations to institute a "basic research study" of chemists' salaries in comparison with those of other occupational groups. Finally, in the fall of 1970, with the nation's economy faltering and the scope of the proposed activities significantly limited, the Council agreed to create not only an ad hoc Committee on Economic Status but also an ad hoc Committee on Public Relations. These were made continuing "other committees" of the Council in the fall of 1972 and the fall of 1973, respectively.

There exists in the ACS membership (and the Council) always a full spectrum of opinion not only on the types of activity deserving the greatest weight—scientific to professional or beyond—but even on whether there ought to be multiple roles rather than a single purpose for the Society. It seems clear, however, that during times of economic hardship the spectrum shifts strongly in favor of increases in member assistance, employment protection, and related professional activities. This was the case in the early 1970s, when the pressures on the Society's leadership for sharply increased professional activity appeared to exceed those of any previous period. In the spring of 1971, the President, Melvin Calvin of the California Section, announced to the Council that he had created an ad hoc Committee on Professionalism to "consider all possible actions the ACS might take. . . to fulfill the professional needs of the membership." During its two years of existence, the committee made a number of useful recommendations relating to vesting of industrial pensions, employment of female chemists, increasing undergraduate chemistry standards, collecting and publicizing data on employment conditions by company, and establishing a legislative counselor program.

Another force on the Society's leaders during this period was the disenchantment of many younger members and the fact that relatively fewer younger chemists were seeking to join than in the past. In July of 1970, the Board of Directors created the Younger Chemists' Task Force (to become a joint committee with the Council in the fall of 1973) and charged it with advising the Society on ways "by which the Society might utilize most effectively the thinking, energy, and enthusiasm" of young chemists. The initial report of this group was chilling for many members (particularly older ones) who traditionally had regarded the Society primarily as the citadel of a great science, worthy of support and devotion on those grounds alone. In a "white paper" in 1973, the Task Force stated: "The young people of today . . . do not respond as strongly to the intellectual beauty of chemistry as a science. . . They are approaching chemistry more as a craft than as an intellectual activity." Moreover, "Young chemists do not at present look upon the Society's permission to join as a great favor or honor. The ACS does not now visibly affect their lives in a way to command allegiance." It seemed clear, and the white paper so stated, that the Society would have to undertake programs more meaningful to chemists of such persuasion to attract and retain their support. Younger members, for example, were largely responsible for organizing the beginnings of the Division of Professional Relations, approved by Council in 1972, and still predominate in its leadership.

Public Affairs

The second major nontechnical area in which many but not all members felt the Society should be involved was public affairs, particularly attempting to influence national legislation. Because the Board of Directors is charged with speaking externally for the Society, it has been the source of most ACS activities in public affairs. On occasion, the Council has played a role, particularly through the joint Board/Council Committee on Chemistry and Public Affairs.

Curiously, the Hancock report did not overtly recognize public affairs as an activity of the Society, although attempts at such activity predated the reorganization of 1947. In the spring of 1946, for example, the Council asked the Board to be concerned with legislation "which affects chemists or scientists" and stated that the Board, with the advice of the Council, should try to take stands on such issues in the name of the Society. In the spring of 1948, the Council asked the Board to seek legislation on the licensing or registration of chemists.

On rare occasions, the Council has intervened directly in matters of federal appropriations. One such case came in the fall of 1951, when the Council (and Board) called on Congress to pass the full National Science Foundation appropriation requested by the administration.

The early 1960s were marked by heightened concern over the possibility of war, and members of the scientific community particularly reflected that concern. In the spring of 1963, the Eastern New York Section introduced in Council a resolution urging the ACS "to participate in programs designed to reduce international tensions and develop nonmilitary alternatives to war." This request nonplussed many Councilors, who wondered in particular what the ACS appropriately could do. Thus, the motion was tabled. In the fall of that year, the Council agreed to form a Committee on Nonmilitary Alternatives to War which would hold open hearings and recommend appropriate action. The Eastern New York resolution was removed from the table and given to the committee. The following fall (1964), the committee summarized a lengthy report on proposed activities of the Society in this area and asked the Council to endorse the proposed plan in principle. The Council tabled the motion until after a poll showed that 58% of a membership sample favored having a permanent committee devoted to exploring the question. After limited additional discussion, the Council nevertheless instructed the committee to confine its deliberations to the question of whether the ACS "should be active in the area of nonmilitary alternatives to war." The following spring (1965), the committee reported that present organs of the Society could undertake whatever the ACS might do appropriately to reduce tensions and provide nonmilitary alternatives to war. With this, the committee requested and received discharge.

In the spring of 1968, the Council first noted formally the impact of social disadvantage on opportunities for technical training and a career in science. It adopted a resolution on the "lack of training and employment evidenced in certain segments of our population." The Council urged that appropriate officers of the Society encourage local sections to support the development of local industrial programs for training and employing the disadvantaged and to assist educational institutions in disadvantaged areas with offers of tutorial assistance and chemical supplies.

At the same meeting, with the Vietnam War drawing large numbers of young people into the service, the Council adopted a resolution, for transmittal to the executive and legislative branches of government, expressing concern that Selective Service requirements would act to reduce substantially the flow of trained technical persons into industry.

In 1925, the Council had adopted a resolution opposing United States ratification of the Geneva Protocol (prohibiting the use of chemical warfare). In the fall of 1970, the Council (and the Board) reversed that position and urged the Senate to ratify the protocol. In the fall of 1973, the Council (and Board) adopted a further resolution saying that it had no reservations about the protocol (the national administration had expressed reservations about banning the use of herbicides and other nontoxic agents).

In the 1970s, with the growth in the number of issues of national policy affecting, or affected by, chemistry, the opportunities and obligations of the Society to provide the Congress with information and advice became difficult to manage. It became clear not only that staff would have to be augmented (which it was), but that many more members than served on the Committee on Chemistry and Public Affairs would have to be enlisted. In the fall of 1972, the President's Committee on Professionalism proposed to the Committee on Chemistry and Public Affairs that selected members of local sections be appointed "Legislative Counselors to the President" (of the ACS), so that each could initiate and develop a continuing dialogue with his or her Senator or Representative. Each counselor would be available to that member of Congress, both as a constituent and as a scientist, to supply technical information on request; each also, when appropriate, could offer information on the position of the Society on certain issues. This proposal was implemented in the spring of 1974 by the President-Elect, William Bailey of the Washington Section, under the auspices of the Committee on Chemistry and Public Affairs. In mid-1975 the program was renamed Congressional Science Counselor Program; by the end of that year at least one ACS member had been assigned to nearly all members of Congress.

Division Governance and Operations

Of the organizational elements of the governance of the ACS, the divisions can fairly be said to be the primary custodians of the science of chemistry within the Society. It is through the divisional programs at ACS national meetings and interim symposia and conferences that members present their research results orally to the chemical community. These presentations frequently are the prelude to publication in the Society's journals. Furthermore, the ACS journals select the members for their advisory boards largely from the divisions.

The Society formed its first five divisions during 1908; the number reached the current 28 in 1974. To form a division, at least 50 ACS members must petition the Council which, upon the recommendation of its Standing Committee on Divisional Activities, grants the proposed new division probationary status. Then, within three years of satisfactory operation, the division may request permanent status. Under the ACS Constitution, each division is represented by two Councilors. The Society's Bylaws require that each division shall have a chairman, a vice-chairman and/or chairman-elect, a secretary, a treasurer, and such other officers as the bylaws of the division may specify. The offices of secretary and treasurer may be combined. The ACS Bylaws further specify that each division shall have an executive committee which shall consist of the officers as members ex officiis and such other members, appointed or elected, as required by the bylaws of the division, and that all officers of a division shall be elected in accordance with its bylaws.

The divisions, as the ACS Bylaws suggest, have enjoyed considerable autonomy in organization and, historically, have guarded jealously against any attempt (real or presumed) to encroach upon it. Many active division members contend further that, inasmuch as the Society does not allocate a portion of ACS members' dues to divisions (as it does for local sections), it should not attempt to dictate how the divisions should conduct their affairs. Thus the trend of the mid-1970s toward "democratizing" the Society met mixed reactions among the divisions. A manifestation of the trend was the Constitutional amendment, effective Jan. 1, 1976, requiring that divisional (and local section) Councilors and Alternate Councilors be elected by mail ballot. As of 1975, the bylaws of 18 divisions required election by mail ballot, but did not necessarily require more than one candidate for each office (nor did the amendment). Thus the amendment affected only the 10 divisions that elected their officers at business meetings held at ACS fall national meetings. Still, regardless of their present practices, the divisions generally feel that, in principle, they know best how to serve their own needs. They feel also that if the Society wishes to impose costly procedures upon them, it should provide the necessary funds. The divisions' memberships are geographically dispersed, which complicates communication and the maintenance of continuity. They think it especially important, therefore, that their officers be individuals of proven interest and reliability, not the winners of popularity contests.

Regardless of their modes of operation, almost all the divisions, year in and year out, have mounted outstanding scientific programs and managed their own administrative affairs as well. Divisional officers must meet seemingly endless series of deadlines in developing meeting programs and divisional publications. Many divisions include the duties of program chairman with those of the office of chairman-elect. Others assign this responsibility to the office of the secretary; still others have a separate, appointive position of program chairman (or secretary). The acceptance of nominations for these offices frequently is related directly to the availability of secretarial help to the prospective nominees.

The divisions felt the need to interact with one another long before the Society established any formal committees to consider their requirements. Thus in 1924 was formed the precursor to the present Divisional Officers Group, whose members (the DOGs) are past and present divisional officers. Although DOG is primarily an informal group that gathers socially at each ACS national meeting, from time to time over the years it has pinpointed divisional needs that the Society ultimately has met by appropriate official action. In this manner, the Divisional Officers Group has served an important function as a forum over its more than 50 years of existence.

The longstanding divisional concern with ACS national meetings is evident in the following excerpt from the Council Minutes of Sept. 7, 1946:

There was extended discussion of future meeting plans. It was generally agreed that national meetings have become so big that they can be held in only a very few places, thus depriving many members of the opportunity for participation. While there were many proponents of regional meetings, support was far from unanimous. It was moved, seconded, and carried that the President appoint a committee of divisional officers to develop possible solutions to the problem of excessively large national meetings and that it report its findings at the next meeting of the Council.

A Committee on National Meetings was so appointed, ad interim, in November 1946. It was discharged in April 1947, at which time a Committee on Divisional Activities was appointed as a temporary advisory committee of the Council. This committee, in turn, was replaced in March 1948 by the Council Committee on National Meetings and Divisional Activities, a standing committee. In 1971, the Committee on National Meetings and Divisional Activities was split into the Committees on Divisional Activities and on Meetings and Expositions. These two committees work together closely, with many joint subcommittees, in looking after the interests of the divisions per se and as they relate to the Society's national meetings.

More than 14,000 registered for the ACS national meeting in September 1966 in New York City, where 1971 papers were given in sessions arranged by 25 divisions.

For many years, the office of the Executive Secretary was responsible for providing full-time staff assistance to the divisions. But this arrangement became inadequate as the divisions' technical programs—and therefore ACS national meetings—grew larger and more complex. Consequently, in 1949 a Meetings Department was established under the Membership Secretary of the staff structure to handle the physical and financial aspects of divisional meetings. Divisional activities were not identified as such, but were given staff support either as part of membership activities or in direct relation to participation in national meetings.

In 1965, headquarters established the Office of National Meetings and Divisional Activities, whose name and functions related directly to the Council Committee on National Meetings and Divisional Activities. At about the same time, that committee began scheduling interviews with the officers of six divisions at each national meeting. These interviews began to reveal varying degrees of lack of information and communication, even though an Annual Divisional Officers' Digest was published commencing in 1966.

The upshot was the First Divisional Officers Conference, sponsored by the Committee on National Meetings and Divisional Activities and held in Boston in October 1968. This conference was successful in bringing together the divisional officers, the committee, and the staff to ascertain the common concerns of the majority of the divisions, as opposed to individual situations and personal preferences. The second such conference was held at the spring national meeting in Minneapolis in 1969 and the third at the fall meeting in New York in 1969. Since 1971 the conferences have been held once a year, at the fall meeting; it has become customary to encourage the divisions to send their chairmen-elect, who stand to benefit the most from the exchange of information. Upon its establishment, in 1971, the Committee on Divisional Activities assumed the sponsorship of these conferences, but the Committee on Meetings and Expositions continues to be represented because of the continuing mutual concern with divisional activities.

In late 1969, headquarters established an Office of Divisional Activities and Informal Communications to coordinate divisional activities and promote divisional participation in overall Society programs. Budgetary problems reduced the scope of this office to Divisional Activities alone in 1971, but it continues to increase its services to the divisions. Through 1975 it was in the Department of Membership Activities, but commencing with 1976 it will be in the Department of Meetings and Expositions (until 1971 the Office of National Meetings and Divisional Activities).

In 1971, the ACS gave the divisions the option of having the Society bill and collect their dues without charge; the option included the maintenance of membership records and the related generation of lists, address labels, etc. As of 1976, 26 of the 28 divisions (including one probationary division) had taken up the option. The change resulted from the combined efforts of the Divisional Officers Group, the Committee on National Meetings and Divisional Activities (and its successor committees), the Divisional Officers Conferences, and staff. The Division of Chemical Literature (now Chemical Information) spearheaded the effort, and served as the experimental division in 1967 and 1968.

For many years, divisional program chairmen were urged by the Committee on National Meetings and Divisional Activities to meet at each national meeting to exchange information about their plans for future symposia and avoid overlaps in subject matter. The effectiveness of this mechanism was limited, however. For this and other reasons the Committee on Meetings and Expositions decided to sponsor an Annual Program Coordination Conference, the first of which took place in May 1973. These day-long conferences not only serve as a means of coordinating future programming, but are designed also as a forum for developing new concepts in programming and meeting techniques. The delegates include those responsible for programming at both regional and national meetings. Through 1975 the conferences had proved an unqualified success in improving program coordination, but had made limited progress in developing new programming and meeting techniques. The Society underwrites travel expenses for program chairmen who cannot be funded by any other source.

In 1974, the establishment of an Experimental Program Development Fund for Divisions Programming at National Meetings was proposed jointly by the Committees on Divisional Activities and on Meetings and Expositions to the Board of Directors. The Board approved the proposal on an experimental basis for 1975–76. The Fund for the first time gives the divisions a direct monetary subsidy. Of the $52,000 made available annually, $32,000 is derived from income at national meetings and $20,000 from ACS general funds. As of 1975, the Fund had provided outlays to all divisions except the Rubber Division, Inc., which prefers not to arrange technical programs at national ACS meetings. The Committees on Divisional Activities and on Meetings and Expositions were gathering data preparatory to proposing in 1976 that the Board of Directors put the program on a permanent basis.

Local Section Governance and Operations

The concept of local sections in the American Chemical Society was born of a dissident movement within the Society soon after it was formed and was based largely on the example of the Society of Chemical Industry, in Great Britain, which had organized a number of sections in different parts of the United Kingdom. The leaders of the movement included F. W. Clarke of Cincinnati, Ohio, and Harvey W. Wiley of Lafayette, Ind. The two were among the dissidents who resigned early from the struggling ACS, Clarke in 1877 and Wiley in 1881. Both were strongly interested in establishing a truly American chemical society, rather than an organization of chemists in and around New York City.

Sometime after they resigned, Dr. Clarke became chief chemist of the U.S. Geological Survey, in Washington, D.C., and Dr. Wiley became chief chemist of the U.S. Department of Agriculture in the same city. They collaborated in organizing the Chemical Society of Washington, established in 1884 and a distinct rival of the New York-based American Chemical Society. In 1888, Wiley and Clarke, after consulting with other chemical societies, such as the Association of Official Agricultural Chemists, recommended the organization of a national chemical society to cover all of North America; in 1889 they issued a circular letter to chemists in North America in which they proposed to form such a society by affiliating as far as possible existing local organizations. They suggested that the Society meet annually at such time and place as agreed upon from year to year; local sections, like the sections of the British Society of Chemical Industry, would meet regularly in as many scientific centers as possible, all publishing their work in one official journal.

The ACS, meanwhile, was plagued by indifference and mass resignations. Prof. C. E. Munroe of Newport, R.I., an ACS member, proposed a reorganization that, among other changes, would provide for local sections of the ACS. It is probable that Munroe, Clarke, and Wiley were cooperating to reform the troubled Society. The upshot, at any rate, was the Constitution of June 6, 1890, which established local sections as autonomous units, controlling their own operations and joined together under the aegis of the ACS. Chemists in Rhode Island, led by Prof. Munroe and J. H. Appleton, petitioned to form the Rhode Island Section, and approval came Jan. 21, 1891. The New York Section was chartered on Sept. 30, 1891, and the Cincinnati Section on March 29, 1892. On April 13, 1893, the Chemical Society of Washington became the fourth local section, but under the name "The Chemical Society of Washington, a Local Section of the American Chemical Society". That name still exists in 1976. From 1893 through 1897, 10 ACS members could petition to establish a local section. In 1898 the number was increased to 20 members, and in 1948 to 50 members, the requirement still in effect in 1976.

Society Bylaws require each local section to have a chairman, a vice-chairman and/or chairman-elect, a secretary, a treasurer, and such other officers as its bylaws may specify; the offices of secretary and treasurer may be combined. The Constitutions of 1890, 1892, and 1897, together with the ACS Bylaws, adopted Oct. 27, 1893, established the principle of allocating some fraction of ACS member dues to local sections to support their operations. These documents also established the local section in the governance of the Society: until 1912, the chairman of each local section was an ACS Vice-President; beginning in 1898, as we have seen, each local section was allowed one Councilor for each 100 members or principal fraction thereof. By 1946, 125 local sections were electing almost 450 Councilors. The Constitution of 1948, as we have seen, reduced the number of local section Councilors by establishing the concept of the "divisor" and, as amended effective in 1964, limited the ultimate number of local section Councilors. Under these provisions in 1976, 359 Councilors represented 175 local sections.

The principles of dues allocation and proportional representation for local sections, established in the 1890s, remained unchanged as the Society entered its 100th year. But local section autonomy, which ACS statutes had carefully protected during much of that period, was being eroded in several ways. One of these was the Constitutional amendment, effective in 1976, requiring mail-ballot election of local section (and division) Councilors and Alternate Councilors. A Bylaw amendment, also effective in 1976, required local sections to submit their annual reports to the Society as a condition of receiving their dues allocations for the succeeding year. And critics were asking whether local sections of low activity should even continue to exist.

The allocation of ACS member dues to local sections began under the Constitution of 1890. In 1912 came the first apportionment of funds on a sliding scale based on the membership of the section, a concept still in effect in 1976. That first scale provided a minimum of $50 for any local section and a per capita payment ranging from $1.50 for sections with less than 50 members to 75 cents for those with more than 200 members. This schedule was revised upward in 1946, 1962, and 1970, when it reached the level still in effect in 1976: a base allotment of $450, plus $4.00 for each of the first 500 members, $3.00 for each of the next 2000, and $2.00 for each over 2500. Since 1946, local section treasurers have been required to make a complete financial report yearly to the Society, and local sections have been allowed to carry over allocated funds for use in the following year. Since 1948, local sections have been allowed to accrue allocated moneys in reserve funds. The total apportionment of ACS member dues to local sections in 1975 exceeded $420,000. In 1976, discussions were under way on the need for increased allotments created by inflation. To justify the need, local sections were being asked for reviews of the cost-effectiveness of their programs.

The initial function of the ACS local sections was to disseminate chemical information. Over the years, as the science grew and fractionated into subdisciplines, scientific journals were established to effectively discharge this function. At the same time, the proliferation of subdisciplines made dissemination of chemical information via the local section meeting more difficult, because proportionately fewer members were interested in any given topic. The needs of members changed, and the activities of the sections changed accordingly.

Before 1946, local sections sponsored the Society's two annual national meetings. As these meetings grew larger, fewer cities could handle them, and in 1946 the ACS assumed responsibility for all phases of national meetings—except expositions held in connection with them—including the site and all physical and financial aspects. Local sections continued to help with social functions, plant tours, and public relations activities.

Regional meetings of groups of ACS local sections have a long history, although the present structure of such meetings dates only from World War II. Prewar series included 15 Midwest Regional Meetings (1908–40), 15 Northern New York Intersectional Meetings (1923–36), a dozen Ohio-Michigan Regional Meetings (1923–39), and five Pacific Intersectional Meetings (1928–32). World War II brought a hiatus in regional meetings, and it became more and more difficult for members to attend national meetings. In 1945, in fact, no national meetings were held because of wartime travel restrictions, and 12 local sections held "meetings-in-miniature" as substitutes. Two such meetings were held in 1946, nine in 1947, and six in 1948, and some local sections were still holding them in 1975.

In 1944, again because of wartime travel restrictions, came the first of the modern series of regional meetings. It took place in Baton Rouge, La., was called the Southwest Regional Meeting, and in 1976 was still being held annually. The spring of 1945 saw the first Northwest Regional Meeting, in the Puget Sound area. By 1975 the country had been divided

TABLE IX

Regional Meetings in 1976

Year of Establishment	*Name*
1944	Southwest
1945	Northwest
1948	Southeastern
1963	Middle Atlantic (formerly Delaware Valley Regional Meeting)
1964	Great Lakes
1964	Midwest
1964	Western
1968	Central
1968	Northeast
1972	Rocky Mountain (held in alternate years)

into 10 regions, so that every local section could cosponsor a regional meeting annually or biennially. For the centennial year of 1976, groups of ACS local sections scheduled 10 regional meetings (Table IX). The Society staff provides technical advice for these meetings and sometimes helps to operate registration facilities, but the primary responsibility rests with the host local section.

To recognize and honor exceptional contributions in the practice of chemistry, local sections over the years have established major awards: the William H. Nichols Medal, New York Section, 1902; the Willard Gibbs Medal, Chicago Section, 1910; the Hillebrand Award, Chemical Society of Washington, 1925; the Edgar Fahs Smith Memorial Lecture, Philadelphia Section, 1929; the Theodore William Richards Medal, Northeastern Section, 1929; the Jacob F. Schoellkopf Medal, Western New York Section, 1930. Succeeding years saw many additional awards, and, with increasing emphasis on the teaching of high school chemisty, many local sections established awards for high school teachers. Other local section awards include those to high school and college students. The 1974 edition of the Review of Local Section Annual Reports indicated that 75 local sections presented a total of 254 separate awards.

In 1936, a Committee on Problems of Local Section Officers suggested that the Society establish a Speakers Tour service, operated by the headquarters staff. The service would be a cooperative program that would decrease costs and provide speakers of note. In 1976, this service was still in existence, expanded and refined, but operating under the initial precepts of the recommendation. The 1976 list of prospective speakers, published by the Local Section Speakers Service, listed 431 individuals covering 25 distinct fields of chemistry. The local sections that participate in the service are divided into 26 touring groups so that tours are one week long, an advantage to the busy speaker. Speakers' expenses and part of the operating overhead are borne by the cooperating local sections.

In 1967 a new project for local sections, Operation Interface, was established. This project was designed to bring teachers from colleges and universities together with representatives of industrial organizations to discuss matters of mutual interest at the interface between academe and industry. The Chicago Section sponsored the first such project. Several such Operation Interface programs were scheduled for 1976, with grants from the ACS Corporation Associates fund.

ACS local sections conduct a variety of other activities. Well over 1000 local section meetings are conducted annually, and eight of the larger local sections and one small one have established subsections where member population and distribution indicate the need. In an attempt to serve needs of small segments of members for scientific information, some 20 local sections have established more than 75 topical groups, many of which meet monthly. More than 110 local sections are served by a local section publication or a newsletter. To assist with growing financial problems, 39 local sections levy dues in addition to the apportionment of national dues. Many local sections have formed committees on public affairs, which are concerned with environmental improvement and with providing counsel on matters of chemical science to state and federal legislators. Other sections have established committees concerned with questions of chemical safety. Still others have committees on professional relations, one function of which is to provide employment aids to local section members. A vigorous activity of many local sections is continuing education for their members; such efforts include sponsored symposia, continuation courses, lecture series, and audio and film courses provided by ACS headquarters.

In 1956, the Society organized a staff Office of Local Section Activities, whose manager serves as staff liaison to the Council Committee on Local Section Activities. In 1965, the Office of Local Section Activities initiated a series of conferences for local section officers to facilitate exchange of ideas, programs, and techniques. In the mid-1970s, six such conferences

Delegates from 30 local sections in Ohio and the surrounding states assembled in May 1965 for the second ACS regional conference of local section officers, at the Chemical Abstracts Service building in Columbus.

were being held annually. In 1969, based on staff study and recommendation, the Committee on Local Section Activities established an ad hoc Committee for Section Action to study the concept and operation of local sections and suggest changes to make them a more vigorous unit of the Society. The Committee for Section Action issued its report in 1970; many of its suggestions for effective activities in local sections have been implemented in succeeding years.

The Society entered its centennial year with 175 local sections, whose memberships ranged from 53 to 6392 members and totaled more than 102,000. Local section officers were facing challenges to the historical policy of autonomy that had encouraged initiative and otherwise helped to train thousands of them for further service to the ACS. There was no reason to doubt, however, that local sections would adjust their methods and programs to the changing needs of members and remain a significant force in the affairs of the Society.

The ACS Governance Study

In the spring of 1974, the Board of Directors announced to the Council that it had authorized an expenditure of $50,000 (later increased to $100,000) to retain a consultant "to conduct a preliminary study of ACS structure, governance, and business management." The Council endorsed this action and also the establishment of a joint committee, composed of members of the Board and of the Council Policy Committee, to oversee the study. Arthur D. Little, Inc., was chosen later as the consultant.

There were two principal reasons for the A. D. Little study. The first was a growing need perceived by the Board to revise and improve the business management function of the ACS, and this alone would probably have led to an outside study. The second reason was the proposal to restructure the governance of the Society that had been offered, for discussion, by the Council Policy Committee's Subcommittee on Long-Range Planning. The presence of this petition on the spring 1974 Council agenda led the Board to realize that major changes in governance might result and that they had best come after an independent study. Thus, the original plan to study business management only was expanded to include structure and governance as well.

The subcommittee's restructuring proposal itself had several basic causes. One factor, certainly, was the increasing number of petitions for fundamental changes in election procedures and Council representation that were being filed in the early 1970s. It was becoming clear also that the committee system within the Society was cumbersome and duplicative and needed major revision. Even more fundamental was the growing sense of responsibility in ACS affairs that the Council had acquired during the late 1960s and early 1970s. These were the times when the Council had been motivated to increase sharply the Society's concern for the professional needs of members and to move deeply into the funding of member activities by the Board. It has always been the Council that the membership, particularly the local sections, has pressured for reforms and improvements in Society affairs, and during this period such pressures increased steeply.

Finally, and most fundamentally, the restructuring proposal was the product of a profound feeling that the Society's deliberative assembly, the Council, was ready for, wanted, and was entitled to a larger voice in the ultimate policy decisions of the Society and, indeed, that as the body most directly representative of the members, the Council should bear the ultimate responsibility. In this, the subcommittee was proposing to take the Society full circle to the governance of its earliest years.

Dr. Robert W. Cairns, the second Executive Director of the Society, assumed the post in December 1972.

CHAPTER IX

HEADQUARTERS STAFF AND OPERATIONS

THE headquarters staff and operations of the American Chemical Society at the end of 1975 (Figure 1) attested well to the long road the Society had traveled in its first 100 years. Reporting to Executive Director Robert W. Cairns in Washington, D.C., were three divisions with a total staff of some 300; reporting to him from Columbus, Ohio was a fourth division, Chemical Abstracts Service, with a staff of more than 1100. The combined administrative staff in 1975 was larger than the membership of the Society until its 24th year and was serving more than 110,000 ACS members on annual expenditures of $35.2 million (1975 estimate).

The first administrative office of the Society was manned by Dr. Albert C. Hale. He served the ACS from 1889 through 1902, first as Corresponding Secretary and later as the first General Secretary. Although Hale worked for the Society only part-time, the membership increased from 200 to 2000 during his tenure (he received a 10% commission for collecting members' dues, then $5.00 annually). Because of this growth, Dr. Hale's secretarial duties began to interfere seriously with his full-time position as professor of chemistry and head of physical sciences at Boys High School in Brooklyn, N.Y. In 1902, therefore, he found it necessary to resign.

With Hale's resignation, the Society combined the office of Secretary with the editorship of the *Journal of the American Chemical Society*. On Jan. 1, 1903, the post of Secretary devolved on Prof. William A. Noyes, who had become Editor of *JACS* in 1902. At this time the administrative office of the ACS moved to Terre Haute, Ind., where Noyes was professor of chemistry at Rose Polytechnic Institute. The ACS office moved again in mid-1903 when Prof. Noyes was named chief chemist of the National Bureau of Standards in Washington, D.C. The building for the newly formed NBS was not ready, so Prof. Noyes and ACS moved into temporary quarters in the chemical laboratory of his alma mater, Johns Hopkins University. In January 1905, the South Building of the National Bureau of Standards became the first Washington address of the Society. From 1902 to 1907, ACS membership increased from 2000 to 3400, and with it Prof. Noyes' workload also increased. During that time he had not only created a new NBS laboratory and led its research and service work, but managed also to lay plans for and serve as the first Editor of *Chemical Abstracts*. In June 1907, Prof. Noyes resigned his secretarial duties when he became chairman of the chemistry department at the University of Illinois. At that time the Society separated the positions of Secretary and Editor again so Prof. Noyes could continue as Editor of *JACS* and of *CA*.

The next move for the ACS offices was to Durham, N.H. In September 1907, Prof. Charles L. Parsons, who was chairman of the department of chemistry at New Hampshire College, became the third General Secretary of the Society. In 1912, ACS once again found its way back to Washington after Dr. Parsons, in 1911, was named chief chemist of the newly organized Bureau of Mines. The Society changed its address several times from 1912 to 1919, but always enjoying the hospitality of the Bureau of Mines.

In 1917, ACS membership topped 10,000, and once again the workload became too much for a part-time position. This time, however, the Secretary opted for the Society. Dr. Parsons resigned his position as chief chemist of the Bureau of Mines, but continued as chief of the Bureau's Division of Mineral Technology until 1919. During his period of dual employment, he received no salary from the ACS, but did receive a set commission and was allowed funds for secretarial assistance. The following is an excerpt from Dr. Parsons' business report in January 1923:

The Secretary's first stenographer in Durham, N.H., when the Society had only one journal, was the only employee in the office, and was paid $45.00 a month, and, as her time was far from occupied, she claimed she received too much. The "good old days" have passed.

The Secretary's own salary was increased from $3500 to $5000 when he left the government employ and agreed to give half his time to the Society's work. He actually gives more than three-fourths. The Society can reduce this item any time it thinks it wise to do so. During the stress of late 1921 the Secretary proposed to the Directors that it be reduced to $4000. During 1922 the President has twice requested him to allow him to propose that it be increased. This the Secretary has declined to do, for he prefers, in his own consciousness, to feel that he is rendering service to chemistry without reference to financial reward.

Upon his resignation from the Bureau of Mines in 1919, Dr. Parsons and his four ACS employees moved into rented space in the Mills Building at 1709 G Street, N.W., in the District of Columbia. This remained the Society's headquarters until 1941.

During the period from 1907 to 1922, the Society's publications program was experiencing the same growth and office relocations as the Secretary's office. By 1922, the office of the *Journal of Industrial & Engineering Chemistry* under its new Editor, Harrison E. Howe, had arrived in Washington and located at 810 18th Street, N.W. In 1927, however, the Society's secretarial and publications activities once again became centralized when Mr. Howe and his staff moved into the Mills Building.

During the subsequent decade, the ACS continued to grow—financial activities, local section services, and other activities and services required for the Society and its members. In December of 1940, therefore, the American Chemical Society purchased a building at 1155 16th Street, N.W., for $150,000. In April of 1941, Dr. Parsons, his Assistant Manager, Alden H. Emery, Howe, and 26 other employees moved into the remodeled apartment house that was now ACS headquarters. There was so much room in the five-story building that more than half the space could be and was rented. Over the next 10 years, however, the Society's activities and staff grew to the point that it not only needed all the building space but had to rent additional space as well.

Dr. Parsons retired at the end of 1945. During his more than 38 years of service, he had guided the ACS through a membership growth from 3400 to 43,000 with a $1.5 million budget. In January of 1946, Alden H. Emery was elected Secretary and Business Manager for the Society. Like Dr. Parsons, he had resigned his position with the Bureau of Mines and joined the ACS staff as Parsons' assistant in 1936.

The Treasurer of the Society, who had been located in New York from 1941, moved to Washington in 1948. The move not only concentrated

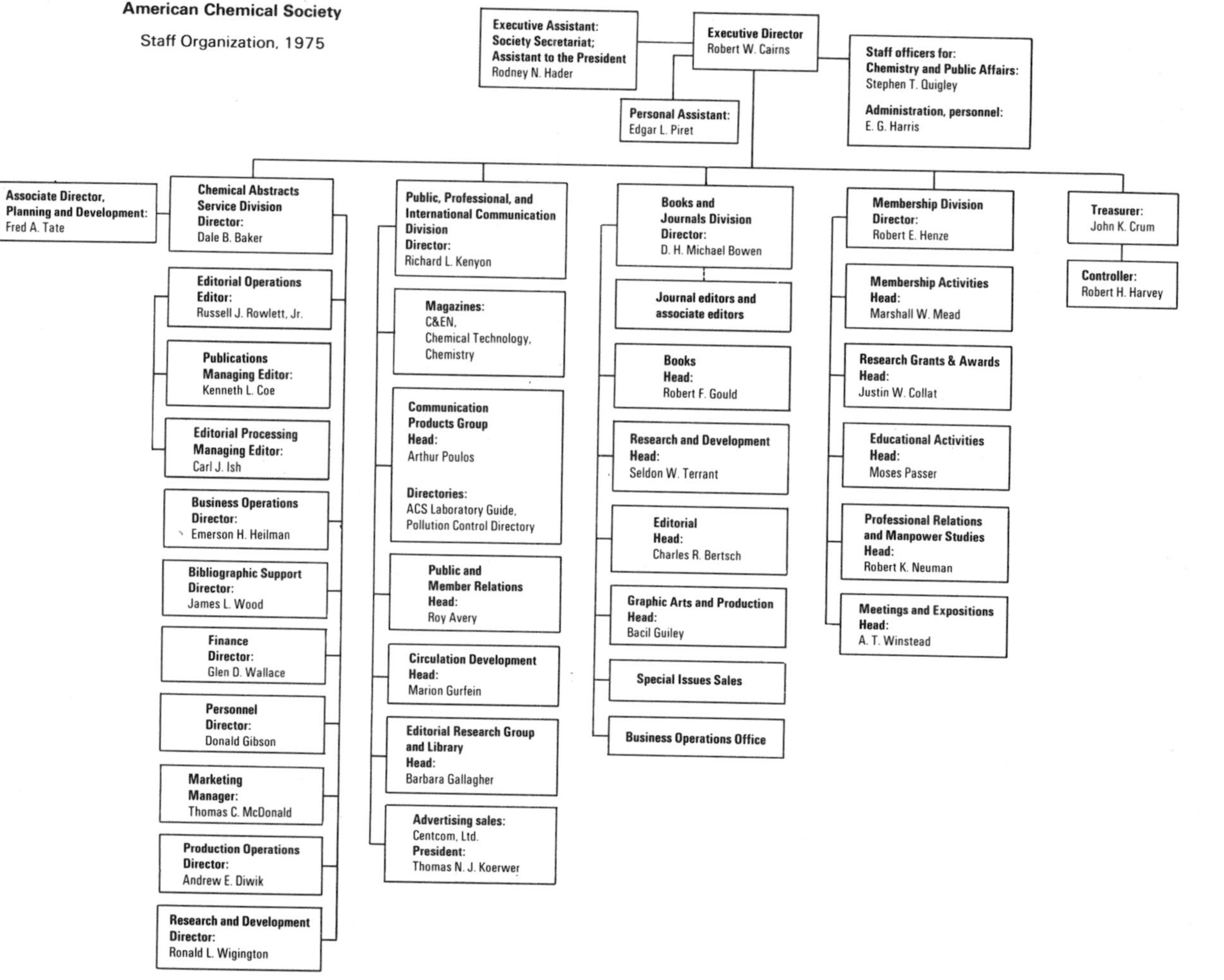
American Chemical Society
Staff Organization, 1975
Executive Assistant:
Society Secretariat;
Assistant to the President
Rodney N. Hader
Executive Director
Robert W. Cairns
Staff officers for:
Chemistry and Public Affairs:
Stephen T. Quigley
Administration, personnel:
E. G. Harris
Personal Assistant:
Edgar L. Piret
Associate Director,
Planning and Development:
Fred A. Tate
Chemical Abstracts
Service Division
Director:
Dale B. Baker
Editorial Operations
Editor:
Russell J. Rowlett, Jr.
Publications
Managing Editor:
Kenneth L. Coe
Editorial Processing
Managing Editor:
Carl J. Ish
Business Operations
Director:
Emerson H. Heilman
Bibliographic Support
Director:
James L. Wood
Finance
Director:
Glen D. Wallace
Personnel
Director:
Donald Gibson
Marketing
Manager:
Thomas C. McDonald
Production Operations
Director:
Andrew E. Diwik
Research and Development
Director:
Ronald L. Wigington
Public, Professional, and
International Communication
Division
Director:
Richard L. Kenyon
Magazines:
C&EN,
Chemical Technology,
Chemistry
Communication
Products Group
Head:
Arthur Poulos
Directories:
ACS Laboratory Guide,
Pollution Control Directory
Public and
Member Relations
Head:
Roy Avery
Circulation Development
Head:
Marion Gurfein
Editorial Research Group
and Library
Head:
Barbara Gallagher
Advertising sales:
Centcom, Ltd.
President:
Thomas N. J. Koerwer
Books and
Journals Division
Director:
D. H. Michael Bowen
Journal editors and
associate editors
Books
Head:
Robert F. Gould
Research and Development
Head:
Seldon W. Terrant
Editorial
Head:
Charles R. Bertsch
Graphic Arts and Production
Head:
Bacil Guiley
Special Issues Sales
Business Operations Office
Membership Division
Director:
Robert E. Henze
Membership Activities
Head:
Marshall W. Mead
Research Grants & Awards
Head:
Justin W. Collat
Educational Activities
Head:
Moses Passer
Professional Relations
and Manpower Studies
Head:
Robert K. Neuman
Meetings and Expositions
Head:
A. T. Winstead
Treasurer:
John K. Crum
Controller:
Robert H. Harvey

the financial operations, but made it possible to centralize all of the functions of the office of the Executive Secretary. Dr. Emery's title had been changed to Executive Secretary effective in 1948. By 1958, the headquarters staff had grown to 225, with a space requirement of 35,000 sq ft. To increase efficiency, the Secretary's office was reorganized into four administrative divisions: financial, business, membership affairs, and Petroleum Research Fund. It was also in 1958 that the headquarters staff moved to temporary quarters at 1801 K Street, N.W., while its new national headquarters was being erected.

In 1955, what had been the *Industrial & Engineering Chemistry* group of publications was formally given the title "Applied Journals" with a Director of Publications. In 1960 the business affairs of the fundamental journals were brought together under a Director of Planning. Then, in January 1965, the entire group of publications became the ACS Publications under a single Director, who answered to the Board of Directors.

On Feb. 1, 1960, the ACS headquarters staff, which had grown to 300, returned to 1155 16th St. to occupy some 45,000 sq ft of the 80,000 sq ft available in the new headquarters. The structure's total cost was very near the original estimate of $3 million. All this to serve a membership that had grown to more than 90,000 on annual expenditures of more than $8 million.

In July 1965, Executive Secretary Alden H. Emery was named Honorary Secretary after serving the Society for 29 years; he remained in the latter post for one year, retiring in July 1966. Dr. B. R. Stanerson, Deputy Secretary since 1960, was elected to succeed Dr. Emery as Executive Secretary. During Dr. Emery's more than 19 years as Secretary and then Executive Secretary, ACS membership more than doubled, from 43,000 to 100,000; annual expenditures rose from almost $1.7 million in 1946 to $15.3 million in 1965; local sections increased from 109 to 166 and divisions from 18 to 25; publications went from four to 18 (not including *Chemical Abstracts*).

It was also at this time that the Board of Directors voted that the Director of Chemical Abstracts Service and the Treasurer of the Society should report directly to the Board. Therefore, as of July 1, 1965, the heads of each of the major areas of ACS activity—Executive Secretary's office, finance, Publications, and CAS—were directly responsible to the Board.

Due to growing needs within the Society, the Board of Directors in 1969 named Dr. Frederick T. Wall as the first Executive Director of ACS. On Oct. 1, 1969, he took over as the full-time chief executive officer responsible directly to the Board. In turn, the directors of the four staff operations—CAS, Publications, Executive Secretary, and Treasurer—reported to the Executive Director rather than individually to the Board

***Dr. Frederick T. Wall,** first Executive Director, ACS, served in the post from October 1969 until September 1972.*

as they had been doing. Then, in 1970, Dr. Wall announced the restructuring and renaming of three of the four major staff units within the Society. The office of the Executive Secretary became the Membership Division, ACS Publications became the Publications Division, and Chemical Abstracts Service became the Chemical Abstracts Service Division. Also, a new staff position—Secretary of the Society—was created. A further reorganization was effected in March 1971. The Publications Division was divided: a Books and Journals Division took over the books and research journals; a newly created Division of Public Affairs and Communication took over Public Affairs, Public Relations, the magazines, and part of the publishing business operations. In 1975, the Division of Public Affairs and Communication became the Division of Public, Professional, and International Communication as the Society moved to strengthen its international activities. At that time the Department of Chemistry and Public Affairs was shifted into the office of the Executive Director.

Dr. Wall retired from his ACS post in September 1972. He was succeeded by Dr. Robert W. Cairns on Dec. 12, 1972.

CHAPTER X

ACS DIVISIONS AND THEIR DISCIPLINES

THE lifeblood of any science and technology is the knowledge and data that circulate and diffuse through the technical community. In the American Chemical Society, 27 divisions (Table I) bear the main burden of fostering face-to-face technical interchange among members and other scientists and engineers. A 28th division, the Division of Professional Relations, fosters interchange on topics such as the economic status of chemists, professional ethics, and the chemist's duties in regard to the public interest. Each division provides a forum for scientists and engineers with mutual interests and goals in a particular field of chemistry. The divisions discharge their duties primarily by arranging the technical programs for ACS national meetings and for meetings they hold independently or cooperatively with other scientific bodies. Papers from these meetings may appear later in one or another of the Society's publications, which are primarily responsible for printed technical interchange; five divisions preprint the papers presented at their sessions at the Society's national meetings.

TABLE I

The Divisions of the American Chemical Society

Name	*Founded*[a]	*Members July 1, 1975*	*Affiliates*[b] *July 1, 1975*
Agricultural and Food Chemistry	1908	1,245	191
Analytical Chemistry	1938	3,728	229
Biological Chemistry	1913	3,085	15
Carbohydrate Chemistry	1921	587	17
Cellulose, Paper and Textile Division	1922	589	48
Chemical Education, Inc.	1924	3,068	310
Chemical Information	1948	851	140
Chemical Marketing and Economics	1952	699[c]	173[c]
Colloid and Surface Chemistry	1926	1,157	80

TABLE I (Continued)

Name	Founded[a]	Members July 1, 1975	Affiliates[b] July 1, 1975
Computers in Chemistry (probationary)	1974	251	29
Environmental Chemistry	1915	1,475	83
Fertilizer and Soil Chemistry	1908	461	7
Fluorine Chemistry	1963	423	6
Fuel Chemistry	1925	590	5
History of Chemistry	1927	238	22
Industrial and Engineering Chemistry	1908	3,552	–
Inorganic Chemistry	1956	2,071	50
Medicinal Chemistry	1909	3,054	365
Microbial Chemistry and Technology	1961	545	84
Nuclear Chemistry and Technology	1963	781	90
Organic Chemistry	1908	4,840	–
Organic Coatings and Plastics Chemistry	1927	2,826	231
Pesticide Chemistry	1969	816	86
Petroleum Chemistry, Inc.	1922	1,811	108
Physical Chemistry	1908	2,070	33
Polymer Chemistry	1950	4,042	396
Professional Relations	1972	549	–
Rubber Division, Inc.	1919	2,372[c]	1,664[c]

[a] Year of founding, if later than 1948, marks beginning of probationary period; divisions were granted full status within three years.
[b] "Affiliates" are associated with the division, but are not members of the Society, either because they do not qualify for membership or for other reasons do not wish to join.
[c] As of Dec. 31, 1974.

The number of divisions in the Society has increased steadily over the years, a trend that reflects clearly the explosive growth of the past century in man's understanding and use of chemistry. Expanding knowledge perforce brought specialization. As a given subdiscipline of the science emerged and blossomed, its practitioners discerned at some point a need for specialized communication among their peers, the kind of need that ACS divisions were designed specifically to fill.

The orientation of individuals toward particular subdisciplines of chemistry began in the 19th century. An early indicator came in 1893 in Chicago at the World's Congress of Chemists, which was organized in part by the ACS. The 76 papers given at the meeting were grouped into nine sections, including analytical, inorganic, organic, and physical chemistry as well as bibliography (chemical literature). In June 1908 the Society formed its first division, the Division of Industrial Chemists and Chemical Engineers, now the Division of Industrial and Engineering Chemistry. By the end of that year the ACS had added four more divisions: Agricultural and Food Chemistry, Fertilizer Chemistry, Organic Chemistry, and Physical and Inorganic Chemistry. The number of divisions had grown to 18 by 1940 and to the present 28 by April 1974, when the probationary Division of Computers in Chemistry came into being.

To achieve and maintain permanent status under Society regulations, a division must, among other requirements, arrange at least one technical session annually. Most, but not all divisions schedule sessions at the two ACS national meetings held each year, in spring and fall; many hold additional, interim symposia or conferences. Divisions elect their own officers and may charge dues, publish newsletters and preprints of technical papers, and pursue additional activities subject to Society regulations. A division may establish subdivisions within its structure, and these commonly have been the precursors of new divisions. Members of divisions must be members of the Society, but an individual may become an affiliate of a division without joining the Society. This provision is designed to accommodate persons who have a professional interest in a division's field of chemistry but do not qualify for Society membership or for other reasons do not wish to join. Despite the degree of specialization in modern chemistry, two or more subdisciplines often will overlap to a significant degree. Accordingly, divisions of the Society regularly arrange joint symposia at national meetings on topics of mutual interest.

Although the existence and nature of ACS divisions has resulted largely from the manner in which chemistry has evolved during the past century, the stimulus has not been in one direction only. Over the years the divisions' technical activities have contributed markedly to the vitality of chemical science and technology in this country and, indeed, worldwide. It is thus very much to the point to pair these brief biographies of the divisions of the American Chemical Society with historical sketches of the fields of chemistry that occupy their energies.

AGRICULTURAL AND FOOD CHEMISTRY

The founding of the Division of Agricultural and Food Chemistry came in 1908, but interest had been apparent four years earlier when the Society began to group papers into topical sections. One of these was the Agricultural, Sanitary and Physiological Chemistry Section, later the Agricultural and Food Section and the immediate precursor of the division, one of the first five to be formed in the ACS. The division's first chairman was W. D. Bigelow, and the secretary was W. B. D. Penniman. The first program was held in 1909, and, although the first paper dealt with whiskey, the time nevertheless marked the beginning of an exciting era in food chemistry. In 1906 Hopkins proposed that scurvy was due to a nutritional deficiency, and in 1911 Funk coined the term "vitamine" to describe essential ingredients present in trace amounts in foods. Despite the heavy involvement of agricultural and food chemists in the search for the new nutritional substances, the division does not appear to have provided a forum for accounts of the pursuit. Programs continued to emphasize food analysis, preparation, and processing. Interest in the division dwindled until 1927, when a group of members made a special effort to reorganize it. A symposium on insecticides at the St. Louis ACS meeting in 1928 attracted much attention and new members, and the vitality of the division since has never been in doubt.

Programing in specialized areas has long been important to the division. Its first symposium, in 1922, was on edible oils and fats. The first vitamin symposium was held in 1935. This topic was to be repeated at least 20 times up to 1960, including five symposia on vitamin B_{12} between 1949 and 1951. Subdivisions also have played an important part in the division's activities; two of them evolved into the independent Divisions of Microbial Chemistry and Technology and of Pesticide Chemistry. Currently the division has a Flavor Subdivision, founded in 1965, and a Protein Subdivision, founded in 1970. To enhance communication among its members the division publishes the quarterly newsletter, *Cornucopia,* which first appeared in 1960.

Discussion of a given subject before the division often has presaged much wider interest. A paper in 1952 discussed "Analysis of Human Fat for DDT and Related Substances." Another, in 1958, examined "Control of Physiological Processes in Plants by Chemicals." The year 1960 saw symposia on "Carcinogenic Hazards of Food Additives" and "Production and Use of Enzymes in Agricultural and Food Processing"; each topic was later to attract widespread interest and concern.

WHEN the American Chemical Society was founded, in 1876, agricultural chemistry and food chemistry were barely discernible as areas of scientific study. A book on farming published 16 years earlier stated: "Scientific agriculture stands today with phrenology and biology and magnetism. No farmer ever yet received any benefit from any analysis of the soil and it is doubtful if anyone ever will."

Fertilizer had long been used, but Paris green had not been used to control the ruinous Colorado potato beetle until 1870, and Bordeaux mixture would not come along until 1885. The Connecticut legislature established the first agricultural experiment station in the western hemisphere only in 1875 as a result of the prodding of Samuel W. Johnson, professor of analytical chemistry at Yale and the acknowledged leader in agricultural science. The full effect of the Morrill Act of 1862, providing for land-grant colleges, was yet to be felt in 1876; the Hatch Act, setting up agricultural experiment stations in every state, would not be passed until 1887. The Babcock test for butterfat was introduced in 1890, and George Washington Carver did not begin his renowned work until 1896. Congress would not pass the Food and Drugs Act until 1906.

Studies of the composition of soils, manure, feed, and plants were under way by 1876, but we knew practically nothing of the essentials of nutrition. Scientists recognized the major classes of food constituents—proteins, fats, carbohydrates, and minerals—but they appreciated only slightly that these alone did not account fully for animal growth. Frankland and Atwater were emphasizing the concept of the caloric value of food, but 30 years would pass before F. G. Hopkins proposed that certain debilitating diseases might be due to nutritional deficiencies.

The existence of very low levels of certain essential substances in foods had become firmly enough fixed by 1911 that Casimir Funk coined the term "vitamine," later shortened to vitamin, to describe them. The role of the agricultural scientist in the pursuit of these elusive substances should not be underestimated. The discovery of most of the vitamins was sparked by nutritional research in agricultural laboratories. Progress was steady and sure. E. V. McCollum demonstrated the existence of vitamin A in 1913, and Steenbock discovered vitamin D in 1922. C. G. King characterized ascorbic acid in 1932, and R. R. Williams synthesized thiamine in 1937. In the latter year the preventive factor for the pellagra-like "black tongue" disease was identified as nicotinic acid by Elvehjem, Madden, Strong, and Wooley. In 1939 vitamin K was synthesized independently by Almquist and by Doisy, and in 1940 pantothenic acid was synthesized by Karl Folkers (ACS President in 1962). The essential nature of trace elements in nutrition has been recognized since 1928, and the nature of the essential fatty acids since 1932. Between 1932 and 1934, W. C. Rose and his co-workers discovered the amino acid threonine, and within a decade they had determined the approximate requirement of laboratory animals and man for the essential amino acids.

Fermentation processes were important even in ancient civilizations in producing foods and beverages and in utilizing agricultural crops. In 1906 the Free Industrial and Denatured Alcohol Act opened the way for industrial alcohol, and during World War I acetone and butanol were produced by fermentation of grain. During World War II fermentation technology was stimulated markedly as a result of its use in producing penicillin. The continued discovery and development of antibiotics ever since has done a great deal to put our knowledge and use of fermentation on a scientific basis. Antibiotics have proved important also in preventing disease in meat-producing animals and have become valuable feed additives in animal husbandry.

Chemists and chemical engineers contributed significantly to the progress in food technology that came during the 1940s. Wartime requirements spurred advances in dehydrated foods. The successful preparation of frozen orange juice and its commercialization after 1947 opened the door to the modern frozen food industry. Perhaps the most important factor in the almost complete elimination of nutritional deficiency diseases has been the advent of refrigeration, dehydration, and the other storage techniques that today give urban populations access to protective foods throughout the year. (Preservation of foods by radiation sterilization was investigated widely during the period just after World War II, but is not yet used generally.) Other notable developments in food technology in recent years have included noncaloric sweeteners, instant coffee, potato flakes, fish flour, and high-protein soft drinks and snack foods. Such items bring convenience or improved nutrition or both to the consumer; often they extend the usefulness of nutritious but otherwise unattractive food sources.

Biochemists have used radiotracers and other new investigative tools to improve sharply our understanding of the physiology of plants. Melvin Calvin (ACS President in 1971) and his co-workers have detailed the carbon pathways in photosynthesis; Daniel Arnon and others have clarified the energy pathways. Other studies have upgraded our knowledge of the chemistry of nitrogen fixation. Scientists have identified various plant hormones with the ability to control growth and development in plants. (The term "hormone" was used first in 1909 by Starling, an animal physiologist.) Uses in agriculture have followed the discovery of hormones such as auxin, the gibberellins, and kinetin and of the role of ethylene. Genetic variation of plants has been pursued and, with improved methods of analysis and better understanding of nutritional needs, has led to the development of higher-yielding, more nutritious cereal crops. E. Mertz and co-workers at Purdue University, for example, identified a type of corn, Opaque II, which is rich in the amino acid lysine.

Food chemistry and flavor chemistry continue to produce new information about the constituents of food. Because food and flavor chemistry use modern instrumental techniques of analysis to the fullest and often, as with chromatography, lead the way, they have been responsible for important progress in analytical chemistry in general. Flavor chemistry has contributed much to the acceptability of new foods and has achieved particular notice in the area of new sweeteners to replace sucrose.

Feeding the peoples of the world is one of the most important and perplexing problems facing humanity. The study of protein—its production, analysis, and use—currently is a key area in agricultural and food chemistry. New and attractive protein foods can be fabricated economically from soybeans and other diverse materials. Such foods undoubtedly will play a significant role in preserving human life.

The past decade or so has brought many questions on the impact of agricultural chemicals and food additives on our external and internal environments. Numerous studies are currently measuring the side effects of scientific agriculture on our ecology and health. Problem chemicals are being withdrawn from use or, if retained, must be used in a controlled manner. Agriculture-related pollutants, such as feedlot manure and food wastes, also can create serious difficulties. Processes are under development to convert such wastes into food and energy.

Research in agricultural and food chemistry always has been marked by close cooperation among governmental, academic, and industrial scientists. From such interactions have come the major accomplishments of yesterday, and from such scientists will come the research breakthroughs of tomorrow. Charles L. Flint has said:

> . . . the present is but the dawn of a new era—an era of improvements of which we cannot yet form an adequate conception. . . . [the improvements] show that a greater application of mind to the labors of the hand is to distinguish the future over all past generations, for the large number of young [people] who will go forth every year . . . , many of them thoroughly instructed in chemistry and kindred sciences, will give us, at least, the conditions for new discoveries which will open the way to higher triumphs, and so lead on to the golden age of American agriculture.

Mr. Flint, later president of the Massachusetts Agricultural College, was speaking in 1876 as the nation was preparing to celebrate its Centennial. That his words were prophetic for the 100 years now past is abundantly clear; agricultural and food chemists are in a sound position, scientifically, to make them prophetic for the next 100 years.

ANALYTICAL CHEMISTRY

The ACS Division of Analytical Chemistry was formed in the fall of 1940 in a culmination of events of the previous five years. In the fall of 1935, the section on analytical chemistry in the Division of Physical and Inorganic Chemistry had held a symposium on microchemistry, which comprised relatively new techniques that could cope with samples too small for the classical methods of analysis. In September 1937, a secton on microchemistry was formed; a year later this group became the Division of Microchemistry, with W. A. Kirner as chairman and L. T. Hallet as secretary-treasurer. In the fall of 1940, the analytical section of the Division of Physical and Inorganic Chemistry merged with the new division to form the Division of Analytical and Microchemistry; G. E. F. Lundell was elected chairman and F. W. Power secretary-treasurer. In 1949 the division assumed its present name, a change that reflected the rapid broadening of analytical chemistry to embrace a number of disciplines besides microchemistry.

In 1948, the Division of Analytical Chemistry initiated its annual summer symposia. These symposia, cosponsored by the ACS journal *Analytical Chemistry,* have continued without a break. In 1966, the division established its Summer Fellowship awards and in 1970 its Full-Year Fellowship awards. The awards go to full-time graduate students who are working toward a Ph.D. in analytical chemistry; they consist of $1000 summer fellowships and $5000 full-year fellowships and are supported by individuals and industrial and other organizations. Through 1974, the division had awarded 38 Summer Fellowships and 13 Full-Year Fellowships. The division's *Newsletter* for members is published four times yearly. In addition, the division is one of seven organizations cooperating in the Federation of Analytical Chemistry and Spectroscopy Societies, which held its first annual meeting in November 1974.

CHEMISTRY, like any physical science, relies strongly on measurement to obtain the data indispensable to understanding. It is thus not surprising that analytical chemistry was a leading field of scientific endeavor by 1876. A century earlier, in fact, Lavoisier's explanation of the process of combustion was based largely on his use of an analytical balance. The structure of chemical analysis in 1876, however, was essentially empirical rather than being based on true understanding. Most methods of the day were gravimetric. The analytical balance had reached a high degree of precision and reliability, and many chemists tended to rely on it instead of venturing into the volumetric techniques advocated by Mohr in his 1855 textbook. Their reluctance is perhaps understandable when one realizes that precalibrated glassware was not available. Reliance on gravimetry could lead to trouble, however, because of coprecipitation and other sources of error; and volumetric methods can save time, so the trend in this direction soon took hold.

Only a small number of reactions were available for use in titrimetry by 1876. The only acid-base indicator in use was litmus. Redox titrations were limited to self-indicating reagents like permanganate and iodine. Redox indicators were restricted to indigo, introduced by Gay-Lussac for hypochlorite determination, and ferricyanide and iodide-starch, the latter two used only as external indicators. The first synthetic indicator, phenolphthalein, appeared in 1887 and was followed soon by tropeolin and methyl orange. By 1893 some 14 synthetic acid-base indicators were known, but not until 1925 was a synthetic redox indicator (diphenylamine) introduced. Fluorescein had been proposed as a fluorescent indicator as early as 1876, but it was too far ahead of its time and was not a success.

About a dozen color-forming reactions were in use in 1876, according to M. G. Mellon, and the number would double by the turn of the century. Chemists still use some of these reactions, including the determination of Fe^{+3} by KSCN and of NH_4^+ by Nessler's reagent. Observations were made simply by comparing hand-held test tubes or, for greater precision, by using Nessler tubes. The Duboscq (color) comparator had been invented only in 1870; this important instrument was so well conceived that it is used today to some extent.

Chemists of the period relied on the classical or "wet-chemical" methods of analysis that, in refined form, are still very much alive. By 1876, however, they could discern the possibility of exploiting other methods for analytical purposes. The spectroscope of Bunsen and Kirchhoff, introduced in 1860, permitted a significant improvement in the qualitative analysis of the readily excited metals. The use of electrochemical techniques had been suggested in 1867 for evaluating copper ores. These developments presaged the strong trend toward instrumental methods of analysis that began to emerge around 1930 and is still accelerating.

1876–1900

During the last quarter of the 19th century, to a large extent, order was brought into chaos in analytical chemistry. The trend was spurred no doubt by the introduction of the periodic table, in 1869, but it followed mainly from the work of the great physicochemical theoreticians, J. Willard Gibbs, van't Hoff, Arrhenius, Ostwald, and Nernst. Previous books on analytical chemistry (Rose, 1829; Fresenius, 1841; Mohr, 1855) were essentially "cookbooks," useful in passing on the arts learned over the years. But now a new kind of book appeared, in which the theory of analytical procedures was for the first time developed and used to improve methodology. Such books were exemplified by Ostwald's "Die wissenschaftlichen Grundlagen der analytischen Chemie" (1894). This lucid treatise explained such matters as the functioning of acid-base indicators in terms of ionic and molecular equilibria. Ostwald's treatment was perhaps oversimplified in the light of later developments, but at the time it exerted a great unifying force. Analytical chemistry, meanwhile, was served by two journals, Fresenius' *Zeitschrift für analytischen Chemie* (started in 1862) and *The Analyst* (1875), a British publication.

Among the significant general innovations during 1876–1900 were the Gooch crucible (1878); the first specific organic reagent (mixed α-naphthylamine and sulfanilic acid), a test for the nitrite ion (1879); the Kjeldahl procedure for nitrogen (1883); the Winkler reaction for dissolved oxygen (1888); and the Jones reductor (1889). The first mention of the use of a centrifuge to separate precipitates appeared in 1894 (Behrens), and in 1896 Harm reported using ion exchange to separate sodium and potassium from sugar-beet residues.

Spectroscopy provided a number of developments. A rapid series of improvements in visual comparators and photometers culminated in the Pulfrich photometer (1894) and the photometers based on the attenuation of polarized light. The latter were invented independently by Glan and by Hüfner (1877) and improved by Krüss (1895). The König–Martens visible spectrophotometer, which appeared in 1894, was of this type; it was so well designed and constructed that a König–Martens instrument was in use at the National Bureau of Standards for high-precision measurements at least as late as 1949.

In other branches of optics, work of the highest importance was in progress. In 1880, S. P. Langley introduced photographic plates into spectroscopy. This innovation freed the spectroscopist for the first time from the difficult, tedious, and too-subjective task of observing spectra visually. The following year, Langley described a sensitive bolometer that permitted him to study spectra well into the infrared region; by 1890 he reported measurements as far into the region as 5.5 μm. Prof. Rowland of the Johns Hopkins University physics department constructed a highly sophisticated engine for ruling diffraction gratings. He became the world's principal producer of gratings from 1893 until his death, in 1901. Rowland's work effectively established the grating as the primary optical element of spectroscopic technology.

The progress of electroanalytical techniques in the last quarter of the 19th century was not so spectacular. Electrogravimetric analysis was the first to grow, largely through the influence of Classen's 1882 book on the subject. Freudenberg (1891) and others used the method to separate metals through control of potentials, an endeavor facilitated by LeBlanc's introduction of the hydrogen electrode in 1893. Both conductometric and potentiometric titrations were described in 1893. The first electroanalytical work in the United States apparently was done by Edgar Fahs Smith and reported in 1890.

1900–1930

Theoretical understanding of the various types of titrimetric analysis expanded dramatically in the early decades of the 20th century. Salm (1907) explained the distinction between the neutral point and the equivalence point in acid-base reactions and showed how to select appropriate indicators. This and similar calculations were greatly simplified by Sørenson's introduction of the pH concept (1909).

Many practical innovations appeared at about this time. Vorländer (1903) reported the first use of a nonaqueous solvent in titration. Mulliken's text on the classification of organic compounds for qualitative identification (1904) profoundly influenced the future of that branch of analysis. The year 1905 saw the characterization of four samples of cast iron at the National Bureau of Standards, the first work there on a series of certified reference materials. In 1918, Feigl began his work on analysis by means of spot tests, work that culminated eventually in a book (1954).

A number of firms on both sides of the Atlantic developed large spectrographs during the period 1900–30. These included Adam Hilger, Ltd., and Kynoch, Ltd., in Great Britain and Gaertner and Bausch & Lomb in the U.S. Qualitative identification by spectroscopy soon far exceeded the capability of the technique in quantitative work (by intensity measurement). Thus the famous spectroscopist H. Kayser wrote in 1910: "Summarizing the results of all the experiments carried out, I conclude that quantitative spectral analysis is impracticable." Nevertheless, ultraviolet-visible spectrophotometers were available as early as 1910 in the form of a rotating sector attachment for a spectrograph; valid absorption spectra could be obtained, albeit at the cost of considerable time and patience.

C. V. Raman, in 1928, reported his observation of a weak scattering of light by certain substances, accompanied by a change in the wavelength. The effect had been predicted by Smekal five years earlier, but Raman apparently was unaware of this. Raman spectroscopy soon became useful in research, but the effect was too weak to become really successful as an analytical tool until the advent of laser excitation in the 1960s.

The year 1923 saw the publication of two parallel books important in electroanalytical chemistry: E. Muller on potentiometric titrations and I. M. Kolthoff on conductometric titrations. Biamperometric titration was originated as far back as 1897 by Salomon; the procedure was clarified and corrected by Nernst and Merriam (1905) and then forgotten, only to be rediscovered by Foulk and Bawden in 1926.

The two most important developments of the period in electroanalytical chemistry were the discovery of the pH-sensitivity of a glass membrane and the invention of the polarograph. Cremer in 1906 observed a spontaneous difference in potential across a thin glass membrane separating solutions identical except for acidity. Three years later came a report by Haber and Klemensiewicz, but nothing more was heard on the subject until 1929, when McInnes and Dole showed that a practical glass electrode for pH measurement was a distinct possibility. The high electrical resistance of the glass membrane made accurate measurement of potential difficult, so further work had to await the development of suitable electronics. In the interim, scientists investigated the measurement of pH by other means, particularly the quinhydrone electrode (Biilmann, 1920) and the antimony electrode (Uhl and Kestrauck, 1923).

The first paper on polarography, published by Heyrovský in 1922, dealt with a modified application of the Lippmann capillary electrometer. Investigation of certain anomalies in the response of the electrometer led to a remarkably useful correlation between the current-voltage curves and the concentration of reducible species in the electrolyte. Recording and plotting the data was a considerable task, so an instrument was designed (1925) to perform both functions simultaneously. This was the earliest analytical instrument with its own built-in recording device.

Nowhere was the parallel development of theory and instrumentation more in evidence during the first quarter of the century than in nuclear chemistry. Largely under the leadership of Ernest Rutherford, the theory of nuclear transformations was worked out; at the same time a series of new instruments was designed for measuring the radiations. The Geiger counter (1908) and spinthariscope (Crookes, 1903) were useful immediately. The latter soon relinquished its popularity, only to regain it in greater measure in the form of the scintillation counter when electronic detection became available (1948).

The first unique application of nuclear methods to analysis was made by Hevesy and Paneth (1913) with the introduction of tracer techniques. Tracer methods were used later by Ehrenberg (1925) to follow the precipitation of $PbCrO_4$. Significant also in respect to atomic theory was the mass spectrometer. The first machine to merit the name was constructed by J. J. Thomson (1910), who used it to observe the first direct evidence of the existence of an isotope, neon-22. Aston (1919) improved on Thomson's design and by 1924 had determined isotopic ratios in about 50 elements.

1930–1960

The year 1930 marked the first significant use of electronic techniques in analytical instrumentation. Photoelectric devices had been tried sporadically, but the first successful one was the recording spectrophotometer invented by Hardy at the Massachusetts Institute of Technology (1930) and later manufactured by the General Electric Co. This instrument, like Heyrovský's Polarograph, had its own recording mechanism linked directly to the scan drive. The pen motor was controlled by a thyratron operated by the detecting photocell; the motor not only drove the pen, but also a null-restoring polarizer in the Hüfner photometer.

The success of the Hardy-GE spectrophotometer encouraged other manufacturers to enter the field. Central Scientific Co. used a photovoltaic (barrier-layer) photocell with an Eagle-mounted concave grating in its manually operated "Spectrophotelometer" (1941). In the same year, National Technical Laboratories (later Beckman Instruments) produced the first photoelectric spectrophotometer to extend into the ultraviolet. This was the famous Beckman Model DU, which soon became the workhorse of spectrophotometry laboratories in this country and abroad. The DU had no recording capability, but the null principle used in its electronic design made it nearly independent of the idiosyncrasies of vacuum tubes and other components, so it was capable of high photometric precision. The DU was followed (1947) by the Uvispek of Adam Hilger, Ltd., which had about the same specifications, and by the first of the Cary recording spectrophotometers (Applied Physics Corp., 1946).

In 1951 the American Optical Co. produced a rapid-scanning spectrophotometer in which the absorption spectrum of a sample was scanned 60 times per second and displayed on the screen of a cathode-ray oscilloscope. Unfortunately the device was ahead of its time and had to be withdrawn from the market. Twenty years later an instrument very similar in concept, but using far more sophisticated electronics, was manufactured by Tektronix, Inc.

In electrochemistry, the first impact of electronics was seen in a glass-electrode pH meter (Beckman, 1935), which soon found its way into all sorts of laboratories. Other electronic developments of the time consisted principally of improving the sensitivity and convenience of existing instruments or procedures. Thus a number of automatic titrators appeared, as did a portable conductivity apparatus.

The potentiometric servo-actuated, strip-chart recorder was a most significant advance, felt in nearly all branches of instrumentation. Leeds & Northrup Co. pioneered in this field in 1932 with its first electronic recorder.

Apart from electronics, the major development in analytical chemistry was the rapid evolution of chromatography during the 1940s and 1950s. The trend started in 1941 when Martin and Synge described partition chromatography with a liquid carrier. Three years later, Consden, Gordon, and Martin extended the method to utilize a paper support. In 1952, James and Martin reported the use of a gas as carrier. The first commercial gas chromatograph appeared in 1955.

Other developments included the thermobalance (Duval, 1953) and the derivative thermobalance (Erdey, Paulik, and Paulik, 1954) which opened up an important new area. One of the most far-reaching non-instrumental discoveries was the chelating power of ethylene diamine tetraacetic acid (EDTA) and similar compounds, as first reported by Schwartzenbach (1946) and in the treatise by Welcher (1958).

Since 1960

The "electronics revolution" that began in analytical chemistry in 1930 was followed by a second one in about 1960, when active devices made from semiconducting materials such as germanium and silicon began to prove advantageous. The new "solid-state" units used power and space far more efficiently than their vacuum-tube counterparts, and they could be produced at much lower cost as well. At first, transistors merely replaced vacuum tubes, with the minimum necessary changes in circuitry, a process called 'transistorization.' Soon, however, entirely new and improved circuitry was devised to utilize the active devices to the best advantage. These advances made it more reasonable to expand the data-processing and control functions within instruments, on an automatic or semiautomatic basis. Thus a mass spectrometer built in the 1960s or 1970s typically includes several racks of electronics, with hundreds of transistors.

In the end uses of analytical instrumentation, one can discern several design trends, some of which seem mutually exclusive. On the one hand are many special-purpose instruments such as an NMR spectrometer that can be used only to measure moisture, or a titrator designed specifically for the Karl Fischer titration. On the other hand are the large multipurpose instruments such as the electroanalytical instrument of Princeton Applied Research Corp., which can be operated in many modes to accommodate nearly all known electro techniques. Another trend is toward modularization, which has two major advantages: the customer need invest only in the modules applicable to his special needs, and servicing and repair are greatly facilitated.

A very important nonelectronic development has been the invention of the laser (1960). Within a decade, a gas laser as exciting source has revolutionized Raman spectroscopy. Lasers also can be used for intense heating of a very small area of the surface of a sample, which has been a need in mass spectroscopy and other analytical techniques.

The Future

As the ACS enters its centennial year, what can we look forward to in analytical chemistry? The second electronics revolution is continuing, with the introduction of medium- and large-scale integrated circuits. These circuits incorporate many hundreds or thousands of transistors and other semiconductor components on a single chip of silicon. Thus a small number—say one to six—of such chips, combined in a so-called "microprocessor," will constitute all the electronics needed for handling data in a major instrument. Clearly, developments along such lines will figure prominently in the future of analytical chemistry.

Another possibility lies in adapting the laser, especially the dye laser, to optical instruments. It will be possible, for example, to build a tunable dye laser such that a wide spectral range can be scanned by changing the frequency at which the laser resonates. This will eliminate the exceedingly inefficient combination of a continuous light source and a high-resolution monochromator.

In 1962, Liebhafsky commented, "Like it or not, the chemistry is going out of analytical chemistry." Presumably he meant that a valid analysis frequently can be obtained by a trained operator manipulating the controls of an automated instrument and knowing nothing of the chemical processes involved. It is clear on the other hand that the tremendous advances in analytical instrumentation are greatly improving our chances of finding out what is involved in innumerable chemical processes, both animate and inanimate. At the same time, the classical or noninstrumental aspects of analysis involve unsolved, fundamental problems whose solutions will benefit both instrumental analysis and chemistry as a whole.

Without doubt it can be said that the need for reliable methods of chemical analysis—often at levels of a few parts per million and even lower—in many cases has outrun the state of the art. The need is pressing now and will become more pressing in the future across the full spectrum of human activity—in medicine, in the environment, in industry, and in many other areas.

BIOLOGICAL CHEMISTRY

The history of the ACS Division of Biological Chemistry can be divided into three periods: from its founding in 1913 to World War II; the "Brand" years, including those years when Dr. Brand's immediate associates followed him and continued his leadership until the middle 1950s; and the 20 years up to 1976. Of special note is the interplay of the division with the American Society of Biological Chemists, founded seven years earlier, in 1906. These two major societies representing biochemists in the United States have worked hand in hand to further the interests of biochemistry not only in this country but throughout the rest of the world.

The events leading up to the formation of the Division of Biological Chemistry (DOB), in 1913, are vague. The first chairman, Dr. Carl L. Alsberg, a Columbia University M.D., had two years' training in biochemistry at the Universities of Strassburg and Berlin. Alsberg also was a founding member of the American Society of Biological Chemists (ASBC). After serving four years as DOB chairman, he became president of the ASBC in 1917. This set a precedent that has been followed throughout the history of the division. Biochemists of prominence, such as R. J. Anderson, H. B. Lewis, C. G. King, J. T. Edsall, E. E. Snell, H. Lardy, P. Boyer, John Buchanan, F. Putnam, Alton Meister, and D. E. Koshland, were chairmen of the division and also presidents or secretaries of the sister society. Many were active in other capacities as well, thus offering biochemistry a most effective forum for communications. DOB and ASBC have had a consistent one-third overlap in membership, which overlap in 1975 totaled about 1000.

Before World War II, it is unlikely that the division ever had more than 200 members, and interest in biochemistry in the American Chemical Society was restricted to the more chemical and physical rather than the biological aspects of the field. But in 1941, a dominant figure and personality appeared; Dr. Erwin Brand took over the leadership, and interest in the division suddenly increased. Dr. John T. Edsall, a close friend and associate, outlined Dr. Brand's accomplishments and his contributions to the division in an address at the Fiftieth Anniversary Banquet of the division in 1963.

Following Brand's death, in 1953, the division established the Erwin Brand Travel Award, which by 1975 was still being supported by contributions of $1.00 from the annual dues. This fund is used to defray the expenses of travel by young investigators to international biochemical congresses. A travel committee of three, appointed by the chairman, meets with a similar committee of the ASBC, which also has a travel fund. This combined Travel Award Committee accepts applications from all biochemists and makes the awards. As much as $10,000 was distributed by the division in 1973, when more than 30 young biochemists received travel support from the Brand Fund.

John T. Edsall succeeded Brand as secretary of the division in 1948 and was chairman in 1948, 1949, and 1951. He was succeeded as chairman in 1952 by Richard H. Barnes, who not only served two terms, but later, as a member of the ACS Council Committee on Publications, played an important role in establishing the new journal, *Biochemistry*.

The division's membership had grown from 263 in 1941 to 1785 by 1955. The secretary's job of recordkeeper, program chairman, and treasurer had become too much for one person; with Dr. Carl Baker's term, a treasurer was included, the first being Mr. Robert Harte. It was the custom of the division to select secretaries and treasurers who had available without cost the clerical, mailing, and stenographic services necessary for the job. The team of Baker and Harte was succeeded by Julius Schultz, a professor of biochemistry at Hahnemann Medical College in Philadelphia, and Dr. Art Heming, department chairman at Smith, Kline & French Co. By 1975, Dr. Schultz had been a member of the division's executive committee and, alternately, secretary or program chairman for 20 years.

After World War II, biochemists began to press for their own international union. The course of this development, in which Erwin Brand played an important role, has been described in a statement by Dr. R. Barnes:

> A peppery, aggressive and fascinating personality, Erwin Brand was the driving force that changed the Division from primarily a forum for presenting scientific papers to an influential force in the development of organized biochemistry internationally as well as nationally. The first International Congress of Biochemistry was held at Cambridge, England in August, 1949. On that occasion, a group of twelve biochemists from eight countries established an International Committee for Biochemistry under the chairmanship of Sir Charles Harington. The British Biochemical Society had polled biochemists around the world and concluded that there was strong feeling for the establishment of an International Union of Biochemistry.

International biochemistry was already represented in a section of biochemistry within the International Union of Pure and Applied Chemistry (IUPAC) and influential biochemists in the U.S. were supporting the continued representation within IUPAC and opposed the establishment of an independent I.U.B. Representation on the Harington Committee from the U.S. included appointees from both the ACS Division of Biological Chemistry and the American Society of Biological Chemists. The same held true for representation in the Section on Biochemistry of IUPAC. Controversy concerning the merits of one international union representing biochemistry versus two continued, but finally through the efforts of the Harington Committee ICSU [International Council of Scientific Unions] approved the establishment of the independent IUB, and country membership and mechanisms of adherence to the new Union were initiated in 1953.

In the U.S., adherence to IUB was established through the National Academy of Sciences and the National Committee was set up so that representation of both the ACS Division of Biological Chemistry and the American Society of Biological Chemists would be maintained. Thus internationally as well as in the U.S., biochemistry found two avenues through which it functions; one through affiliation with Chemistry (IUPAC and ACS) and the other independently (IUB and ASBC). The dual representation internationally apparently has not been disorienting as had been predicted by a number of U.S. biochemists. Certainly, in this country, the dual representation of biochemists has been beneficial in several respects.

Throughout the critical period in the development of IUB, Erwin Brand served biochemistry well. His political savvy and aggressive personality were of extreme importance in aligning U.S. biochemists with the establishment of an independent union, and after this was accomplished he played an important role in assuring representation of both the ACS Division and ASBC in the U.S. National Committee on Biochemistry.

Robert Harte writes of those times that "in the forefront at the time of organizing the IUB were people like Severo Ochoa, Elmer Stotz, John Edsall, Erwin Brand, Jesse Greenstein, Dick Barnes, and Carl Baker, and I recall many Saturdays and Sundays spent at the Academy building where pros and cons were argued."

Another great event was the founding of the journal *Biochemistry*. As a member of the ACS Council Committee on Publications, Dr. Barnes brought to the division's executive committee the information that the ACS was interested in a journal of biochemistry. This lay dormant, especially since the other executive committee members close to the ASBC were not sure of the reaction of that society to competition with the highly popular *Journal of Biological Chemistry,* the only society publication in the field in the U.S. The other journal in biochemistry was the *Archives of Biochemistry,* published by Academic Press. Action came when Dr. E. Stadman of the National Institutes of Health, a member of the executive committee, was appointed chairman of the committee to poll the membership on the feasibility of and the need for a new journal. A complicated questionnaire designed by the committee was returned by more than 1000 members: an overwhelming majority favored another journal. With this information, the ACS asked the division's chairman, Paul Boyer, followed by Dr. D. E. Green, to participate in selecting an editor. Dr. Boyer and Dr. Jack Buchanan visited Prof. Hammet of Boston and with good fortune succeeded in finally selecting Dr. Hans Neurath as Editor. Although the division has no responsibility for the publication, Dr. Neurath presents a report on *Biochemistry* at each meeting of the executive committee. He became chairman of the division in 1961–62.

The great surge of federal support for research brought the rumblings of the revolution in biology in the 1950s; in the 1960s this became a full-fledged revolt that made biochemistry one of the glamor sciences in the popular press. It was strongly reflected in the output of papers and presentations at meetings, and the division's membership reached 3000 before the end of the 1960s. Many recipients of the ACS awards in biochemistry and enzyme chemistry, the Eli Lilly and Paul-Lewis awards, presented their original work at symposia organized by the division. During Dr. Max Lauffer's chairmanship in 1950 a policy was established that the Eli Lilly and Paul-Lewis awardees present addresses on their fields of activity. This policy remained unchanged in 1975. Many of the awardees and others who participated in the symposia, such as S. Ochoa, K. Block, B. Stein, S. Moore, F. Lipman, V. DuVigneaud, C. Anfinsen, M. Nirenberg, and J. D. Watson, went on to win the Nobel prize.

These programs brought great crowds and boosted the membership, but they created a new problem. The awards symposia were held in the spring, a week before the meeting of the Federation of American Societies for Experimental Biology, at which ASBC met. Dr. F. Putnam, secretary of ASBC, asked the DOB executive committee if the division could restrict its meetings to the fall and drop the meeting in the spring. The ASBC then would consider the DOB's fall meeting as its own. The straw that broke the camel's back was the ACS Denver meeting in the spring of 1964 where Nirenberg gave his award address with Ochoa as chairman of the symposium. Alton Meister, D. Koshland, S. Moore, W. Stein, and Fritz Lipman were all participants as chairmen and speakers. It was pointed out that such an assembly of biochemists should be heard by much greater numbers than attended the spring meeting; the fall meeting offered an opportunity for much greater crowds. After an exchange of mutual meetings of the ASBC council and the DOB executive committee, the division decided to drop its spring meeting.

Public awareness became a byword in scientific societies as the Government's need for technical knowledge grew along with the dollar-volume of appropriations affecting both science and society. By the 1970s, many biochemists had advanced to positions of political influence. Dr. William McElroy of Johns Hopkins had become Director of the National Science Foundation; Dr. Philip Handler of Duke, who had been Chairman of the National Science Board, became President of the National Academy of Sciences; Dr. H. E. Carter of Illinois became Chairman of the National Science Board; Dr. Carl Baker was appointed Director of the National Cancer Institute. In 1971, six months before Congress began hearings on the new multimillion dollar National Cancer Attack Program, a symposium was organized to take place at the national meeting in Washington on the Role of Basic Research on the Current Status of the Cancer Problem. The subject was important, because legislation was aimed principally at clinical treatment rather than research. Dr. F. Putnam, Dr. E. Frei, now Director of the Sidney Farber Cancer Institute in Boston, and Dr. Emmanuel Farber, President of the Association for Cancer Research, participated in the program. Following the symposium, William McElroy, Philip Handler, and Carl Baker were to hold a panel discussion. Congressman Paul Rogers, who had planned to attend, could not because he had opened his hearings on the same day, but Congressman Claude Pepper did appear and offered his support. More than 1000 members crowded the room.

Because biochemists in Latin America lacked adequate communications with their North American colleagues, the Pan American Association of Biochemical Societies (PAABS) was founded in 1970. Among the organizers of this group were B. Horecker, a member of the division's executive committee for several years and a former president of ASBC, Dr. P. P. Cohen, also a former member of the executive committee, and Dr. W. J. Whelan. The first president of PAABS was Nobel laureate and biochemist L. Leloire; the second was B. Horecker. William Whelan became Secretary-Treasurer, followed by R. Estabrook; currently the Secretary-General is former DOB chairman F. Putnam.

CARBOHYDRATE CHEMISTRY

The Division of Carbohydrate Chemistry was an outgrowth of a Section of Sugar Chemistry formed within the ACS in 1919. The interest in forming the section had been provided by an article published by C. A. Browne in *Industrial & Engineering Chemistry*. In it, he urged American companies to manufacture polariscopes—important to sugar chemists—

and other optical instruments in order to relieve shortages created during World War I and to forestall similar shortages in the future. Browne and other sugar chemists, in discussing the polariscope problem, conceived a plan for a permanent organization devoted to sugar chemistry and technology. When it developed that those who would form the nucleus of the group were already members of the Society, they sought and received permission to form the Section of Sugar Chemistry. Two years later, in 1921, the section became the Division of Sugar Chemistry and Technology, with S. J. Osborn as chairman and F. J. Bates as secretary. Over the years, the division's interests broadened gradually from sugar to carbohydrates in general, and in 1952 the name was changed to the Division of Carbohydrate Chemistry.

CELLULOSE, PAPER AND TEXTILE

The Cellulose, Paper and Textile Division, established in 1922, grew out of the Cellulose Chemistry Section formed in 1920 within the ACS Division of Industrial and Engineering Chemistry. The first chairman of the division was G. J. Esselen, and the Secretary-Treasurer was L. F. Hawley. Since 1923, divisional programs have included many papers on the structure and composition of wood, lignin, extractives, and related topics. Because "cellulose chemistry" did not fully reflect the division's interests, the name was changed in 1961 to Cellulose, Wood and Fiber Chemistry. In 1973, to recognize the still-changing nature of the field, the division assumed its present name.

In 1962, with financial donations from industry, the division established the Anselme Payen Award. This annual award consists of a scroll, a plaque, and an honorarium and can be given to any scientist worldwide for outstanding research in the division's fields of interest (before 1973, only residents of North America were eligible for the award).

THE word "cellulose" was coined just 37 years before the American Chemical Society was formed. It appeared first in 1839 in the French Academy's assessment of Anselme Payen's research on "ligneous matter," the then-current term for the combination of lignin and cellulose that forms the woody cell walls of trees and other plants. In 1838 Payen reported that ligneous matter, then considered a single substance, contained two chemically distinct materials. One of these, the French chemist declared, had the same chemical composition as starch, but differed in structure and properties. Whether Payen himself named this substance "cellulose" is uncertain.

Cellulose technology advanced steadily in the 19th century. The Fourdrinier papermaking machine was introduced in France and England around 1800. These machines, and the expanding interest in newspapers, bred a serious shortage of rags, then the major source of cellulose pulp for papermaking. The response was a marked effort to obtain cellulose from straw, grass, and wood. Mechanical pulping, in limited use by about 1850, has become the source of most of the world's newsprint today. Between about 1850 and 1885 came the soda, sulfite, and sulfate processes for pulping wood chemically. A major economic development since 1950 has been the perfection and widespread use of semichemical pulping, conceived initially in the late 1920s by scientists at the U.S. Forest Products Laboratory in Madison, Wis. Semichemical pulping converts wood into fiber with a higher yield than does chemical pulping, and it produces fibers that are stronger than those from mechanical pulping. It has become the leading process for making the center ply of the corrugated board used in shipping containers, the largest consumers of pulp in the world today.

During the past 35 years, the soda process for pulping wood has declined in importance, but it is still used to purify cotton linters. The sulfite process has held its position because of improvements such as replacing the calcium-based liquor with soluble bases. Now, sodium, ammonium, or magnesium liquors permit pulping chemicals to be recovered and by-products such as lignin to be converted into energy (steam) by burning and thus solve the major pollution problem of the sulfite process. These developments were pioneered by George Tomlinson and others in Canada. Despite these improvements, the more efficient kraft (sulfate) pulping process, which can pulp all species of wood, has become the dominant chemical process; it has had a major economic impact on much of the South, where fast-growing Southern pines provide much of the world's fibermaking raw material. Two developments in the early 1940s contributed heavily to the success of the kraft process. The first was a relatively simple and economical method for bleaching the product to a very high level of stable whiteness. The second was the discovery that a simple process step (prehydrolysis) would provide pulp that, when fully bleached, was suitable for making cellulose derivatives.

Before the use of chlorine dioxide became routine, only sulfite pulps were considered "bleachable." The use of this chemical resulted from research to circumvent two problems that threatened the existence of a dissolving pulp mill in Canada. They were a combination of wartime restrictions on the use of chlorine and the need to pulp hardwoods because the supply of the preferred softwood was almost exhausted. W. Howard Rapson and Morris Wayman, working with the chemists and engineers of the Canadian International Paper Co. research center at Hawkesbury, Ont., devised a practical system for producing on-site the chlorine dioxide that solved both problems. Since then, commercial processes for producing and using chlorine dioxide have been devised by Frank Dole of the Mathieson Chemical Co.; T. G. Holst; S. H. Persson; J. Schuber and W. A. Kraske (Solvay Process); G. A. Day and E. F. Fenn

(Brown Co.); and E. Kesting. The most modern systems are based on the later work by Rapson and his co-workers at the University of Toronto; these systems are the result of design studies and equipment development by Hooker Chemical Co. and by Erco Industries, Ltd.

The paper products available in 1975 differ markedly from those of 25 years ago. In the paper industry, the thrust of process development has been to make products faster and more uniform. Engineers have had a major role in this effort. The thrust of the chemists' work has been to control the properties of paper to make it more versatile. Tensile strength, tear strength, and burst strength, both wet and dry, have been improved; totally new forms include flame retardant papers, dimensionally stable papers, stiffer papers, and softer papers. As a result, the use of paper has increased tremendously: in 1950 this country consumed 29 million tons of paper; in 1975 it was expected to consume 68 million tons.

Paper is prepared basically by forming a sheet from a slurry of pulp fibers in water, dewatering, pressing, and drying. On the surface the process seems simple, but in fact it is complex and still not completely understood. Still, our knowledge of papermaking has increased tremendously during the past 25 years. As a result, we now have newsprint machines 335″ wide that operate at 3300 ft/min. We also have linerboard machines that produce 1500 tons of paper daily.

Very few papers are produced only from wood fibers and water. Many additives and coatings have been developed that have had a major impact on the paper industry. In the late 1950s, wet-strength resins were developed that double or triple the strength of paper when wet. Since then a wide range of other wet-end additives has appeared; they include dry-strength resins based on starch or acrylamide and retention aids, for fines and pigments, based on polyacrylamide.

The introduction of roll coating about 1950 permitted high-speed coating of paper. High-speed coaters, however, created rheological problems so that existing formulations became unsatisfactory. The problem was solved by developing starch and starch derivatives as binders, as well as more finely divided pigments. These also resulted in smoother, brighter coatings that improved appearance and printability. Not all coating is done for aesthetic or graphic purposes. Much of it is done to improve the functional properties of paper, usually to increase its barrier or adhesive properties. In most end uses of this nature, the strength of paper is combined with the barrier or adhesive properties of the added plastic material.

Basic studies on fibers, their behavior in water, their reaction to heat and pressure, and the mechanism of bonding have combined in the past 30 years with pulping and bleaching developments to give us the huge assortment of paper we use in the mid-1970s. Some of the major and highly technological types are communication papers (printing, computer, Xerox); packaging products (boxes, bags, combinations with plastics); building and construction materials; health and hygiene papers; and nonwoven fabric. No single dramatic event revolutionized the paper industry; progress has resulted from many, many small developments: improvements in understanding, in printing, in binders, and in plastics.

Advances in papermaking were paralleled by progress in cellulose chemical technology. Cellulose nitrate (guncotton) was discovered in 1833 and combined with nitroglycerine in about 1865 to make double-base smokeless powder. In 1872 the Celluloid Co. was formed in the U.S. to exploit the invention of the first plastic, celluloid (cellulose nitrate plasticized with camphor). Mercerization of cellulose fibers—by treatment with caustic soda—became an important industrial process in the second half of the 19th century. Also commercialized during that period were several processes for making rayon, a regenerated cellulose fiber, by the viscose process. In 1923 the first automobile finished with cellulose nitrate lacquer appeared, and within four years the finish was being applied to every automobile manufactured. The Swiss chemists Camille and Henri Dreyfus developed a process for producing cellulose acetate yarn. In 1919 they founded Celanese, Ltd. in England and in 1924 the U.S. predecessor company to Celanese Corp. to make acetate fibers and fabrics. The Celluloid Co. and Eastman Kodak Co. introduced cellulose acetate photographic film in the 1920s. In 1927 Celluloid and Celanese joined forces to produce cellulose acetate plastic in the form of sheets, rods, and tubes. The move marked the birth of a major new industry, the custom molding of plastics. By the 1920s chemists also had learned to make a number of other cellulose derivatives, such as carboxymethylcellulose. This compound was first described in 1921; eventually, along with methyl- and ethylcellulose, it would help to replace the widely used natural gums from Asia and the Near East.

Cellulose technologists have learned to make dissolving pulps, for rayon and other products, by the kraft woodpulping process, which for many years could not be used to produce dissolving pulps of satisfactory reactivity. The problem was solved in the 1940s by the discovery that hemicelluloses in the pulp were converted to a form that resisted removal and that the conversion could be avoided by an acid-treatment step prior to pulping. Using this knowledge, International Paper Co. started up the first kraft dissolving pulp mill in the U.S. in 1950 at Natchez, Miss. Since then, a major fraction of the new sources of chemical cellulose has been based on the prehydrolysis kraft process.

Cellulose chemistry has long played a vital role in textile development. Mercerization, rayon, and cellulose acetate fibers have been mentioned already. In 1954, cellulose triacetate became a commercial fiber. Textiles made of it dry quickly, have good easy-care properties, and can be permanently pleated. The modern nylons, acrylics, polyesters, and other synthetic fibers owe much to the understanding of polymers that stemmed initially from research on cellulose.

Fiber development in terms of structure and of methods for producing fabrics and garments has been the result of much chemical research. The production of fibers from natural materials was suggested as early as 1664 by Robert Hooks, but not until 1857 was cellulose dissolved in cuprammonium solution by Schweizer. This discovery led to Bemberg rayon and the general rayon industry that began in the latter 1800s. Chardonnet patented nitrocellulose, which formed fibers when he squirted a dilute solution of it in ethyl alcohol and ethyl ether through fine holes under water—wet spinning—or discharged the solution upward in air to evaporate the solvent—dry spinning. Thus one man introduced two of the three techniques used today to spin fibers.

Bleaching of fibers began in this country as early as 1804; it employed a process, patented in 1798, for generating chlorine–water solution. Before, fibers had been bleached by exposing them to sunlight. Chlorine bleaching was used first with papermakers' rags and recovered paper stock. It was soon learned that, for minimum fiber damage, an alkaline chlorine bleach solution was preferable to chlorine–water. By 1830 the practice was almost universal in this country for papermaking, and it appears that the textile industry must have followed the lead of the papermakers, although the exact date does not seem to have been recorded.

The efforts of many large chemical and textile companies, private laboratories, and government laboratories have yielded tremendous advances in textiles. It all began with the recognition that the major problem with cellulosic fibers was their lack of dimensional stability. In this area, knowledge of the basic reaction of cellulose with formaldehyde was a beginning that has led us to resin-finishing reactions that impart easy-care/durable-press properties. The first mention of the formaldehyde reaction is in a 1904 German patent issued to Blumer; the first U.S. patent was issued in 1917 to Block–Pimental. Formaldehyde alone produces many effects that seriously limit the uses of the modified fibers, and this spurred further research. The introduction in the 1950s of cyclic urea derivatives—dimethylol ethyleneurea was the first—has led to today's many finishes. Later, dimethylol dihydroxyethyleneurea was developed by Badische Anilin-& Soda-Fabrik and was accepted in Europe by the late 1950s. It was popularized in this country when postcure techniques came into use, and new methods of producing the compound were discovered in the early 1960s. Postcuring allowed the fabric to be shaped into a garment before it was crosslinked to make it resistant to deformation. The results far surpassed those of all previous finishes. High levels of resiliency could be used, now that it was not necessary to limit resiliency to permit tailoring.

Interest in flame resistance of textiles dates back several hundred years, the first application being to canvas scenery used in the theater. These early treatments were based on water-soluble salts and were largely nondurable. The development of semidurable and durable flame retardants became prominent in the 1940s. For example, a commercial semidurable process, the Banflame process, based on urea and phosphoric acid, was introduced in 1948. Clayton and Heffner developed a class of flame retardants that were important in military uses during World War II. These materials were based on insoluble salts or oxides such as antimony oxide in combination with chlorinated paraffin. Later in the 1940s it was generally recognized that a combination of nitrogen and phosphorus compounds was probably most suitable for treating textiles. In 1954 Reeves and Guthrie published their discovery of a new flame retardant based on tetrakis(hydroxymethyl)phosphonium chloride. Since then a multitude of durable flame retardants for cotton has appeared based on this compound and on nitrogen-containing compounds.

Soil-release finishes for textiles may be placed in four broad classes: fluorocarbon agents; co-crystallizing agents; acrylic emulsions; and water-soluble acrylics. Each of the four classes serves a more or less specific purpose. Among the first generation were fluorinated alcohols, which created almost as many problems as they solved. Second generation fluorocarbon finishes have "dual action" and exhibit true soil-release properties as well as the ability to repel oily stains, an ability imparted by incorporating hydrophilic groups into the fluoropolymer. By 1937, the technology of water-repellent finishing had advanced to the point where durability to laundering and dry cleaning was a reality. The polysiloxanes, or silicones, introduced about 1950, are noted for the durable, soft, smooth hand they impart to fabrics. They have the further advantage of imparting a superior initial water-repellency to a wide range of fibers, especially the synthetics. Thus they hold an important segment of the market.

In 1976, by applying our present knowledge of chemical finishes and fabric construction, it is possible to produce comfortable, attractive, and durable textiles that will serve the user well through many a day over very wide ranges of temperature, humidity, radiation, and abrasion.

In 1922, when the ACS Division of Cellulose Chemistry was formed, fundamental research on cellulose was just entering the highly productive era that has continued to this day. A major step came in 1920 with the discovery that celluloses in delignified wood pulp, flax, ramie, and cotton showed practically the same x-ray pattern. Such studies would do much eventually to clarify the structure of cellulose, the mechanical properties of fibers, and the physical basis for the fiber-forming properties of polymeric systems. A second major step was the "primary-valence chain" theory of cellulose polymers. The theory grew out of the work of a number of chemists, including Nobelists Walter Haworth in England and Hermann Staudinger in Germany; it led to the proposition that cellulose is a linear polymer of glucose. In the same period, K. H. Meyer and Herman Mark in Germany determined the dimensions of the unit cell of cellulose. This and x-ray information led, in the early 1930s, to the formulation of the modern fibrillar theory of fiber structure. The concepts of polymer structure and behavior that originated in studies on cellulose have since played a critical role in the development of synthetic polymers.

Studies on the nature, location, and biogenesis of the hemicelluloses illustrate the extent to which organic, physical, and biological chemistry are being applied in fundamental research on cellulose and its sources in nature. This effort to understand a basic world resource (woody and nonwoody plants) has led to the pioneering development of many techniques and instruments that have become useful in other fields. In 1950, of the three major components of wood, only cellulose and lignin were readily identified and characterized. The hemicelluloses, the third component, were considered a mysterious mixture. The situation was due not only to the complex nature of hemicelluloses but also to controversial definitions of them. Since about 1962, publications dealing with hemicelluloses have increased at an explosive rate.

Our knowledge of the xylan- and mannan-containing hemicelluloses has moved forward so rapidly that few realize how little we knew in the early 1950s. At that time, the hemicelluloses could be isolated only after the wood had been delignified, and it was often assumed that only those substances that survived this rough treament were truly hemicelluloses. The complex nature of carbohydrate-containing extracts from hardwoods and especially from softwoods suggested that many components were present. The inadequacies of the classical methods of analysis generally employed at the time presented an insurmountable barrier to research. The advent of chromatography broke this barrier, but the basic concepts of the technique had to be adapted to the requirements of wood and polysaccharide chemistry. Among the many workers who helped to obtain the basic data and to make sense of the conflicting, confusing results were Aspinall, Bishop, Bouveng, Carter, Croon, Dutton, Gustafsson, Hagglund, Hirst, Hamilton, Isherwood, J. K. N. Jones, Lindberg, McPherson, Mahomed, Meirer, Mitchell, Ranby, Ritter, Saarnio, F. Smith, Timell, Thompson, Wathen, Whistler, Wise, Yamamori, and Yundt. These workers in Japan, England, and the U.S. were doing research on plant gums and starches, and they developed the techniques of polysaccharide chemistry that were necessary to elucidate the structures of these complex carbohydrates.

In the early 1950s, several groups of scientists, despite the uncertainties of their studies, proposed an unusual branched structure for the xylan molecule. This radical concept met much opposition, but the data on xylans from many sources were overpowering. Further, as a result of systematic research, others proposed a more filiform configuration for the xylan molecule, with all the terminal units of the xylan being either xylose or glucuronic acid. From these concepts emerged the presently accepted configuration consisting of a 1,4-xylose backbone, uronic acid side groups, and acetyl groups on certain positions in the molecule. This structure has been upheld through the years and found to be common to xylans from many plants.

Pioneering research by Bertrand in 1899 and by Sherrard and Blanco in 1923 indicated that a mannose-containing polysaccharide was present in wood; because of its resistance to extraction, they thought it was part of a "manocellulose," that is, a linear molecule composed only of mannose but otherwise identical in structure to cellulose. The adverse effect of mannose on the characteristics of dissolving pulps provided much incentive to study this hemicellulose. In the mid-1950s, five, almost simultaneous publications described the detection and isolation of a low-molecular-weight glucomannan from coniferous holocelluloses and sulfite pulps. The discovery that these glucomannans do not dissolve readily in potassium hydroxide explained the persistence of the polymer in certain types of alpha-cellulose; this destroyed the concept of a "mannocellulose."

Bishop and his co-workers demonstrated that traces of terminal galactose units were part of the so-called coniferous glucomannan. Timell combined this evidence and postulated that conifers contain no glucomannan—only a family of galactoglucomannans whose galactose contents differ and range from 1 or 2% up to 25%. Subsequent research has shown that the galactoglucomannan polymers contain acetyl groups and have different ratios of anhydrogalactose components and different distributions of components in different trees of the same species.

This effort to identify and understand the hemicelluloses has been both academic and practical. Industrial products include arabinogalactan, xylitol, furfural, furfuryl alcohol, and yeast from wastes containing hemicelluloses. The effect on the properties of pulp has been mentioned before, and from this knowledge have come sophisticated processes that, by the early 1960s, provided high-purity woodpulps, rivaling those from cotton linters, as a common article of commerce. The effect of hemicelluloses on the properties of paper is significant and complex; they are necessary for papermaking, but too much is harmful. It has become common, therefore, to blend fibers from different species and processes in order to obtain the desired product performance.

Knowledge of hemicelluloses, together with experience gained in biochemistry, has led to a deeper understanding of the plant cell wall and its formation during plant growth. This research will lead ultimately to an understanding of the mechanism of growth of a plant, of the structure of the cell wall, and of the functions of specialized cells within the plant. Future developments can be predicted to be at least as exciting as those of the past 25 years. Genetic control of composition and structure may well be the key to providing mankind with all the forms of cellulose needed for everyday life.

CHEMICAL EDUCATION

A Section of Chemical Education met at the 39th general meeting of the ACS, in 1909 in Baltimore, and papers were presented at the meetings in 1910 and 1911. Thereafter the section seems to have been inactive until 1921, when it was revived at the instigation of Neil E. Gordon with the support of Edgar Fahs Smith, ACS President in 1895 and in 1921–22. In 1924, the section became the Division of Chemical Education, with Dr. Gordon as chairman and Wilhelm Segerbloom as secretary.[1]

ONE of the first projects of the ACS Division of Chemical Education was the Standard Minimum High School Course Outline. Work on the outline was started by a Committee on Chemical Education appointed by President Edgar Fahs Smith in 1922 as a committee of the then-Section of Chemical Education. By early 1924, the committee had drawn up the outline, circulated it around the country, and collected some 30,000 criticisms. The outline was revised in April 1924 at the first meeting of the newly formed division. The result was accepted widely as a standard of excellence; moreover, it gave teachers useful leverage in obtaining additional facilities and help from their administrators and school boards. The outline remained in use until 1936, when the division revised and republished it as "An Outline of Essentials for a Year of High School Chemistry."

The content of the high school chemistry curriculum was a matter of some moment by the 1920s, although the issues were hardly the same as in 1975. In 1924, vigorous debate resulted from an article, "Should the Electron Theory Be Included in High School Chemistry," in the *Journal of Chemical Education*. Virginia Bartow, treasurer of the division 1931–41, has reported that the staff at the University of Illinois had "long

[1] Other ACS activities in education are described in Chapter II.

and sometimes furious" discussions before they decided to tell their freshmen about the electron.

Also in 1924, again at the division's first meeting, a committee was set up to evaluate the papers submitted in the ACS Prize Essay Contest, which the Society sponsored from 1923 to 1931. Before this pledge was redeemed, some five million secondary- and college-level students had submitted essays. The division's Committee on Prize Essays, with 50 members (not all of them members of the division) and 987 members of state committees, became involved. The contest was a great success. It had a healthy impact on chemical education and on public knowledge of the role of chemists and chemistry; at the same time, it uncovered a serious shortage of authoritative publications on chemistry and other sciences in school and public libraries. The Prize Essay Contest was financed by Mr. and Mrs. Francis P. Garvan. As a memorial to their daughter, the Garvans gave an estimated $1 million to finance the contest and to help the Chemical Foundation, of which Garvan was president, supply sets of chemical references to libraries throughout the nation.

The Chemical Foundation played an important role in the 1920s and early 1930s in the fortunes of the Division of Chemical Education and in chemical education in general. The foundation was established in 1919 by executive order of President Woodrow Wilson. Its primary purpose was to license to U.S. industry the chemical processes, apparatus, and products covered by enemy patents (mostly German) taken over by the U.S. Alien Property Custodian during World War I. The foundation, under Mr. Garvan, was to use the proceeds from licensing to stimulate the development of the nation's chemical industry, a function that necessarily involved support for chemical education.

At the ACS fall meeting of 1923, Neil Gordon outlined to the Section of Chemical Education the problems of getting papers on chemical education, with their particular specialization, published in ACS or other journals. Gordon then polled 750 men and women interested in chemical education and found considerable support among them for an independent journal devoted to their field. To start such a journal he sought, but failed to obtain, help from the ACS, whose policy was not to supply funds to sections or divisions. Gordon did, however, get permission to publish the journal (with the Society accepting no financial responsibility) and raised the money by selling $2000 worth of advertising on the basis of a "dummy" magazine. The first issue of the *Journal of Chemical Education,* with Neil Gordon as Editor, appeared in 1924.

The Chemical Foundation took an immediate interest in the new journal, which it subsidized from 1924 until 1932, when the depression compelled it to withdraw its support. The foundation also paid Dr.

Gordon's full salary during those nine years and allowed him to stay at the University of Maryland, where he was then teaching. With the end of the Chemical Foundation's aid, in 1932, the division signed an agreement with its printer, Mack Printing Co. of Easton, Pa., under which Mack assumed the business management and financial responsibility for the journal, while the division retained the ownership and editorial control. During the depression, Harvey F. Mack personally carried the division's printing bill, which sometimes was as much as $20,000; without his help, the journal would have folded. As of Jan. 1, 1945, the long-standing debt of the *Journal of Chemical Education* was paid in full. By the end of 1975, the journal had completed 52 years as "A Living Textbook of Chemistry."

Another long-time activity of the Division of Chemical Education has been its testing program for high school and college students. The work began in 1932 with studies by a Committee on Examinations and Tests. In 1934, the committee was authorized to cooperate with the American Council on Education to develop tests at all levels of chemical education. Cooperative Chemistry Tests were given in 1934–36, although only about 20% of the colleges and universities involved returned reports on a total of 1465 students. The program continued to grow, and in the summer of 1941 the Committee on Examinations and Tests, under chairman Otto M. Smith, worked for a week at the University of Chicago on the problems of testing chemistry students. In the same year, T. A. Ashford joined the committee; in 1946, Dr. Ashford became chairman, succeeeding Dr. Smith, who by then had spent 16 years in the post

In 1975, the testing program was still in existence, and Dr. Ashford, at the University of South Florida, was still chairman of the division's Examinations Committee. The committee makes tests available for colleges—at both the undergraduate and graduate levels—and, in cooperation with the National Science Teachers Association, for high schools. All are designed to give teachers an objective means of measuring individual and group achievement in their courses. In 1974, the committee offered some 40 different tests. In that year it sold, at nominal cost, about 150,000 tests and 250,000 answer sheets.

In 1942, the division held its first book exhibit at an ACS fall meeting, an idea conceived by Ben Gould, advertising manager for the *Journal of Chemical Education* from 1933 to 1967, when he retired. At the Society's fall meeting in Chicago in 1975, the 33rd Annual Chemical Education Book Exhibit displayed a variety of audiovisual materials and about 1000 books from 70 publishers. The books included 12 published by the journal. Mainly these were collections of articles from the journal; the first such collection, combining 20 articles on "The Discovery of the Elements," was written by Mary Elvira Weeks and appeared in 1933.

The division sponsored its first Conference on General Chemistry at Oklahoma A&M, June 12–23, 1950. The event was a precursor of many similar sessions. Such conferences were costly, and in 1952 the division began to solicit funds from the Ford Foundation, the Edison Foundation, and others to support the conferences and other operations. In 1954, the summer program was expanded. Workshops were held at North Carolina State and Kenyon, and the National Science Foundation (NSF) supported the First Chemistry Institute, at the University of Wyoming. The latter was the first major cooperative effort involving the division, NSF, and a university chemistry department. In 1956, the division's Committee on Institutes and Conferences announced NSF support for summer programs at Indiana, Oregon State, and Michigan State. In 1957, NSF sponsored 10 Institutes especially for chemistry teachers, with many division members taking part.

The same year, 1957, saw the Reed College Conference on the Teaching of Chemistry, sponsored by the division and the Crown–Zellerbach Foundation. At the conference, which owed much of its success to the division's Harry F. Lewis, a group under the direction of L. E. Strong wrote the first 1000 pages of rough draft for the Chemical Bond Approach Project, which in the next few years would develop a new kind of introductory chemistry course. The CBA Project received continuing support from NSF. In January 1960, preliminary work was started on the Chemical Education Material Study (CHEM Study), a second new approach to introductory chemistry that also was supported by the National Science Foundation. This project was not sponsored by the division. It was directed by J. Arthur Campbell and chaired by Glenn T. Seaborg. The CBA curriculum was the more sophisticated of the two; it made no concessions in the interest of understanding. CHEM Study, on the other hand, was prepared to oversimplify to clarify a subject. The two projects did not compete for prominence, but comparisons were inevitable and eventually CHEM Study was accepted more widely.

In 1956, a subcommittee of the division's Committee on Institutes and Conferences reported 125 invitations for "Visiting Scientists." During the year, seven division scientists visited 22 institutions in 14 states. Their purpose was to stimulate students and faculty through lectures and conversations directed mainly to the subject matter of chemistry. This was the beginning of a coordinated, continued move to improve undergraduate instruction in chemistry. In 1962 came the Advisory Council on College Chemistry, which had generally the same mission. This mammoth communications enterprise was funded by NSF, and many division members became involved. In 1968, NSF decided to phase out the council owing to lack of funds, but the College Chemistry Consultants Service, a unit of the council, remained in being. In 1970, the divsion took over the direction of the service, with funding from NSF. The service was due to expire in 1974, again for lack of funds, but in mid-year NSF decided to provide funds for about 100 further campus visitations by the service's consultants. Unlike the visiting scientists, the consultants advise a chemistry department on its total program and its relationship to the institution as a whole.

In 1969, the division's William B. Cook initiated the planning for the International Conference on Education in Chemistry, which took place at Snowmass-at-Aspen, Colo. in July 1970. In addition to the efforts of the division itself, manpower and funds for this massive event were provided by the Society, NSF, the ACS Petroleum Research Fund, and the Research Corp. The week-long conference was designed to clarify the many problems facing chemical educators; it produced a lengthy report, including many recommendations for strengthening chemical education at all levels.

In addition to the many special projects the Division of Chemical Education has worked on during the 52 years 1924–75, it has regularly arranged technical sessions at ACS national meetings. Frequently these sessions have included joint symposia with other divisions, a means of blending the scientific and educational aspects of chemistry to the benefit of both. The effects of the division's long-time efforts would appear to have been predicted accurately at the first meeting of the revived Section of Chemical Education in September 1921 in New York City. At that meeting, chairman Edgar Fahs Smith remarked in part, "A Section on [sic] Chemical Education is just as important as a Section on any division of Science. I think that in the end it will create greater unity in our teachings and lead to improvement in the teaching of chemists."

CHEMICAL INFORMATION

The Division of Chemical Information was formed in late 1948—originally as the Division of Chemical Literature—with Norman C. Hill as chairman. Almost from the beginning, ACS national-meeting programs had included papers on various aspects of the chemical literature. In the late 1930s and early 1940s, as the literature expanded and chemical information became essential during World War II, papers on the chemical literature increased in number and enjoyed better-than-average attendance. It was apparent that many chemists were vitally concerned with operations in the chemical literature. In 1943 a number of them organized the Chemical Literature Group of the Division of Chemical Education. On attaining divisional status, in 1948, these chemists became a community with a forum for presenting papers that have played an important role in advancing the art and science of chemical documentation, the discipline practiced by literature chemists. In 1975, the division assumed its present name. Like other divisions, the Division of Chemical Information is a diversified group of chemists with work-related titles: editor, writer, translator, indexer, abstractor, librarian, bibliographer, historian, market analyst, information system designer, computer programmer, patent chemist, patent attorney, and others.

Since 1949, its first full year, the Division of Chemical Information has issued a newsletter, *Chemical Literature,* to inform members of divisional plans and activities. An annotated bibliography of current publications of interest to the members ran regularly in *Chemical Literature.* Since this feature was duplicated by others, the division directed its publications committee to initiate a cooperative program with other organizations. This effort led to *Documentation Abstracts,* launched in 1966 as a cooperative venture with the Special Libraries Association and the American Documentaton Institute (now the American Society for Information Science). *Documentation Abstracts* became self-supporting in its second year, 1967, and in 1969 changed its name to *Information Science Abstracts.*

As happens in all ACS divisions, the Division of Chemical Information recognized an acute need for a suitable medium for publishing papers of interest to its members. The division appointed a publications committee late in 1957 to seek a solution to this problem. ACS staff and officers were sympathetic, and the outcome was the *Journal of Chemical Documentation,* introduced by the ACS in 1961. Two issues per year were planned, but the flow of papers was such as to warrant three issues the first year and four per year since. Recent years have seen increasing involvement of chemists with computers, especially for information systems, data acquisition and analysis, and correlations. In 1975, in harmony with this trend, the *Journal of Chemical Documentation* changed its name to *Journal of Chemical Information and Computer Sciences.* In late 1974, the Division of Computers in Chemistry was formed, and the *Journal* serves both that division and the Division of Chemical Information, as well as others in the evolving areas of chemical information and computer sciences.

Besides participating actively in most ACS national meetings, the division has held two meetings on its own: Jan. 19–21, 1958, in Pittsburgh, Pa., with 157 attending; and Mar. 14–17, 1973, in Columbus, Ohio, with 147 attending. The division also has encouraged its members to plan programs for ACS regional meetings.

WHEN the American Chemical Society celebrated its 25th anniversary, in 1901, chemists tended to take a proprietary interest in all of their science. Chemistry had not yet entered the age of specialization, although it was beginning to break into analytical, organic, physical, industrial, etc., to emphasize what one did the most. The literature of chemistry was evolving slowly from a prolonged infancy. Only a handful of journals was in existence; the more important were published in Germany, France, and England. American chemists read these journals, and some contributed to them almost exclusively. The few domestic journals of note to the founding members of the ACS were: Chandler's *American Chemist; Druggists' Circular and Chemical Gazette; American Gas Light Journal and Chemical Repertory;* Silliman's *Journal of Science; Journal of the Franklin Institute;* Remsen's *American Chemical Journal;* and the *Proceedings of the American Chemical Society,* which became the *Journal of the American Chemical Society* in 1879. A five-foot shelf was more than large enough to hold the books essential to a chemist in the latter part of the 19th century.

From the time chemistry became a science, knowing the chemical literature has been an essential obligation of the professional chemist. Practitioners of no other science have been as involved as chemists in working with their literature, establishing journals, setting up indexing and abstracting systems and services, systematizing nomenclature, and correlating data. Significant chemical journals began to appear abroad in the 19th century. They included Justus Liebig's *Annalen der Chemie* in 1832 (Germany); *Journal fuer Praktische Chemie* in 1834 (Germany); *Comptes Rendus* in 1835 (France); *Journal of the Chemical Society, London* in 1848 (England); *Berichte der Deutschen Chemischen Gesellschaft* in 1868 (Germany); and *Gazzetta Chimica Italiana* in 1871 (Italy).

Even though the chemical literature was relatively small during the 19th century and until the end of World War I, chemists considered it large enough to warrant the abstract journal. Abstracts appeared first in regular journals, a service that was reappearing in some specialized journals by 1975. Then came the abstracts-only journals—*Chemisches Zentralblatt* in 1830, *Chemical Abstracts* in 1907, and *British Abstracts* in 1926.

Beginning in the second quarter of the 19th century, research in organic chemistry began to expand rapidly, and by 1900 chemists had synthesized and studied hundreds of thousands of compounds. Friedrich Beilstein (1838–1906), concerned with maintaining ready access to the proliferating knowledge of organic chemicals, assembled the available information over many years for his own use. He published the material in 1880 as a handbook, with subsequent editions. The fourth edition of Beilstein's *Handbuch der Organischen Chemie* covered the literature through 1909; it was published beginning in 1919 and has been updated with supplements covering subsequent decades. A similar dedication to the literature was marked by Leopold Gmelin's *Handbuch der Anorganischen Chemie,* first published in 1817. It was taken over later by the Deutsche Chemische Gesellschaft and since 1945 has been published by the Gmelin Institut.

Until the 19th century, chemistry was essentially an art whose lack of a universally accepted language made communication extremely difficult. But such a language began to emerge with the Avogadro hypothesis of 1811, the introduction of symbols by Berzelius in 1813, the elucidation of atomic and equivalent weights by Cannizzaro in 1858, and Kekulé's concepts of quadrivalent carbon in 1858 and of the unsaturated benzene ring in 1865. Chemists have been concerned intimately with developing their language, specifically its symbols and nomenclature; they have made it second only to the language of mathematics in semantic content and in ease and conciseness of communication. The chemist's interest in nomenclature has been evident in the existence of international nomenclature committees, the ACS Council Committee on Nomenclature, ACS divisional nomenclature committees, and the nomenclature activities of the Chemical Abstracts Service.

Correlation of chemical information has contributed much to both the understanding and the teaching of chemistry. The introduction of the concept of homology by Dumas and the classification of chemical elements in the periodic table by Mendeleev are outstanding examples of how the chemical literature can yield new knowledge.

Thus, like Molière's *bourgeoise gentilhomme,* who spoke prose without knowing it, chemists have been working as literature chemists without knowing it since chemistry has been a science.

Emergence of Literature Chemists

The 20th century has seen rapid growth in the chemical industry, the numbers of professional chemists and chemical engineers, and the chemical literature throughout the world. The number of abstracts in *Chemical Abstracts* rose from 12,000 in 1907 to about 350,000 in 1975; the number of journals monitored increased from fewer than 400 to some 14,000.

As the literature of chemistry expanded and the number of chemists increased, the educational process and the conduct of research in industry accelerated the fragmentation of chemistry into subdisciplines. What chemists did began to define the disciplines of chemistry. Thus by 1940 the ACS had 18 divisions that defined general areas in which chemists and chemical engineers worked. Also by 1940, the individual chemist or chemical engineer no longer could hope to maintain a current, working knowledge of the literature without aid. In response to this need came a new specialist, the literature chemist. Some literature chemists, such as the chemical librarian, the translator, and the patent chemist, emerged slowly in the chemical industry during the 1920s and the 1930s. Others, such as indexers and abstractors, emerged with the introduction of information services, such as *Chemical Abstracts.* Not until the 1940s however, were these specialists numerous enough to form a group within the Division of Chemical Education and, finally, an ACS division.

The accomplishments of literature chemists prior to the 1940s can be summarized in a number of concepts or paradigms:

- Chemical libraries and document files.
- The science of definition and classification.
- Chemical symbols, structural formulas, and nomenclature.
- Chemical structure correlations.
- Chemical journals, abstracting services, and compendia.
- Indexes: dictionary, formula, classified, heterocyclic, etc.

Over the past 25 years, literature chemists have been using and extending these concepts and paradigms and introducing new ones.

Hand-sorted (with a needle) punched cards and the many ingenious indexing and classification schemes for their use were introduced by literature chemists in the 1940s; many papers on this subject appeared in the programs of the Division of Chemical Literature. This high activity and the apparent potential of punched cards prompted the ACS Board of Directors to establish in 1946 a special committee, with financial support, whose efforts culminated in 1951 with the publication of the book, "Punched Cards," by J. W. Perry and R. S. Casey (two early chairmen of the Division of Chemical Literature).

The intrinsic value of the work on hand-sorted punched cards was realized most fully in the 1960s, when the computer began to be an important tool for literature chemists. Many of the concepts introduced for hand-sorted punched cards were compatible with the logic inherent in computers. Consequently, the transition from manual to computerized information systems was relatively easy and rapid.

With the advent of computerized information systems, literature chemists became involved in new activities, such as system design, programming, and data acquisition and analysis. Practically all literature chemists today have been affected by the use of computers for information storage and retrieval, either directly, as on-line users of computerized services, or indirectly, as intermediaries between their clients and the computer. Where previously the literature chemist's desk was covered with 3 x 5 index cards, then by hand-sorted punched cards, and then by the 80-column tab card, it is common now to find only a computer terminal.

Although literature chemists remain actively interested in nomenclature, linear notation systems have been introduced since the 1940s to solve problems associated with the naming of chemicals. In the early 1950s, members of the Division of Chemical Literature participated in the National Research Council's evaluation of four notation systems. Since then, many papers on new or revised notation systems have been and continue to be presented before the division and published in the *Journal of Chemical Information and Computer Sciences.* Another active area has been the representation of chemical structures by topological systems. Topological representation has achieved major importance in storing and retrieving chemical structures by computer and has played an important role in the development by Chemical Abstracts Service of its CAS Registry System for identifying, naming, and registering chemical structures.

The chemical literature has been doubling at about a 12-year interval. In the face of such growth, important information services like *British Abstracts* and *Chemisches Zentralblatt* became too costly and had to be terminated, thus increasing the burden on Chemical Abstracts Service. Fortunately, technology had advanced by the early 1960s to the point where computers could be used effectively in many chemical information operations. Conversion to computers, particularly for operations as massive as those at CAS, is highly involved and requires considerable R&D. The successful conversion at CAS was achieved by an expanding staff of dedicated literature chemists and computer scientists and their assistants.

Chemical information services by now have become big business. A number of organizations in addition to CAS have been performing R&D for processing large data bases. They include the BioSciences Information Service, the Engineering Index, the National Library of Medicine, the Institute for Scientific Information, the National Technical Information Service, Derwent, Ltd., and many others.

What the future holds is anybody's guess. But projecting the past through the present, it is not unreasonable to visualize a one-world total information system that every scientist and engineer can use via a personal terminal to tap man's store of chemical knowledge.

CHEMICAL MARKETING AND ECONOMICS

The ACS Division of Chemical Marketing and Economics was formed in September 1952, with Fred A. Soderberg as chairman and Hal G. Johnson as secretary-treasurer. The beginnings of the division go back to the early 1940s, when a group of chemical engineers formed the Technical Service Group of the Chemical Industry (later the Commercial Chemical Development Association and by 1975 the Commercial Development Association). During 1946, members of the Technical Service Group tried to get the membership standards lowered to bring in younger men working in field service and chemical marketing. The move failed, but the interested members then explored the possibility of setting up a parallel group in the Society, with the Technical Service Group to provide guidance. The result was the formation, in 1947, of the chemical marketing section of the ACS Division of Industrial & Engineering Chemistry. The section became a subdivision of the I&EC division in 1950 and, finally, an independent division.

Among other objectives, the Division of Chemical Marketing and Economics strives to encourage communication and understanding "between the technical function and the commercial functions of the chemically oriented industrial community" and "between the academic community and the commercial functions of the chemical industry." The ACS began to publish preprints of CM&E papers at national meetings in 1960, but for some years the division has been publishing its own reprints.

CHEMICAL marketing and economics is concerned essentially with the commercial aspects of chemically oriented industry and thus relies heavily on technical and business information obtained by marketing research. The concepts of marketing research are surprisingly old. In R. Ferber's "Handbook of Marketing Research," L. C. Lockley cites the Fugger family as the first to use information from direct observation of business affairs:

In 1380 Johann Fugger left his native Swabian village of Graben to settle in Augsburg, for he engaged in the international sale of textiles. Family members settled in strategic capitals, moving into finance, trade, manufacturing, and mining. They helped finance the rulers of the Western European countries and dominated international finance until the mid-17th century.

Apparently the Fugger family regularly exchanged detailed letters on trade conditions and finance in the localities of their branches, so that all partners could base their decisions on current knowledge of supply and demand for money and goods. Later, the Rothschild family in Germany, England, and France appears to have practiced the same kind of transfer of information. Mr. Lockley concludes that there was a need to collect, analyze, and disseminate commercial information although the original business was largely local and there was little need for information from distant parts. The Industrial Revolution sharply increased the need for information on which to base marketing decisions.

Techniques for gathering business information, whether chemically oriented or otherwise, have evolved in haphazard fashion. In 1948 the American Marketing Association defined marketing research as, "The gathering, recording, and analyzing of all facts about problems relating to the transfer and sale of goods and services from producer to consumer." In the same period, Lockley stated that marketing research "requires formulation of problems; location of information sources; obtaining of information by inquiry, observation, or inference; tabulation and analysis; and presentation of research results in a way which helps management make wiser marketing decisions."

In the United States the first instances of marketing research techniques came from attempts to forecast elections. In 1879, in working on a proposed schedule for the Nichols–Shepard Co., manufacturers of agricultural machinery, N. W. Ayer and Sons wired state officials and publishers throughout the country for information on expected grain production. With the resulting data, the agency was able to construct a crude but formal market survey by states and counties. The next recorded marketing research occurred at E. I. du Pont de Nemours & Co., Inc. Even today, Du Pont has a companywide service division, "The Trade Analysis Division," that analyzes salesmen's call reports monthly. Some believe that this effort was started by a Mr. Patterson, beginning in 1892, first alone and then within sales groups totaling 65 people by 1902. Sales are reported to have doubled, tripled, and then quadrupled during 1892–1902, when Patterson is credited with having conquered the explosives market. In 1915 the U.S. Rubber Co. established a department of commercial research, as did Swift & Co. at about the same time.

Not until after World War I did market research in the chemical process industries become appreciably important. Until the mid-1920s most of it was conducted with little support from general statistics. In 1926 the Domestic Commerce Division of the U.S. Department of Commerce scheduled a conference to discuss possible services that the department could perform for marketing research. A committee of marketing research people discussed the question and reported its belief that a national census of distribution was needed badly. As a result of the pressures generated by that conference, the first Census of Distribution was taken in 1929; today the Census of Business is taken regularly and covers retailing, wholesaling, and the service industries. This basic material has produced many background publications useful to marketing research.

One of the early textbooks in marketing research was "Commercial Research: An Outline of Working Principles," copyrighted in 1919 by C. S. Duncan of the University of Chicago. Other early texts were J. George Frederick's "Business Research and Statistics," published in 1920, and Z. Clark Dickenson's "Industrial and Commercial Research," published in 1928. In 1929 the Ronald Press published "Marketing Investigations" by William J. Reilley, a marketing professor at the University of Texas and formerly on the research staff of Procter and Gamble. In 1921, Percival White's "Market Analysis" was published by McGraw–Hill. The latter text was one of the first to go through a series of editions. In 1937, "The Technique of Marketing Research" appeared under the sponsorship of the American Marketing Association. It had been written cooperatively by a committee of experts. Also in 1937 "Market Research and Analysis" by Lindon O. Brown was published by the Ronald Press. It is probably fair to say that marketing research was presented infrequently in business schools before 1930 and omitted infrequently after 1937.

The role of sampling in marketing research became more important in later years, bringing more interest in the use of mathematics and statistics. The importance of sampling and the proper application of statistics is evident in the discrediting of the *Literary Digest* presidential poll, which used an invalid sampling method. The *Digest's* forecasts for the 1936 presidential election underestimated Franklin Roosevelt's vote by 19%.

In the early and middle 1930s statistical training began to improve greatly in the colleges, especially in England. Technical preparation in this country improved, and quota sampling—or some approach to quota sampling—was under heavy attack. Two concepts emerged rapidly. The first was that bias could be decreased or eliminated if individuals or cases were chosen randomly—that is, with a table of random numbers or by some method of drawing from an urn. The second concept was that the areas could be chosen randomly and could provide the basis for a suitable sample.

Leaders in sampling techniques were George H. Gallup, Arthur C. Nielsen, and perhaps Louis B. Harris, although his techniques apply to public relations, not marketing. George Gallup did some readership studies for the DeMoines *Register and Tribune* in 1928 and developed readership research methods for both newspapers and magazines. In about 1935 he founded The American Institute of Public Opinion in Princeton, N.J., and attracted attention by predicting that the *Literary Digest* presidential poll in 1936 would be inaccurate. Gallup applied his poll techniques in areas beyond marketing and politics. His work in poliomyelitis, for example, uncovered symptoms that medicine had not yet isolated.

Arthur C. Nielsen developed a marketing research organization beginning in 1923. His business suffered during the early 1930s because of the depression, but studies for two important clients helped to reestablish it, primarily because they turned it sharply from industrial research to consumer research. One of these was a study for General Electric on timing gears for appliances, and the other was for Du Pont on Duco coatings for automobiles. In 1933 Nielsen established the concept of monitoring retail inventories on a continuing rather than a spot basis. In 1934 he introduced the concept in the form of the Nielsen Drug Index and later the Nielsen Food Index. These examples are significant in being related to consumer items. The sampling technique they embody is used to test advertising and to test product, package, and price changes and the application of marketing techniques.

In late 1940, new approaches to sampling, quotas, and so-called probability sampling emerged. In a probability sample, all respondents or cases are selected by random methods. Such a sample is acceptable for testing for size and extent of sampling error. One great drawback is that the technique is hard to administer in the field and requires close and costly supervision. Theoretically, however, the probability sample answers many questions that have arisen on problems of sampling.

The "Chemical Business Handbook," edited by John H. Perry and published by McGraw-Hill in 1954, contains a section (Part 5) entitled "History, Development and Functions of Marketing Research in the Chemical Process Industries" by Robert B. Wittenberg. He reports that marketing research has become an essential tool of management. It provides and evaluates market and marketing information for management guidance in the following fields: policy establishment; program establishment; product studies; market studies; method studies; and various areas related to government, institutions, and advertising. Wittenberg states that until after World War I, the chemical process industries were concerned largely with "heavy chemicals"—those made and sold in large tonnages to relatively few customers—so that the need for marketing research was not very great or at least not very apparent. After 1925, however, chemical marketing research began to develop the standard practices and professional standing that distinguish it today, when a given company often may wish to monitor markets for a particular chemical all the way to the end-use level, which may comprise a variety of products. Thus, organized chemical marketing research is only about 50 years old.

Survey design is still not entirely satisfactory, as is true of various other techniques used in marketing research, but it has made considerable progress. This was true especially in the 1960s, as interest turned to Bayesian sampling and greater emphasis on subjective probabilities. New emphasis on marketing research has come also with the advent of new concepts in marketing information systems, concepts that have been implemented by the use of the computer.

The nature of modern marketing research in the chemical and other industries is evident in the titles and authors of chapters in Ferber's "Handbook of Marketing Research." Chapter II, "The Functions of Marketing Research," is by George Fisk, Department of Marketing, Syracuse University. Chapter III, "Marketing Information Systems," is by Kenneth P. Yule, Department of Business Administration, University of Illinois. Chapter IV, "Marketing Decision Information Systems: Some Design Considerations," is by David B. Montgomery, Graduate School of Business, Stanford University. This chapter discusses the information systems and the design of models and computer usage. Last but not least is Chapter V, "The Organization of the Marketing Research Function," by Richard D. Crisp of Richard D. Crisp and Associates, Inc., Pasadena, Calif.

Computer simulation is likely to witness considerable development during the 1970s. The emphasis in marketing research recently has been on the design of models that simulate real situations and from which important management decisions can be made. At this stage in the development of the art it is difficult to build the total complexity of reality into the simulation models and computer games that have been designed. As a result, some of the models have had to include estimates, which lead to deviations that are open to criticism. The economists, however, have not fared as well with their models and systems for prognosticating the U.S. or the world economy. Certain segments of the economy have been modeled successfully, but the complexity of the total system has made the models of it inaccurate.

During the past 50 years many chemical companies have become consumer oriented as opposed to only industry oriented. Many have become multinational, others have become large, and the need for marketing information certainly has become more apparent. One of the common problems is the need for more communication and understanding among top management, research and development groups per se, and marketing people.

COLLOID AND SURFACE CHEMISTRY

The Division of Colloid and Surface Chemistry was established in September 1926 in Philadelphia at the Society's Golden Jubilee meeting. The first chairman was H. B. Weiser, the first secretary F. E. Bartell. Initially the name was the Division of Colloid Chemistry; it was changed to the present name in 1960. The division grew out of the activities of the National Research Council's Committee on the Chemistry of Colloids, formed in 1919. The committee sponsored the first four of the annual National Colloid Symposia, which began in 1923; participants in the first two of these meetings laid the plans that led to the new division. Since 1927 the division has been the principal sponsor of the National Colloid Symposium, which has taken place every year excepting 1933 and two years during World War II. In 1976 the event was renamed the Colloid and Surface Science Symposium.

In 1970, the division established the Victor K. LaMer Award, which is given annually to the author of an outstanding doctoral thesis accepted by a U.S. or Canadian university. The award, $1000, is supported by funds made available to the division by the friends and family of the late Dr. LaMer. The division awards its Certificate of Merit to recognize service to the division and to colloid and surface chemistry. This award consists of a certificate and a lifetime membership in the division; it is given annually (if warranted) and may go to more than one recipient in the same year. The division administered an Undergraduate Award until 1973, when Lever Bros. Co. withdrew as sponsor. The award was made to students for research performed before receipt of the bachelor's degree; the division expects to resume it with funds from another sponsor.

IT is becoming more and more evident that colloid and surface science is interdisciplinary. Chemists, physicists, biologists, physiologists and others are involved with problems in the field. This range of disciplines is remarkable for a science whose name, "colloid," was first suggested by Graham in 1861 based on the Greek work *κολα* meaning "glue," and whose foundations span no more than 150 years.

Brown (1828) described the movement of pollen grains suspended in liquids, but not until 1905 did Einstein discuss these phenomena theoretically and propose that the mean square displacement per interval of time equals twice the diffusion coefficient. Three Nobel prizes (R. Millikan, physics, 1923; J. Perrin, physics, 1926; T. Svedberg, chemistry, 1926) were awarded at least partly for attempts to prove experimentally this manifestation of molecular motion.

In 1903, Zsigmondy and Siedentopf developed the ultramicroscope and studied the Brownian motion of colloidal particles. For this and work that proved the heterogeneous nature of colloid solutions, Zsigmondy was awarded the Nobel Prize for Chemistry in 1925. Hillyer, Burton, and others perfected and used the electron microscope (1935–45) to observe and study details of particles and surfaces. The early limits of resolution, about 100 Å, have been extended to 5–10 Å, and resolutions better than 5 Å have been reported for ion and proton microscopes.

Advances in various types of spectroscopy have made it possible to study molecules both at and adsorbed on surfaces. Vibrational spectroscopy, both transmission and reflection, has yielded much information on adsorbed molecules, and multiscanning techniques have enhanced these results. Now available are spectroscopic instruments that make it possible to study the electronic nature of surfaces. These include x-ray photoelectron spectroscopy, electron spectroscopy for chemical analysis (ESCA), uv photoelectron spectroscopy (UPS), low energy electron diffraction (LEED), Auger electron spectroscopy (AES), and electron-stimulated desorption (ESD). These techniques, coupled with ultraclean surfaces prepared by high vacuum techniques, enhance markedly our understanding of many solid-state problems and find uses in fields such as catalysis, adhesion, and corrosion

The studies of Brown indicated that colloidal particles exist in a size range where the energy of random impingement by solvent molecules is adequate to keep them distributed uniformly in the medium, apparently independently of the action of gravity. The observations of the scattering of light by small particles by Tyndall (1869) and the ultramicroscopic observations of Zsigmondy supported this hypothesis. Lord Rayleigh (1881) developed an electromagnetic theory of light-scattering that was extended by Lorenz (1890) to light reflection by transparent spheres and by Maxwell–Garnett (1904) and Mie (1908) to turbid media. Debye (1943) extended the work of Smoluchowski and of Einstein (1906) and simplified the light-scattering experimental techniques so that they could be used to measure the molecular weight or the aggregate weight as well as the shape of many colloidal systems. LaMer and Sinclair (1949) applied Rayleigh–Mie scattering to a number of aerosol preparations. Kerker and his associates, principally Matijevíc and Kratohvil (1965), continued this work and prepared aerosols with narrow size distribution. By using digital computers with their light-scattering data they were able to follow the particle-size distribution of a coagulating system and compare the results with Smoluchowski's (1917) theory.

Debye and Scherrer (1916) recognized the crystalline nature of many lyophobic colloids–rigid particles, either amorphous or crystalline, or small liquid droplets. Kruyt suggested that the stability of lyophobic colloids is governed almost solely by the electric charge of the particles and that of lyophilic colloids by both charge and solvation. Lyophilic colloids are either large single molecules or clusters of such molecules dissolved in the dispersion medium so that each link or group of the molecule is in contact with the medium. This is not true of lyophobic colloids, which are therefore never stable thermodynamically. Also, lyophobic sols may be precipitated with small amounts of electrolyte, which is not usually the case with lyophilic sols. Schulze (1882) and Hardy (1900) showed that the flocculation value is determined mainly by the valency of the ions, which are charged oppositely to the particles of the lyophobic sols. Smoluchowski (1917) developed a theoretical, diffusion-limited kinetic equation for the aggregation process. London (1930) developed a quantum-mechanical explanation of the nonpolar van der Waals forces, those weak attracting forces acting between all molecules. Kallman and Willstätter (1932) and de Boer and Hamaker (1936) suggested that the force acting between colloidal particles might be due to the double-layer interaction coupled with the van der Waals–London interaction.

Until about 1940, the literature displayed little agreement on the role of various forces involved with the stability of lyophobic sols or even on the question of whether the double-layer interaction produces attraction or repulsion. However, the work and subsequent publications of Langmuir (1938), Derjaguin/Landau, and Verwey/Overbeek (1939–45) showed that the stability of lyophobic colloids and suspensions was due to the mutual repulsion arising from the interaction of two electrochemical double layers and the attraction by the London–van der Waals forces. The DLVO theory, as it is now known, has been generally accepted and used extensively, but it does have limitations, particularly as a quantitative predictive theory for many systems of practical importance. It has little or no relevance to the stability of colloidal dispersions in nonaqueous, nonionic solvents, where surface-charge and steric effects arising from adsorbed polymers are important.

The studies on Brownian movement, diffusion, light-scattering, and mobilities of charged colloidal particles and their theoretical interpretations resulted from the interaction of mathematicians, physicists, and colloid scientists. During this same period, physical chemists began to use thermodynamic principles in colloid and surface chemistry. Colloid chemistry expanded rapidly within the framework of physical knowledge, and surface chemistry began to emerge. Notable concepts advanced were the Laplace (1806) and Young (1855) theories of pressure drop across a curved surface and Gibbs' (1875) showing how solutes that concentrate at the surface of a liquid lower the surface tension. In this same period, the Kelvin (1871) equation relating vapor pressure to the curvature of a liquid interface was derived and used subsequently to describe the behavior of water in the capillaries of a porous solid. Interest grew rapidly in various interfacial phenomena, particularly at gas–liquid and liquid–liquid interfaces; the number of papers published in this field has doubled every 6–6.5 years to the present day.

Agnes Pockels (1891) and Lord Rayleigh (1899) predicted monolayer formation and measured surface tension with a Wilhelmy plate. Devaux (1904) conducted the first spreading experiments and measured areas using movable barriers. The force measurement technique was introduced by Langmuir (1917), who established the basic science of adsorption on surfaces. Harkins followed Langmuir and made a systematic thermodynamic development of the work of Young and Dupré (1869) on contact angles and surface forces. During this same period, Adam (1922) and Rideal (1931) extended the study of insoluble monolayers to many interfacial systems, particularly those involving biological interactions and membranes.

The concept of colloids as glue-like materials had to be revised, for it became clear at the beginning of the century that a fine state of subdivision is more pertinent in defining colloids. Freundlich's (1909) experimental data showed the importance of large surface area and the concept of adsorption. These properties found much use in the fields of pigments, heterogeneous catalysis, and separations based on adsorption. The development of preparative chromatography enabled scientists to work with high purity materials. It is customary to divide adsorption of vapors and gases on solids into two large classes, physical adsorption and chemical adsorption or chemisorption. Physical adsorption is rapid and, except for hysteresis, is reversible, and the adsorbate may be readily removed and recovered unchanged chemically. Chemisorption may be rapid or slow and may be difficult to reverse, and desorption may involve chemical changes. Equations for adsorption isotherms were suggested initially by Freundlich (1926) and by Langmuir (1918), but a major increase in interest in physical adsorption evolved when Brunauer, Emmett, and Teller (1938) presented their (BET) theory of multimolecular adsorption, which made it possible to estimate surface areas of adsorbents. Hill (1949) showed how thermodynamic information from experiments could be used to test statistical thermodynamic theories of adsorption.

Culver and Tompkins (1949) used surface potential measurements and Gomer (1955) and Boeker (1955) used the field-emission and field-ionization microscopes in chemisorption studies. Eischens (1958) and Kiselev (1962) used spectroscopic methods to determine the state of adsorbed molecules, and Selwood (1962) used magnetic measurements for similar problems. MacRae (1963) used low energy electron diffraction to study surface atoms. Ever since Taylor (1931) introduced the concept of active sites in catalysis, surface chemists have been concerned with detecting the heterogeneity of surfaces. Roginsky and Keier (1946) partially covered a surface with an adsorbent of one isotopic composition followed by further coverage of the same adsorbent with a different isotopic composition. Partial desorption would show the degree of surface heterogeneity from analysis of isotope mixtures. Eley (1955) covered the surface with one gas and studied the kinetics of its exchange with its gas–phase isotope. Thomas and Cranstoun (1963) utilized a displacement technique, using mixed isotopes of mercury to displace hydrogen and hydrogen–tritium previously adsorbed.

The marked development of computers during the past decade has permitted the application of EHMO, CNDO, and SCF-X_a calculations about the character of bonding adsorbates to surfaces, which should make it possible shortly to calculate the energies of transition states in catalytic reactions. Porous glasses with high surface areas were prepared by Nordberg (1944) and used by numerous groups to study the state of adsorbed molecules. In the past 10 years or so, molecular sieve catalysts have been almost the only type used in cracking petroleum. This goes back to studies by Weigel and Steinhoff (1925), who discovered that a zeolite, chabazite, had selectively adsorbed water, methanol, ethanol, and formic acid, but had excluded from its pores acetone, benzene, and ether. McBain called these materials molecular sieves because of their ability to separate molecules on the basis of size. Milton and Breck (1961) synthesized a family of zeolitic materials of industrial importance.

Surface states, depending on their origin either intrinsic or extrinsic, are those electronic levels of a crystalline solid whose wave functions have finite, nonvanishing values only near the surface. The existence of intrinsic surface states has been surmised from experiments with photoconducting cadmium sulfide crystals by Somorjai and Lester (1965) and Levine and Mark (1966). Mark (1964) used photocurrent decay to show the interaction between intrinsic surface states and various adsorbates. Weisz (1952) used "chemisorption" to describe the formation of adsorbed ions by the exchange of electrons between the adsorbate and the band structure of the adsorbent. From consideration of extrinsic surface states of a chemisorbed ion, Mark (1968) concluded that higher index surfaces furnish greater stability to the formation of chemisorbed lattice constituents during growth, and that the dominant surfaces of dispersed adsorbents such as powders and polycrystalline layers are naturally more stable and it is precisely these that offer the least stability for chemisorption. The marked interest in this field and in the general subject of clean surfaces is shown by the series of symposia, "Molecular Processes at Surfaces," which have been part of the Division of Colloid and Surface Chemistry's program since 1973.

Although Svedberg was much concerned with the preparation of hydrophobic colloids by dispersion and condensation (1912), and with his studies on Brownian movement, his Nobel address (1926) covered only his results on the ultracentrifugation of proteins. This so-called sedimentation velocity technique demonstrated that many common proteins had measurable and reproducible molecular weights. Coupled with the development of the moving-boundary electrophoresis apparatus by Tiselius (1930), which earned him the Nobel prize in 1948, the Svedberg technique had a profound influence on macromolecular chemistry and biochemistry; in fact many colloid chemists devoted the rest of their careers mainly to proteins and other biological macromolecules. Recent improvements in instruments and experimental techniques, coupled with newer concepts of the structure of aqueous solutions in general, have markedly extended the application of ultracentrifugation to complex biological systems.

The reversible formation of micelles, bimolecular layer aggregates, was used by McBain (1925) and by Hartley (1936) to explain abrupt changes in various properties of solutions that occur at "critical micelle concentrations." McBain had postulated the presence of many-layered aggregates in dilute soap solutions, whereas Harley proposed that micelles were two-layered and essentially spherical, a concept that has proven to be reasonably correct. Debye (1949) showed that micelles result from the van der Waals attraction between paraffin-chain tails and the electrical repulsion of the charged heads and that their size depends on the length of the paraffin chain and the ionic strength of the solution. Since these are dynamic systems, a degree of micellar size distribution may exist, but only some high speed, possibly pulsed, techniques will resolve this problem. At higher soap concentrations, liquid crystals are present; depending on composition, they can cause marked changes in various properties of the solution. Precise measurements of the enthalpy and free energy of micelle formation have yielded reliable values of the entropy. These results suggest that solvent molecules rearrange at the surface of the micelle. Lawrence (1937) and Hartley (1938) showed that the solubility of organic compounds increases markedly above the critical micelle concentration. With the addition of various amounts of hydrocarbons to soap solutions, more or less stable systems resulted because of processes that have been termed emulsification, microemulsification, and solubilization. Characteristics of emulsions have been reviewed by Clayton (1954) and Becher (1957), of microemulsions by Schulman (1955), and of solubilized systems by Klevens (1951). Lawrence (1958) justifiably objected to the term solubilization, which he said was first used by McBain, on the premise that these systems are quite ordinary ones of three partly miscible components.

The stability of emulsions and foams is related to the mechanical properties of surface films. Although methods for measuring the viscosity of surface films have been known for some time, there is at present no coherent picture of the viscoelastic properties of surface films. Emulsifying agents were classified numerically by Griffin (1949) on an HLB scale, a hydrophilic–lipophilic balance of the emulsifier molecule. HLB values are related to the distribution of surface active agents between water and oil under certain conditions. Griffin obtained these HLB values from an empirical numerical correlation of the emulsifying and solubilizing properties of different surface active agents, and Davies (1957) calculated them directly from chemical formulas, using empirically determined group numbers.

Harkins (1938) demonstrated the existence of marked effects of phase and temperature on surface viscosity as well as the existence of non-Newtonian viscosity behavior in monolayers. Joly (1946) studied the viscosity of monolayers under various spreading pressures and concluded that there were many phase changes in the systems investigated. The orientation of polar molecules at the oil–water interface in water emulsions would lead to electrical double layers in emulsions and foams, which would tend to stabilize emulsion droplets. Van den Tempel (1953) extended the concept of lyophobic colloid stability to aggregation of oil-in-water emulsions and found that observed flocculation rates agreed reasonably well with theory.

Schulman and Cockbain (1940) showed that mixed emulsifiers, such as soap plus long-chain alcohols, would result in films of unusual strength and emulsions of considerable stability. These complex mixtures were shown to enhance hydrocarbon solubility in micellar solutions, were utilized to produce microemulsions, and would produce foams with unusual drainage properties. LaMer (1944) showed that such mixed soap–long chain monomolecular films markedly decrease the evaporation of water from reservoirs.

Reuss (1808) showed that the electrical charge carried by many colloidal particles is important in their behavior and explains their movement in an electrical field. This phenomenon, initially called cataphoresis and now electrophoresis, was first observed by Linder and Picton (1894). Helmholtz (1879) and Smoluchowski (1903) interpreted it in terms of an electrical double layer. Gouy (1910) and Chapman (1913) treated the case of a plane charged surface and the resulting diffuse double layer. Stern (1924) suggested that the region near the surface should be considered in two parts, an inner compact double adsorbed layer and a diffuse Gouy–Chapman layer. Lyklema and Overbeek (1961) proposed the presence of an inner layer of desolvated, potential determining ions; they regarded the Stern layer as a second layer of at least partially solvated ions, which have a rather long lifetime in the layer and thus are somewhat immobile. Sparnaay (1958) used the concept of a free volume fraction to introduce ionic radii in the form of a correction term in the Gouy–Chapman treatment. Grahame (1947) reviewed the possible structuring of the Stern layer into an inner Helmholtz plane (IHP) and an outer Helmholtz plane (OHP) that marks the beginning of the diffuse layer. Bockris et al. (1963) showed that surface charge densities indicated small degrees of surface coverage by adsorbed ions and that highly oriented water molecules occupied most of the solid surface. Levine and Bell (1962) introduced a discrete charge effect, a form of repulsion between adsorbed ions, which leads to a prediction that the potential at the OHP goes through a maximum as the potential at the surface is increased.

Various electrokinetic phenomena result from relative motion between a charged surface and the bulk solution. Smoluchowski (1903) developed equations that clarified various electrokinetic effects such as electrosmosis and streaming potentials. Lippmann (1895) comprehensively investigated the electrocapillary effect and, following this work, it was possible to measure the free energy of the interface and to calculate surface excesses and surface charge density. Frumkin (1923) and Grahame and Whitney (1942) treated these phenomena in more detail by considering the nature of the adsorbed layer and the general framework of the derivation of the Gibbs equation.

With changes in solvent, the electrical aspects of solutions become more complicated. For mixed solvents, Butler (1929) and Frumkin and Damaskin (1954) explained the observed changes in terms of polarizability of the displaced water by the organic molecule. For nonaqueous media, van der Minne and Hermanie (1952) put the electrophoresis technique on a reasonably sound basis. Koelmens and Overbeek (1954) and Parfitt, Lewis, and McGown (1966) showed that it was possible to compare experimental data with DLVO theory, at least for dilute dispersions; in concentrated dispersions, Albers and Overbeek (1959) indicated that the comparison is not valid.

The contact angle of a liquid on a solid surface has been shown to be a function only of the forces exerted on the liquid. The most basic factor in wetting is the free energy of interaction between the liquid and solid phases across the interface. The force that promotes wetting is adhesion, which is the sum of forces resulting from a number of kinds of interfacial attraction—dispersion forces, hydrogen bonds, other polar interactions, pi bonds, and electrostatic interactions—all of which can be estimated. Young (1805) linked contact angle to surface energetics, and Gibbs (1878) described the basic thermodynamics of wetting of smooth, homogeneous, nondeformable solids. Buff (1960) verified this latter treatment from a rigorous statistical–mechanical point of view. Bikerman (1950) pointed out that contact angles and drop shapes depend very strongly on anisotropics in the surface. For various solvents and surfaces, Bartell (1953) and Johnson and Deltra (1964) showed that the advancing angles increased and receding angles decreased with increasing roughness.

Fox and Zisman (1952) developed an important empirical relationship of contact angles with critical liquid surface tension of wetting for the surface. Good and Girifaleo (1957) used statistical mechanics to develop a theory of contact angles; Fowkes (1962) and Good (1963) used the theory to explain the deviation of interfacial behavior from ideality. Subsequently, Good (1967) attempted to interpret Zisman's critical surface tension of wetting in terms of a back-to-back bilayer film separating two bulk regions of a solid.

From contact angle measurements of various liquids on high energy surfaces, Hare and Zisman (1955) found that certain liquids were nonspreading because the molecules adsorbed on the solid formed a film when critical tension of wetting was less than the surface tension of the liquid itself. These they termed "autophobic" liquids. The work on the wetting properties of organic liquids on high energy surfaces introduced new concepts of the mechanisms of the reactions between solids and liquids.

A surface is called hydrophobic when water does not spread on it—the contact angle is high between a drop of water and the surface and the heat of immersion is low. Because of numerous heterogeneities in surfaces, contact angle measurements are subject to a number of interpretations. Leslie (1802) showed that heat was evolved upon immersion of some solids in liquids. Parks (1902) tried to obtain such data on a unit area basis, but not until the BET (1938) method was achieved could this be accomplished.

For perfect surfaces, adhesive strengths should be very high, and the force necessary to separate the surfaces by a molecular dimension could be 600 kg/cm^2. According to Bowden (1931), the usual values are much smaller because of plastic and viscous flow that may precede fracture and the presence of surface flaws and dislocations.

Many colloid and surface chemical processes are not in equilibrium in a strict thermodynamic sense. Defay, Prigogine, and Bellemans (1951) extended the concepts of the thermodynamics of irreversible processes developed by Prigogine (1947) to problems in surface tension and adsorption. Mechanical and thermal equilibrium are always supposed to be established, but no equilibrium with respect either to adsorption processes or to chemical changes is assumed in the system.

The Freundlich and Langmuir isotherms have been applied widely to adsorption from solution, the former as an empirical equation, the latter limited to a monolayer. Freundlich (1926) extended Traube's (1891) rule and stated that the adsorption of organic substances from aqueous solution increases strongly and regularly as we ascend the homologous series. Holmes and McKelvey (1928) showed that this rule is reversed for a polar adsorbent and a nonpolar solvent. Bartell and Fu (1929) demonstrated that, even in aqueous solutions, strength of adsorption would decrease with decreasing degree of hydration. Hansen and Craig (1954) observed that an inverse relationship generally exists between the extent of adsorption of a compound and its solubility in the solvent used. Strain (1942) and Zechmeister (1950) indicated that similar constitutive effects must be considered in chromatographic experiments. The Langmuir equation was found to hold when dilute solutions were involved, but Hansen (1949) found a marked increase in adsorption as the saturation concentration was approached, indicating multilayer adsorption. Bartell and Donahue (1952) found adsorption that was characteristic of capillary condensation.

Surface areas have been estimated from solution adsorption studies if the isotherms were found to fit the Langmuir equation. Smith and Hurley (1949) used a fatty acid in cyclohexane, and data obtained by van der Waarden (1951) showed that aromatics lay flat on the adsorbing surface. Adsorption of electrolytes may involve the electrolyte's being adsorbed in toto, or ions of one sign may be held very strongly and those of opposite sign as a diffuse layer. The ions of the diffuse layer could be replaced by others of the same sign. This essentially is an ion exchange process similar to that reported by Way (1850). Zeolites, because of their ion exchange capabilities, were used extensively as water softeners. During the period 1935–50, many synthetic organic cation exchangers were developed, often for special purposes, and these are now more important than the zeolites.

COMPUTERS IN CHEMISTRY

The formation of the ACS Division of Computers in Chemistry, in April 1974 upon petition by almost 700 ACS members, reflects the large and growing impact of the electronic machines on chemical science and technology. The division's interests span the use of computers in both theory and experiment—from dealing with the complex mathematics of quantum chemistry to manipulating and displaying data. The division also has interests that overlap those of other ACS divisions whose programs have a significant computer component. These include the Divisions of Analytical Chemistry, Chemical Education, Chemical Information, Chemical Marketing and Economics, Organic Chemistry, and Physical Chemistry.

THE growth in the use of computers in chemistry has paralleled the development of computers themselves in the past quarter century. Theoretical chemists pioneered the application of computers to chemical problems primarily because of their need to evaluate the difficult integrals that arise in calculating the energy of molecules. The initial steps came in the late 1940s with the calculation of atomic self-consistent fields, but another decade passed before the development of new computational algorithms and procedures designed to exploit the capabilities of computers in dealing with such problems.

The basic elements of quantum chemistry were established in the late 1920s and the 1930s. However, desk calculators were slow, and the few available differential analyzers often were prone to serious error; computations usually were undertaken only to test some fundamental principle. In the early 1950s, calculation of the properties of a simple diatomic molecule, or a Hückel calculation of a hydrocarbon with about a dozen carbon atoms, could take a year on a desk calculator. For these reasons, many chemists viewed theoretical chemistry as an esoteric discipline having little to do with the real problems of chemistry.

This attitude was to change with the availability of high-speed computers and efficient computer programs. Also contributing to the change were certain advances in the laboratory. New experimental methods like nuclear magnetic resonance (NMR) and electron spin resonance (ESR), for example, could not be explained by classical concepts, so that quantum ideas became perforce an integral part of the new theory, and the calculations dictated by the quantum approach were feasible as a rule only on the high-speed computers that were becoming available. By now the computer has become so vital to theoretical chemistry that the National Academy of Sciences—National Research Council and other interested parties in 1975 were seeking to establish a National Resource for Computation in Chemistry. Such a center would be a vital step toward widespread scientific resource-sharing via computer networks, including steps such as certification of computer programs for quantum chemical calculations.

The computer has influenced profoundly not only NMR and ESR but the development of other experimental methods and their associated apparatus. These include nuclear quadrupole resonance, optical spectroscopy, chromatography, x-ray crystallography, infrared spectroscopy, mass spectroscopy, kinetics, and microwave spectroscopy. The role of the computer in these areas takes two general forms. One is the direct (on-line) automation of experiments, in which the computer governs the real-time activity of the apparatus. The other form is the indirect (off-line) mode, in which the computer primarily stores and refines data. In the latter mode, electronic computers greatly facilitate the use of sophisticated techniques such as time-averaging, curve-fitting, and Fourier transforms. Indeed, fast Fourier transforms are among the most important of the innovations that have occurred in instrumentation because the technique permits simultaneous sampling of a number of quantities (energy, absorbance, power dissipated, etc.). The method allows one to transform reversibly between time and frequency and, therefore, to select preferred fitting algorithms for maximizing signal-to-noise ratio in the shortest time. Carbon-13 NMR spectroscopy could not have been applied to macromolecular research without the computer, which is an essential element of the Fourier transform techniques used in such work.

Another important use of computers in chemistry is "phenomenology" —the processing of a mass of data to extract correlations and thus to build a predictive model of the behavior of the compound(s) or process to which the data pertain. Still other uses include information storage and retrieval and nonnumerical operations in general.

The rapid growth of the minicomputer industry has provided small and powerful machines that are relatively inexpensive. Present integrated circuit module technology enables the chemist to design and construct his own interfaces between his experiment and the small control computer. Minicomputers offer a comparatively high degree of flexibility and freedom from direct intervention in the experiment. In chemical research laboratories, in consequence, they have become as common as infrared spectrometers. Somewhat larger computers, furthermore, can handle a number of experiments at the same time with the flexibility of the single-purpose minicomputers.

The use of computers has been facilitated significantly by flexible graphical displays that present results rapidly in a form that is easy to assimilate. These devices range from digital plotters to dynamic visual displays that the user can interact with. The latter capability, for example, allows the scientist to compare calculated and experimental spectra rapidly and to modify the pertinent theory and the resulting computed spectra until the latter come into accord with the experimental spectra. This practice is common in NMR work.

Departments of chemistry in colleges and universities have recognized the increasing importance of the computer in the day-to-day activities of professional chemists. Computer courses are becoming steadily more common in chemical curricula; chemical educators are requiring that competence in using the machines become integral to the training of chemists. An important reason for the trend is the fact that the practicing chemist more and more will have to develop his own specialized algorithms and related procedures for applying the computer to particular problems.

ENVIRONMENTAL CHEMISTRY

The ACS Division of Environmental Chemistry was formed in 1915, originally as the Division of Water, Sewage, and Sanitation Chemistry. A section by the latter name, formed in 1913 at the instigation of Edward Bartow, was the forerunner of the new division; the first divisional chairman was Prof. Bartow (ACS President in 1936), and the secretary was H. P. Corson. In 1959 the division's name was changed to Water and Waste Chemistry. Beginning in 1954, the division regularly held symposia in cooperation with the ACS Committee on Air Pollution, and these activities culminated in 1964 in a second change of name, to Water, Air, and Waste Chemistry. In 1973 the division assumed its present name.

The Division of Environmental Chemistry's several name changes reflect the evolving interests not only of its members but of society at large. The division's focus today is on sound chemical approaches to natural water quality, air pollution phenomena and their control, and the technology of domestic and industrial water and waste treatment. The emphasis in these areas is on research, as opposed to operating data and routine tests. The division recognizes the complex nature of environmental problems by organizing multidisciplinary programs of its own as well as joint symposia with other ACS divisions.

The division has preprinted extended abstracts of its meeting papers since 1961–62. Members of the division helped to instigate and contributed significantly to "Cleaning Our Environment: the Chemical Basis for Action," published in September 1969 under the aegis of the Society's Committee on Chemistry and Public Affairs. By mid-1975, the Society had distributed more than 20,000 complimentary copies of this 250-page paperback on environmental chemistry and had sold more than 50,000. The division also was active in the launching of the ACS journal *Environmental Science & Technology,* which has appeared monthly since January 1967.

The Division of Environmental Chemistry's Bartow Award, named after its first chairman, is given annually for the divisional paper most outstanding in content and presentation; the award was first given in 1952. Also in 1952 the division awarded its first Certificate of Merit for a notable first appearance before the division; the certificate is designed to encourage the presentation of papers by new and younger members. In 1957 the Division of Environmental Chemistry gave the first of its Distinguished Service Awards, which recognize individuals who have performed outstanding service for the division over a relatively long period.

THAT interest in the chemistry of the environment is not new is evident in the fact that the ACS Division of Environmental Chemistry, under its various names, is more than 60 years old. Indeed, the division was formed to provide a more satisfactory forum for chemists who were working actively on water supply, sewage disposal, and related problems.

In 1939, Dr. Edward Bartow, the first chairman of the division, looked back on 25 years of water chemistry. He noted that the period had seen large-scale application of the lime and soda water-softening processes, as well as the use of ion exchange, for softening and purifying water for household, laundry, industrial, and municipal purposes. Water sterilization methods had been extended from bleaching powder to liquid chlorine and the chloramines; pH control methods had been introduced for coagulation and softening. The activated sludge sewage-treatment process had been developed from the first U.S. experiments (reported at the division's first meeting) to the completion of the world's largest installation of that type by the Chicago Sanitary District. Studies of stream pollution had shown the need to treat sewage and industrial wastes. Studies of in-house treatment of industrial wastes had proved profitable for the factories concerned. Naturally radioactive waters had been found. Surveys of the fluoride content of waters throughout the nation had been made as a result of the suspicion that fluoride in drinking water was the cause of mottled enamel on teeth. Means of removing tastes and odors from water with activated carbon and chloramines had been developed.

In September 1963, division chairman Henry C. Bramer summed up another 25 years of progress in water chemistry. Treatment methods for sewage and industrial wastes, he noted, had evolved from the elimination of gross pollution under special circumstances to sophisticated methods in nationwide use. At the ACS national meeting in January 1963, the division had devoted its entire program to a symposium on wastewater renovation (held jointly with the Division of Industrial and Engineering Chemistry). The symposium concentrated on treatments that would yield water suitable for various uses and that would include environmentally sound means of disposing of separated contaminants. It moved beyond the classical biological approaches to physical-chemical methods such as ion exchange, adsorption, and electrodialysis. Other symposia of the period covered conversion of saline to fresh water, water for nuclear power generation, boiler water chemistry, and water for television picture-tube production.

At the time of these symposia, interest in environmental chemistry had begun to spread well beyond the realm of the specialist. Congress had enacted the first identifiable federal program for water pollution control in 1948 and the first for air pollution control in 1955. The resulting activity was reflected in the programs of the Division of Environmental Chemistry. The number of papers in those programs in the 35 years ending in 1947 was equaled by the number in the 15 years 1948–63.

Regulation of the environment since 1963 has intensified steadily at all levels of government, and the consequent demands on environmental chemists have intensified in like measure. It has become increasingly clear that sound environmental control involves a delicate balance of many factors. It is clear also that such a balance cannot be struck without extensive interdisciplinary research and development in which chemistry plays a vital role. The goals of such work include alternative sources of energy; catalysts, scrubbers, and other means of controlling emissions to the environment; and economical recycling processes for many materials. Goals in the chemistry of the environment itself include deeper understanding of the behavior, interactions, and effects of contaminants as they move from source to sink or receptor.

Progress in environmental chemistry relies heavily on analysis—the ability to measure the constituents of the environment and to determine their chemical and physical forms. Analysts have been hard put to keep up with the demands of environmental control in the past decade or so, but they have made headway nevertheless. Potential problems with long-lived organic compounds, for example, would never have come to light without analytical methods that can detect such compounds at levels as low as a few parts per trillion. The measurement of air contaminants has been bolstered by the advent of permeation tubes and other devices that make it possible to generate standard atmospheres for testing analytical methods and instrumentation. Despite these and other advances, however, environmental analysis poses many problems. A general one is standardization of methods and instruments so that data obtained by different scientists in different laboratories or geographical areas will be comparable. More specific needs include methods for determining the chemical forms of sulfur and particulate matter in air and of elements such as phosphorus in water.

Advances of the past decade in atmospheric chemistry include better understanding of the mechanisms of smog formation, of the fate of carbon monoxide in the air, and of the sources of ozone in urban atmospheres. An important need is improved simulation models that can be used to devise least-cost strategies for controlling air pollution in metropolitan areas. An unprecedented attack on the problem was under way in 1975 in St. Louis, Mo. under the direction of the Environmental Protection Agency. The theme of the effort—the St. Louis Regional Air Pollution Study—is coordinated measurement of many interrelated variables of urban air pollution within the same time frame.

Progress with the water environment has included the development and use of physicochemical treatment processes for removing substances such as nitrogen, phosphorus, and trace organic contaminants from wastewaters. Other developments include the use of liquid oxygen to upgrade the efficiency of the classical biological treatment of wastewater and of ozone in disinfecting the effluent from treatment plants. The behavior of certain substances, such as mercury, in ambient waters has been elucidated to a degree, but the behavior and fate of many others remains partly or wholly a mystery. Notable examples include radionuclides and particulate matter. The already complex chemistry of natural waters, moreover, is complicated further by the many biological processes involved.

These few examples of progress and problems illustrate very roughly the scope of modern environmental chemistry. The full scope of the field, as well as its evolution over the six decades ending with 1976, are evident in the programs of the ACS Division of Environmental Chemistry during that period. It is clear that the division and similar forums are growing steadily more essential to the proper dissemination and use of interdisciplinary knowledge of the environment in all of its aspects.

FERTILIZER AND SOIL CHEMISTRY

The Division of Fertilizer and Soil Chemistry was one of the first five divisions in the Society, all formed in 1908. The division's first chairman was F. B. Carpenter; the secretary-treasurer was J. E. Breckenridge. The programs were concerned initially with analytical and control procedures, but in the 1940s the emphasis shifted toward fertilizer technology. Originally, the division's name was the Division of Fertilizer

Chemistry; the current name was adopted in 1952 as part of an attempt to broaden the interests of the division to include fertilizer use as well as manufacture. Specific efforts were made, with considerable success, over the ensuing years to encourage agronomists to present papers on fertilizer usage and to join the division. However, divisional meetings have continued to serve mainly as a forum for discussion of fertilizer processes and properties.

EVER-INCREASING demands for food and fiber have spurred worldwide growth in chemical fertilizers, which in 1975 were improving crop yields enough to feed one billion people—one quarter of the world's population. Fertilizer chemistry is based on simple reactions that have been known for many years. The major ones are the reaction of hydrogen and nitrogen to form ammonia and the liberation of phosphate from phosphate rock by heat or acid. The application of these reactions yields a variety of fertilizers containing nitrogen and phosphorus, two of the three primary plant nutrients. Large natural reserves of potash have given us fertilizers containing potassium, the third primary nutrient. Carbon, hydrogen, and oxygen, as well as the secondary nutrients, calcium, magnesium, and sulfur, generally have been widely available. The recognition of plants' needs for the micronutrients that make up the remainder of the 17 essential elements completed the development of basic fertilizer chemistry.

The growth in understanding of soil chemistry has paralleled and sometimes overlapped the development of fertilizer chemistry. Perhaps the most important principle established was the cation- or base-exchange properties of soil. Soil holds cations such as ammonium, calcium, magnesium, sodium, potassium, and hydrogen. One cation may replace another, and as basic elements are taken up by the plant they are replaced by the hydrogen ion, which leaves the soil acid. These ion-exchange properties were correlated with the fine-particle-size or clay fraction of soil and also with its organic fraction. Specific chemical and physical reactions that cause soil to bind phosphate and potassium also were established. Proper use of fertilizers, and soil management in general, have thus been shown to depend on fertilizer-soil interactions.

Nitrogen

The critical importance of chemistry to the fertilizer industry is typified by the synthetic ammonia process. Fritz Haber in Germany in 1902 began seeking a practical synthesis; by 1909, using an osmium catalyst, he had produced ammonia at a concentration of 6%. The German firm Badische Anilin-& Soda-Fabrik (BASF) acquired Haber's patents and joined the work. Carl Bosch of BASF played a key role with his low-cost catalyst of iron and small amounts of other metals. In 1913, BASF put a 30-ton-per-day plant into operation; by the end of World War I, a second plant was producing 240,000 tons of ammonia per year.

In the U.S., General Chemical patented an ammonia process in 1918, and the first successful plant was built in 1921. Production grew relatively slowly in this country until World War II, when 10 new plants were built with a total annual capacity of 800,000 tons (of nitrogen). The main stimulus was the need for ammonia to make nitric acid for use in making munitions. In 1951, the Government launched a rapid amortization program to help industry expand its capacity for ammonia; by the end of 1958, 58 plants were operating with a capacity of 4,254,200 tons of nitrogen annually. This, and later, even more spectacular growth, made low-cost ammonia available for other fertilizer materials such as ammonium nitrate, urea, and ammoniated phosphates.

Steam reforming of natural gas (methane) is the most widely used source of hydrogen for synthetic ammonia. The process was developed in 1920–30, with variations aimed originally at using methane from coke-oven gas, but large-scale use of methane had to await the availability of natural gas. In 1940–60, a network of transcontinental pipelines was laid to move gas from large new reserves in the Texas–Oklahoma–Louisiana area to industrial centers in the eastern and north central U.S. Until 1939, 90% of the hydrogen for ammonia was obtained from coke and steam; by 1957, natural gas accounted for 72% of the hydrogen for ammonia in the U.S. The future source of hydrogen for ammonia is critical in view of the world energy shortage. Some foresee that pipeline gas made from coal in large, central plants will replace natural gas. Naphtha, heavy oil, and electrolysis of water also can readily be visualized as alternative sources of hydrogen.

Anhydrous ammonia was first used directly as a commercial fertilizer in 1943. It accounted for 2–3% of total nitrogen fertilizer use in 1945; by 1972, it accounted for 39%. It is remarkable that ammonia, a gas boiling at —33°C, a corrosive base, and a strong irritant to the eyes, nose, throat, and skin, has found everyday use in farming. Its direct use as a fertilizer was pioneered by Leavitt at Shell Development and Andrews at Mississippi State and their collaborators. Leavitt investigated the addition of ammonia to irrigation water in the 1930s and recognized the need for applying ammonia directly to soil; this work led to a patent in 1942. Andrews in 1947 published principles for direct application of anhydrous ammonia. He showed that when applied to soil, it binds firmly to clay and organic matter. This early work demonstrated that direct application of ammonia was practical and overcame the initial skepticism. The subsequent progress in fittings and in metering devices has made direct application a practical and economical method.

The major commercial outlet for nitric acid in the U.S. is fertilizers. Before 1920 most nitric acid was made from Chilean sodium nitrate (caliche) by reaction with sulfuric acid, although the oxidation of ammonia with air to produce nitric acid was known as early as 1839. As the new synthetic ammonia industry gained momentum, oxidation of ammonia became the commercial method. The trend toward use of nitric acid in fertilizers began in the U.S. in 1928 with production of sodium nitrate, which reached 500,000 tons per year in the middle 1930s and continued to expand until about 1945. Since then, sodium nitrate has lost ground steadily to ammonium nitrate and ammonia.

During World War II, the rise in ammonia production in the U.S. supported a corresponding expansion in nitric acid for making explosives. After the war, the output of the new plants went to meet the great demand for fertilizer nitrogen. In 1943, some 30,000 tons of ammonium nitrate was used as fertilizer in this country; by 1950 the figure had reached 559,000 tons. From 1945 to 1955, ammonium nitrate was the largest-volume source of fertilizer nitrogen in the U.S. By 1975 it remained second to ammonia, although urea eventually may challenge this position.

Urea was first synthesized commercially from ammonia by I. G. Farbenindustrie in Germany in 1920 and by Du Pont in the early 1930s at Belle, W. Va. In 1956–57, urea represented only 4% of world nitrogen consumption; by 1962–63 its share was 9%. The growth stemmed from advances in urea production technology, the compound's unique combination of properties, and better understanding of its use. Large plants are now commonplace, and unit costs of production have declined sharply. Urea is a high-analysis product (45% nitrogen) with relatively low transportation and distribution costs; it can be used as a solid, in solutions, or as an intermediate in producing compound fertilizers—properties that give it advantages over most other sources of fertilizer nitrogen.

Phosphorus

In 1875, annual consumption of fertilizer materials in the U.S. was approaching one million tons, about 20% of it domestic phosphate rock. Phosphate rock is a complex fluorophosphate in which fluorine binds the phosphorus in a form that makes it difficult for crops to use. Chemical or thermal means can be used to put the phosphate in a soluble, more readily available form. Phosphate rock was treated commercially with sulfuric acid to give "superphosphate" as early as 1849 in this country. It remains the basic raw material for all phosphate fertilizers. South Carolina deposits were important initially, but were surpassed by those in Florida in about 1900. Florida deposits, still the mainstay in 1975, are supplemented by significant amounts from Tennessee and from Idaho and other western states.

The original superphosphate process consisted simply of mixing phosphate rock and sulfuric acid, holding the slurry in a container until it solidified, and then storing the material for some weeks to allow the chemical reactions to go to completion. This process leaves a high proportion of unusable calcium sulfate in the product. The concentration of useful phosphate can be increased by substituting phosphoric acid for sulfuric acid to yield "concentrated" or "triple" superphosphates.

Improved methods for manufacturing phosphoric acid were required to make production of concentrated superphosphate practical. Elemental phosphorus was produced by heating phosphate rock in an electric furnace as early as 1891 and in a blast furnace somewhat later. The importance of elemental phosphorus as a source of pure phosphoric acid was first demonstrated by the Bureau of Soils, U.S. Department of Agriculture, between 1918 and 1922. A number of commercial plants for furnace acid were built later. Improved processes for making wet-process phosphoric acid—by treating phosphate rock with sulfuric acid—were developed still later, particularly between 1927 and 1932. By 1940, both furnace and wet-process phosphoric acid were in full-scale commercial production. Either may be used for triple superphosphate, with cost and markets generally favoring wet-process acid.

The availability of both furnace and wet-process phosphoric acids permitted significant output of triple superphosphates, which approached 300,000 tons in 1939 and approximately doubled in the next 10 years. This growth has continued, while production of ordinary superphosphate has declined. Process improvements by the Tennessee Valley Authority (TVA) and promotion of the agronomic and economic advantages of triple superphosphate combined to make it the largest-volume phosphorus fertilizer in the U.S. by 1965.

The second largest source of fertilizer phosphorus in this country is ammonium phosphates, made by ammoniation of phosphoric acid. Ammonium phosphates emerged as important fertilizers, starting in the 1950s, as a result of the availability of low-cost ammonia and wet-process phosphoric acid. The trend was favored by the need for soluble phosphates in alkaline soils of the Midwest and stimulated by the development of the TVA drum ammoniator.

Potassium

Before World War I, the U.S. depended on potash deposits in Germany for its potassium fertilizers; the war spurred the establishment of a domestic potash industry. Fractional crystallization processes were worked out to obtain potash from the brines of Searles Lake, Calif., and the salt flats near Wendover, Utah. Between World Wars I and II, processes were developed for recovering and concentrating potash from the solid, low-grade ores near Carlsbad, N.M. The discovery of vast potash deposits in Canada has been the most important development in the potash industry in the past two decades. With the possible exception of recent discoveries in Russia, these deposits contain the greatest reserves of high-grade ore in the world. Canadian reserves are estimated at 37% of world supply, Soviet reserves at 49%.

Conversion of potassium ores to fertilizer generally involves little chemical processing. The ore is mined, beneficiated, and either applied to the soil directly or in fertilizer mixtures. Many of the U.S. deposits (sylvite, containing potassium chloride) are not competitive with Canadian material except in the Southwest, where shipping charges are low. Potassium chloride accounts for more than 95% of the fertilizer potassium consumed. Other potassium salts are preferred in certain agronomic situations. Potassium sulfate is recovered in the U.S. from langbeinite ($K_2SO_4 \cdot 2MgSO_4$), or the mineral may be used as such. Potassium nitrate and potassium phosphates have attractive features, but are not now made commercially.

Micronutrients

The finding that microamounts of certain elements are essential to plants was one of the most important discoveries in agriculture. In a classical experiment in 1860, plant physiologists Sachs and Knop grew plants with roots immersed in a solution of nutrient salts. By excluding first one element and then another, they demonstrated the need for specific elements. Until about 1920, plant physiologists believed that only the 10 elements first reported by Sachs and Knop were essential to normal plant life. These were carbon, hydrogen, oxygen, nitrogen, phosphorus, potassium, sulfur, calcium, magnesium, and iron. Additional essential elements established over the following 30 to 40 years included boron, zinc, copper, manganese, molybdenum, chlorine, and sodium.

In 1976, in many crop situations, micronutrients are as critical as primary nutrients. The Florida citrus industry, for example, could never have developed without using trace elements. Zinc is used widely; in some parts of the country it is the most common nutrient added, after nitrogen. Sorghum frequently shows iron deficiency, and sugar beets will not grow without traces of boron. Micronutrients may be applied to the soil directly, with fertilizer, or as foliar sprays. They will become even more critical in the future as heavy crop production depletes the natural supplies in the soil.

FLUORINE CHEMISTRY

The Division of Fluorine Chemistry, established in 1963, evolved from a symposium on fluorine chemistry sponsored by the Division of Industrial and Engineering Chemistry at the ACS fall meeting in 1946. Speakers at that symposium reported on the extensive and then newly declassified work on fluorine that had been undertaken during World War II in the development of the atomic bomb. Subsequent symposia led in 1950 to the formation of a Subdivision of Fluorine Chemistry in the I&EC division and, 13 years later, to full divisional status. The first chairman of the division was Paul Tarrant; the secretary-treasurer was J. E. Castle.

The division administers the Fluorine Division Award for Creative Work in Fluorine Chemistry, sponsored by PCR, Inc. The award was established in 1971 and consists of a plaque and an honorarium of $1000. It has been presented annually beginning in 1972.

THE beginnings of fluorine technology are lost in antiquity, but almost certainly they date from man's first use of naturally occurring compounds of fluorine in ceramics and metallurgy. Ceramic glazes and the worthless gangue associated with metallic ores had to be melted, but the effectiveness of such processes was limited by the peak temperature attainable with the fires in use at the time. An unknown technologist discovered that melting points could be lowered and melting eased by adding calcium fluoride (fluorite, fluorspar). This newly useful material, "the rock that flows," was an important spur to technology and gave fluorine its name. Fluorspar and cryolite, a sodium-aluminum fluoride, are still used widely as fluxes in the metallurgy of iron, as electrolytes in making aluminum and magnesium, and as additives in ceramics.

Fluorine chemistry began with the liberation of hydrogen fluoride from fluorspar and sulfuric acid by Marggraff (1768) and Scheele (1771). Their attempts to isolate the acid in all-glass apparatus failed, but Gay-Lussac and Thénard (1809) succeeded by using inert lead or silver apparatus. Hydrogen fluoride, an important intermediate for making most other fluorine compounds, is still to a major extent made industrially from fluorspar and sulfuric acid.

Not until 1886 did the French chemist Moissan isolate elemental fluorine, by electrolysis of hydrogen fluoride. As is now known, fluorine is a most unusual element. Many of its compounds are stable, but some are highly reactive. The fluorine molecule (F_2) itself is extraordinarily reactive, and the reactions of elemental fluorine are exceptionally difficult to control. For example, Moissan and Chavanne (1905) tried to react solid methane with liquid fluorine at $-187°C$, but notwithstanding the very low reaction temperature and the skill of the investigators, the result was a disastrous explosion. Even "inert" gases such as xenon have been found recently to form compounds with fluorine. Thus research on fluorine itself is difficult and was undertaken by only a handful of chemists from the time Moissan isolated the element until about 1920. The leaders were Moissan himself, Ruff in Germany, and Swarts in Belgium.

The foundations of modern organofluorine chemistry were laid by the Belgian chemist Swarts, a contemporary of Moissan. Swarts found that organofluorine compounds could be prepared by halogen exchange of organic chlorides, bromides, and iodides with inorganic fluorides, particularly the fluorides of antimony. Swarts began his work about 1890 and was the only worker publishing in this field for about 25 years after 1900. By 1920, the scientific literature on fluorine was contained almost entirely in a 128-page book by Ruff that was published in that year.

Interest in fluorine chemistry began to grow more rapidly in the early 1920s. This was especially true in the U.S., where G. H. Cady, J. H. Simons, and others were beginning their research in the field. Many fluorine compounds were synthesized and studied; their characteristics often were unusual and they sometimes challenged the chemical theories of the day, thus stimulating further research. The work of Swarts led Thomas Midgley, Albert Henne, and others at General Motors to conclude that a compound of chlorine, fluorine, and carbon (dichlorodifluoromethane) should make a superior refrigerant. This proved to be the case, and Midgley announced the discovery at the national meeting of the American Chemical Society in April 1930. A company formed by General Motors and Du Pont began to produce the new refrigerant, the first of the Freons, in 1931, a step that marked the beginning of today's thriving fluorochemicals industry. The Freon process was based on hydrogen fluoride, derived from fluorspar; when World War II erupted, chemists still could generate fluorine itself only in gram quantities, and the element could not be purchased at any price. Nevertheless, workers in the field by then were predicting great utility for fluorine chemistry, which in the end would contribute to the war effort in a major way in three areas: poison gas, aviation gasoline, and the atomic bomb.

The Government did not know whether poison gas would be used during the war, but prudence dictated preparedness. Of the toxic agents discovered during the period of concern, several were fluorophosphates, products of fluorine chemistry.

The relationship of fluorine chemistry to aviation gasoline began in the 1930s when Simons and his colleagues found that hydrogen fluoride is an excellent catalyst for many organic chemical reactions, including alkylation, the source of an important component of high-octane gasoline. V. N. Ipatieff and Herman Pines then developed the HF alkylation process, first used commercially in 1942. The HF process, and the sulfuric acid alkylation process first used commercially in 1938, were the sources of the high-octane aviation fuel so essential during the war; both processes remain in use in 1976.

Early in the atomic bomb project, uranium hexafluoride was found to be the only volatile compound of uranium that could be used to separate uranium-238 and the fissionable uranium-235 by gaseous diffusion, the main means of separation. Fluorine was required in large amounts to make uranium hexafluoride, but before 1940 production facilities for the element did not exist. The necessary techniques were developed in extensive work by Harshaw Chemical, Hooker Electrochemical, Du Pont, and Union Carbide, all under government contract. Fluorine units with multiton daily capacity sprang up at the three wartime gaseous diffusion plants; they used electrolytic cells and a fused salt electrolyte of potassium fluoride and hydrogen fluoride. This process, too, is still in use.

Uranium hexafluoride is highly reactive, so that materials of construction of uncommon stability were required for use in the gaseous diffusion plants. Simons and Block (1937) studied the direct fluorination of carbon in the presence of mercury salt explosion inhibitors and reported the formation of CF_4, C_2F_6, C_3F_8, C_4F_{10} (two isomers), C_5F_{10}, C_6F_{12}, and C_7F_{14}. Simons found these compounds to be thermally and chemically very stable and suggested, correctly, that they might be resistant to uranium hexafluoride. The youthful organic fluorochemicals industry still was making only the simple one- and two-carbon chlorofluorocarbons used as refrigerants. Thus development was started on more complex fluorocarbon liquids and solids for use as lubricants, coolants, plastics, and elastomers in the processing of UF_6. Tetrafluoroethylene had been obtained by Ruff and Bretschneider in 1933 by decomposing CF_4 in an electric arc, and by Locke, Brode, and Henne (1934) by dehalogenating $ClCF_2CF_2Cl$ with zinc; in 1940 the compound was discovered to polymerize. At about the same time, chlorotrifluoroethylene, $CF_2=CFCl$, was oligomerized and polymerized to materials of outstanding stability. Wartime demand led to the commercial production of fluorocarbons, polytetrafluoroethylene (Teflon), and polyfluorotrichloroethylene (Kel–F). The pressures of the gaseous diffusion work greatly accelerated work on fluoropolymers, and research in many industrial and university laboratories created numerous new fluorocarbons and processes for making them. Many of these materials are used in 1976 as plastics, oils, and greases.

The end of the war brought the disclosure of extensive new knowledge of fluorine chemistry held back until then by security restrictions. Teflon and the Freons became widely known and used. The Freon-based insecticide bomb, first used early in the war, presaged the aerosol container industry. Fluorocarbons came into use as: fire extinguishants; insulators, both liquid and gaseous, in electrical condensers and transformers; solvents; lubricants and hydraulic fluids; and blowing agents in plastics. Fluoride ion was recognized as a normal constituent of foods and natural waters in very low concentrations. Some fully fluorinated fluorocarbons now show promise as artificial blood in experimental animals, and others appear exceptionally useful as radio-opaque materials.

Despite their chemical inertness, fluorocarbons may not be free of side effects. Chlorofluorocarbon aerosol propellants, because of their stability and extensive use, have built up to measurable, though still very low, levels in the atmosphere; some scientists hypothesize that products of their photochemical breakdown in the stratosphere are depleting the ozone layer that shields the earth from harmful wavelengths of ultraviolet radiation. Intensive research on the problem was under way in 1975.

Inorganic fluorides also find many uses: in fluoridating water to protect against dental caries and as a dentifrice additive for the same purpose; in etching glass; as a flux in metallurgy; and as an alkylation catalyst (HF) for making detergent intermediates as well as high-octane gasoline.

The broad commercial development of fluorine chemistry in the past 30 years has been accompanied by great experimental and theoretical progress. For example, medical research on fluorinated steroids provided strong impetus for the development of selective fluorination methods, and the search for new fluorinated drugs and anesthetics, as well as for prosthetics and agrichemicals, continues apace. In addition, the availability of fluorocarbons and many derivatives and the special properties associated with fluorinated groups have led to their widespread use as stable electronegative probes in organic and inorganic chemistry. The resulting broadened interest in fluorine chemistry, both academic and industrial, promises many more advances in the future.

FUEL CHEMISTRY

The ACS Division of Fuel Chemistry had its beginnings in 1922 with the formation of the Gas and Fuel Section of the Division of Industrial and Engineering Chemistry. In 1925 the section became the Division of Gas and Fuel Chemistry and, in 1960, the Division of Fuel Chemistry. The first chairman of the division was S. W. Parr, then the dean of American coal chemists; the secretary-treasurer was O. O. Malleis. The division concentrates today on the chemical problems and the engineering and economic aspects of the conversion of fossil fuels to energy, chemicals, or other forms of fuel. The group's primary interests do not include petroleum, the basic concern of the Division of Petroleum Chemistry.

In the early years, divisional symposia dealt with topics such as the properties of coal, combustion, carbonization, and the use of gaseous fuels. The post-World War II years saw greater emphasis on synthetic fuels. Subsequent symposia have covered fuel cells, high-energy fuels, advanced power cycles, pollution control, mine safety, automotive emission control, low-sulfur fuels from coal, fuels from wastes, fluidized-bed combustion, and energy storage. In 1957 the division began to bind preprints of its symposia papers into volumes. Since 1972, current and back issues of preprint volumes have been available on microfilm from the ACS.

The Division of Fuel Chemistry administers two awards. The Henry H. Storch Award, established in 1964, is given annually to a citizen of the U.S. who has contributed most to fundamental or engineering research on the chemistry and utilization of coal or related materials in the preceding five years. The award consists of a plaque and an honorarium of $500. The BCR–Richard A. Glenn Award, established in 1955, is sponsored by Bituminous Coal Research, Inc. The award recognizes excellence in quality and presentation of papers relating to coal or coal-derived products. It is given before each semiannual meeting of the division and consists of a certificate and an honorarium of $100. The award need not be given if no paper meets the standard of excellence.

THE development of fuel science and technology may be said to have started in 1350 A.D., when coal first became a commercial commodity. Gas was first recognized as a state of matter in 1620, and coal was first distilled in the laboratory to yield gas in 1660. The carbonization of coal to produce metallurgical coke was known in the late 1600s, but was not practiced on a large scale until 1730. Coke was a by-product of a process, developed in 1792, in which coal was distilled in an iron retort to produce illuminating gas. The first coke ovens to collect gas as a by-product were built in France in 1856. By the early 1900s it had become apparent that the by-product coke oven produced gas more economically than the gas retort; by the late 1930s in this country, the by-product oven had almost entirely displaced the beehive coke oven, in which the gases are burned and vented to the atmosphere.

In 1855 Bunsen invented the atmospheric gas burner, which paved the way for gas as a source of heat as well as of light. The need for gas of high illuminating power was met by Welsbach's invention of the incandescent mantle, in 1884, and the advent of Edison's incandescent electric lamp a few years later presaged the end of gas lighting. At the same time, electric lighting created a huge new use for coal in electric power plants. Pulverized coal was tried in such a plant in 1876 but failed owing to insufficient knowledge of the combustion process. Four decades later, in 1917, the U.S. Bureau of Mines reported that efficient combustion required long flame travel unless air and coal were mixed vigorously. On this basis, Bumines and the Milwaukee Electric Railway and Light Co. cooperated in the first successful use of pulverized coal with steam boilers.

The first unit to make low Btu gas for heating was built in 1832. The principle was to burn coal or coke with a restricted supply of air to yield a gas containing about 25% carbon monoxide and 13% hydrogen. By 1900 this so-called "producer gas" had become an important source of heat, but its use declined thereafter. In 1780 Fontana had discovered "water gas" (blue gas), made by decomposing steam over a bed of incandescent coke. The resulting gas contained about 40% carbon monoxide and 50% hydrogen. The first successful water gas process was developed by Lowe in 1875; the first serious attempt to study the chemistry of the process was the work of Clement, Adams, and Haskins, published in 1911 by the U.S. Bureau of Mines. The use of this technology in the U.S. increased steadily until about 1926, when it began to decline slowly.

Work on the direct hydrogenation of coal to make liquid hydrocarbons began in Germany in 1913, and a 31 ton-per-day pilot plant was built there in 1921. Also in 1913, Fischer and Tropsch in Germany conceived the idea of making liquid hydrocarbons from water gas; the first experimental results were published in 1923. In 1928–30, Smith, Hawk, and Golden at the Bureau of Mines, in studying the Fischer–Tropsch synthesis, discovered the chemical basis of the OXO process used commercially to make long-chain alcohols.

By 1925, the major areas of application of fuel technology—carbonization, combustion, gasification, and hydrogenation—were well defined and active. The work rested on a broad and growing scientific base, supported in significant degree by the enormous progress in organic chemistry that had occurred since 1850, primarily in Germany.

Between 1925 and World War II came many developments of lasting importance. Means were devised for classifying coals and their constituents. In the U.S. a standard method was developed for determining a carbonizing coal's yield of coke, gas, and tar and for evaluating their utility. Steady progress was made in the understanding of the basic combustion reactions and of the combustion of coal in beds as well as in pulverized form; indeed, the work on pulverized coal remains the most widely applied advance of any of the period.

Work on coal gasification increased our understanding of the basic reactions and produced a number of innovations. The Lurgi process—high-pressure gasification of coal with oxygen and steam—was commercialized in 1936. It is being used more and more abroad, and units may soon be built in this country. The process yields a gas whose heating value is about half that of natural gas. Dent and his co-workers, studying the direct gasification of coal and coke with hydrogen, laid the foundation for modern hydrogasification processes.

Hydrogenation of coal to liquid fuel was practiced first in Germany in 1925 and later in France, Great Britain, and Japan. In 1936, H. H. Storch and his colleagues at the Bureau of Mines began their studies of direct hydrogenation of domestic coals. Coal hydrogenation was used commercially in Germany in World War II. The Fischer–Tropsch synthesis of hydrocarbons from carbon monoxide and hydrogen was studied intensively in several countries, including the U.S. In 1936–40, the process was commercialized by Ruhrchemie in Germany, where it was used during World War II to make gasoline, diesel oil, and other products.

Through the end of World War II, research in fuel chemistry focused largely on the uses of coal as such. Major interests included the ranking, components, and structure of coals; means of converting coal to a smokeless or less smoke-producing fuel; ways to use blends to produce better metallurgical coke; and methods of exploiting less costly, high-volatile coals. The behavior of coals under heat and pressure was studied; processes were developed for separating valuable chemical products from the gas, oil, tar, and pitch resulting from carbonization of coals. In large measure the research was basic, and most of it was performed by a relatively few organizations—the Bureau of Mines, schools of mineral industries in various universities, and a number of State Geological Surveys or Bureaus of Mineral Industries. Large users of coal concentrated on carbonization, mainly to develop more economical processes or better coke.

As World War II drew to a close, this country's consumption of oil was skyrocketing, and the rate of discovery of new domestic reserves had begun to decline. The consequent concern over petroleum supplies led in 1944 to the passage of the Synthetic Liquid Fuels Act. The programs of the Division of Fuel Chemistry soon showed a sharp rise in papers on coal gasification, coal hydrogenation, reactions of coal, and Fischer–Tropsch synthesis, all directed at producing liquid fuels. This burst of activity persisted from about 1945 to 1953 and then vanished—along with the federal Office of Synthetic Liquid Fuels—as quickly as it had appeared, submerged by the cheap and abundant petroleum found in the Middle East.

The U.S. Bureau of Mines built a coal hydrogenation demonstration plant based on German know-how obtained at the end of the war. It was designed to charge just under two tons of coal per hour and produce 200 barrels of gasoline per day. By 1953, when the plant was closed, it had processed 2000–4000 tons each of four types of coal and one lignite. The emphasis was on operability; not all of the problems were solved, but much was learned. The Bureau of Mines also built a research and development laboratory at Bruceton, Pa., where it extended the work on coal hydrogenation that it had started in 1936. The new laboratory's work on catalysts, process variables, reaction mechanisms, and other elements of the process contributed a good deal to the construction and operation of the demonstration plant.

Interest in synthetic liquid fuels also stimulated the construction of two Fischer–Tropsch plants, primarily to make gasoline. One of these was completed in 1950 and put into operation by the Carthage Hydrocol Co. at Brownsville, Tex. Little was published of the technical results, but the process proved unable to compete economically with the production of gasoline from petroleum. The Bureau of Mines, using a somewhat different approach to the Fischer–Tropsch synthesis, built a demonstration plant at Louisiana, Mo. Unsolved problems remained when the plant was shut down, in 1953, but there is reason to believe that the process was capable of reliable, sustained operation.

Both coal hydrogenation and the Fischer–Tropsch process required large amounts of gas—about 8000 cubic feet of hydrogen per barrel of liquid product for hydrogenation; about 30,000 cubic feet of synthesis gas (carbon monoxide and hydrogen) per barrel of liquid for the Fischer–Tropsch approach. The only practical source of the gas in 1946 was coal, so that coal gasification came under very intensive study. A basic problem was that, to minimize distribution costs, synthetic liquid fuels would have to be made in most parts of the country and, if possible, from local coals. The consequent need to cope with a variety of coals puts very stringent restrictions on the gasification process. Of the variety of processes tried—almost all of them using oxygen and steam to gasify the coal—only two went beyond a throughput of 100–150 pounds of coal per hour. The Bureau of Mines plant at Louisiana, Mo. processed one ton of coal per hour and was primarily an experimental unit (synthesis gas for the companion Fischer–Tropsch unit came from another source). The Du Pont gasifier at Belle, W. Va. handled 17 tons of coal per hour and produced 25 million cubic feet per day of synthesis gas. The design was based on work by Du Pont, Babcock and Wilcox Co., and the Bureau of Mines. The unit operated for more than two years, primarily to supply hydrogen for ammonia synthesis, but was closed down when the cheaper natural gas replaced coal as a raw material.

By 1960, research in fuel chemistry had shifted back to the areas of major interest in 1940. This is not to say that work in those areas had come to a halt in the intervening 20 years. Despite the diversion of effort to synthetic liquid fuels, knowledge of the composition, character, and use of fossil fuels had expanded markedly. The years since have seen continued progress in elucidating the structure of coal, particularly with the help of modern instrumentation, and in clarifying the factors involved in the chemical reactivity of coal. In the past decade, moreover, work has intensified steadily on minimizing the environmental impact of coal, a problem compounded in the past few years by renewed fears of a shortage of petroleum and natural gas. These pressures have led once again to strong interest in coal conversion processes, but the goals this time are clean-burning fuel oil and clean-burning gas of the same energy content as natural gas.

The sulfur in coal, upon combustion, enters the air as sulfur dioxide unless preventive measures are taken. One such measure is to scrub the sulfur dioxide from the stack gases, and much effort was being expended by 1975 to develop processes that do this. A second measure, precombustion cleaning of coal, is not generally applicable because it can remove only that fraction of the sulfur—up to perhaps 50%—that is not built into the structure of the fuel. The third approach is the various coal conversion processes now in development.

Four large coal gasification pilot plants were operating in the U.S. in 1975. The processes involved are the Hygas process of the Institute of Gas Technology, the CO_2–acceptor process of Consolidation Coal Co., the

Synthane process of the Bureau of Mines, and the Bigas process of Bituminous Coal Research, Inc. Each process is based on technology developed initially in the 1930s and 1940s, but each has novel variations. These processes are designed to produce fuel gas high in methane, the main constituent of natural gas. In each, sulfur, nitrogen, and oxygen in the coal are converted to gaseous compounds that can be removed from the product gas. In addition to these four approaches to gasification, a number of others are being explored.

Also in the pilot plant are processes for converting coal to liquids. In the COED (Coal Oil Energy Development) unit of FMC Corp., powdered coal is pyrolyzed to liquid hydrocarbons in a series of fluid beds. A second unit, operated by Consolidation Coal Co., treats coal with a "hydrogen-donor" solvent; the resulting liquid hydrocarbons are extracted, and the solvent is rehydrogenated and recycled. The PAMCO process, developed by Pittsburg and Midway Coal Mining Co., produces "solvent-refined" coal. Pulverized coal, mixed with a coal-derived solvent, is hydrogenated at moderate pressure, the resulting hydrogen sulfide and light hydrocarbons are removed, and the solvent is removed by flash distillation to recover the solidified coal product.

A PAMCO pilot plant that processes six tons of coal per day began operating in 1974. The federal Office of Coal Research is sponsoring a second pilot plant that also began operating in 1974; in addition to testing the process it will assess the combustion characteristics of the product. In another approach, the H–Coal process, coal is hydrogenated in an ebulliated bed reactor system developed by Hydrocarbon Research, Inc. In one test, the process converted each ton of coal to 3.6 barrels of butane and low-sulfur liquid hydrocarbons, representing 93% coal conversion. Besides the coal-to-liquids processes already in pilot plant, a number of new conversion systems are under study in the laboratory.

In parallel with the progress in coal conversion, fuel scientists in recent years have continued to clarify the chemistry of combustion. Environmental concerns have bestowed added importance on such research. It is now possible, for example, to control the combustion of coal in power plants so as to minimze—but not eliminate—the formation of oxides of nitrogen. Considerable work has been done also on the behavior of limestone added to a coal combustion system as a means of removing sulfur dioxide.

Trends in fuel chemistry in the past decade or two have reflected a rapidly shifting pattern of fuel use and supply and, more recently, the effort to control environmental degradation. The scientific and technological breakthroughs in energy supply in recent years have been mainly in petroleum chemistry, where much of the research has been concentrated.

Concern over energy in the years ahead, however, has spurred sharply expanded research on coal and oil shale, and there is no reason to doubt that fuel chemists will provide their share of the knowledge required to exploit these vast domestic sources of energy after 1976.

HISTORY OF CHEMISTRY

The Division of History of Chemistry offers a meeting ground for all ACS members interested in any phase of the history of chemistry, domestic or foreign. Between its founding (as a section) in 1921 and April 1975, division members presented 1414 papers before division sessions at 96 national meetings of the ACS.

Two historians of chemistry were instrumental in founding the division. They were Edgar Fahs Smith, professor of chemistry and later provost of the University of Pennsylvania, and Charles Albert Browne, chemist with the Bureau of Chemistry and Soils in Washington, D.C. At the fall national meeting in New York City in 1921, while Smith was ACS President, these two and others organized a Section of History of Chemistry with Browne as chairman and L. C. Newell as secretary. At the national meeting in Birmingham, Ala., the following spring, the section presented its first formal program and arranged an exhibit of rare books, autograph letters, and apparatus. In 1927 the section became a division.

One of the main objects of the division has been to take an active interest in the preservation of chemical landmarks. Beginning about the time of the division's founding, Browne and other division members labored to restore Joseph Priestley's chemical apparatus and return it to his home in Northumberland, Pa. They completed their work in time for the dedication of the Priestley house and museum, Sept. 5, 1926, as part of the Golden Jubilee meeting of the ACS. Another divisional activity has been to commemorate important events and discoveries in the history of chemistry with appropriate symposia; a few examples follow:

- The 100th Anniversary of Liebig's Epoch-Making Address Before The British Association for the Advancement of Science (1940).
- The Kekulé–Couper Centennial Symposium (1958).
- The Benzene Centennial Symposium on Aromatic Character and Resonance (1965).
- The Werner Centennial Symposium (1966).
- Recent Landmarks in the DNA Revolution (1968).
- A Century of Chemical Periodicity (1969).
- The Tswett Centennial Symposium (1972).
- The LeBel–van't Hoff Centennial Symposium (1974).

The division also administers the Dexter Award in the History of Chemistry; Dexter Chemical Corp. established the award in 1956 to recognize outstanding service to the history of chemistry. The award consists of a plaque and $1000 and is given not more than once a year.

THE German poet and dramatist Goethe once wrote that "the history of science is science itself." His observation seems to be borne out by the experiences of Kekulé, Ramsay, and Rayleigh, among many others. Kekulé, the chemist who proposed the cyclohexatriene structure of benzene, read widely in the classics of chemistry before making any original scientific discoveries of his own. Sir William Ramsay and Lord Rayleigh carefully read Henry Cavendish's 1785 paper on nitrogen, which led them to the discovery of the inert gas argon in 1894, more than a century later.

Thus it is not surprising that several founders of the American Chemical Society were keen and active students of the history of chemistry. Most notable among them were Benjamin Silliman, Jr., Henry Carrington Bolton, and Charles F. Chandler. Silliman was professor of chemistry at Yale University and editor of the *American Journal of Science.* Bolton, a widely traveled man of independent means, retired in 1887 at the age of 44 to devote himself to research in the history and bibliography of chemistry. His "Select Bibliography of Chemistry" is consulted frequently today by historians and practicing scientists. In a letter in the April 1874 issue of Chandler's journal, *American Chemist,* Bolton proposed that the year 1774, with its discovery of oxygen by Priestley and of chlorine by Scheele, be considered the starting point for modern chemistry and that its centennial be celebrated in the current year by a meeting of chemists "at some pleasant watering-place to discuss chemical questions, especially the wonderfully rapid progress of chemical science in the past hundred years." His suggestion was greeted enthusiastically and led eventually to the formation of the American Chemical Society.

Despite its many members with an active interest in chemical history, the ACS in its early days had no outlet or provision for research in the field, and many of the older chemists who were interested in history eventually dropped out of the ACS. When Chandler's *American Chemist* stopped publishing, in 1877, historical articles found no alternate outlet, and work in the history of chemistry entered a latent period.

Interest in the subject did not vanish entirely, however. Among the younger chemists active in the Society at the beginning of the 20th century were the two outstanding historians of chemistry, Edgar Fahs Smith (1854–1928) and Charles Albert Browne (1870–1947). Both had studied in Germany and were interested particularly in the history of chemistry in the United States. Smith, a three-time President of the Society (1895, 1921–22) and the author of a book, "Chemistry in America," approached the subject from a study of chemical developments in and around Philadelphia. Browne, who wrote several histories of the ACS as well as 148 historical articles, approached the subject in terms of agricultural chemistry. Strangely enough, these two avid collectors of early chemical books and manuscripts, despite their common interests, did not meet until the 60th national meeting of the Society, in Chicago in September 1920.

Dr. Smith had long planned to found his own journal in the history of chemistry. He hoped that it "would be the most effective means of emphasizing the great cultural value of chemistry and of resisting its present exceedingly materialistic trends." As early as the Birmingham ACS meeting in April 1922, Dr. Smith and Dr. Browne discussed the prospects for founding such a journal. At the New Haven meeting in April 1923, Smith reported that he had approached W. H. Nichols of New York (ACS President in 1918–19) for a grant of $100,000 to be used for the publication of an *American Journal of Historical Chemistry.* This arrangement and others of later date did not materialize, and the project became a frequent subject of discussion at divisional meetings.

Finally, in 1948, 20 years after Edgar Fahs Smith's death, his cherished idea of an American journal devoted exclusively to the history of chemistry became a reality with the appearance of the first volume of *Chymia,* a journal in book format subtitled *Annual Studies in the History of Chemistry.* The journal was sponsored jointly by the Division of History of Chemistry and the Edgar Fahs Smith Memorial Collection in the History of Chemistry at the University of Pennsylvania and had the generous financial support of Denis I. Duveen. *Chymia* was published by the University of Pennsylvania Press, and 12 volumes, containing a total of 139 papers and 2485 pages, appeared before it ceased publication, in 1967. Articles in English, German, French, and Spanish were contributed by scholars from the U.S., the U.S.S.R., England, France, the Netherlands, Brazil, Peru, Latvia, Argentina, Bulgaria, India, Canada, and Scotland. For the first two volumes (1948 and 1949), Tenney L. Davis was the editor-in-chief; the remaining 10 volumes (1950–1967) appeared under the editorship of Henry M. Leicester.

Dr. Smith's widow presented his collection of historical records on chemistry, along with an endowment, to the University of Pennsylvania. The collection comprises some 15,000 volumes of primary source material ranging from 16th century alchemy tracts to the "newer alchemy" of radioactivity and atomic physics. Portions of the libraries of Charles A. Browne and Tenney L. Davis have been added to the collection as well.

Courses in the history of chemistry once were offered at almost every college and university, but by 1975 their number had been reduced drastically. In most schools their demise has been the result of the increasing requirements of other courses in the typically crowded curriculum of modern chemistry. Those who believe it is impossible to *educate* a chemist without teaching the history of chemistry deplore the decline of these courses.

In recent years, too, the history of chemistry has emerged as an independent discipline, which has led to competition between historians of chemistry and chemical historians. The historians of chemistry, who see themselves as historians first and chemists second, seem to be in the ascendancy in 1975; the chemical historians, who were trained primarily in chemistry and entered the history of chemistry through a back door, appear to have been eclipsed. These chemists and amateur historians are the intellectual descendants of the original authors of the history of chemistry, the practicing chemists who sought to add an extra and human dimension to their laboratory research.

With the increasing specialization of modern life, the tasks of the history of chemistry have largely been taken over by the professional historians with their emphasis on historical method rather than on a detailed knowledge of the intricacies of the individual science. Ideally, the practitioner of the history of science, in this case chemistry, should have a comprehensive and profound knowledge of history and the social sciences as well as of the subject matter of the individual sciences, together with a reading ability in a number of languages. Such individuals are rare if not nonexistent. Without them, one must be content with the two approaches —the one stressing history and the other science.

C. P. Snow, especially in his book "The Two Cultures," has focused attention on the split between the practitioners of the humanities and those of the sciences. The history of chemistry, by treating chemistry as a human activity carried out in a milieu of other human activities, can bridge the chasm that splits these twin fields of human endeavor. In this service, the historians of chemistry and the chemist–historians each have crucial roles to play.

INDUSTRIAL AND ENGINEERING CHEMISTRY

The Division of Industrial and Engineering Chemistry, formed in 1908, was the first division of the American Chemical Society. At the end of 1974 it was the third largest division, despite the fact that since its founding it had been the progenitor of 11 other existing divisions. The division initially was the Division of Industrial Chemists and Chemical Engineers and assumed its present name in 1919; the first chairman was Arthur D. Little (ACS President in 1912–13), and the secretary was B. T. B. Hyde.

The few years leading up to the birth of the new division were marked by growing pains within the Society and by the appointment of a committee, in 1906, to consider measures that would avert fragmentation into more specialized groups. In 1907 President Marston T. Bogert added to this committee a subcommittee charged with examining the feasibility of publishing a journal for industrial chemists, by then the Society's largest membership group. In December 1907 the subcommittee recommended to the Council not only that ACS launch the *Journal of Industrial and Engineering Chemistry*, whose first issue appeared in 1909, but that it form the new division as well. Division and journal were closely associated until the latter ceased publication, in response to changing times and needs, in December 1970. In the period 1962–70, in fact, any U.S. or Canadian subscriber could become a member of the division simply by validating his ACS membership card.

In January 1971 the Society launched a new journal, the monthly *Chemical Technology*. The charge to the journal in part was that it treat not only chemistry and engineering but also the sister disciplines that must be brought to bear to bring any industrial innovation to commercial fruition. To help meet *ChemTech's* objectives, its Editor is advised by a panel of divisional representatives chaired by a past-chairman of the I&EC division; in 1975, 16 divisions were represented on the panel.

Because of the highly interdisciplinary nature of its subject matter, the Division of Industrial and Engineering Chemistry traditionally has been one of the Society's most prolific cosponsors of symposia with other divisions. The division also has long sponsored symposia in addition to its program activities at ACS national meetings. In 1934, the division sponsored the first of its Chemical Engineering Symposia, an event that since has been held annually, usually during the last week of December. In June 1963 came the division's first State of the Art Symposium; these events, too, have been held annually. At the request of the Society, the division also has cosponsored symposia with other organizations. Typical examples are "The Planning of Experiments" (1963), with the American Society for Quality Control; a symposium of 12 papers (1967) at Frankfurt on Main, with the Deutsche Gesellschaft fur Chemisches Apparatewesen; and the First International Symposium on Chemical Reaction Engineering (1970), with the American Institute of Chemical Engineers and the European Federation of Chemical Engineers.

The division presents its Joseph E. Stewart Award annually to division members who have rendered distinguished and faithful service to ACS and to the Division of Industrial and Engineering Chemistry (the award was the I&EC Honor Scroll, 1933–70, and was renamed in 1971).

INDUSTRIAL and engineering chemistry embraces a wide range of activities involved primarily in the chemical and other manufacturing industries, but also in such disparate fields as saline water conversion, nuclear power generation, and even fluid flow in the human circulatory system. When the ACS Division of Industrial Chemists and Chemical Engineers was formed, in 1908, the people it served worked mainly in or in direct support of manufacturing; research had not yet emerged as an organized industrial endeavor, and the "industrial chemist" and "chemical engineer" often were considered more or less synonymous in terms of their daily activities. This situation would change over the years, one signpost being the division's change in name, to Industrial and Engineering Chemistry, in 1919.

Today one rarely hears "industrial chemist" used in the original sense; chemists who work in industry are much more likely to be identified by field of endeavor—polymer chemist, medicinal chemist, petroleum chemist, and so on. The industrial chemist tends to have a strong interest in seeing ideas reduced to practice; the chemical engineer is interested in chemical reaction vessels and the physical processes that must be applied to reactor products to convert them to useful materials. Both chemists and chemical engineers are found in research and development, although the chemist is more likely to be working out new reactions and the engineer to be building them into scaled up processes and designing the necessary equipment. Both are found in manufacturing, although the engineer is more likely to be tending to plant operations and the chemist to be troubleshooting reaction upsets and providing analytical and other support services. Both are found in a variety of other industrial activities, including technical sales and service, purchasing, industrial health, and pollution control.

The evolution of industrial and engineering chemistry is inseparable from the growth of chemical engineering, which has yet to see its 100th year as a distinct profession. By the time the American Chemical Society was organized, manufacturing chemists had long been aware that their success depended not only on the basic chemistry of their processes, but also on the design of reaction vessels, the selection of materials of construction, and other factors that came under the heading of "engineering." Thus in 1887 a committee was appointed at Massachusetts Institute of Technology "to consider instruction in engineering as relating especially to applied chemistry." A curriculum was established in 1888, and the first class in chemical engineering ever graduated received its diplomas from MIT in 1891. By the turn of the century, only one other school, the University of Michigan, had adopted such a curriculum, but the basic idea had taken root firmly.

The two fundamental concepts of classical chemical engineering are "unit operations" and "unit processes." They were introduced in 1915 by Arthur D. Little in a report to the Corporation of Massachusetts Institute of Technology. Little pointed out that, "Any chemical process, on whatever scale conducted, may be resolved into a coordinated series of . . . 'unit actions,' as pulverizing, mixing, heating, roasting, absorbing, condensing, lixiviating (extracting a soluble constituent from a solid mixture), precipitating, crystallizing, filtering, dissolving, electrolyzing, and so on." The number of these basic "unit operations" is not very large, Little argued, and any given process uses relatively few of them. Thus a clear, quantitative understanding of such unit actions should enable the chemical engineer to cope with the much larger number of processes in which one or more of them might be involved.

Little's approach caught on rapidly and soon would evolve into the conceptual cornerstone of chemical engineering. He did not himself draw a clear line between unit operations and unit processes, but that distinction was emerging by 1921 when the Division of Industrial and Engineering Chemistry held its first formal symposium on unit operations. The distinction began to crystallize in 1923 with the appearance of the classical textbook, "Principles of Chemical Engineering," by W. H. Walker, W. K. Lewis, and W. H. McAdams. Those authors defined "unit operations" implicitly to include primarily physical processes—fluid flow, crushing and grinding, filtration, distillation, and the like. (Each author, appropriately enough, served as chairman of the division, Walker in 1918, Lewis in 1922, and McAdams in 1927.) In 1935 came a second book, "Unit Processes in Organic Synthesis," in which editor P. H. Groggins treated unit processes as those involving primarily chemical phenomena, such as nitration, alkylation, and halogenation. The division's first formal symposium under the title "unit processes" came in 1937, although for some years its programs had included individual papers and symposia on more or less specific processes.

As chemical engineering evolved in theory and practice, chemical science was moving with equal speed. The two combined to fuel the growth of an already sophisticated chemical industry. In the two decades 1920–40, the value of chemicals produced in the U.S. climbed more than eightfold to exceed $3.7 billion, and similar expansion occurred abroad. At the same time, chemists and chemical engineers were contributing markedly to the technology and operations of other industries, including petroleum, food, pulp and paper, pharmaceuticals, and primary metals. Research and development had become an established feature of the industrial scene; during the 1930s, especially, new products and processes were appearing regularly. A few examples: the first freon refrigerant announced (1930); neoprene in commercial production (1931); vitamin C in commercial production (1934); 100-octane gasoline available commercially (1935); polystyrene offered commercially (1937); pilot plant production of synthetic gylcerol announced (1938); catalytic cracking of petroleum in operation on a commercial scale (1939).

Practitioners of industrial and engineering chemistry were pushed to unprecedented lengths by the demands of World War II. This country's production of the general purpose synthetic elastomer, styrene–butadiene rubber, for example, rose from essentially zero in 1940 to more than 719,000 long tons in 1945, the peak year. The atomic bomb project required the design and construction of gaseous diffusion plants, on a scale never before attempted, to separate fissionable uranium-235 from uranium-238. A penicillin production process was developed and put to use. A large synthetic ammonia industry sprang up. And to these and many similar developments should be added the tremendous wartime expansion of productive capacity in chemically-based industry in general.

In the years since World War II, the industrial applications of chemical science and technology have expanded steadily both in this country and abroad. During 1950–70, the number of chemists working in industry in the U.S. rose some 250%, to about 93,000. Just under half of these were working in the chemical industry; the remainder were employed across the full spectrum of manufacturing and nonmanufacturing industry. Paralleling the growth in industrial chemical manpower has been a stream of new, chemically based products and processes. Among the products have been synthetic detergents, water-thinned latex paints, synthetic diamonds, oral contraceptives, and a variety of plastics and synthetic fibers for both consumer and industrial use. Among the new processes have been zone refining for making transistor-grade silicon, stereospecific polymerization for making, among other materials, polyisoprene or synthetic "natural" rubber, and synthetic ammonia plants that produce

1000–1500 tons/day at costs much lower than could be achieved in earlier, smaller plants.

Industrial chemistry does not claim an unsullied record. With its benefits have come penalties, both real and potential, such as environmental damage and the toxicity of certain materials, especially in long-term, low-level exposure. The work of the ACS Division of Industrial and Engineering Chemistry in providing a forum for the exchange of information on both benefits and penalties is apparent in some of its recent or planned symposia:

- Environmental impacts of chemical engineering.
- Catalytic conversion of coal (to other energy sources, such as pipeline fuel gas).
- Chemical engineering in medicine (such as the use of liquid membranes to oxygenate blood in artificial lungs).
- Occupational safety.
- Single cell protein (as made by feeding microorganisms on hydrocarbons or cellulosic wastes).
- Polymeric membranes for separation and purification (as in purifying water by reverse omosis).
- Fire safety aspects of polymeric materials.

A few examples such as these cannot indicate fully the scope of the division's interests. They do, however, portray its long-time objective—"to support programs that promote the science, techniques, and technology of chemical process and product development" in the broadest sense.

INORGANIC CHEMISTRY

The ACS Division of Inorganic Chemistry enjoyed its first full year in 1957 under chairman John C. Bailar, Jr., and Secretary–Treasurer Larned B. Asprey. Until 1956, inorganic and physical chemists in the Society had operated within the Division of Physical and Inorganic Chemistry, founded in 1908. The dominance of physical chemists in that division, however, and the growing numbers of inorganic chemists led the latter to seek and achieve independent divisional status.

At every national meeting since the fall of 1959 the Division of Inorganic Chemistry has sponsored two or three symposia (excepting the spring of 1973 when only one was sponsored). The division also has sponsored special biennial symposia in the summers of 1968 (Banff, Alberta), 1970 (Blacksburg, Va.), and 1972 (Buffalo, N.Y.) and in January 1975 (Athens, Ga.). In 1967 the division established an Organometallic Subdivision and in 1973 a Solid State Subdivision.

INORGANIC chemistry, defined loosely as the chemistry of all elements except carbon, has had a long and venerable history. Most chemistry during the 18th and 19th centuries was primarily inorganic chemistry, with a sizable contribution from analytical chemistry, which had not yet come into its own as a separate branch of the science. During the first half of the 19th century, however, attention was focused increasingly on the new discipline of organic chemistry, which by the second half of the century was moving much more rapidly than the once dominant inorganic chemistry. A century ago, inorganic chemistry was primarily preparative chemistry; its descriptive aspects were confined to chemical formulae, melting and boiling points, and elementary optical properties such as color and refractive index. Into the early 1900s inorganic chemistry remained sluggish, largely because of the absence of a suitable theory of chemical bonding. Progress was made, however, in a few areas, of which four arbitrarily selected examples are the Periodic Law, the noble gases, the rare earths, and the coordination theory of bonding.

Dmitri Mendeleev's first periodic table was published in 1869. Despite its imperfections, it did demonstrate the Periodic Law—undoubtedly the greatest generalization in all of inorganic chemistry–under which the properties of the elements repeat themselves at periodic intervals. That first table also incorporated several principles that contributed to the later acceptance of the Periodic Law; these included the listing of the elements by increasing atomic weight and the blank spaces for undiscovered elements. Lothar Meyer discovered the Periodic Law independently at about the same time as Mendeleev, and John Newlands several years earlier had noted the repetition of the properties of the elements when they are arranged in order of increasing atomic weight. Among the contributors to the evolution of the Periodic Law were a number of Americans, including Oliver Wolcott Gibbs (1822–1908), Josiah Parsons Cooke, Jr. (1827–1894), and Gustavus Hinrichs (1836–1923). Lewis Reeves Gibbes, who published his "Synoptical Table of the Elements" in 1886, was completely unaware of the work of Mendeleev, Meyer, and Newlands some years earlier.

The unexpected discovery of argon in 1894 by Ramsay and Rayleigh opened a new and exciting chapter in inorganic chemistry. In 1888, William Francis Hillebrand, the American mineralogical chemist and President of the ACS in 1906, noticed that treatment of the mineral uraninite with acid yields an inert gas, which he believed to be nitrogen. This observation led directly to the discovery of helium by Ramsay in 1895. In 1907 H. P. Cady of the University of Kansas, a pioneer in research with liquid ammonia, together with D. F. McFarland of the same university, discovered helium in the natural gases of Kansas, thus initiating a new industry.

During the late 19th century much progress was made in the exceedingly difficult separation and identification of the lanthanides or rare earths. J. Lawrence Smith (1818–83), the Amercian mineralogical and analytical chemist, investigated the rare earths in samarskite and verified Mosander's conclusions on the complex nature of yttria. B Smith Hopkins (1873–1952) did considerable work on methods of separation, atomic weights, and spectra of the rare earths. British-born Charles James (1880–1928), working in the U.S., discovered lutetium, the last of the lanthanides but because of his caution and delay in publication, the French chemist Georges Urbain was awarded priority. Another American chemist who did research on the rare earths was Herbert Newby McCoy (1870–1945), who in 1907 with W. H. Ross recognized the existence of the chemically inseparable elements that Frederick Soddy named isotopes. McCoy also gave the first quantitative proof that the α-ray activity of uranium compounds is directly proportional to their uranium content.

In inorganic chemistry a long stalemate had resulted from overdependence on organic structural concepts, particularly Kekulé's dogma of constant valence. Not until the end of the 19th century, as a consequence of Werner's revolutionary coordination theory, was structural inorganic chemistry placed on a sound theoretical footing. Before Werner's work, however, research on what were then known as "molecular compounds" or "complex compounds" was pursued in the United States. In 1856, Oliver Wolcott Gibbs and Frederick Augustus Genth (1820–93) published a 67-page memoir, "Researches on the Ammonia–Cobalt Bases," marking "the first distinct recognition of the existence of perfectly well defined and crystallized salts of cobalt–ammonia bases." The German-born Genth, who was President of the American Chemical Society in 1880, was one of the foremost mineralogical chemists in the United States, while Gibbs devoted his later years to a vast, extremely complicated, and almost unexplored field, the heteropoly compounds, which he called the complex inorganic acids. Gibbs also was a pioneer in the chemistry of the platinum metals as was James Lewis Howe (1859–1955), who began his research in the 1890s and became the outstanding American authority on and bibliographer of the platinum metals in general and the undisputed world authority on the chemistry of ruthenium in particular.

Alfred Werner (1866–1919), with his ideas on bonding (coordination number, isomerism, etc.), probably did more than any scientist in the past 100 years to bring order into inorganic chemistry. Werner won the Nobel prize in 1913, and in the early 1920s his ideas began to take hold. These ideas, along with the development of valence bond theory by Heitler and London and others and the introduction of quantum theory led to rapid advances in the description of chemical bonding.

In the next two decades up through World War II, remarkable advances were made by "physical chemists" in the development of methods for studying chemical compounds to gain a better description of their structure, bonding, etc. These methods included visible and infrared absorption spectroscopy, x-ray crystallography, and electron diffraction. The years to follow saw the introduction of new methods of structural study, such as nuclear magnetic resonance, electron paramagnetic resonance, Raman spectroscopy, photoelectron spectroscopy, etc. At the same time came giant steps in the theory required to describe the bonding in inorganic compounds. The major impetus in this field was R. S. Mulliken's theories of molecular orbitals.

Nevertheless, despite the research in progress, inorganic chemistry was in the doldrums during the first four decades of the 20th century. As the late Ronald S. Nyholm observed, it was widely regarded as a dull and uninteresting part of the undergraduate curriculum, seemingly consisting chiefly of unconnected facts with no system comparable to that found in organic chemistry and with none of the rigor and logic characteristic of physical chemistry. In 1928, when the noted coordination chemist John C. Bailar, Jr., began his research at the University of Illinois, only half a dozen or so universities were offering doctorates in inorganic chemistry or publishing much in the field.

By 1939, however, the stage was set for a resurgence that has become known as the renaissance in inorganic chemistry. According to Nyholm, this modern inorganic chemistry, defined as the integrated study of the preparation, qualitative and quantitative composition, structure, and reactions of compounds, most of which do not contain carbon, was reborn from the old chemistry primarily as a result of three factors: "the purposeful application of physicochemical theories, especially the ideas of quantum mechanics . . . the use of new preparative techniques, such as ion exchange, and new physical methods, such as infrared spectroscopy and electrical and magnetic measurements for solving old problems . . . [and] the atomic energy program, coming just at a time when inorganic chemistry was ready for a big advance, focused attention on the need and opportunities for research in the field." It was probably the urgency of the atomic energy program that catalyzed the development so vigorously. For example, when many of the products of nuclear fission were found to lie between elements 57 and 71 in the periodic table, research on the rare earths, once the prerogative of a few, became of interest to numerous academic and industrial scientists. Also, as a direct result of the wartime atomic bomb project, an entirely new family of elements, the transuranium elements, was discovered by nuclear bombardment of uranium and heavier elements. This country seems to have had and still has a virtual monopoly on the synthesis of the transuranium elements. The 11 elements from neptunium (element 93) to lawrencium (element 103) were all prepared at the University of California, Berkeley. For their work in the field at Berkeley, Glenn T. Seaborg and Edwin M. McMillan shared the Nobel Prize for Chemistry in 1951.

Werner's work, ironically enough, was so complete and all-encompassing that many felt that he had already answered all the important questions in inorganic chemistry and that few advances were left to be made. The renaissance in inorganic chemistry has changed this attitude, and coordination chemistry is one of the most active fields in chemistry today. In fact, "complex chemistry" is an approach to the entire field of inorganic chemistry and not just a specialized topic. Some of the more active areas of current inorganic research include valence theory, stabilization of unusual oxidation states, unusual coordination numbers, metal–metal cluster compounds, heteropoly compounds, the rarer elements, macrocyclic ligands, organometallic compounds including π-bonded sandwich compounds, rates and mechanisms of reactions, relationships between structure and reactivity, and complexes containing oxygen and nitrogen molecules. Many coordination compounds have served as model compounds for biological systems, and the interdisciplinary field of bioinorganic chemistry is one of the most exciting fields of contemporary chemical research.

The Organometallic Subdivision of the ACS Division of Inorganic Chemistry deals with a field that a century ago was just making its first important impact on chemistry as a whole.

Mendeleev and Meyer were both attempting to work out a periodic table by writing the elements in the order of their atomic weights, but were having difficulty in deciding between the combining weights and true atomic weights. The problem was that no one knew which were the characteristic or principal valences of the elements of variable oxidation state. A long search showed that such elements gave only a single organometallic derivative, R_nM. So the new branch of organometallic chemistry, discovered by Robert Bunsen 30 years before, and developed by his student Edward Frankland in England through a long series of researches beginning in 1849, bore fruit in Mendeleev's classic paper establishing the concept of periodicity in 1871.

In 1875, chemists would have known of the work of Sir Edward Frankland in preparing from his diethylzinc reagent alkyl derivatives of the other elements. They would have known also of Cadet's fuming arsenical liquid, first obtained in 1760 by heating arsenic oxide with potassium acetate, which had been reinvestigated by a host of chemists from Berzelius, Thénard, Bunsen, and Cahours to Baeyer, Dumas, Frankland, and Kolbe. Completely forgotten, by contrast, would have been Zeise's pioneering paper in 1827 on the ethylene complex of platinum, only rediscovered in the 1930s. The dimeric nature of trimethylaluminum also had been known for a decade by 1875, but had attracted little interest, and no explanation was available.

The next 100 years were to prove very fruitful. Mond explored the action of carbon monoxide on nickel metal in 1890 and discovered the first metal carbonyl. Gosio had the first glimmer of recognition of biological methylation in 1897 when investigating poisonous vapors from arsenic-containing wallpapers. Dimroth carried out the first aromatic mercuration in 1898, and this was followed in 1900 by Grignard's discovery of the alkylating abilities of magnesium halide preparations. Organic chemists for the first time were beginning to look toward organometallic chemistry for new synthetic methods.

Dilthey produced the first silicone in 1905, and Kipping began a long series of researches into the differences between carbon and silicon at about the same time; neither chemist recognized the usefulness of the silicone polymers produced on hydrolysis of their compounds. Ehrlich, on the other hand, was pursuing a purposeful search for an effective organoarsenical chemotherapeutic agent, and his efforts were crowned with success in 1907 with the *para*-aminophenylarsenicals. The next decade saw the flowering of the work of Shlenk on the action of the alkali metals on unsaturated organic compounds, the reaction of potassium with naphthalene having been explored by Berthelot in 1867. Shlenk's work on the aromatic radical anions in solution in 1928 was very far ahead of its time, and so was that of Stock, who produced the first carborane from diborane and acetylene in 1923. In 1928, Ziegler laid the groundwork for the Nobel prize he shared with Natta, in 1963, in papers on the polymerization of olefins; in 1929, Paneth produced free radicals for the first time from the pyrolysis of organometallic derivatives, which much stimulated chemical bonding theory.

Ziegler, Gilman, and Wittig began to publish convenient reactions for the preparation and use of organolithium reagents in the 1930s, demonstrating the great power of these compounds in organic synthesis. In 1926, meanwhile, Midgley had turned up tetraethyllead as the most effective antiknock known, allowing much more energy to be derived from petroleum.

The transition metal chemistry that had lain fallow through this period began to reemerge with Hieber, who prepared the first carbonyl hydrides in 1931, and Hein, who prepared transition metal–nonmetal bonds a decade later. Organotin compounds first were used as stabilizers for polyvinyl chloride in 1936, and Rochow discovered his direct synthesis for the economical production of silicone polymer precursors in 1940. Calingaert subjected the redistribution reaction to detailed study at the same time.

The end of World War II saw the hydrosilylation reaction published by Sommer in 1947 and the beginnings of the perfluoroalkyl derivation of the elements in 1949 by Emeléus. Structural studies and theoretical contributions by Rundle established the electron-deficient methyl bridge in 1953, and theoretical and experimental studies by Craig, Maccoll, Nyholm, Orgel, and Sutton and experimental contributions by Hedberg on the planarity of trisilylamine and the acidity order of silyl-substituted benzoic acids by Chatt established the $(p \rightarrow d)-\pi$ bonding concept in 1954 and 1955. The first announcement of the hydroboration reaction was made by Brown in 1956, and the first resolution of optically active silicon compounds by Sommer in 1959.

The organotransition metal area had in the meantime exploded into activity. The first patent on the oxo- or hydroformylation process had been issued in 1944, and Reppe's wartime German work on the oligomerization of olefins was published beginning in 1948. Dewar and Chatt were proposing explanations for the formation of olefin and acetylene complexes in 1951 and 1953. It seemed that a reinvestigation of Zeise's early work would have been in order. Instead, the discovery of ferrocene was published by Pauson in 1951 and followed quickly by independent discoveries by Miller, Tebboth, and Tremaine in 1952, and by Fischer and by Wilkinson also in 1952. Woodward recognized the compound's aromatic properties and named it ferrocene. Antecedent compounds containing carbon–transition metal σ-bonds had been made in 1907 by Pope and Peachey, using platinum, and by Gilman in his trimethylgold in 1948, but the discovery of ferrocene was the opening shot in the renaissance of organometallic chemistry. Fischer prepared dibenzenechromium in 1955; Longuet–Higgins' and Orgel's postulated cyclobutadiene–stabilization on a transition metal atom in 1956 was effected by Criegee for the tetramethyl derivative in 1959 and for the parent compound in 1965 by Petit. Zeiss reinvestigated the polyaromatic chromium compounds of Hein and reinterpreted them in terms of π–complexation in 1957. Stone prepared the first cyclooctatetraene complexes in 1959, and Streitweiser synthesized the analogous uranium derivatives in 1968.

Piper and Wilkinson in 1956 recognized with remarkable intuition that the α–cyclopentadienyliron and copper compounds they had made were undergoing a rapid shifting of metal–carbon bond sites, and Davison, Cotton, and Meutterties postulated the detailed explanations.

Crowfoot–Hodgkin in 1961 had determined the structure of vitamin B_{12} which was found to contain a cobalt–carbon σ–bond; Hawthorne had explored the chemistry of the many diffferent types of carborane cage derivatives, including in 1965 the carborane analogs of ferrocene.

Transition metal salts together with organoaluminum compounds had been used by Ziegler in 1955 to polymerize α–olefins, the mechanism of the hydroformylation of olefins had been elucidated by Heck in 1960, and the production of acetic acid from methanol by organo–rhodium catalysts was demonstrated in 1971. Fischer synthesized carbene complexes of the transition metals in 1967, and Wilkinson and Lappert codiscovered in the early 1970s the α–elimination-stabilized silylmethyl ligand that allows derivatives of metals in low oxidation states to be isolated. Homogeneous catalysis by transition metal complexes continues to be an area holding the hopes and active attention of a number of prominent chemists, including Wilkinson, Halpern, Chatt, Collman, Basolo, Parshall, King, Heck, Holm and Wojcicki.

MEDICINAL CHEMISTRY

The ACS Division of Medicinal Chemistry was organized in 1909, originally as the Division of Pharmaceutical Chemistry, with A. B. Stevens as chairman and B. L. Murray as secretary. Early programs of the division reflected largely interests in drug assay methods and formulation improvements. Because of growing interest in the chemical structure of drugs, however, the division's name was changed in 1920 to the Division of Chemistry of Medicinal Products and in 1928 to its present name. The division's primary goal is to stimulate progress in medicinal chemical research. One of its chief means of doing so has been through programs and symposia at ACS national meetings. In addition, in 1948 the division organized the First National Symposium in Medicinal Chemistry, an event that has been held biennially ever since. The division recognizes excellence in medicinal chemical research through its Award in Medicinal Chemistry, which it has given biennially since 1966.

The division supports and participates actively in the *Journal of Medicinal Chemistry*, which ACS has published since 1962 (in its first year as the *Journal of Medicinal and Pharmaceutical Chemistry*). In 1965 the division itself began publishing "Annual Reports in Medicinal Chemistry." This review of research and discoveries in medicinal chemistry is now recognized as the most significant review of its kind.

IN 1951, a distinguished team of medicinal chemists reviewed the achievements of medicinal chemistry from 1876 to 1951.[1] At the time, a decade of creation of almost unbelievable "miracle drugs" had just passed. The major vitamins had been synthesized and were being added to foods, raising the hope that nutritional deficiency diseases would soon be a thing of the past, at least in the developed countries. Many hormones, including various steroid hormones, had been synthesized, and their use to correct a variety of disorders was near. Several of the major antibiotics had been isolated and were being produced commercially; among them was chloramphenicol, the first one to be manufactured by chemical synthesis. The peak of research and development of the antihistaminics had passed, and a few cardiovascular and autonomic drugs were being introduced into medicine. Potent synthetic analgetics had

[1] Moore, M. L., *Ind. Eng. Chem.*, **43**, 577 (1951).

been discovered, and the effort to separate dependence-liability from the useful pharmacological properties of such drugs was well advanced.

The decade preceding 1951 had covered the years of World War II, when almost all available medicinal chemical manpower had been channeled into urgent military work, such as the development of antimalarial drugs, penicillin, and the sulfanilamides. In spite of the massive efforts of American medicinal chemists and biological scientists, the fundamental discoveries in these areas came from abroad. The clinically useful antimalarials of today, excepting primaquine, a U.S. development, were discovered in Germany and England. Penicillin, a British discovery, was brought to the point of commercial fermentation and structural elucidation at Oxford University. Sulfanilamide came from France, after a German discovery that a related red dye killed bacteria. Spectacular successes in the discovery and development of useful drugs have occurred since in many lands, including the U.S., where many of the major developments in modern medicinal chemistry have been made. Many members of the ACS Division of Medicinal Chemistry have contributed to the theoretical foundations that barely existed in 1950 and to the ever-increasing impact of drug discovery, drug development, and the explanation of drug action on public health and the happiness of living in our age.

Antiinfectious Agents

In 1951 isoniazid was introduced, capping a successful campaign against tuberculosis whose first success was the discovery of the clinical antituberculous activity of *p*-aminosalicylic acid only a few years earlier. These drugs have removed this ancient white plague from the list of major causes of death.

During the decade 1948–58, many of the major naturally occurring antibiotics now in use came into being. A few of them were synthesized by ingenious methods in which novel reactions and stereoselective steps played a major role. New fermentation procedures brought decisive changes in manufacturing processes. Some of these fermentations could be guided to give important fragments of the molecules of the natural antibiotics. Using these fragments as synthetic intermediates, a vast array of semisynthetic antibiotics has been assembled. A few of these have reached clinical utility and possess many advantages over their natural prototypes. For example, of the 10,000 or more penicillins synthesized from 6-aminopenicillanic acid, several are orally active, less allergenic, and tolerated better than benzylpenicillin and have a wider or differently useful spectrum of antibacterial activity. Similar conditions hold for the many semisynthetic cephalosporins. The most active areas of antibiotics research are the tetracyclines, the aminoglycosides, and the polyene and macrolide antibiotics. The 70–odd antibiotics in clinical use in 1975 have reduced mortality rates from serious infections by 80–90%. There is still great need for new broad-spectrum antibiotics for gram-negative infections and for strains of all bacteria that have become resistant to the existing drugs.

The antibacterial and antiprotozoal action of 2-substituted 5-nitrofuran derivatives has been known for 30 years. The clinical importance of the drugs in genito-urinary and alimentary tract infections was established only gradually, however, after the traditional distrust of nitro compounds, dating back to the toxic weight-reducing nitrophenols, had been overcome. Similarly, the nitroimidazole derivative, metronidazole, developed at Rhone–Poulenc, in France, has become the prototype antitrichomonal chemotherapeutic.

The war in Southeast Asia (1962–72) exposed millions of U.S. military personnel to tropical infections and infestations. Research on such diseases had been lagging for two decades. More urgent domestic health problems had occupied biomedical scientists, government funding was unavailable, and the pharmaceutical industry could see no way to recover research investments from markets in underdeveloped tropical societies. This picture was changed by the rapid emergence of plasmodial strains resistant to the antimalarial drugs developed during World War II. Many organic chemists devoted a decade of synthesis efforts to tenuous leads in antimalarial research under the sponsorship of the Walter Reed Army Medical Research Center. More than 150,000 new compounds were screened, mostly against *Plasmodium berghei* in mice, by Leo Rane of the University of Miami. The difficulty of this program is emphasized by the fact that only a few drugs have been found that can cure or suppress resistant plasmodial infections in vivax and falciparum malarias, and these have not yet been developed for broad clinical use. With U.S. personnel no longer endangered by resistant malarias, funding of antimalarial research again is declining.

Medicinal research on antifungals, anthelmintics, antiamebics, and other antiparasitic agents has advanced only slowly. Its visible successes have been more often in veterinary infestations than in clinical situations. Even so, several of the veterinary drugs developed during 1950–75 have had a major impact on sheep and cattle raising. However, the increasing ecological restrictions on such drugs, and on insecticides and other pesticides, have posed a serious and continuing threat not only to veterinary and tropical medicine but also to agriculture and crop improvement in a starving world.

Many infections became treatable with antibiotics and other antimicrobial agents by the mid-1950s, and mortality, at least from bacterial diseases, fell sharply. The lives of many children could now be saved by these drugs, and the statistical longevity of the population increased correspondingly. Only serious virus infections continue to defy chemotherapy. Antiviral agents have been tried clinically, but none has helped to stem the periodic epidemics of viral diseases. Fortunately, effective preventive vaccines have been developed for measles, mumps, rubella, and poliomyelitis. Nevertheless, the frequent mutations of influenza and common cold viruses have dimmed the prospects of successful immunization for these diseases and place high priority on chemotherapeutic research in these areas.

Antihypertensive and Diuretic Drugs

As the longevity of the population increased, chronic diseases of advanced age moved to the fore as targets of medicinal research. This trend was augmented by the research support these diseases received from governmental sources.

Hypertension is a major cause of death through cardiac and cerebrovascular accidents. The role of catecholamines in hypertension is well documented, and many attempts have been made to block these amines by hindering their biosynthesis or their action at the receptors. One of the earlier antihypertensive drugs was reserpine. Its principal biochemical effect is the inhibition of the active transport of catecholamines and serotonin into tissue storage sites, thereby enabling intraneuronal monoamine oxidase to destroy these amines.

Another approach was through inhibition of dopa decarboxylase, the enzyme that catalyzes the biosynthesis of dopamine from dopa. The action of this enzyme is inhibited by α-methyldopa, although this homolog of dopa is also a substrate of the decarboxylase. Competitive antagonism of dopamine formation, joined by antagonism of biosynthetic α-methylnorepinephrine to norepinephrine, reduces the effects of dopamine and norepinephrine and lowers the blood pressure.

The most widely used antihypertensive drugs are the thiazides and their congeners. They were first synthesized by Novello and Sprague at Merck Sharp and Dohme Laboratories and found by Beyer and Baer to be potent, orally active diuretics. Even more pronounced diuretic effects, with less inhibition of carbonic anhydrase, have been achieved with the hydrothiazides, in which the 3,4–double bond is saturated. In hypertensive individuals, several hydrothiazides also exert a mild antihypertensive effect.

Because the thiazides combine antihypertensive and diuretic properties, they have been used extensively to relieve edema caused by cardiac insufficiency. Their diuretic activity may have to be supplemented by other agents, because the thiazides promote kaliuresis as well as the desirable natriuresis. This drawback is minimzed in 2,4,7-triamino-6-phenylpteridine (triamterene), developed by Wiebelhaus, Weinstock, and their associates at Smith Kline and French Laboratories.

Among other cardiovascular disorders, much progress has been made in the therapy of cardiac arrhythmias. Both atrial arrhythmia and the dangerous ventricular arrhythmias respond well to propranolol, a β-adrenergic blocking agent introduced in Britain by Black, et al., in 1964. Propranolol also effectively abolishes symptoms of angina pectoris.

Antihyperglycemics

A side effect of the sulfanilamides led to an additional important class of drugs, the orally active antihyperglycemic agents. The sulfanilamides had been known to reduce the concentration of blood glucose. A series of thiadiazolyl sulfanilamides causing marked hypoglycemia in man were studied in France, and modification of their molecules led Franke and Fuchs in Germany in 1955 to test arylsulfonylureas for the same activity. Several of these compounds are now used clinically to manage maturity-onset or stable diabetes. Their oral activity gives them one advantage over insulin, which must be administered by injection at frequent intervals.

Nonsteroidal Antiinflammatory Agents

Among functional disorders that are aggravated by advancing age, arthritis and other inflammatory conditions have presented a stubborn and unresolved obstacle to medicinal chemical research. The lack of understanding of the underlying causes of these diseases remains the main deterrent in devising meaningful animal model systems against which new compounds can be tested. It is no exaggeration to call the past quarter of a century the age of the biochemistry of physiological processes. There is hope that as biochemists turn to the explanation of pathological abnormalities, a new wave of progress in medicinal thought will follow. An indication of this could be seen in the discovery by John R. Vane at the Royal College of Surgeons, in London, that nonsteroidal antiinflammatory drugs, including the 85-year-old aspirin as well as indomethacin, mefenamate, phenylbutazone, and acetaminophen, inhibit the biosynthesis of prostaglandin $F_{2\alpha}$. This prostaglandin is known to induce inflammation in rats and in man.

Since the full tide of steroid research, 20–30 years ago, there has been little to match the excitement offered by the organic chemical, biochemical, and medicinal modification of the prostaglandins after about 1960. One pharmaceutical company is reputed to have invested more than $100 million in such work. This concentrated effort was given an additional spark by the great unifying discovery of Nobel laureate (1971) Earl Sutherland that the adrenergic hormones, the prostaglandins, and, apparently, the steroids and other hormones act on the adenylcyclase system mediating the release of cyclic AMP and GMP. It is a bit too early to say whether any really important new drugs will repay this monumental effort in a practical way.

Hormones

In 1953, the synthesis of two polypeptide hormones of the posterior lobe of the pituitary gland was announced by Vincent du Vigneaud, an achievement crowned in 1955 by a Nobel prize. Not only did this work make oxytocin and vasopressin available in pure form, it opened the door to molecular modification of these and other peptide hormones. Improved methods of peptide synthesis, especially the Merrifield solid phase automated procedure, made possible the synthesis of ever larger polypeptides, including such hormones as insulin, MSH, ACTH, somatotropin, the angiotensins, and several enzymes.

It had been deemed axiomatic that nature had evolved in the animal body the most active metabolic catalysts, as seen in the uncontested position of the vitamins and essential coenzymes. This was held also for the many steroid hormones. However, Fried and Borman were among the first to recognize that certain modifications of steroid structure produce predictable biological effects. This served as a starting point in the molecular modification of glucocorticoids and other steroid hormones. Methylated, halogenated, unsaturated, and ring-modified analogs began to compete with hydrocortisone and other natural therapeutic hormones. The most widely known result was the development of synthetic congeners of progesterone by Carl Djerassi, which made possible the compounding of antiovulatory contraceptives by Gregory Pinkus. The effects of these "pills" on population control and the sexual liberation of women cannot yet be assessed fully.

Psychopharmacological Agents

The first neuroleptic drug, chlorpromazine, appeared in France in 1952. Its effect and that of some of its congeners on the treatment of psychotic diseases and on the conversion of overcrowded and restrictive mental institutions to psychiatric hospitals has been recorded as a medical and social epic of drug therapy. An alternative to the tricyclic neuroleptics was found later in the butyrophenone derivatives, whose first representative, haloperidol, was launched clinically by Paul Janssen in Belgium in 1959.

Psychopharmacology came also to include drugs for neurotic disorders and endogenous depressions, which became amenable to drug therapy during the next decade. Minor tranquilization of neurotic patients was achieved with meprobamate through the work of Frank Berger of Carter–Wallace Laboratories. Anxiety and other neurotic conditions were suppressed with the benzodiazepines, discovered by Leo Sternbach and Lowell O. Randall of Hoffmann–LaRoche, Inc. Depressed states became treatable with monoamine oxidase inhibitors and the tricyclic drugs imipramine, amitriptyline, and their congeners. Simultaneously, it was found that the serious and widespread neurological disorder parkisonism was caused in part by a deficiency of dopamine in certain brain tissues, such as the substantia nigra, and that this defect could be ameliorated by administration of L-dopa, the biogenic precursor of dopamine.

Antineoplastic Agents

The group of diseases covered by the collective term cancer remained, until recently, irreversible; cures were provided only by surgical removal of the primary tumor if it had not metastasized. There is little hope for tumors that have migrated to additional, remote tissues. Destruction of tumors by irradiation can delay death, but the effective and the toxic doses of radiation lie too closely together to permit acceptable long-term therapy. The same holds for the chemotherapy of tumors. Medicinal efforts along these lines go back to antiquity. Every known plant and plant product has been tested for its ability to inhibit the growth of tumors. The combined screening programs of the Cancer Chemotherapy National Service Center (CCNSC) of the National Cancer Institute, the Sloan–Kettering Institutes for Cancer Research, the Chester–Beatty Research Institute in London, the Pasteur Institutes, and a few minor screening contractors have examined a quarter million compounds for inhibition of animal tumor growth. Of these, several hundreds have advanced beyond the primary and secondary screening criteria. Among them are inhibitors of specific enzymes involved in tumor cell metabolism, inhibitors of mitosis, and many structures that do not lend themselves to classification on biochemical grounds.

Antineoplastic drug research began in the 1940s with the discovery of the antileukocytic properties of military mustard gases. Isosteric expansion of this "lead" to the nitrogen mustards furnished the first nucleophilic alkylating agents that could prolong significantly the lives of patients with certain types of leukemia. Under the aegis of the CCNSC hundreds of more sophisticated alkylating agents were screened, and a few of them were nontoxic enough to become useful in treating leukemias. Other active compounds were designed as antimetabolites, either as structural analogs of nucleosides or protein constituents or of substrates of anabolic enzymes. The most useful drugs of this kind are linked to the names of George Hitchings, Charles Heidelberger, John A. Montgomery, and Roland K. Robins.

Of the alkylating agents, nitrogen mustard itself, chlorambucil, L-sarcolysin, and, especially, cyclophosphamide have proved themselves clinically, although their toxicity limits the period of treatment. Cyclophosphamide gives a 20–30% cure rate in Burckitt's lymphoma, and busulfan has become the agent of choice against chronic myelocytic leukemia. Chemotherapy combined with radiotherapy has effected cures in retinoblastoma. The antifolate drug methotrexate offers a cure rate of 70% in choriocarcinoma in women and 30–50% remissions in acute childhood leukemia. Combination therapy of methotrexate, vincristine, 6-mercaptopurine, and prednisone is used to produce prolonged remission rates in acute leukemia.

The lesions most responsive to 5-fluorouracil are tumors of the gastrointestinal tract, breast, and female genital organs, where 15–30% of often prolonged remissions have been observed. 6-Mercaptopurine, another antimetabolite, is about as effective against chronic myelocytic leukemia as is busulfan, and the *Vinca* alkaloids vinblastin and vincristine show promise against acute leukemias and some solid tumors. The quite toxic antibiotic actinomycin D has been used to treat renal lesions of Wilms' tumor in children and has achieved survival rates in excess of 80%.

These and a few other successes in the chemotherapy of malignant tumors have raised hopes for further cures of these diseases. There have been some fallouts from such work, catalyzed by studies under the National Cancer Act. Many antitumor agents also act as immunosuppressants and have found use in preventing rejection of transplanted tissue. Because some animal tumors, and perhaps even some forms of human breast cancer, have been shown to be caused by viruses containing reverse transcriptase, many antitumor drugs have been tested in viral diseases. Herpes simplex responds to 5-iodouridine and to nucleosides containing arabinose instead of ribose, especially Ara-A.

Theoretical Developments

It has been clear for some time that the external pressures and restrictions typified by federal laws and regulations are not the only causes of the current slowing of drug discovery and development. Long before a drug is selected for study in depth, an enormous waste of scientific manpower, materials, and overhead occurs because of ignorance of the mechanisms of drug action. Much of the search for new drugs is still done by methods reminiscent of those in use 50 years ago, although a small amount of such work is done now by more directed and sophisticated pathways.

The discovery of drugs by starting from scratch is still an unaccountable empirical adventure. One can screen compounds for a given disease entity. One can follow up therapeutic folklore in natural products research and, if one is lucky, one such hint in several thousands will lead to a novel active material. All other methods of drug discovery require an available "lead" or prototype compound. This holds even for the two most sophisticated methods of searching for novel drugs: the biological observations of side actions of an experimental or clinically used drug; and the molecular modification of recognized biochemical metabolites.

In the first method, the side effects are usually undesirable or at best unnecessary, but may, in structurally derived substances, become a valuable asset in an often quite different branch of medicine. Pharmacologists watch for such opportunities routinely in the pharmacological laboratory and in experimental clinical medicine. In fact the demands of the Food and Drug Administration for uniqueness of action and absence of side effects in drugs have sharpened the alertness of experimental biologists to side effects in tissues unrelated to the main thrust of the investigative drug. Such properties in turn have motivated medicinal chemists to undertake molecular modifications to suppress the original principal activity and enhance a potentially useful side effect.

For the molecular modification of normal biochemical metabolic products, again, a prototype structure, that of the metabolite, is needed as a starting point. Many of the thousands of antimetabolites found active in vitro fail to affect disease conditions in vivo. In spite of experimental support for several hypotheses to explain this failure, the causes are still obscure. Explanations of receptor hypotheses are still in their infancy.

In spite of these uncertainties, the medicinal chemist can base his designs on whatever knowledge exists about the general biology of a given disease and the structure-activity relationships of families of chemical compounds. A totally empirical approach is no longer the primary method of drug development as it was of necessity 25 or 30 years ago. Medicinal chemists in the 1970s are in a state of "enlightened empiricism." The present advances in biochemistry and pharmacology should permit them to discard the "empiricism" altogether in another quarter century. The emphasis on the role of drug metabolism initiated by B. B. Brodie, and the calculations of the effects of substituents on partition coefficients by Corwin Hansch, exemplify the hopes for broader theoretical foundations.

MICROBIAL CHEMISTRY AND TECHNOLOGY

The Division of Microbial Chemistry and Technology was formed in 1961. Since 1946, a group of chemists, microbiologists, and engineers with common interests in fermentation research and technology had been arranging programs at ACS national

meetings. This group, known initially as the Fermentation Section and later as the Fermentation Subdivision of the Division of Agricultural and Food Chemistry, was the precursor of the new division, whose first officers were Gilbert M. Shull, chairman, and Henry J. Peppler, secretary-treasurer. The division's programs have focused on industrial, academic, and governmental research in applied microbiology, chemical engineering, and bioengineering. Among the major topics of concern are ethanol and distillery processes; antibiotics, vitamins, and pharmaceuticals; yeast propagations; organic acids and solvents; and cultures for legume inoculation, cheese making, and other food fermentations. By 1975, developments in the biotechnologies of enzymology and tissue cultures had prompted a proposal that the division be renamed the Division of Microbial and Biochemical Technology.

The division reports its program planning and other activities to the members thrice yearly in a mimeographed publication, *the fermento(e)r*. (The dual spelling of the name recognizes both the fermentation vessel and the biocatalyst in it.) The division encourages participation in international meetings. In 1968 it cosponsored the 3rd International Fermentation Symposium, at Rutgers University. The other cosponsor was the Fermentation Industries Section of the International Union of Pure and Applied Chemistry.

The division's Microbial Chemistry and Technology Distinguished Service Award recognizes extraordinary service to the division and the fermentation industry or exceptional achievement in international professional relations. The first award was in 1966; six members of the division have been so honored. The division's William H. Peterson Award recognizes promising students in biochemistry, bioengineering, and microbiology. Recipients are selected on the basis of their presentation of resarch results at a divisional annual meeting. Thus far, two students have received the award, in 1972 and 1974.

FERMENTATION as a practical art has been practiced for millennia in processing foods and beverages: bread, cheese, sauerkraut, vinegar, pickles, olives, milk, beer, wine, distillates. In this country industrial fermentations go back to the start-up of a commercial process for making lactic acid in 1881. Until about 1920, industrial fermentations generally lagged behind European discoveries and applications in chemistry, microbiology, and manufacturing technology, but in the past half century large-scale, microbially-dependent processes have evolved steadily in the U.S. (Table II).

For many years commercial fermentation was used mainly to convert carbohydrates to products such as citric acid, gluconic acid, lactic acid, acetone, butanol, and ethanol. By 1940 a few companies were using fermentation to make these chemicals as well as bakers' yeast and a few crude enzymes with protease, invertase, or amylase activity. The industry grew gradually during this period, stimulated in particular by the advent of a riboflavin (vitamin B_2) process and the booming antibiotics era that began with penicillin in 1942. Major expansion in the fermentation industries began around 1960 and is still in process. By 1973, microbiological processes were turning out some 60 antibiotics and a variety of other products, including enzymes, organic acids and solvents, amino acids, vitamins, pharmaceuticals, and yeast.

Some fermentation processes, meanwhile, have been replaced partly or wholly by chemical processes. Prominent among these are the fermenta-

TABLE II

Evolution of Fermentation Processes in the U.S.

1870–1880	Bakers' yeast, lactic acid.
1900–1910	Ethanol, amylases, glycerol.
1920–1930	Acetone, butanol, citric and gluconic acids.
1930–1940	Proteases, riboflavin, sorbose, food and feed yeast, phenylacetylcarbinol.
1940–1950	Penicillin, streptomycin, neomycin, aureomycin, vitamin B_{12}, pectinases, cellulases.
1950–1960	More antibiotics, steroid oxidation, invertase, glucose oxidase, kojic and 2-ketogluconic acids, glutamic acid, lysine.
1960–1970	Tetracyclines, penicillins and other antibiotics; gibberellic acid, dihydroxyacetone, milk coagulants, lipases, catalase, glucoamylase, lactase, single cell protein, industrial waste utilization.

tions for acetone, butanol, lactic acid, riboflavin, ethanol (nonbeverage), gluconic acid, and glutamic acid. Restrictions on petrochemical feedstocks, coupled with an adequate supply of starchy grains, could bring fermentation back into the picture for some of these products.

Research on microbiological processes today is directed in part at making complicated molecules that cannot readily be produced by purely chemical methods. Such molecules include enzymes, vitamin B_{12}, and certain antibiotics and amino acids. For some of the amino acids, research is under way on replacing carbohydrates with methanol, ethanol, acetic acid, and cheap hydrocarbons as sources of energy for the microorganisms. Yeast is being used commercially to convert hydrocarbons to citric acid, and other hydrocarbon fermentations may soon become economical. Also in use on a limited basis are microbial insecticides. They are highly selective toward the target pests, readily degradable, and may replace certain chemical insecticides that are less selective and less readily degradable. Production of single-cell protein by growing microorganisms on hydrocarbons is well along in development; current problems include the cost of the hydrocarbons and questions about nutritional value and toxicity. Research and development are under way also on conversion of waste cellulose to single-cell protein by bacteria and to glucose by fungi.

Fermentation seems likely to be involved for some years in meeting the antibiotic needs of the future. These needs include a nonallergenic penicillin; an antibiotic that, taken orally, is effective against systemic Gram-negative infections; an antifungal agent that is effective systemically and is less toxic than current compounds; and a growth-promoting antibiotic for animal feeds that is not cross-resistant with antibiotics used against animal infections. Research was under way in 1975 on all of these needs.

Another likely development is the use of animal cell cultures for antiviral vaccines or to produce hormones and other medicinals. Current problems include the selection of cells for study and producing them in large amounts. Promising targets include the polypeptide hormones of the pituitary gland and other organs and tissues.

More broadly, fermentation processes and products have certain characteristics that should adapt them particularly well to meeting the demands of the future. In brief:

- Fermentations may be designed to process renewable raw materials such as cellulose.
- Fermentations can be used to process a variety of waste materials.
- Fermentation products are the result of biological metabolism and are thus by nature readily biodegradable.
- Industrial fermentation need not depend on large reserves of minerals and fossil fuels and thus should favor underdeveloped or resource-poor countries.
- Fermentation can be used to convert wastes and other materials to foods and feeds.

NUCLEAR CHEMISTRY AND TECHNOLOGY

The Division of Nuclear Chemistry and Technology was formed in 1963, after several years as a subdivision of the Division of Industrial and Engineering Chemistry. The first chairman was Morton Smutz; the secretary-treasurer was Premo Chiotti. The division's interests extend from fundamental studies of nuclei to practical applications of radioactivity, and its members come in about equal proportions from academic, governmental, and industrial institutions. The use of radioactive nuclides has expanded greatly over the past 30 years in all branches of science. Thus the division's activities encompass a broad spectrum of science and technology. Divisional symposia at ACS national meetings, for example, often are sponsored jointly with other divisions, and the topics have ranged from analyses of the moon to reactor safety.

The division sponsors the Charles D. Coryell Undergraduate Award in Nuclear Chemistry. Up to two awards of $500 each are presented annually, and the first two were presented in 1970. Awardees are selected for the ingenuity, novelty, and potential usefulness of their completed nuclear or nuclear-oriented research projects in chemistry or chemical technology.

NUCLEAR chemistry[1] originated in the early work of Pierre and Marie Curie that began shortly after Becquerel's discovery of radioactivity in 1896. The Curies characterized and concentrated the new elements polonium and radium, both announced in 1898, depending strongly on chemical manipulations and chemical reasoning. In 1900, the collaboration between physicist Ernest Rutherford and chemist Frederick Soddy culminated in recognition of radioactive disintegrations as subatomic changes and thus revolutionized the concept of atoms as immutable entities. Soddy's chemical insight led him eventually to recognize that several radioelements, each with its own distinctive characteristics, could occupy a single position in the periodic table. He arrived thereby at the concept of "isotopes," a term he used in a 1913 paper. Soddy also was the first to propose that the occurrence of isotopes was not confined to radioelements, but "that each known element may be a group of nonseparable elements occupying the same place [in the periodic table], the atomic weight not being a real constant, but a mean value."

There were other pioneers in nuclear chemistry. K. Fajans, independently of Soddy, contributed much to the understanding of the connection between radioactivity and atomic structure and in 1913 formulated some of the rules that govern the chemical behavior of trace quantities of materials. O. Hahn discovered a number of naturally occurring radioactive species, cleared up the decay sequence in the ^{235}U series (1918), and contributed importantly to knowledge of the behavior of substances at extremely low concentrations (1926). G. von Hevesy, with F. A. Paneth (1913–14), originated the idea of radioactive isotopes as tracers and later (1935) pioneered their application to biological problems. Following the discovery of the neutron by Chadwick in 1932 and of artificial radioactivity by I. Curie and F. Joliot in 1934, the scope of nuclear chemistry expanded to encompass the entire periodic table. The current era began with two closely related discoveries, of fission in 1938 and of the first two transuranium elements in 1940.

O. Hahn and F. Strassmann painstakingly developed the firm chemical evidence that lent credibility to nuclear fission, which ran counter to then-accepted ideas of nuclear physics. Since then, nuclear fission has been a special concern of chemists, primarily because of the particular suitability of the techniques of chemistry in studying the phenomennon. Chemists established the gross features of the asymmetric mass-split in low-energy fission and the quantitative yields of approximately 100 fission products.

[1] This review is taken from a report by the Panel on Nuclear Chemistry of the Committee for the Survey of Chemistry, Division of Chemistry and Chemical Technology. Publication 1292-C, National Academy of Sciences–National Research Council, Washington, D.C., 1966. The original material has been modified and edited for use here.

Likewise, the dependence of these yields on the nature of the fissioning nucleus and on the amount of excitation has been determined by radiochemical techniques. In 1962–63, Soviet nuclear chemists and physicists discovered transuranium nuclear isomers, with very short half-lives, that decay primarily by spontaneous fission. This work led to the concept of a second minimum in the fission barrier caused by shell effects in highly deformed nuclei.

In 1940, neptunium was discovered by E. McMillan and P. Abelson and plutonium by G. Seaborg, E. McMillan, J. Kennedy, and A. Wahl. Since then, 11 other transuranium elements have been produced, largely through the efforts of nuclear chemists at Berkeley, Calif., and at Dubna in the U.S.S.R. The discovery of element 103, reported in 1961, completed the identification of elements in the actinide series. Element 104, reported in 1964, and element 105, reported in 1968, are expected to have chemical properties similar to those of hafnium and tantalum, respectively; element 106, the most recent to be reported (in 1974), is expected to resemble tungsten.

The synthesis of elements 93 through 101 was achieved by neutron and cyclotron bombardments of target materials of lower atomic number. By 1955, enough einsteinium had been produced in the Materials Testing Reactor in Idaho to serve as a target for helium ions, which led to the discovery of element 101. Elements 102 through 106 were produced by bombardment of heavy elements (americium, curium, californium) with heavy ions, that is, ions heavier than helium. The synthesis of new elements via the heavy-ion approach, however, becomes increasingly difficult with increasing atomic number: reaction yields become extremely small as a result of strong competition from fission reactions, and the relatively neutron-deficient isotopes attainable by heavy-ion bombardment are short-lived. Synthesis via slow buildup by neutron capture in reactors is essentially blocked at fermium-258, which is so unstable with regard to spontaneous fission that its half-life is only 0.3 millisecond. A possible approach to the synthesis of superheavy elements is to employ highly endothermic reactions between heavy nuclei in an inverse fission process, such as reactions induced by xenon ions in which the compound nuclei would be formed with relatively low-excitation energy. Research groups in Orsay, France; Dubna, U.S.S.R.; and Berkeley, U.S., are now using heavy-ion accelerators to study such reactions.

More than 100 isotopes of transuranium elements have been identified. In addition, more than 1400 radioactive nuclides have been identified throughout the periodic table. Detailed measurements of the nuclear properties of these nuclides by chemists and physicists contributed significantly to the formulation in 1949 of the independent-particle or shell model of the nucleus by Maria Goeppert-Mayer and J. H. D. Jensen and in 1953 of the collective model by A. Bohr and B. Mottelson. The latter model combines with the shell model the important concept of collective motions, such as rotations and vibrations of the nucleus as a whole. The past two decades have seen much progress in testing, modifying, and refining these basic models through studies of radioactive decay, particle emission in nuclear reactions, and Coulomb excitation.

Nuclear Dynamics

The dynamics of nuclear reactions is an area that attracts many nuclear chemists. It is concerned with the details of processes that occur when nuclear species are transformed into other nuclear species on interaction with bombarding particles. These may be elementary particles (such as photons, neutrons, protons, and mesons) or complex nuclei (such as deuterons, helium nuclei, carbon-12, neon-20, and the like).

Nuclear dynamics may be divided into three parts: studies of low-energy nuclear fission, already mentioned, and of low- and high-energy nuclear reactions, which are distinguished somewhat arbitrarily by bombarding energies of less than and more than 100 million electron volts (MeV), respectively. Nuclear chemists and physicists have done much work in attempts to characterize particular low-energy reactions as proceeding either via compound-nucleus or via direct-interaction mechanisms. Most reactions induced by neutrons, photons, or alpha particles with energies of up to 30 or 40 MeV appear to be interpretable as compound-nucleus processes, but considerable evidence exists also for direct interactions even at these low energies, and some reactions appear to exhibit intermediate behavior. Reactions proceeding by a direct-interaction mechanism hold additional interest in that they provide information on nuclear energy levels complementary to that obtained by conventional nuclear-spectroscopic techniques.

The study of high-energy reactions has depended on the availability of high-energy accelerators. Thus nuclear chemists have explored the energy range of bombarding particles provided by each successive generation of accelerators: the synchrocyclotrons (100–450 MeV) in the late 1940s; the proton synchrotrons (1000–6000 MeV) in the 1950s; the strong-focusing synchrotrons (up to 30,000 MeV) in the 1960s, and up to 400,000 MeV in the 1970s. Various phenomena have been discovered in each energy region, and many require further detailed investigation.

Spallation reactions, characterized by emission of one to several dozen small fragments, have received much attention, both experimentally and theoretically. The complex patterns of these reactions and their dependence on bombarding energy, on mass, and on the charge of the target nucleus are accounted for in terms of the cascade-evaporation model. In the cascade stage the incident particle sets off a rapid sequence of collisions between individual particles in the nucleus, leading to ejection of some particles involved in this cascade and to formation of a more-or-less highly excited intermediate nucleus. This excited intermediate, like the compound nucleus in low-energy reactions, is then deexcited on a slower time scale by the evaporation of additional particles. Fission, pictured as the breakup of an intermediate excited nucleus, also occurs in high-energy bombardments.

A third process, fragmentation, has been postulated to account for reaction products not easily ascribable to fission or spallation, particularly low-mass fragments formed in large yield from heavy-element targets. Effects related to details of nuclear structure are observed most readily in relatively simple reactions and have been searched for in (p,pn) and (p,2p) reactions. A small number of experiments with incident pi-mesons and mu-mesons have been reported. The number of meson experiments promises to increase dramatically with the availability of the meson factories at Los Alamos, N.M., and Würenlingen in Switzerland.

Since the discovery of nuclear fission and the first transuranium elements, nuclear chemists have played indispensable roles in both military and peaceful applications of nuclear energy. The technological requirements of these programs are often related so closely to fundamental research that it becomes difficult to separate the two. A prime example is the transuranium elements. Their chemical properties were first established by following their radiations when the elements were available only in unweighably small quantities; on the basis of this information the original large-scale separation processes were devised. Similarly, the development of processes for isolating weighable amounts of the elements neptunium through einsteinium from neutron-irradiated material has depended on nuclear chemical investigations on a tracer scale. A further example is that chemists, particularly active in identifying, characterizing, and determining the relative production of the many products of nuclear fission, sought this information originally to gain a better understanding of the fission process. The data turned out to be essential for designing proper chemical processes for purifying plutonium, because fission products are the main radiation hazard in purification plants.

Studies of fission products, covering the middle third of the periodic table, also focused attention on the chemical behavior of many unfamiliar elements—zirconium and niobium, selenium and tellurium, yttrium and the rare earths, and the new element technetium. Interest in these elements contributed strongly to the development of ion-exchange and solvent-extraction procedures, which now are used widely.

Without doubt, nuclear fission and the development of processes for large-scale production of some transuranium elements (especially plutonium) have had far-reaching consequences for mankind. They are not, however, the only examples of advances in other disciplines that are traceable directly to developments in nuclear chemistry. The vast field of isotopic (radioactive and stable) tracers, so important in chemistry, biology, medicine, agriculture, and industry, sprang from nuclear chemistry. Not only did nuclear chemists invent, develop, and first use the tracer method itself, they also discovered and characterized some of the most important radioactive tracer isotopes, such as carbon-14 and iodine-131. Nuclear chemists were responsible in addition for discovering most of the radioactive nuclides, such as cobalt-60 and cesium-137, used as powerful radiation sources in industry and medicine.

With the advent of tracer methods came two other fruitful developments. One was the study of the chemical effects of isotopic substitutions; these provide a most useful tool in the study of chemical reactions and bonds and form the basis for the most successful method of separating the isotopes of hydrogen and other light elements. The second development was the invention of techniques for labeling complex molecules with radioactive atoms; noteworthy among these is labeling with energetic atoms—so-called hot atoms—produced in nuclear reactions.

In analytical chemistry, radioactivation has opened up previously inaccessible ranges of sensitivity. In this method, characteristic and easily detected radioactive elements are generated by bombarding the sample with neutrons or other radiation. The achievement of the extremely high purities required for transistor materials depended on such a method for measuring trace impurities. Isotope dilution and the Mössbauer effect (announced in 1958) are other important analytical methods developed and used by nuclear chemists.

Geology and geochemistry were revolutionized by the introduction of radioactivity clocks, which established absolute and accurate measurements of geologic time. By this means, geologists can determine the time elapsed since the solidification not only of various terrestrial rocks but also of meteorites and lunar materials. Another radioactivity clock, different in character and applicable to more recent dates, was provided by the ^{14}C-dating method of W. Libby (1947). It has given significant impetus especially to archaeology and anthropology. A final example of cross-fertilization is the impact that physicists' and chemists' investigations of nuclear reactions have had on recent thinking in astrophysics. One may cite two notable milestones: the discovery (1952) of spectral lines of the relatively short-lived element technetium in stellar spectra, which stimulated the development of theories of element synthesis in stellar interiors;

and the discovery (1952), by chemical means, of the process of many successive neutron captures on a rapid time scale in thermonuclear explosions. The latter discovery led to the realization that such processes may be important in supernovae and thus to new ideas about the mechanisms of the gigantic stellar explosions and their roles in element synthesis.

Instrumentation

Effective research in nuclear chemistry depends on proper instrumentation. The availability of specialized (and costly) equipment often determines the direction of research in this area; the periodic technological breakthroughs that make possible instruments of higher quality present opportunities to do new types of research and to reexamine old problems.

Research on induced nuclear transformations is possible only with a nuclear reactor or accelerator. Accelerators usually have been designed primarily with the purposes of physicists in mind; nuclear chemists have had to adapt their work to the type of accelerator available. As high-energy particles of first hundreds and then thousands of MeV were produced, their interactions with complex nuclei were investigated; as beams of the new particles became sufficiently intense, the reactions of pi- and mu-mesons were studied; with the development of machines to accelerate heavy ions, the new types of reactions produced by such particles were characterized.

This dependence on the characteristics of available accelerators will, no doubt, continue to influence the course of much nuclear chemical research. As the beams of K-mesons and antiprotons become more intense, their reactions with complex nuclei will be studied. Yet the technological developments that made reactor and accelerator design more standard and efficient have made possible the construction of machines that have as their design goal the requirements of nuclear chemists. The High Flux Isotope Reactor at Oak Ridge, Tenn., for example, has as its prime aim the production of high-neutron fluxes for manufacturing transuranium elements; the heavy-ion linear accelerator at Berkeley, Calif. was designed for synthesizing new transuranium isotopes; cyclotrons at Argonne, Ill., and at Berkeley were built and are operated primarily for research in nuclear chemistry.

In detectors for nuclear radiations and in associated electronic equipment, significant advances in the past 35 years have made it possible to reexamine abandoned problems and test developing theories. Advances in detectors have provided increased sensitivity, improved resolving power, and decreased resolving time. For detecting electrons, the Geiger counters of the 1940s have been superseded by proportional counters and scintillation counters with much shorter response time. Unlike Geiger counters, they deliver an output electrical impulse that is proportional to the energy loss in the detector. In determining the energies of gamma rays, an essential part of the study of nuclear energy levels, the development of thallium-activated sodium iodide scintillation crystals made possible the enormous amount of research in nuclear spectroscopy in the 1950s. The resolution of these detectors, although remarkably improved, is still poor compared to that obtainable in other regions of the electromagnetic spectrum and compared to that needed to check theoretical predictions of nuclear models.

The newest development in detectors has been the solid-state semiconductor devices made of silicon or germanium. They improve resolution by an order of magnitude over scintillation counters in determining the energy of charged particles or gamma radiation. The gamma-ray spectroscopic possibilities are particularly exciting and call for restudy of nearly all known radioactive isotopes with solid-state detectors.

Radiation detectors of higher resolving power would have meant an intolerable increase in the time required for measurements had there not been parallel development of electronic multichannel analyzers. These instruments process electrical impulses of various sizes and store them in computer-type memory systems. Multichannel analyzers make possible simultaneous measurements over an entire energy spectrum instead of at isolated points. Recent developments in this field resulted in instruments that can provide energy-correlated distributions for the outputs of detectors that measure radiations emitted in coincidence by a radioactive substance or as a result of a nuclear transformation.

This increased efficiency of producing data has made much nuclear chemical research dependent on access to high-speed computational facilities. In fact, raw data from many experiments are never examined by the experimentalist in the traditional visual manner but are fed directly into a computer.

While nuclear chemists are the beneficiaries of advances in instrumentation, they have also contributed to this field. In particular, they led in the development of techniques for measuring extremely low levels of radioactivity. Such developments were inspired by some of the most challenging problems of nuclear research: establishment of the radiocarbon-dating method; detection of a few atoms of a new element; measurement of cosmic-ray-induced nuclear transformations. The development of new techniques for measuring low-level radioactivity undoubtedly will make possible many new investigations of physical and chemical processes taking place on a large scale.

ORGANIC CHEMISTRY

A section on organic chemistry, chaired by James Flack Norris, met for the first time at the ACS Philadelphia meeting in December 1904. In 1908 the section became the Division of Organic Chemistry, with E. C. Franklin as chairman and R. H. McKee as secretary. Specialties within "organic chemistry" are legion, and many of them over the years have found specialized forums in other ACS divisions, including Medicinal Chemistry, Carbohydrate Chemistry, Petroleum Chemistry, and Polymer Chemistry. Still, despite this diversion of membership, the Division of Organic Chemistry by the end of 1975 was the largest in the Society.

The growth of the division led to the recurring event now called The National Organic Chemistry Symposium, first held in 1925, which was planned to permit more intimate scientific exchange than could be achieved at the much larger ACS national meetings. The division held these symposia biennially during the Christmas holiday through 1941, when the series was interrupted by wartime travel restrictions, and since 1947 has held them in alternate years in June. Continued growth has eroded the intimacy of these affairs, so that the division lately has begun to sponsor additional, topical meetings of restricted attendance. The first such event, on anion activation, was held in the fall of 1973 and the second, on nucleophilic substitution, in the spring of 1975.

Beginning in 1959, a highlight of The National Organic Chemistry Symposium has been the presentation of the biennial ACS Roger Adams Award in Organic Chemistry and the recipient's award address. The award, initiated in 1959, consists of a gold medal, a sterling silver replica of the medal, and $10,000. It is sponsored jointly by Organic Syntheses, Inc., Organic Reactions, Inc., and the Division of Organic Chemistry.

The division has long been associated intimately with the *Journal of Organic Chemistry,* which the American Chemical Society acquired in 1954. The journal was conceived by Morris Kharasch, who obtained the backing of the University of Chicago and the Williams and Wilkins publishing house. At the ACS national meeting in the spring of 1935, Kharasch revealed his plans to the division, which, though enthusiastic, could not support the venture officially because the Society would not be the publisher. Members of the division became members of the editorial board, and it was agreed unofficially that the board of the journal would always include the secretary of the division. Although the journal is now published by the Society, the divisional secretary still is an ex officio member of the editorial board, and the division's executive committee retains the privilege of nominating candidates for the editorship when a new editor is to be selected.

A second publishing venture began in the early 1960s, when the division's executive committee recognized the practicing chemist's need for an authoritative source of organic chemical nomenclature. New kinds of compounds were being produced that caused great difficulties for writers of articles in the field. In fact, the dynamic nature of the science foredoomed the project to constant updating, and some thought that it would never reach fruition. Nevertheless, John H. Fletcher, who had undertaken to assemble and edit the book, completed it in 1973 with the help of coauthors Otis C. Dermer and Robert B. Fox, and an enormous amount of volunteer labor. "Nomenclature of Organic Compounds" (Advances in Chemistry Series No. 126) was published by the Society in 1974 and offered to members of the division at very modest cost.

THIS brief, historical account of progress in organic chemistry is intended to show the most significant advances in the field and to emphasize the key role that organic chemistry has played in the development of related fields that have become distinct disciplines. These include, among others, the chemistry of polymers, both natural and synthetic, biochemistry, molecular biology, medicinal chemistry, organic photochemistry, and the various fields of spectroscopy (mass, ultraviolet, infrared, nuclear magnetic resonance, electron spin resonance).

Organic chemistry as a discrete discipline began in this country around 1875; this corresponds with the founding of the Johns Hopkins University with its graduate program (1876) and with the beginning of systematic research at several centers other than Hopkins. The few general principles on which structural organic chemistry is based had been laid down by 1875. As G. N. Lewis said in 1931, "Structural organic chemistry, although developed without mathematics, except of the most elementary sort, is one of the very greatest scientific achievements. An enormous mass of information was reduced to a well-ordered system through the aid of a few simple principles."

Before 1914, the leading centers of research in organic chemistry in this country were Hopkins, Yale, Harvard, Columbia, Chicago, and Michigan; Hopkins Ph.D.'s were the most numerous single group in teaching institutions and government laboratories. In general, the work published in the *American Chemical Journal* and the *Journal of the American Chemical Society* was competent or better. Michael, Stieglitz, Nef, Acree, Noyes, Kohler, and others published excellent work on syntheses and reaction mechanisms; Stieglitz, and later Norris, suggested carbonium ions as reactive intermediates.

The entire development of reaction mechanism studies in this country is epitomized in four papers by chemists whose active careers together span the full century 1875–1975. Acree (1907) presented a partial picture of acid-catalyzed oxime formation. Michael (1919) illustrated the effects of structural changes in carbonyl compounds on the rate of semicarbazone formation, a paper rather antiquated in viewpoint. The 1932 Conant–Bartlett publication on the kinetics of semicarbazone formation, thoroughly modern in viewpoint and technique, gave a sharply defined illustration of the thermodynamics–kinetics dichotomy, which since has become a fundamental concept in physical organic chemistry. In 1958, using spectrophotometric instrumentation not available in 1932, Jencks showed that a tetrahedral addition product was the intermediate in semicarbazone formation.

The years 1914–32 saw the development of the electron-pair concept of valence, principally through the insight of G. N. Lewis. Lewis' generalized theory of acids and bases was to become extraordinarily important in later experimental and theoretical work. In many problems, including the major one of elucidating the structures of proteins and nucleic acids, hydrogen bonding was to play a fundamental role. The quantum mechanical theory of valence was developed by Pauling and by Hückel. These ideas for the first time gave a theoretical basis for the structural theory of organic chemistry, particularly for conjugated and aromatic systems. A qualitative anticipation of these theories of conjugated systems had been developed independently by C. K. Ingold in England. The application of valence bond and molecular orbital theory in theoretical calculations and in suggesting synthetic problems has had a profound and increasing effect on organic chemistry up to the present.

Research in structural organic chemistry became increasingly expert in this country. The work of Jacobs on the cardiac aglycones, of Roger Adams on chaulmoogric acid, of Hudson on sugars, of Levene on nucleic acids and stereochemistry, and of Sumner on the isolation of the first crystalline enzyme, urease (1926), was as important as any similar work being done anywhere.

Equally prophetic for the future of organic chemistry was the continuing development of methods in organic synthesis. The annual publication of tested synthetic procedures, "Organic Syntheses," originated in the University of Illinois laboratory with Adams and C. S. Marvel and, in its second half century, remains a vital influence. Even more important may have been the series of monumental papers (1929–36) on polymerization by Wallace Carothers (1896–1937) of Du Pont. Carothers' work not only cleared up the chemical nature of polymers, eliminating vague, incorrect speculations, it also laid the foundation for the development of high polymers for textile fibers, elastomers, and plastics of many kinds.

The development of physical organic chemistry during 1914–32, including reaction mechanisms and relations among chemical structure, physical properties, and reactivity, clearly foreshadowed the very prominent place the field was to assume in this country. Although important work in this area had been done by Lapworth, Ingold, and Robinson in England and by Brønsted in Sweden, the American development was largely independent. Conant and Lucas developed outstanding schools in this area. Hammett's "Physical Organic Chemistry" (1940) organized the significant available knowledge and provided a viewpoint comprehensive and prescient enough to include many future findings. It remains an American chemical classic, comparable to those of Gibbs, Lewis, Pauling, and Flory.

In the period 1933–45 the trends mentioned above became clearer. The total synthesis of equilenin, with all four possible stereoisomers, by W. E. Bachmann marked the opening of an era of total synthesis of complex natural products that is still going on. The synthesis of retronecine by Adams and Leonard and of quinine by Woodward and Doering are further examples.

The structures of complex natural products, some of them of great biochemical importance, were unraveled. Among these were vitamin B_1 (R. R. and R. J. Williams), in which uv spectroscopy proved very useful, gossypol (Adams), marijuana constituents (Adams), the cardiac aglycones

(Jacobs), vitamin E (Smith), vitamin B_6 (pyridoxal, Folkers), vitamin K (Doisy), and many others.

A very important development was the recognition of free radical chain reactions, pioneered by M. S. Kharasch and his students, Walling and Mayo. Kharasch became the recognized world leader in free radical studies. The free radical path of vinyl polymerization was established by Flory, Price, and others and soon developed into a field of its own; Marvel was the first university chemist in this country to make a major study of polymerization processes.

Hammett developed a broadly successful linear free energy equation that connected the rate of reaction of aromatic compounds with the dissociation constants of substituted benzoic acids. The determination of Hammett "rho and sigma constants" was to become a standard characterization of aromatic substituents.

The chemical research projects supported by the Office of Scientific Research and Development during World War II produced useful results on antimalarial drugs and many other problems. More significant, however, was the pattern of government support for research that grew out of the wartime organizations. The National Science Foundation (established 1950) and numerous other government agencies supported chemical research on a large scale, which enabled American organic chemistry and many other sciences to achieve world preeminence.

The wartime work on penicillin yielded the first widely successful antibiotic and started the intensive searches for other mold metabolites that have continued to the present. This work has produced large numbers of compounds of novel chemical structure and some with very useful clinical properties. Further, the penicillin problem emphasized the power of x-ray crystallography in establishing chemical structures, a fact that was to become a major aspect of the "Instrumental Revolution."

Before 1940, the instrumental techniques available to organic chemists had not changed markedly from those of 1890, except perhaps for polarography and column chromatography. Ultraviolet spectroscopy was used in the 1870s, and ir and uv measurements on organic compounds were published from 1920 on. The instruments were individual models, however, built and operated by specialists; they were too cumbersome to use and too limited in number to permit general use in the determination of spectra. The Instrumental Revolution may be said to have started in the 1930s with pH meters equipped with glass electrodes; in the 1940s came the Beckman DU uv spectrophotometer, a sturdy instrument of good resolving power for qualitative and quantitative use in the determination of electronic spectra. After 1945, commercial ir spectrometers became available, allowing routine qualitative or quantitative determination of vibrational spectra for characterization of organic compounds. Optical rotatory dispersion and circular dichroism measurements have been brought to a new peak of usefulness by new instrumentation and detailed investigations (Djerassi), permitting the determination of absolute configurations.

Mass spectrometers were available to organic chemists after 1945. The resolution and range of molecular weights accessible increased steadily. The combination of high resolution mass spectroscopy with a computer can yield possible empirical formulas for each ion formed. Extensive studies are being directed to computer programs that can give structural formulas for samples from analysis of the mass spectrometric cracking pattern.

The combination of computer and x-ray diffraction has made the determination of structure almost routine, even for compounds not containing an atom of high scattering power. For molecules of low symmetry properties, the calculations necessary before computers were formidably laborious, but complete structures can now be determined for enzymes and proteins. A recent striking example of a simpler compound is the steroid batrachotoxin, a compound of extraordinary toxicity and interesting neuropharmacological properties, whose structure was established by x-ray without using a heavy atom as a marker (I. Witkop and J. Karle).

Since 1940, separations of complex mixtures by various partition methods have developed rapidly along with instrumental advances. Chromatography by paper, ion exchange, vapor phase, thin layer, affinity, and high pressure liquid, as well as countercurrent distribution, have been used widely in organic chemistry and in biochemistry. Many fundamental advances in these two fields since 1945 have depended on partition methods, combined with labeling by radioactive or stable isotopes.

The most versatile and important of all the new instrumental methods undoubtedly is nuclear magnetic resonance (NMR). It was discovered in 1946 (Purcell, Bloch) and commercial instruments were soon available. No technique has shown such a variety of important applications and such apparently limitless potential for further development. Its application to organic problems was pioneered by Roberts. It has become a highly developed field in itself, with its own journals and reviews and many excellent monographs.

At first, NMR was used mainly for ^{1}H, for structural determinations; by considering chemical shifts and spin-spin coupling and decoupling, detailed structural and stereochemical information could be obtained in a very short time. Improvements in instrumentation have made ^{13}C NMR routine for compounds containing the natural abundance of ^{13}C; extensive structural information results. The usefulness of NMR in measuring rates of reaction was recognized early; rates of rotation around single bonds, as in dimethyl formamide, can be obtained and energy barriers to rotation calculated. An example is the measurement of the rate of flipping of the alkyl group on the oxygen in the oxonium salt (**1**) (Lambert). Rates of conformational changes can frequently be measured, and the impact of NMR on stereochemistry has been profound. The "shift reagents" for extending the range of chemical shifts have greatly increased

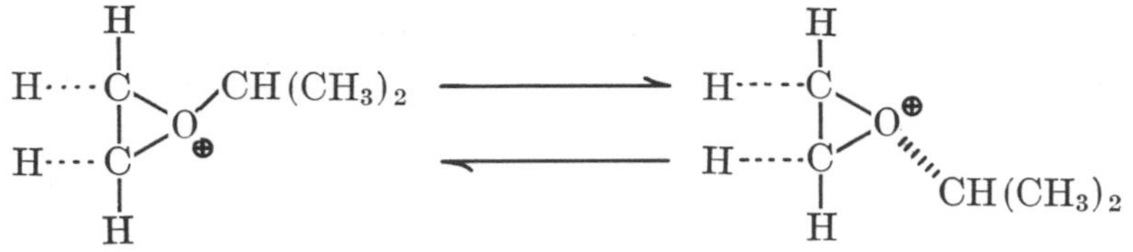

1.

the information available from NMR. The chemical shift of protons on a ring system is the best indication of aromatic or nonaromatic character. The NMR technique has developed a number of variations, such as spin-echo and the nuclear Overhauser effect.

By combining NMR, ir, and mass spectroscopy, structural problems that would have taken months or years before 1940 can frequently be solved in days. These techniques, plus x-ray crystallography, have made classical chemical degradations for structure determination obsolete in most cases.

Electron spin resonance (ESR), which shows the presence of an unpaired electron, has revolutionized the study of free radicals. It is a powerful tool in studying structures of biochemically important compounds by attaching a "spin label," usually an N–O group with an odd electron (McConnell).

The problem of aromatic character has shown striking advances since 1945, and it is clear that E. Hückel's condition of $(4n + 2)\pi$ electrons for aromatic character is well followed for monocyclic systems. The synthesis and studies of cyclooctatetraene (COT) and related compounds by Cope, and the compound's subsequent ready availability by the German process for polymerizing acetylene, confirmed Willstätter's report that COT was nonaromatic.

The lower analog of benzene, cyclobutadiene, was obtained by Pettit as an iron carbonyl complex and was found to be a highly reactive nonaromatic compound. The tropylium ion (**2**), first recognized by Doering and Knox, obeys Hückel's rule and shows aromatic stabilization. The pyrene derivatives (**3** and **4**) of Boekelheide show the most spectacular change from aromatic to nonaromatic character, as determined by the NMR shielding of the protons in the ring plane, and of the CH_3 protons in the π cloud above and below the aromatic ring. Compound **3** has 14π electrons and is aromatic, while **4** is nonaromatic (16π electrons). Breslow has shown that several ring systems containing 4π electrons, such as the

H H H H H H H *2.*

CH_3 CH_3 *3.* 2K [CH_3 CH_3]$^{\ominus}$ $2K^{\oplus}$ *4.*

cyclopropenyl anion, cyclobutadiene, and the cyclopentadienyl cation, are "antiaromatic," that is, destabilized compared to the corresponding open chain compounds.

The new aromatic system ferrocene (**5**), which was recognized by Woodward and Wilkinson, is the parent compound of a family of aromatic "sandwich compounds" with typical aromatic stability and substitution reactions. The iron atom is bound to the carbocyclic rings by sharing of the π electrons. Aromatic character has been assigned to such structures

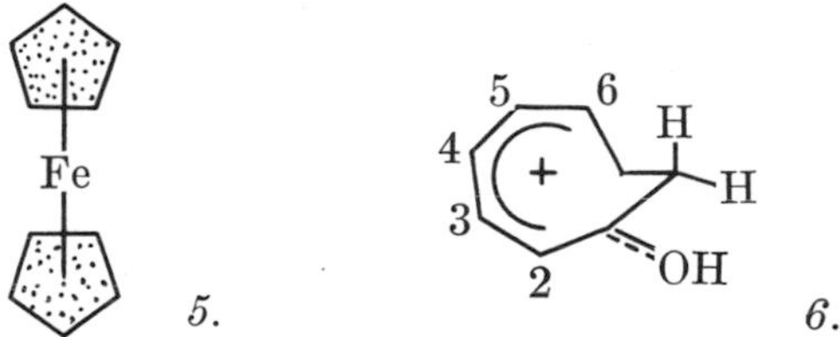

5. *6.*

as **6**, the hydroxyhomotropylium ion, by Winstein, from the NMR of protons 2–6.

Various "Dewar benzenes" (**7**) have been prepared by Van Tamelen and others, one method being the cyclobutadiene addition to an acetylene. Other isomers of benzene such as the prismane (**8**) and benzvalene (**9**,

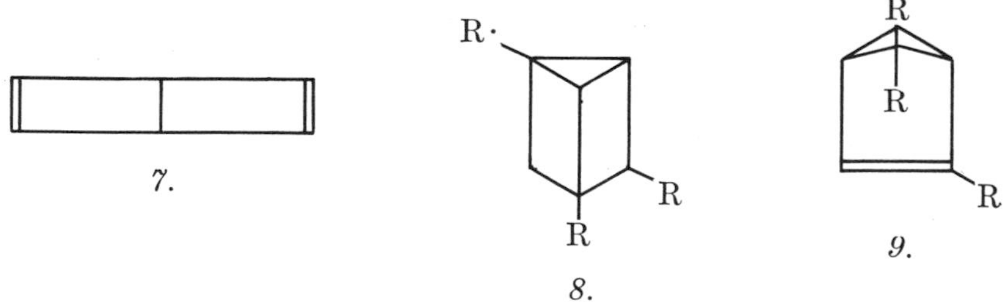

7. *8.* *9.*

R=t–Bu) have been synthesized photochemically from tri-*t*-butylbenzene. The preparation of cyclophanes, such as **10**, by Cram, Boekelheide, and others has shown the effect of the π electron cloud on spectra, reactivity, and stereochemistry.

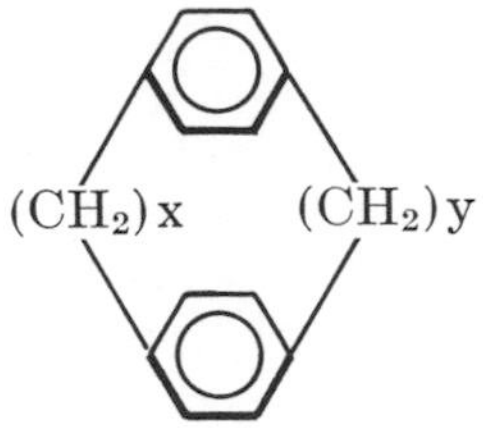

(CH$_2$)x (CH$_2$)y R R H_3C CH_3 O

10.  *11.* 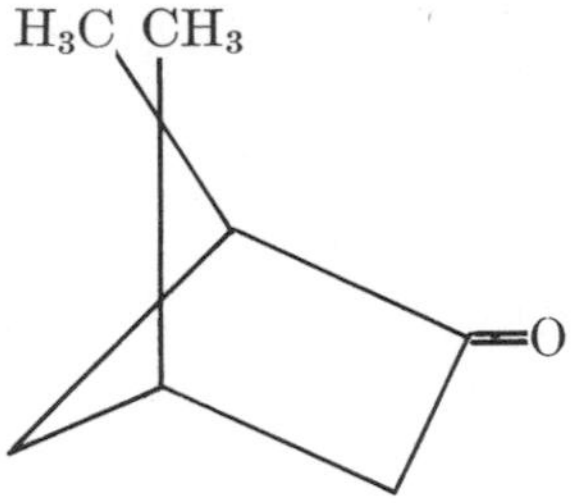 *12.*

Many highly strained saturated ring systems have been prepared and their reactions and thermodynamic properties studied. Representative are **11–14.**

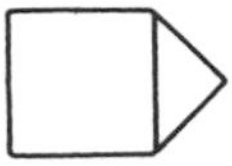

CH_3 CH_3 O

13. 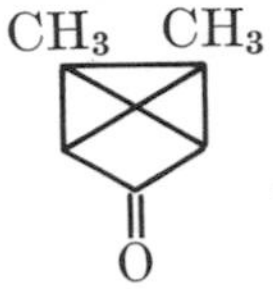 *14.*

The recent development of very strongly acidic and basic agents, and of dipolar aprotic solvents, has aided organic synthesis and extended ideas of physical organic chemistry alike. Hauser's preparation of dicarbanions has developed into important synthetic procedures. Now available for synthesis are bases of very high proton-removing power but of low nucleophilicity to carbon; they include lithium diisopropylamide, or butyllithium complexed with ligands such as tetramethylethylenediamine. The combination of a base like potassium *t*-butoxide with a dipolar aprotic solvent like dimethyl sulfoxide increases the basic properties of the alkoxide ion by many powers of ten. The action of strong bases on aromatic halides leads to intermediates ("arynes") with a "triple" bond in the ring; Roberts has furnished the most conclusive proof for benzyne (**15**) and its congeners.

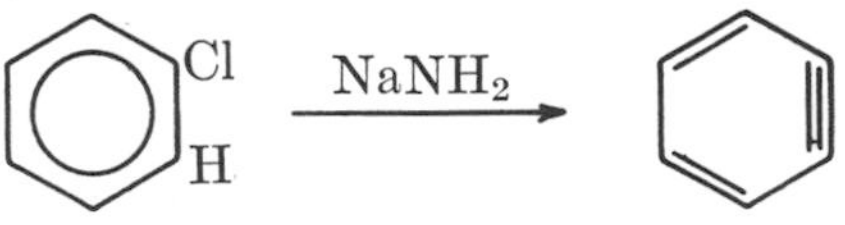

Cl H $\xrightarrow{NaNH_2}$ *15.*

On the acidic side, the range has been increased equally; very strong protonic or Lewis acids, such as F_3CSO_3H, FSO_3H, and SbF_5 have been developed; a mixture of the latter two ("magic acid") is capable of protonating even saturated hydrocarbons. The positive ions formed can be identified by NMR and in some cases by isolation as salts (Olah).

Compounds capable of transferring alkyl groups to almost any nucleophile, such as trialkyloxonium salts, have found wide usefulness and development. One of the most powerful is FSO_3CH_3.

The dipolar aprotic solvents such as dimethylformamide (DMF), dimethyl sulfoxide (DMSO), and hexamethylphosphoramide (HMP) increase rates of nucleophilic displacements by many powers of ten (Sheehan, Zaugg) and now are utilized routinely in synthesis. Their effect is due to their preferential solvation of cations.

Nef's idea of methylenes has been revived by the kinetic demonstration of dichlorocarbene, CCl_2, as a reaction intermediate (Hine, Skell); its synthetic usefulness is due to Doering and Seyferth. Carbenes are also readily produced by photochemical decompositions of diazoalkanes. The synthesis of cyclopropanes by addition of methylene to a double bond by the Simmons–Smith procedure has been very useful; similar syntheses utilize DMSO derivatives (Corey).

The total synthesis of natural products has attained structures of increasing complexity and has required methods of increasing stereospecificity and sophistication. It is clear that future synthetic work will rely on computer-planned sequences. Worthy successors to Bachmann's equilenin synthesis include the total syntheses of morphine (Gates), cortisone (Sarett, Johnson), strychnine, chlorophyll, and vitamin B_{12} (Woodward), colchicine and squalene epoxide (Van Tamelen), griseofulvin (Stork), penicillin (Sheehan), aflatoxins (Buchi), tetracycline (Muxfeldt), and the prostaglandins (Corey, Fried, and other groups). Many programs on synthesis of potential anticancer and antimalarial agents contained excellent synthetic work, as did synthetic activities in laboratories of pharmaceutical companies.

Structural work on insect hormones and pheromones (Meinwald) yielded results of much chemical interest and of potential importance in pest control. The study of the biogenesis of steroids, alkaloids, and other natural products developed rapidly both here and abroad; the elucidation of the biochemical pathways of conversion of acetate to polyisoprenoids and steroids, which is still going on, has been one of the great scientific achievements of the past few decades.

Detailed mechanisms of enzyme action have been probed by many physical organic chemists, including Bender, Bruice, Breslow, Jencks, and others. The work of Westheimer and Vennesland on the mode of action of NAD as coenzyme in enzymatic oxidation–reduction remains unsurpassed in elegance and usefulness.

The invention of the automated amino acid analyzer by Moore and Stein and of the Edman sequencer has made the determination of amino acid sequences in proteins and polypeptides practical even for hundreds of units. Synthesis of polypeptides has developed steadily, with du

Vigneaud's synthesis of oxytocin the first achievement; the synthesis of ribonuclease, the first synthetic enzyme, is a recent landmark. These syntheses have required many ingenious investigations of suitable blocking groups, acylating agents, and methods of avoiding racemization. The solid phase method of peptide synthesis developed by Merrifield represents not only an important procedure, but also the application of polymer chemistry to synthetic problems in organic chemistry.

Another specialized field, the nucleosides, nucleotides, and polynucleotides, presents structural and synthetic problems of great difficulty, but of fundamental importance for molecular biology. Holley first solved the sequencing problem for alanine transfer ribonucleic acid, which has more than 75 nucleotide units. Khorana's monumental synthetic studies, culminating in the synthesis of a gene, are based on synthetic organic chemical skill of the first rank.

The application of conformational analysis to ring systems is based on Barton's classical paper of 1951. The fundamental idea of a potential barrier preventing free rotation even around the C–C bond of ethane, with the eclipsed arrangement of hydrogens that of highest energy, is due to Pitzer. American organic chemists have developed conformational analysis, in acyclic and cyclic systems, in many directions. Only a few examples are addition to a double bond to form diastereoisomers (Cram), elimination and rearrangement reactions (Curtin, Saunders), the classical Winstein–Holness use of a *t*-butylcyclohexane derivative, where the *t*-butyl group stays equatorial and determines stereochemistry at other centers, carbonyl reactions in cyclohexanes (Eliel), and Turner's determination of energy differences in cyclohexanes and Decalins. Modern steroid chemistry and a large proportion of stereospecific synthetic methods are mainly based on, or are rationalized by, conformational reasoning. The advent of NMR has given a quantitative as well as a qualitative probe for preferred conformations and barriers to conformational changes.

Conformational studies imply a fundamental concern of recent years, the nature of the transition state of a reaction and its relation to starting materials and products. The Curtin–Hammett principle says that the ratio of rates of formation of two products does not depend on the ratio of conformational isomers most closely resembling the transition state to a given product, but on the difference in the free energies of the two transition states. Hammond suggested that a real molecule will approximate in geometry a transition state that it approximates in energy. One can thus speak of a transition state that resembles the starting materials rather more closely than it resembles the products, or vice versa, depending on the correspondence of the energy of the transition state with that of starting materials or products. This qualitative view is frequently useful in correlating rates and equilibria and is well illustrated by Bartlett's work on decompositions of peresters.

A fundamental addition to the theory of organic chemistry is the principle of conservation of orbital symmetry (Woodward and Hoffman). The symmetry of the molecular orbitals in the nonbonding and excited (antibonding) states determines the possibility of a thermal reaction or of a photochemical reaction (via an excited state). The principle is applicable to a variety of addition and isomerization reactions, such as the Cope rearrangement; the stereochemistry here and in related cases can be predicted from the rules of orbital symmetry.

Consideration of the possibilities of the Cope rearrangement led to the discovery by Doering of fluxional or degenerate compounds, which rearrange at room temperature to isomers indistinguishable except by isotopic labeling. Thus bullvalene cannot be represented by a single structural formula at room temperature, but continuously undergoes rearrangements; three forms, all equivalent, are shown.

Free radical reactions (C. Walling, "Free Radical Reactions in Solution," Wiley, 1957) have been investigated intensively for their industrial, preparative, and mechanistic importance; ESR measurements show the intermediacy of free radicals in vinyl polymerization. No attempt to review this field adequately will be made; outstanding work has been done by M. S. Kharasch, Walling, Mayo, Bartlett, Hammond, Brown, Kochi, and many more.

Recent work has developed elegant synthetic procedures that involve radical anion chain reactions (Kornblum, Russell, Bunnett). The very interesting Lewis acid, tetracyanoethylene (TCNE), studied by the Cairns group at Du Pont, undergoes one electron transfer very readily. Some of the congeners of TCNE form complexes that show very high electrical conductivity (Perlstein).

Two of the most valuable synthetic contributions of the past 30 years are due to H. C. Brown. The complex metal hydrides, such as $LiAlH_4$, $NaBH_4$, and their many variants, as convenient and specific reagents for reduction of carbonyl and other unsaturated groups, are almost limitless in application to synthesis, even including the determination of mechanisms of action of enzymes. Brown has shown further that alkylboranes are extraordinarily versatile synthetic intermediates, convertible under ordinary laboratory conditions to a variety of organic compounds—alcohols, carbonyl compounds, amines, and many more.

Organic photochemistry has developed dramatically in the past 20 years, both mechanistically and preparatively; four annual reviews now appear in the field. Hammond's determination of the mechanism of triplet sensitization and Chapman's technique of low temperature photolysis have been milestones; Chapman's technique has been used to study arene oxides, which are important biochemical intermediates, illustrating again the influence of advances in one field on progress in another, apparently quite unrelated. Photochemical cycloadditions frequently yield four-membered rings, this being a key step in the synthesis of cubane (Eaton). Organic photochemistry has been investigated actively by Leermakers, Turro, Zimmerman, Bartlett, Dauben, Wagner, and many others.

The prevailing ideas of steric hindrance were successfully altered by Brown after 1945; one of his striking results was the loss of basic character in 2,6-di-*t*-butylpyridine.

The role of solvent and neighboring groups in displacement reactions was examined both qualitatively and quantitatively by Winstein. This problem has been an active one ever since Bartlett's demonstration of the minimal reactivity of bridgehead halogens. Early ^{14}C measurements by Roberts showed unexpected rearrangements in carbonium ions, but the exact importance of nonclassical carbonium ions has not yet been clearly defined.

After 100 years, American organic chemistry is a vigorous and prolific science, with enthusiastic workers in a host of industrial and academic laboratories, with a continual stream of new ideas, problems, and techniques, and with a continuing influence on related fields.

ORGANIC COATINGS AND PLASTICS CHEMISTRY

The ACS Division of Organic Coatings and Plastics Chemistry grew out of efforts by chemists in the paint and varnish industry, in the early 1920s, to create a forum where they could discuss their work and problems. After two successful meetings—in Washington, D.C., in June 1922 and May 1923—they decided to form a permanent group, either independent or affiliated with an existing society. As a result, about 100 chemists petitioned the ACS to form a Paint and Varnish Section, which came into being in December 1923 with H. A. Gardner as chairman and W. T. Pearce as secretary. In 1927 the section became the Division of Paint and Varnish Chemistry; it has renamed itself three times to reflect changes in members' interests and progress in coatings chemistry. In 1940 the name became Paint, Varnish and Plastics Chemistry; in 1952 it became Paint, Plastics and Printing Ink Chemistry; and in 1960 the division assumed its present name.

At the ACS spring meeting in St. Louis in 1941, the division set out to see if it could compile preprints of its national meeting papers in time to send them to all of its members about two weeks before the meeting. This was done for the fall meeting of 1941. The preprint booklet was not quite complete—the papers were condensed, and some arrived too late to be included—but the members' reaction was good, and the division has published preprints without interruption ever since.

The Division of Organic Coatings and Plastics Chemistry administers the Arthur K. Doolittle Award, which is presented at each spring meeting for the best paper from the division's two previous programs. The award, a certificate and a cash honorarium, was established by Union Carbide Corp. in 1953. It is financed from royalties from the book, "Technology of Solvents and Plasticizers," by Dr. Doolittle, a past chairman of the division and a retired research executive at Carbide and Carbon Chemicals, a predecessor company of Union Carbide.

THE following historical sketch is biased somewhat toward "organic coatings," but the division nevertheless retains its long-time strong interest in "plastics," which is covered more extensively under Polymer Chemistry.

The paint industry in the United States is generally agreed to have started about 1810 and to have changed very little during the subsequent century. The advent of phenol–formaldehyde resins, in 1909, marked a turning point. Varnish had always been made by incorporating natural gums and resins into a limited number of drying oils, but the similarity of the phenol–formaldehyde materials to certain of the natural resins spurred coatings chemists into an eventually successful search for synthetic resins for their products. Scientific progress was slow at first, but began to accelerate after World War I. Physical chemists, for example, began increasingly to contribute to basic knowledge of pigments, which, with vehicles, are the major components of coatings. Of the types of coatings available in World War I, only three were not displaced in the succeeding 50 years. The three were exterior house paints, asphaltic coatings, and red-lead oil paints for structural steel, and even they had been modified extensively.

About 60 years ago, chemists began working to improve the properties of linseed, tung, castor, soybean, and other drying oils used as vehicles and film-formers in coatings. First they established the molecular structures of the oils; then they postulated drying mechanisms to guide the modification of the structures to achieve the desired properties. A typical result of such research was the dehydration of castor oil to make it behave more like tung oil. In a relatively short period this effort produced a variety of drying oils adapted to specific types of coatings. After a coating is applied, the drying oils it contains react with oxygen in the air and polymerize to form a resinous film; polymerization may be accelerated and controlled by driers—usually metallic salts of organic acids—in the coating. Current understanding of the mechanisms of drying-oil polymerization, particularly in alkyd coatings, owes much to the research of R .H. Kienle and A. G. Hovey. The work of T. F. Bradley, W. B. Johnson, and D. H. Wheeler did much to establish the mechanisms of polymerization and drying in drying oils per se.

In the early 1920s, chemists developed nitrocellulose finishes that broke the production-line bottleneck caused by the slow-drying finishes then used on automobiles, which had entered a time of explosive growth. The new nitrocellulose coatings, whose properties included high solids (pigment) and low viscosity, allowed autos to be finished in hours instead of days. The brittleness of the new coatings was eliminated at first by modifying them with plasticizers, such as castor oil and dibutyl phthalate, and later by adding alkyd resins, which also enhanced resistance to ultraviolet light from the sun and to outdoor conditions in general. Alkyds eventually displaced the nitrocellulose materials entirely for use on autos.

The 1920s saw the emergence of alkyd resins as vehicles and film formers in protective coatings. Smith had shown in 1901 that phthalic anhydride and glycerol react to form a hard, resinous film; more than a decade later, M. J. Callahan and then W. C. Arsem showed that the reaction could be modified by oleic and stearic acids, which improve the drying and flexibility of alkyds. Today, alkyds are made of a variety of combinations of polyhydric alcohols, such as glycerol, polybasic acids or anhydrides, such as phthalic anhydride, and monobasic fatty acids, such as oleic and stearic. Some alkyds are polymerized by oxygen in the air, others by heat. Their outstanding features include toughness and retention of gloss when exposed out-of-doors. The alkyds were especially well suited to automobile finishes—although they were displaced by acrylic resins after about 1957—and made possible the "wrinkle" and "crackle" finishes used widely on metal products. By the 1950s the technology had advanced to the point where emulsions of alkyds in water were being used extensively to finish interior walls.

Vinyl resins appeared in the plastics industry in the early 1930s and soon were playing a major role in protective coatings. The vinyls displayed properties unavailable in previous coatings materials—natural resins, alkyds, nitrocellulose, and most phenolics. Vinyls could be made, for example, that were colorless, odorless, tasteless, nontoxic, and inert to most chemicals. These properties made them particularly useful in food packaging; citrus juices, beer, and many other items normally corrosive to metal could now be packaged in cans with vinyl finishes. During World War II, vinyl finishes and polydisulfide rubber linings permitted the construction of large-scale aviation fuel facilities from lined concrete, releasing critical steel for other purposes.

Since the advent of the vinyls, a stream of new and improved polymeric materials has been formulated into protective coatings for many purposes. Copolymer resins of vinylidene chloride and acrylonitrile are used widely in coatings for petroleum fuel storage tanks. The resistance of silicones to heat has solved several problems: silicone insulating varnishes have allowed the design of electric motors much smaller than previously required for a given horsepower; silicone coatings now protect many surfaces from corrosion at temperatures that would destroy other finishes. Epoxy resins provide coatings of unusual adhesion and inertness for use in particularly rugged conditions. Resin-bonded tetrafluoroethylene coatings improve the corrosion resistance of surfaces and add lubricating properties as well. Urethane coatings display among other properties the abrasion resistance and durability that make them useful in applications such as varnishes for gymnasium floors.

Research on pigments, which give coatings their color and hiding power, has paralleled that on vehicles in that it did not really begin to accelerate until World War I. Work by chemists on the light absorption, index of refraction, particle diameter, and general surface chemistry of pigments has contributed much to the improvement of coatings. At the first meeting of the Paint and Varnish Section of the ACS, in 1923, for example, W. D. Harkins and his associates dealt with topics such as the flocculation of pigments as a result of the low heat of wetting of hydrophilic oxides in nonpolar liquids. F. E. Bartell used the contact angle between liquids and solids to measure the preferential absorption of liquids by solids and the displacement of one liquid by another when both were in contact with a solid. At a divisional meeting in 1955, W. A. Zisman and his colleagues described the wetting properties of organic liquids on high-energy surfaces, thus introducing new concepts of the mechanisms of reactions between solids and liquids. These and many related efforts have led to marked improvements in the dispersion stability, spectral properties, and durability of commercial pigments.

A notable development in pigments was the phthalocyamines, announced in 1933 by R. P. Linstead and his associates in England and introduced in the U.S. in 1936. These pigments, which resulted from a good deal of both basic work and development, were the first completely new chemical type to appear for many years. The tinting strength of copper phthalocyamine, for example, is more than twice that of the iron blues used for many years and at least 20 times that of ultramarine. The phthalocyamines, moreover, are far more durable than pigments used previously, which has made possible the production of green and blue colors of high quality.

A second notable development in pigments was the emergence of titanium dioxide, which was first made commercially in this country in 1920 and, starting in about 1940, has gradually become the dominant white pigment in the coatings industry. The current utility of titanium pigments is a result of years of research and development both on their properties and on means of making them economically. All but the darkest colored paints must contain white pigment to supply opacity and brightness; most, in fact, contain more white than colored pigment. Titanium dioxide, in addition to its other advantages, has markedly greater hiding power than other white pigments. In consequence it has largely displaced all earlier white pigments except certain lead compounds, and even they are limited to specific applications to which they are particularly well suited.

Perhaps the most striking phenomenon of the past 25 years in coatings has been the spiraling growth of water-thinnable paints. Water-base paints of the casein type were available before World War II, but had only limited application. The war produced shortages in coatings materials and, from the synthetic rubber program, a supply of water emulsions of butadiene–styrene latexes. From this combination sprang an entirely new family of paints. Pigments that were compatible with the emulsions materialized quickly, and the resulting coatings proved easy to apply and clean, and performed well in many applications. They promptly revolutionized interior finishing, in the process creating a large do-it-yourself market. The water-base paints have been improved substantially since they first appeared and are used widely today for both interior and exterior applications. The three main latex systems are butadiene–styrene, polyvinyl acetate, and acrylics. Alkyd systems, as mentioned earlier, also are available in water–emulsion form.

The water-base paint phenomenon has been succeeded in recent years by a variety of other developments in coatings science and technology. These have included both new and improved coatings and more effective means of applying them. Alkyds, for example, have continued to improve and today are the single most important class of resin in the coatings industry. Acrylic resins have progressed rapidly and probably come closer than any others to matching the performance and utility of the alkyds. Polyvinylidene fluoride finishes are virtually immune to ultraviolet radiation and may last up to 30 years out-of-doors. These materials are now used widely on coil- and spray-coated aluminum and galvanized steel. Among the newer methods of applying coatings is electrodeposition, which in some ways resembles electroplating. The coating is electrodeposited from a water bath and fused to final form in an oven. The process offers a number of advantages, especially in large-scale operations, and its use has been growing rapidly. Among many other developments in coatings in the past few years have been water-based industrial enamels, electron beam and ultraviolet curing of paints, replacements for mercury and lead in pigments, and new formulations that reduce the emission of organic vapors to the air.

PESTICIDE CHEMISTRY

The Pesticides Subdivision formed in 1951 in the Division of Agricultural and Food Chemistry achieved divisional status in 1969 as the Division of Pesticide Chemistry. The first officers were D. G. Crosby, chairman, and W. F. Phillips, secretary-treasurer. Papers presented before the division have spanned a wide range of subjects, including chemical syntheses, structure-activity relationships, macro- and microanalytical techniques and procedures, pesticide formulation, metabolism of pesticides in plants and animals, and mode of action and fate of pesticides in the environment.

At the time of its organization, the division established an International Career Award, first given in 1970, that it presents annually for outstanding contributions to research in the chemistry of pesticides. Later, to recognize the sponsor, the award was renamed the Burdick and Jackson International Award for Research in Pesticide Chemistry. In May 1970, the division established a fellowship awards program. A maximum of three such awards are made annually to individuals who have contributed their time, talent, and services to the division for a period of at least six years.

In June 1972, the Division of Pesticide Chemistry sponsored a conference in Fargo, N.D., on "Pesticide Metabolites, Enzymology, Isolation, and Characterization." A similar conference was held in Vail, Colo., in June 1975 on "Bound and Conjugated Pesticide Residues in the Environment." The division of Pesticide Chemistry publishes a semiannual newsletter, the *Picogram,* which carries information on divisional activities and other items of interest.

IN the 1860s, farmers in the newly cleared midwestern and western farmlands of the U.S. faced a serious economic problem with the Colorado potato beetle, an insect native to the eastern slopes of the Rockies, where it fed on wild nightshade. The beetle was spreading eastward, finding a new food in the potato plant, whose culture was spreading westward. Paris green (copper acetoarsenite), then in use as a paint pigment, was found to be highly effective against the beetle and by 1870 was in wide use as an agricultural insecticide. Within a few years London purple, a mixture of calcium arsenite, calcium arsenate, purple dye, and miscellaneous materials, was introduced as an insecticide. Paris green and London purple were the most commonly used insecticides in the U.S. until 1900. They had many disadvantages, however, not the least of which was that they were highly toxic to man and other animals.

In 1892, lead arsenate was used for the first time. By 1910 it had largely replaced London purple and Paris green, and it remained the leading insecticide against orchard pests until the end of World War II. Calcium arsenate (a mixture of several calcium arsenates, calcium arsenite, and an excess of calcium hydroxide) became available commercially about 1919. It was less costly than lead arsenate and become the top insecticide for controlling the boll weevil on cotton. However, calcium arsenate never supplanted lead arsenate for use in orchards because of its toxicity to both fruit and foliage.

In those pre-DDT days, farmers also had at their disposal inorganic fluorine compounds, inorganic mercury compounds, and petroleum oils. Plant-derived materials, such as nicotine from tobacco, were also common.

Pyrethrum, from a species of chrysanthemum, provided a general-purpose insecticide. The insecticidally active material, found in the pyrethrum flower, is unique in the speed with which it paralyzes insects. Its structure was determined by 1938; four separate and complex organic chemicals were identified. In 1949, one of the compounds, allethrin, was synthesized and produced in commercial quantities, but was not as effective as the natural pyrethrins as a general purpose insecticide. Closely related compounds have been made commercially since.

Synthetic Organic Insecticides

The early 1940s saw the first of the synthetic organic insecticides that soon would dominate the field. The first synthetic organic insecticide had been the potassium salt of dinitro-*o*-cresol, which was marketed in Germany in 1892 for use against aphids and grasshoppers. Forty years later, several organic thiocyanates were marketed on a sizable scale in the U.S. for use as fly sprays and household insecticides. But the advent of DDT marked a watershed in synthetic organic insecticides, of which soon there were three major classes—the organochlorines (including DDT), the organophosphorus materials, and the carbamates.

DDT was first synthesized in Germany in 1874. Biologist Paul Muller at J. R. Geigy, S.A., in Switzerland, discovered its insecticidal properties in 1939, an achievement that won him the 1948 Nobel Prize for Physiology or Medicine. The first samples of DDT arrived in the U.S. in 1942. The compound was easily produced from inexpensive raw materials—chloral, chlorobenzene, and sulfuric acid. The U.S. armed forces used DDT widely during World War II for mosquito and fly control. Its first agricultural use was in Europe, against the Colorado potato beetle, whose eastward migration by then had carried it across the Atlantic.

DDT's success was immediate and spectacular. The compound has been estimated during its first eight years of use to have prevented 100 million illnesses and five million deaths by controlling the insect vectors of malaria, yellow fever, and gastrointestinal diseases. Between 1952 and 1964 in India alone, DDT reduced the cases of malaria from 75 million to 100,000. Similar triumphs followed in rapid succession as the compound was found to control a variety of insect species. Besides its low cost, DDT was easy to apply and was far less toxic to man and higher animals than were the less effective arsenates it replaced. The compound is chemically stable, so that its residues remain in treated areas to kill insects migrating from untreated areas. Ironically, it would be DDT's stability that would cause it so much difficulty in the 1960s and 1970s as environmental awareness increased. The compound's use in this country began to decline in that period, in part because of the development of insect resistance; by 1975, a series of court decisions had imposed a near-total ban on DDT, although it is still used abroad, primarily in public health.

DDT was the first of a long series of chlorinated hydrocarbon insecticides, including the chemically similar methoxychlor and TDE as well as benzene hexachloride, chlordane, heptachlor, dieldrin, aldrin, and toxaphene. With the success of all of these materials, basic knowledge of organic chemical syntheses was applied in an all-out search for insecticides and other pesticides. Thousands of organic chemicals were synthesized in the laboratories of U.S. chemical companies and screened for pesticidal properties; scientists tried to correlate structure with activity in the hope of being able to design more effective compounds. But by 1975, the organochlorine insecticides themselves, like DDT, the first of the group, were either banned, under severe restriction, or threatened by one or the other on environmental grounds.

Organophosphorus insecticides, the second of the three major classes, resulted from research that began in the early 1930s in Germany. Approximately 50 organophosphorus pesticides were registered in the U.S. in 1975. They encompassed a range of characteristics—from highly toxic to relatively safe, from moderately persistent to rapidly degradable, from specific to broad spectrum.

Gerhard Schrader in Germany synthesized the organophosphorus compound TEPP (tetraethyl pyrophosphate) in 1938; it was the first synthetic insecticide to be as active as nicotine. Research on organophophorus compounds continued during the early 1940s in Germany, the United Kingdom, and the U.S. The goal shifted to war gases (nerve gases), but much was learned that later was advantageous in insect control.

After World War II, Allied scientists learned of the German development, parathion, useless as a war gas but promising as a broad spectrum insecticide. By 1948 U.S. chemical companies were manufacturing parathion and methyl parathion commercially. Both compounds are highly toxic to mammals. Thus the introduction of malathion in the early 1950s was important, because this organophosphorus compound exhibited not only broad insecticidal activity but relatively low toxicity to mammals as well.

The organophosphorus compounds also provided, in the systemic members of the class, an important new route of application. Systemic compounds are absorbed by the plant through its roots, stems, or leaves and move through the plant's vascular system. The plant thus becomes temporarily lethal to insects that ingest its juices in the process of feeding on it.

The organophosphorus insecticides generally degrade more rapidly in the environment than the more persistent organochlorines. Some, in fact, degrade more rapidly than is desirable for fully effective insect control; this property limits their usefulness to the control of insects that threaten just before harvest, but at the same time avoids toxic residues in the harvested crop. Organophosphorus compounds are used widely not only on crops but to control animal pests such as cattle grubs, horn flies, fleas, and ticks.

The first carbamate insecticides, the third of the three major classes, were developed in the late 1940s in Switzerland, where scientists made a series of insecticides based on the carbamate structure of the alkaloid physostigmine, which occurs naturally in the poisonous Calabar bean. In the early 1950s research at the University of California at Riverside established the insecticidal activity of another series of carbamates. This research contributed to the development of carbaryl, the first major carbamate insecticide to be applied commercially to a wide range of crops. More recently, carbamate insecticides such as Temik, Furadan, Baygon, Zectran, and Bux have been introduced for commercial use.

The carbamates exhibit a wide range of toxicity to mammals. Most of the compounds act topically, but one new type shows a high degree of systemic activity. Chemists may build subtle structural variations into the carbamates to render them more specific than the parent generations of insecticides. Consequently, carbamates may offer economically sound means of protecting certain crops with minimal ecological hazard.

The early 1950s saw growing concern in the U.S. over the widespread use of the new organic pesticides and the possible effects, on humans and animals, of long-term, low-level exposure to their residues. In 1954, Congress passed the Miller amendment to the Food, Drug, and Cosmetic Act which established broad, new controls on pesticides. The legislation requires the manufacturer to prove that the pesticide is useful for its intended purpose, that it is not a hazard to public health or wildlife, and that its uses are in accord with good agricultural practice. If the pesticide is to be used on food or feed, a tolerance for it must be established (usually in parts per million of pesticide residue in or on the raw food or feed). Before establishing a tolerance, the government requires manufacturers to provide data on the pesticide's toxicity to specified laboratory test animals, the amount and timing of application to the specific crops covered, and the amount of residue that will remain in or on the food or feed at varying intervals after the pesticide is applied. The manufacturer also must develop a practical and sensitive analytical method that will identify and measure the pesticide and any of its toxic metabolites in or on the food or feed in question. Concern over pesticide residues has continued to increase since the Miller amendment was passed, and federal controls on pesticides and their uses have grown increasingly stringent.

Herbicides

Large-scale application of chemicals to control weeds did not begin until after World War II. Discovery of the selective herbicidal properties of 2,4-D and 2,4,5-T during the war and their commercial introduction shortly thereafter set off rapid acceptance of the practice. For the first time, man was able to destroy only unwanted vegetation; earlier chemical herbicides, such as petroleum oils and sodium arsenite, killed all vegetation on contact, not just the weeds. With the newly discovered and selective herbicidal compounds came wide recognition that weeds reduce the yield of food and fiber crops, serve as havens for harmful insects and plant diseases, and produce pollen that causes widespread discomfort and even suffering among allergic humans.

The herbicidal activity of the phenoxy compound 2,4-D, the most widely used of all herbicides, was discovered in England in the early 1940s. 2,4,5-T, silvex, and MCPA are other important phenoxy herbicides. MCPA and 2,4-D control broadleaf weeds in grasslands and rangelands and in major crops such as wheat, oats, flax, rice, and sugarcane. 2,4-D, 2,4,5-T, and silvex help to increase the livestock-carrying capacity of rangelands infested with noxious brush and poisonous plants. The same chemicals also are used widely to maintain rights-of-way for public utilities, railroads, and highways. 2,4-D is made from three low-cost commodity chemicals: phenol, acetic acid, and chlorine. This keeps its cost low and contributes to the economy of using it. The same is generally true of the other phenoxy herbicides. All of them show high ratios of benefiits to risks, low toxicity to mammals, and relatively rapid biodegradability by soil microorganisms.

The herbicidal activity of the carbamate as well as of the phenoxy herbicides was discovered in England, and two of them, IPC and CIPC, were introduced commercially in England and the U.S. after World War II. Subsequently, a series of highly selective thiocarbamates was developed. Another carbamate, Carbyne, can distinguish between two species of grasses—wheat and wild oats. This capability has been very useful on wheat, where the introduction of the combine intensified the wild oat problem by causing wild oat seeds to be distributed in the field at harvest time instead of being confined to thrashing areas. Sutan and Eptam are carbamates developed for selective grass control in corn and other important field crops. Other important herbicides introduced in the early 1960s in the U.S. include substituted ureas, amides, triazines, and dinitroanilines.

Other Pesticides

In addition to insecticides and herbicides, chemists and other scientists have developed pesticides that fight plant diseases, nematodes, and rodents. The best of these are organic materials introduced since about 1940. They include:

• Captan, the first fungicide to offer the possibility of eradicating a plant disease once it was established.

• D–D soil fumigant, a mixture of dichloropropene and dichloropropane that was introduced in 1943 and began the modern era of nematocides.

• Warfarin, an anticoagulant that causes fatal internal bleeding in rodents and is the most effective and widely used of the rodenticides.

The increasing environmental awareness of recent years has brought pesticides generally into poor repute in the public mind. Nevertheless, the need for pest control will always exist and, indeed, can only grow stronger in the light of the world's problems with food. Pesticide chemistry and chemists have a vital role to play in the development of increasingly effective and ecologically desirable means of meeting that need.

PETROLEUM CHEMISTRY

The Division of Petroleum Chemistry was formed in 1922 with T. G. Delbridge as chairman and W. A. Gruse as secretary. The nucleus of the division was the petroleum section formed in 1921 within the Division of Industrial and Engineering Chemistry. Interest in petroleum chemistry was already brisk when the division was formed—in 1920 the nation consumed some 450 million barrels of oil—and it has grown steadily in the subsequent half century. Thus the division has become one of the larger ones in the Society; its technical sessions at national meetings usually require a full week, often with one or more days of simultaneous sessions.

The Division of Petroleum Chemistry has been preprinting its papers almost since it came into being. Since 1955, the preprints have been a serialized document with a quota of four issues yearly; many public and institutional libraries have standing orders for the preprints as they appear. Members are informed of divisional activities, including future programs, through newsletters issued three or four times annually.

THE petroleum industry was 60 years old when the Division of Petroleum Chemistry was formed, and it had adapted already to the automobile's voracious demand for gasoline. In 1910, when automobile registrations in the U.S. reached 500,000, gasoline still accounted for only 13% of the products made from petroleum—other products included kerosene and lubricants. These materials were obtained from crude oil by simple distillation, and refiners realized that they could not keep up with the growing demand for cheap gasoline unless they could learn to break the larger hydrocarbon molecules into the size range of those in gasoline. The first commercially important result was the thermal cracking process invented by chemist William Burton and his colleagues at Standard Oil Co. (Ind.) and introduced in 1913. Since then, the use of liquid and gaseous hydrocarbons in this country has grown phenomenally: petroleum consumption climbed 13-fold in the five decades ending in 1973, and natural gas consumption rose almost 30-fold:

Energy Use in the U.S.
(Quadrillion [10^{15}] Btu[a])

	1920	*1930*	*1940*	*1950*	*1960*	*1970*	*1973*
TOTAL	19.8	22.3	23.9	34.0	44.6	67.4	75.6
PETROLEUM	2.7	5.9	7.7	13.5	20.1	29.6	34.7
GAS	0.8	2.0	2.7	6.2	12.7	22.0	23.6

[a] One barrel of oil is equivalent to about 6 million Btu.

These numbers are so large as to be almost incomprehensible. The magnitude of our fossil fuel consumption is brought home, however, by the realization that each day in the U.S. each person on the average uses 50 pounds of fossil fuels—23 of oil, 13 of gas, and 14 of coal. In 1975, petroleum accounted for about 43% of our energy supply; gas, 33%; coal, 18%; hydroelectric, 4%; and nuclear, 2%.

The fuel-use pattern in this country has altered greatly over the past half-century. The greatest growth in energy consumption has been for transportation (now about 25%) and for generation of electricity (also about 25%). This changing pattern has markedly influenced the trends in research and development in petroleum chemistry, which has created a remarkable series of new refining processes. These processes have been essentially all catalytic and are used to make high octane gasoline from gaseous hydrocarbons (by polymerization and alkylation), from naphtha (by reforming), and from heavier petroleum fractions (by cracking and hydrocracking). A new and important factor has been the need to remove sulfur from oil, both for environmental reasons and to facilitate certain catalytic refining reactions in which sulfur poisons (inactivates) the catalyst. This need is being met by hydrodesulfurization processes.

Discoveries in petroleum chemistry have led sometimes to new industrial processes and products. This has been true particularly in petrochemicals, which now consume nearly 10% of the petroleum used and are the main source of raw materials for the plastics industry. Many discoveries in petroleum chemistry, moreover, create opportunities not known before—for example, the production of single-cell protein from paraffins.

For its 50th anniversary meeting, the Division of Petroleum Chemistry arranged seven symposia on progress of petroleum chemistry in the previous 50 years.[1] Following are selected highlights of these symposia, which outline important innovations in a number of areas in addition to those mentioned below. These areas include hydrocarbon alkylation, polymerization, isomerization, and dehydrogenation (as well as petrochemical manufacture, which is beyond the scope of this review).

Hydrocarbon Chemistry

Perhaps the most important advance in theoretical hydrocarbon chemistry during the past 50 years was the postulation that highly reactive intermediates are involved. Such intermediates can be free radicals, carbonium ions, carbanions, or others such as carbenes. Catalytic surface complexes should also be included. In the 1930s, F. C. Whitmore, describing carbonium ions, postulated the existence of an electron-deficient carbon atom; the deficiency would induce migration of electrons to yield a rearranged product. More recently, G. A. Olah has suggested the existence of two types of positive carbon ion: the classical trivalent ion and a pentacoordinated nonclassical ion. The pentacoordinated ion is considered to consist of five atoms bound to a carbon atom by three single bonds and a two-electron three-center bond. Postulation of the pentacoordinated ion has many advantages in explaining reaction mechanisms.

In 1964, R. L. Banks and G. C. Bailey showed that alkenes disproportionate to homologs of higher and lower molecular weight in the presence of alumina-supported molybdenum oxide (and similar catalysts) at 100°—200°C. Propylene, for example, disproportionated to ethylene and n-butylenes with 94% efficiency at 43% conversion. The reaction may be considered to proceed via a four-center ("quasicyclobutane") intermediate involving the catalyst and the four doubly bonded carbon atoms of two molecules of olefin. The Banks–Bailey reaction has been studied by many laboratories throughout the world and has been shown to be general for olefins. It is important theoretically and is used commercially to convert propylene to high-purity ethylene and butenes and to synthesize specialty olefins.

Considerable progress has been made in alkali catalyst hydrocarbon reactions. The best known application perhaps is the polymerization of butadiene, catalyzed by metallic sodium, to make synthetic rubber. The behavior of carbanions in such reactions differs markedly from that of carbonium ions. Alkyl groups in carbanions do not migrate; primary carbanions are more stable than secondary ions which are more stable than tertiary ions. The reverse is true with carbonium ions, so that

[1] *Preprints, Div. Pet. Chem.*, ACS, **17**, 3, 4 (1972).

catalytic reactions involving them yield much different products than do those involving carbanions.

Catalytic Reforming

Catalytic reforming of petroleum has taken great strides since the first unit was built in this country, in 1939. Free World catalytic reforming capacity totals more than 7 million barrels per day, about 50% of it in the U.S. and Canada. The bulk of the U.S. capacity has been built during the past two decades in response to sharply increased demand (beginning about 1952) for high-octane-number gasolines needed to fuel high-compression-ratio, large-displacement, automotive engines.

Aromatics (benzene, toluene, and xylenes) are important components of high-octane reformate. The hydrogen produced as a by-product in reforming has become increasingly valuable for use in the associated hydrotreating and hydrocracking operations.

A significant advance was the discovery and recognition of the dual function (isomerization and hydrogenation) of reforming catalysts. The mechanism of hydrocarbon reactions over these dual-function catalysts has been established and is illustrated for the isomerization of *n*-pentane:

$$n\text{-pentane} \rightarrow \text{1- or 2-pentene} + H_2 \text{ (on Pt site)}$$

$$\text{1- or 2-pentene} \rightarrow \text{branched pentenes (on acid sites)}$$

$$\text{Branched pentenes} + H_2 \rightarrow \text{isopentane (on Pt site)}$$

The amount of olefin in the gas phase is extremely small because of the large excess of hydrogen used in the process and the moderate temperature and high pressure. A related advance was the recognition that a dual-function catalyst, in a single reactor, can accomplish more than is achieved by passing reactants in sequence through two reactors, each filled with a different, single-function catalyst.

Platinum-containing reforming catalysts dominated the field throughout the 1960s. Late in the 1960s, several catalyst suppliers announced new catalysts which have come to be known as bimetallic. The early 1970s continued to show rapid development of additional bimetallic catalysts. These catalysts offer much higher activity and maintain their selectivity much longer than do earlier materials. The improved selectivity and activity permit reformers to operate at lower pressure and at lower gas-recycle ratios or higher severity or both. The development of bimetallic reforming catalysts capable of drastically improving performance represents potentially the most far-reaching improvement in catalytic reforming. These recent developments represent a "quantum jump" in catalyst performance.

The addition of rhenium to a platinum reforming catalyst results primarily in a more stable catalyst. Data on Rheniforming (which uses a platinum–rhenium catalyst) a heavy naphtha at lower pressures, for example, show large increases in yield as operating time progresses. It has been suggested that some kind of platinum–rhenium couplet is involved, although it is not known how this complex acts. Other metals and also sulfur are believed to have a beneficial effect. An important point is that platinum–rhenium does not, per se, lead to higher yields; the initial yields are the same as those obtained with conventional catalysts. However, higher yields can be achieved with platinum–rhenium catalysts because they can operate at lower pressures, where conventional reforming catalysts are not stable, but at otherwise equivalent conditions.

Catalytic Cracking

A "quantum jump" improvement in catalytic cracking has been achieved by the introduction of so-called crystalline aluminosilicate or zeolitic catalysts. The crystalline aluminosilicate catalysts are more active, more stable, and produce considerably more of the desired products, gasoline and light fuel oil, compared with amorphous silica–alumina catalysts. At equal conversion levels, they also produce considerably less of the undesirable products, coke and light gas.

Durabead 5, the first commercial zeolitic cracking catalyst, was introduced in March 1962. Since then a variety of other zeolitic catalysts have been developed for use in both moving and fluid beds. The zeolitic catalysts now are used in more than 90% of cracking units. The use of x and y zeolites with a port size of about 10 angstroms has provided for even further improvements, including very high catalytic activity. The ion-exchanged faujasites can have cracking activities many orders of magnitude greater than those of silica–alumina. In turn, the use of highly active catalysts has permitted process variations, including riser cracking, which gives the most efficient contact between catalyst and reactants.

Crystalline zeolites are "molecular sieves" that have pores of uniform dimension— for example, 4 Å—in the size range of the molecules of interest. Early concepts of molecular sieves prompted their use as "shape selective" catalysts with pores that could accommodate molecules of certain molecular dimensions but exclude others. They would accommodate straight-chain molecules, for example, but not branched-chain molecules. Actually, selective reactions were found possible, although crystalline zeolites with pores in the range of 10-Å diameter have become the most significant in catalytic cracking. While attention has been given to the unsuual electronic charge fields in the pores of the zeolites and their effects on reactant molecules, the scientific basis of the catalytic action and selectivity of zeolites has not been resolved.

Hydrocracking, Hydrodesulfurization

The decade of the 1960s witnessed the establishment of modern hydrocracking processes in petroleum refining. From a single demonstration unit of 1000 barrels-per-day capacity in 1960, commercial installations have grown to more than 50 units with a total capacity of about 800,000 barrels per day. The demand for gasoline and middle distillates and the availability of low-cost hydrogen have played major roles in this growth. The most significant factor has been the development of superior catalysts having excellent, long-lived activity at operating conditions far less severe than those used in the older hydrocracking processes.

Research is creating increasingly active and economical hydrocracking catalysts whose composition can be tailored to meet particular feed and product objectives. These materials are dual-function catalysts having a critical balance of functions.

The "Paring" reaction is one of the most interesting new reactions discovered. It involves hydrocracking hexamethylbenzene over a dual-function catalyst to yield light isoparaffins and C_{10} and C_{11} methyl benzenes. Essentially no ring cleavage occurs, and hydrogenolysis to form methane occurs to only a small degree. To account for the formation of isoparaffin products, a process of isomerization leading to side-chain growth, followed by cracking of side chains of four or more hydrocarbons, has been proposed.

Hydrodesulfurization is one of the most significant petroleum-refining operations. With lighter petroleum fractions, desulfurization is important for subsequent catalytic refining steps, such as reforming, because sulfur can poison the catalysts. New emphasis on hydrodesulfurization, particularly of heavy fuels, has stemmed from the strict limits imposed on the content of boiler fuels to control emission of sulfur dioxide to the air. Current regulations call for as little as 0.8 lb SO_2 per million Btu and even less in some cities.

In 1960, about 2¼ million barrels per day of hydrotreating and hydrocracking capacity was installed in the U.S. By 1971, capacity had grown to 5½ million barrels per day or one half the 11 million barrels of oil processed daily. Thus, one of the remarkable features of hydrotreating and hydrocracking is the rapid growth rate and high capacity reached. A second significant feature of commercial hydrotreating has been its application to heavier petroleum fractions, including residuals, which are more difficult to desulfurize than are the lighter fractions. The heavier end of a heavy distillate with an end point of 1050°F may have a molecular weight of 800–900. Much of the sulfur in these cuts is present in substituted benzo- and dibenzothiophenes. The residual fractions from such cuts contain asphaltenes, in highly dispersed colloidal form, with an average molecular weight of 5000—10,000. (The asphaltene molecule has been postulated to have a layer structure in which each stack contains four or five aromatic sheets about 14 Å across.) To remove the sulfur from these large, complex molecules requires severe hydrodesulfurization. This means raising the temperature to a point where bonds that do not involve sulfur-removal will begin to break, which leads to undesirable side reactions.

The chemistry of hydrodesulfurization is only partially understood. Most researchers believe that for thiophene the reaction proceeds in two steps: the first step produces butylene and hydrogen sulfide, and the second is a hydrogenation of butylene to butane. However, the first step must be complex and must proceed through a butadiene intermediate.

Cobalt molybdate and nickel molybdate catalysts are used extensively in hydrodesulfurization. Relatively little is known of how they act, although in operation the active components exist as lower-valence sulfides (German processes have used nickel–tungsten sulfide catalysts). Diffusional effects are important in these catalytic processes. Important also is the tolerance of the catalyst for the nickel and vanadium in oil. These metals tend to inactivate the catalyst, shortening its effective life and increasing costs. Current techniques can desulfurize oil to any desired level; the need is for catalysts and processes that are more efficient and less costly.

PHYSICAL CHEMISTRY

The Division of Physical Chemistry was formed in 1908, originally as the Division of Physical and Inorganic Chemistry, one of the first five to be established by the Society. The first chairman was Charles H. Herty, a future ACS President (in 1915–16); the first secretary was Wilder D. Bancroft, who had founded the *Journal of Physical Chemistry* in 1896. In the division's early years, its officers came from the fields of analytical, colloid, inorganic, and physical chemistry, all of which were within its purview. Growth in each of these fields, however, resulted in the splitting off of a separate Division of Colloid Chemistry in 1926 and a Division of Microchemistry—now the Division of Analytical Chemistry—in 1938. By 1939, the methods of physical chemistry had helped to set the stage for a resurgence in inorganic chemistry, which had lain dormant for several decades. The resurgence gained strength steadily, and in 1956 a separate Division of Inorganic Chemistry was split off from the Division of Physical and Inorganic Chemistry, which at that time assumed its present name.

POLYMER CHEMISTRY

No unit in the American Chemical Society dealt solely with polymers until 1946, when the intense interest in the field spurred the formation of a High Polymer Forum by several interested divisions: Paint, Varnish and Plastics; Physical and Inorganic; Rubber; Cellulose; Colloid; Organic; and Petroleum. From 1946 to 1949, 162 papers were presented at 27 sessions of the High Polymer Forum. This high level of activity led to the formation in 1950 of the *Division of Polymer Chemistry* (initially the Division of High Polymer Chemistry), with C. S. Marvel as chairman and Herman Mark as secretary-treasurer.

In 1960 the division started *Polymer Preprints* to publish meeting papers just before they are presented. The journal now has the largest circulation of any in the polymer field; its size and cost, in fact, determine the maximum number of papers that can be accepted for presentation at division meetings. In 1962 the division sponsored its first Biennial Polymer Symposium, an event that has been continued without a break. The early 1960s also saw the formation of a divisional education committee, which led to a speakers bureau and several ACS short courses for updating members' knowledge of the polymer field. And in 1968, with strong encouragement from the division, the American Chemical Society began publishing the bimonthly journal, *Macromolecules,* with F. H. Winslow as editor.

MANKIND has used polymeric materials for centuries, but until about 100 years ago, all such materials were used in their natural state or in slightly modified form. Naturally occurring polymers were commercially important long before their structures were understood.

A case in point is rubber, which has been used in its natural form for hundreds of years. As early as 1826 the empirical structure of rubber was known, and in 1860 it was shown that destructive distillation of rubber produced isoprene, which would be found later to be the basic, repeating unit in the polymer. The structure of rubber, however, remained unknown. In 1839, Charles Goodyear discovered vulcanization. The process was the first major modification of rubber and greatly extended its utility. In 1851, the usefulness of the material was extended further—to the production of rigid articles—when a patent was issued on a means of making hard rubber or ebonite from natural rubber. In 1879, G. Bouchardat polymerized isoprene to a rubberlike material. In the years 1911–13, S. V. Lebedev and C. Harris managed independently to polymerize butadiene. Attempts to synthesize rubber did not become commercial until World War I, when Germany produced methyl isoprene as a substitute for natural rubber. The material displayed poor quality, but its use was necessitated by the Allies' blockade.

Until this period, scientists felt that the properties of polymers, especially their apparently high molecular weights, could be explained by the presence of large ring structures or aggregates of small molecules called colloids. In 1920, Hermann Staudinger proposed a different explanation. He argued that materials of high molecular weight actually are composed of very long molecules or macromolecules. The chemists of the day opposed this idea strongly, but Staudinger persisted. In 1953, after 644 publications in the field and long after his basic ideas had been confirmed, he received the Nobel Prize for Chemistry. Staudinger was not the only early contributor to the understanding of polymers. In 1928, for example, K. H. Meyer and Herman Mark postulated that crosslinks exist in natural rubber. Staudinger followed up this idea in 1929 by differentiating linear from network polymers.

In the late 1920s and early 1930s questions on polymer structure were generated as fast as answers. An important link between polymer structure and properties was established in 1934 when Guth and Mark, as well as Kuhn, treated the problem of the different configurations a polymer molecule might assume. This work created a basis for solving problems that deal, for example, with the elasticity of rubber and the high viscosity of polymer solutions.

Perhaps the single most important synthesis in polymer chemistry was reported in 1955 by Prof. Guilio Natta, who had made natural rubber, *cis*-polyisoprene, from isoprene using a Ziegler catalyst. At last, chemists had matched nature's feat of producing a controlled stereoisomerism in rubber. The technique—heterogeneous polymerization—was applied also to polyethylene and polypropylene and is used today to control the properties of many polymers. For their work in the field, Natta and Karl Ziegler shared the Nobel Prize for Chemistry in 1963.

Although *cis*-polyisoprene was one of the first polymers used by man, it was one of the last to be synthesized. In the interim, chemists produced and studied a number of other polymers.

One of the first natural polymers to be heavily modified was cellulose. Industry based on the chemical modification of cellulose seems to have been started by Alexander Parkes, an English metallurgist, in 1862. In that year he displayed articles molded from a material he called Parkesine and won a medal for "excellence of quality" at the International Exposition in London. The material was made by treating cotton waste with nitric and sulfuric acids and was softened with camphor and castor oil. Parkes embarked on a commercial venture, but it soon failed because of technical problems. In 1870, however, John Wesley Hyatt started the Celluloid Corp. in New Jersey, and celluloid—cellulose nitrate plasticized with camphor—was soon a widely used plastic. Cellulose acetate, another modification of cellulose, was discovered in 1865, but was not used extensively until the early 1900s.

The high flammability of nitrocellulose spurred a search for a substitute. One result was the phenolic resins patented in 1899 in England by Arthur Smith for use as electrical insulation. Smith's process did not succeed commercially, but his material was the first high polymer made from small molecules. In 1908, the Belgian-born chemist Leo Baekeland (President of ACS in 1924) worked out a process for controlling the extremely fast reaction between phenol and formaldehyde so that objects could be molded or shaped from the resulting phenol-formaldehyde resin. Baekeland's material, which he named Bakelite, assured the still-continuing commercial success of the phenolics.

From that point on, the polymer industry blossomed. By the early 1920s, large-scale production of vinyl chloride–acetate resins was under way. In 1839, E. Simon had reported the conversion of styrene to a gelatinous mass, but not until 1930 did Germany begin making polystyrene commercially; the U.S. followed in 1937. Also in 1937, Wallace Carothers at Du Pont synthesized the polyamide nylon 6,6 and started a multimillion dollar industry in the U.S. Carothers' work led Otto Bayer in Germany in 1938 to produce a polyurethane, Perlon U, designed to compete with nylon 6,6.

The early development of the polymer industry was paralleled by progress in polymer science. In 1860–63, Lourenco reported the synthesis of polyethylene glycols by condensing ethylene glycol in the presence of ethylene dihalides. The complexity of the product was recognized, but not its structure. In fact the basic nature of condensation polymers was not recognized until 1910–20.

At the same time as Lourenco was reporting his work, Thomas Graham recognized that, in solutions, certain materials diffused much more slowly through membranes that did others. This insight led to a classification scheme in which slowly diffusing materials were called colloids and those that could be crystallized were called "crystalloids."

The year 1872 in Germany saw the first polymerizations of two important materials, polyvinyl chloride (PVC) by E. Baumann and phenolic resins by Adolf Bayer. Eight years later, R. Fitlig and F. Englehorn prepared polymethacrylic acid.

Until the 1880s, scientists had no way to measure the molecular weights of polymeric materials. With the formulation of the laws of Raoult and van't Hoff in this period, techniques became available. Chemists did not believe the extremely high values obtained and ascribed them to molecular aggregation or to inapplicability of Raoult's and van't Hoff's laws to colloidal materials. The problem was not resolved until 1926, when

Staudinger pointed out that, in contrast to association colloids, polymer molecules in solution always exhibit colloidal properties.

In the 1920s, along with Staudinger, investigators such as Mark, Herzog, Katz, Kurtmeyer, Sponsler, and Dore studied the x-ray diffraction patterns of natural polymers such as cellulose. They found that polymers consisted not of ring structures or colloidal aggregates, but of long chains that passed from one unit cell to another.

In the late 1920s and early 1930s, Staudinger pointed out that the molecular weight of synthetic polymers is polydisperse—that is, such polymers contain chains of differing lengths. At the same time, he developed the solution viscosity technique for determining the molecular weights of polymers rapidly.

In 1929, Carothers began a systematic study of the organic chemical reactions that lead to polymers of definite structure. Carothers hoped also to establish certain structure-property relationships. This research led to his major contributions in the areas of polyamides, polyesters, and other condensation polymers.

In 1930, Kuhn published the first application of statistical methods to the study of polymers. Paul J. Flory then applied these methods to polymerization reactions and the configurations of polymer chains. This and other contributions led to Flory's receipt of the Nobel Prize for Chemistry in 1974.

The second World War accelerated polymer development markedly among both friends and foes. In this country, the conversion from natural rubber to the general purpose styrene–butadiene rubber (GR-S or government rubber–synthetic) was a major chemical achievement. Germany also relied heavily on synthetic elastomers. Polyethylene, discovered in England, was used in radar and a variety of other applications because of its superior electrical-insulating properties. Nylon found uses in parachutes and elsewhere. These and other developments laid the base for the huge polymer industry of 1976 and hastened the progress of polymer science as well.

The synthetic polymer industry emerged as an important segment of this country's economy shortly after World War II. Before then, production was limited by wartime shortages as well as by the newness of the materials themselves. About the only synthetic polymers available for production in the early 1950s were nylons, polyethylene, polystyrene, phenolics, cellulosics, PVC, acrylics, silicones, ABS, polyesters, and rubbers. By 1975, some 40 major classes of polymeric materials were in production in commerial quantities as were some rather exotic materials developed for specific uses. Plastics and resins production in this country in 1974 approached 25 billion pounds.

Since the early 1950s, new polymers, new methods of producing already important polymers, and new ways to use these materials have appeared at a steady pace. The most important materials developed and used in large amounts during this period include epoxies, fluoropolymers, polycarbonates, polyethers, polypropylene, and polyurethanes. The areas of free-radical, cationic, and anionic polymerization grew concurrently. The two most important developments in synthesis perhaps were heterogeneous and stereoregular polymerization; the first is used to make polyethylene, for example, and the second to make polypropylene and *cis*-polyisoprene.

It has been suggested that the past quarter century of polymer history can be subdivided into three overlapping periods. The 1940s and 1950s saw the major research thrust in synthesis. The physics of polymeric materials was studied vigorously in the 1950s and 1960s with areas such as crystallization, rheology, thermal analysis, and configurational statistics receiving attention. The 1960s and 1970s appear to be decades of engineering applications. Areas receiving considerable support by the mid-1970s were composites and medical applications of polymers. Exceptions to this chronological scheme include the study of the solution properties of polymers. Polymer scientists recognize also that polymers rarely can be considered usefully from the standpoint of their chemistry or physics alone. An understanding of structure–property relationships demands consideration of the chemistry, physics, and engineering application of the polymer in question.

The foregoing chronology does not mean that research and development in each area has ceased. It would be more accurate perhaps to describe each period as a time when interest in each area accelerated rapidly and plateaued. What lies ahead cannot be predicted, but the tremendous and growing use of polymers almost guarantees their continued technical importance.

The evolution of education in polymer science and technology reflects both the relative youth of the field—understanding of the chemistry and physics of polymers began to emerge only 50 years ago—and its demand for a blend of chemistry, physics, and engineering. When the Division of Polymer Chemistry was formed, a specific program in polymers in this country was offered only at one school: the Polytechnic Institute of Brooklyn with its Polymer Research Institute (headed by Herman Mark). Polymer studies elsewhere were largely isolated efforts by dedicated teachers in the classical area of organic chemistry. The succeeding 25 years have seen a notable change in this picture. Today (1975), slightly more than 100 institutions in the U.S. offer about 1840 credits in polymer science and engineering. These schools include 32 that offer at least specific programs in the field and, at most, degrees ranging from B.S. to Ph.D.

PROFESSIONAL RELATIONS

The Division of Professional Relations came into existence in April 1972 amid a crisis in employment and controversy over the formation of the division itself. Only seven months earlier, at the ACS national meeting in September 1971, the Council had voted 156 to 142 to reject the petition to form the division, but this time the Council approved by a vote of 125 to 89. The division's first officers were Thomas Fitzsimmons, chairman, Norman Pinkowski, secretary, and Myron Linfield, treasurer.

The formation of such a division was not a new idea. In 1947 it had been proposed that the American Institute of Chemists become a division of the Society responsible for professional matters, AIC's primary interest. The plan was approved by the ACS Directors and by mail ballot of AIC members; it was presented to the Council in April 1947, but withdrawn after prolonged debate. A committee assigned to study the issue reported in September 1947 its unanimous belief "that formation of a professional division is not desirable at this time." The stated objections were largely organizational, as were those expressed 25 years later during the Council session that finally brought the division into being. Some Council members in 1972 objected on the grounds that the proposed division apparently sought to represent the general membership rather than only its own members. Others foresaw conflict and overlap with the Council Committee on Professional Relations. Still others thought the proposed division might become a union. In the end these objections proved insufficient to overcome the endorsements of the Committee on Professional Relations and the executive committees of several large local sections.

The memberships of the endorsing local sections were among those most affected by the employment crisis or threat of crisis that then was overtaking the chemical profession. Many chemists and chemical engineers in the previous few years had been subjected to "mass layoffs" or "multiple employment terminations." Research and development laboratories around the country were cutting back sharply, and the employment situation for chemical professionals was critical, as it was for the nation in general.

For many years, chemists and chemical engineers had been hearing that their knowledge and skills were in short supply and in great demand. The sudden shift in the economic winds was jolting. Many complained that the layoffs or terminations were mishandled or unnecessary and that the American Chemical Society was "not doing anything."

The petition to form a Division of Professional Relations was drafted by Thomas Fitzsimmons. At the ACS national meeting in Chicago in September 1970, he sought petitioners among the job-seeking chemists around the Employment Clearing House and in the audience at a Ph.D. job-market symposium. In four hours he had the 50 signatures necessary to bring the petition legally before the Council.

The objectives of the Division of Professional Relations are to represent its own membership and to inform the general membership on professional rather than scientific matters; to increase the awareness of members and to influence Society policies on professional matters through the organization of appropriate programs, conferences, and discussion groups; to assess member opinion on professional matters and to make this information available to Society members through appropriate means.

"Professional matters" almost always refers to conflicts between chemists and institutions and the resolution of those conflicts—usually by compromise—in a satisfactory manner. With an employer, for example, the conflict may involve salary or working conditions; with a scientific society it may involve "return " on dues investment; with society at large it may involve public health or environmental protection; and so on. The Division of Professional Relations' range of interests is evident in its national meeting symposia: Government Incentives for Research and Development; Unionization and the Professional Chemist; Guidelines for Employers—Fact or Fantasy?; New Venture Businesses—How It's Done; Legal Rights and Professional Responsibility; Development of Professional Attitudes; Professional Relations—Members in Action; Social Significance for the Chemist; Scientists and the Legislature; Licensing, Regulation, and Certification; Insurance.

RUBBER CHEMISTRY

An India Rubber Chemistry Section (later the Rubber Chemistry Section) was formed in the American Chemical Society in 1909 and, despite early difficulties, operated successfully until 1919, when it became the Division of Rubber Chemistry. The first chairman of the new division was J. B. Tuttle; the secretary was A. H. Smith. The early problems of the section had stemmed from the intense secrecy among companies in the rubber industry, but World War I and the growth of the automobile spurred technical activity and bred a more enlightened attitude in the industry.

The Rubber Division is unique in that, beginning in 1969, it held its two yearly meetings separately from the Society and was scheduled to do so through 1980. The division did, however, schedule sessions for the ACS Centennial meeting in New York in April 1976. In 1927, the division established its first four local groups, in Akron, Boston, Los Angeles, and New York. In 1975 it had 21 such groups, including two in Canada; these Rubber Groups or Rubber and Plastics Groups had a collective membership of 11,000, of which 35% were members or affiliates of the division.

In 1928, the Rubber Division launched the journal *Rubber Chemistry & Technology*, which in 1975 it still published (bimonthly except January–February) and believed to be the most comprehensive publication in rubber chemistry and related subjects in the world. Scientifically, the journal is the division's largest single undertaking. The division also publishes the annual "Bibliography of Rubber Literature," which was started in 1935 by *Rubber Age* and taken over by the division in 1940.

In 1939, the division established its Charles Goodyear Medal, which is given annually to recognize valuable contributions to the science or technology of rubber or related subjects; it consists of a gold medal, a scroll, and an honorarium of $500. In 1974, in seeking the best means of spending money in support of rubber chemistry, the division initiated a Research Project Award on a trial basis. The award consisted of financial support for a research project in polymer chemistry selected by the division from proposals solicited from academic institutions. The first two awards were made in 1974 and 1975.

Since the fall of 1962 the Rubber Division has offered a correspondence course in rubber chemistry. The course is given twice yearly and is handled out of the University of Akron. Registration is limited to 300 for each course, and each has always been filled. The original textbook, "Introduction to Rubber Technology," was succeeded in the fall of 1973 by an updated version, "Rubber Technology." In 1975 the division was expanding this work to include an advanced course, and the textbook, "Science and Technology of Rubber," was to be published in 1976.

PROGRESS in the rubber industry necessarily has involved many scientific disciplines and commercial enterprises as well as associated industries like chemicals, carbon black, steel, petroleum, agriculture, machinery, plastics, and fabrics. The industry began in a primeval tropical jungle of South America along a river, perhaps the Amazon of Brazil, when early man became aware of a liquid material that oozed from a fallen tree. In time this material, now known as rubber latex, became useful to those Stone Age men as an adhesive, for making balls, and for other applications. A Spanish historian, Antonio de Herrera y Tordesillas, first wrote of this gummy and elastic substance in 1725, but not until 1770 did the English chemist Joseph Priestley give it the name of rubber. Priestley reportedly had found that the gum was most useful in rubbing out errors made by the lead pencil. Michael Faraday, another English scientist, used rubber to make a fabric-reinforced hose. He also established that rubber consisted mainly of carbon and hydrogen with the empirical formula C_5H_8.

In 1839 Charles Goodyear, in his home outside of Boston, discovered that incorporating sulfur into rubber and heating it greatly improved the physical properties of the end product. He delayed until 1841 in filing for a patent, which did not issue until June 15, 1844. Goodyear spent the remaining 19 years of his life successfully fighting court battles with infringers of his patent. His discovery, called vulcanization, is considered the most important in the history of rubber. Thomas Hancock, an Englishman working on rubber at about the same time as Goodyear, recognized the benefits of sulfur in rubber and has shared some of the credit for this great discovery.

In 1877 came an important event, the transfer of seeds of the rubber tree (*Hervea brasiliensis*) from Brazil to London and then to the Far East. The move is credited to Sir Henry Wickham. The seeds, from the best stands of Para, started the great plantations of the Far East that were needed to supply the increasing demands for rubber during the early growth of the automobile. Not until 1910, however, did the first quantity of plantation rubber appear on the market.

Another important event of the period was the invention of the pneumatic tire by the Englishman Robert Thomson, who in 1845 was granted a patent on an "air tube device." The device was not too successful, and not until 1884 did the Scotsman John B. Dunlop reinvent the pneumatic tire by using thin layers of fabric impregnated with rubber and subsequently vulcanized. Alexander Parkes found in 1846 that raw rubber immersed in a solution of sulfur chloride, without heating, resulted in vulcanization; this is known as "cold vulcanization." In 1851, Nelson Goodyear, a brother of Charles, discovered that adding large quantities of sulfur to rubber and later vulcanizing it produced a tough, horny, durable material that became known as hard rubber or ebonite. Charles MacIntosh and his English firm, meanwhile, were producing double-textured garments consisting of two layers of fabric with a layer of rubber between.

During this early period many scientists, including the Germans C. O. Weber and R. N. Henriques, experimented with rubber and its vulcanizates to determine the mechanism by which sulfur brought about these beneficial changes. They found that sulfur actually combined chemically with rubber. Charles Goodyear used inorganic lead components to enhance or accelerate vulcanization, and in 1881 T. Rowley of England discovered that ammonia greatly accelerated the vulcanization of rubber. It was found also that salts and oxides of copper, vanadium, silver, and manganese were deleterious to vulcanizates. These latter two events were the beginning of the great advances that were to be made in developing organic accelerators and antioxidants in the years ahead.

In 1899, A. H. Marks received U.S. and British patents on a process for reclaiming rubber with alkali. Not only was this process a great technical achievement, it became very important commercially some years later when the Stevenson plan was enacted to restrict natural rubber production in order to obtain a stable and higher price for the product. In 1927, within five years after the plan was enacted, reclaim accounted for 33% of the rubber used, and it was the chief weapon used to destroy the plan.

The first year of the 20th century saw a total world production of 54,000 tons of raw rubber, all of it from wild natural sources, mostly in South America. Ten years of the new century would elapse before plantation rubber made its debut, and then only 11,000 tons entered world markets. This same year 83,000 tons of wild rubber, accounting for 87% of the market, would be produced, but from then on wild rubber declined steadily. By 1929, 863,000 tons of natural rubber was produced to satisfy the automobile's terrific appetite, and wild rubber accounted for only 3% of it. The development of high-yielding plantation rubber trees and subsequent bud grafting of trees with selected clones was one of the most important events of this period. It was started in 1917 by P. J. S. Cramer in Indonesia. During the next 50 years, the annual yield of rubber per acre of trees increased from less than 300 lb to more than 2000 lb. This is most significant in today's world with an energy crisis that surely will increase our dependence on natural rubber. During the past few years, moreover, scientists in Malaysia have found that treating the *Hevea* tree with 2-chloroethyl phosphonic acid (Ethrel), which slowly releases ethylene, can increase the tree's output of rubber by 50% or more.

A new era in rubber chemistry was born in 1906, when George Oenslager and A. H. Marks discovered that aniline was an improved accelerator of vulcanization. From this development emerged the industry's first commercial organic accelerator, thiocarbanilide, an aniline derivative that was less toxic and had better processing characteristics than aniline itself. I. Ostromyslenski, a Russian scientist, in 1915 was issued a patent on the use of zinc alkylxanthate as an accelerator; the compound was significant in that it was the first accelerator to contain no nitrogen. Until then it was thought that nitrogen was a necessary ingredient of any organic accelerator. In 1921, L. B. Sebrell and C. W. Bedford made the most important accelerator discovery of all, 2-mercaptobenzothiazole. The compound had many advantages, including fast action at elevated temperatures but practically no activity at lower temperatures, where mixing and processing are done. It resulted in a plateau-type vulcanization condition that decreased the tendency toward overcure and subsequent inferior physical properties. It also greatly improved the resistance of rubber vulcanizates to aging. Today, thiazole derivatives account for 65% of total accelerator usage, even though the field continues to be well plowed, with many new types of accelerators available. Diphenylguanidine was patented by M. L. Weiss in 1922, but litigation went up to the U.S. Supreme Court, which declared the patent invalid. This accelerator is still in use today, usually as a secondary type. The Italians G. Bruni and E. Romani in 1920 developed zinc salts of dithiocarbonic acids, which are called ultra accelerators because of their high speed of vulcanization. These are not normally used in dry rubber because of scorch, but they became very important for latex applications, an industry that was just about to be born.

Work on antioxidants, important organic chemicals for building resistance to aging into rubber products, started about the same time as that on accelerators. Although aniline was discovered as an accelerator by Oenslager and Marks, W. Ostwald was issued a German patent on it as an antioxidant in 1908. Many excellent antioxidants based on amines, secondary amines, and condensation products of amines with carbonyl compounds were developed subsequently. Although these chemicals enjoy acceptance as excellent antioxidants in today's market, their one great deficiency is staining. This would not be detrimental in all-black stocks, but in many products, including white sidewall tires, staining is a handicap. Shortly after World War II, U.S. chemists experimented with reaction products of cresols and olefins, which the Germans had used to stabilize synthetic rubber, and developed a whole new group of nonstaining antioxidants called hindered phenols.

In 1904 S. C. Mote of England discovered that carbon black made by the incomplete burning of natural gas greatly increased the mechanical strength of rubber. Lampblacks, which had been used since the days of Goodyear, were lacking in this respect. Diamond Rubber Co. in 1912 produced tire tread compounds made with reinforcing channel black that wore 10 times better than those made previously with lampblack, clay, or zinc oxide. In that first year, approximately 10 million pounds of the new carbon black was used. Channel blacks were succeeded by thermal furnace blacks, which became commercial in 1922. Since 1942, high structured furnace blacks have been made, and these are produced by both oil-furnace and gas-furnace methods. Parallel with the improvement in the quality of blacks was the development of pelletizing of blacks for easier handling and processing. The technical progress of carbon black has played a most important part in making the high mileage tires we enjoy today. In 1939 the first silica reinforcing pigments were produced, but not until 1950, when silica was made by the pyrogenic high temperature process, did it approach the quality of carbon black as a reinforcing pigment for rubber.

Work on the fundamental chemical nature of rubber occupied many scientists during the 1920s and 1930s. They included Nobel laureate (1953) Herman Staudinger, who proposed the bold concept that rubber was made up of macromolecules of colloidal size and containing 1000 or more isoprene groups. The first synthetic rubber, called methyl rubber, was made in Germany during World War I by polymerizing dimethylbutadiene. Polysulfide rubber, called Thiokol, was discovered by J. C. Patrick in 1927. And in 1931 a team of Du Pont scientists, including J. A. Nieuwland, W. H. Carothers, and Arnold Collins, discovered polychloroprene, called neoprene. German chemists, including E. Tschunker and W. Bock, developed butadiene–styrene polymer, SBR (Buna S), to the experimental factory stage in 1930; it proved to be the first synthetic rubber for tires, following additional improvement and experience with it in tire building. Buna N, a copolymer of butadiene and acrylonitrile developed by German chemists E. Tschunker and E. Konrad about the same time, was an oil-resistant synthetic that became known in this country as NBR. R. M. Thomas and W. J. Sparks of Standard Oil (N.J.)—now Exxon—discovered butyl rubber, a copolymer of isobutylene with a small quantity of butadiene for vulcanization sites.

Although natural rubber is taken from the rubber tree as a 30% colloidal dispersion in a serum consisting mostly of water, and called latex, latex itself did not become a commercial product until the late 1920s. W. L. Utermark in 1923 patented the centrifuge process for concentrating latex to 60–67% total solids. At about the same time Paul Schidowitz found that rubber particles could be vulcanized and still remain in the latex form. In the early 1930s at Dunlop Rubber a group of scientists headed by E. A. Murphy developed the first commercial process for making foam cushioning material, which consisted of beating air into soap-stabilized latex and gelling the liquid mass with salts of hydrofluorosilicic acid. With the subsequent development of emulsion-type synthetic rubbers, such as SBR and NBR, systems were developed to concentrate these into useful, commercial latices. Products made from latices today are dipped goods, such as gloves, adhesives, paper applicators, carpet backings including foam, additives for asphalt to improve road construction, and many others. About 8% of the world's rubber is consumed in latex form.

The advent of World War II set the U.S. Government and private companies, including the major rubber products producers, on a path that was to lead to a great technological triumph, the wartime synthetic rubber program. All of the major rubber companies and many chemical companies had experience in making one or more synthetic rubbers. But no one company or combination of companies had the knowledge, experience, and capital needed for an enterprise that could make the type and quantity of synthetic rubber needed for the war effort. In 1941, the Government, four of the major rubber companies, and Standard Oil Co. (N.J.) agreed to pool their information on styrene–butadiene rubber and develop a new all-purpose rubber that could be processed on standard equipment and made into satisfactory tires. An emulsion recipe with a monomer charge ratio of 75 parts butadiene and 25 parts styrene was established. By June 1942, one SBR rubber plant was operating. By the end of the war, the nation had built 15 SBR plants, two butyl rubber plants, 16 plants for producing butadiene, and five plants for making styrene. Some years after the war the Government sold these plants to private companies, which continued to improve the quality and processing of the rubber.

On Feb. 21, 1951, the federal Reconstruction Finance Corporation (RFC) announced that oil-extended GR–S (government rubber–synthetic) masterbatch was being made. In this process, a tough, high-molecular-weight styrene–butadiene rubber, unsuitable for making tread stock because of its inferior processing properties, could be extended with high boiling petroleum hydrocarbons, to as high as 50% of the weight of the rubber, while still in the latex stage. The resulting rubber had better processing and extrusion properties, gave a softer ride and longer wear, and cost much less than regular SBR. Tires made of the oil-extended SBR, however, were subject to degradation by ozone, which had not previously been a serious problem; the degradation showed up as severe cracking in tire sidewalls. At about the same period, ambient ozone concentrations began to rise in several metropolitan areas, such as Los Angeles. The problem created a need for antiozonants, which were first developed in 1954 by chemists at the Rock Island Laboratory. In the end, several types of antiozonants were developed, including dialkyl–phenylene diamines, alkyl–aryl–phenylene diamines, and waxes.

In the mid-1950s, a new era in synthetic rubber dawned when three large competitors, Goodrich, Firestone, and Goodyear revealed within a short period the discovery of synthetic *cis*-polyisoprene, a rubber that matched natural rubber not only in chemical configuration but in physical properties as well. Karl Ziegler of Germany and Giulio Natta in Italy in the early 1950s had shown how to polymerize olefins with regular configurations, for which they shared the Nobel prize in 1963. Goodrich scientists, licensed by Ziegler, used a trialkyl aluminum/titanium tetrachloride catalyst in making *cis*-polyisoprene rubber; Firestone used lithium metal; Goodyear used aluminum triethyl and a cocatalyst. Trade names for the products are Ameripol SN (Goodrich), Coral rubber (Firestone), and Natsyn (Goodyear). The first commercial production of synthetic natural rubber using a lithium catalyst was in 1960 by Shell; Goodyear started production of Natsyn in 1962. Some years later Goodrich began making Ameripol SN.

Ethylene–propylene rubber was introduced by Standard Oil Co. (N.J.) in 1951. Then came an ethylene–propylene nonconjugated diene terpolymer, now called EPDM, introduced by Du Pont in 1963, by Uniroyal in 1964, by Copolymer Rubber and Chemical in 1967, and by Goodrich in 1971. EPDM at first seemed capable of replacing SBR in many applications because of its improved aging and relatively low cost raw materials. However, it remains a specialty rubber with good potential growth.

Butyl rubber, developed before World War II but reaching maturity during the war, was in high demand for auto and truck tire tubes. It was thought that this rubber could become, like SBR, a second general purpose rubber. The tubeless tire, pioneered by Goodrich, reduced demand for butyl, but it is now finding increasing demand in tubes for truck tires, in water barrier units, and in other applications. Chlorobutyl rubber, made by subjecting a solution of butyl rubber to chlorine vapor, has increased the possibility of covulcanization with other rubbers and is now available from Enjay.

Professor Dr. Otto Bayer in May 1975 received the Charles Goodyear Medal for his discovery, starting in the mid-1930s, of polyurethane. This is a radically different type of rubber, in that products may be made by mixing in liquid form a polyol and a polyisocyanate, along with necessary catalysts and extenders, and injecting the liquid mixture into a mold. The chemicals react quickly and exothermally, and within a minute or two the product may be taken from the mold. Polyurethane foam has taken over and expanded the foam seating market. It is used in adhesives, coatings, and tough, off-the-road solid tires. During the past two years it has been used in fabricating the front ends of automobiles to make them flexible and less subject to damage on impact. Companies such as Goodyear, Firestone, and recently the Polyair Co. of Austria have announced progress in making cast, cordless, pneumatic urethane tires. No such tires are in commercial use, but many groups are working toward that end.

Cis-1,4-polybutadiene, another stereo solution polymer, is used in tire treads as a blend with SBR and *cis*-1,4-polyisoprene to improve tread wear. It has the highest resilience of any rubber. Stereo styrene–butadiene rubber made by the solution process has found increasing demand in the past few years because of its higher cured modulus and superior abrasion resistance. This material also could develop into a general purpose rubber.

In the 1960s came a new family of thermoplastic block copolymers that require no cure and can be remolded. They consist of butadiene and styrene copolymerized with alkyl lithium and having a chain consisting of several butadiene units followed, by several styrene units, followed by more butadiene units, and so on. Several companies, including Firestone, Shell, and Phillips, are manufacturing these thermoplastic rubbers. At room temperature they behave like cured rubber, except for high temperature properties. These polymers can be injection molded and have found utility in shoe soles and heels.

Of the many finished products of rubber, tires are the largest consumer of raw rubber, at 70% of the total. They are followed by mechanical goods, which include conveyor belts, motor mounts, etc.; latex products, largely carpet backing and foam; and footwear, which accounts for 5.0% of the raw rubber. All of these prodcts have climbed the development ladder in the past several years, but none so dramatically as tires, which have changed from a natural rubber/cotton base in 1935, to a natural rubber/rayon base in 1938, to a synthetic rubber/nylon system in 1947, to synthetic/wire in 1955, to synthetic/polyester in 1962, and to synthetic/fiberglass in 1967. During this period, tires had to support increasingly greater loads and travel at higher speeds. The bias tire gave way to the belted/bias tire which now is giving way to the radial tire. This story of the automobile tire could be repeated for the tires used on larger and more demanding trucks and earthmoving vehicles.

The rubber industry in 1975 is entering a period beset with many problems: pollution, energy, smaller tires and less miles of driving, profitability, and the large capital investment needed to keep up with the many advances. Nearly 500,000 Americans earn their livings in the industry, whose annual volume of business is $18 billion. During 1974, the U.S. rubber industry consumed 719,079 tons of natural rubber and 2,388,974 tons of synthetic, accounting for 30% of all the raw rubber consumed in the world. The industry produced a total of 2,516,531 tons of synthetic rubber of all types. Scientists, engineers, and technologists have played important roles in the progress of this industry, and they can be expected to continue to do so.

CHAPTER XI

THE RECORD

The Presidents*

1876
John W. Draper
1811–1882

1877
J. Lawrence Smith
1818–1883

1878
Samuel W. Johnson
1830–1909

1879, 1888
T. Sterry Hunt
1826–1892

1880
Frederick A. Genth
1820–1893

1881, 1889
Charles F. Chandler
1836–1924

1882
John W. Mallet
1832–1912

1883–85
James C. Booth
1810–1888

1886
Albert B. Prescott
1832–1905

1887
Charles A. Goessmann
1827–1910

1890
Henry B. Nason
1831–1895

1891
George F. Barker
1835–1910

1892
George C. Caldwell
1834–1907

1893–94
Harvey W. Wiley
1844–1930

1895, 1921–22
Edgar Fahs Smith
1854–1928

1896–97
Charles B. Dudley
1842–1909

1898
Charles E. Munroe
1849–1938

1899
Edward W. Morley
1838–1923

1900
William McMurtrie
1851–1913

1901
Frank W. Clarke
1847–1931

1902
Ira Remsen
1846–1927

1903
John H. Long
1856–1918

1904
Arthur A. Noyes
1866–1936

1905
Francis P. Venable
1856–1934

1906
William F. Hillebrand
1853–1925

1907–08
Marston T. Bogert
1868–1954

1909
Willis R. Whitney
1868–1958

1910
Wilder D. Bancroft
1867–1953

1911
Alexander Smith
1865–1922

1912–13
Arthur D. Little
1863–1935

1914
Theodore W. Richards
1868–1928

1915–16
Charles H. Herty
1867–1938

1917
Julius Stieglitz
1867–1937

1918–19
William H. Nichols
1852–1930

1920
William A. Noyes
1857–1941

1923
Edward C. Franklin
1862–1937

1924
Leo H. Baekeland
1863–1944

1925–26
James F. Norris
1871–1940

1927
George D. Rosengarten
1869–1936

1928
Samuel W. Parr
1857–1931

1929
Irving Langmuir
1881–1957

1930
William McPherson
1864–1951

1931
Moses Gomberg
1866–1947

1932
L. V. Redman
1880–1946

1933
Arthur B. Lamb
1880–1952

1934
Charles L. Reese
1862–1940

1935
Roger Adams
1889–1971

1936
Edward Bartow
1870–1958

1937
Edward R. Weidlein
1887–

1938
Frank C. Whitmore
1887–1947

1939
Charles A. Kraus
1876–1967

1940
Samuel C. Lind
1879–1965

1941
William Lloyd Evans
1870–1954

1942
Harry N. Holmes
1879–1958

1943
Per K. Frolich
1899–

1944
Thomas Midgley, Jr.
1899–1944

1945
Carl S. Marvel
1894–

1946
Bradley Dewey
1887–1974

1947
W. Albert Noyes, Jr.
1898–

1948
Charles A. Thomas
1900–

1949
Linus Pauling
1901–

1950
Ernest H. Volwiler
1893–

1951
N. Howell Furman
1892–1965

1952
Edgar C. Britton
1891–1962

1953
Farrington Daniels
1889–1972

1954
Harry L. Fisher
1885–1961

1955
Joel H. Hildebrand
1881–

1956
John C. Warner
1897–

1957
Roger J. Williams
1893–

1958
Clifford F. Rassweiler
1899–1976

1959
John C. Bailar, Jr.
1904–

1960
Albert L. Elder
1901–

1961
Arthur C. Cope
1909–1966

1962
Karl Folkers
1906–

1963
Henry Eyring
1901–

1964
Maurice H. Arveson
1902–1974

1965
Charles C. Price
1913–

1966
William J. Sparks
1905–

1967
Charles G. Overberger
1920–

1968
Robert W. Cairns
1909–

1969
Wallace R. Brode
1900–1974

1970
Byron Riegel
1906–1975

1971
Melvin Calvin
1911–

1972
Max Tishler
1906–

1973
Alan C. Nixon
1908–

1974
Bernard S. Friedman
1907–

1975
William J. Bailey
1921–

1976
Glen T. Seaborg
1912–

* Biographies of all ACS presidents who died prior to 1974 are in the book, "American Chemists and Chemical Engineers," Wyndham D. Miles, ed., ACS (1976).

Chairmen of the Board

1931–33
Charles L. Reese
1862–1940

1934–44
Thomas Midgley, Jr.
1889–1944

1945–49
Roger Adams
1889–1971

1950–53
Charles A. Thomas
1900–

1954–55
Ernest H. Volwiler
1893–

1956–58
Ralph Connor
1907–

1959–60, 1962–66
Arthur C. Cope
1909–1966

1961
Louis P. Hammett
1894–

1966–70
Milton Harris
1906–

1971
Byron Riegel
1906–1975

1972
Robert W. Cairns
1909–

1974–76
Herman S. Bloch
1912–

Elected Members of the Board of Directors

Leason H. Adams 1940-45 (R)
Roger Adams 1932-34 and 1941-49 (R)
George P. Adamson 1931-33 (AL)
Meinhard Alsberg 1880-89
John H. Appleton 1895-96
Maurice H. Arveson 1948-59 and 1961-64 (AL)
Peter T. Austen 1893-97
Norman C. Babcock 1948-52
William J. Bailey 1973-75 (AL)
Wilder D. Bancroft 1920-31 (R)
Harry E. Barnard 1917-21
Edward Bartow 1934-36 (R)
Charles Baskerville 1906-07
Lawrence W. Bass 1946-49 (R)
Willard D. Bigelow 1913-22 and 1925-33 (R)
Erle M. Billings 1932-40 (R)
Herman S. Bloch 1971-76 (R)
Marston T. Bogert 1909-20
Elmer K. Bolton 1936-38 and 1940-43 (R) (AL)
Ralph W. Bost 1949-51 (R)
William Brady 1913-16
Abraham A. Breneman 1885-98
David S. Breslow 1973-75 (R)
Wallace R. Brode 1949-60 (R)
Arthur M. Bueche 1966-67 (R)
Robert W. Cairns 1961-66 and 1970-72 (R) (AL)
George C. Caldwell 1892
Herbert E. Carter 1965-67 (R)
Paul Casamajor 1879-87
Charles F. Chandler 1879-1905
Ralph Connor 1954-65 (AL)
Lloyd M. Cooke 1968-70 (R)
Arthur C. Cope 1951-66 (R)
Bryce L. Crawford 1969-77 (AL)
Paul C. Cross 1968-71 (AL)
Farrington Daniels 1950-52 (R)
Louis M. Dennis 1908-10
Charles Doremus 1893-1904
Willard H. Dow 1937-48 (AL)
Charles B. Dudley 1898-1908
C. Tessie DuMotay 1880
Lawrence T. Eby 1963-65 (R)
Robert C. Elderfield 1960-65 (R)
Arthur H. Elliott 1880
Herman Endemann 1879-90
Gustavus J. Esselen 1934-41 (AL)
Henry Eyring 1949-51 (R)
W. Conard Fernelius 1951-59 (R)
Patricia A. M. Figueras 1975-77 (R)
C. Harold Fisher 1969-71 (R)
Paul J. Flory 1960-62 (R)

**(R) Regional, (AL) At Large*

Edward C. Franklin 1926-30
L. H. Friedburg 1889-92
Albert H. Gallatin 1881
Joseph F. Geisler 1890-92
William E. Geyer 1882-83
Thomas S. Gladding 1883 and 1886
J. Goldmark 1879-81
S. A. Goldschmidt 1879
Mary L. Good 1972-77 (R)
William M. Habirshaw 1979-84
Albert C. Hale 1889-92
Albert P. Hallock 1891-92
R. W. Hall 1891-92
Louis P. Hammettt 1956-61 (R)
William E. Hanford 1968-70 (R)
Milton Harris 1966-72 (AL)
Anna J. Harrison 1976 (R)
John B. F. Herreshoff 1882
Arthur J. Hill 1963-44 (R)
Henry A. Hill 1971-75 (R)
William Hoskins 1923-26
T. Sterry Hunt 1889
Walter H. Kent 1888-90
F. T. King 1890-91
Raymond E. Kirk 1950-55 (R)
Charles A. Kraus 1945-50 (R)
A. R. Ledoux 1881
Albert R. Leeds 1879-94
Morton Liebschutz 1886-87
Townes R. Leigh 1937-39 (R)
Samuel C. Lind 1946-48 (R)
Arthur D. Little 1910-12 and 1914-22
Edward G. Love 1910-18
Randolph T. Major 1955-60 (AL)
Raymond P. Mariella 1974-78 (AL)
Robert F. Marschner 1966-70 (R)
Charles F. McKenna 1890-92
William McMurtrie 1893-99, 1906-10
Thomas Midgley, Jr. 1931-42 (AL)
G. M. Miller 1879
Winfred O. Milligan 1961-66 (R)
Henry Morton 1880-84
William A. Mosher 1967-72 (R)
Charles E. Munroe 1894-97
Charles E. Munsell 1884-87
John H. Nair 1964-67 (AL)
Pauline Newman 1973-78 (R)
William H. Nichols 1879-81
William D. Niederhauser 1976-78 (R)
James F. Norris 1927-35 (R)
W. Albert Noyes, Jr. 1967-69 (R)
T. D. O'Connor 1886-90
Emil Ott 1948-50 (R)

Charles G. Overberger 1962-67 and 1969-71 (R)
T. J. Parker 1905-14
Samuel W. Parr 1926-29
Charles A. Parson 1946-48 (R)
Robert W. Parry 1974-76 (AL)
Albert B. Prescott 1896-97
George A. Prochazka 1882-83
H. M. Rau 1884-85
Clifford F. Rassweiler 1955-58 (AL)
Charles L. Resee 1927-32 (R)
E. Emmet Reid 1934-36 (R)
P. deP. Ricketts 1882
Byron Riegel 1959-68 (R)
George D. Rosengarten 1918-27
William Rupp 1884-91
A. H. Sabin 1893-94
Walter A. Schmidt 1931-33 (R) 1935-36 (R) 1943-50 (AL)
Charles H. Schultz 1879
John C. Sheehan 1966-68 (R) and 1971-76 (AL)
T. O'Connor Sloane 1883-86
Alexander Smith 1912-18
Ernest E. Smith 1900-05
William J. Sparks 1959-62 (AL)
Edward R. Squibb 1879-81
Gardner W. Stacy 1970-78 (R)
James H. Stebbins, Jr. 1882-92
Bradford R. Stanerson 1972-77 (AL)
Robert E. Swain 1937-45 (R)
Henry P. Talbot 1921-27
Charles A. Thomas 1942-46 and 1950-53 (AL)
Charles L. Thomas 1955-68 (AL)
Glenn E. Ullyott 1972 (R)
Emerson Venable 1972-74 (R)
Ernest H. Volwiler 1944-55 (AL)
Martin E. Waldstein 1884-89
Frederick T. Wall 1962-64 (R)
Elwyn Waller 1879-87 and 1890-92
John C. Warner 1949-56 (AL)
George W. Watt 1969-73 (AL)
Edward R. Weidlein 1939-47 (R)
Roy L. Whistler 1956-58 (R)
Milton C. Whitaker 1931-36 (AL)
Frank C. Whitmore 1929-31 and 1933-36 (R)
Willis R. Whitney 1916-25
Harvey W. Wiley 1895-98
Hobart H. Willard 1934-40 (R)
Robert E. Wilson 1931-39 (AL) 1941-45 (R)
Durand Woodman 1883-93
William G. Young 1952-60 (R)

Administrative Officers

Recording Secretaries

1876-77 **Isidor Walz**
1878 **Meinhard Alsberg**
1879 **S. A. Goldschmidt**
1880 **Arthur H. Elliott**
1881 **Albert H. Gallatin**
1881-82 **James H. Stebbins, Jr.**
1883 **Thomas S. Gladding**
1884-86 **Charles E. Munsell**
1887-88 **T. D. O'Connor**
1889 **Durand Woodman**
1889-90 **Charles F. McKenna**
1891-92 **Durand Woodman**

Corresponding Secretaries

1876-77 **George F. Barker**
1878 **Henry Morton**
1879-87 **Paul Casamajor**
1888-89 **Meinhard Alsberg**
1890-92 **Albert C. Hale**

Secretaries

1893-02 **Albert C. Hale**
1903-07 **William A. Noyes**
1907-45 **Charles L. Parsons**

Executive Secretaries

1946-65 **Alden H. Emery**
1965-70 **Bradford R. Stanerson**

Executive Directors

1969-72 **Frederick T. Wall**
1972- **Robert W. Cairns**

Treasurers

1876-78 **William M. Habirshaw**
1879-80 **William H. Nichols**
1881-86 **T. O'Connor Sloane**
1886-89 **James H. Stebbins, Jr.**
1890-91 **F. T. King**
1892-99 **Charles F. McKenna**
1899-16 **Albert P. Hallock**
1917-19 **Edward G. Love**
1919-31 **John E. Teeple**
1931-47 **Robert T. Baldwin**
1948-73 **Robert V. Mellefont**
1973-75 **Milton Harris**
1975- **John K Crum**

Chairmen of Committees of the Board of Directors

Chemical Abstracts Service (S)*

1965-69 Byron Riegel
1970-76 Bryce L. Crawford, Jr.

Education and Students (S)

1953-56 W. Conard Fernelius
1957-58 Roy L. Whistler
1959-60 Wallace R. Brode
1961 Byron Riegel
1962-65 Robert C. Elderfield
1966-67 Herbert E. Carter
1968-71 William A. Mosher
1972-76 Gardner W. Stacy

Finance (S)

1893-95 A. P. Hallock
1895-99 Durand Woodman
1899-00 Elwyn Waller
1901-09 J. Howard Wainwright
1909-10 P. C. McIlhiney
1910-19 Edward G. Love
1919-30 John E. Teeple
1931-50 Robert T. Baldwin
1953 Ernest H. Volwiler
1954 John C. Warner
1955 Ralph Connor
1956-59 Maurice H. Arveson
1960-63 Charles L. Thomas
1964-66 Robert W. Cairns
1967-68 Charles L. Thomas
1969-70 William E. Hanford
1971-72 Milton Harris
1973-74 Henry A. Hill
1975-76 John C. Sheehan

Awards and Recognitions (S)

1953-56 Wallace R. Brode
1957-59 W. Conard Fernelius
1960 John C. Bailar
1961-66 Winfred O. Milligan
1967-70 Robert F. Marschner

Petroleum Research (S)

1956-58 Louis P. Hammett

Grants and Fellowships (S)

1959-60 William G. Young
1961-65 Robert C. Elderfield
1966-69 John C. Sheehan
1970 George W. Watt

Grants and Awards (S)

1971 George W. Watt
1972-74 John C. Sheehan
1975 Henry A. Hill
1976 Pauline Newman

Public, Professional, and Member Realtions (S)

1950 Emil Ott
1951-52 Walter A. Schmidt
1953 Edgar C. Britton
1954 Farrington Daniels
1955-59 Clifford F. Rassweiler
1960-61 Albert L. Elder
1962 William J. Sparks
1963 Robert W. Cairns
1964-65 Byron Riegel
1964-67 John H. Nair, Jr.
1968 William E. Hanford
1969-70 Lloyd M. Cooke
1971 Charles H. Fisher
1972 Henry A. Hill
1973-74 Bradford R. Stanerson
1975-76 Raymond P. Mariella

Publications (S)

1950-51 John C. Warner
1952 Arthur C. Cope
1953 John C. Warner
1954-58 Arthur C. Cope
1959-60 Louis P. Hammett
1961-62 Paul J. Flory
1963-71 Charles G. Overberger
1972 William A. Mosher
1973-76 Mary L. Good

Board of Trustees, Group Insurance Plans for ACS Members (Sp)*

1967-73 Robert V. Mellefont
1974-76 Robert E. Henze

Civil Defense and Disaster (Sp)

1956-66 Conrad E. Ronneberg
1967 Walter A. Lawrence
1968-69 Frederick Bellinger
1970-72 Clayton E. Matthews

Chemical Disaster (O)*

1973-76 Russell M. Bimber

Corporation Associates (Sp)

1952 Robert B. Semple
1953-65 Maurice H. Arveson
1966-67 Robert W. Cairns
1967-70 John A. Leermakers
1971-72 Richard W. Roberts
1973-74 Alexander Ross
1976 Blaine C. McKusick

Committee on Investments (Sp)

1950-54 Robert T. Baldwin
1955-73 Robert V. Mellefont
1974-76 Milton Harris

Chemists Club Library Oversight Committee (O)

1973 Pauline Newman
1974-76 Arthur D. F. Toy

Clinical Chemistry (O)

1950-54 Warren M. Sperry
1954-61 John G. Reinhold
1962-66 William B. Mason
1967-68 Richard J. Henry
1969-71 William B. Mason
1972-75 Morton K. Schwartz
1976 Kurt M. Dubowski

Exchanges (O)

1917-51 Evan J. Crane
1954-60 Dale B. Baker
1961-76 James L. Wood

Frasch Foundation Awards (O)

1950-58 Bernard E. Proctor
1959-76 Philip K. Bates

Paper (Sp)

1936-57 Byron L. Wernhoff
1958-68 Marvin C. Rogers
1968-73 Alex Glassman
1975-76 William D. Schaeffer

Pensions (Sp)

1946-65 Alden H. Emery
1965-72 Robert V. Mellefont
1973-75 Milton Harris
1976 John C. Crum

Petroleum Research Fund Advisory Board (O)

1954-62 Cary R. Wagner
1963-69 Arthur L. Lyman
1970-71 Frank G. Ciapetta
1972-76 Charles A. Walker

(S) Standing, (SP) Special, (O) Other

Chairmen of Committees of the Council

Committee on Committees

1975-76 Edwin R. Shepard

Council Policy

Established 1923. The Chairman of this Committee is the ACS President, ex-officio. Starting in 1948 the Committee has elected a Vice-Chairman, who is spokesman for the Committee before the Council.

Vice Chairman

1948-50 John C. Warner
1951-53 Clifford F. Rassweiler
1954-56 John H. Nair, Jr.
1957-58 Byron Riegel
1959-61 Frank T. Gucker
1962 Henry Eyring
1963 J. Harold Perrine
1964 Charles C. Price
1965-67 Lloyd M. Cooke
1968-70 Albert C. Zettlemoyer
1971-73 Pauline Newman
1974-75 David C. Young
1976 David M. Wetstone

Chairmen of Committees of the Council (continued)

Nominations & Elections

1947 Arthur C. Cope
1948 Carl S. Marvel
1949 William J. Sparks
1950 Beverly L. Clarke
1951-52 Ralph A. Connor
1953 Wallace R. Brode
1954-55 Winfred O. Milligan
1956-58 Robert C. Elderfield
1959-61 Charles C. Price
1962-63 William A. Mosher
1964-66 George W. Watt
1967-68 Frank T. Gucker
1969-71 William J. Bailey
1972 Harry E. Whitmore
1973-75 Robert B. Carlin
1975-76 Ellsworth E. McSweeney

Chairmen of Council Committees

Chemical Education (S)

1947 William G. Young
1948-50 Henry E. Bent
1951 Albert F. McGuinn
1952-54 William von Fischer
1955 Byron Riegel
1956 L. Reed Brantley
1957-59 Sherman S. Shaffer
1960-62 Gardner W. Stacy
1963-64 Moses Passer
1965 Robert H. Lindquist
1966-67 Donald L. Swanson
1968 Edward N. Wise
1969-71 J. Trygve Jensen
1972-73 Patricia A. M. Figueras
1974-75 Stanley Kirschner
1976 James J. Hazdra

Constitution & Bylaws (S)

1948 Edward Mack, Jr.
1949-50 Frederick D. Rossini
1951-52 Preston L. Brandt
1953-55 Charles L. Thomas
1956-58 Lloyd M. Cooke
1959-61 Bruce L. Ritz
1962 Lloyd M. Cooke
1963-65 Albert C. Patterson
1966-68 David C. Young
1969-70 Lester C. Krogh
1971-72 Robert B. Fox
1973-74 Joe A. Adamcik
1975-76 Paul V. Smith, Jr.

National Meetings & Divisional Activities (S)

1946 Laurence L. Quill
1947 Francis J. Curtis
1948 Gustav Egloff
1948-50 Milton Harris
1951 Ellsworth E. McSweeney
1952-54 William A. Pardee
1955-57 John C. Bailar, Jr.
1958-60 Winfred O. Milligan
1961 Frank B. Johnson
1962-64 Thurston E. Larson
1965-67 Ambrose G. Whitney
1968-69 John W. LeMaistre
1970 Edward M. Fettes

Divisional Activities (S)

1971-72 J. Wade Van Valkenburg, Jr.
1973-74 Allen L. Alexander
1975 Carlos M. Bowman
1976 James D. Idol, Jr.

Local Sectional Activities (S)

1943-44 Lawrence M. Henderson
1944-46 Carl F. Prutton
1946-47 Robert C. Swain
1948 Clifford F. Rassweiler
1949-50 Joseph S. McGrath
1951-53 Carl F. Graham
1954 Jules D. Porsche
1955-57 J. Harold Perrine
1958-60 Bernard M. Sturgis
1961 David W. Stewart
1962-63 Glenn E. Ullyot
1964-65 Leonard V. Sorg
1966-68 George W. Campbell
1969-71 Arthur H. Hale
1972 Dale N. Robertson
1973-74 Andrew J. Frank
1975-76 M. Wayne Hanson

Meetings and Expositions (S)

1971-72 Edward M. Fettes
1973 Glenn E. Ullyot
1974-76 Bruno J. Zwolinski

Membership Affairs (S)

1947-49 Nathan L. Drake
1950-51 Frank E. Brown
1952-53 Britton A. Shippy
1954-56 Ralph E. Silker
1957-58 Betty Sullivan
1959-60 George W. Watt
1961 William E. McEwen
1962-64 LeRoy W. Clemence
1965-66 Charles O. Gerfen
1967 Ellsworth E. McSweeney
1968-69 John C. Edwards
1970-71 David A. Shirley
1972-74 Edwin R. Shepard
1975-76 Phillip S. Landis

Professional Relations (S)

1947-48 William A. Mosher
1949-50 Wayne W. Hilty
1951 Gordon A. Alles
1952-53 Herman S. Bloch
1954 Donovan J. Salley
1955-56 Herman S. Bloch
1957-59 John K. Taylor
1960-61 Frederick C. Nachod
1962-64 George H. Morse
1965-67 Joseph E. Stewart
1968-69 Henry A. Hill
1970-71 Raymond P. Mariella
1972 Samuel M. Gerber
1973-75 Albert C. Zettlemoyer
1976 Ilmari F. Salminen

Future ACS Dues Requirements

1967-69 Paul N. Craig

Program Review (S)

1969-72 Jack A. Carr
1973-74 Donald L. Swanson
1975-76 Clayton F. Callis

Publications (S)

1947 Edward R. Weidlein
1948 Ernest H. Volwiler
1949 Calvin S. Fuller
1950-52 Abraham L. Marshall
1953 Arthur Rose
1954-55 Howard S. Nutting
1956-57 Otis C. Dermer
1958-59 Robert F. Marschner
1960-62 Edward R. Atkinson
1963-65 John S. Ball
1966-68 Blaine C. McKusick
1969 William G. Dauben
1970-71 Norman Rabjohn
1972 Ernest L. Eliel
1973-75 Arthur Fry
1976 Ernest L. Eliel

Admissions (O)

Established 1876

1917-38 W. D Bigelow
1939-47 W. D. Collins
1947-52 Benjamin D. Van Evera
1953 Norman Bekkedahl
1954-56 Robert C. Vincent
1957 John L. Torgesen
1958-59 Howard W. Bond
1960-67 Calvin F. Stuntz
1968-70 Earl L. Meyers
1971-76 David A. Rowley

Standard Apparatus

1925-52 William D. Collins
1953-55 W. Stanley Clabaugh

Analytical Reagents (O)

1917-19 W. D. Bigelow
1920-42 William D. Collins
1943-55 Edward Wichers
1956-66 W. Stanley Clabaugh
1967-73 Vernon A. Stenger
1974-76 Samuel M. Tuthill

Hazardous Chemicals & Explosives

1923-27 Charles E. Munroe
1928-30 George St.J. Perrott
1931-55 George W. Jones
1955-58 Mathew M. Braidech

Chemical Safety (O)

1961 Herman S. Bloch
1962-66 Herbert K. Livingston
1967-69 Mark M. Chamberlain
1970-72 Ernest I. Becker
1973-76 Howard H. Fawcett

Economic Status (O)

1970-71 Samuel M. Gerber
1972-74 Alan L. McClelland
1975-76 Madeleine M. Joullié

Chairmen of Council Committees (continued)

Nomenclature (O)

1894 Edward Hart
1922 Arthur B. Lamb
1924 Evan J. Crane
1925 Harrison E. Howe
1927-57 Evan J. Crane
1958-63 Leonard T. Capell
1964-76 Kurt L. Loening

Public Relations (O)

1970-74 Arnet L. Powell
1975-76 Helen M. Free

To Study Technician Training

1964-66 William G. Young

To Study Technician Affiliation with the ACS

1965 LeRoy W. Clemence

Technician Curriculum

1966-69 Carleton W. Roberts

Technician Activities (O)

1966-71 LeRoy W. Clemence
1972-73 Carleton W. Roberts
1974-76 J. Fred Wilkes

Women Chemists (O)

1927-35 Glenola B. Rose
1936-39 Lois W. Woodford
1940-43 May L. Whitsitt
1943-47 Cornelia T. Snell
1947-48 Hoylande D. Young
1949-51 Marjorie J. Vold
1951-52 H. Marjorie Crawford
1953-57 Gladys A. Emerson
1958-61 Essie White Cohn
1962-66 H. Gladys Swope
1966-70 Florence H. Forziati
1970-72 Helen M. Free
1973-75 Susan S. Collier
1976 Nina M. Roscher

(S) Standing Committee
(O) Other Committee
(Sp) Special Committee

Chairmen of Joint Board-Council Committees

ACS Centennial Coordinating

1972-76 Bradford R. Stanerson

Education Liaison and Advisory Panel

1967 Herbert E. Carter
1968-70 William A. Mosher

Chemical Education Planning and Coordinating

1971-72 William A. Mosher
1973 Patricia A. M. Figueras
1974-76 Peter E. Yankwich

Chemistry and Public Affairs

1965-68 Charles C. Price
1969-70 Joseph E. Stewart
1971-72 Frank A. Long
1973-76 Charles G. Overberger

Copyrights

1969-70 Ernest E. Campaigne
1971-76 Ben H. Weil

Committee on Air Pollution

1952-58 H. Fraser Johnstone
1959-62 Richard D. Hoak
1963-64 James P. Lodge
1965-68 Aubrey P. Altshuller

Environmental Improvement

1968-76 Thurston E. Larson

International Activities

1963 W. Albert Noyes, Jr.
1964-66 Robert C. Elderfield
1967-68 W. Albert Noyes, Jr.
1969-73 John C. Sheehan
1974-76 Robert W. Parry

Patent & Related Legislation

1917 Leo H. Baekeland
1918-22 E. J. Prindle
1922 C. P. Townsend
1923-38 Henry Howard

Patent Matters and Related Legislation

1964-69 Pauline Newman
1970-74 John T. Maynard
1975-76 Willard Marcy

Professional Programs Planning and Coordinating

1973-76 Albert C. Zettlemoyer

Professional Training

1936-37 Frederick W. Willard
1937-38 Roger Adams
1938-39 Erle M. Billings
1939-40 Robert E. Swain
1941-46 W. Albert Noyes, Jr.
1947-48 Samuel C. Lind
1949-57 William G. Young
1958-59 Louis P. Hammett
1960-65 Herbert E. Carter
1966-73 Cheves Walling
1974-76 Herbert S. Gutowsky

Younger Chemists Task Force

1970-72 Connie Hoiness
1973 Robert G. Linck

Younger Chemists

1973-75 Robert G. Linck

Chairmen of Joint Board-Council Policy Committees

ACS Governance, Structure, and Business Management

1974-76 Robert W. Cairns

Task Force to Carry Forward a More Extensive Consideration of the ADL Report

1975-76 Paul V. Smith, Jr.

Editors of Society Publications

Proceedings of the American Chemical Society

Committee
1876-77 **Hermann Endemann**
Walden Shapleigh
Elwyn Waller

Committee
1878 **Arno Behr**
Hermann Endemann
Walden Shapleigh

Journal of the American Chemical Society

1879 **Hermann Endemann**
1880 **Gideon E. Moore**
1881 **Hermann Endemann**

Committee
1882 **Elwyn Waller**
Charles A. Doremus
L. H. Friedburg

Committee
1883 **Martin E. Waldstein**
Charles A. Doremus
Elwyn Waller
1884-92 **A. A. Breneman**
1893-01 **Edward Hart**
1902-17 **William A. Noyes**
1918-49 **Arthur B. Lamb**
1950-62 **W. Albert Noyes, Jr.**
1963-69 **Marshall Gates**
1970-74 **Martin Stiles**
1975- **Cheves Walling**

Journal of Physical and Colloid Chemistry

(Published to 1927 at Cornell University; published 1927-51 under the auspices of the American Chemical Society, the Faraday Society, and (until 1936) the Chemical Society. In the period 1947–50 it carried the name *Journal of Physical and Colloid Chemistry*. Acquired by American Chemical Society in 1952.)

1896-27 **Wilder D. Bancroft** and **Joseph E. Trevor**
1927-32 **Wilder D. Bancroft**
1933-51 **Samuel C. Lind**
1952-64 **W. Albert Noyes, Jr.**
1965-69 **Frederick T. Wall**
1970- **Bryce Crawford**

Chemical Abstracts

1907-09 **William A. Noyes**
1010-14 **Austin M. Patterson**
1915-58 **Evan J. Crane**
1959-61 **Charles L. Bernier**
1961-66 **Fred Tate**
(Acting Editor)
1967- **Russell J. Rowlett, Jr.**

Industrial and Engineering Chemistry

1909-10 **William D. Richardson**
1911-16 **Milton C. Whittaker**
1917-21 **Charles H. Herty**
1922-42 **Harrison E. Howe**
1943-56 **Walter J. Murphy**
1957-63 **Will H. Shearon, Jr.**
1964-70 **David E. Gushee**

ACS MONOGRAPHS

1921-42 **William A. Noyes**
(Scientific Series)
1921-23 **John Johnston**
(Technologic Series)
1924-41 **Harrison E. Howe**
(Technologic Series)
1942 **Harrison E. Howe**
1943-47 **Frederick W. Willard**
1948-61 **William A. Hamor**
1962-72 **F. Marshall Beringer**

Chemical and Engineering News

(Established as News Edition of Industrial and Engineering Chemistry; name changed in 1942)
1923-43 **Harrison E. Howe**
1943-56 **Walter J. Murphy**
1956-62 **Richard L. Kenyon**
1963-68 **Gordon H. Bixler**
1969-73 **Patrick P. McCurdy**
1974 **Richard L. Kenyon**
(Acting Editor)
1974- **Albert E. Plant**

Chemical Reviews

1924-29 **William A. Noyes**
1930-38 **Gerald Wendt**
1939-49 **W. Albert Noyes, Jr.**
1950-66 **Ralph L. Shriner**
1967- **Harold Hart**

Chemistry

(Founded in 1927 as **Chemistry Leaflet,** became **Science Leaflet** in 1933, and again **Chemistry Leaflet** in 1941, renamed **Chemistry** in 1944. Published by Pennsylvania State College until 1944 then by Science Service. Acquired by American Chemical Society in 1962.)

1927-43 **Pauline Beery Mack**
1944-57 **Helen Miles Davis**
1958-61 **Watson Davis**
1962-63 **Rodney N. Hader**
(Acting)
1964- **O. Theodor Benfey**

Analytical Chemistry

(Established as **Analytical Edition** of **Industrial and Engineering Chemistry** in 1929; name changed in 1948.)

1929-43 **Harrison E. Howe**
1944-56 **Walter J. Murphy**
1957-65 **Lawrence T. Hallett**
1966- **Herbert A. Laitinen**

Journal of Organic Chemistry

(Published 1936-54 by Williams & Wilkins Co., from 1955 by the American Chemical Society. Edited 1936-38 by its policy committee: Marston T. Bogert, Henry Gilman, John R. Johnson, Maurice S. Kharasch, and Ralph L. Shriner. Edited 1938-51 by Lyndon F. Small.)

1952-61 **George H. Coleman**
1962- **Frederick D. Greene**

ADVANCES IN CHEMISTRY SERIES

1950-59 **Walter J. Murphy**
1960- **Robert F. Gould**

Journal of Agricultural and Food Chemistry

1953-56 **Walter J. Murphy**
1956-64 **Rodney N. Hader**
1965- **Philip K. Bates**

Journal of Chemical and Engineering Data

1956-63 **Will H. Shearon, Jr.**
1963-64 **Rodney N. Hader**
(Acting)
1965-70 **Bruce H. Sage**
1971- **Bruno J. Zwolinski**

Journal of Medicinal Chemistry

(Published as **Journal of Medicinal and Pharmaceutical Chemistry,** 1959-61, by Interscience Publishers, edited by Arnold H. Beckett and Alfred Burger.)

1962-71 **Alfred Burger**
1972- **Philip S. Portoghese**

Journal of Chemical Information and Computer Sciences

(Formerly **Journal of Chemical Documentation;** name changed February 1975)
1961- **Herman Skolnik**

Inorganic Chemistry

1962-63 **Robert W. Parry**
1964-68 **Edward L. King**
1969- **M. Frederick Hawthorne**

Biochemistry

1962- **Hans Neurath**

Industrial & Engineering Chemistry Product Research & Development

1963-64 **Byron M. Vanderbilt**
1965-67 **Rodney N. Hader**
(Acting Editor)
1968- **Howard L. Gerhart**

Editors of Society Publications (continued)

Industrial & Engineering Chemistry Fundamentals

1963- **Robert L. Pigford**

Industrial & Engineering Chemistry Process Design and Development

1963- **Hugh M. Hulburt**

Environmental Science and Technology

1967-74 **James J. Morgan**
1975- **Russell F. Christman**

Accounts of Chemical Research

1969- **Joseph F. Bunnett**

Chemical Technology

1971- **Benjamin J. Luberoff**

Clinical Lab Digest

1972-76 **Arthur Poulos**

Journal of Physical and Chemical Reference Data

Published by the American Chemical ical Society and the American Institute of Physics for the National Bureau of Standards.

1972- **David R. Lide, Jr.**

Editors of Divisional Publications

Journals

Journal of Chemical Education

1924-32 **Neil E. Gordon**
1933-40 **Otto Reinmuth**
1940-55 **Norris W. Rakestraw**
1955-67 **William F. Kieffer**
1967- **William T. Lippincott**

Rubber Chemistry & Technology

1928-57 **Carroll C. Davis**
1957-64 **David Craig**
1964 **Samuel D. Gehman** (acting)
1965-68 **Edward M. Bevilacqua**
1969- **Earl C. Gregg, Jr.**

(Other than journals and newsletters)

Cornucopia 1960
Agricultural & Food Chemistry

Chemical Information Bulletin 1948
(formerly *Chemical Literature*)
Chemical Information

EnvirofACS 1951
(formerly *AquafACS* and *Aero and AquafACS*)
Environmental Chemistry

The Fibril Angle 1975
Cellulose, Paper and Textile Chemistry

Picogram 1975
Pesticide Chemistry

Preprints of Meeting Papers 1958
Fuel Chemistry

Preprints of Meeting Papers 1946
Organic Coatings & Plastics Chemistry

Preprints of Meeting Papers 1938
Petroleum Chemistry

Preprints of Meeting Papers 1960
Polymer Chemistry

Reprints of Meeting Papers 1960
Chemical Marketing & Economics

Local Section Publications

(With publishing section and year founded or span)

Accelerator 1916
Indiana (R)*

Alembic 1947
Trenton

Amalgamator 1945
Milwaukee

Bayou Boilings 1947
Southwest Louisiana

Blue Ridge Chemist 1948
Virginia Blue Ridge

Borinchem 1964
Puerto Rico

Branched Chain 1945-67
East Tennessee (R)

Bulletin 1923
Virginia

Capital Chemist 1951
Washington

Catalyst 1915
Philadelphia

Chemical Bond 1951
St. Louis

Chemical Bulletin 1914
Chicago (R)

Chemical Reactions 1941-65
Joliet

Chemical Record 1953
Columbus

Chesapeake Chemist 1945
Maryland

Cintacs 1964
Cincinnati

Condensate 1946-59
Sacramento

Condenser 1946-68
Texas-Louisiana Gulf (Sabine-Neches)

Crucible 1918
Pittsburgh

Dayton Section Monthly Bulletin 1967

DelChem Bulletin 1945
Delaware

Detroit Chemist 1928

Eastern New York Chemist 1955

Filter Press 1946
Georgia

Fission Product 1955
Rhode Island

Flacs 1946
Florida

Free Radical 1952
Kanawha Valley

Genesee Valley Chemunication 1949
Rochester

Indicator 1919
New York and North Jersey (R)

Isotopics 1925
Cleveland (R)

Kansas City Chemist 1950

LaCrosse-Winona Newsletter 1970

Louisville Newsletter 1969

Mid-Hudson Chemist 1970

Midland Chemist 1964

Minnesota Chemist 1949

Mississippi Newsletter 1959

NF=B Double Bond 1928
Western New York

North Carolina Newsletter 1970

Northeast Tennessee Section Newsletter 1972

Nucleus 1922
Northeastern

Octagon 1918
Lehigh Valley

Pipet 1945
Binghamton

Puget Sound Chemist 1940

Resonator 1947
South Jersey

Retort 1948
Northeast Indiana

* Regional, with one or more other sections.

Local Section Publications (continued)

Publication	Local Section	Year
Scalacs	Southern California (R)	1945
Southeaster	Southeastern Pennsylvania	1940
Southern Chemist	Memphis (R)	1941-72
Southwest Retort	Dallas-Fort Worth (R)	1944
Susquehanna Valley Newsletter		1970
Syracuse Chemist		1908
Test Tube	Northern New York	1953
Toledo Section Newsletter		1970
Tunacshro	Hampton Roads	1974
Valchemist	Connecticut Valley	1947
Vapor Pressure	Northesast Oklahoma (R)	1931
Vortex	California (R)	1940
Western Maryland Bulletin		1964-67

Chairmen of Divisions

(Founding year of division is that of first chairman unless indicated otherwise; earlier chairmen were of the originating section or, after 1948, of the division during its probationary period.)

Agricultural and Food Chemistry

(Established 1908)

Year	Chairman
1909	**Willard D. Bigelow**
1910	**Charles D. Woods**
1911-13	**Harry E. Barnard**
1914-15	**Floyd W. Robinson**
1916	**L. M. Tolman**
1917-18	**T. J. Bryan**
1919	**William D. Richardson**
1920-21	**Charles E. Coates**
1922	**T. J. Bryan**
1923	**H. A. Noyes**
1924-25	**C. H. Bailey**
1926-27	**E. F. Kohman**
1928-29	**Fred C. Blanck**
1930	**R. C. Roark**
1931	**J. S. McHargue**
1932-33	**Henry A. Schuette**
1934-35	**Donald K. Tressler**
1936	**John H. Nair, Jr.**
1937-38	**Henry R. Kraybill**
1939	**Roy C. Newton**
1940	**Charles N. Frey**
1941	**Gerald A. Fitzgerald**
1942-43	**Ellery H. Harvey**
1944-46	**Nollie B. Guerrant**
1947	**Bernard L. Oser**
1948	**Paul Logue**
1949	**Carl R. Fellers**
1950	**L. E. Clifcorn**
1951	**Bernard E. Proctor**
1952	**Asger Funder Langlykke**
1953	**Arthur N. Prater**
1954	**Clair S. Boruff**
1955	**Walter O. Lundberg**
1956	**Albert L. Elder**
1957	**Delbert M. Doty**
1958	**Herbert L. J. Haller**
1959	**Frank M. Strong**
1960	**Lloyd W. Hazleton**
1961	**Leonard S. Stoloff**
1962	**John C. Sylvester**
1963	**Herbert E. Robinson**
1964	**Fred E. Deatherage, Jr.**
1965	**J. Wade Van Valkenburg, Jr.**
1966	**John F. Mahony**
1967	**Louis Lykken**
1968	**F. Leo Kauffman**
1969	**Daniel MacDougall**
1970	**Kenneth Morgareidge**
1971	**Irwin Hornstein**
1972	**Emily L. Wick**
1973	**Stanley J. Kazeniac**
1974	**George E. Inglett**
1975	**Richard J. Magee**
1976	**Roy Teranishi**

Analytical Chemistry

(Authorized 1938 as Microchemistry Division; name changed 1940 to Analytical & Microchemistry; changed 1949 to present name.)

Year	Chairman
1936	**A. A. Benedetti-Pichler**
1937	**Frank Schneider**
1938	**Walter R. Kirner**
1939	**Lawrence T. Hallett**
1940	**Clyde W. Mason**
1941	**G. E. F. Lundell**
1942	**George L. Royer**
1943	**Harvey C. Diehl**
1944	**Edward W. D. Huffman**
1945-46	**William M. MacNevin**
1947	**Mary L. Willard**
1948	**Phillip J. Elving**
1949	**Wayne A. Kirklin**
1950	**Grant T. Wernimont**
1951	**Hobart H. Willard**
1952	**Beverly L. Clarke**
1953	**Herbert A. Laitinen**
1954	**G. Frederick Smith**
1955	**William G. Batt**
1956	**Jesse W. Stillman**
1957	**R. P. Chapman**
1958	**John H. Yoe**
1959	**Warren W. Brandt**
1960	**H. A. Libhafsky**
1961	**Charles N. Reilley**
1962	**J. Mitchell, Jr.**
1963	**Lockhart B. Rogers**
1964	**Howard V. Malmstadt**
1965	**Donald H. Wilkins**
1966	**Donald Cooke**
1967	**John K. Taylor**
1968	**Edward C. Dunlop**
1969	**Fred Warren McLafferty**
1970	**Sidney Siggia**
1971	**James Carl White**
1972	**Henry Freiser**
1973	**Richard S. Juvet, Jr.**
1974	**George H. Morrison**
1975	**Robert A. Osteryoung**
1976	**James D. Winefordner**

Biological Chemistry

Year	Chairman
1913-17	**Carl L. Alsberg**
1918	**Winthrop John V. Osterhout**
1919	**Isaac K. Phelps**
1920	**Ross A. Gortner**
1921	**Arthur W. Dox**
1922	**Howard B. Lewis**
1923	**Josiah Simpson Hughes**
1924	**William T. Bovie**
1925	**R. Adams Dutcher**
1926	**Rudolph J. Anderson**
1927	**John R. Murlin**
1928	**Paul E. Howe**
1929	**Michael Xavier Sullivan**
1930	**D. Breese Jones**
1931	**Icie G. Macy**
1932	**Leonard Amby Maynard**
1933	**John Boyer Brown**
1934	**Robert C. Lewis**
1935	**Roe E. Remington**
1936	**Charles Glen King**
1937	**Conrad A. Elvehjem**
1938	**Walter C. Russell**
1939	**Joseph J. Pfiffner**
1940	**George O. Burr**
1941	**Herbert O. Calvery**
1942	**Ben Harry Nicolet**
1943	**Horace A. Shonle**
1944	**Elmer M. Nelson**
1945-46	**Arthur Knudson**
1947-48	**Erwin Brand**
1949	**John T. Edsall**
1950	**Max A. Lauffer**
1951	**John T. Edsall**
1952	**Richard R. Barnes**
1953	**Esmond E. Snell**
1954	**Jesse P. Greenstein**
1955	**Otto Schales**
1956	**Sidney Weinhouse**
1957	**Henry A. Lardy**
1958	**Sidney W. Fox**
1959	**Paul D. Boyer**
1960	**David E. Green**
1961	**Hans Neurath**
1962	**Felix Haurowitz**
1963	**Earl R. Stadtman**
1964	**John N. Buchanan**
1965	**Alton Meister**
1966	**Frank W. Putnam**
1967	**Otto Karl Behrens**
1968	**Daniel E. Koshland**
1969	**Frank M. Huennekens**
1970	**Harold A. Scheraga**
1971	**Robert A. Alberty**
1972	**P. Roy Vagelos**
1973	**Mildred Cohn**
1974	**Robert M. Bock**
1975	**Mary J. Osborn**
1976	**Bernard L. Horecker**

Chairmen of Divisions (continued)

Carbohydrate Chemistry

(Authorized 1921 as Sugar Chemistry Division; name changed 1939 to Sugar Chemistry and Technology; changed 1952 to present name.)

1919-21 Charles A. Browne
1922 S. J. Osborn
1923 William Dodge Horne
1924 Frederick W. Zerban
1925-26 William B. Newkirk
1927 Charles E. Coates
1928 Frederick Bates
1929 H. C. Gore
1930 William L. Owen
1931 Julian K. Dale
1932 W. L. Howell
1933 H. W. Dahlberg
1934 Otto A. Sjostrom
1935 Herbert C. Gore
1936 Julian K. Dale
1937 W. R. Fetzer
1938 Horace S. Isbell
1939 A. R. Nees
1940 James M. Brown
1941 R. Max Goepp, Jr.
1942 Sidney M. Cantor
1943 George H. Coleman
1944 Guido E. Hilbert
1945-46 Robert C. Hockett
1947 Wendell W. Moyer
1948 Melville L. Wolfrom
1949 Ward W. Pigman
1950 William Z. Hassid
1951 George T. Peckham, Jr.
1952 Roy L. Whistler
1953 Thomas R. Gillett
1953-54 Nelson K. Richtmyer
1954-55 Carl C. Kesler
1955-56 John Sowden
1956-57 Norman F. Kennedy
1957-58 Victor R. Deitz
1958-59 Dexter French
1959-60 John L. Hickson
1960-61 John W. LeMaistre
1961-62 John T. Goodwin, Jr.
1962-63 Frederic R. Senti
1963-64 Rex Montgomery
1964-65 John E. Hodge
1965-66 Eugene L. Powell
1966-67 Elizabeth M. Osman
1967-68 Roger William Jeanloz
1968-69 Arthur G. Holstein
1969-70 James Noble BeMiller
1970-71 Louis Long, Jr.
1971-72 Theodore H. Haskell
1972-73 Fraidoun Shafizadeh
1973-74 Hans Helmut Baer
1974-75 Wendell W. Binkley
1975-76 Milton S. Feather

Cellulose, Paper, and Textile Chemistry

(Authorized 1922 as Cellulose Chemistry Division; name changed 1961 to Cellulose, Wood, and Fiber Chemistry; changed 1974 to present name.)

1919-22 Harold Hibbert
1923-24 Gustavus J. Esselen, Jr.
1924-26 Harry LeB. Gray
1926-27 Bjarne Johnsen
1927-28 Louis E. Wise
1928-29 John L. Parsons
1929-30 Earl C. Sherrard
1930-32 Fred Olsen
1932-33 Harold Hibbert
1933-34 Carleton E. Curran
1934-35 Harry F. Lewis
1935-36 George J. Ritter
1936-37 Emil Heuser
1937-38 William F. Henderson
1938-39 George L. Clark
1939-40 Melville L. Wolfrom
1940-41 William O. Kenyon
1941-42 Elwin E. Harris
1942-43 Harold M. Spurlin
1943-44 Ernst Berl
1944-46 Clifford B. Purves
1946-47 Milton Harris
1947-48 Charles R. Fordyce
1948-49 Rollin F. Conaway
1949-50 John S. Tinsley
1951 Wayne A. Sisson
1952 Joseph L. McCarthy
1953 Kyle Ward, Jr.
1954 Alfred J. Stamm
1955 Eugene D. Klug
1956 Reid L. Mitchell
1957 Carlton M. Conrad
1958 Leo B. Genung
1959 Herman F. Mark
1960 Orlando A. Battista
1961 George C. Daul
1962 Roy L. Whistler
1963 Tore Erik Timell
1964 Malcolm Chamberlain
1965 Jeremiah W. Weaver
1966 Wilson A. Reeves
1967 Donald F. Durso
1968 James Russell
1969 Jerome F. Saeman
1970 Mary E. Carter
1971 Joseph H. Dusenbury
1972 Fraidoun Shafizadeh
1973 Albin F. Turbak
1974 Stanley P. Rowland
1975 Robert F. Schwenker, Jr.
1976 John J. Willard

Chemical Education, Inc.

(Section of Chemical Education granted divisional status 1924.)

1921-23 Edgard F. Smith
1923-24 Neil E. Gordon
1924-25 William A. Noyes
1925-26 Wilhelm Segerblom
1926-28 B. Smith Hopkins
1928-29 William McPherson
1929-31 John Nesbit Swan
1931-32 Owen L. Shinn
1932-33 Lyman C. Newell
1933-34 Ross A. Baker
1934-35 Robert E. Swain
1935-36 Harrison Hale
1936-39 Otto M. Smith
1937-38 B. Clifford Hendricks
1938-40 Martin V. McGill
1940-41 Rufus D. Reed
1941-42 Frank E. Brown
1942-43 Arnold J. Currier
1943-46 Lawrence L. Quill
1946-47 John C. Bailar, Jr.
1947-48 Edward L. Haenisch
1948-49 Douglas G. Nicholson
1949-50 Otis C. Dermer
1950-51 James A. Campbell
1951-52 Robert B. Alyea
1952-53 Paul H. Fall
1953-54 Alfred B. Garrett
1954-55 Calvin A. VanderWerf
1955-56 Harry F. Lewis
1956-57 Norris W. Rakestraw
1957-58 Harry H. Sisler
1958-59 Grant W. Smith
1959-60 Leallyn B. Clapp
1960-61 John F. Baxter
1961-62 L. Carroll King
1962-63 Edward C. Fuller
1964 Luke E. Steiner
1965 Robert C. Brasted
1966 William E. Morrell
1967 Wendell H. Slabaugh
1968 William G. Kessel
1969 William B. Cook
1970 Laurence E. Strong
1971 Anna J. Harrison
1972 Robert W. Parry
1973 Shelton Bank
1974 Samuel P. Massie
1975 Robert C. West, Jr.
1976 Gilbert P. Haight, Jr.

Chemical Information

(Authorized 1948 as Division of Chemical Literature; name changed 1975 to present name.)

1948-49 Norman C. Hill
1950 Evan J. Crane
1951 James W. Perry
1952 Julian F. Smith
1953 Robert S. Casey
1954 Byron A. Soule
1955 Milburn P. Doss
1956 Melvin G. Mellon
1957 John H. Fletcher
1958 Ben H. Weil
1959 Hannah Friedenstein
1960 Karl F. Heumann
1961 Herman Skolnik
1962 Fred R. Whaley
1963 Dean F. Gamble
1964 Carleton C. Conrad
1965 Harriet A. Geer
1966 Howard T. Bonnett
1967 Helen R. Ginsberg
1968 Joe Haller Clark
1969 Carlos Morales Bowman
1970 Frederick Bennett Broome
1971 Robert E. Maizell
1972 Stephen J. Tauber
1973 James E. Rush
1974 Charles Granito
1975 Barbara A. Montague
1976 Bruno M. Vasta

Chemical Marketing and Economics

1952-53 Frederick A. Soderberg
1954 Carl A. Setterstrom
1955 Hal G. Johnson
1956 Lawrence H. Flett
1957 Carl A. Sears, Jr.
1958 Ambrose G. Whitney
1959 Malcolm M. Renfrew
1960 John J. Glover
1961 Robert S. First
1962 John W. Slaton
1963 David S. Alcorn
1964 Frenklin W. Wedge
1965 Stanley T. Pender
1966 Robert I. Stirton
1967 Ramon A. Mulholland
1968 Newman H. Giragosian
1969 Robert R. Burns
1970 Russell C. Kidder
1971 Nelson E. Thornton Eby
1972 Richard P. Germann
1973 R. M. Hull
1974 J. Kenneth Craver
1975 Henry F. Whalen, Jr.
1976 Ely Balgley

Chairmen of Divisions (continued)

Colloid and Surface Chemistry

(Authorized 1926 as Division of Colloid Chemistry; name changed 1960 to present name.)

1926-28 **Harry B. Weiser**
1928-30 **Floyd E. Bartell**
1931 **Ross A. Gortner**
1932 **Elmer O. Kraemer**
1933 **Elroy J. Miller**
1934 **Wesley G. France**
1935 **Alfred J. Stamm**
1936 **Samuel S. Kistler**
1937 **Richard Bradfield**
1938 **John W. Williams**
1939 **Lloyd H. Reyerson**
1940 **Ernst A. Hauser**
1941 **Arthur M. Buswell**
1942 **Fred Olsen**
1943 **Winfred O. Milligan**
1944 **James W. McBain**
1945-46 **Geoffery E. Cunningham**
1947 **C. Edmund Marshall**
1948 **Robert C. Vold**
1949 **Desiree S. LeBeau**
1950 **Sydney Ross**
1951 **Miroslav W. Tamele**
1952 **George E. Boyd**
1953 **Desiree S. LeBeau**
1954 **Harold T. Byck**
1955 **John D. Ferry**
1956 **Ralph A. Beebe**
1957 **Albert C. Zettlemoyer**
1958 **William A. Zisman**
1959 **Victor K. LaMer**
1960 **Stephen Brunauer**
1961 **Donald P. Graham**
1962 **Hendrick Van Olphen**
1963 **Frederick R. Eirich**
1964 **Frank H. Healey, Jr.**
1965 **Stanley G. Mason**
1966 **Milton Kerker**
1967 **B. Roger Ray**
1968 **Frederick M. Fowkes**
1969 **Errol Desmond Goddard**
1970 **Egon Matijevic**
1971 **Arthur W. Adamson**
1972 **Robert S. Hansen**
1973 **Howard B. Klevens**
1974 **Paul Becher**
1975 **Gabor A. Somorjai**
1976 **William H. Wade**

Computers in Chemistry

(Authorized 1974)

1975 **Peter G. Lykos**
1976 **Edward C. Olson**

Dye Chemistry

(Authorized 1920)

1919 **Charles L. Reese**
1920-21 **A. B. Davis**
1922-24 **William J. Hale**
1925-26 **R. Norris Shreve**
1927-29 **Moses L. Crossley**
1930-31 **Harold W. Elley**
1932-35 **William D. Appel**

(Merged 1935 with Division of Organic Chemistry.)

Environmental Chemistry

(Authorized 1915 as Division of Water, Sewage, & Sanitation Chemistry; name changed 1959 to Water & Waste Chemistry; name changed 1964 to Water, Air, & Waste Chemistry; changed 1973 to present name.)

1913-16 **Edward Bartow**
1917 **R. B. Dole**
1917 **E. H. S. Bailey**
1918-19 **Robert Spurr Weston**
1920 **J. W. Ellms**
1921 **W. P. Mason**
1922-23 **A. M. Buswell**
1924 **W. W. Skinner**
1925-26 **F. W. Mohlman**
1927 **W. D. Collins**
1928-29 **Stuart E. Coburn**
1930-31 **W. D. Hatfield**
1932-33 **A. S. Behrman**
1934-35 **E. S. Hopkins**
1936-37 **Richard C. Bardwell**
1938-39 **A. P. Black**
1940 **O. M. Smith**
1941 **C. R. Hoover**
1942 **Charles S. Howard**
1943 **Louis F. Warrick**
1944 **Willem Rudolfs**
1945-46 **Clarence C. Ruchhoft**
1947 **Willard W. Hodge**
1948 **H. Gladys Swope**
1949 **William Stericher**
1950 **S. Kenneth Love**
1951 **John J. Dwyer**
1952 **John F. Wilkes**
1953 **H. C. Marks**
1954 **Thurston E. Larson**
1955 **Richard D. Hoak**
1956 **J. Carrell Morris**
1957 **W. Allan Moore**
1958 **Frederick K. Lindsay**
1959 **William L. Lamar**
1960 **Robert S. Ingols**
1961 **Hilding B. Gustafson**
1962 **John J. Maguire**
1963 **Henry C. Bramer**
1964 **A. Curtis Reents**
1965 **Frank M. Middleton**
1966 **Louis F. Wirth, Jr.**
1967 **James P. Lodge, Jr.**
1968 **Sebastian C. Caruso**
1969 **Benjamin F. Willey**
1970 **Aubrey P. Altshuller**
1971 **Frances L. Estes**
1972 **Robert A. Baker**
1973 **C. Ellen Gonter**
1974 **Leslie B. Laird**
1975 **John I. Teasley**
1976 **Donald F. Adams**

Fertilizer and Soil Chemistry

(Established 1908 as Division of Fertilizer Chemistry; name changed 1952 to present name.)

1909-10 **F. B. Carpenter**
1911-13 **Paul Rudnick**
1914-18 **J. E. Breckenridge**
1919-27 **F. B. Carpenter**
1928-39 **Egbert W. Magruder**
1940-46 **H. B. Siems**
1947 **Charles A. Butt**
1948-49 **Jackson B. Hester**
1950-51 **Vincent Sauchelli**
1952 **Samuel F. Thornton**
1953 **Arnon L. Mehring**
1954 **Jesse D. Romaine**
1955 **George H. Serviss**
1956 **Grover L. Bridger**
1957 **Stacy B. Randle**
1958 **Kenneth G. Clark**
1959 **M. Dwight Sanders**
1960 **Travis P. Hignett**
1961 **John O. Hardesty**
1962 **Lawrence B. Hein**
1963 **David R. Boylan, Jr.**
1964 **William J. Tucker**
1965 **William J. Hanna**
1966 **Alvin B. Phillips**
1967 **Charles E. Waters**
1968 **F. J. L. Miller**
1969 **Archie V. Slack**
1970 **George Burnet, Jr.**
1971 **Richard L. Gilbert**
1972 **David W. Bixby**
1973 **Charles H. Davis**
1974 **John T. Hays**
1975 **John G. Getsinger**
1976 **Maurice A. Larson**

Fluorine Chemistry

(Authorized 1963)

1963 **Paul Tarrant**
1964-65 **Ogden R. Pierce**
1966 **Alan Mathieson Lovelace**
1967 **John E. Castle**
1968 **Joseph D. Park**
1969 **Charles Buford Colburn**
1970 **Jeanne M. Shreeve**
1971 **Christ Tamborski**
1972 **Neil Bartlett**
1973 **William A. Sheppard**
1974 **Eugene C. Stump**
1975 **Carl G. Krespan**
1976 **Robert Filler**

Fuel Chemistry

(Gas & Fuel Section authorized 1925 as division; name changed 1960 to present name.)

1922 **Arno C. Fieldner**
1923 **Samuel W. Parr**
1924 **Robert T. Haslam**
1925-26 **Samuel W. Parr**
1927 **George G. Brown**
1928 **Arno C. Fieldner**
1929 **Stephen P. Burke**
1930 **Horace C. Porter**
1931 **George St. J. Perrott**
1932 **Joseph D. Davis**
1933 **Alfred W. Gauger**
1934 **Harold J. Rose**
1935 **Wilbert J. Huff**
1936 **Otto O. Malleis**
1937 **Alfred R. Powell**
1938 **Homer H. Lowry**
1939 **Frank H. Reed**
1940 **Henry H. Storch**
1941 **Henry C. Howard**
1942-43 **Orin W. Rees**
1944 **Gilbert Thiessen**
1945-47 **Selah S. Tomkins**
1948 **Calvert C. Wright**
1949 **G. Robert Yohe**
1950 **Ralph E. Brewer**
1951 **Arthur A. Orning**
1952 **John F. Foster**
1953 **George D. Creelman**
1954 **Corliss R. Kinney**
1955 **Martin B. Neuworth**
1956 **Harlan W. Nelson**
1957 **Charles C. Russell**
1958 **Howard R. Batchelder**
1959 **R. T. Eddinger**
1960 **Richard A. Glenn**
1961 **Joseph Grumer**
1962 **Joseph H. Wells**

Chairmen of Divisions (continued)

Fuel Chem. (cont.)

1963 Robert S. Montgomery
1964 Raymond Friedman
1965 James W. Eckerd
1966 J. Donald Clendenin
1967 Henry Robert Linden
1968 Everett Gorin
1969 Frank Rusinko
1970 Irving Wender
1971 Johnstone S. MacKay
1972 Martin D. Schlesinger
1973 Robert T. Struck
1974 Joseph Freid
1975 Frank C. Schora, Jr.
1976 Wendell H. Wiser

History of Chemistry

(Section authorized as division 1927.)

1922-26 Frank B. Dains
1927-32 Lyman Newell
1933-34 Frank B. Dains
1935-39 Tenney L. Davis
1940-41 James F. Couch
1942-46 Harrison Hale
1947-51 Henry M. Leicester
1952-54 Virginia Bartow
1955-56 Eduard Farber
1957-59 Wyndham D. Miles
1960-61 Virgil R. Payne
1962-64 Aaron J. Ihde
1965 Wyndham D. Miles
1966 O. Theodor Benfey
1967 Martin M. Levey
1968 Melville Gorman
1969 Jack J. Bulloff
1970 George B. Kauffman
1971 June Z. Fullmer
1972 Florence E. Wall
1973 Peter Oesper
1974 Robert M. Hawthorne
1975 O. Bertrand Ramsey
1976 Carl Alper

Industrial and Engineering Chemistry

1908-10 Arthur D. Little
1911 George C. Stone
1912-13 George D. Rosengarten
1914-15 George P. Adamson
1916-17 Harrison E. Howe
1918 William H. Walker
1919 Harlan S. Miner
1920-21 Harry D. Batchelor
1922 Warren K. Lewis
1923-24 D. R. Sperry
1925-26 W. A. Peters, Jr.
1927 William H. McAdams
1928-29 Robert J. McKay
1930-31 Robert E. Wilson
1932-33 Donald B. Keyes
1934-36 Walter G. Whitman
1937 Thomas A. Boyd, Jr.
1938-39 Walter L. Badger
1940-41 Barnett F. Dodge
1942 Lawrence W. Bass
1943-44 R. Norris Shreve
1945-46 Thomas H. Chilton
1947 Francis J. Curtis
1948 Henry F. Johnstone
1949 Joseph C. Elgin
1950 Lincoln T. Work
1951 William A. Pardee
1952 Melvin C. Molstad
1953 J. Henry Rushton
1954 Charles J. Krister
1955 Edward W. Comings
1956 Charles M. Cooper
1957 Edmond L. d'Ouville
1958 DeWitt O. Myatt
1959 James M. Church
1960 Otto H. York
1961 Joseph E. Stewart
1962 G. R. Seavy
1963 Brage Golding
1964 Arthur R. Rescorla
1965 Arthur Rose
1966 Robert B. Beckmann
1967 Robert Landis
1968 Merrell R. Fenske
1969 Leo Friend
1970 Robert N. Maddox
1971 James D. Idol
1972 William E. Hanford
1973 Vernon A. Fauver
1974 James R. Couper
1975 Peter K. Lashmet
1976 David E. Gushee

Leather and Gelatin

(Authorized 1921 as Division of Leather Chemistry; name changed 1923.)

1919 Edwin E. Marbaker
1920-27 John Arthur Wilson
1928-29 August C. Orthman
1929-30 Edwin R. Theis
1930-31 Frank Shephard Hunt
1931-33 Henry B. Merrill
1933-37 J. Harold Hudson

(Division discontinued 1938.)

Inorganic Chemistry

(Divisional status granted 1958.)

1956-57 John C. Bailar, Jr.
1958 John F. Gall
1959 Henry Taube
1960 Eugene O. Brimm
1961 Therald Moeller
1962 Donald R. Martin
1963 George H. Cady
1964 W. Conard Fernelius
1965 Robert W. Parry
1966 Winston M. Manning
1967 Daryle H. Busch
1968 Earl L. Muetterties
1969 Eugene G. Rochow
1970 Fred Basolo
1971 Jack Halpern
1972 Riley Schaeffer
1973 Theodore L. Brown
1974 Llewellyn H. Jones
1975 Alan G. MacDiarmid
1976 Harry B. Gray

Medicinal Chemistry

(Established 1909 as Division of Pharmaceutical Chemistry; name changed 1920 to Chemistry of Medicinal Products; name changed 1927 to present name.)

1909-10 Alviso Burdett Stevens
1911-13 Benjamin L. Murray
1914-15 Frank R. Eldred
1916 John H. Long
1917 Lyman F. Kebler
1918 Frank O. Taylor
1919-21 Charles E. Caspari
1922-23 Edgar B. Carter
1924-25 Ernest H. Volwiler
1926-27 Horace A. Shonle
1928 Arthur W. Dox
1929 Arthur J. Hill
1930 Arnold E. Osterberg
1931 Arthur D. Holmes
1932 Oliver Kamm
1933 Harold C. Hamilton
1934 Paul N. Leech
1935 John H. Waldo
1936 R. Norris Shreve
1937 Donalee L. Tabern
1938 George D. Beal
1939 Walter H. Hartung
1940 Frederic Fenger
1941 Russel J. Fosbinder
1942 John H. Gardner
1943-44 John H. Speer
1945-46 Edward F. Degering
1947 Maurice L. Moore
1948 Frederick F. Blicke
1949 Leon A. Sweet
1950 Kenneth N. Campbell
1951 Richard O. Roblin, Jr.
1952 Chester M. Suter
1953 Marcus G. Van Campen, Jr.
1953-54 Alfred Burger
1954-55 Martin T. Leffler
1955-56 Thomas P. Carney
1956-57 A. Wayne Ruddy
1957-58 Robert M. Herbst
1958-59 Chester J. Cavallito
1959-60 Edward E. Smissman
1960-61 John H. Biel
1961-62 Richard V. Heinzelman
1962-63 Kenneth E. Hamlin, Jr.
1963-64 Joseph H. Burckhalter
1964-65 Cornelius K. Cain
1965-66 James E. Gearien
1966-67 Barry M. Bloom
1967-68 Ernest E. Campaigne
1968-69 Edward Lewis Schumann
1969-70 Raymond E. Counsell
1970-71 Arthur Allan Patchett
1971-72 Joseph G. Cannon
1972-73 Warren J. Close
1973-74 Lester A. Mitscher
1974-75 Irwin J. Pachter
1976 Allan P. Gray

Microbial Chemistry and Technology

(Authorized 1963)

1960-61 Robert K. Finn
1961-62 Gilbert M. Shull
1962-63 Marvin J. Johnson
1963-64 George Edward Ward
1964-65 David Perlman
1965-66 Henry J. Peppler
1966-67 Arthur Earl Humphrey
1967-68 William Everett Brown
1968-69 Peter Hosler
1969-70 Ralph F. Anderson
1970-71 Henry Robert Bungay
1971-72 Rudy J. Allgeier
1972-73 William D. Maxon
1973-74 Daniel I. C. Wang
1974-75 Jerome S. Schultz
1975-76 Ira D. Hill

Chairmen of Divisions (continued)

Nuclear Chemistry and Technology

1963 Archie E. Ruehle
1964 Morton Smutz
1965 Charles E. Stevenson
1966 George E. Boyd
1967 Gerhart Friedlander
1968 W. Wayne Meinke
1969 Anthony Turkevich
1970 George A. Cowan
1971 Grover D. O'Kelley
1972 John R. Huizenga
1973 F. Sherwood Rowland
1974 Ellis P. Steinberg
1975 David A. Shirley
1976 Gregory R. Choppin

Organic Chemistry

(Established 1908)

1910 Edward C. Franklin
1911 George B. Frankforter
1912-13 Treat B. Johnson
1914-15 F. F. Allen
1916 C. G. Derick
1917 J. R. Bailey
1918 William J. Hale
1919 L. W. Jones
1920 E. Emmett Reid
1921 Roger Adams
1922 Hans T. Clarke
1923 Frank C. Whitmore
1924 R. R. Renshaw
1925 Julius A. Nieuwland
1926 Marston T. Bogert
1927 Frank B. Dains
1928 William Lloyd Evans
1929 Edward C. Franklin
1930 Frank C. Whitmore
1931 James B. Conant
1932 Homer Adkins
1933 Carl S. Marvel
1934 Claude S. Hudson
1935 Arthur J. Hill
1936 Henry Gilman
1937 L. Charles Raiford
1938 Lyndon F. Small
1939 Werner E. Bachmann
1940 Cliff S. Hamilton
1941 Nathan L. Drake
1942 Lee I. Smith
1943 Louis F. Fieser
1944 Ralph L. Shriner
1945-46 Samuel M. McElvain
1947 Arthur C. Cope
1948 Paul D. Bartlett
1949 William G. Young
1950 Ralph W. Bost
1951 William S. Johnson
1952 Robert C. Elderfield
1953 Walter N. Lauer
1954 Max Tishler
1955 A. Harold Blatt
1956 Nelson J. Leonard
1957 John D. Roberts
1958 Melvin S. Newman
1959 Karl A. Folkers
1960 John C. Sheehan
1961 William E. Parham
1962 Stanley J. Cristol
1963 William G. Dauben
1964 Delos F. De Tar
1965 Theodore L. Cairns
1966 Herbert O. House
1967 Gilbert J. Stork
1968 Harry H. Wasserman
1969 Jerrold Meinwald
1970 Frederick G. Bordwell
1971 Ronald Breslow
1972 Jerome A. Berson
1973 Andrew Streitwieser, Jr.
1974 Ernest L. Eliel
1975 Jeremiah P. Freeman
1976 Howard E. Simmons, Jr.

Organic Coatings and Plastics Chemistry

(Paint & Varnish Section given divisional status 1927; name changed 1940 to Paint, Varnish, & Plastics Chemistry; name changed 1952 to Paint, Plastics, & Printing Ink Chemistry; name changed 1960 to present name.)

1924 Harry A. Gardner
1925-26 John R. MacGregor
1927-28 W. T. Pearce
1929 P. E. Marling
1930 J. S. Long
1931 P. R. Croll
1932 Harley A. Nelson
1933 Floyd E. Bartell
1934 Robert J. Moore
1935 Wayne R. Fuller
1936 E. W. Boughton
1937 Roy H. Kienle
1938 E. E. Ware
1939 William Howlett Gardner
1940 Edwin J. Probeck
1941 G. G. Sward
1942 Shailer L. Bass
1943 William W. Bauer
1944 Calvin S. Fuller
1945-46 Adolf C. Elm
1947 Ralph H. Ball
1948 Paul O. Powers
1949 Malcolm M. Renfrew
1950 Charles R. Bragdon
1951 Ellsworth E. McSweeney
1952 Francis Scofield
1953 Arthur K. Doolittle
1954 John K. Wise
1955 Albert C. Zettlemoyer
1956 Russell B. Akin
1957 J. Kenneth Craver
1958 L. Reed Brantley
1959 Allen L. Alexander
1960 Walter A. Henson
1961 Edward G. Bobalek
1962 Edward R. Mueller
1963 William O. Bracken
1964 George R. Somerville
1965 Raymond R. Myers
1966 Robert F. Helmreich
1967 Frank P. Greenspan
1968 John C. Cowan
1969 Alfred E. Rheineck
1970 Kenneth N. Edwards
1971 Carlton W. Roberts
1972 Louis J. Nowacki
1973 L. H. Princen
1974 George E. F. Brewer
1975 Clara D. Craver
1976 Lieng Huang Lee

Pesticide Chemistry

(Authorized 1970)

1969 Donald G. Crosby
1970 Elvins Y. Spencer
1971 Wendell Francis Phillips
1972 Philip C. Kearney
1973 Roger C. Blinn
1974 C. H. Van Middelem
1975 Henry F. Enos, Jr.
1976 Julius J. Menn

Petroleum Chemistry

1922-23 Thomas G. Delbridge
1924-27 Ralph R. Matthews
1928-30 J. Bennett Hill
1931-32 Cary R. Wagner
1933-34 Frederick W. Sullivan, Jr.
1935-36 Frank W. Hall
1937-38 Jacque C. Morrell
1939-40 Per K. Frolich
1941-42 Joseph K. Roberts
1943-44 Cecil L. Brown
1945-46 Stewart S. Kurtz, Jr.
1947-48 Gustav Egloff
1949 Wayne E. Kuhn
1950 Arlie A. O'Kelly
1951 Bernard H. Shoemaker
1952 Fred E. Frey
1953 Arthur L. Lyman
1954 Frederick D. Rossini
1955 Everett C. Hughes
1956 Alex G. Oblad
1957 Sherman S. Shaffer
1958 Harold M. Smith
1959 William E. Bradley
1960 Wheeler G. Lovell
1961 Charles L. Thomas
1962 Robert F. Marschner
1963 Bernard M. Sturgis
1964 Augustus H. Batchelder
1965 John S. Ball
1966 William J. Coppoc
1967 Howard L. Yowell
1968 George H. Denison, Jr.
1969 Grant C. Bailey
1970 Vladimir Haensel
1971 Charles E. Moser
1972 Harold Beuther
1973 Charles P. Brewer
1974 George A. Mills
1975 Joe W. Hightower
1976 A. R. Vander Ploeg

Physical Chemistry

(Established 1908 as Division of Physical & Inorganic Chemistry; name changed 1958 to present name).

1909 Charles H. Herty
1910 Edward C. Franklin
1911 Henry P. Talbot
1912 W. Lash Miller
1913 Samuel L. Bigelow
1914-15 George A. Hulett
1916 Irving Langmuir
1917 Henry T. Talbot
1918 Samuel L. Bigelow
1919 William E. Henderson
1920 William D. Harkins
1921 Harry N. Holmes
1922 Samuel E. Sheppard
1923 Robert E. Wilson
1924 Edgar Graham
1925 Arthur E. Hill
1926 Harry B. Weiser
1927 George Shannon Forbes
1928 George L. Clark
1929 Victor K. LaMer
1930 Ward V. Evans
1931 Farrington Daniels
1932 Hobart H. Willard
1933 W. Albert Noyes, Jr.
1934 Donald H. Andrews
1935 N. Howell Furman
1936 John W. Williams
1937 Herbert L. Johnston
1938 Harold S. Booth
1939 George Scatchard
1940 G. Frederick Smith
1941 John G. Kirkwood
1942 W. Conard Fernelius

Chairmen of Divisions (continued)

Physical Chem. (cont.)

1943 Ralph E. Gibson
1944 Oscar K. Rice
1945-46 Thomas F. Young
1947 Paul M. Gross
1948 Henry Eyring
1949 Martin Kilpatrick
1950 John C. Bailar, Jr.
1951 Milton Burton
1952 Glenn T. Seaborg
1953 Frank A. Long
1954 Joseph W. Kennedy
1955 Frank T. Gucker
1956 Pierce W. Selwood
1957 John E. Willard
1958 David P. Stevenson
1959 Robert L. Burwell, Jr.
1960 Benjamin P. Dailey
1961 Joseph O. Hirschfelder
1962 Richard M. Noyes
1963 Bryce L. Crawford, Jr.
1964 Paul C. Cross
1965 Rudolph A. Marcus
1966 Richard B. Bernstein
1967 Herbert S. Gutowsky
1968 Walter Kauzmann
1969 B. Seymour Rabinovitch
1970 Max T. Rogers
1971 David Warren McCall
1972 Peter E. Yankwich
1973 John W. Ross
1974 F. Sherwood Rowland
1975 Sidney W. Benson
1976 Bernard Weinstock

Polymer Chemistry

(Authorized 1951)

1950-51 Carl S. Marvel
1952 William E. Hanford
1953 Paul J. Flory
1954 Raymond M. Fuoss
1955 Herman F. Mark
1956 Raymond F. Boyer
1957 Thomas G. Fox
1958 Arthur V. Tobolsky
1959 Frank R. Mayo
1960 Charles G. Overberger
1961 Turner Alfrey, Jr.
1962 Maurice Morton
1963 Arthur M. Bueche
1964 Field H. Winslow
1965 Robert Simha
1966 Edward M. Fettes
1967 William J. Bailey
1968 Walter H. Stockmayer
1969 John R. Elliott
1970 Jack B. Kinsinger
1971 William E. Gibbs
1972 Joseph P. Kennedy
1973 Jesse C. H. Hwa
1974 Otto Vogl
1975 John K. Stille
1976 Frederick E. Bailey

Professional Relations

1972-74 Thomas Fitzsimmons
1975 Gordon L. Nelson
1976 Warren D. Niederhauser

Rubber Chemistry

(India Rubber Chemistry Section authorized 1919 as a division.)

1919 John B. Tuttle
1920 Warren K. Lewis
1921 W. W. Evans
1922 Clayton Wing Bedford
1923 William B. Wiegand
1924 Elwood B. Spear
1925 Charles R. Boggs
1926 John M. Bierer
1927 Ray P. Dinsmore
1928 Harry L. Fisher
1929 Arnold H. Smith
1930 Stanley Krall
1931 Herbert A. Winkelmann
1932 Ernest R. Bridgwater
1933 Lorin B. Sebrell
1934 Ira Williams
1935 Sidney M. Cadwell
1936 Norman A. Shepard
1937 Harlan L. Trumbull
1938 Archie R. Kemp
1939 George K. Hinshaw
1940 Ernest B. Curtis
1941 Roscoe H. Gerke
1942 John N. Street
1943 John T. Blake
1944 Harold Gray
1945-46 Willis A. Gibbons
1947 Walter W. Vogt
1948 Harry E. Outcault
1949 Howard I. Cramer
1950 Fred W. Stavely
1951 John H. Fielding
1952 Waldo L. Semon
1953 Seward G. Byam
1953-54 James C. Walton
1954-55 John M. Ball
1955-56 Arthur E. Juve
1956-57 Benjamin S. Garvey, Jr.
1957-58 Raymond F. Dunbrook
1958-59 E. H. Krismann
1959-60 William J. Sparks
1960-61 Wesley S. Coe
1961-62 George E. Popp
1962-63 Gilbert H. Swart
1963-64 James D. D'Ianni
1964-65 Edwin B. Newton
1965-66 Norman S. Grace
1966-67 Dale F. Behney
1968 Glen Alliger
1969 Thomas H. Rogers
1970 Paul G. Roach
1971 John H. Gifford
1972 Albert E. Laurence
1973 Eli Dannenberg
1974 Francis M. O'Connor
1975 Benjamin Kastain
1976 Earl C. Gregg, Jr.

Divisional Officers Group

1924 R. Rex Renshaw
1926-30 Erle M. Billings
1931-33 John N. Swan
1933-36 John H. Nair, Jr.
1936-37 Herrick L. Johnston
1937-38 Donalee L. Tabern
1938-39 George Scatchard
1939-41 Cliff S. Hamilton
1941-42 Charles R. Hoover
1942-43 Howard I. Cramer
1943-44 Arthur C. Cope
1944-46 Chester M. Alter
1946-47 Cary R. Wagner
1947-48 Milton Harris
1948-49 John C. Bailar, Jr.
1949-50 Winfred O. Milligan
1950-51 Milton Burton
1951-52 Winfred O. Milligan
1952-53 Alexander Oblad
1953-54 Albert C. Zettlemoyer
1954-55 G. Frederick Smith
1955-56 Joseph E. Stewart
1956-57 William E. Parham
1957-58 L. Reed Brantley
1958-59 Warren W. Brandt
1959-60 Hilding B. Gustafson
1960-61 Brage Golding
1961-62 Malcolm M. Renfrew
1962-63 John F. Baxter
1963-64 Fred R. Whaley
1964-65 J. Wade Van Valkenburg, Jr.
1965-66 Malcolm Chamberlain
1966-67 Carleton C. Conrad
1967-68 Howard L. Yowell
1968-69 Robert I. Stirton
1969-70 Edward C. Dunlop
1970-72 Elvins Y. Spencer
1972-73 Lambertus M. Princen
1973-74 Donald F. Durso
1974-75 J. Kenneth Craver
1975-76 Clara D. Craver

Local Section Chairmen

(Sections chartered year of first listing unless noted otherwise. Location of section headquarters in parentheses unless title is location.)

Akron (Ohio)

1923 Walter W. Evans
1924 Norman A. Shepard
1925 Harry L. Fisher
1926 Raymond P. Dinsmore
1927 James W. Schade
1928 Henry F. Palmer
1929 Lorin B. Sebrell
1930 Webster N. Jones
1931 Hezzelton E. Simmons
1932 George Oenslager
1933 George K. Hinshaw
1934 Harlan L. Trumbull
1935 Arthur W. Carpenter
1936 A. Evan Boss
1937 Charles R. Park
1938 Joseph P. Maider
1939 Raymond F. Dunbrook
1940 William I. Burt
1941 Winfield Scott
1942 Arthur W. Sloan
1943 Arthur E. Warner
1944 John N. Street
1945 Ben S. Garvey, Jr.
1946 Paul J. Flory
1947 Edwin B. Newton
1948 John H. Bachmann
1949 Edward A. Willson
1950 George E. P. Smith, Jr.
1951 Guido H. Stemple, Jr.
1952 James D. D'Ianni
1953 W. Glenn Mayes
1954 Thomas L. Gresham
1955 Elbert E. Gruber
1956 Nelson V. Seeger
1957-58 Ralph F. Wolf
1959 Harold P. Brown
1960 Henry A. Pace
1961 Walter C. Warner

Local Section Chairmen (continued)

Akron (Ohio) (cont.)

1962 **Leora E. Straka**
1963 **Kenneth W. Scott**
1964 **Harold Tucker**
1965 **Glenn H. Brown**
1966 **Larry E. Forman**
1967 **Henry J. Kehe**
1968 **Vernon L. Folt**
1969 **William E. Bissinger**
1970 **Otto C. Elmer**
1971 **Joginder Lal**
1972 **Ronald W. Smith**
1973 **Richard M. Wise**
1974 **Joseph A. Beckman**
1975 **Charles P. Rader**
1976 **Roger A. Crawford**

Alabama

(Birmingham)

1913-14 **Bennett B. Ross**
1915 **Stewart J. Lloyd**
1916 **Alfred H. Olive**
1916-17 **Jack Percival Montgomery**
1918 **C. N. Wiley**
1919-20 **Wallace L. Caldwell**
1921 **James Forrest Carle**
1922-23 **Augustus Guy Overton**
1924 **Jesse R. Harris**
1925 **Roger Williams Allen**
1926 **Harry Burn**
1927-28 **David Hancock**
1929-30 **George J. Fertig**
1931 **John Richard Sampey**
1932 **Ernest V. Jones**
1933 **Willard M. Mobley**
1934 **Emmett B. Carmichael**
1935 **John Xan**
1936 **Benjamin F. Clark**
1937 **A. Richard Bliss, Jr.**
1938 **Samuel S. Heide**
1939 **Cleburne Ammen Basore**
1940 **J. Thompson Vann**
1941 **George D. Palmer**
1942 **Harold E. Wilcox**
1943 **Simon R. Dean**
1944 **C. S. Whittet**
1945-46 **Thomas Lafayette McWaters**
1946 **Russell L. Jenkins**
1947 **Wilbur A. Lazier**
1948 **David H. Thompson**
1949 **Edgar E. Hardy**
1950 **Carl Bordenca**
1951 **David H. Chadwick**
1952 **James L. Kassner**
1953 **Robert E. Burks, Jr.**
1954 **Warner W. Carlson**
1955 **Locke White, Jr.**
1956 **James A. Johnson, Jr.**
1957 **Samuel Booth Barker**
1958 **William J. Barrett**
1959 **Kenneth M. Gordon**
1960 **Charles E. Feazel**
1961 **Don B. Griffin**
1962 **William S. Wilcox**
1963 **Leonard L. Bennett, Jr.**
1964 **William C. Coburn, Jr.**
1965 **Robert H. Garner**
1966 **Oscar L. Hurtt, Jr.**
1967 **Wynelle D. Thompson**
1968 **William Niedermeier**
1969 **Robert Glaze**
1970 **Harmon Leslie Hoffman, Jr.**
1971-72 **Ervin R. Van Artsdalen**
1973 **Dean Calloway**
1974 **Carrol G. Temple, Jr.**
1975 **John A. Montgomery**
1976 **B. W. Ponder**

Ames (Iowa)

1915 **John Anderson Wilkinson**
1916 **Chester Charles Fowler**
1917 **Rex R. Renshaw**
1918 **Charles A. Mann**
1919 **William G. Gaessler**
1920 **John Hall Buchanan**
1921 **Frank E. Brown**
1922 **Henry Gilman**
1923 **Anson Hayes**
1924 **Winfred F. Coover**
1925 **Victor E. Nelson**
1926 **Norman A. Clark**
1927 **Ralph M. Hixon**
1928 **Ellis I. Fulmer**
1929 **John Anderson Wilkinson**
1930 **Nellie Naylor**
1931 **Henry Gilman**
1932 **Frank C. Vilbrandt**
1933 **John Hall Buchanan**
1934 **Victor E. Nelson**
1935 **Norman A. Clark**
1936 **Iral B. Johns**
1937 **Leland A. Underkofler**
1938 **Walter Bernard King**
1939 **Harley A. Wilhelm**
1940 **Emerson W. Bird**
1941 **Henry A. Webber**
1942 **Harvey C. Diehl**
1943 **Rachel H. Edgar**
1944 **Burrell F. Ruth**
1945 **Lester Yoder**
1946 **Winfred F. Coover**
1947 **Byron H. Thomas**
1948 **Sidney W. Fox**
1949 **Lionel K. Arnold**
1950 **George F. Stewart**
1951 **Adolph F. Voigt**
1952 **Joseph F. Foster**
1953 **George S. Hammond**
1954-55 **Charles V. Banks**
1956 **Robert S. Hansen**
1957 **Dexter French**
1958 **Ernest Wenkert**
1959 **Adrian H. Daane**
1960 **Chas. H. DePuy**
1961 **Robert S. Allen**
1962 **David E. Metzler**
1963 **John D. Corbett**
1964 **Orville L. Chapman**
1965 **Robert E. McCarley**
1966 **Harry J. Svec**
1967 **John G. Verkade**
1968 **William C. Wildman**
1969 **Robert A. Jacobson**
1970 **Walter S. Trahanovsky**
1971 **Bernard J. White**
1972 **Robert J. Angelici**
1973 **Thomas J. Barton**
1974 **Jon C. Vardy**
1975 **Harvey Diehl**
1976 **Gerald J. Small**

Arizona

(Tucson)

1925 **Ernest Anderson**
1926 **Theophil F. Buehrer**
1927 **Lathrop E. Roberts**
1928 **Oscar C. Magistad**
1929 **John D. Sullivan**
1930 **I. Grageroff**
1931 **Lila Sands**
1932-33 **Frank S. Wartman**
1934 **Ernest Anderson**

(*See* Southern Arizona)

Arkansas

(Fayetteville)

1921-22 **Charles F. Robinson**
1923-24 **J. W. Johnson**

(*See* University of Arkansas)

Arkansas-Louisiana Border

(Chartered 1946 as Ark-La-Tex Section; name changed 1970.)

1970 **Rudolph B. Horstmann**

(Name changed again 1971 to Northwest Louisiana.)

Ark-La-Tex

1946 **John B. Entrikin**
1947 **Guy S. Mitchell**
1948 **D. N. Barrow**
1948 **Harry J. Sheard**
1949 **Edward C. Greco**
1950 **Aubrey W. Trusty**
1951 **Justin O. Griffin**
1952 **Harold E. Abbott**
1953 **Willard M. Dow**
1954 **Marvin A. Smith**
1955 **Harry Karam**
1956 **Donald W. Emerich**
1957 **William F. Cummer**
1958 **Harold E. Hammar**
1959 **David E. Sullenberger, Jr.**
1960 **Alan H. Crosby**
1961 **Carl B. Sutton**
1962 **A. Adler Hirsch**
1963 **Glenn L. Shepherd**
1964 **Robert E. Pitts**
1965 **Leonard E. Savory**
1966 **Marshall R. Kesling**
1967 **Marvin W. Hanson**
1968 **John B. Sardisco**
1969 **Charles H. Whiteside**

(Name changed 1970 to Arkansas-Louisiana Border.)

Auburn (Ala.)

1950 **James E. Land**
1951 **William T. Miller**
1952 **Samuel H. Nichols, Jr.**
1953 **William B. Bunger**
1954 **Edwin Hove**
1955 **Edwin O. Price**
1956 **Frank J. Stevens**
1957 **Paul F. Ziegler**
1958 **Parker P. Powell**
1959 **Paul F. Ziegler**
1960 **J. Marshall Baker**
1961 **Herman D. Alexander**
1962 **Donald L. Vives**
1963 **Paul Melius**
1964 **Robert H. Dinius**
1965-66 **Leslie E. Baker**
1967 **Curtis H. Ward**
1968 **Frank J. Stevens**
1969 **Earle C. Smith**
1970 **William C. Neely**
1971-72 **William R. Mountcastle, Jr.**
1973 **James H. Hargis**
1974 **Lawrence F. Koons**
1975 **Joseph L. Greene**
1976 **Michael E. Friedman**

Local Section Chairmen (continued)

Baton Rouge (La.)

1939 **Arthur R. Choppin**
1940 **Maurice B. Amis**
1941 **Raoul Louis Menville**
1942 **Arthur G. Keller**
1943 **David F. Edwards**
1944 **Roger W. Richardson**
1945 **O. Edward Kurt**
1946 **Sumner B. Sweetser**
1946 **Ernest A. Fieger**
1947 **Cecil L. Brown**
1947 **Robert S. Asbury**
1948 **Charles E. Starr, Jr.**
1949 **Philip W. West**
1950 **C. Gruman Steele**
1951 **George W. Beste**
1952 **Arthur L. LeRosen**
1952 **James P. McKenzie**
1953 **John L. Porter**
1954 **Frank B. Johnson**
1955 **Martin B. Smith**
1956 **Louis L. Rusoff**
1957 **Charles F. Gray**
1958 **Henry G. Allen**
1959 **Arvid A. Anderson**
1960 **Franklin Conrad**
1961 **Earl B. Claiborne**
1962 **David H. Campbell**
1963 **James G. Traynham**
1964 **Hulen B. Williams**
1965 **Tillmon H. Pearson**
1966 **Paul E. Koenig**
1967 **Julian B. Honeycutt, Jr.**
1968 **Harry V. Drushel**
1969 **Fred J. Impastato**
1970-71 **Gene C. Robinson**
1972 **Charles Boozer**
1973 **Wm. A. Baddley**
1974 **Wm. H. Daly**
1975 **G. C. Gaeke, Jr.**
1976 **Harold J. Wahlborg**

Binghamton (N.Y.)

1941-42 **August H. Brunner, Jr.**
1942 **Harold E. Pletcher**
1942-43 **Keith Famulener**
1943-44 **Frank J. Kaszuba**
1944-46 **Harold C. Harsh**
1946-47 **Benjamin R. Harriman**
1947-48 **Thomas R. Thompson**
1948-50 **F. W. H. Mueller**
1950 **Ralph A. Copeland**
1951 **Elton L. Beavan**
1952 **Francis H. Gerhardt**
1953 **Martin A. Paul**
1954 **Lee C. Hensley**
1955 **Fritz H. Dersch**
1956 **Raymond Walford**
1957 **Charles H. Benbrook**
1958 **Charles A. Clark**
1959 **C. Max Hull**
1960 **Michael T. Orinik**
1961 **Joseph E. P. Apellaniz, Jr.**
1962 **Harold A. Levine**
1963 **Donald E. Trucker**
1964 **Wallace L. Bostwick**
1965 **John Kushner**
1966 **Emil Bruno Rauch**
1967 **Stanley P. Popeck**
1968 **Guenther H. Klinger**
1969 **Bruce McDuffie**
1970-71 **Frank J. Loprest**
1972 **Felix Viro**
1973 **John A. Welsh**
1974 **Lawrence P. Verbit**
1975 **Carl E. Johnson**
1976 **John M. Para**

Boulder Dam

(Henderson, Nev.)

1944 **Hillard L. Smith**
1944-46 **Carl K. Stoddard**
1946 **N. E. McDougal**
1947 **Thomas A. Sullivan**
1947 **Fred E. Littman**
1948 **Larry E. Trumbull**
1949 **Bruce J. Boyle**
1950 **Frank G. Radis**
1951 **Harold J. Wurzer**
1952 **Harold H. Houtz**
1953 **Raynard V. Lundquist**
1954 **Leonard J. Edwards**
1955-56 **Alford L. Andersen, Jr.**
1957 **Maurice J. Miles**
1958 **George R. Stewart**
1959 **Harold Leitch**
1960 **Daniel D. Walker, Jr.**
1961 **Edgar E. Millaway**
1962 **Duncan W. Cleaves**
1963 **Linden E. Snyder**
1964 **Clarence Lee Boyd**
1965 **Meade A. Stirland**
1966 **Morgan S. Seal**
1967 **Nathan B. Coe**
1968 **Robert B. Smith**
1969-70 **Paul Howard Norton**
1971 **Walter Chester Nowak**
1972 **Richard Titus**
1973 **Donald K. Pennelle**
1974-75 **Richard R. Renner**
1976 **James W. Mullins**

Brazosport

(Freeport, Tex.)

1958 **Vincent A. Thorpe**
1959 **R. Bruce LeBlanc**
1960 **Paul E. Muehlberg**
1961 **Julius C. Sanders, Jr.**
1962 **John C. Smith**
1963 **Sherman Kottle**
1964 **J. M. Leathers**
1965 **George E. Ham**
1966 **Brad H. Miles**
1967 **Clarence R. Dick**
1968 **Simon Miron**
1969 **John Philip Buettner**
1970 **Oscar L. Hollis**
1971 **Basil C. Doumas**
1972 **Paul D. Ludwig**
1973 **Evan A. Mayerle**
1974 **Harry L. Spell**
1975 **Thomas F. LaGest**
1976 **John L. Massingill**

California

(1901; Berkeley)

1902 **Edmond O'Neill**
1903 **E. C. Burr**
1904 **J. M. Stillman**
1905 **Felix Lengfeld**
1906 **Alonzo E. Taylor**
1907 **S. W. Young**
1908 **Arthur Lachman**
1909 **M. E. Jeffa**
1910-11 **Edward C. Franklin**
1911 **Ralph A. Gould**
1912 **Franklin T. Green**
1913 **T. J. Wrampelmeir**
1914 **Abbott A. Hanks**
1915 **Edward C. Franklin**
1916 **Fred G. Cottrell**
1917 **Joel H. Hildebrand**
1918 **John W. Bockman**
1919 **L. H. Duschak**
1920 **Robert E. Swain**
1921 **Wm. C. Bray**
1922 **Bryant S. Drake**
1923 **C. L. Alsbery**
1924 **C. W. Porter**
1925 **Roy R. Rogers**
1926 **Herman A. Spoehr**
1927 **W. H. Sloan**
1928 **George S. Parks**
1929-30 **Charles G. Maier**
1931 **Russell W. Millar**
1932 **C. L. Baker**
1933 **James H. C. Smith**
1934 **Merle Randall**
1935 **Robert E. Swain**
1936 **Ludwig Rosenstein**
1937 **Carl L. A. Schmidt**
1938-39 **Wilhelm Hirschkind**
1940 **Theodore K. Cleveland**
1941 **Thomas D. Stewart**
1942 **Paul F. Bovard**
1943 **Norman N. Gay**
1944 **Troy C. Daniels**
1945 **Wm. E. Vaughan**
1946 **Louis B. Howard**
1947 **Robert Matteson**
1948 **Gerhard K. Rollefson**
1949 **Richard Wister**
1950 **Robert G. Larsen**
1951 **Melvin Calvin**
1952 **A. H. Batchelder**
1953 **Alva C. Byrns**
1954 **Alan C. Nixon**
1955 **Harry S. Mosher**
1956 **James O. Clayton**
1957 **Kenneth J. Palmer**
1958 **Wendell M. Stanley**
1959 **Theodore W. Evans**
1960 **William T. Stewart**
1961 **Richard M. Lemmon**
1962 **Glenn L. Allen, Jr.**
1963 **Fred Stitt**
1964 **Fred H. Stross**
1965 **T. K. Cleveland**
1966 **John Y. Beach**
1967 **Donald S. Noyce**
1968 **William L. Stanley**
1969 **Walter B. Petersen**
1970-71 **Fred Rust**
1972 **Louis R. Pollack**
1973 **Don R. Baker**
1974 **Stephen A. Rodemeyer**
1975 **Nylen L. Allphin**
1976 **Alan C. Nixon**

Canton, Mo.

Chartered 1950

1951 **Benjamin I. Lyon**
1952 **Edward C. Tarpley**
1953 **Wallace P. Elmslie**
1954 **William A. Hensley**

(Name changed 1954 to Quincy-Keokuk.)

Carolina-Piedmont

(Charlotte, N.C.)

1943 **Charles H. Stone**
1944-46 **Charles H. Higgins**
1946 **Cecil Waltham Gilchrist**
1946 **Dave E. Truax**
1947 **Brooks S. Liles**
1948-50 **Norman D. Doane**
1950 **John B. Gallent**
1951 **Robert H. Baker, Jr.**
1952 **David F. Mason**
1953 **Arthur H. Noble, Jr.**
1954 **Laura T. Hall**
1955 **Charles R. Holtzclaw**

Local Section Chairmen (continued)

Carolina-Pied. (cont.)

1956 Carle W. Mason
1957 George K. Kologiski
1958 Robert D. Williams
1959 R. Henry Teeter
1960 Betty J. Livingstone
1961 William L. Milheim
1962 Homer R. Ketchie
1963 John F. Luther
1964 Thomas E. Lesslie
1965 James M. Fredericksen
1966 Preston E. Grandon
1967 Howard P. Belue
1968 Sherman L. Burson
1969 Paul L. Weinle
1970 William B. Robertson
1971 James R. Kuppers
1972 Blair W. Drum
1973 Frank B. Tutwiler
1974 Gerald W. Davis
1975 Joseph C. Hubball
1976 Joe B. Davis

Central Arizona

(Phoenix)

1961 J. Smith Decker
1962 Michael J. Sullivan
1963 Alice L. Mathis
1964 Gordon D. Perrine
1965 Castle O. Reiser
1966 George U. Yuen
1967 Morton E. Munk
1968 Bernard Van Pul
1969 Theodore M. Brown
1970 Arnold E. Bereit
1971 Ray A. Cattani
1972 Michael L. Parsons
1973 Samuel R. Lewis
1974 Charles J. Horn
1975 William E. Bidleman
1976 Patricia M. Waeschle

Central Arkansas

(Little Rock)

1950 Paul L. Day
1951 Carroll F. Shukers
1952 Clyde A. Broyles
1953 Alfred G. Hewitt
1953-54 Joseph E. Pryor
1955 Howard L. Leventhal
1956 Carl D. Douglass
1957 Thomas E. Shook
1958 Lavert A. Adams
1959 Robert M. Atchley
1960 Charles R. Nony
1961 Wilson J. Broach
1962 Daniel M. Mathews
1963 William D. Williams
1964 Charles A. Mazander, Jr.
1965 Robert W. Shideler
1966 Bernard F. Armbrust, Jr.
1967 John E. Stuckey
1968 Wendell L. Fortner
1969 James O. Wear
1970 Joe C. Wright
1971-72 Don England
1973 J. Lyndoll York
1974 Damon V. Royce, Jr.
1975 Donald C. De Luca
1976 Bryan D. Palmer

Central Massachusetts

(Worcester)

1947 John R. Erickson
1948 Samuel S. Kistler
1949 Harry B. Feldman
1950 Howard A. Whittum
1951 Jesse L. Bullock
1952 Rev. B. A. Fiekers
1953 Alexander L. Gordon
1954 Harold Levy
1955 Wilmer L. Kranich
1956 Robert H. Haberstroh
1957 Arthur E. Martell
1958 Carleton P. Stinchfield
1959 Andrew Van Hook
1960 Lowell H. Milligan
1961 Harris Rosenkrantz
1962 Robert E. Wagner
1963 Walter G. Dahlstrom
1964 Richard B. Bishop
1965 Edward N. Trachtemberg
1966-67 Neal L. McNiven
1968 Robert C. Plumb
1969 John J. Killoran
1970 Edward L. Eagan
1971 William Andruchow, Jr.
1972 Philip E. Stevenson
1973 Elihu J. Aronoff
1974 Leo W. Ziemlak
1975 Harry C. Allen
1976 John J. Falvey

Central New Mexico

(Albuquerque)

1947 Melvin Leon Brooks
1948 Richard Dean Baker
1949 Anthony R. Ronzio
1950 Wright H. Langham
1951 Charles F. Metz
1952 Morris F. Stubbs
1953 Joe Fred Lemons
1954 R. H. Miller
1955 Jesse L. Riebsomer
1956 Robert A. Penneman
1957 Karl S. Bergstresser
1958 John F. Suttle
1958-59 Joseph A. Schufle
1960 Joseph A. Leary
1961 Sherman W. Rabideau
1962 Charles E. Holley, Jr.
1963 Galen W. Ewing
1964 Eldon L. Christensen
1965 Joseph A. Schufle
1966 Adam F. Schuch
1967 Alan E. Florin
1968 Thomas T. Castonguay
1969-70 Vincent C. Anselmo
1971 R. Gillette Bryan
1972 Al Zerwekh
1973 Clifford R. Keizer
1974 Marvin C. Tinkle
1975 Thomas R. Henderson
1976 Donald Hoard

Central North Carolina

(Greensboro)

1950 Philip H. Latimer, Jr.
1951 Harvey A. Ljung
1952 Paul M. Ginnings
1953 Edmund O. Cummings
1954 Charles M. Sprinkle
1955 Guita Marble
1956 Andrew Miga
1957 Paul H. Cheek
1958 John H. Whiteside
1959 Robert H. Cundiff
1960 Harry B. Miller
1961 James E. Greer
1962 Albert W. Cagle
1963 William W. Menz
1964 Phillip J. Hamrick, Jr.
1965 Eugene W. Jones
1966 James T. Dobbins, Jr.
1967 Fred J. Schultz
1968 Cornelius F. Strittmatter
1969 Henry L. Anderson
1970 Marjorie Newell
1971 Claude I. Lewis
1972 Herbert W. Baird
1973 Edward R. Epperson
1974 Donald H. Piehl
1975 Sherri Forrester
1976 John L. McKenzie

Central Ohio Valley

(Huntington, W.Va.)

(Chartered 1947 as Tri-State Section; name changed 1949.)

1949 George C. Meredith
1950 Harry E. Tschop
1951 Joseph S. Beddall
1952 Allen W. Scholl
1953 Blaine C. Mays
1954 Francis J. Gibson
1955 Stanley C. Church
1956 William H. Toller
1957-58 Francis A. Koehler
1959 Oliver J. Zandona
1960 George S. Brown
1961 J. Stephen Ogden
1962 Jane W. Mittendorf
1963 Roland O. Meyer
1964 Kenneth P. Fuller
1965 J. F. Bartlett
1966 Stuart H. Morgan
1967 J. Holland Hoback
1968 James T. Corcoran
1969 James Edward Douglass
1970 Kenneth R. Robinson
1971 Orlando Antony Vita
1972 James T. Corcoran
1973 William G. Lipscomb
1974 Melvin Mosher
1975 Thomas F. Lemke
1976 Tharol L. McClaskey

Central Pennsylvania

(State College)

(Chartered 1924 as State College Section; name changed 1925.)

1926 Gerald L. Wendt
1927 Ernest B. Forbes
1928 R. Adams Dutcher
1929 Walter Thomas
1930 Harry H. Geist
1931 Arthur K. Anderson
1932 J. Harris Olewine
1933 John G. Aston
1934 Howard O. Triebold
1935 Alfred W. Gauger
1936 Arnold J. Currier
1937 Donald Stevens Cryder
1938 Nelson W. Taylor
1939 Albert Witt Hutchison
1940 Russell C. Miller
1941 Samuel T. Yuster
1942 Arthur Rose
1943 Frank C. Whitmore
1944 Nollie Burnham Guerrant
1945 Helen M. Davis
1946 Thomas S. Oakwood
1947 Calvert C. Wright
1948 Robert V. Boucher
1949 Corliss R. Kinney
1950 Robert W. Schiessler
1951 Paul M. Althouse

Local Section Chairmen (continued)

Central Pa. (cont.)

1952 Theodore S. Polansky
1953 Ralph P. Seward
1954 Gordon H. Pritham
1955 Lester Kieft
1956 Harold J. Read
1957 John R. Hayes
1958 Donald M. Rockwell
1959 Andrew A. Benson
1960 Joseph A. Dixon
1961 Harold L. Lovell
1962 Howard B. Palmer
1963 Carl O. Clagett
1964 James J. Fritz
1965 Ralph O. Mumma
1966 E. Erwin Klaus
1967 Laxman N. Mulay
1968 Richard L. McCarl
1969 Thomas V. Long, II
1970 Robert L. Kabel
1971 John D. Sink
1972 Jennings H. Jones
1973-74 William A. Steele
1975 Edward J. Tracey
1976 Gregory J. McCarthy

Central Texas

(1917; Austin)

1918-19 George S. Fraps
1920-21 Wilby T. Gooch
1922 Charles C. Hedges
1923-24 Nicholas C. Hamner
1925-28 Wilby T. Gooch
1929 John Campbell Godbey
1930 W. S. Mahlie
1931 Henry R. Henze
1932 James Laurence Whitman
1933 C. W. Burchard
1934 May L. Whitsett
1935 Harry Louis Lochte
1936 Harry Marshall Bulbrook
1937-38 Henry R. Henze
1939 William A. Felsing
1940 James E. Adams
1941 Harry Louis Lochte
1942 William A. Cunningham
1943 Lewis F. Hatch
1944 William A. Felsing
1945 Kenneth A. Kobe
1946 Norman Hackerman
1947 Robbin C. Anderson
1948 John J. McKetta
1949 Beverly Guirard
1950 Gilbert H. Ayres
1951 William Shive
1952 Philip S. Bailey
1953 J. David Malkemus
1954 Leon O. Morgan
1955 Royston M. Roberts
1956 Stanley H. Simonsen
1957 William J. Peppel
1958 Howard F. Rase
1959 James E. Boggs
1960 J. David Malkemus
1961 Philip H. Moss
1962 Harry Louis Lochte
1963 Jefferson C. Davis, Jr.
1964 George P. Speranza
1965 Hugo Steinfink
1966 Joseph J. Lagowski
1967 Floyd Edward Bentley
1968 David M. Himmelblau
1969 Patrick Edward Cassidy
1970-71 John Michael White
1972 Robert M. Gipson
1973 Douglas B. Manigold
1974 Robert L. Soulen
1975 Robert L. Soulen
1976 Peggy W. Glass

Central Utah

(1969; Provo)

1970 Jerald S. Bradshaw
1971 James J. Christensen
1972 J. Bevan Ott
1973 Raymond N. Castle
1974 Byron J. Wilson
1975 Francis Nordmeyer
1976 Lee D. Hansen

Central Wisconsin

(Steven Point)

1972 C. Marvin Lang
1973 Diane A. Tewksbury
1974 John W. Hollis, Jr.
1975 Ronald Roberts
1976 Douglas Radtke

Chattanooga (Tenn.)

(Chartered 1937 as Southeast Tennessee Section; name changed 1945.)

1946 Ivine W. Grote
1947 Hans Thurnauer
1948 Raymond C. Adams
1949-51 Milton Gallagher
1952 T. J. McIntosh
1953 Kendol C. Gustafson
1954 William F. Luther, Jr.
1955 Donald H. Gunther
1956 Frederick W. Hayward
1956 Murray Raney
1957 Thomas G. Street, Jr.
1958 Harold L. Elmore
1959 Harold H. Goslen
1960 Hugh J. Bronaugh
1961 John Christensen
1962 Ralph E. Berning
1963 Benjamin H. Gross
1964 Carl Cain, Jr.
1965 Ross F. Russell
1966 Raymond W. Ingwalson
1967 James S. Galbraith
1968 Donald E. Bruce
1969 Horst Wolfgang Schmank
1970 Norman Peek
1971 Franklin L. Boyer
1972-73 John W. Musser
1974 Julian Glasser
1975 Marie S. Black
1976 Robert L. McNeely

Chicago (Ill.)

1895-96 Frank Julian
1897 William Hoskins
1898 Joseph P. Grabfield
1899 C. E. Linebarger
1900 Warren Rufus Smith
1901 Felix Lengfeld
1902 Gustav Thurnaur
1902 Edward Gudeman
1903 William August Puckner
1904 Julius A. Stieglitz
1905 William Brady
1906 Herbert N. McCoy
1907 Warren Rufus Smith
1908 William D. Richardson
1909 Willard A. Converse
1910 Thomas J. Bryan
1911 Stephen T. Mather
1912-13 Arthur Lowensteing
1913 Harry McCormack
1914 Otto Eisenschiml
1915 William D. Harkins
1916 Austin Van Hoesen Mory
1917 Lucius M. Tolman
1918-19 Lawrence V. Redman
1920-21 W. Lee Lewis
1922 Carl S. Miner
1923 George Albert Menge
1924-26 Dudley K. French
1926 Paul Nicholas Leech
1927 Sterling L. Redman
1928 Benjamin B. Freud
1929 Ward V. Evans
1930 Herman I. Schlesinger
1931 Bernard E. Schaar
1932 David Klein
1933 Lee Francis Supple
1934-35 Walker M. Hinman
1935-36 Arthur Guillaudeu
1936-37 Paul Van Cleef
1937 Jacque C. Morrell
1938 Charles D. Hurd
1939 Cary R. Wagner
1940 William F. Henderson
1941 George L. Parkhurst
1942 Warren C. Johnson
1942-43 Roy C. Newton
1943-44 Lawrence M. Henderson
1944 Robert E. Zinn
1945 Maurice H. Arveson
1946 Robert K. Sommerbell
1947 Herbert E. Robinson
1948 Charles L. M. Thomas
1949 Walter M. Urbain
1950 Byron Riegel
1951 Robert F. Marschner
1952 Jules D. Porsche
1953 Marvin C. Rogers
1954 Herman S. Bloch
1955 Lloyd M. Cooke
1956 Hoylande D. Young
1957 Gordon T. Peterson
1958 LeRoy W. Clemence
1959 Bernard S. Friedman
1960 Raymond P. Mariella
1961 Wayne Cole
1962 F. Leo Kauffman
1963 Gifford W. Crosby
1964 Edmund Field
1965 J. Frederick Wilkes
1966 Charles K. Hunt
1967 Roy H. Bible, Jr.
1968 Fred Klepetar
1969 Thomas H. Donnelly
1970 Thomas J. Kucera
1971 Ellis K. Fields
1972-73 Walter S. Guthmann
1974 Richard W. Mattoon
1975 Louis J. Sacco
1976 James P. Shoffner

(Cincinnati (Ohio)

(Chartered 1892)

1893 Chauncey R. Stuntz
1894 John Uri Lloyd
1895 Karl Langenbeck
1896 E. Twitchell

Local Section Chairmen (continued)

Cincinnati O. (cont.)

1897 E. C. Wallace
1897 William L. Dudley
1898 O. W. Martin
1899 William Simonson
1900 Thomas Evans
1901 William H. Crane
1902 R. W. Hochstetter
1903 H. E. Newman
1904 J. F. Snell
1905 Sigmund Waldbatt
1906 A. Springer
1907 Harry S. Fry
1908 L. W. Jones
1909 Fred Hamburg
1910 Joseph W. Ellms
1911 Max H. Goettsch
1912 Archibald Campbell
1913 Frank C. Broeman
1914 Archibald Campbell
1915 Charles T. P. Fennell
1916 Clarence P. Bahlmann
1917 Carlos P. Long
1918-19 M. B. Graff
1920 Hugh M. Campbell
1921-22 George K. Elliott
1923 Albert P. Mathews
1924 Louis W. Bosart
1925 Geo. D. McLaughlin
1926 A. O. Snoddy
1927 Everett R. Brunskill
1928 C. R. Bragdon
1929 Albert S. Richardson
1930 R. F. Reed
1931 W. M. Burgess
1932 Howard Ecker
1933 John T. R. Andrews
1934 Clifford J. Rolle
1935 Francis F. Heyroth
1935 C. H. Allen
1936 Frederick E. Ray
1937 R. H. Ferguson
1938 Arthur W. Broomell
1939 Walter H. McAllister
1940 Procter Thomson
1941 Frederic E. Holmes
1942 W. C. Gangloff
1943 Ronald C. Stillman
1944 Clarence C. Ruchhoft
1945 Daniel J. Kooyman
1946 Ralph E. Oesper
1947 Fred H. Snyder
1948 Hoke S. Greene
1949 Joseph F. Treon
1950 Henry R. Kreider, Jr.
1951 Donald S. Hirtle
1952 Oscar T. Quimby
1953 Charles E. Frank
1954 Willard I. Upson
1954 Nathan N. Crounse
1955 C. Austin Sprang
1956 Thomas B. Cameron, Jr.
1957 Carl J. Opp
1958 Elton S. Cook
1959 Keith W. Wheeler
1960 Herbert Nordsieck
1961 Milton Orchin
1962 Francis M. Middleton
1963 Robert A. Harris
1964 Alfred Richardson
1965 Ted J. Logan
1966 Floyd L. James
1967 Edwin R. Andrews
1968 Irving L. Mador
1969-70 Norman A. Bates
1971-72 Sally A. Vonderbrink
1973 David H. Gustafson
1974 David E. O'Connor
1975 R. Marshall Wilson
1976 Thomas W. Gibson

Cleveland (Ohio)

1909 Franklin T. Jones
1910 A. W. Smith
1911 J. W. Brown
1912 H. Gruener
1913 Edward C. Holton
1914 C. B. Murray
1915 A. T. Baldwin
1916 W. C. Moore
1917 O. F. Tower
1918 A. W. Smith
1919 H. D. Bachelor
1920 F. M. Dorsey
1921-22 L. C. Drefahl
1923 William R. Veazey
1924 Harry N. Holmes
1925 Harold S. Booth
1926 John D. Morron
1927 M. J. Rentschler
1928 N. K. Chaney
1929 A. F. O. Germann
1930 W. L. Reinhardt
1931 Norbert A. Lange
1932-33 Edwin G. Pierce
1934 William R. Veazey
1935 C. G. Schluderberg
1936 Robert E. Burk
1937 Lauchlin M. Currie
1938 Carl F. Prutton
1939 E. C. Hughes
1940 Herman P. Lankelma
1941 Arthur S. Weygandt
1942 Eric A. Arnold
1943 W. J. Bartlett
1944 Glenn H. McIntyre
1945 Frank Hovorka
1946 M. J. Rentschler
1947 M. R. Hatfield
1948 Gerald M. Juredine
1949 A. G. Bowers
1950 Wm. Von Fischer
1951 Lester L. Winter
1952 Oliver J. Grummitt
1953 Karl S. Willson
1954 Robert L. Savage
1955 George C. Whitaker
1956 Thomas J. Walsh
1957 Thomas W. Mastin
1958 Harrison M. Stine
1959 Charles V. Mitchell
1960 Melvin J. Astle
1961 Clark O. Miller
1962 Harry J. Dietrick
1963 Ralph L. Dannley
1964 Albert A. Arters
1965 J. Reid Shelton
1966 Samuel M. Darling
1967 Henry Z. Sable
1968 George E. Blomgren
1969 Lester Friedman
1970 Allen P. Arnold
1971 James A. Walsh
1972 Norman Standish
1973 Roger E. Stansfield
1974 Joan P. Lambros
1975 John P. Fackler
1976 Glenn R. Brown

Coastal Empire

(Savannah, Ga.)

1958-60 Wyndham E. Priddle
1961 Don Martin
1962 James W. Miles
1963 Robert T. Lukat
1964 Seldon Page Todd
1965 Edgar E. Sellers
1966 Charles K. Clark
1967 Roger M. McKinney
1968 Robert Wm. Johnson
1969 Albert S. Perry
1970 Cedric Stratton
1971 Gordon O. Guerrant
1972 Paul E. Robbins
1973 Martha Tootle Cain
1974 Norman M. Dennis
1975 Bruce Anderson
1976 Frank E. Taylor

Colorado

(Denver)

1920 Samuel C. Lind
1921 John B. Ekeley
1922-23 Sidney J. Osborn
1924-25 Larry D. Roberts
1926-27 Reuben G. Gustavson
1928-30 A. R. Nees
1930-31 Clarence M. Knudson
1932 Leon S. Ward
1933 Howard L. Wiley
1934 Leon S. Ward
1935-36 Frank E. E. Germann
1937 Henry W. Dahlberg
1938 Alfred C. Nelson
1939 William L. Conrad
1940 Warren W. Howe
1941 Scott A. Powell
1942 H. B. Van Valkenburg
1943 Robert T. Phelps
1944 Lewis H. Chernoff
1945 Rev. T. Louis Keenoy
1946 Emil G. Swanson
1947 Bernard B. Longwell
1948 Edward W. D. Huffman
1949 Robert A. Baxter
1950 George O. G. Lof
1951 Rex E. Lidov
1952 Stanley J. Cristol
1953 Arthur L. Fowler
1954 John S. Meek
1955 S. Barney Soloway
1956 Rudolph M. Anker
1956-57 Harold F. Walton
1958 Thomas S. Chapman, Jr.
1959 Henry J. Richter
1960 Louis C. Gibbons
1961 Walter H. Dumke
1962 Norman F. Witt
1963 John L. Ellingboe
1964 Dale N. Robertson
1965 William C. Stickler
1966 William T. Miller, S.J.
1967 Alice O. Robertson
1968 John A. Beel
1969 Douglas G. Heberlein
1970 Larry L. Miller
1971 Richard L. Taber
1972 Alice O. Robertson
1973 Alvin L. Schlage
1974 Ed. L. King
1975 Lowell A. King
1976 Dwight M. Smith

Columbus (Ohio)

(Chartered 1897)

1898-06 Henry A. Weber
1907 William McPherson
1908 George O. Higley
1909 William L. Evans
1910-11 Charles W. Foulk

Local Section Chairmen (continued)

Columbus, O. (cont.)

1912-13 James R. Withrow
1914-16 Charles W. Foulk
1917 George F. Weida
1918-20 Cecil E. Boord
1921 Charles P. Hoover
1922 W. E. Henderson
1923 George O. Higley
1924 Evan J. Crane
1925 Jesse E. Day
1926 Clyde S. Adams
1927 Cecil E. Boord
1928 William McPherson
1929 O. L. Barneby
1930 Charles B. Morey
1931 Edward Mack
1932 John Brown
1933 W. Clarence Ebaugh
1934 W. A. Manuel
1935 Wallace R. Brode
1936-37 Wesley G. France
1937 Richard Bradfield
1937 H. V. Moyer
1938 Herbert L. Johnston
1939 H. A. Depew
1940 Melville L. Wolfrom
1941 Richard S. Shutt
1942 Alfred B. Garrett
1943 Robert C. Williams
1944 Melvin S. Newman
1945 John S. Crout
1946 Joseph H. Koffolt
1947 Wm. MacNevin
1948 Frank H. Verhoek
1949 Paul O. Powers
1950 Roy G. Bossert
1951 Randall G. Heiligmann
1952 Fred E. Deatherage, Jr.
1953 Elsworth E. McSweeney
1954 Lawrence P. Biefeld
1955 Kenneth W. Greenlee
1956 John N. Pattison
1956-57 Palmer B. Stickney
1958 James V. Robinson
1959 Richard D. Morin
1960 John W. Clegg
1961 Samuel A. Woodruff
1962 Randall G. Rice
1962-63 Frank C. Croxton
1964 Jack G. Calvert
1965 Paul K. Glasoe
1966 Richard C. Himes
1967 Robt. I. Leininger
1968 Joyce B. Healy
1969 Ferd R. Wetsel
1970 Paul G. Gassman
1971 William R. Dunnavant
1972 H. Dale Hannan
1973 William A. Hoffman
1974 Edward L. King
1975 Eugene J. Mezey
1976 Theodore Kuwana

Connecticut Valley

(Hartford, Conn.)

1911-12 F. W. Farrell
1912 K. R. Sternberg
1912-15 Hervert E. Emerson
1915-18 Moses L. Crossley
1918-19 B. H. Smith
1919-20 Charles R. Hoover
1920-22 Joseph S. Chamberlain
1922-24 G. Albert Hill
1925 George B. Hogaboon
1926-27 Lewis B. Allyn
1928 V. K. Krieble
1929 A. R. Lincoln
1930 E. M. Shelton
1931-32 Mary Lura Sherrill
1933-35 John E. Cavelti
1936 Joseph B. Ficklen
1937 Evald L. Skau
1938 Ralph A. Beebe
1939 Philip S. Barnhart
1940 Emma P. Carr
1941 C. A. Peters
1942 Samuel E. Q. Ashley
1943 S. B. Smith
1944 H. S. Schwenk
1945 Harry F. Miller
1946 Walter S. Ritchie
1947 Robert D. Dunlop
1948 M. Gilbert Burford
1949 Leonard C. Flowers
1950 William C. Lee
1951 Harold W. Mohrman
1952 Ernest R. Kline
1953 Jane L. Hastings
1954 David C. Grahame
1955 John M. DeBell
1956 George E. Hall
1957 Rolf Buchdahl
1958 Lawrence H. Amundsen
1959 Edgar J. Page
1960 George W. Cannon
1961 Cecil E. Johnson
1962 John W. Sease
1963 Edgar E. Hardy
1964 Norbert Platzer
1965 Peter L. Costas
1966 Jane L. Maxwell
1967 Morton Kramer
1968 James Bobbitt
1969 Speros P. Nemphos
1970 Richard H. Groth
1971 Donald D. Donermeyer
1972 Maris C. Markham
1973 Edwin K. Mahlo
1974 Otto Vogl
1975 Jeremy W. Gorman
1976 John Fletcher

Cornell

(1902; Ithaca, N.Y.)

1903 Louis M. Dennis
1904 Wilder D. Bancroft
1905 Emile M. Chamot
1906 Louis M. Dennis
1907 Wilder D. Bancroft
1908 A. W. Browne
1909 Louis M. Dennis
1909 R. C. Snowdon
1909 Gustav E. F. Lundell
1910 H. W. Redfield
1911 L. J. Cross
1912 R. P. Anderson
1913 Burton J. Lemon
1914 F. W. B. Welch
1915-16 T. R. Biggs
1917 T. L. Lyon
1918 Frank E. Rice
1919 Louis M. Dennis
1920 Emile M. Chamot
1921 T. R. Briggs
1922 Fred H. Rhodes
1923 Melvin Nichols
1924 Frederick R. Georgia
1925 R. B. Corey
1926 A. E. McKinney
1927 Albert W. Laubengayer
1928 Fred H. Rhodes
1929 John R. Johnson
1930 Leonard A. Maynard
1931 M. L. Nichols
1932 James B. Sumner
1933 Wilder D. Bancroft
1934 James M. Sherman
1935 David B. Hand
1936 Clive M. McCay
1937 William F. Bruce
1938 P. F. Sharp
1939 John R. Johnson
1940 Richard Bradfield
1941 Albert W. Laubengayer
1942 James B. Sumner
1943 Peter Debye
1944 Barbour L. Herrington
1945 Milicent L. Hathaway
1946-48 Gordon H. Ellis
1948 William T. Miller, Jr.
1949 R. L. Von Berg
1950 Harold H. Williams
1951 Simon H. Bauer
1952 J. Eldred Hedrick
1953 Michell J. Sienko
1954 Walter L. Nelson
1955 Harold A. Scherago
1956 Herbert F. Wiegandt
1957 James L. Hoard
1958 V. N. Krukovsky
1959 Jerrold Meinwald
1960 Robert K. Finn
1961 W. Donald Cooke
1962 Robert W. Holley
1963 Robert A. Plane
1964 Charles F. Wilcox
1965 Ferdinand Rodriguez
1966 Arthur L. Neal
1967 Gordon G. Hammes
1968 Robert C. Fay
1969 Victor H. Edwards
1970 Richard J. Guillory
1971-75 William R. Bergmark
1976 Harold C. Mattraw

Corning (N.Y.)

1953-55 Robert A. Cameron
1955-56 Samuel R. Scholes, Jr.
1956-57 David F. Fortney
1957-58 Leon R. Schlotzhauer
1958 Ralph B. Elliott
1959 Donald E. Campbell
1960 Charles B. Rutenber
1961 Robert H. Dalton
1962 Clifford E. Myers
1963 William A. Plummer
1964 Charles W. W. Hoffman
1965 Harmon M. Garfinkel
1966 Clayton D. Spangenberg, Jr.
1967 Augustus M. Filbert
1968 Vaughn K. Gustin
1969 David R. Rossington
1970 Ivan E. Lichtenstein
1971 A. Stuart Tulk
1972 Roger F. Bartholomew
1973 Richard D. Sands
1974 Martin B. MacInnis
1975 Francis P. Fehlner
1976 James L. Brown

Dallas-Fort Worth (Tex.)

1937 N. C. Hamner
1938-39 Willis H. Clark
1940 John T. Murchison
1941 H. G. Whitmore
1942 M. H. Gorin
1943 James L. Carrico
1944 Harry M. Bulbrook
1945 Thomas S. Bacon
1946 Maurice Moren Barr

Local Section Chairmen (continued)

Dallas-Ft. Worth (cont.)

1947 Harold A. Jeskey
1948 H. Truehart Brown
1949 Madison L. Marshall
1950 Morris Brown
1951 T. S. McDonald
1952 James J. Spurlock
1953 Robert James Speer
1954-55 E. Wilson Nance
1956 Ogden Baine
1957 Gordon K. Teal
1958 Price Truitt
1959 Elmer R. Alexander
1960 Thom. S. Burkhalter
1961 Robert W. Higgins
1962 Edgar F. Meyer
1963 Morton F. Mason
1964 Russel O. Bowman
1965 William R. Foster
1966 Russell C. Walker
1967 John J. Banewicz
1968 Norman E. Foster
1968-69 William H. Glaze
1970 Morton D. Prager
1971 William H. Watson
1972 Herman C. Custard
1973 Peter R. Girardot
1974 Kenneth A. Hansen
1975 E. Thomas Strom
1976 Manfred G. Reinecke

Dayton (Ohio)

1930-31 Charles A. Thomas
1932 Edgar W. Fasig
1933 Carroll A. Hochwalt
1934 Paul E. Marling
1935 Benjamin C. Morris
1936 Fred L. Chase
1937 James H. Lum
1938 Clyde S. Adams
1939 Charles E. Waring
1940 Mathias E. Haas
1941 Nicholas N. T. Samaras
1942 John D. Coleman
1943 Edward N. Rosenquist
1944 Paul W. K. Rothemund
1945 John W. Wright
1946 Carlyle J. Stehman
1947 Richard S. Gauger
1948 James F. Corwin
1949 Howard K. Nason
1950-51 Joseph J. Burbage
1952 William S. Emerson
1953 Vincent J. Wottle, S.M.
1954 Edward Orban
1955 Bernard S. Wildi
1956 John L. Lucier
1957 Robert A. Ruehrwein
1958 Clay A. Aneshansley
1959 Allen S. Kenyon
1960 Carl I. Michaelis
1961 William H. Yanko
1962 Marjorie C. Lyon
1963 J. Brennan Gisclard
1964 Ralph V. Montello
1965 Lerroy V. Jones
1966 Joseph A. Pappalardo
1967 Glen R. Buell
1968 William C. Jenkin
1968-69 Julius G. Villars
1970 David J. Karl
1971 John J. Fortman
1972 William G. Scribner
1973 J. Theodore Brown
1974 Bertran C. Blanke
1975 George G. Hess
1976 Don B. Sullenger

Decatur-Springfield

(1972; Decatur, Ill.)

1973 Kenneth B. Moser
1974 Joseph A. Empen
1975 Frank Verbanac
1976 Akiva Pour-El

Delaware

(1917; Wilmington)

1918 Lamont du Pont
1919 Charles L. Reese
1920 George M. Norman
1921 Fletcher B. Holmes
1922 Charles M. Stine
1923 Frederick C. Zeisberg
1924 C. D. Porch
1925 Arthur P. VanGelder
1926 James L. Bennett
1927 Jesse W. Stillman
1928 Hamilton Bradshaw
1929 Arthur W. Kenney
1930 Ernest M. Symmes
1931 Herbert A. Lubs
1932 Wallace H. Carothers
1933 George G. Lahr
1934 Marshall T. Sanders
1935 Emmett F. Hitch
1936 James W. Hill
1937 J. Merrian Peterson
1938 Miles A. Dahlen
1939 Rollin F. Conaway
1940 George H. Scheffler
1941 Samuel Lenher
1942 J. Harrel Shipp
1943 Harry R. Dittmar
1944 Robert W. Cairns
1945 Swanie S. Rossander
1946 George E. Holbrook
1947 Raymond F. Schultz
1948 Richard A. Schreiber
1949 Aubrey O. Bradley
1950 Vernal R. Hardy
1951 Arnold L. Lippert
1952 Hugh W. Gray
1953 William A. Mosher
1954 Walter E. Mochel
1955-56 Harold M. Spurlin
1957 Albert V. Willett, Jr.
1958 George R. Seidel
1959 Gilbert P. Monet
1960 Herbert K. Livingston
1961 Arthur C. Stevenson
1962 Herman Skolnik
1963 John T. Maynard
1964 Robert H. Varland
1965 Arthur H. Hale
1966 Blaine C. McKusick
1967 David S. Breslow
1968 John J. Drysdale
1969 John Mitchell
1970 Eugene E. Magat
1971 Harold C. Beachell
1972 Benjamin C. Repka
1973 John R. Schaefgen
1974 Wm. H. Calkins
1975 Allan Cairncross
1976 E. J. Vandenberg

Detroit (Mich.)

1912 R. W. Perry
1913 L. D. Vorse
1913 Niels C. Ortved
1914 Arthur B. Connor
1915 Chas. T. Bragg
1916 Howard T. Graber
1917 Edward J. Gutsche
1918 B. J. Rivett
1919-20 E. E. Follin
1921 J. C. Moore
1922-23 Frank O. Taylor
1924 W. G. Nelson
1925 S. R. Wilson
1926 G. W. Winchester
1927 L. W. Rowe
1928 D. Segal
1929-30 Edward Lyons
1930-31 Icie G. Macy
1932 H. M. Merker
1933 Robert L. Jones
1934 Thomas A. Boyd
1935 Sidney M. Cadwell
1936 Oliver Kamm
1937 Joseph J. Jasper
1938 Arthur W. Bull
1939 Arthur Rautenberg
1940 Harvey M. Merker
1941 O. Edward Kurt
1941-42 Horace H. Bliss
1943 Ralph D. Hummel
1944 George Calingaert
1945 H. H. Williams
1946 Everette L. Henderson
1947 Carl F. Graham
1948 Charles K. Hunt
1949 George Rieveschl, Jr.
1950 T. H. Vaughn
1951 J. R. Bright
1952 John L. Eaton
1952-53 William G. Frederick
1954 Calvin L. Stevens
1955 George W. Moersch
1956 Albert G. Gassman
1957 Ralph L. Seger, Jr.
1958 Phelps Trix
1959 Stanley Kirschner
1960 George E. F. Brewer
1961 George H. Coleman
1962 Thomas H. Coffield
1963-64 Leon Rand
1965 Leon A. Sweet
1966 Thomas O. Morgan
1967 Gordon O. Knapp
1968 Charles R. Harmison
1969 Herbert K. Livingston
1970 Gilbert J. Mains
1971 Bernard E. Nagel
1972 John P. Oliver
1973 H. Harry Szmant
1974 Richard L. Lintvedt
1975 Salvatore A. Fusari
1976 Bernard Weinstock

East Tennessee

(Knoxville)

1930-40 Charles O. Hill
1940 M. F. Stubbs
1941 J. H. Robertson
1942 R. R. Ralston
1943 W. H. MacIntire
1944 Arthur L. Davis
1945 V. C. Henrich
1946 J. H. Wood
1946 James L. Gabbard
1947 Calvin A. Buehler
1948 Charles D. Susano
1949 Fred Griffitts
1950 Harvey A. Bernhardt
1951 Carl T. Bahner

Local Section Chairmen (continued)

E. Tenn. (cont.)

1952 Robert H. Lafferty, Jr.
1953 Harry A. Smith
1954 Ellison H. Taylor
1955 Arthur D. Melaven
1956 Myron T. Kelley
1957 Wm. T. Smith, Jr.
1958 James C. White
1959 David A. Shirley
1960 Karl E. Rapp, Jr.
1961 Jerome F. Eastham
1962 Clair J. Collins
1963 Albert J. Myers
1964 Eugene J. Barber
1965 William E. Bull
1966 Raymond W. Stoughton
1967 Newell S. Bowman
1968 Robert Mayo McGill
1969 Gleb Mamantov
1970 Wallace Davis, Jr.
1971 Carl Boggess Honaker
1972 Ralph Livingston
1973 W. Alexander Van Hook
1974 James H. Junkins
1975 Earl L. Wehry, Jr.
1976 Robert L. Farrar, Jr.

East Texas

(1970; Longview)

1971 Charles H. Whiteside
1972 Donald E. Gwynn
1973 Harrold E. Abbott
1974 James D. Wicks
1975 Ben R. Condry
1976 Bennie F. Walker

Eastern New York

(Schenectady)

1908 Willis R. Whitney
1909 Matthew A. Hunter
1910 Edward Ellery
1911 John Hurley
1912 C. Casper Zapf
1913 Azariah T. Lincoln
1914 A. D. Carrier
1915 Irving Langmuir
1916-17 William C. Arsem
1918 Arthur Knudson
1919-20 Saul Dushman
1921-22 G. M. Johnstone Mackay
1923-24 Gorton R. Fonda
1925-26 Wheeler P. Davy
1927 Thomas A. Wilson
1928 Henry S. Van Klooster
1929 Charles B. Hurd
1930 Arthur J. Sherburne
1931-32 Louis Navias
1933-34 Neil T. Gordon
1935-37 Arthur W. Davison
1938 Abraham Lincoln Marshall
1939 Harold M. Faigenbaum
1940-42 Francis J. Norton
1943-45 Nicholas E. Oglesby
1945 Wolfgang Huber
1946 Winton I. Patnode
1947 Oscar E. Lanford
1948-50 Leon E. Hoogstoel
1950 Robert O. Sauer
1951 Frederick C. Nachod
1952 Lewis S. Coonley
1953 William E. Cass
1954 Alexander R. Surrey
1955 Floyd W. Green
1956 Albert C. Titus
1957 Jason E. Dayan
1958 Walter H. Bauer
1959 Leonard W. Niedrach
1960 Robert K. Bair
1961 Stanley C. Bunce
1962 Arthur E. Newkirk
1963 Arthur O. Long
1964 Stephen E. Wiberley
1965 Heinz G. Pfeiffer
1966 Sydney Archer
1967 Robert L. Strong
1968 Howard A. Vaughn, Jr.
1969 Edward D. Homiller
1970 Fred F. Holub
1971 William B. Martin, Jr.
1972 Harry F. Herbrandson
1973 Joseph G. Wirth
1974 John G. Lanese
1975 Ronald E. Brooks
1976 Matthew T. Gladstone

Eastern North Carolina

(Kinston)

1954 Edward R. Kane
1955 Wm. A. Bridgers
1956 John V. Flanagan
1957 Robert E. Kitson
1958 John M. Griffing
1959 John M. Christens
1960 Robert J. Collins
1961 Grover W. Everett
1962 Tomlinson Fort, Jr.
1963 Lawrence W. Kendrick, Jr.
1964 Edward A. Haseley
1965 Sidney B. Maerov
1966 Jack O. Derrick
1967 Harold J. E. Segrave
1968 James D. Hodge
1969 Donald Faull Clemens
1970 William B. Bond
1971 R. Tilden Burrus
1972-73 Robert C. Lamb
1974 John D. Jernigan
1975 Phillip P. Burks, Jr.
1976 Robert C. Morrison

El Paso (Tex.)

(Chartered 1960)

1961 Jesse A. Hancock, Jr.
1962 John N. Abersold
1963 Louis E. Kidwell, Jr.
1964 Floyd B. O'Neal
1965 Samuel M. Schwartz
1966 William H. Rivera
1967 Harold E. Alexander
1968 T. David Ling
1969 William R. Cabaness, Jr.
1970 Walter J. Decker
1971 Harold K. Myers

(Merged 1972 into Rio Grande Valley.)

Erie, Pa.

1923 Paul Henkel
1924-25 Maxamillian A. Krimmel
1926 I. R. Valentine
1927 Warren W. Hilditch
1928 Richard E. Lee
1929 C. H. Reese
1930 G. McClelland
1931 W. E. Coon
1932 John L. Parsons
1933 Leo A. Goldblatt
1934 John L. Parsons
1935 W. H. Powers
1936 F. M. Truesdale
1937 E. T. Weibel
1937 D. T. Jackson
1938 John E. Cavelti
1939 A. B. Evans
1940 C. H. Steinford
1941 William F. Reichert
1942 John C. Tongren
1943-44 H. J. Morris
1945 Herbert S. Rhinesmith
1946 John E. Toppari
1947 H. M. State
1948 W. K. Mohney
1949 Henry E. Obermanns
1950 Edward C. Boyer
1951 Carl Anderson
1952 Louis W. Balmer
1953 Lewis N. Pino
1954 James P. Coffman
1955 Robert W. Finholt
1956 Calvin M. Dolan
1957 John E. Cavelti
1958 Frank M. Precopio
1959 Morris J. Danzig
1960 Lewis W. Pyle
1961 Robert M. Lukes
1962 William M. De Crease
1963 Laurance R. Webb
1964 Morton M. Rayman
1965 Geza Gruenwald
1966 Louis W. Balmer
1967 David P. Spalding
1968-69 Robert H. Becker
1970 Randolph D. Sites
1971 Robert F. Leifield
1972 Clifford H. Cox
1973 Mary Charles Weschler
1974 Frederick M. Sitter
1975 Paul R. Guevin, Jr.
1976 William P. Fornof, III

Florida

(1924; Gainesville)

1925 Townes R. Leigh
1926 R. W. Ruprecht
1927-28 Alvin P. Black
1929 Robert S. Bly
1930 Vestus Twiggs Jackson
1931 Leland J. Lewis
1932 John F. Conn
1933 Perry A. Foote
1933 Walter H. Beisler
1934 J. J. Taylor
1935 Cash B. Pollard
1936-38 Bailey F. Williamson
1938 Burton J. Otte
1939 Gertrude Vermillion
1940 Fred H. Heath
1941 Allan T. Cole
1942 Isabel McKinnell
1943 J. Erskine Hawkins
1944 Joseph P. Bain
1945 George P. Shingler
1946 Rowland B. French
1947 Louis Gardner MacDowell, Jr.
1948 Ralph A. Morgan
1949 Vincent E. Stewart
1950 Lawrence R. Phillips
1951 Seymour S. Block
1952 Robert Fuguitt
1953 John V. Vaughen
1954-55 George B. Butler

Local Section Chairmen (continued)

Florida (cont.)

1956 Karl Dittmer
1957 Paul Tarrant
1958 George K. Davis
1959 Richard W. Wolford
1960 Armin H. Gropp
1961 Robert J. Dew, Jr.
1962 Harry H. Sisler
1963 J. Allen Brent
1964 Harry P. Schultz
1965 Ray V. Lawrence
1966 Wallace S. Brey, Jr.
1967 Carl Bordenca
1968 Werner Herz
1969 Theodore W. Beiler
1970 Paul C. Maybury
1971 Alfred P. Mills
1972 Arthur N. Baumann
1973 William R. Dolbier, Jr.
1974 Richard D. Dresdner
1975 Sik Vung Ting
1976 Thomas M. Willard

Fort Wayne, Ind.

1926 R. T. Bohn
1927 Eskel Nordell
1928 C. D. Dilts
1929 Paul H. Adams

(*See* Northeastern Indiana)

France

(Authorized 1918 but never functioned.)

Georgia

(Atlanta)

1904 H. C. White
1905 William Henry Emerson
1906 J. F. Sellers
1907 J. M. McCandless
1908 H. B. Arbuckle
1909 William Henry Emerson
1910 R. E. Stallings
1911 W. P. Heath
1912 F. B. Porter
1912 William C. Dumas
1913-14 H. C. Moore
1915 Ray C. Werner
1916 V. H. Bassett
1917 T. C. Law
1918 A. M. Muckenfuss
1919 Ray F. Monsalvatge
1920 L. Blockhart
1921 J. Sam Guy
1922 Charles A. Butt
1923 J. F. Sellers
1924 J. S. Brogdon
1925-27 John L. Daniel
1928 Andrew M. Fairlie
1929-31 Paul Seydel
1932 Osborne R. Quayle
1933 Harold B. Friedman
1934 William H. Jones
1935 Alfred W. Scott
1936 Luther B. Lockhart
1937 Paul Weber
1938 Elise C. Shover
1939 Everett E. Porter
1940 Howard M. Waddle
1941 John J. McManus
1942 Wyatt C. Whitley
1943 James R. Hall
1944 W. M. Spicer
1945 Charles J. Brockman
1946 Hershell H. Cudd
1947 Charles T. Lester
1948 H. L. Edwards
1949 F. Homer Bell
1950 Waldemar T. Ziegler
1951 W. Joe Frierson
1952 Lee W. Blitch
1953 T. H. Whitehead
1954 Gene M. Roberts
1955 Eugene P. Cofield, Jr.
1956 William H. Eberhardt
1957 K. T. Holley
1958 Marion T. Clark
1959 W. Herbert Burrows
1960 Preston H. Williams
1961 John C. Simms
1962 William H. Waggoner
1963 Henry M. Neumann
1964 Ivy M. Parker
1965 Grover Dunn
1966 Leon Mandell
1967 Robert S. Ingols
1968 William G. Trawick
1969 Robert M. Sims
1970 Eugene P. Cofield, Jr.
1971-72 James P. Kinney, Jr.
1972 Donald G. Hicks
1973-74 Sol Cohen
1975 Alice J. Cunningham
1976 Raymond D. Kimbrough

Hampton Roads

(Norfolk, Va.)

1944 Frank W. Wilder
1945 Samuel F. Thornton
1946 Paul Caldwell
1947 Herbert G. Wall
1948 James H. Zwemer
1949 Walter B. F. Randolph
1950 Calder S. Sherwood, III
1951 George I. Earnest, Jr.
1952 Peter Eustis
1953 Karl Ellingson
1954 William J. Francis
1955-56 Mearl A. Kise
1957 David C. Spence
1958 Carlyle T. McCloud
1959 William E. Perry
1960 William F. Hullibarger
1961 Ramon A. Morano
1962 Jesse Lunin
1963 Robert S. Hufstedler
1964 Parker M. Baum
1965 Allen M. Murphy
1966 Allen K. Clark
1967 Edgar E. Linekin
1968 Charles E. Bell, Jr.
1969 Donald M. Oglesby
1970 Alvin M. Olson
1971-72 Theron L. Moore
1973 Wavell W. Fogleman
1974 Samuel Anderson
1975 Lawrence Sacks
1976 Gary B. Hammer

Hawaii

(Honolulu)

1922 J. C. James
1923-24 Guy R. Stewart
1925 Frank T. Dillingham
1926-28 William T. McGeorge
1929-31 Harold Johnson
1932 H. Ennis Savage
1933 Ronald Q. Smith
1934 Samuel S. Peck
1935 John H. Payne
1936 Edward L. Campbell
1937 John Norman Spencer Williams
1938 H. Darwin Kirschman
1939 Harold E. Clark
1940 John M. Lowson
1941 Earl M. Bilger
1942 O. C. McBride
1943 Ronald Q. Smith
1944 Robert F. Gill
1945 Carl A. Farden
1946 Hugo P. Kortschak
1947 George E. Felton
1948 G. Donald Sherman
1949 George O. Burr
1950 Willis A. Gortner
1951 Leonora N. Bilger
1952 Melvin Levine
1953 Arthur S. Ayres
1954 Paul J. Scheuer
1955 Ralph M. Heinicke
1956 Leon J. Rhodes
1957 Ernest H. Thomas
1958 John J. Naughton
1959 Charles E. Mumaw
1960 George E. Sloane
1961 Wai Y. Young
1962 John W. Hylin
1963 Robert W. Leeper
1964 Raymond I. Mori
1965 Richard G. Inskeep
1966 David H. Miles
1967 Louis G. Nickell
1968 Lawrence H. Piette
1969 L. Reed Brantley
1970 Edgar F. Kiefer
1971 David R. V. Golding
1972 Julian Adin Mann
1973-74 Kerry Yasunobu
1975 Ray L. McDonald
1976 George M. Richards

Heart O' Texas

(1968; Waco)

1969 Thomas Chester Franklin
1970 Charles H. Burnside
1971 Malcolm Dole
1972 Robert D. Krienke
1973 Albin G. Pinkus
1974 Bob Ford
1975 David E. Pennington
1976 Sam H. Hastings

Idaho

(Idaho Falls)

1952 Ernest C. Wadsworth
1953 Junius Larsen
1954 Cyril M. Slanksy
1955 George V. Beard
1956 Albert E. Taylor
1957 Ralph C. Shank
1958 Charles E. Stevenson
1959 James E. Rein
1960 Elton H. Turk
1961 George E. Heckler
1962 John R. Huffman
1963 Artell G. Chapman
1964 Earl R. Ebersole
1965 Cyril M. Slansky
1966 Kenneth L. Rohde
1967 George A. Huff
1968 Glenn L. Booman
1969 Robert P. Schuman

Local Section Chairmen (continued)

Idaho (cont.)

1970 Donald W. Rhodes
1971 Stanley Satoshi Yamamura
1972 Fred O. Cartan
1973 Kenneth Faler
1974-75 Marven A. Wade
1976 Robert C. Girton

Illinois-Iowa

(1923; Davenport, Iowa)

1924-25 C. H. Christman
1926 Harold L. Parr
1927 H. A. Geauque
1928 Wilfred Hinde
1929 O. K. Smith
1930-32 Adrienne E. Anderson
1932-33 Robert E. Clayton
1933-35 H. C. Goldschmidt
1936 Jeremiah F. Goggin
1937 Guy Williams
1937 A. L. Fuhrman
1937 Gerald O. Inman
1938 Arthur C. Hanson
1939 G. A. Lillis
1940 Fred Cook
1941 Arthur N. Johnson
1942-43 Charles L. Guettel
1944 John P. Magnusson
1945 Harry L. Faigen
1946 Carl E. Ekblad
1947 George T. Peckham, Jr.
1948 Donald R. Larson
1949 Lee D. Ough
1950 Garrett W. Thiessen
1951 Manley R. Hoppe
1952 W. R. Fetzer
1953 Edward L. Hill
1954 S. J. Vellenga
1955 Robert G. Rohwer
1956 Allen M. Varney
1957 Benjamin T. Snawver
1958 Thomas Rice
1959 Robert A. Berntsen
1960 Harry C. Muffley
1961 William J. Nelson
1962 John Thomas Garbutt
1963 Bernard J. Bornong
1964 Robert W. Haack
1965 Julian Corman
1966 Lynn G. Wiedenmann
1967 Robert Dworschack
1968 E. Daniel Hubbard
1969 Morton A. Eliason
1970 Vincent C. O'Leary
1971 Donald L. Kiser
1972 Melbert E. Peterson
1973-74 William S. Shore
1975 Emil H. Carlson
1976 David G. DeWit

Indiana

(Indianapolis)

1906 Harry E. Barnard
1907 Robert H. Lyons
1907 Harry E. Barnard
1908 Robert H. Lyons
1909-10 Richard B. Moore
1911 Frank R. Eldred
1912 John White
1913 Albert D. Thorburn
1913 William M. Blanchard
1914-15 Frank B. Wade
1916-18 Frederick C. Atkinson
1919 H. W. Rhodehamel
1920 Frank B. Wade
1921 Edgar B. Carter
1922-23 Harry E. Jordan
1924 Ivy L. Miller
1925 Horace A. Shonle
1926 Paul Smith
1927 Cecil K. Calvert
1928 Robert M. Lingle
1929 William Higburg
1930 John H. Waldo
1931 Charles T. Harman
1932 Rolla N. Harger
1933 Auburn A. Ross
1934 Carl E. Stone
1935-36 Norman J. Harrar
1936 Norman T. Shideler
1937 Asa N. Stevens
1938 Edward J. Hughes
1939 Edward H. Niles
1940 Lawrence L. Newburn
1941 Neil Kershaw
1942 John A. Leighty
1943 James W. Meek
1944 John R. Kuebler
1945 Imer H. Stuart
1946 Wayne W. Hilty
1947 Robert James Kryter
1948 Wallace F. Benson
1949 John M. Goodyear
1950 Malcolm Mitchell
1951 William E. Schaefer
1952 W. Brooks Fortune
1953 Robert B. Forney
1954 Theodore G. Delang
1955 Thomas P. Carney
1956 Edward E. Kennedy
1957 Albert E. Jarvis
1958 Dale T. Wilson
1959 Robert M. Brooker
1960 Edwin R. Shepard
1961 Granville B. Kline
1962 Harold A. Nash
1963-64 Maurice E. Clark
1965 LeRoy A. Springman
1966 Sidney A. Kilsheimer
1967 Robert E. Staten
1968 Norbert R. Kuzel
1969 Carl L. Hake
1970 Richard T. Rapala
1971 Wm. A. Nevill
1972 William A. Nevill
1973 Jesse L. Bobbitt
1974 Peter W. Rabideau
1975 Frank E. Gainer
1976 Joseph Kirsch

Indiana-Kentucky Border

(Evansville, Ind.)

1954-55 Norman O. Long
1956 Herbert P. Sarett
1957 Frederick A. Grunwald
1958 Homer E. Stavely
1959 Frederick H. Pfarrer
1960 Lowell E. Weller
1961 Coy W. Waller
1962 Philip Kinsey
1963 John R. Corrigan
1964 James C. Hickle
1965 Peter Tavormina
1966 Stanley J. Kykstra
1967 Robert E. Carnahan
1968 Aubrey A. Larsen
1969 Kiyoshi Hattori
1970 Wao Hua Wu
1971-72 Billy J. Fairless
1973 Tellis A. Martin
1974 Charles M. Combs
1975 Thomas J. Gerteisen
1976 Robert F. Majisvski

Inland Empire

(Spokane, Wash.)

1952 Frank L. Howard
1953 Warren S. Peterson
1954 Herbert L. Redfield
1955 Donald H. Thompson
1956 Lydia G. Savedoff
1957 Glade M. Wilson
1958 Richard V. Paulson
1959 Harold A. Page
1960 James R. Brathovde
1961 Gilbert B. Manning
1962 Vincent T. Stevens
1963 Henry J. Wittrock
1964 Raphael B. Penland
1965 George H. Stewart
1966 John E. Douglas
1967 Saul Kessler
1968 Robert S. Winniford
1969 Emil Fattu
1970 Dennis J. Kelsh
1971 Edward L. Foubert, Jr.
1972 Roy K. Behm
1973 M. Ronald Johns
1973 O. Jerry Parker
1974 John M. Rancour
1975 Frank Doolittle
1976 Dennis McMinn

Iowa

(Iowa City)

1905-06 W. S. Hendrixson
1907-08 A. A. Bennett
1908-09 Nicholas Knight
1909 Elbert W. Rockwood
1910 C. N. Kinney
1911 Launcelot Winchester Andrews
1912 C. O. Bates
1913 W. J. Karslake
1914-16 J. Newton Pearce
1917 Perry A. Bond
1918 J. A. Baker
1919 R. Monroe McKenzie
1920 W. S. Hendrixson
1921 Perry A. Bond
1922 Edward Bartow
1923 L. Charles Raiford
1924 Harry F. Lewis
1925 George H. Coleman
1926 Jacob Cornog
1927 N. O. Taylor
1928 J. L. Whitman
1929 H. L. Olin
1930 George H. Coleman
1931 W. G. Eversole
1932 L. J. Waldbauer
1933 C. P. Berg
1934 A. P. Hoelscher
1935 L. Smith
1936-37 James B. Culbertson
1938 P. A. Bond
1939 L. J. Waldbauer
1940 H. H. Rowley
1941 Henry A. Mattill
1942 L. Charles Raiford
1943 George Glockler
1944 Joseph I. Routh
1944 John H. Arnold
1945 Joseph I. Routh
1946 Robert M. Featherstone
1947 Robert S. Casey
1948 Stanley Wawzonek
1949 Walter F. Edgell
1950 G. Kalnitsky

Local Section Chairmen (continued)

Iowa (cont.)

1951 W. H. Montgomery
1952 John Philip Hummel
1953 R. Thomas Sanderson
1954 Charles Tanford
1955 LeRoy Eyring
1956 Alexander I. Popov
1957 J. C. Culbertson
1958 Gene F. Lata
1959 Norman C. Baenziger
1960 Robert E. Buckless
1961 Ronald T. Pflaum
1962 John R. Doyle
1963 Rex Montgomery
1964 Willis B. Person
1965 John K. Stille
1966 William E. Bennett
1967 Richard D. Campbell
1968 Charles A. Swenson
1969 Donald John Pietrzyk
1970 E. David Cater
1971 Edward B. Buchanan, Jr.
1972 M. Wayne Forsyth
1973 Wm. A. Deskin
1974 Robert E. Coffman
1975 Bruce Friedrich
1976 Jeffrey E. Keiser

Joliet (Ill.)

1945-46 Harry P. Kramer
1946-47 Alfred Long
1948 Nicholas A. Manfred
1949 Eugene A. Kazmark
1950 Charles S. King
1951 Henry L. Rohs
1951 Byron White
1952 John E. Kennedy
1953 Maurice P. Novak
1954 Emil M. Stolz, Jr.
1955 William E. Bunting
1956 James H. Carnett
1957 S. Bruce Humphrey
1958 Joseph V. Karabinos
1959 Paul Carus
1960 J. O. Philip Lindahl
1961 Donald T. Lurvey
1962 Eunice M. Moore
1963 James J. Hazdra
1964 Gerald G. Wilson
1965 Robert J. Westfall
1966 David C. Knoderer
1967 Robert Z. Muggli
1968 John Robert Samlin
1969 Robert J. Rolih
1970 Jerome R. Wiley
1971 John L. Hughes
1972 David C. Knoderer
1973 Wallis G. Hines
1974 Jerold L. Armstrong
1975 Agnes T. Fekete
1976 John E. Hanson

Kalamazoo (Mich.)

1942 George R. Laure
1943 Gerald Osborn
1944-45 Fred L. Chappell, Jr.
1946 Harry F. Meier
1947 D. Roberts Erickson
1948 Eugene H. Woodruff
1949 Allen B. Stowe
1950 Theodore W. Conger
1951 Jack W. Hinman
1952 Laurence E. Strong
1952-53 Richard Heinzelman
1954 Lawrence G. Knowlton
1955 Douglas A. Shepherd
1956 Fred Kagan
1957 John B. Wright
1957 Robert H. Reitsema
1958 Peter D. Meister
1959 David H. Gregg
1960 Hugh V. Anderson
1961 Kurt D. Kaufman
1962 George B. Whitfield, Jr.
1963 Ross Robert Herr
1964 Robert H. Anderson
1965 Paul F. Wiley
1966 Arch B. Spradling
1967 Don C. Iffland
1968-69 Brian Bannister
1970 Fred J. Bassett
1971 Robert C. Nagler
1972 Robert E. Harmon
1973 Daniel Lednicer
1974 Robert C. Kelly
1975 Thomas O. Oesterling
1976 Dean W. Cooke

Kanawha Valley

(Charlestown, W.Va.)

1927-28 J. R. MacMillan
1929 Clyde L. Voress
1930 Carl G. Campbell
1931 Granville A. Perkins
1932 John S. Beekley
1933 Chester S. Heath
1934 Bernard H. Jacobson
1935 Ralph L. Dodge
1936 H. Chester Holden
1937 W. Elwood Vail
1938 Henry L. Cox
1939 Thomas W. Bartram
1940 William T. Nichols
1941 Ashby C. Blackwell
1942 George Vincent Schofield
1943 Victor R. Thayer
1943 Jacob N. Wickert
1944 Edward S. Blake
1945 John R. McConnell
1946 Dwight Williams
1947 Ray W. McNamee
1948 Richard O. Zerbe
1949 Sidney D. Smith
1950 Thomas R. Miller
1951 George S. Haines
1952 P. Ernest Roller
1953 Wendel P. Metzner
1954 William J. Tapp
1955 Willard A. Payne
1956 Walter H. Walker
1957 Nelson R. Eldred
1958-59 Davenport Guerry, Jr.
1960 Samuel C. Harris
1961 Benjamin Phillips
1962 Harry W. Kilbourne
1963 Victor A. Yarborough
1964 Fred W. Stone
1965 Paul J. Moore
1966 Abraham N. Kurtz
1967 Donald L. Heywood
1968 Phil H. Miller
1968-69 William C. Kuryla
1970 Leonard O. Moore
1971 Herbert P. Kagen
1972 Robert A. Bleidt
1973 Harry T. Zika
1974 Clyde A. Pentz
1975 Bernard R. Krabacher
1976 Jonathan J. Kurland

Kansas City (Mo.)

(Chartered 1900)

1901-03 Edgar H. S. Bailey
1903-07 J. Robert Moechel
1908-09 Hamilton P. Cady
1910 F. W. Bushong
1911-12 Frank B. Dains
1913 L. S. Bushnel
1914 L. D. Havenhill
1915 Roy Cross
1916-19 W. A. Whitaker
1920 Rudolph Hirsch
1920 Carl J. Patterson
1921 Harry C. Allen
1922 W. Bradford Smith
1923 Ray Q. Brewster
1924-25 Carl F. Gustafson
1925 H. M. Elsey
1926-27 Henry E. Hancock
1928 G. Harry Clay
1929 Arthur W. Davidson
1930-31 James E. Wildish
1931-32 George W. Stratton
1932-33 Herbert M. Steininger
1933 Hamilton P. Cady
1934-35 Rodney K. Durham
1935-36 Frank B. Dains
1936-37 Royce H. LeRoy
1937-38 Ray Q. Brewster
1938-39 Harold P. Brown
1940 James A. Austin
1941 Herbert M. Steininger
1942 Leonard V. Sorg
1943 Erskine S. Longfellow
1944 Willard M. Hoehn
1945 Charles J. Boner
1946 G. Harry Clay
1947 Calvin A. Vander Werf
1948 Charles L. Shrewsbury
1949 Lindley S. DeAtley
1950 Milton P. Puterbaugh
1951 Jacob Kleinberg
1952 Ralph G. O'Brien
1953 Max H. Thornton
1954 Perry L. Bidstrup
1955 Lidwig C. Krchma
1956 Louis H. Goodson
1957 Leonard V. Sorg
1958 Delta W. Gier
1959 Henry F. Woodward, Jr.
1960 Francis V. Morriss
1961 John C. Lamkin
1962 James R. Costello, Jr.
1963 Buell W. Beadle
1964 Myron L. Wagner
1965 Donald M. Coyne
1966 Edwin T. Upton
1967 J. Earl Barney, II
1968 James D. Wheeler
1969 Laurence W. Breed
1970 Edward R. Levy
1971 William C. Doyle
1972 Robert L. Stutz
1973 Albert D. McElroy
1974 Ivan C. Smith
1975 Gaylord R. Atkinson
1976 Franklin W. Pogge

Kansas State University

(Manhattan)

1928 Josiah Simson Hughes
1929 Howard W. Brubaker
1929-31 Ernest Baker Keith
1932 Carrell H. Whitnah
1933 Harold N. Barham
1934 Charles W. Clover
1935 W. Lawrence Faith
1936 James L. Hall
1937 Benjamin L. Smits
1938 Mendel E. Lash

Local Section Chairmen (continued)

Kansas State (cont.)

1939 Hubert Whatley Marlow
1940 John W. Greene
1941 John H. Shenk
1942 Arthur C. Andrews
1942-43 Ralph Conrad
1944 Ralph E. Silker
1945 William Alexander Van Winkle
1945-46 Maynard L. McDowell
1947 William G. Schrenk
1948 Howard L. Mitchell
1949 Carl Alfred Dorf
1950 John E. Devries
1951 Robert E. Clegg
1952 Henry T. Ward
1953 Alfred T. Perkins
1954 R. Kenneth Burkhard
1955 Max Milner
1956 G. Williams Leonard
1957 Jack L. Lambert
1958 Scott Searles, Jr.
1959 Williom H. Honstead
1960 Donald B. Parrish
1961 Byron S. Miller
1962 Willard S. Ruliffson
1963-64 Herbert C. Moser
1965 Adrian H. Daane
1966 R. Kenneth Burkhard
1967 Robert M. Hammaker
1968 Clifton E. Meloan
1969 Kenneth Conrow
1970 Herbert C. Moser
1971 Richard N. McDonald
1972 Wayne C. Danen
1973 Geneva S. Hammaker
1974 R. Kenneth Burkhard
1975 James L. Copeland
1976 Larry E. Erickson

Kentucky Lake

(Murray, Ky.)

1958 Walter E. Blackburn
1959 Robert W. Levin
1960 Otis W. Fortner
1961 Howard P. Huyck
1962 Harry S. Weglicki
1963 Peter Panzera
1964 Andrew S. Wood
1965 Norman Campbell
1966 J. Channing Hale
1967 Robert L. Siegmann
1968 Lewis E. Barbre
1969 Maurice P. Christopher
1970 Robert E. Simmons
1971 Ernest E. Atkins
1972 Melvin B. Henley
1973 Lloyd A. King
1974 Clifford A. Powell
1974-75 Annette W. Gordon
1976 Gary Boggess

LaCrosse-Winona

(1968; LaCrosse, Wis.)

1969 Gerald Rausch
1970 Brother I. Ambrose
1971 Paul J. Lynch
1972 Robert G. Doerr
1973 C. Richard Kistner
1974 Geo. E. Knudson
1975 John L. Allen
1976 Gary T. Bender

Lake Superior

(Duluth, Minn.)

1943 Otto H. Johnson
1943-44 Gerald Chapman
1945 E. R. Beachtel, Sr.
1946 Albert H. Pedler
1947 John C. Cothran
1948-49 Ole Forsberg
1950-51 Odin A. Sundness
1951 Alfred D. Ludden
1952 Albert H. Pedler
1953 Harlan D. Fayle
1954 Alfred D. Ludden
1955 Moses Passer
1956 L. Keith Coad
1956-57 Jack C. Holland
1958 Otis R. Videen
1959 Francis J. Glick
1960 F. B. Moore
1961 Edward J. Cowles
1962 Howard M. Thomas
1963 Nathan Allen Coward
1964 James C. Nichol
1965 Larry C. Thompson
1966 Donald K. Harriss
1967 Ronald K. Roubal
1968 Thomas Joseph Bydalek
1969 Ronald Caple
1970 Robert M. Carlson
1971 Gary E. Glass
1972 Vincent R. Magnuson
1973-74 Mary A. Riehl
1975 Gilman D. Veith
1976 Herbert L. Kopperman

Lehigh Valley

(Bethlehem, Pa.)

1894 William H. Chandler
1895 Edward Hart
1896-97 Albert Ladd Colby
1897-03 Joseph W. Richards
1904 Porter W. Shimer
1905 William B. Newberry
1906 Harry Drew
1907 George P. Adamson
1907 Harry Drew
1908 George P. Adamson
1909 Harry M. Ullmann
1910-11 John T. Baker
1912-16 Harry M. Ullmann
1917 Frank G. Breyer
1918-19 Eugene C. Bingham
1920 Richard Boyd
1921 Henry J. Wisor
1922 Paul R. Croll
1923 Eugene C. Bingham
1924 Walter O. Snelling
1925 James S. Long
1926-27 Charles C. Nitchie
1927 Emory F. Marsiglio
1928 Roy N. Young
1929 Harvey A. Neville
1930 Harley A. Nelson
1931 Luther F. Witmer
1932 Merl H. Meighan
1932 Carl D. Pratt
1933 Warren W. Ewing
1934 Fred A. Steele
1935 James H. DeLong
1936 Walter O. Snelling
1937 Charles W. Simmons
1938 Howard M. Cyr
1939 Charles H. Love
1940 Esther A. Engle
1941 R. Graham Cook
1942 Thomas H. Hazlehurst
1943 Adolf C. Elm
1944 J. Hunt Wilson
1945 George H. Brandes
1946 Grey F. Rolland
1947 Earl J. Serfass
1948 James H. Gardner
1949 Britton A. Shippy
1950 Saul R. Buc
1951 Walter J. McCoy
1952 C. W. Siller
1953 Albert C. Zettlemoyer
1954 Franklin S. Eisenhouer
1955 Thomas B. Lloyd
1956 Alfred M. Sadler
1957 Clayton Thomas Kleppinger
1958 Edward C. Truesdale
1959 William F. Hart
1960 William J. Wiswesser
1961 Paul M. Leininger
1962 Peter P. Prichett
1963 Raymond R. Myers
1964 Velmer B. Fish
1965 Fred W. Cox, Jr.
1966 Robert D. Billinger
1967 James F. Feeman
1968 Robert Fredericks
1969 Stuart S. Kulp
1970 Malcolm L. White
1971 Carl Theodore Kleppinger
1972 Ned D. Heindel
1973 David N. Stehly
1974 Clarence J. Murphy
1975 Paul E. Maleskey
1976 Harold F. Bluhm

Lexington (Ky.)

1912-13 Alfred M. Peter
1914 Franklin E. Tuttle
1915 Ralph N. Maxson
1916 C. A. Nash
1917 Leslie A. Brown
1918 G. D. Buckner
1919 James S. McHarque
1920 Mary E. Sweeny
1921 O. M. Shedd
1922 Saxe D. Averitt
1923 Leslie A. Brown
1924 J. M. Saunders
1925 G. D. Buckner
1926 V. F. Payne
1927 A. Lloyd Meader
1928 Franklin E. Tuttle
1929 J. Stanton Pierce
1930 Charles Barkenbus
1931 R. I. Rush
1932 John R. Mitchell
1933 Franklin E. Tuttle
1934 Olus J. Stewart
1935 M. Hume Bedford
1935 David W. Young
1936 Harry R. Allen
1937-38 Charles F. Krewson
1938 Wayne H. Keller
1939-40 Fred W. Oberst
1941 W. H. Hensley
1942 James L. Gabbard
1942 Julian H. Capps
1943 Martin E. Weeks
1944 Stacy B. Randle
1945 Simon H. Wender
1946 Robert N. Jeffrey
1947 Albert G. Geiser
1948 Lyle A. Dawson
1949 Jacob R. Meadow
1949 Julian H. Capps
1950 Reedus R. Estes
1951 James M. Schreyer
1951 Valva C. Midkiff
1952 Joseph B. Beard, III

Local Section Chairmen (continued)

Lexington, Ky. (cont.)

1953 William F. Wagner
1954 Rodney E. Black
1955 Gerrit Levey
1956 Ellwood M. Hammaker
1956 William K. Plucknett
1957 Robert M. Boyer
1958 John F. Steinback
1959 Arthur W. Fort
1960 John M. Patterson
1961 Walter T. Smith, Jr.
1962 Paul G. Sears
1963 Wm. D. Ehmann
1964 James E. Douglas
1965 George W. Pope
1966 Stanford L. Smith
1967 Donald E. Sands
1968 Ellis Vincent Brown
1969 Henry Hermann Bauer
1970 Thomas R. Beebe
1971 Robert D. Guthrie
1972 Harold N. Hanson
1973-74 Paul L. Corio
1975 Joseph W. Wilson
1976 John R. Bladeburn

Lima (Ohio)

1960-61 David F. Wright
1962 Karl L. Shull
1963 James A. Bright
1964 Wilbur H. Huber
1965 J. Richard Weaver
1966 Darrel G. Dock
1967 Donald J. Bettinger
1968 Morris J. Groman
1969 Gred E. Dresher
1970 Robert C. McConnel
1971 Byron L. Hawbecker

(Name changed 1971 to Northwest Central Ohio.)

Louisiana

(New Orleans)

(*See also* New Orleans)

1906-08 W. R. Betts
1908-10 F. C. Johnson
1910-12 Philip Asher
1912-13 B. P. Caldwell
1914-15 W. L. Howell
1916 John L. Porter
1917-18 O. S. Williamson
1919-20 Frank W. Liepsner
1921-22 Hal W. Moseley
1923 Cassius L. Clay
1924-25 S. A. Mahood
1926 Frank W. Liepsner
1927 John L. Porter
1928-29 W. O. Griffen
1930-31 Harold A. Levey
1932-33 Hal W. Moseley
1934-35 R. P. Walton
1936-37 James J. Ganucheau
1938-39 Raymond Freas
1940-41 William M. Lehmkuhl
1941-42 A. Watson Chapman
1943-44 E. F. Pollard
1945 Thomas B. Crumpler
1946 Kyle Ward
1947 Harry P. Newton
1948 Hans B. Jonassen
1949 Wesley H. Sowers
1950 Leo A. Goldblatt
1951 John M. Scott
1952 Sheldon J. Haneman
1953 James A. Kime
1954 Winston R. deMonsabert
1955 Rivers Singleton
1956 Carroll L. Hoffpauir
1957 Homer R. Jolley
1958 Lawrence R. Collins
1959 Carl M. Conrad
1960 Calvin C. Rolland
1961 Jules A. Lorio, Jr.
1962 Robert T. O'Connor
1963 Jack K. Carlton
1964 Charles L. Jarreau
1965 Leon Segal
1966 Jack H. Stocker
1967 Antoine F. Alciatore
1968 Robert L. Ory
1968-69 Mary L. Good
1970 Boleslaus M. Kopacz
1971 Tadeusz K. Wiewiorowski
1972 Ruth R. Benerito
1973 Manie K. Stanfield
1974 Ralph J. Berni
1975 Darrell J. Donaldson
1976 David J. Miller

Louisville (Ky.)

1908-09 R. M. Parks
1910-12 Richard C. Lord
1913-20 Charles E. Martin
1921-22 George A. Goodell
1923 Cecil E. Bales
1924 Frank M. Shipman
1925 Cecil E. Bales
1926-27 Frank M. Shipman
1928 Martin L. Degavre
1929 Louis F. Heitz
1930 Andrew J. Snyder
1931 Robert Roger Bottoms
1932 W. L. Clore
1933 Robert Craig Ernst
1934-35 Martin Louis Degavre
1936 Clarence C. Vernon
1937 Theo. Hubbuch
1938 Harry Eaton Carswell
1939 Francis C. Huber
1940 Edward E. Litkenhous
1941 Paul A. Kolachov
1942 Grover L. Corley
1943 Theodore E. Field
1944 Gordon C. Williams
1945 Otto J. Mileti
1946 Martin R. Broadbooks
1947 Wilson R. Barnes
1948 Thomas R. Linak
1949 Ronald E. Reitmeir
1950 Robert M. Reed
1950 Sigfred Peterson
1951 Maurice C. Brockmann
1951 Melvin R. Arnold
1952 Robert L. McGeachin
1953 J. Richard Goertz
1954 Hamilton W. Putnam
1955 Gradus L. Shoemaker
1956 Clarence M. Woodworth, Jr.
1957 Harold G. Cooke, Jr.
1958 Samuel C. Spalding, Jr.
1959 David Apotheker
1960 Milton F. V. Glock
1961 Hans Spauschus
1962 Gerald W. Recktenwald
1963 John M. Daly
1964 Charles H. Jarboe
1965 Robert F. Vance
1966 Daniel M. Sweeny
1967 Samuel L. Cooke, Jr.
1968 William M. Keely
1969 Algerd F. Zavist
1970 William G. Bos
1971-72 Gordon R. Coe
1973 Thomas E. Kargl
1974 Alexander F. Rosenberg
1975 Irwin W. Tucker
1976 John E. Kennedy, Jr.

Maine

(Orono)

1912-13 Arthur B. Larcher
1914 Martin L. Griffin
1915 Walter V. Wentworth
1916-19 Arthur B. Andrews
1920 Lucus H. Merrill
1921-22 Charles A. Brautlecht
1923 Ralph H. Price
1924-25 Bertrand F. Brann
1926-27 Carl Otto
1928 Joseph F. Kolouch
1929 Frederick W. Adams
1930 William L. Gilliland
1931-32 Bertrand L. Brann
1933 William L. Gilliland
1934 Lynwood S. Hatch
1935 John Darrah
1936 Walter A. Lawrence
1937-38 William C. Root
1939 George F. Parmenter
1940 Fred C. Mabee
1941 Frederick T. Martin
1942 Samuel E. Kamerling
1943 George F. Parmenter
1944 William B. Thomas
1945 Irwin B. Douglass
1946 William C. Root
1947 George F. Parmenter
1948 William A. Lawrence
1949 Frederic T. Martin
1950 Samuel E. Kamerling
1951 Wendall A. Ray
1952 A. B. Andrews
1953 Frederic T. Martin
1954 John L. Parsons
1955 James S. Coles
1956 Lyle C. Jenness
1957 Evans B. Reid
1958 William B. Thomas
1959 Samuel E. Kamerling
1960 Malcolm L. Jewell
1961 James L. Wolfhagen
1962 Evans B. Reid
1963 Walter A. Lawrence
1964 Gordon L. Hiebert
1965 Robert D. Dunlap
1966 Paul E. Machemer
1967 William B. Thomas
1968 Samuel E. Kamerling
1969 Charles R. Russ
1970 James R. Young
1971 Evans B. Reid
1972 James G. Boyles
1973 Dana W. Mayo
1974 Brian Green
1975 Alan G. Smith
1976 John L. Gordon

Maryland

(Baltimore)

1914 William B. D. Penniman
1915 C. P. Van Gundy
1916-17 E. Emmet Reid
1918-19 Albert E. Marshall
1920 Frederick C. Blanck
1921 F. M. Boyles
1922 Shepard T. Powell
1923 Carl Haner
1924 Neil E. Gordon
1925 Henry A. B. Dunning
1926-27 Leslie H. Ingham

Local Section Chairmen (continued)

Maryland (cont.)

1927 Edward S. Hopkins
1928 Joseph C. W. Frazer
1929-30 Arthur A. Backhous
1931-32 William B. D. Penniman
1933 Duncan MacRae
1934-36 Fitzgerald Dunning
1936 Donald H. Andrews
1937-38 John C. Krantz, Jr.
1939-40 Channing W. Wilson
1941 Howard H. Lloyd
1942 Paul H. Emmett
1943 Wilton Cope Harden
1944 William H. Hartung
1945 William F. Reindollar
1946-47 Giles B. Cooke
1947-49 John A. Herculson
1950 Alsoph H. Crowin
1951 Charles E. Brambel
1952 Leslie Hellerman
1953 Winslow H. Hartford
1954 William H. Summerson
1955 Belle Otto
1956 Harry C. Freimuth
1957 Raymond M. Burgison
1958 Edward A. Metcalf
1959 Edward M. Hoshall
1960 Lloyd C. Felton
1961 Richard L. Hall
1962 George L. Braude
1963 Samuel L. Goldheim
1964 Arthur J. Emergy, Jr.
1965 George M. Steinberg
1966 F. Marion Miller
1967 William H. Stahl
1968 Frank T. Parr
1969 Harold Delaney
1969-70 Joseph A. Cogliano
1971 Richard J. Kokes
1972 Joyce J. Kaufman
1973 Yale H. Caplan
1974 Donald E. Jones
1975 Anthony A. Bednarczyk
1976 John L. Kolbe

Memphis (Tenn.)

1939 Clarence B. Weiss
1940 Paul D. Bowers
1941 Thomas P. Nash, Jr.
1942 Linwood N. Rogers
1943 Edward R. Stevens
1944 E. Foster Williams
1945 Eldon H. Ruch
1946 E. Foster Williams
1947 Ernest E. Hembree
1948 Vincent C. O'Leary
1949 Joseph N. Pless
1950 Ignatius Leo
1951 Frank A. Anderson
1952 Earl W. Gutliph, Sr.
1953 Elmore Holmes
1954 Arthur F. Johnson
1955 Samuel F. Clark
1956 Vincent C. O'Leary
1957 M. Foster Moose
1958 James L. A. Webb
1959 Peter F. Marchisio
1960-61 Dempsie B. Morrison
1962 Elton Fisher
1962-63 Jesse W. Fox
1964 Daniel R. Marks
1965 George R. Payne
1966 J. Edward Doody, FSC
1967 Charles C. Irving
1968 Don P. Claypool
1969 James Gordon Beasley
1970 V. Marie Easterwood
1971 Charles H. Smith
1972 Carl D. Slater
1973 Kenneth A. Kulken
1974 Hemuth M. Gilow
1975 John D. Pera
1976 Martin Morrison

Michigan State University

(East Lansing)

1917-19 Dwight T. Ewing
1920 Andrew J. Patten
1921 Ralph C. Huston
1922 David Randall
1923 Charles S. Robinson
1924 H. S. Reed
1925 Bruce E. Hartsuch
1926 Charles D. Ball, Jr.
1927 Clark S. Robinson
1928 Dwight E. Ewing
1929 R. C. Huston
1930 Arthur J. Clark
1931 Henry E. Publow
1932 O. B. Winter
1933 Carl A. Hoppert
1934 Wilfred C. Lewis
1935 Arthur J. Clark
1936 Philip J. Schaible
1937 R. C. Huston
1938 Dwight T. Ewing
1939 Arthur J. Clark
1940 Carl A. Hoppert
1941 Charles D. Ball, Jr.
1942 Erwin J. Benne
1943 Elmer E. Leininger
1944 Clifford W. Duncan
1945 Dwight T. Ewing
1946 Bruce E. Hartsuch
1947 Lawrence L. Quill
1948 Charles D. Ball, Jr.
1949 Ralph L. Guile
1950 Richard L. Bateman
1951 Frederick B. Dutton
1952 Harold M. Sell
1953 Robert M. Herbst
1954 Maurice G. Larian
1955 Charles N. McCarty
1956 Kenneth C. Stone, Jr.
1957 Max T. Rogers
1958 Harold Hart
1959 Richard U. Byerrum
1960 Carl H. Brubaker
1961 Andrew Timnick
1962 James L. Dye
1963 Richard H. Schwendeman
1964 James L. Fairley, Jr.
1965 Jack B. Kinsinger
1966 Harry A. Eick
1967 William H. Reusch
1968 Frederick H. Horne
1969 Richard S. Nicholson
1970 Thomas J. Pinnavaia
1971 Stanley R. Crouch
1972 Donald G. Farnum
1973-74 James F. Harrison
1975 Loran L. Bieber
1976 Frederick M. Bernthal

Mid-Hudson

(Poughkeepsie, N.Y.)

1942 Frank H. Bruner
1943 Carrol W. Griffin
1944 Norman McBurney
1945 Bradford R. Stanerson
1946 Mary L. Sague
1947 Charles E. Moser
1947 Ernest A. Rodman
1948 Charles E. Moser
1949 George B. Hatch
1950 Mary A. Plunkett
1951 George B. Arnold
1952 Fidele J. Pira
1953 John A. Patterson
1954 H. Marjorie Crawford
1955 Herbert E. Vermillion
1956 Donald Stewart Allen
1957 Harry C. Becker
1958 Ellsworth K. Holden
1959 Morford C. Throckmortan
1960 Gene A. Silvey
1961 Morris A. Wiley
1962 Curt W. Beck
1963-64 Joseph C. Komyathy
1965 Donald G. Hulslander
1966 James G. Dadura
1967 Sidney L. Phillips
1968 Malvin J. Michelson
1969 Robert E. Rehwoldt
1970 George S. Saines
1971 Burton J. Masters
1972 John M. Larkin
1973 Harry F. Bell
1974 Wm. M. Cummings
1975 Klaus D. Beyer
1976 Rodney L. Sung

Midland (Mich.)

(Chartered 1919)

1920 Herbert H. Dow
1921-22 E. O. Barstow
1923-24 Ross Sanford
1925 Mark E. Putnam
1926 Charles J. Strosacker
1927 Willard H. Dow
1928-29 Edgar C. Britton
1930 Thomas Griswold, Jr.
1931-32 Lawrence F. Martin
1933 Shailer L. Bass
1934 Howard S. Nutting
1935 Wilfred M. Murch
1936 Joseph W. Britton
1937 Don LeLance Irish
1938 George H. Coleman
1939 Noland Poffenberger
1940 Raymond H. Boundy
1941 J. Donald Hanawalt
1942 John J. Grebe
1943 H. Avery Stearns
1944 Arthur J. Barry
1945 Ralph P. Perkins
1946 Lewis R. Drake
1947-48 Laurence L. Ryden
1949 Julius E. Johnson, Jr.
1950 Ezra Monroe
1951 Stevens S. Drake
1952 Melvin J. Hunter
1953 William C. Bauman
1954 Donald E. Pletcher
1955 Eldon L. Graham
1956 Fred W. McLafferty
1956-57 David C. Young
1958 Turner Alfrey, Jr.
1959 Etcyl H. Blair
1960 John W. Gilkey
1961 Rodney D. Moss
1962 Ethan C. Galloway
1963 Ogden R. Pierce
1964 Carleton W. Roberts
1965 Robert M. Wheaton
1966 Malcolm Chamberlain
1967 Donald R. Weyenberg
1968 Douglas A. Rausch
1969 Kenneth L. Burgess
1970 Gordon E. Hartzell
1971 C. Elmer Wymore
1972 Jack F. Mills
1973 Donald R. Peterson
1974 Gary E. LeGrow
1975 Alvin E. Bey
1976 John A. Schneider

Local Section Chairmen (continued)

Milwaukee (Wis.)

(Chartered 1908)

1909 **Martin M. Rock**
1910 **John M. Thomas**
1910 **Hugo W. Rohde**
1911 **Frank S. Low**
1912 **Clarence Herbert Hall**
1913-14 **Alfred J. Schedler**
1915 **George N. Prentiss**
1916 **Robert W. Bauer**
1917 **C. Bartlett Dickey**
1918 **Geo. Kemmerer**
1919 **Ben L. Saloman**
1920 **John A. Wilson**
1921 **Clarence A. Nash**
1922 **Ralph M. Kibbe**
1923 **Clare H. Hall**
1924 **Thomas Harry Cochrane**
1925 **Henry B. Merrill**
1926 **Paul R. Croll**
1927 **Mahlon E. Manson**
1928 **August C. Orthman**
1929 **William W. Bauer**
1930 **William R. Pate**
1931 **Thomas R. Moyle**
1932 **Walter D. Kline**
1933 **C. R. McKee**
1934 **Walter McCrory**
1935 **Curtis E. Norton**
1936-37 **William M. Higby**
1937 **George M. Buffett**
1938 **John R. Koch**
1939 **Harold H. Tucker**
1940 **Norton A. Thomas**
1941 **Robert R. Austin**
1942 **Howard L. Gerhart**
1943 **Robert O. Guettler**
1944 **Carroll E. Imhoff**
1945 **E. Leon Foreman**
1946 **Lohi A. Burkardt**
1947 **Melvin E. Ellertson**
1948 **Merle G. Farnham**
1949 **Herbert Heinrich**
1950 **James F. McKenna**
1951 **Frank J. Rudert**
1952 **Harold C. Krahnke**
1953-55 **Enos H. McMullen**
1955 **Hamilton A. Pinkalla**
1956 **Clark O. Miller**
1957 **Herbert L. Ellison**
1958 **John G. Surak**
1959 **John D. Jenkins**
1960 **Kenneth D. Brown**
1961 **Henry J. Peppler**
1962 **John H. Biel**
1963 **William K. Miller**
1964 **Stephan E. Freeman**
1965 **Clarence F. Peterman**
1966 **Allan G. Boyes**
1967 **Kenneth E. Miller**
1968 **Howard E. Mann**
1969 **Glenn R. Svoboda**
1970 **Frederick J. Kohls**
1971 **John T. Suh**
1972 **Albert Jache**
1973 **John R. Rogers**
1974 **Richard Bayer**
1975 **John A. Bauer**
1976 **Elizabeth M. Kramer**

Minnesota

(Minneapolis)

1906-14 **George B. Frankforter**
1915 **Roscoe W. Thatcher**
1916-17 **Harry Snyder**
1918 **Arthur D. Hirschfelder**
1919-20 **Lauder W. Jones**
1921 **Ross A. Gartner**
1922 **F. H. MacDougall**
1923 **Charles H. Mann**
1924 **LeRoy S. Palmer**
1925 **Oscar E. Harder**
1926 **Victor H. Roehrich**
1927 **J. J. Willaman**
1928 **M. Cannon Sneed**
1929 **Samuel C. Lind**
1930 **Albert A. Schaal**
1931 **R. C. Sherwood**
1932 **George H. Montillon**
1933 **Clyde J. Bailey**
1933 **Ralph E. Montonna**
1934 **S. G. Stoltz**
1935 **George Glockler**
1936 **Clayton O. Rost**
1937-38 **Halvor Orin Halvorson**
1938 **Charles A. Mann**
1939 **W. M. Sandstrom**
1940 **C. G. Ferrari**
1941 **George O. Burr**
1942 **Harold P. Klug**
1943 **Lee I. Smith**
1944 **William F. Geddes**
1945 **Henry N. Stephen**
1946 **Richard T. Arnold**
1947 **Bryce Low Crawford, Jr.**
1948 **William E. Lundquist**
1949 **William N. Lipscomb**
1950-51 **Betty Sullivan**
1952 **Lloyd H. Reyerson**
1953 **Paul D. Boyer**
1954 **Matthew W. Miller**
1955 **Harold A. Wittcoff**
1956 **Walter M. Lauer**
1957 **Donald H. Wheeler**
1958 **Courtland L. Agre**
1958-59 **James O. Hendricks**
1960 **Robert C. Brasted**
1961 **Michael H. Baker**
1962 **William D. Larson**
1963 **Lester C. Krogh**
1964 **Stephen Prager**
1965 **Richard W. Fulmer**
1966 **Olaf A. Runquist**
1967 **Richard Nicholsen**
1968 **Martin Allen**
1969 **William S. Friedlander**
1970 **Wayland E. Noland**
1971 **Stuart A. Harrison**
1972 **Stuart W. Fenton**
1973 **Carl A. Dahlquist**
1974 **Earl R. Alton**
1975 **Julianne H. Prager**
1976 **Mary E. Thompson**

Mississippi

(Jackson)

1959 **Louis L. Sulya**
1960 **Charles E. Lane**
1961 **Lyell C. Behr**
1962 **Jimmie B. Price**
1963 **Charles A. Payne**
1964 **Mahlon P. Etheredge**
1965 **Harold B. White, Jr.**
1966 **Marshall E. Propst, Jr.**
1967 **Donald W. Emerich**
1968 **Charles L. Dodgen**
1969 **Georgia L. Hatcher**
1970 **Paul A. Hedin**
1971 **Louis Gunning**
1972 **Charles R. Brent**
1973 **Thomas Fisher**
1974 **Roy A. Berry, Jr.**
1975 **Shelby F. Thames**
1976 **Delbert H. Miles**

Mobile (Ala.)

(Chartered 1948 as Mobile-Pensacola Section; Pensacola dropped when that section was formed in 1960.)

1961-62 **Elwood B. Trickey**
1963 **Philip G. McCracken**
1964 **Edward Klein**
1965 **William J. Rimes**
1966 **Kurt Tauss**
1967 **Paul D. Cratin**
1968 **Robert Curry**
1969 **William Standford Durrell**
1969-70 **George Saul**
1971 **Thomas G. Jackson**
1972 **James L. Lambert**
1973 **Kurt Tauss**
1974 **Richard P. Gideon, Jr.**
1975 **Ernest Spinner**
1976 **Russell S. Andrews**

Mobile-Pensacola

(Mobile, Ala.)

1948-49 **Robert B. Reynolds**
1949-50 **Reid H. Leonard**
1950 **Edwin M. Trigg**
1951 **Bernard B. Bond**
1952 **Charles E. Lane, Jr.**
1953 **Joseph H. Stump, Jr.**
1954 **William D. McNally**
1955 **Edward M. Milner**
1956 **Richard D. Meyer**
1957 **John Wharton**
1958 **Edwin G. Rothbauer**
1959 **Kleim Alexander**
1959 **Francis J. Kearley**
1960 **Edward F. Rehm**

Mojave Desert

(Ridgecrest, Calif.)

1946 **Edward H. Spreen**
1947 **Ross W. Moshier**
1948 **Leo W. Briggs**
1949 **John H. Shenk**
1950 **Harvey S. Eastman**
1951 **Vincent Morgan**
1952 **Robert W. Van Dolah**
1953 **George T. Deck**
1954 **E. St. Clair Gantz**
1955 **James V. Wiseman**
1956 **Donald S. Villars**
1957 **Jack W. Walker**
1958 **Ronald A. Henry**
1959 **Harold J. Gryting**
1960 **George G. Gale**
1961 **William G. Finnegan**
1962 **Donald S. Arnold**
1963 **Raymond T. Merrow**
1964 **Wallery M. Sergy**
1965 **Gerald C. Whitnack**
1966 **Peter G. Cortessis**
1967 **Arnold T. Nielsen**
1968 **William W. Merk**
1969 **Alvin S. Gordon**
1970 **Arnold Adicoff**
1971 **Gerald C. Whitnack**
1971 **Peter R. Hammond**
1972-73 **Donald M. Moore**
1974 **Harold J. Gryting**
1975 **Russell Reed, Jr.**
1976 **Jack M. Pakulak**

Local Section Chairmen (continued)

Monmouth County

(Red Bank, N.J.)

1945 Virgil Francis Payne
1946 Meredith F. Parker
1947 Ernst T. Franck
1948 John N. Mrgudich
1948 Hugh V. Alessandroni
1949 Virgil Francis Payne
1950 Keith G. Misegades
1951 Erik G. Linden
1952 John P. Wadington
1953 William F. Nye
1954 Walter Raymond Wooley
1955 Edward R. Scheffer
1956 Arthur F. Daniel
1957 Theodore S. Williams
1958 Charles A. Russell
1959 Vaughan C. Chambers
1960 Lawrence S. White
1961 Marvin S. Fink
1962 Vincent J. Webers
1963 John N. Mrgudich
1964 Thomas J. Engelbach
1965 Raymond J. LeStrange
1966 W. C. Pfefferle, Jr.
1967 John V. Teutsch
1968 Charles Wendell Beggs
1969 Robert E. Fuguitt
1970 Glen A. Thommes
1971 Gerald L. Weiss
1972 Edward D. Cohen
1973 Thomas S. Harvey
1974 Nina M. Roscher
1975 Marilyn Parker
1976 Eugene W. Seitz

Montana

(Bozeman)

1928 A. J. Johnson
1928 Oden E. Sheppard
1929 Edmund Burke
1930 Oden E. Sheppard
1931 Alfred E. Koenig
1932 Oden E. Sheppard
1933 J. W. Howard
1934 Birger L. Johnson
1935-37 Guy E. Sheridan
1938 P. C. Gaines
1939-41 Don C. Evans
1941 John F. Suchy
1944-45 A. R. Johansson
1946 Alfred E. Koenig
1947-49 Birger L. Johnson
1949 Earl C. Lory
1950-52 Charles N. Caughlan
1952 Edward W. Anacker
1953 John M. Stewart
1954 George D. MacDonald
1955-56 Kenneth J. Goering
1957 John W. Howard
1958 Edwin G. Koch
1959 Ray Woodriff
1960 Richard E. Juday
1961 Kenneth M. McLeod
1962 Graeme L. Baker
1963 Leland M. Yates
1964 Edgar Milter Layton, Jr.
1965 Robert K. Osterheld
1966 Joseph I. Murray
1967 Edward W. Anacker
1968 Forrest D. Thomas II
1969 Frank Enri Diebold
1970 Earl H. Hoerger
1971 Ronald E. Erickson
1972 Alvin Fitzgerald
1973 Donald R. Beuerman
1974 Wayne P. VanMeter
1975 Kenneth Emerson
1976 Ralph J. Fessenden

Nashville (Tenn.)

1911 William L. Dudley
1912-13 R. Wilfred Balcom
1914-15 John I. D. Hinds
1916-17 James Flack Norris
1918-19 Edsel A. Ruddimann
1920-22 James M. Breckinridge
1923-24 James W. Sample
1925-26 Louis J. Bircher
1927-29 O. W. Boies
1930 James M. Breckinridge
1931-37 J. C. Carlin
1938-41 James M. Breckenridge
1942 Waite P. Fishel
1943 Demetrius Franklin Farrar
1944 George M. Smith
1945 Ward C. Sumpter
1946 Edward E. Litkenhous
1947 H. L. Kickison
1948 James K. Witt
1949 Donald E. Pearson
1950 Glenn Dooley
1951 Hanor A. Webb
1952 Word B. Bennett, Jr.
1953 Carl P. McNally
1954 Alvis M. Holladay
1955 Lamar Field
1956 Walter E. Cole
1956-57 Oscar Touster
1958 Arthur W. Ingersoll
1959 Milton T. Bush
1960 J. Eldred Wiser
1961 Kenneth K. Innes
1962 Larry C. Hall
1963 Howard G. Ashburn
1964 Exum D. Watts
1965 Thomas W. Martin
1966 Donald E. Pearson
1967 James K. Witt
1968 John E. Bishop
1969 Gordon Wilson, Jr.
1970 I. Wesley Elliott
1971 Howard E. Smith
1972 Leland Estes
1973 David O. Johnston
1974 Robert V. Dilts
1975 Martha Andrews
1976 Melvin D. Joesten

Nebraska

(Lincoln)

1895-05 H. H. Nicholson
1906-08 Samuel Avery
1908-09 F. J. Alway
1909-10 Wilson H. Low
1910-13 Herbert A. Senter
1914-15 Benton Dales
1916-17 F. W. Upson
1918 George Borrowman
1919 Horace G. Deming
1920 Donald J. Brown
1921 Ernest Henderson
1922 B. Clifford Hendricks
1923 M. J. Blish
1924-25 Cliff S. Hamilton
1926 Clarence J. Frankforter
1927 Roscoe C. Abbott
1928-29 Samuel Avery
1930-31 E. Roger Washburn
1932 H. Armin Pagel
1933-34 M. J. Blish
1935 H. A. Durham
1936 Clifton W. Ackerson
1937 B. Clifford Hendricks
1938 M. D. Weldon
1939 Walter E. Militzer
1940 Deton J. Brown
1941 Joseph Bell Burt
1942 Carl E. Georgi
1943 Rudolph M. Sandstedt
1944 T. J. Thompson
1945 H. Armin Pagel
1946 Lewis E. Harris
1947 Carl E. Georgi
1948 Donald E. Fox
1949 Herbert T. Bates
1950 Bennett Dale Hites
1951 Henry F. Holtzclaw
1952 Raymond L. Borchers
1953 Norman H. Cromwell
1954 Cecil E. Vanderzee
1955 John H. Pazur
1956 Henry E. Baumgarten
1957 James H. Looker
1958 Robert H. Harris
1959 J. M. Brim
1960 Robert M. Hill
1961 Robert H. Glazier
1962 Gordon A. Gallup
1963 Arnold A. Alberts
1964 John J. Scholz
1965 B. B. Johnston
1966 Robert C. Larson
1967 Richard Dam
1968 Richard E. Gilbert
1968-69 George Sturgeon
1970 Christopher J. Michejda
1971 James D. Carr
1972 Norman E. Griswold
1973 Michael L. Gross
1974 Robert M. Grossman
1975 Daniel B. Howell
1976 Anne Denise George

New Haven (Conn.)

1912 J. P. Street
1913 William Buell
1914 B. W. McFarland
1915 Trent B. Johnson
1916 Ralph Gibbs Van Name
1917 C. H. Mathewson
1918 Horace T. Smith
1919 Bertram B. Boltwood
1920-21 Harold Hibbert
1922 Merrill C. Burt
1923 Treat B. Johnson
1924 John L. Christie
1925 Arthur J. Hill
1926 H. W. Foote
1927 N. Saxton
1928 Stuart R. Brinkley
1929 C. H. Mathewson
1930 Harry A. Curtis
1931 John J. Donleavy
1932 Arnold H. Smith
1933 Bernard F. Dodge
1934 Harold G. Dietrich
1935 Robert D. Coghill
1936 John A. Timm
1937 Clifford C. Furnas
1938 Erwin B. Kelsey
1939 H. B. Vickery
1940 Dewitt T. Keach
1940-41 Clinton Doede
1942 Herbert S. Harned
1943 Werner Bergmann

Local Section Chairmen (continued)

New Haven, Ct. (cont.)

1944 Harding Bliss
1945 Bingham J. Humphrey
1946 Henry C. Thomas
1947 Wesley S. Coe
1948 James English, Jr.
1949 Joseph Fleischer
1950 D. Loren Schoene
1951 Joseph S. Fruton
1952 W. F. Bruckach
1953 Charles A. Walker
1954 Herman A. Bruson
1955 Harold G. Cassidy
1956 Charles D. McCleary
1956-57 Henry C. Thomas
1957-58 Dean E. Peterson
1959 Randolph H. Bretton
1960 Clinton W. Mac Mullen
1961 Richard D. Gilbert
1962 J. Vozas R. Vaisnys
1963 Emmanuel G. Kontos
1964-65 Orville J. Sweeting
1965-66 Walter Lwowski
1966-67 John Philip Faust
1967-68 Frank E. Lussier
1968-69 William R. Andrews
1969-70 William T. Neville
1970 Richard J. Ferrari
1970-72 Harold Greenfield
1972-73 Stephen E. Cantor
1973-74 Maurice Raymond
1974-75 Maurice R. Chamberland
1975-76 Martin I. Jacobs

New Orleans (La.)

(Chartered 1894)

1895-98 Abraham Louis Metz

(*See* Louisiana)

New York (N.Y.)

(Chartered 1891)

1892-93 Alvah Horton Sabin
1893-96 Peter T. Austen
1896-99 William McMurtrie
1899 Charles F. McKenna
1900 Charles A. Doremus
1901 Marston T. Bogert
1902 Thomas J. Parker
1903 Edmund H. Miller
1904-05 William J. Schieffelin
1906 A. A. Breneman
1907 Henry C. Sherman
1908 Leo H. Baekeland
1909 Morris Loeb
1910 Charles Baskerville
1911-12 Arthur C. Langmuir
1912 Arthur B. Lamb
1912 Herbert R. Moody
1913 B. C. Hesse
1914 Allen Rogers
1915 T. B. Wagner
1916 J. Merritt Methews
1917-18 Charles H. Herty
1919 David W. Jayne
1920 Ralph H. McKee
1921 John E. Teeple
1922 Martin H. Ittner
1923-24 Clarke E. Davis
1925 James Kendall
1926 Benjamin T. Brooks
1927 Arthur W. Thomas
1928 Charles R. Downs
1929 Rex R. Renshaw
1930 J. G. Davidson
1931 Arthur E. Hill
1932 Walter S. Landis
1933 Victor K. LeMer
1934 John M. Weiss
1935 Arthur W. Hixson
1936 Lawrence W. Bass
1937 David P. Morgan
1937 William C. Mac Tavish
1938-39 William W. Winship
1939 Louis P. Hammett
1940 Robert Calvert
1941 Ralph H. Muller
1942 Charles N. Frey
1943 Vincent du Vineaud
1944 Beverly L. Clarke
1945 Ross Allen Baker
1945 Cornelia T. Snell
1946 Raymond E. Kirk
1947-48 Hans T. Clarke
1949 Robert M. Burns
1950 John H. Nair, Jr.
1951 H. Burton Lowe
1952 Edward J. Durham
1953 Emmett S. Carmichael
1954 William M. Sperry
1955 Alvan H. Tenney
1956 Henry B. Hass
1957 Thomas I. Taylor
1958 Nathan Weiner
1959 Charles G. Overberger
1960 S. Fisher Gaffin
1961 George L. McNew
1962 Adolph J. Stern
1963 Peter P. Regna
1964 Kenneth Morgareidge
1965 Arthur B. Kemper
1966 George B. Brown
1967 Charles R. Dawson
1968-69 Herman Gershon
1970-72 Hervert Meislich
1972-73 J. Trygve Jensen
1974-75 Arthur Fon Toy
1976 Gary W. Sanderson

North Alabama

(Huntsville)

1952 John M. Geisel
1953 Carlyle J. Stehman
1954 Frank W. James
1955 Madison L. Marshall
1956 William F. Arendale
1957-58 Harry E. Anschutz
1958-59 William R. Lucas
1960 Patrick H. Hobson
1961 Charles B. Colburn
1962 John Lomartire
1963 Thomas A. Neely
1964 Wilbur A. Riehl
1965 Harry W. Hamme
1966 Chester W. Huskins
1967 David A. Flanigan
1968 Travis E. Stevens
1969 Samuel P. McManus
1970 Harold Dwain Coble
1971 William D. Stephens
1972 Ronald C. McNutt
1973 Martin C. Manger
1974 Barry D. Allen
1975 Clifford J. Webster
1976 Gary Workman

North Carolina

(1895; Raleigh)

1896-98 Francis P. Venable
1899 Charles Baskerville
1900 Benjamin W. Kilgore
1901 William Alphonso Withers
1902-03 Charles E. Brewer
1904-05 Irvin Sawter Wheeler
1906 Charles H. Herty
1907 William H. Pegram
1908 James E. Mills
1909 William Anderson Syme
1910 William Myron Allen
1911 George M. MacNider
1912 Luther B. Lockhart
1913-14 Leon F. Williams
1914-15 James M. Bell
1915-16 John William Nowell
1916-17 James Kemp Plummer
1917-18 Howard Bell Arbuckle
1918-20 John William Nowell
1920-21 James M. Bell
1921-22 Paul Gross
1922-23 Alvin Sawyer Wheeler
1923-24 John O. Halverson
1924 Francis Weber Sherwood
1925 Frank Elmore Rice
1926 James T. Dobbins
1927 Paul M. Ginnings
1928 L. G. Willis
1929 Frank K. Cameron
1930 Lucius A. Bigelow
1931 Francis Weber Sherwood
1932-33 Horace D. Crockford
1934 John H. Saylor
1935 Ralph W. Bost
1936 Edward Mack, Jr.
1937 Warren C. Vosburgh
1938 Nevill Isbell
1939 Edwin C. Markham
1940 Walter E. Jordan
1941 Ivan D. Jones
1942 Charles S. Black
1942-43 Frank Houston Smith
1944 Willis A. Reid
1945 Marcus E. Hobbs
1946 Oscar K. Rice
1946-47 Charles S. Black
1948 Douglas G. Hill
1949 Samuel B. Knight
1950 Charles K. Bradsher
1951 John W. Nowell
1952 Frank Houston Smith
1953 Arthur Roe
1954 Walter J. Peterson
1955 Pelham Wilder, Jr.
1956 Frances C. Brown
1957 George O. Doak
1958 Frederick R. Darkis
1959 John W. Dawson
1960 S. Young Tyre, Jr.
1961 Richard H. Loeppert
1962 Howard A. Strobel
1963 William F. Little
1964 Ralph C. Swann
1965 J. Keith Lawson, Jr.
1966 J. Charles Morrow, III
1967 E. Clifford Toren, Jr.
1968 Monroe E. Wall
1969 G. Gilbert Long
1970 Robert Ghiradelli
1971 Vivian T. Stannett
1972 Peter Smith
1973 Halbert C. Carmichael
1974 Richard J. Thompson
1975 Maurice M. Bursey
1976 Monica Nees

North Central Oklahoma

(Ponca City)

1955 Elbert L. Hatlelid
1956 James C. Kirk
1957 Robert S. Munger
1958 William C. Hamilton
1959 Harold H. Eby
1960 George C. Feighner
1961 William L. Groves, Jr.

Local Section Chairmen (continued)

No. Cent. Okla. (cont.)

1962 Flynt Kennedy
1963 Pat W. K. Flanagan
1964 Wayne R. Sorenson
1965 W. Dean Leslie
1966 Richard E. Laramy
1967 Billy J. Williams
1968 Richard L. Every
1969 Gerald Perkins, Jr.
1970 Charles M. Starks
1971 Albert M. Durr, Jr.
1972 Allan Lundeen
1973 Stephen E. McGuire
1974 Donald L. Whitsill
1975 Paul Washecheck
1976 John L. Riddle

North Jersey

(Newark, N.J.)

1925-27 David Wesson
1928-29 Augustus Merz
1930 Lawrence V. Redman
1931 George M. Maverick
1932 Allan F. Odell
1933 John H. Schmidt
1934 Per K. Frolich
1935-36 William T. Read
1936 Robert B. Sosman
1937-38 Moses L. Crossley
1938 Roscoe H. Gerke
1939 Robert J. Moore
1940 Randolph T. Major
1941 Delmer L. Cottle
1942 Calvin S. Fuller
1943 Edward R. Allen
1944 Horace E. Riley
1945 Robert E. Waterman
1946 William J. Sparks
1947 Ira D. Garard
1948 Harold F. Wakefield
1949 Frank R. Mayo
1950 George L. Royer
1951 Burnard S. Biggs
1952-53 John Lee
1954 Cecil L. Brown
1955 Karl A. Folkers
1956 Vlon N. Morris
1956-57 Oscar P. Wintersteiner
1958 Ellis V. Brown
1959 H. Herbert Fox
1960 William Rieman, III
1961 Ernst T. Theimer
1962 Lawrence T. Eby
1963 Albert W. Meyer
1964 John L. Lundberg
1965 Paul V. Smith
1966 W. Lincoln Hawkins
1967 Samuel M. Gerber
1968 Julian J. Leavitt
1969 Neil M. Mackenzie
1970 Howard Eugene Heller
1971 George E. Heinze
1972 Gerald Smolinsky
1973 Richard W. J. Carney
1974 Arnold D. Lewis
1975 Galen W. Ewing
1976 Benjamin J. Luberoff

Northeast Georgia

(Athens)

1968 S. William Pelletier
1969 Winfield Morgan Baldwin, Jr.
1970 Joseph Paul LaRocca
1971 George William Bailey
1972 C. Harry Neufeld
1973 Richard K. Hill
1974 Charles H. Stammer
1975 Charles D. Blanton
1976 Douglas B. Walters

Northeast Oklahoma

(Bartlesville)

1947 Cecil C. Ward
1948 James R. Owen
1949 John E. Mahan
1950 Guy Waddington
1951 Emory W. Pitzer
1952 Richard W. Blue
1953 Barton H. Eccleston
1954 R. Vernon Jones
1955 James L. Hart
1956 Delbert Pidgeon
1957 Kenneth J. Hughes
1958 James E. Pritchard
1959 John P. McCullough
1960 Howard W. Bost
1961 John C. Hillyer
1962 Dan E. Smith
1963 C. Wayne Moberly
1964 Harry T. Rall
1965 Robert P. Zelinski
1966 Wm. T. Nelson, Jr.
1967 Carl W. Kruse
1968 Robert T. Johansen
1969 Henry L. Hsieh
1970-71 J. Paul Hogan
1972 Donald R. Douslin
1973 Paul S. Hudson
1974 Clifford W. Childers
1975 Roland H. Harrison
1976 William B. Hughes

Northeast Tennessee

(Kingsport)

1932-34 Herbert G. Stone
1934 Horace B. Huddle
1935 Martin Wadewitz
1936 Leonidas Rosser Littleton
1937 Paul T. Jones
1938 Louis J. Figg, Jr.
1939 Jack J. Gordon
1940 Albert C. Adams
1941 Matthew Weber, Jr.
1942 Chester H. Penning
1943 Lester W. A. Meyer
1944 Joseph H. Brant
1944-46 James E. Magoffin
1946 Robert B. Hickey
1947 Edward M. McMahon
1948 George E. Cronheim
1949 Robert H. Hasek
1950 Milton E. Lubs
1951 William D. Kennedy
1952 William M. Gearhart
1953 Paul V. Brower
1954 Roger M. Schulken, Jr.
1955 Norman H. Leake
1956 Clyde A. Glover
1957 John Frank Fuzek
1958 Wayne V. McConnell
1959 Harold Reynolds
1960 Richard L. McConnell
1961 Gaylord K. Finch
1962 Hugh W. Patton
1963 Kermit B. Whetsel
1964 Vinton A. Hoyle, Jr.
1965 James E. Poe
1966 Louis E. Mattison
1967 Vernon W. Goodlett
1968 James H. Lady
1969 David L. Nealy
1970 Peter E. Morrisett
1971 Buford L. Poet
1972 John M. McIntire
1973 Gerald P. Morie
1974-75 Dale E. Van Sickle
1976 Linda Jane Adams

Northeast Wisconsin

(1930; Appleton)

1931 Harry F. Lewis
1931 L. A. Youtz
1932 Louis C. Fleck
1933 John R. Fanselow
1934 Stephen F. Darling
1935 A. Lewenstein
1936 J. Glenn Strieby
1937 Herbert L. Davis
1938 D. Romund Moltzau
1939 Kenneth A. Craig
1940 Rev. Peter P. Pritzl
1941-42 Joseph O. Frank
1943 Emil Heuser
1944 Louis E. Wise
1945 John W. Green
1946 Nelson E. Rodgers
1947 Paul E. Truttschel
1948 Abbott Byfield
1949 Linton E. Simerl
1950 Ara O. Call
1951 Paul F. Cundy
1952 John C. Bletzinger
1953 James O. Thompson
1953-54 James H. Dinius
1955 Herbert T. Peeler
1956 Lawrence A. Boggs
1957-58 John P. Butler
1959-60 Harold A. Swenson
1961 Robert A. Gorski
1962 George D. Stevens
1963 George A. Lauterbach
1964 Lawrence L. Motiff
1965 Dale G. Williams
1966 Robert E. Weber
1967 Edward A. Baetke
1968 Gilbert F. Pollnow
1969 Arild J. Miller
1970 Donald C. Johnson
1971 Robert E. Anderson
1972 Jerrold P. Lokensgard
1973 William D. Guither
1974 Fredric N. Miller
1975 Dwight B. Easty
1976 Orval Rautmann

Northeastern

(Boston, Mass.)

1898-99 Arthur A. Noyes
1900 Arthur D. Little
1901 John Alden
1901 Leonard Parker Kinnicutt
1902 Augustus H. Gill
1903 William H. Walker
1904 James F. Norris
1905 Charles L. Parsons
1906 Louis A. Olney
1907 Frank G. Stantial
1908 Gilbert N. Lewis
1909 S. W. Wilder, Jr.
1910 Walter L. Jennings
1911 William C. Bray
1912 Robert Spurr Weston
1913 J. Russell Marble
1914 William K. Robbins
1915 Frank H. Thorp

Local Section Chairmen (continued)

Northeastern (cont.)

1916 Henry P. Talbot
1917 Elwood B. Spear
1918 Homer J. Wheeler
1919 George S. Forbes
1920 Robert W. Neff
1921 James B. Conant
1922-23 Gustavus J. Esselen
1924 Hermann C. Lythgoe
1925 Lyman C. Newell
1926 Kenneth L. Mark
1927 Arthur D. Holmes
1928 Allan Winter Rowe
1929 David E. Worrall
1930 Reid Hunt
1931 William P. Ryan
1932 Raymond Stevens
1933 Lester A. Pratt
1934 Kenneth E. Bell
1935 Grinnell Jones
1936 Hervey J. Skinner
1937 Michael J. Ahearn, S.J.
1938 Ernest H. Huntress
1939 Thorne M. Carpenter
1940 John J. Healy, Jr.
1941 Avery A. Ashdown
1942 Frederick S. Bacon
1943-44 Harold A. Iddles
1945 Stuart B. Foster
1946 Chester P. Baker
1947 Chester M. Alter
1948 Ernest C. Crocker
1949 Avery A. Morton
1950 John T. Blake
1951 John A. Timm
1952 Thomas R. P. Gibb, Jr.
1953 Paul D. Bartlett
1954 C. Richard Morgan
1955 Arthur C. Cope
1956 Edward R. Atkinson
1957 Lockhart B. Rogers
1958 Howard H. Reynolds
1959 John L. Oncley
1960-61 Lloyd H. Perry
1962 M. Kent Wilson
1963 Henry A. Hill
1964 Lawrence J. Heidt
1965 Arnet L. Powell
1966 Robert A. Shepard
1967 George E. Kimball
1968 Arno H. A. Heyn
1969 Russell T. Werby
1970 Robert E. Lyle
1971 Edward F. Levy
1972 Ernest I. Becker
1973 Edward J. Modest
1974 Phyllis A. Brauner
1975 Janet S. Perkins
1976 Harry E. Keller

Northeastern Indiana

(Fort Wayne)

(*see also* Fort Wayne)

1946-48 Maurice M. Felger
1948 Ralph Edward Broyles
1949 Joseph H. LaFollette
1950 Judson West, Jr.
1951 Paul L. Brunner
1952 Harry B. Bolson
1953 R. Conway Knapp
1954 H. Landon Thomas
1955 R. Conway Knapp
1956 Earl L. Smith
1957 Rod G. Dixon
1958 Paul Fulkerson
1959 Harry Boyko, Jr.
1960 Wendell D. Mason
1961 Ernst C. Koerner
1962 Harold Rothchild
1963 Warren E. Hoffman
1964 Marvin A. Peterson
1965 William E. Donahue
1966 George J. Cocoma
1967 Robert B. Young
1968 John J. Flynn, Jr.
1968-69 Sharon K. Slack
1970 Ronald R. McVay
1971 Adrian J. Good
1972 Stephen P. Coburn
1973 Richard A. Pacer
1974 Thomas A. Shelby
1975 Royce L. Hutchinson
1976 Chester A. Pinkham, III

Northeastern Ohio

(Painesville)

1944-45 Carol L. Campbell
1946 Marvin Achterhof
1947 George F. Rugar
1948 Howard W. Davis
1949 Maxwell J. Skeeters
1950 Archie Hill
1951 John E. Brothers
1952 Clifford A. Neros
1953 Leslie P. Seyb
1954 Tom S. Perrin
1955 Alfred Hirsch
1956 Edward J. Amann
1957 Clayton L. Dunning
1958 Howard E. Everson
1959 Clemens J. Urbanski
1960 Robert G. Banner
1961 Frank B. Slezak
1962 Charles E. Entemann
1963 Frank W. Hengeveld
1964 Clifford J. Burg
1965 Alex Hlynsky
1966 Russell M. Bimber
1967 Thomas A. Magee
1968 H. Norman Benedict
1969 Lawrence A. Retallick
1970 Joseph J. Dietrich
1971 William O. Emrich
1972 James M. Kolb
1973 Lester E. Coleman
1974 Norman M. Pollack
1975 John S. Staral
1976 James A. Scozzie

Northern Indiana

1922 Milber M. MacLean
1923 Edgar N. Weber
1924 V. C. Bidlack

(Name changed 1925 to St. Joseph Valley.)

Northern Intermountain

(Pullman, Wash.)

1912 J. Shirley Jones
1913-15 Elton Fulmer
1916 Carl M. Brewster
1917 Clare C. Todd
1918-23 John A. Kostalek

(*See* Washington-Idaho Border)

Northern Louisiana

(1925; Shreveport)

1926 Fred J. Mechlin

(*See* Northwest Louisiana)

Northern New York

(Potsdam)

1951-52 Francis William Brown
1953 William K. Viertel
1954 Herman L. Shulman
1955 Gilbert E. Moos
1956 Milton Kerker
1957 Clarke L. Gage
1958 Nelson F. Beeler
1959 Paul E. Merritt
1960 James B. Reed
1961 Charles H. Stauffer
1962 Egon Matijevic
1963 Gilbert E. Moos
1963 Charles A. Howe
1964 Charles H. Stauffer
1965 Clarke L. Gage
1966 Josip Kratohvil
1967 Gordon L. Galloway
1968 Hans H. G. Jellinek
1969 Samuel S. Stradling
1970 Donald Rosenthal
1971 Lauri Vaska
1972 George Sheats
1973 Kenneth West
1974 Nicholas Zeuos
1975 Peter Zuman
1976 Charles A. Blood, Jr.

Northern West Virginia

(1921; Morgantown)

1922-24 Friend E. Clark
1925 Willard W. Hodge
1926 C. Alfred Jacobson
1927 Samuel Morris
1928 Earl C. H. Davies
1929 Robert B. Dustman
1930 Armand R. Collett
1931 Clarence E. Garland
1932 Hubert Hill
1933 John A. Gibson, Jr.
1934 Stephen P. Burke
1935 Charles L. Lazzell
1936 Audrey H. Van Landingham
1937 Lilly Bell Deatrick
1938 Gordon A. Bergy
1939 Walter A. Koehler
1940 Timothy J. Cochrane
1941 Ira J. Duncan
1942 A. J. W. Headlee
1943 Homer A. Hoskins
1944 P. Lloyd MacLachlan
1945 Virgil G. Lilly
1946 Howard P. Simons
1947 Harvey D. Erickson
1948 David F. Marsh
1949 Lawrence D. Schmidt
1950 James L. Hall
1951 Don C. Iffland
1952 Kenneth L. Temple
1953 Harold V. Fairbanks
1954 James B. Hickman
1955 John J. S. Sebastian
1956 A. J. W. Headlee
1957 Chester W. Muth
1958 George L. Humphrey
1959 Reginald F. Krause
1960 Frederick J. Lotspeich
1961 Peter Popovich
1962 George A. Hall
1963 Damon C. Shelton
1964 Leslie J. Kane
1965 Armine D. Paul
1966 Alexander D. Kenny
1967 Elizabeth D. Swiger
1968 Anthony Winston

Local Section Chairmen (continued)

No. W. Virginia (cont.)

1969 William T. Abel
1970 Robert V. Digman
1971 William D. Ruoff
1972 Gabor Fodor
1973 Ralph Booth
1974 William R. Moore
1975 Charles G. McCarthy
1976 Vincent J. Traynelis

Northwest Central Ohio

(Chartered 1960 as Lima Section; name changed 1971.)

1972 Donald J. Harvey
1973 Dale L. Wilhelm
1974 Howard L. Haight
1975 Harold D. Knierieman
1976 Robert W. Suter

Northwestern Utah

(Salt Lake City)

1924 Walter D. Bonner
1925 Elton L. Quinn
1926 Charles E. Maw
1927 Thomas B. Brighton
1928-29 Antoine M. Gaudin
1930 Ludvig Reimers
1931 Nephi E. McLachlan
1932-33 Sherwin Maeser
1934 Hugh Peterson
1935 Alpha Johnson
1936 Moyer D. Thomas
1937 M. Elmer Christensen
1937 Albert C. Titus
1938 Loren C. Bryner
1939 Corliss R. Kinney
1940 Arthur Fleischer
1941 Joseph K. Nicholes
1942 John R. Lewis
1943 Walter D. Bonner
1944 Russell H. Hendricks
1945 Glen C. Ware
1946 Theodore M. Burton
1947 Moyer D. Thomas
1948 George V. Beard
1949 Delbert A. Greenwood
1950 John H. Wing

(Name changed 1951 to Salt Lake.)

Northwest Louisiana

(Shreveport)

(Chartered 1946 as Ark-La-Tex Section; name changed 1970 to Arkansas-Louisiana Border; changed again 1971 to above.)

1971 Charles B. Lowrey
1972 Larry G. Spears
1973 Robert D. Schwartz
1974 David B. McCullough
1975 Joseph W. Goerner
1976 Anita C. Olson

Norwich (N.Y.)

(Chartered 1970)

1971 R. Norman Johnson
1972 Donald H. Kelly
1973 Dyral C. Fessler
1974 Mario J. Cardone
1975 George S. Denning
1976 Raymond E. Dann

Ohio Northern

(Ada, Ohio)

1927-28 Jesse R. Harrod
1929-30 L. C. Sleesman

(*See* Lima)

Oklahoma

(Stillwater)

1919 Hilton Ira Jones
1920 Edwin DeBarr
1921-24 Guy Y. Williams
1925 Fred W. Padgett
1926 Charles K. Francis
1927 Bruce Houston
1928 Frederick W. Lane
1929 Lloyd E. Swearingen
1930 Harvey M. Trimble
1931 Cecil T. Langford
1932-34 Charles L. Nickolls
1935-36 Fred E. Frey
1937 Richard L. Huntington
1937 Cecil T. Langford
1938 Jerry R. Marshall
1939 Ralph J. Kaufmann
1940 J. Paul Jones
1941 Gordon D. Byrkit
1942 Harold M. Smith
1943-44 Ralph W. Boyd
1945 Leonard F. Sheerar
1946 Grant C. Bailey
1947 Paul H. Horton
1948 Bruce Houston
1949 Wyman H. Meigs
1950 Otto M. Smith
1951 Bernard O. Heston
1952 Clyve Allen
1953 James E. Webster
1954 Howard H. Rowley
1955 Simon H. Wender
1956 Henry P. Johnston
1957 Horace H. Bliss
1958 Jack O. Purdue
1959 Paul F. Kruse, Jr.
1960 Marvin R. Shetlar
1961 Ernest M. Hodnett
1962 Alfred J. Weinheimer
1963 Marvin T. Edmison
1964 William E. Neptune
1965 Roger E. Koeppe
1966 Charles L. Cahill
1967 Reagan Bradford
1968 George Gorin
1969 Francis J. Schmitz
1969-70 George R. Waller, Jr.
1971 Bill B. Arnold
1972 Mary P. Carpenter
1973 Thomas W. Clapper
1974 Donald I. Hamm
1975 John A. Burr
1976 George V. Odell

Omaha (Nebr.)

1920-22 Herbert A. Senter
1923-24 Charles F. Crowley
1925 Victor E. Levine
1926 Llewellyn B. Parsons
1927 Samuel A. Rice
1928 John A. Krance, S.J.
1929 J. A. Land
1930 Raymond S. Burnett
1931 John J. Guenther
1932 Christopher L. Kenny
1933 W. T. Bailey
1934 M. D. Mize
1935 Nicholas Dietz, Jr.
1936-37 William G. Haynes
1937 George F. Stewart
1938 J. P. Maxfield
1939 S. James O'Brien
1940 Otto C. Johnson
1941 Wm. Dean Yohe
1942 William K. Noyce
1943 Walter R. Urban
1944-45 Donald M. Findlay
1946 James T. Smith
1947 J. Leslie Hale
1948-49 Philippos E. Papadakis
1949 Eugene T. Drake
1950 Herbert P. Jocobi
1951 Marinus P. Bardolph
1951 Leslie A. Brown
1952 Wallace L. Rankin
1953 Howard J. Anderson
1954 Arthur L. Dunn
1955 John C. McMillan
1956 Michael J. Carver
1957 Violet M. Wilder
1958 Dawn N. Marquardt
1959 Hugh L. Davis
1960 Donald J. Baumann
1961 Phillip J. Stageman
1962 Fred J. Mleynek
1963 Walter W. Linstromberg
1964 Philip L. Stageman
1965 Julia D. Buresh
1966 Carl Chin
1967 Dale P. J. Goldsmith
1968 C. Robert Keppel
1969 George C. Phelps
1970 Kazuo H. Takemura
1971 Arley L. Goodenkauf
1972 Francis M. Klein
1973 Mary M. Hill
1974 James K. Wood
1975 H. V. Subbaratnam
1976 Anne Maria Coverdell

Orange County

(Anaheim, Calif.)

1962 Clayton McAuliffe
1963 Wendell G. Markham
1964 Sol W. Weller
1965 Vernon E. Stiles
1966 Edgar W. Fajans
1967 Rodger W. Baier
1967-68 Andrew F. Montana
1969 Lyman L. Handy
1970 Frederic J. Kakis
1970-71 Lloyd R. Snyder
1972 Don C. Atkins, Jr.
1973 Claude P. Coppel
1974 Margil W. Wadley
1975 Vance Gritton
1976 John C. Middleton

Oregon

(Corvallis)

1912-13 A. L. Knisely
1914 Orin F. Stafford
1915-16 William Conger Morgan
1917-20 F. A. Olmstead
1921-22 Ralph K. Strong
1923-25 Harry G. Miller
1926 Roger J. Williams
1927 J. R. Harrod
1928-29 Francis H. Thurber
1930 F. Leonard Cooper
1931 Orin F. Stafford
1932 Earl C. Gilbert
1933 Charles H. Johnson
1934 F. Leonard Cooper
1935 J. Shirley Jones

Local Section Chairmen (continued)

Oregon (cont.)

1936 Roger J. Williams
1937 Walter R. Carmody
1938 Leo E. Friedman
1939 Bert E. Christensen
1940 Edward T. Luther
1941 Charles S. Pease
1942 Glen C. Ware
1943 Joseph S. McGrath
1944 Edward G. Locke
1945 Pierre J. Van Rysselberghe
1946 Leo Friedman
1947-48 Joseph Schulein
1949 Vernon Cheldelin
1950 Albert W. Stout
1951 Harold O. Ervin
1952 William E. Caldwell
1953 Max B. Williams
1954 Arthur H. Livermore
1955 Harry Freund
1956 Robin E. Moser
1957 J. Dean Patterson
1958 Lewis A. Thayer
1959 Francis J. Reithel
1960 Donald W. Turnham
1961 James C. Anderson
1962-63 Dale W. Richardson
1963-64 Anton Postl
1964-65 Allen B. Scott
1965-66 H. Darwin Reese
1966-67 John L. Kice
1967-68 Richard M. Noyes
1968-69 Elliot N. Marvell
1969-70 Lloyd J. Dolby
1970-71 Theran D. Parsons
1971-72 John F. W. Keana
1972-73 William J. Fredericks
1973-74 Charles E. Klopfenstein
1974-75 H. Hollis Wickman
1975-76 Tom W. Koenig

Ouachita Valley

(Chartered 1949 as South Arkansas Section; name changed 1966.

1966 Robert H. Colvard
1967 Bobby D. La Grone
1968 James P. Berry
1969 Robert L. Holt
1970 Morley T. Johnston
1971-72 Harry E. Moseley
1973-74 Kenneth J. Miller
1975 Sally E. Cauthen
1976 Gerald L. Ellis

Panhandle Plains

(Amarillo, Tex.)

1937 C. W. Seibel
1938 Chester A. Pierle
1939 Elmer G. Hammerschmidt
1940 George L. Heller
1941 Roy E. King
1942 W. M. Deaton
1943 Ira Williams
1944 Chester A. Pierle
1945 Emmett B. Reinbold
1946 Elmer B. Hammerschmidt
1947 Rector P. Roberts
1948 E. M. Frost, Jr.
1949 Charles F. Fryling
1950 Jack S. Skelly, Jr.
1951 L. O'Brien Thompson
1952 Lawrence R. Sperberg
1953 Clyde H. Mathis
1954 J. Frank Svetlik
1955 Ross Van Volkenburgh
1956 Curt B. Beck
1957 Andries Voet
1958 Roger D. Whealy
1958-59 Walter B. Polk
1960 James C. Word, Jr.
1961 William N. Whitten, Jr.
1962 Murl B. Howard
1963 Charles L. Klingman
1964 Franklin W. Baer
1965 Muerner S. Harvey
1966 Richard N. Cooper, Jr.
1967 John Wesley Balentine
1968 Gene A. Crowder
1969 Elmer T. Suttle
1970 Peter Aboytes
1971 Kenneth L. Ladd, Jr.
1972 James D. Woodyard, Jr.
1973 Anil K. Sircar
1974 Zach L. Estes
1975 Philip C. Tully
1976 Joel D. Oliver

Penn-Ohio Border

(1950; Youngstown)

1951 John O. Collins
1952 Wm. A. Beckman
1953 Donald E. Babcock
1954 Creig S. Hoyt
1955 Irwin Cohen
1956 Henry M. McLaughlin
1957 Joseph B. Littman
1958 R. Warren Leisy
1959 Ralph F. Lengerman
1960 Hobson D. DeWitt, Jr.
1961 Robert James Grubb
1962 William H. McCoy
1963 Richard A. Hendry
1964 Edward W. Naegele, Jr.
1965 Leonard B. Spiegel
1966 Philip A. Shreiner
1967 Perry Warrick, Jr.
1968 Inally Mahadeviah
1969 Edward L. Safford
1969-70 Kenneth M. Long
1971 Elmer Foldvary
1972 Evelyn Halpern
1973 Andrew Pronay
1973-74 Peter W. Von Ostwalden
1975 Ralph E. Yingst
1976 Thomas N. Dobbelstein

Penn-York

(Warren, Pa.)

1939 Ogden Fitz Simons
1940 Bennett S. Ellefson
1941 Russell E. Palmateer
1942 Chas. W. Cable
1943 Harold A. Krantz
1944 Everett J. Schneider
1945 Robert E. Dunham
1945 Robert G. Capell
1945 Robert E. Dunham
1946 Ralph W. Hufferd
1947 Robert C. Amero
1947 Chas. W. Cable
1948 Milton C. Hoffman
1949 Verlin L. Miller
1950 Joseph N. Breston
1951 Wm. T. Granquist
1952 John Milglarese
1953 Robert L. Lambert
1954-55 Roy L. Overcash
1956 Ralph A. Johnson
1957 A. Jerome Miller
1958 Peter W. Parsons
1959 Richard M. Smith
1960 Theodore W. Blickwedel
1961 Malcolm R. Rankin
1962 Reverdy E. Baldwin
1963 Calaldo Cialdella
1964 George I. Beyer
1965 Luther D. Dromgold
1966 Frank Rusinko, Jr.
1967 Robt. W. Lambdin
1968 Gilbert E. Moos
1969 James D. McCoy
1970 Melvin B. Redmount
1971 Marvin Boskin
1972 William Turek
1973 Joseph G. Free
1974 Charles P. Buhsmer
1975 John F. Rakszawski
1976 Robert L. Mintz

Pensacola (Fla.)

(Chartered 1960)

(*see also* Mobile-Pensacola)

1961 Robert Johnson
1962 Douglas L. Johnson
1963 Frederick E. Detoro
1964 Karl Alfred Kubitz
1965 Edwin Earl Royals
1966 Jerry G. Morrison, Jr.
1967 Raymond E. Kourtz
1968 Bill H. Daughdrill
1969 Ralph K. Birdwhistell
1970 John J. Hicks, Jr.
1971 Jerome E. Gurst
1972 Ralph W. Smith
1973 Stephen P. Tanner
1974 Charles E. Cutchens
1975 Clifford W. J. Chang
1976 Herman P. Benecke

Peoria (Ill.)

1939-40 Geo. C. Ashman, Sr.
1941 Robert D. Coghill
1942 Gerald C. Baker
1943 Reid T. Milner
1944 John H. Shroyer
1945 John C. Crown
1946 G. Rockwell Barnett
1947 Cecil T. Langford
1948 Harold J. Deobald
1949 Howard M. Teeter
1950 Ralph I. Claassen
1951 Carl E. Rist
1952 Leonard Stone
1953 Cyril D. Evans
1954 H. Orville Bensing
1955 Allen K. Smith
1956 Mark C. Paulson
1957 Charles R. Russell
1958 Fred O. Hale
1959 Robert J. Dimler
1960 Clair S. Boruff
1961 Herbert J. Dutton
1962 Karl L. Smiley
1963 Frederick R. Senti
1964 Thomas F. Cummings
1965 Leonard L. McKinney
1966 James M. Van Lanen
1967 Ivan A. Wolff
1968 Kenneth E. Kolb
1969 Felix H. Otey
1970 William H. Tallent
1971 Rex D. Steinke
1972 Everett H. Pryde
1973 Wilbur C. Shaefer
1974 Doris K. Kolb
1975 Joseph S. Wall
1976 Kenneth D. Carlson

Local Section Chairmen (continued)

Permian Basin

(1959; Midland, Tex.)

1960 Glen B. White
1961 Jesse Looney
1962 Thomas L. McGinnis
1963 James Leonard Hutson
1964 Raymond B. Seymour
1965 Birt Allison
1966 Wallace Wilhelm
1967 Lehman G. Richardson
1968 Billy Joe Reynolds
1969 Ronald G. Howell
1970 Henry B. Dirks,, Jr.
1971 Daniel J. Kallus
1972 Willis Gunn
1973 Edward D. Taylor
1974 Norbert F. Cywinski
1975 James M. Watson
1976 Edwin B. Kurtz

Philadelphia (Pa.)

1899-01 Harry W. Jayne
1902 J. Merritt Matthews
1903 John Marshall
1904 Owen L. Shinn
1905 Samuel P. Sadtler
1906 David W. Horn
1907 Gellert Alleman
1908 Daniel W. Fetterolf
1909 Charles E. Vanderkleed
1910 Arthur M. Comey
1911 George C. Davis
1912-13 Walter T. Taggert
1914 Harry F. Keller
1915 Harlan S. Miner
1916 Clement S. Brinton
1917 Abraham Henwood
1918 Charles L. Reese
1919 Harlan S. Miner
1920 William A. Pearson
1921 George E. Barton
1922 James G. Vail
1923 J. Howard Graham
1924 Hiram S. Lukens
1925 Horace C. Porter
1926 J. Bennett Hill
1927 J. Spencer Lucas
1928 William Stericker
1929 Elmer C. Bertolet
1930 Lawrence M. Henderson
1931 William J. Kelly
1932 Carl Haner
1933 Horace M. Weir
1934 P. Edward Rollhaus
1935 Martin Kilpatrick
1936-37 Wesley R. Gerges
1938 Alexander G. Keller
1939 Newcomb K. Chaney
1940 Henry J. M. Creighton
1941 S. W. Ferris
1942 Harry A. Alsentzer, Jr.
1943 Arthur Osol
1944 John M. McIlvain
1945 Melvin C. Molstad
1946 Floyd T. Tyson
1947 Arthur B. Hersberger
1948 C. Harold Fisher
1949 Willard A. LaLande, Jr.
1950 Allan R. Day
1951 Felix C. Gzemski
1952 Joseph W. E. Harrisson
1953 O. Davis Shreve
1954-55 J. Harold Perrine
1956 Glenn E. Ullyot
1957 Waldo C. Ault
1958 Claude K. Deischer
1959 Joseph N. Bartlett
1960 James W. Wilson, III
1961 Ellington M. Beavers
1962 James L. Jezl
1963 J. Hartley Bowen, Jr.
1964 Adalbert Farkas
1965 Benjamin S. Garvey, Jr.
1966 Paul N. Craig
1967 Newman M. Bortnick
1968 Daniel Swern
1969 Donald H. Saunders
1970 Marian F. Fegley
1970-71 Grafton D. Chase
1972 F. William Kirsch
1973 David J. Cooper
1974 John F. Gall
1975 Frederick H. Owens
1976 Lyle H. Phifer

Pittsburgh (Pa.)

1903 Alexander G. McKenna
1904 Harry E. Walters
1905 George P. Maury
1906-08 Joseph H. James
1909 Horace C. Porter
1910 James O. Handy
1911 John K. Clement
1912 Walter O. Snelling
1913 George D. Chamberlain
1914 Raymond F. Bacon
1915 Arno C. Fieldner
1916 Karl F. Stahl
1917 George A. Burrell
1918 Samuel R. Scholes
1919 Rufus E. Zimmerman
1920 E. Ward Tillotson
1921 Henry C. P. Weber
1922 James O. Handy
1923 Edward R. Weidlein
1924 Alexander Silverman
1925 Warren F. Faragher
1926 Clarence J. Rodman
1927 William A. Hamor
1928 Alexander Lowy
1929 Edward E. Marbaker
1930 J. Clyde Whetzel
1931 Leonard H. Cretcher
1932 Harry V. Churchill
1933 William P. Yant
1934 T. George Timby
1935 Charles S. Palmer
1936 Chester G. Fisher
1937 Gerald J. Cox
1938 Charles Glen King
1939 Lloyd H. Almy
1940 Earl K. Wallace
1941 James N. Roche
1942 John C. Warner
1943 Helmuth H. Schrenk
1944 William A. Gruse
1945 Harold K. Work
1946 Robert N. Wenzel
1947 Herbert E. Longnecker
1948 Gilbert Thiessen
1949 J. Paul Fugassi
1950 Rob Roy McGregor
1951 Bernard F. Daubert
1952 Edmund O. Rhodes
1953 Earl L. Warrick
1954 Hugh F. Beeghly
1955 Tobias H. Dunkelberger
1956 John R. Bowman
1957 Foil A. Miller
1958 Gordon H. Stillson
1959 Robert B. Anderson
1960 Earl A. Gulbransen
1961 Robert B. Carlin
1962 George W. Gerhardt
1963 Harold P. Klug
1964 Robert A. Friedel
1965 John J. McGovern
1966 W. Conard Fernelius
1967 Paul C. Cross
1968 Kurt Schreiber
1969 Richard Hein
1970 Thomas J. Hardwick
1971 W. Dwight Johnston
1972 Glenn P. Cunningham
1973 Denton M. Albright
1974 James C. Carter
1975 I. Rosabelle McManus
1976 Jack W. Hausser

Portland (Oreg.)

1961-62 James G. Anderson
1963 Storrs S. Waterman, Jr.
1964 William C. Wilson
1965 Cecil K. Claycomb
1966 Herman R. Amberg
1967 Richard T. Van Santen
1968 Elaine Spencer
1969 Harold W. Zeh
1970 Arleigh R. Dodson
1971 Wilbur L. Shilling
1972 Nick J. Mauleg
1973 G. Doyle Daves, Jr.
1974 Chester A. Schink
1975 Joseph W. McCoy
1976 James F. McCormack

Princeton (N.J.)

1927 Robert N. Pease
1928 W. T. Richards
1929 F. B. Stewart
1930 T. J. Webb
1931 Earle R. Caley
1932 Herbert N. Alyea
1933 Henry Eyring
1934 Everett S. Wallis
1935 Wendell H. Taylor
1936 Gregg Dougherty
1937 Chas. P. Smyth
1938 John Turkevitch
1939 Chas. Rosenblum
1940 Richard H. Wilhelm
1941 John Y. Beach
1942 Hubert N. Alyea
1943 R. E. Powell
1944 John F. Lane
1945 Frederick R. Duke
1946 Arthur Tobolsky
1947 Elmer J. Badin
1948 Dean R. Rexford
1949 Clark E. Bricker
1950 Robert H. Goeckermann
1951 Humboldt W. Leverenz
1952 William C. Wildman
1953 Richard O. Steele
1953-54 David Garvin
1955 Howard J. White, Jr.
1956 Wallace H. McCurdy, Jr.
1957 James R. Arnold
1958 Charles Rosenblum
1959 Michael J. Boudart
1960 Richard H. Wilheim
1961 Donald S. McClure
1962 Edward C. Taylor, Jr.
1963 J. R. White
1964 Jackson P. English
1965 Arthur V. Tobolsky
1966 Heinz Heinemann
1967 Kurt M. Mislow
1968 Robert D. Offenhauer
1969 Leland C. Allen
1970 Hugo Stange
1971 Paul V. R. Schleyer
1972 Richard J. Magee
1973 Victor Laurie
1974 Arthur Fontijn
1975 Robert A. Naumann
1976 David F. Ollis

Local Section Chairmen (continued)

Puerto Rico

(Rio Piedras)

1947 Fritz Fromm
1948 Victor Rodriguez-Benitez
1949 Isidoro A. Colon
1950 Juan D. Curet
1951 M. Garcia-Morin
1952 Angel Alberto Colon
1953 Carlos Vincenty
1954 Rafael Pol Mendez
1955 Conrado F. Asenjo
1956 Hector Flores-Gallardo
1957 Marta Cancio-England
1958 Leo B. Lathroum
1959 Luis Amoros-Marin
1960 Herbert F. Wolf
1961 Edwin Roig
1962 Leopold R. Cerecedo
1963 Frederick C. Strong
1963-64 H. H. Szmant
1965 Efrain Toro-Goyco
1966 Ferdinand Sanchez
1967 Gerhard R. Morell
1968 James A. Singmaster, III
1969-70 Rafael N. Infante
1971 Ismael Almodovar
1972 Jose R. Sanchez
1973 Owen H. Wheeler
1973 Jose R. Sanchez-Caldas
1974 Myrian Vargas
1975 Waldemar Adam
1976 Manuel Torrens

Puget Sound

(Seattle, Wash.)

1909 Harry K. Benson
1910 Myrl J. Falkenburg
1911 H. M. Loomis
1912 Albert C. Jacobson
1913 Charles A. Newhall
1914 C. E. Bogardus
1915 Horace G. Byers
1916 Edward A. Pieterle
1917 Rex Smith
1918 Ray W. Clough
1919 Fred H. Heath
1920 A. L. Knisely
1921 Herman V. Tartar
1922 Wm. L. Haley
1923-24 Thomas G. Thompson
1925-26 Ernest D. Clark
1927 H. V. Tartar
1928-30 Walter R. Gailey
1931-32 Bernard B. Coyne
1933 James C. Palmer
1934-36 Bradley MacKenzie
1937 Harvey C. Diehl
1938-39 Earl G. Thompson
1940 Robert S. Roe
1941 John M. Kniseley
1942 Warren L. Beuschlein
1943-44 Sargent G. Powell
1945 Victorian Sivertz
1946 Theodore S. Hodgins
1947 Herbert R. Erickson
1948 Joseph L. McCarthy
1949 Robert D. Sprenger
1950 Collis C. Bryan
1951 Edward C. Lingafelter
1952 Charles V. Smith
1953 Paul C. Cross
1954 Jim C. Drury
1955 George H. Cady
1956 David H. Read
1957 Dirk Verhagen
1958 Benton S. Rabinovitch
1959 Raymond M. Way
1960 Robert W. Moulton
1961 F. Bruce Sanford
1962 Rex J. Robinson
1963 George O. Orth, Jr.
1964 Norman W. Gregory
1965 Otto Goldschmid
1966 Diptiman Chakravarti
1967 Gerald O. Freeman
1968 Verner Schomaker
1969 Edwin E. Barnes
1970 Bernard M. Steckler
1971 Theodore R. Beck
1972 Kyosti Sarkanen
1973 J. Keluin Hamilton
1974 Edward H. Gruger, Jr.
1975 Joseph R. Crook
1976 Lloyd V. Johnson

Purdue

(Lafayette, Ind.)

1922 Louis A. Test
1923 Charles B. Jordan
1924 Ralph E. Nelson
1925 Ralph Howard Carr
1926 Harry C. Peffer
1927 Melvin G. Mellon
1928 Henry R. Kraybill
1929 Harold L. Maxwell
1930 Roy F. Newton
1931 R. Norris Shreve
1932 Henry B. Hass
1933 Arthur R. Middleton
1934 Sigfred M. Hauge
1935 Ralph C. Corley
1936-37 Ed. F. Degering
1937 John L. Bray
1938 Dean C. B. Jordan
1939 Thomas DeVries
1940 Charles L. Shrewsbury
1941 Frank D. Martin
1942 Earl T. McBee
1943 Paul B. Curtis
1944 Fred J. Allen
1945 G. Bryant Bachman
1946-47 Forrest W. Quackenbush
1948 E. St. Clair Gantz
1949 Roy L. Whistler
1950 John E. Christian
1951 Joe M. Smith
1952 Walter F. Edgell
1953 Robert A. Benkeser
1954 Edwin T. Mertz
1955 Herbert C. Brown
1956 Edward L. Haenisch
1957 Nathan Kornblum
1958 Herbert E. Parker
1959 Henry Feuer
1960 Alan F. Clifford
1961 Howard N. Draudt
1962 John F. Foster
1963 Adelbert M. Knevel
1964 Richard A. Sneen
1965 Dale W. Margerum
1966 James W. Richardson
1967 Harry L. Pardue
1968 Michael Laskowski, Jr.
1969 Sam P. Perone
1970 Francis K. Fong
1971 Joseph Wolinsky
1972 Heinz G. Floss
1973 R. Stuart Tobias
1974-75 Albert Light
1976 Otto E. Lobstein

Quincy-Keokuk

(Quincy, Ill.)

(Chartered 1950 as Canton, Mo. Section; name changed 1954.)

1954 William Eaton
1955 Marion J. Caldwell
1956 Chester W. Bennett
1957 Wray M. Rieger
1958 Harold W. Matzke
1959 Kenneth L. Hamm
1960 Albert J. Gehrt
1961 Kenneth H. Goode
1962 Max Q. Freeland
1963 Vernon R. Heaton
1964 Raymond G. Behrensmeyer
1965 Robert W. Shelton
1966 David E. Sims
1967 Richard L. Hardin
1968 Joseph A. Siefker
1969 Benjamin Hughes
1970 William Gasser
1971 John M. Brodmann
1972 Norbert A. Goeckner
1973 John L. Martin
1974 Phillip L. Dillon
1975 Donald E. Walker
1976 Allan T. Kucera

Red River Valley

(Grand Forks, N.Dak.)

1948 George A. Abott
1949 Albert L. Eliason
1950 Albert M. Cooley
1951 Ray T. Wendland
1952 Ernest D. Coon
1953 William B. Treumann
1954 Bernhard G. Gustafson
1955-56 Ralph E. Dunbar
1957 Richard E. Frank
1958 Harold J. Klosterman
1959 W. Eugene Cornatzer
1960 Richard G. Werth
1961 Edward J. O'Reilly, Jr.
1962 Charles W. Fleetwood
1963 Carl A. Wardner
1964 Muriel C. Vincent
1965 Roland G. Severson
1966 Roger B. Meintzer
1967 Virgil I. Stenberg
1968 Gustav P. Dinga
1969 James A. Stewart
1970 William H. Shelver
1971 Francis A. Jacobs
1972 James M. Sugihara
1973 Richard J. Baltisberger
1974 Nicholas Kowanko
1975 Alyn W. Johnson
1976 Dewey O. Brummond

Rhode Island

(1891; Providence)

1893 John Howard Appleton
1894-95 Charles A. Catlin
1896-99 Edward D. Pearce
1900-02 Walter Mills Saunders
1903 Walter E. Smith
1904 Charles M. Perry
1905 A. H. Jameson
1906 R. Clinton Fuller
1907-08 John E. Bucher
1909-10 Frederick A. Franklin
1911-12 Albert W. Claflin
1913-14 William H. Cady
1915-17 Augustus H. Fiske
1918 Robert K. Lyons
1919-20 Robert F. Chambers
1921-24 Samuel T. Arnold
1925 J. C. Hostetter
1926-28 Lucius A. Bigelow
1929-30 Ellery L. Wilson
1931-32 William W. Russell
1933-34 Edward K. Strachan
1935-36 Joseph L. Costa
1937 Charles B. Wooster
1938 Richard B. Earle

Local Section Chairmen (continued)

Rhode Island (cont.)

1939-40 W. George Parks
1941-42 Dana Burks, Jr.
1943 Paul C. Cross
1944-45 Frederick C. Hickey
1946-47 Albert E. Marshall
1948 Leallyn B. Clapp
1949-50 Douglas L. Kraus
1951-52 Karl L. Holst
1953-54 John J. Hanley
1955-56 Mark Weisberg
1957 Harold R. Nace
1958 Charles A. Robinson
1959 Scott Mackenziek, Jr.
1960 Robert H. Elliott
1961 Theodore T. Galkowski
1962 Irving G. Loxley
1963 Edwin H. Beach
1964 Joseph F. Bunnett
1965 Bertram J. Garceau
1966 Mark M. Rerick
1967 Harry Kroll
1968 Harold R. Ward
1969 Melvin A. Lipson
1969-70 Edward F. Greene
1971 Harold Petersen, Jr.
1972-73 Edward L. Shunney
1974-75 Ascanio G. Dipippo
1976 Wilfred H. Nelson

Richland (Wash.)

1948 Robert L. Moore
1949-50 Bernhardt Weidenbaum
1951 Orville F. Hill
1952 Albert H. Bushey
1953 Fred J. Leitz
1954 Harold R. Schmidt
1954-55 Roy C. Thompson
1956 William R. Dehollander
1957 Milton Lewis
1958 Leland L. Burger
1959 Raymond E. Burns
1960 George J. Alkire
1961 Raymond Hugh Moore
1962 Edward W. Christopherson
1963 Merle K. Harmon
1964 Wileber E. Keder
1965 Frederick A. Scott
1966 Charles A. Rohrmann
1967 John R. Morrey
1968 Robert W. Stromatt
1969 Horace H. Hopkins, Jr.
1970 Archie S. Wilson
1970-71 Milton H. Campbell
1972 Earl C. Martin
1973 Glen E. Benedict
1974 Walter D. Felix
1975 Jack Ryan
1976 Wayne H. Yunker

Rio Grande Valley

(Alamagordo, N.Mex.)

(Formed 1971 from El Paso and Southern New Mexico Sections.)

1972 Dennis D. Davis
1973 Marion L. Ellzey, Jr.
1974 Gordon J. Ewing
1975 Michael P. Eastman
1976 Lynford L. Ames

Rochester (N.Y.)

1912-13 Victor J. Chambers
1913 Harrison E. Howe
1914-15 C. E. Kenneth Mees
1916 Charles F. Hutchison
1917 Harry LeB. Gray
1918 Henry H. Tozier
1919 William F. Zimmerli
1920 J. E. Woodland
1921 Hans T. Clarke
1922 Otto Irving Chormann
1923 Edward B. Leary
1924 Lincoln Burrows
1925 Wm. R. Webb
1926 Walter R. Bloor
1927 Florus R. Baxter
1928 Emmett K. Carver
1929 Erle M. Billings
1930 Ivan N. Hultman
1931 Cyril J. Staud
1932 Willard R. Line
1933 Ralph Helmkamp
1934 Harold W. Crouch
1935 Samuel W. Clausen
1936 Gale Nadeau
1937 Charles R. Fordyce
1938 Charles K. Tressler
1939 Edwin O. Wiig
1940 Louis K. Eilers
1941 Harold Hodge
1942 Charles F. H. Allen
1943 Thos. F. Murray, Jr.
1944 Howard S. Gardner
1945 Leslie J. Ross
1946 John F. Flagg
1947 Ethel L. French
1948 Maurice L. Huggins
1949 Winston D. Walters
1950 Edmond S. Perry
1951 Dean S. Tarbell
1952-53 Philip L. Harris
1954 Ralph L. Van Peursem
1955 Gordon D. Hiatt
1956 Marshall D. Gates, Jr.
1957 Paul W. Vittum
1958 Paul W. Aradine
1959 Norbert J. Kreidl
1960 Arnold Weissberger
1961 John H. Dessauer
1962 Stanley R. Ames
1963 Shelby A. Miller
1964 John R. Thirtle
1965 Carl J. Claus
1966 David W. Stewart
1967 William H. Saunder, Jr.
1968 T. Howard James
1969 Carl W. Zuehlke
1970 Robert L. Craven
1971 Lloyd E. West
1972 Clarence G. Heininger, Jr.
1973 Charleton C. Bard
1974 K. Thomas Finley
1975 J. Dolf Bass
1976 Earl Krakower

Sabine-Neches

(Port Arthur, Tex.)

(Chartered 1943 as Texas-Louisiana Gulf Section; name changed 1967.)

1967 James E. Bates
1968 S. B. Crockett
1969 Joe N. Fields
1970 Irvine H. Russell
1971 James L. Hahn
1971-72 Howell R. Brown
1973 William J. Powers, III
1974 Anthony Macaluso
1975 Buford C. Hall
1976 Roy J. Woods

Sacramento (Calif.)

1922 Geo. P. Gray
1923 George H. P. Lichthardt
1924-25 Chester F. Hoyt
1926 William J. Lentz
1927 John Haworth Jonte
1928 H. C. Davis
1929 Ralph Austin Stevenson
1930 J. Gordon Sewell
1931 John H. Norton
1932 Maxwell Adams
1933 George A. Richardson
1934 Frank H. Prittie
1935 Arthur T. Bawden
1936 Geo. W. Sears
1937 Wallace A. Gilkey
1938 Max Kleiber
1939 Roy P. Tucker
1940 Meryl W. Deming
1941 Lincoln M. Lampert
1942 Clarence E. Larson
1943 Harold G. Reiber
1944 Andrew C. Rice
1945 Alfred R. Ebberts
1946 Fred J. Clark
1947 Herbert A. Young
1947-48 Edward L. Randall
1949 Van Paul Entwistle
1950 George H. Morse
1951 Emerson Cobb
1952 Ralph S. Gray
1952 James F. Guymon
1953 J. Ray Schwenck
1954 Frederick P. Zscheile, Jr.
1955 Rex S. Thomas
1956 Charles H. Hughes
1957 Ishmael W. Walling
1958 Lawrence J. Andrews
1959 C. Robert Hurley
1960 Karl Klager
1961 Robert A. White
1962 Edeard P. Painter
1963 Stanley W. Harris
1964-65 Vincent S. DeMarchi
1966 Chester A. Luhman
1967 Albert T. Bottini
1968 John S. Sullivan
1969 Gerhard Blaschczyk
1970 Philip S. Gisler
1971 Major Bard Suverkrop
1972 Joseph B. Di Giorgia
1973 George R. Tichelaar
1974 Richard A. Lungstrom
1975 Theodore J. Hampert
1976 Kenneth G. Hancock

St. Joseph Valley

(South Bend, Ind.)

(Chartered 1922 as Northern Indiana Section; name changed 1925.)

1924 V. C. Bidlack
1925 J. M. Gauss
1926 E. G. Mahin
1927 Marcus W. Lyon, Jr.
1928 Herman H. Wenzke
1929 Leo J. Lovett
1930-31 Henry D. Hinton
1932 M. F. Taggart
1933 Leo J. Lovett
1934 Lawrence H. Baldinger
1935 Marcus W. Lyon, Jr.
1936 Ronald Rich
1937 Glen R. Miller
1938 Geo. F. Hennion
1939 Andrew J. Boyle

Local Section Chairmen (continued)

St. Joseph Valley (cont.)

1940 Ernest J. Wilhelm
1941 Henry J. Klooster
1942 Brother Columba Curran
1943 Harold Goebel
1944 Kenneth N. Campbell
1945 Ernest G. Bargmeyer
1946 P. A. McCusker
1947 Paul L. Bush
1948 Allen S. Smith
1949-50 Alfred H. Free
1951 Edward C. Svendsen
1952 James V. Quagliano
1953 Stephen M. Olin
1954 C. Elton Huff
1955 James P. Danehy
1956 Richard S. Reamer
1957 Glen R. Miller
1957 Paul G. Roach
1958 Milton Burton
1959 Lloyd T. Johnson
1960 Ernest L. Eliel
1961 Henry D. Weaver, Jr.
1962 Vincent J. Traynelis
1963 Helen Free
1964 Francis Lee Benton
1965 Dean F. Gamble
1966 Arthur A. Smucker
1967 George H. Baii
1968 Rudolph S. Bottei
1969 Ernest C. Adams, Jr.
1970 Gerald J. Papenmeier
1971 James M. Day
1972 Thomas P. Fehlner
1972-73 Chauncey Rupe
1973-74 Joseph H. Ross
1975 Robert C. Boguslaski
1976 Geraldine M. Huitink

St. Louis (Mo.)

(Chartered 1907)

1908-09 Launcelot Winchester Andrews
1910-11 Edward H. Keiser
1912-13 W. F. Monfort
1914-15 Leroy McMaster
1916-17 A. C. Boylston
1918 Charles E. Caspari
1919 Gaston DuBois
1920 Charles E. Caspari
1921-22 Frederick W. Russe
1923 Philip A. Shaffer
1924-25 R. R. Matthews
1926 Theodore R. Ball
1927 George S. Robins
1928 Lyn A. Watt
1929 Courtland F. Carrier
1930 Charles N. Jordan
1931 H. Edward Wiedemann
1932 Henry L. Dahm
1933 L. P. Hall
1934 C. H. Swanger
1935 Eugene S. Weil
1936-37 Lyman J. Wood
1937-38 Jules Bebie
1939 John H. Gardner
1940 Franklin D. Smith
1941 Paul A. Krueger
1942 Fredrich Olsen
1943 Adolph H. Winheim
1944 Melvin A. Thorpe
1945 Charles W. Rodewald
1946 Carl E. Pfeifer
1947 Milton L. Herzog
1948 Robert A. Burdett
1949 Ferdinand B. Zienty
1950 August H. Homeyer
1951 Joseph R. Darby
1952 Desiree S. LeBeau
1953 William H. Elliott
1954 Barrett L. Scallet
1955 Roger W. Stoughton
1956 Hal G. Johnson
1957 George K. Robins
1958 Robert D. Rands, Jr.
1959 C. David Gutsche
1960 Charles O. Gerfen
1961 Edward E. Marshall
1962 Clayton F. Callis
1963 Kenneth H. Adams
1964 Samuel M. Tuthill
1965 Thomas P. Lawton, Jr.
1966 Henry D. Barnstorff
1967 Leo J. Spillane
1968 Henry C. Godt, Jr.
1969 George Brooke Hoey
1970 Perry King, Jr.
1971 Franklin E. Mange
1972 John W. Lyons
1973 Thomas P. Layloff
1974 Marvin M. Crutchfield
1975 Roger L. Kidwell
1976 Jordan J. Bloomfield

Salt Lake (City, Utah)

(Chartered 1924 as Northwestern Utah Section; name changed 1951.)

1951 Melvin A. Cook
1952 Ralph S. Gray
1953 Royal C. Mursener
1954 Melvin C. Cannon
1955 H. Smith Broadbent
1956 Sherman R. Dickman
1957 Russell H. Hendricks
1958 Harris O. Van Orden
1959 H. Tracy Hall
1960 George R. Hill
1961 William M. Tuddenham
1962 Carl J. Christensen
1963 Angus U. Blackham
1964 Grant G. Smith
1965 Reed M. Izatt
1966 Robt. J. Heaney
1967 Edward M. Eyring
1968 John W. Holmes
1969 Richard C. Anderson
1970 Ralph E. Wood
1971 Ronald O. Ragsdale
1972 Kenneth W. Nelson
1973 Terry D. Alger
1974 Robert R. Beishine
1975 David M. Bodily
1976 Joseph G. Morse

San Antonio (Tex.)

1947 Alfred R. Rowe, Jr.
1948 Ava J. McAmis
1949 John C. Rentz, Jr.
1950 Earle C. Smith
1951 Oliver P. Smith
1952 Harold E. Weissler
1953 Edward C. Collignon
1954 John W. Allen
1955 Herman Levin
1956 Alden H. Waitt
1957 Jose A. Rivera
1957-58 George T. Deck
1958-59 Herbert C. McKee
1960 Louis Koenig
1961 George R. Somerville
1962 Franklin E. Massoth
1963 Dale A. Clark
1964 John M. Bryant
1965 Charles J. Cummiskey
1966 Russell G. Dressler
1967 Donald E. Johnson
1968 John D. Miller
1969 William J. Wilson, Jr.
1970 A. L. Shackelford
1971 Wanda L. Brown
1972 Charles F. Rodriguez
1973 Larry C. Grona
1974 Charles Howard
1975 Alan D. Elbein
1976 John A. Burke, Jr.

San Diego (Calif.)

1941 Ambrose R. Nichols, Jr.
1942 Paul B. Donovan
1943-44 Arnold B. Steiner
1945 Bernard Gross
1946 Franklin H. Page, Jr.
1947 C. Eldon White
1948 Aaron Miller
1949 Horace H. Selby
1950 Wm. H. McNeely
1951 Arne N. Wick
1952 John O'Connell
1953 Lionel Joseph
1954 Willard H. Wynne
1955 Viola Sommermeyer
1956 George Tohir
1957 Frank J. Riel
1958 Harold Walba
1959 Dudley H. Robinson
1960 Donald A. Hoffman
1961 Charles J. Stewart
1962 Harry N. Barnet
1963 Robert J. Good
1964 Ted G. Traylor
1965 Loy L. Flor
1966 Arthur W. Mosen
1967 Isadore Nusbaum
1968 John Hopperton
1969 Donald D. Myers
1970 Charles J. Stewart
1971 David J. Pettitt
1972-73 Philip Camberg
1974 William Kaplan
1975 Kenneth J. Liska
1976 Ian W. Cottrell

San Gorgonio

(Riverside, Calif.)

1949-50 Francis A. Gunther
1951 Owen K. Burman
1952 W. Conway Pierce
1953 Wm. Gustav Schulze
1954 Ralph B. March
1955 Horton E. Swisher
1956 Reinhold J. Krantz
1957 Edwin F. Bryant
1958 R. Nelson Smith
1958-59 Carleton B. Scott, II
1960 Charles P. Haber
1961 Glenn C. Soth
1962 Roy A. Whiteker
1963 Robert H. Maybury
1964 Willard E. Baier
1965 George Matsuyama
1966 Franklin M. Turrell
1967 Edgar R. Stephens
1968 Grant H. Palmer
1969 Irwin Geller
1970-71 Ernest A. Ikenberry
1972 Oscar P. Hellrich
1973 John E. Simpson
1974 James A. Hammond
1975 Julian L. Roberts
1976 Denny B. Nelson

Local Section Chairmen (continued)

Santa Clara Valley

(Palo Alto, Calif.)

1954 **Irving M. Abrams**
1955 **Douglas A. Skoog**
1956 **Bruce Graham**
1957 **H. Murray Clark**
1958 **Bert M. Morris**
1959 **Richard D. Cadle**
1960 **Erik Heegaard**
1961 **John E. DeVries**
1962 **Worden Waring**
1963 **Eric Hutchinson**
1964 **LeRoy A. Spitze**
1965 **Shirley B. Radding**
1966 **Howard M. Frantz**
1967 **Robert L. Focht**
1968 **M. Floyd Hobbs**
1969 **George W. Shreve**
1970 **John I. Brauman**
1971 **Stephen C. Dorman**
1972 **Donald E. Green**
1973 **Allan Kahn**
1974 **Lois J. Durham**
1975 **Theodore Mill**
1976 **Hubert E. Dubb**

Savannah (Ga.)

1920 **John J. McManus**
1921 **Herbert S. Bailey**
1922 **Eli H. Armstrong**
1923 **J. N. Everson**
1924 **J. C. Howard**

(*See* Savannah River)

Savannah River

(Aiken, S.C.)

1957 **Charles E. Anderson**
1958 **Don G. Ebenhack**
1959 **Steward W. O'Rear**
1960 **Harold M. Kelley**
1961 **Robert L. Folger**
1962 **Daniel S. St. John**
1963 **Robert I. Martens**
1964 **Edward L. Albenesius**
1965 **Donald A. Orth**
1966 **H. John Groh, Jr.**
1967 **W. Knowlton Hall**
1968 **Richard M. Wallace**
1969 **Leon H. Meyer**
1970 **William C. Perkins**
1971 **LeVerne Fernandez**
1972 **Monte L. Hyder**
1973 **Robert S. Dorsett**
1974 **Ned E. Bibler**
1975 **James M. McKibben**
1976 **Jane P. Bibler**

Sierra Nevada

(Reno, Nev.)

1967-68 **Bernard Porter**
1969 **Howard H. Heady**
1970 **Hyung Kyu Shin**
1971 **Robert J. Morris**
1972 **David J. MacDonald**
1973 **Frank G. Baglin**
1974 **Roger A. Lewis**
1975 **Elwood F. Blondfield**
1976 **John H. Nelson**

Sioux Valley

(Sioux Falls, S.Dak.)

1939-40 **James A. Coss**
1941 **James E. Brock**
1942 **Guy G. Frary**
1943 **Lester S. Guss**
1944 **Gregg M. Evans**
1945 **Arthur M. Pardee**
1946 **Emerson Cobb**
1947 **Alvin L. Moxon**
1948 **Edwin H. Shaw, Jr.**
1949 **Elmer R. Johnson**
1950 **John A. Foremke**
1951 **Victor S. Webster**
1952 **Charles R. Estee**
1953 **Eugene I. Whitehead**
1954 **Oscar M. Hofstad**
1955 **Keatha K. Krueger**
1955 **Ralph K. Strong**
1956 **Oscar E. Olson**
1957 **Howard M. Thomas**
1957 **Andrew W. Halverson**
1958 **Harlon L. Klug**
1959 **Andrew W. Halverson**
1960 **George P. Scott**
1961 **John A. Tanaka**
1962 **Robert R. Kintner**
1963 **Donald J. Mitchell**
1964 **Donald E. McRoberts**
1965 **Marvin DeYoung**
1966 **Charles S. Dewey**
1967 **M. Laeticia Lilzer**
1968 **Richard J. Landberg**
1969 **E. Howard Coker**
1970 **Rolland R. Rue**
1971 **Robert J. Peanasky**
1972 **Milton P. Hanson**
1973-74 **James Fries**
1975 **Leo H. Spinar**
1976 **Peter Maldonado**

South Arkansas

(1949; El Dorado)

1950 **Marion D. Barnes**
1951 **Lloyd T. Sandborn**
1952 **Everett E. Cofer**
1953 **John T. Skinner**
1954 **Ralph C. Tallman**
1955 **Leo J. Spillane**
1956 **Neoson B. Russell**
1957 **King I. Glass**
1958 **William B. Stengle**
1959 **Charles N. Robinson**
1960 **Russell E. Koons**
1961 **Dan Nall**
1962 **William K. Easley**
1963 **George L. Heller**
1964 **W. H. Hochstetler**
1965 **David S. Byrd**

(Name changed 1966 to Ouchita Valley)

South Carolina

(1927; Columbia)

1928 **Harry E. Shiver**
1929 **Caleb Archibald Haskew**
1930 **George A. Buist**
1931-32 **James E. Copenhaver**
1933 **Roe E. Remington**
1934 **Coleman B. Waller**
1935 **Glenn G. Naudain**
1936 **John R. Sampey**
1937 **Willard A. Whitsell**
1938 **Mary New**
1939 **Nelson McKaig, Jr.**
1940 **Raymond A. Patterson**
1941 **Harry E. Sturgeon**
1942 **John Albert Southern**
1942 **Maston T. Carlisle**
1943 **J. Kennedy Hodges**
1944 **Jack H. Mitchell, Sr.**
1944-45 **Peter Carodemos**
1946 **Auburn Woods, Jr.**
1947 **Earl Dew Jennings**
1948 **Baynard R. Whaley**
1949 **Ralph M. Byrd**
1950 **Joseph W. Bouknight**
1951 **Caleb Archibald Haskew**
1952 **Robert L. Holmes**
1953 **Harry W. Davis**
1954 **John C. Edwards**
1955 **Thomas E. Wannamaker**
1956 **Clark H. Ice**
1957 **Samuel A. Wideman**
1958 **Peyton C. Teague**
1959 **William M. McCord**
1960 **David D. Humphreys**
1961 **Richard A. Potter**
1962 **Joseph Bayne Doughty**
1963 **Oscar D. Bonner**
1964 **Hood C. Hampton, Jr.**
1965 **Charles E. Durkee**
1966 **John T. Wise**
1967 **John Miglarese**
1968 **Joseph R. Wilkinson**
1969 **Henry D. Reese Page**
1970 **Charles F. Jumper**
1971 **John A. Eberwein**
1972 **Edward E. Mercer**
1973-74 **H. Clyde Osborn, Jr.**
1975 **Spencer R. Arrowood**
1976 **William H. Breazeale, Jr.**

South Central Missouri

(Rolla)

1966-67 **Hector O. McDonald**
1968 **Samir B. Hanna**
1968-69 **Gary L. Bertrand**
1970 **James O. Stoffer**
1971 **B. Ken Robertson**
1972 **Larry M. Nicholson**
1973 **Raymond Venable**
1974 **Samir B. Hanna**
1975 **Donald J. Siehr**
1976 **James T. Wrobleski**

South Jersey

(Gibbstown, N.J.)

1920-21 **H. D. Gibbs**
1922 **C. E. Burke**
1923 **Hermann W. Mahr**
1924 **William F. Twombley**
1925 **Paul C. Bowers**
1926 **M. S. Thompson**
1927 **William A. Douglass**
1928-29 **Paul W. Carleton**
1930 **A. S. Yount**
1931 **William S. Calcott**
1931 **Harold E. Woodward**
1932 **Albert E. Parmelee**
1933 **George H. Schuler**
1934 **Henrietta C. W. Calcott**
1935 **Frederick B. Downing**
1936 **Charles J. Teahan**
1937 **Herbert W. Walker**
1938 **John M. Tinker**
1939 **Donovan E. Kvalnes**
1940 **Charles B. Biswell**
1941 **Harold J. Jones**
1942 **Clifford E. Carr**
1943 **Arlie A. O'Kelly**
1944 **Bernard M. Sturgis**
1945 **Rowland C. Hansford**

Local Section Chairmen (continued)

So. Jersey (cont.)

1946 Henry R. Lee
1947 Lonard C. Drake
1948 Arthur C. Stevenson
1949 Alfred W. Francis
1950 William F. Filbert
1951 John W. Brooks
1952 Arthur D. Gilbert
1952 Charles J. Planm
1953 Willard F. Anzilotti
1953 Phillip D. Caesar
1954 Harrison H. Homes
1955-56 Herbert E. Rasmussen
1957 Robert A. Brooks
1958 Harry G. Doherty
1959 John Sterlong, Jr.
1960 William E. Garwood
1961 Melvin L. Huber
1962 Phillip S. Landis
1963 Charles G. J. Rudershausen
1964 Lyle A. Hamilton
1965 C. Donald Federline
1966 David B. Cox
1967 Edward L. Reilly
1968 Donald M. Nace
1969 William J. Stenger
1970 Leo James McCabe
1970-71 Robert K. Armstrong
1972 Robert W. Barrett
1973 Bernard A. Orkin
1974 M. Girard Bloch
1975 Edwin N. Givens
1976 Abraham O. M. Okorodudu

South Plains

(Lubbock, Tex.)

1964 Joe Dennis
1965 Ulrich Hollstein
1966 Robert J. Thompson
1967 Joe A. Adamcik
1968 William G. Thomas
1969 John A. Anderson
1970 Henry J. Shine
1971 Roy E. Mitchell
1972 Samuel H. Lee
1973 Thomas W. Russell
1974 Jerry L. Mills
1975 John N. Marx
1976 Gary L. Blackmer

South Texas

(Corpus Christi, Tex.)

1945-46 John L. Nierman
1947 Gustave Heinemann
1948 Bruce S. Zinsworth, Jr.
1949 Foster G. Garrison
1950 Fred M. Garland
1951 Oren V. Luke, Jr.
1952 John W. Moore
1953 Arthur W. Schnizer
1954 Darrel M. Jones
1955 Walter E. Heinz
1956 Charles H. Fuchsman
1956 A. Bernard Hoefelmeyer
1957 Frank Marcotte
1958 Howard A. Hoekje
1959 Harold D. Medley
1959-60 Robert E. Scott
1961 Stone D. Cooley
1962 Thomas R. Rogers
1963 Charles C. Hobbs, Jr.
1964 E. Phelps Helvenston
1965 Edward N. Wheeler
1966 Dorothy Stewart
1967 Frank S. Wagner, Jr.
1968 Henry V. Cortez
1969 Frederick R. Hartlage, Jr.
1970 Charles J. Mazac
1971-72 Harold R. Gerberich, Jr.
1973 Jo A. Beran
1974 John S. England
1975 Carroll G. Pate
1976 Alberto M. Olivares

Southeast Kansas

(Pittsburg)

1930 John R. Sheppard
1931 James Henry Davidson
1932 William J. Clapson
1933 P. T. Naudet
1934 J. H. Calbeck
1935 John James Stadtherr
1935 C. V. Miller
1936-37 Howard G. Hodge
1937 H. L. Roscoe
1938 J. H. Calbeck
1939 Donald J. Doan
1939-40 W. B. Parks
1941 Everett J. Ritchie
1942-43 Max E. Colson
1944-46 C. V. Miller
1946 John R. Musgrave
1947-48 Roland G. Holmes
1949 Lawrence W. Braniff
1950 Archie J. Deutschman, Jr.
1951 A. Paul Thompson
1952 Elmer R. Ligon
1953 Neil B. Garlock
1954 James A. Burns
1955 Elton W. Cline
1956 Lorenzo W. Payden
1957 Shirely E. Witherspoon
1958 W. E. Medcalf
1959 James L. Pauley
1960 Fred G. Sprang
1961 Elbert W. Crandall, Jr.
1962 Donald J. Doan
1963 Wendell E. Ferguson
1964 Eric C. Juenge
1965 Eula Ratekin
1966 Joe M. Walker
1967 Thomas M. Medved
1968 Wilbur McPherson
1969 Edward W. Moore
1970 Melvin Potts
1971 Vernon Biamonte
1972-73 Margaret B. Parker
1974 John S. Todd
1975 Donald L. Smith
1976 Elton W. Cline

Southeastern Pennsylvania

(York)

1941 Edmund Claxton
1942 John B. Zinn
1943 Harry E. Bruce
1944 Paul O. Powers
1945 Rollin P. Gilbert
1946 Delos H. Wamsley
1947 John B. Zinn
1948 G. Edward Shubrooks
1949 Robert B. Rohrer
1950 Harold A. Reehling
1951 Ernest A. Vuilleumier
1952 William E. Weisgerber
1953 Harold H. Quickel
1954 Joseph A. Benner
1955 Robert L. Moore
1956 C. Allen Sloat
1957 Horace E. Rogers
1958 John W. deGroot, Jr.
1959 Hugh A. Heller
1960 Edgar S. Long
1961 James F. B. Everett
1962 Donald E. Nichol
1963 Lawrence H. Dunlap
1964 Karl L. Lockwood
1965 Oren Schnee Kaltriter
1966 F. Melvin Sweeney
1967 Robert Griswold
1968 Eugene H. Wells
1969 Ted B. Windsor
1970 Harry M. Bobonich
1971 John M. Kauffman
1972 Harold F. Emitt
1973-74 Stanley J. Chlystek
1975 John R. Snyder
1976 Joseph E. Kunetz

Southeastern Texas

(Houston)

1917 F. W. Bushong
1918 Frank M. Seibert
1919 C. M. Alexander
1920 Felix Paquin
1921 Harry B. Weiser
1922 M. C. Van Gundy
1923 L. S. Bushnell
1924 F. W. Bushong
1925 J. G. Detweiler
1926-28 Walter R. Kirner
1929-32 Allen D. Garrison
1933-35 Byron M. Hendrix
1936-38 Harry B. Weiser
1939-41 Ivan S. Cliff
1942 M. C. Van Gundy
1943 Geo. R. Gray
1944 Geo. H. Richter
1945 C. M. Hickey
1946 Preston L. Brandt
1947 Briggs B. Manuel
1948 Sherman S. Shaffer
1949 W. O. Milligan
1950 Herbert E. Morris
1951 Sam S. Emison
1952 Joseph H. Gast
1953 Harold T. Byck
1953 Loren B. Odell
1954 Rudolph L. Heider
1955 Joe L. Franklin, Jr.
1956 Wm. H. Lane
1957 Simon Miron
1958 Richard B. Turner
1959 Donald R. Lewis
1960 Byron L. Williams
1960-61 Max A. Mosesman
1962 Gordon O. Guerrant
1963 William F. Hamner, Jr.
1964 Albert Zlatkis
1965 James T. Richardson
1966 James V. Cavender, Jr.
1967 Lonnie W. Vernon
1968 Emery B. Miller
1969 Marvin L. Deviney
1970 Robert H. Friedman
1971 Wesley W. Wendlandt
1972 William A. Bailey
1973 Harold A. Palmer
1974 Wm. K. Meerbott
1975 Gerhard G. Meisels
1976 Howard W. Cogswell

Southeast Tennessee

(Chattanooga)

1937-38 J. H. Barnett, Jr.
1939 Walter S. Hude
1940 John W. Edwards
1941 Emerson P. Poste

Local Section Chairmen (continued)

Southeast Tenn. (cont.)

1942 Andrew J. Kelley
1943 Harold S. King
1944 James M. Holbert
1945 William O. Swan

(Name changed 1945 to Chattanooga.)

Southern Arizona

(Tucson)

(*see also* Arizona)

1953-54 Mitchell G. Vavich
1955 Duane Brown
1956 Edward N. Wise
1957 Jacob Fuchs
1958 Douglas S. Chapin
1959 Claude E. McLean, Jr.
1960 Hilding B. Gustafson
1961 J. Smith Decker
1961-62 Winslow F. McCaughey
1963 James W. Berry
1964-65 John P. Schaefer
1966 Robert B. Bates
1967 John V. Rund
1968 James E. Mulvaney
1969 Gordon Tollin
1970 Lee B. Jones
1971 Robert D. Feltham
1972 John H. Enemark
1973 George S. Wilson
1974 Victor J. Hruby
1975 Michael F. Burke
1976 Phillip C. Keller

Southern California

(Los Angeles)

1911-12 L. J. Stabler
1912 D. B. W. Alexander
1912 Henry L. Payne
1913 D. B. W. Alexander
1914 Richard S. Curtiss
1915 Edgar Baruch
1916 Erwin H. Miller
1917 Everette E. Chandler
1918 E. O. Slater
1919 W. L. Hardin
1920 Walter Mark
1921 Stuart J. Bates
1922-23 Walter A. Schmidt
1924-25 William C. Morgan
1926 Henry L. Payne
1927 J. E. Bell
1928 Russel W. Mumford
1929 Chas. J. Robinson
1930 Winfred R. Goddard
1931 Howard J. Lucas
1932 B. A. Stagner
1933 William N. Lacey
1934 Marion Dice
1935 G. Ross Robertson
1936 F. S. Pratt
1937 Raymond B. Stringfield
1938 Arnold O. Beckman
1938 W. A. Bush
1939 Roger W. Truesdail
1940 Gordon Alles
1941 William G. Young
1942 Willard E. Baier
1943 Robert E. Vivian
1944-45 Philip W. Drew
1946 George M. Cunningham
1947 L. Reed Brantley
1948 Harry V. Welch
1950 Charles S. Copeland
1951 William J. Hanson
1952 J. B. Ramsey
1953 Thomas F. Doumani
1954 Anton B. Burg
1955 Paul R. Pariseau
1956 Aries J. Haagen-Smit
1957 M. Elber Latham
1957-58 Max S. Dunn
1959 Loren L. Neff
1960 Don L. Armstrong
1961 Robert L. Pecsok
1962 James L. Bills
1963 Ulric B. Bray
1964 Arthur W. Adamson
1965 Eugene V. Kleber
1966 Kenneth W. Newman
1967 Robert D. Vold
1968 Alvin E. May
1969 Kendrick R. Eilar
1970 Eugene N. Garcia
1971 Morton Z. Fainman
1972 Kenneth L. Marsi
1972-73 Agnes A. Green
1974 Irving R. Tannenbaum
1975 Joseph Quaglino, Jr.
1976 Ralph L. Amey

Southern Illinois

(Carbondale)

1966 James N. BeMiller
1967 Donald H. Froemsdorf
1968 Albert L. Caskey
1969 Russell F. Trimble, Jr.
1970 Juh Wah Chen
1971 David F. Koster
1972 John K. Leasure
1973 Robert L. Smith
1974 Donald W. Slocum
1975 James Tyrrell
1976 Russell F. Trimble, Jr.

Southern Indiana

(Bloomington)

1948 Ernest E. Campaigne
1949 Frederic C. Schmidt
1950 Harry G. Day
1951 Christian E. Kaslow
1952 Lynn L. Merritt, Jr.
1953 John S. Peake
1954 Ralph L. Seifert
1955 Robert B. Fischer
1956 John H. Billman
1957 Ward B. Schaap
1958 William H. Nebergall
1959 Edward J. Bair
1960 Calvin M. Austin
1961 Vernon J. Shiner
1962 Walter L. Meyer
1963 John A. Thoma
1964 Russell A. Bonham
1965 Eugene H. Cordes
1966 Dennis G. Peters
1967 Rupert A. D. Wentworth
1968 Charles S. Parmenter
1969 Joseph J. Gajewski
1970 Kenneth G. Caulton
1971 Gary M. Heiftje
1972 Robert E. Roberts
1973-74 Patrick J. Dolan
1975 Lee J. Todd
1976 Glenn E. Callis

Southern New Mexico

(Alamagordo)

1956-57 William E. Hughen
1958 Alvin D. Boston
1959 W. B. Dancy
1960 Ward W. Repp
1961 Richard E. Witman
1962 Albert E. Richardson
1963 Jack E. Slay
1964 John J. Monagle, Jr.
1965 Edmund A. Schoeld
1966 Irwin D. Smith
1967 Walter H. Kaelin
1968 Floyd W. Neelwy
1969 Walter Lwowski
1970 Eugene R. McGough
1971 Dennis W. Darnall

(Name changed 1971 to Rio Grande Valley in merger with El Paso Section.)

Southwest Louisiana

(1946; Lake Charles)

1947-48 Arthur R. Rescorla
1949 Thomas W. Kirby
1950 Robert W. Rice
1951 E. Clarence Oden
1952 Claude A. Burns
1953 Lowell Betow
1954 Thomas A. McKenna
1955 William R. Jacoby
1956 Quentin J. Trahan
1957 John M. Brierly
1958 Charles L. McGehee
1959 Fred R. Henke
1960 Roland M. Bodin
1961 Millard N. Hudson
1962 John J. Hawkins
1963 Victor Monsour
1964 John S. Tonsley
1965 Jack Newcombe
1966 Raymond J. Carroll
1967 Bobby E. Hankins
1968 Bernherd M. Bekkerus
1969 Herbert J. Alleman
1970 Ronald D. Crain
1971 Colin M. McSwain
1972 Raymond E. Chavanne
1973 Henry J. Hoenes, Jr.
1974 Joseph E. Smith
1975 Robert W. Kline
1976 Catherine C. Christian

State College (Pa.)

1924 Grover C. Chandlee
1925 David F. McFarland

(Name changed 1925 to Central Pennsylvania.)

Susquehanna Valley

(1958; Danville, Pa.)

1959 Lester Kieft
1960 John T. Day
1961 Ben R. Willeford, Jr.
1962 Frank P. Romano
1963 George G. Hazen
1964 Manning A. Smith
1965 John A. Radspinner
1966 Frederick W. Kocher, Jr.
1967 Edward F. Hoover
1968 Jay Alfred Young
1969 James K. Hummer
1970 Hans Venning
1971 James J. Bohning
1972 Gynith Giffin
1973 Neil H. Potter
1974 Constantine C. Vlassis
1975 Charles A. Root
1976 Robert A. Sallavanti

Local Section Chairmen (continued)

Syracuse (N.Y.)

1907 **John A. Mathews**
1908 **William M. Booth**
1909 **Hermon C. Cooper**
1910 **Ernest N. Pattee**
1911 **H. Monmouth Smith**
1912 **Lawrence Carpenter Jones**
1913-15 **Edward S. Johnson**
1916-18 **George M. Berry**
1919 **Raphael S. Fleming**
1920 **Jacob Martin Johlin**
1921 **Louis E. Wise**
1922 **Alexis C. Houghton**
1923 **Ross A. Baker**
1924-25 **Arthur W. Kimman**
1926-27 **John H. Nair, Jr.**
1928 **Washington Platt**
1929 **Carl R. McCrosky**
1930 **Paul A. Keene**
1931 **R. Chester Roberts**
1932 **Clayton C. Spencer**
1933 **Samuel C. Robison**
1934-35 **Neal E. Artz**
1936-37 **Albert L. Elder**
1937 **Dwight C. Bardwell**
1938 **Aden J. King**
1939 **Charles K. Lawrence**
1940 **Gerard M. Edell**
1941 **Richard M. Cone**
1941 **Edwin C. Jahn**
1942 **Frank Porter**
1943 **Paul M. Ruoff**
1944 **William A. Fessler**
1945 **Joseph L. Neal, Jr.**
1946 **Donald J. Saunders**
1947 **Frank W. Staab, Jr.**
1948 **Otto Kay**
1949 **Frank A. Kanda**
1950 **Walter P. Kelly**
1951 **Oliver L. I. Brown**
1952-53 **Wm. B. Wheatly**
1954 **Ewing C. Scott**
1955 **George Oplinger**
1956 **Henry E. Wirth**
1957 **Howard A. Bewick**
1958 **Lee C. Cheney**
1959 **Gerald F. Grillot**
1960 **Robert J. Conan, Jr.**
1961 **William C. Risser**
1962 **Conrad Schuerch**
1963 **John B. Davidson**
1964 **David Aaron Johnson**
1965 **Milton Sach**
1966 **Edward M. Greeley**
1967 **Oliver B. Fardig**
1968 **William A. Baker, Jr.**
1969 **Clarence C. Schubert**
1970 **Daniel J. Macero**
1971 **Charles Masao Ise**
1972 **Donald S. Allen**
1973 **Donald N. McGregor**
1974 **Ray T. O'Donnell**
1975 **Robert J. Conan, Jr.**
1976 **Donald C. Dittmer**

Texas A & M

(College Station, Tex.)

1939 **Charles Winfield Burchard**
1940 **Arthur Russell Kemmerer**
1941 **Fred W. Jensen**
1942 **James D. Lindsay**
1943 **Neil E. Rigler**
1944 **Carl M. Lynman**
1945 **J. Frank Fudge**
1946 **Francis F. Bishop**
1947 **Wm. M. Potts**
1948 **Paul B. Pearson**
1949 **Bryant R. Holland**
1950 **Errol B. Middleton**
1951 **Philip G. Murdoch**
1952 **Luther R. Richardson**
1953 **Royce H. LeRoy**
1953-54 **Wilson W. Minke**
1955 **Robert V. Andrews**
1955-56 **William R. Stephens**
1957 **C. Kinney Hancock**

(Name changed 1958 to Texas A & M-Baylor; changed 1968 back to Texas A & M).

1969 **Edward A. Meyers**
1970 **Alan S. Rodgers**
1971 **Dwight C. Conway**
1972 **Ralph A. Zingaro**
1973 **Arthur Furman Isbell**
1974 **Carlos N. Pace**
1975 **Kenn E. Harding**
1976 **Patrick S. Mariano**

Texas A & M-Baylor

(College Station, Tex.)

(Chartered 1939 as Texas A & M Section; name changed 1958.)

1958 **Raymond Reiser**
1959 **Thomas Jackson Bond**
1960 **Charles D. Holland**
1961 **Virgil Tweedie**
1962 **Roger D. Whealy**
1963 **Charles E. Reeder**
1964 **John M. Prescott**
1965 **John S. Belew**
1966 **Norman C. Rose**
1967 **James L. McAtee**
1968 **Robert B. Alexander**

(Name changed 1968 to Texas A & M).

Texas-Louisiana Gulf

(Port Arthur, Tex.)

1943 **Carl E. Lauer**
1944 **Matthew Weber, Jr.**
1944-45 **Ralph N. Traxler**
1945 **Walter W. Scheumann**
1946 **Vladimir A. Kalichevsky**
1947 **Nelson B. Haskell**
1948 **Harvey N. Frost**
1949 **Carson C. Morrison**
1950 **Earl C. Daigle**
1951 **R. P. Daniels**
1952 **Homer D. Maples**
1953 **Valentine D. Luedeke**
1953 **Harry R. Robinson**
1954 **J. Madison Nelson**
1955 **Gordon H. Miller**
1956 **Robert E. Price**
1957 **Howard A. Young**
1958 **James E. Sampson**
1959 **James B. Hutto**
1960 **G. H. Carter**
1961 **Elmer E. Huffhines, Jr.**
1962 **Richard H. Laird, Jr.**
1963 **J. W. Romberg**
1964 **Harold T. Baker**
1965 **Douglas Ray McKinney**
1966 **Edwin C. Anderson**

(Name changed 1967 to Sabine-Neches.)

Toledo (Ohio)

1917 **Gustavus A. Kirchmaier**
1918-19 **Henry W. Hess**
1920 **Willis B. Holmes**
1921 **Emil A. Schragenheim**
1922-23 **Thomas E. Moore**
1924 **Geo. Anderson**
1925-26 **Walter Eugene Ruth**
1927 **Geo. H. Anderson**
1928-29 **Guy E. Van Sickle**
1930-31 **E. Carleton Mathis**
1932-33 **Harold G. Oddy**
1934-35 **Arthur Rigby**
1936-37 **Nelson W. Hovey**
1938 **Henry R. Kreider**
1939-40 **Julian H. Toulouse**
1941-42 **Harold F. Stose**
1943-45 **John J. Thornton**
1945 **Herman W. Dorn**
1946 **Albert A. Dietz**
1947 **Harold A. Hoppens**
1948 **Lyle E. Calkins**
1949 **Rolland G. Bowers**
1950 **Frank R. Bacon**
1951 **Ben C. Smith**
1952 **Lewis E. Thomas, Jr.**
1953 **Sol Boyk**
1954 **Edgar R. Klinck**
1955 **Thomas R. Santelli**
1956 **Edward Butler**
1956 **Lawrence Kumnick**
1957 **John C. Safranski, Jr.**
1957 **Edward F. Mohler, Jr.**
1958 **Cecil R. Fetters**
1959 **Arthur Herman Black**
1960 **John R. Jones, Jr.**
1961 **Sam W. Wohlfort**
1962 **Daniel E. Cross**
1963 **Albertine Krohn**
1964 **Carl Arnold Johnson**
1965 **Lancelot C. A. Thompson**
1966 **Ell Dee Compton**
1967 **Elmer L. Williams**
1968 **Philip A. Kint**
1969 **David R. Hostetler**
1970 **Robert J. Niedzielski**
1971 **Kenneth Parsons**
1972 **Ivan E. Den Besten**
1973 **Lawrence P. Biefield**
1974 **Charles P. Hablitzel**
1975 **Paul Block, Jr.**
1976 **Ernest L. Lippert, Jr.**

Trenton (N.J.)

(Chartered 1947)

1948 **N. Richard Yorke**
1948 **Carl W. Virgin**
1949 **Newell A. Perry**
1950 **James W. Kemmler**
1951 **Edward M. Fettes**
1952 **John F. Garber**
1953 **Ernest Weiss**
1954 **Franklin O. Davis**
1955-56 **Raymond R. Goldberg**
1957 **Eugene R. Bertozzi**
1957-58 **Dudley W. Thomas**
1959 **John S. Cook**
1960 **Gaylord A. Kanavel**
1961 **Robert Kunin**
1962 **Keith R. Cranker**
1963 **John P. Duffy**
1964 **Herbert Q. Smith**
1965 **Armen D. Yazujian**
1966 **Riad H. Bobran**
1967 **Frederick F. Assmann**
1968 **Edward H. House**
1969 **Willard S. Bundy**
1969-70 **George Bulbenko**
1971 **Frank B. Slezak**
1972 **Leonard R. Darbee**
1973 **Eileen P. Smith**
1974 **Nick S. Semenuk**
1975 **Kenneth D. Bair**
1976 **Rolf F. Foerster**

Local Section Chairmen (continued)

Tri-State

(Huntington, W.Va.)

1947 Frank H. Moser
1948 Frederick C. Wolfe

(Name changed 1949 to Central Ohio Valley.)

Tulsa (Okla.)

1947 Buell O'Connor
1948 Paul H. Cardwell
1949 Wayne E. White
1950 Joel L. Burkitt
1951 Richard N. Meinert
1952 W. Herschel Hopson
1953 Milton O. Denekas
1954 Chas. L. Lunsford
1955 Alvin C. Broyles
1956 J. Vernon Lawson
1957 Arthur L. Draper
1958 James H. Jones
1959 Louis E. Craig
1960 H. Roberts Froning
1961 Charles M. Maddin
1962 Leonard N. Devonshire
1963 Robert W. Provine
1964 Hugh T. Harrison
1965 John T. Loft
1966 Robert A. Van Nordstrand
1967 Myron L. Dunton
1968 Edward S. McKay
1969 James B. Beal, Jr.
1969-70 Richard A. Tomasi
1971 Samuel J. Martinez
1972 Robert B. Earl
1973 Robert B. Rosene
1974 Dayal T. Moshri
1975 Edward B. Butler
1976 Charles B. Lindahl

University of Arkansas

(Fayetteville)

(See *also* Arkansas)

1947 Harrison Hale
1948 Wladimar W. Grigorieff
1949 Wm. K. Noyce
1950 Edward S. Amis
1951 Lyman E. Porter
1952 Granville A. Billingsley
1952-53 Maurice E. Barker
1954 Raymond R. Edwards
1955 Samuel Siegel
1956 Aubrey E. Harvey, Jr.
1957 Jacob Sacks
1958 Arthur Fry
1959 Robert F. Kruh
1960 Paul K. Kuroda
1961 William K. Noyce
1962 Lester C. Howick
1963 James R. Couper
1964 George Donald Blyholder
1965 Samuel Siegel
1966 Arthur Wallace Cordes
1967 Richard N. Porter
1968 Dale A. Johnson
1969 Walter L. Meyer
1970 John A. Thoma
1971 Leslie B. Sims
1972 Lothar Schafer
1973 James F. Hinton
1974 Wm. L. Cairns
1975 Roderic P. Quirk
1976 Francis S. Milett

University of Illinois

(Urbana)

1906 Samuel W. Parr
1907-08 Edward Bartow
1909 Willis B. Holmes
1910-11 Edward Wight Washburn
1912 Philip B. Hawk
1913 Clarence William Balke
1914 Samuel W. Parr
1915 George McPhail Smith
1916 William A. Noyes
1917 Harry Sands Girdley
1918 David Ford McFarland
1919 George Denton Beal
1920 B Smith Hopkins
1921 Roger Adams
1922 Howard B. Lewis
1923 John H. Reedy
1924 Worth H. Rodebush
1925 William C. Rose
1926 Arthur M. Buswell
1927 Carl S. Marvel
1927 Leonard F. Yntema
1928 Thomas E. Phipps
1929 George L. Clark
1930 Donald B. Keyes
1931 John Henry Reedy
1932 Ralph L. Shriner
1933 W. Albert Noyes
1934 Norman W. Krase
1935 Arthur M. Buswell
1936 John C. Bailar, Jr.
1937 G. Frederick Smith
1938 Michael J. Copley
1939 Henry Fraser Johnstone
1940 Louis F. Audrieth
1941 B Smith Hopkins
1942 George L. Clark
1943 Charles C. Price
1944 Frederick T. Wall
1945 Herbert E. Carter
1946 Frank H. Reed
1947 Harold R. Snyder
1948 Therald Moeller
1949 Herbert A. Laitinen
1950 Edward W. Comings
1951 Glenn C. Finger
1952 Virginia Bartow
1953 Carl S. Vestling
1954 Nelson J. Leonard
1955 Herbert S. Gutowsky
1956 Harry G. Drickamer
1957 Reid T. Milner
1958 Peter E. Yankwich
1959 Howard V. Malmstadt
1960 David Y. Curtin
1961 Irwin C. Gunsalus
1962 Robert Forrest Nystrom
1963 Russell S. Drago
1964 Thomas Joseph Hanratty
1965 Theron S. Piper
1966 Theodore L. Brown
1967 Douglas Applequist
1968 Richard S. Juvet, Jr.
1969 Kenneth L. Rinehart, Jr.
1970 Lowell P. Hager
1971 Peter Beak
1972 Iain Paul
1973 Galen D. Stucky
1974-75 Robert M. Coates
1976 Elizabeth P. Rogers

University of Kansas

(Lawrence)

1951-52 Ernest Griswold
1953 John A. Davis
1954 Arthur W. Davidson
1954-55 Joseph H. Bruckhalter
1955-56 Jacob Kleinberg
1957 Russell C. Mills
1957 Frank S. Rowland
1958 William J. Argersinger, Jr.
1959 Philip Newmark
1960 Charles A. Reynolds
1961 Russell Bernard Mesler
1962 Albert W. Burgstahler
1963 Earl S. Huyser
1964 Frank S. Rowland
1965 Mathios P. Mertes
1966 Reynold T. Iwamoto
1967 Ralph N. Adams
1968 Richard L. Middaugh
1969 John A. Landgrebe
1970 Marlin D. Harmony
1971 Robert G. Carlson
1972 Robert Wiley
1973 Judy Harmony
1974 James D. McChesney
1975 Richard S. Givens
1976 Robert P. Hanzlik

University of Michigan

(1899; Ann Arbor)

1900 Albert B. Prescott
1901-02 Paul C. Freer
1903-10 Edward D. Campbell
1911 Moses Gomberg
1912 Edward D. Campbell
1913 Samuel L. Bigelow
1914 A. B. Stevens
1915 L. H. Cone
1916 David M. Lichty
1917 Hobart H. Willard
1918 Edward D. Campbell
1919 William G. Smeaton
1920 Hobart H. Willard
1921 Alfred H. White
1922 Floyd E. Bartell
1923 Chester S. Schoepfle
1924 Alfred L. Ferguson
1925 Clifford C. Meloche
1926 Walter L. Badger
1927 Howard B. Lewis
1928 Philip F. Weatherill
1929 Robert J. Carney
1930 G. Granger Brown
1931 Chester S. Schoepfle
1932 Hobart H. Willard
1933 Malcolm H. Soule
1934 F. F. Blicke
1935 Edwin M. Baker
1936 Roy K. McAlpine
1937 William G. Smeaton
1938 Leigh C. Anderson
1939 J. Hallett Hodges
1940 Werner E. Bachmann
1941 Byron A. Soule
1942 Lee O. Case
1943 Philip F. Weatherill
1944 Lawrence Brockway
1945 Floyd E. Bartell
1946 Joseph O. Halford
1947 Kasimir Fajans
1948 Raymond N. Keller
1949 F. F. Blicke
1950 Peter A. S. Smith
1951 Alfred L. Ferguson
1952 Clifford C. Meloche
1953 Lawrence Brockway
1954 Edgar F. Westrum, Jr.
1955 Christian S. Rondestvdt, Jr.
1956 Philip J. Elving
1957 Robert C. Taylor
1958 Halvor N. Christensen
1959 Robert W. Parry
1960 Lee O. Case
1961 Alfred A. Schilt
1962 Charles L. Rulfs
1963 Wyman R. Vaughan

Local Section Chairmen (continued)

U. of Michigan (cont.)

1964 Milton Tamres
1965 Daniel Thomas Longone
1966 Thomas M. Dunn
1967 Harry B. Mark, Jr.
1968 Paul G. Rasmussen
1969 Jerry H. Current
1970 John R. Wiseman
1971 Arthur J. Ashe III
1972 George W. Moersch
1973 Joseph P. Marino
1974 Donald E. Butler
1975 Ralph W. Rudolph
1976 Robert W. Fleming

University of Missouri

(Columbia)

1909 Perry Fox Trowbridge
1910 Sidney Calvert
1911 Charles Robert Moulton
1912 William G. Brown
1912-13 John A. Gibson
1914 Louis E. Wise
1915 Leonard Dixon Haigh
1916-17 John Wesley Marden
1918-19 Leroy S. Palmer
1920 Herman Schlundt
1921-22 Sidney Calvert
1923 Herman Schlundt
1924-25 Henry D. Hooker, Jr.
1926 Herbert E. French
1927 Richard Bardfield
1928-29 Albert G. Hogan
1930 Allen E. Stearn
1931-32 Sidney Calvert
1933-34 Gerald F. Breckenridge
1935 Addison Gulick
1936 Albert G. Hogan
1937 Gerald F. Breckenridge
1938 Sidney Calvert
1939 Henry E. Bent
1940 Neil E. Gordon
1940 Harry A. Curtis
1941 Leonard Dixon Haigh
1942-43 Luther Ray Richardson
1944 Lloyd B. Thomas
1945 Gerald F. Breckenridge
1946 Albert G. Hogan
1947 Herbert E. Ungnade
1947 Ralph H. Luebbers
1948 Wm. E. Gordon
1949 Allen E. Stearn
1950 C. Edmund Marshall
1951 Dennis T. Mayer
1952 Edward E. Pickett
1953 Norman Rabjohn
1954 Boyd L. O'Dell
1955 Wesley J. Dale
1956 Merle E. Muhrer
1957 James R. Lorah
1958 Charles W. Gehrke
1959 Thomas D. Luckey
1960 Gerhard H. Beyer
1961 George B. Garner
1962 Robert Kent Murmann
1963 Owen J. Koeppe
1964 Edward E. Pickett
1965 Benedict J. Campbell
1966 John Carl Guyon
1967 James A. Rose
1968 John E. Bauman, Jr.
1969 Robert McCreight
1970 David E. Troutner
1971 Robert R. Kuntz
1972 Truman S. Storvick
1973 Willis Byrd
1974 Edwin M. Kaiser
1975 Milton S. Feather
1976 Scott Searles, Jr.

Upper Ohio Valley

(1949; Marietta, Ohio)

1950 Geo. L. Graf, Jr.
1951 Don R. Clippinger
1952 Paul R. Murphy
1953 Ellis L. Krause
1954 Phillip D. Brossman
1955-56 W. Eldon Nees
1957 William B. Corydon
1958 Thomas H. Curry
1959 Edgar E. Hardy
1959-60 Herschel Gene Grose
1961 Robert J. Cote
1962 Paul R. Stockwell
1963 Wilfred B. Howsmon, Jr.
1964 Robert E. Moynihan
1965-66 Hans Georg Gilde
1967 William Duane Huntsman
1968 Lawrence P. Eblin
1969 Harold B. Staley
1970 James Y. Peh Tong
1971 Robert R. Winkler
1972 Laurance A. Knecht
1973 Edward Leon
1974 Donald L. Bailey
1975 Gene A. Westenbarger
1976 Charles L. Myers

Upper Peninsula

(Houghton, Mich.)

1946 Allan F. Olson
1947 Ralph E. Menzel
1948 Alfred A. Camilli
1949 Henry L. Coles
1950 James F. Bourland
1950 Lucian F. Hunt
1951 Royal F. Makens
1952 Walter H. Dieterich
1953 Bart Park
1954 Joseph C. Cameron
1955 Geo. L. Craig
1956 Alfred A. Reiter
1957 Roy E. Heath
1958 Arthur W. Goos
1959 Franklin B. Wittmer
1960 Herbert K. Hedrick
1961 Bart Park
1962 William F. Giuliani
1963 James D. Spain, Jr.
1964 Thomas S. Griffith, Jr.
1965 Teuvo Rudolph Maki
1966 Thomas S. Griffith, Jr.
1967 Laurence G. Stevens
1968 Gerald D. Jacobs
1969 Larry M. Julien
1970 Roger D. Barry
1971 G. Fred Reynolds
1972 Jerome A. Roth
1973 David W. Hubbard
1974 Donald L. Macalady
1975 Laurence B. Hein
1976 David W. Kingston

Vermont

(1916; Burlington)

1917 S. Francis Howard
1918-20 P. Conant Voter
1921-22 George H. Burrows
1923-25 Elbridge C. Jacobs
1926 C. A. Moat
1927 Ben Bennett Corson

(See Western Vermont)

Virginia

(Richmond)

1915 J. Bernard Robb
1916 R. B. Arnold
1917 A. Holmes Allen
1918 Garnett Ryland
1919 E. C. L. Miller
1920 W. Catesby Jones
1921 Graham Edgar
1922-23 Eean W. F. Rudd
1924 Hall Canter
1925 Henry K. McConnell
1926 Wm. Clift
1927 Sidney S. Negus
1928 Roy F. McCrackan
1929 Edwin Cox
1930 William G. Crockett
1931 William J. Nissley
1932 W. R. Cornthwaite
1933 Laureen B. Hitchcock
1934 C. B. Craxton Valentine
1935 Thomas A. Balthis
1936 John H. Yoe
1937 Ira A. Updike
1938 John C. Forbes
1939 Rodney C. Berry
1940 William Ralston
1941 Allen Berne-Allen, Jr.
1942 Hiram A. Hanmer
1943 James W. Cole, Jr.
1944 Robert H. Kean
1945 Wm. R. Harlan
1946 Foley F. Smith
1947 Clifford M. Smith
1948 Edward S. Harlow
1949 Robert E. Lutz
1950 Lynn D. F. Abbott, Jr.
1951 William G. Guy
1952 M. E. Kapp
1953-54 Clinton W. Baber
1955 W. E. Trout, Jr.
1956 Randolph N. Gladding
1957 Guido J. Coli, Jr.
1958 William P. Boyer
1959 Alfred R. Armstrong
1960 Loyal H. Davis
1961 Fred R. Millhiser
1962 Richard M. Irby, Jr.
1963 Warren E. Weaver
1964 Wm. A. Powell
1965 Harold A. Hoffman
1966 Russell J. Rowlett, Jr.
1967 Oscar R. Rodig
1968 James E. York, Jr.
1969 Joseph C. Holmes
1969-70 Preston H. Leake
1971 Lowell V. Heisey
1972 Robert G. Bass
1973 Willard W. Harrison
1974 Herndon Jenkins
1975 Robert T. Kemp
1976 Franklin D. Kizer

Virginia Blue Ridge

(Roanoke, Va.)

1931 Floyd Hamilton Fish
1932 Warren E. Knapp
1933 Lester M. Whitmore
1934 Harry Bucholz Riffenburg
1935 Frederick F. Morehead
1936 Harry I. Johnson
1937 Geo. V. Downing
1938 Robert E. Hussey
1939 Joseph Coxe, Jr.
1940 James F. Eheart
1941 C. Lucian Crockett
1942 John D. Schumacher
1943 Basil W. Waring
1944 Nan V. Thornton
1945 Rollin H. Wampler

Local Section Chairmen (continued)

Va. Blue Ridge (cont.)

1945 Lawrence H. Pownall
1946 Lester M. Whitmore
1947 Frank C. Vilbrandt
1948 Vernon C. Compton
1949 Leslie German
1950 Clarence G. Haupt
1951 Harriett H. Fillinger
1952 Rawie P. Moomaw
1953 John O. Rider
1954 Dorothy D. Thompson
1955 Seldon P. Todd
1956 Robert C. Krug
1957 George A. Lenaeus
1958 Charles W. Smart
1959 Helen L. Whidden
1960-61 Karl E. Balliet
1962 Harold H. Garretson
1963 Gene Wise
1964 Ray C. Jackson
1965 Fred Weymouth Davis
1966 Edw. F. Furtsch
1967 Hugh C. Crafton, Jr.
1968 John H. Wise
1969 Martin H. Gurley
1970 James P. Wightman
1971 Roberta Stewart
1972 Alan F. Clifford
1973 Edwin S. Flinn
1974 Barbara Blair
1975 Sandra Boatman
1976 Harold M. Bell

Wabash Valley

(Terre Haute, Ind.)

1944 Harold V. Fairbanks
1945 Carl William Freerichs
1946 Esther A. Engle
1947 William G. Kessel
1948 Murray Senkus
1949 Wanda L. Campbell
1950 Odon S. Knight
1951 Paul J. Baker, Jr.
1952 Wilbur Louis Keko
1953 J. Raymond McMahn
1954 Edward B. Hodge
1955 John T. Craig
1956 Robert James Harker
1957 John Bascom Tindall
1958 Frank A. Guthrie
1959 William F. Phillips
1960 Jerome L. Martin
1961 Oron M. Knudsen
1962 John A. Riddick
1963-64 Lawrence Rymon Jones
1965 Eugene L. Herbst
1966 John A. Means
1967 Curtis Allan Dhonau
1968 Joseph Roy Siefker
1969 William G. Kammler
1970 E. Cooper Smith
1971 Richard N. Hurd
1972 Theodore K. Sakano
1973 William Arthur Trinler
1974 Melvin L. Druelinger
1975 Jay A. Bardole
1976 Lynn A. Swanson

Washington (D.C.)

1893 F. P. Dewey
1894-95 William Henry Seaman
1895 Charles E. Munroe
1896 Emil Alexander de Schewinitz
1897 Willard D. Bigelow
1898-99 Henry N. Stokes
1900 Henry Carrington Bolton
1901 Victor K. Chesnut
1902 William F. Hillebrand
1903 Frank K. Cameron
1904-05 Eugene T. Allen
1906 Lucius M. Tolman
1907 Peter Fireman
1908 Joseph S. Chamberlain
1909 Percy H. Walker
1910 George H. Failyer
1911-12 William W. Skinner
1912-13 Campbell E. Waters
1914 Michael X. Sullivan
1915 Carl L. Alsberg
1916 Robert B. Sosman
1917 Claude S. Hudson
1918 Francis B. Power
1919 Atherton Seidell
1920 Carl O. Johns
1921 William M. Blum
1922 Roger Clark Wells
1923 William M. Clark
1924-25 L. Heberling Adams
1926 Frederick W. Smither
1927 Edgar T. Wherry
1928 George W. Morey
1929 Raleigh Gilchrist
1930 Horace T. Herrick
1931 Ralph E. Gibson
1932 Edward Wichers
1933 Paul E. Howe
1934 David B. Jones
1935 James F. Couch
1936 James H. Hibben
1936-37 Ben H. Nicolet
1938 Nathan L. Drake
1939 Frank C. Kracek
1940 Raymond M. Hann
1941 Herbert L. J. Haller
1942 Norman Bekkedahl
1943 Sterling B. Hendricks
1944 Edgar R. Smith
1945 Horace S. Isbell
1946 Leo A. Shinn
1947 Benjamin D. Van Evers
1948 Irl C. Schoonover
1949 Chas. E. White
1950 Frederick D. Rossini
1951 James I. Hoffman
1952 James J. Fahey
1953 John K. Taylor
1954 George W. Irving, Jr.
1955 Howard W. Bond
1956 Charles R. Naeser
1957 Bourdon F. Scribner
1958 Joseph R. Spies
1959 William W. Walton
1960 Allen L. Alexander
1961 William J. Bailey
1962 John L. Torgensen
1963 Alfred E. Brown
1964 Leo Schubert
1965 William A. Zisman
1966 Alphonso F. Forziati
1967 Gerhard M. Brauer
1968 Robert B. Fox
1969 Edward O. Haenni
1970 Mary H. Aldridge
1971 Joseph C. Dacons
1972 Fred E. Saalfeld
1973 Harvey Alter
1974 Alfred Weissler
1975 Robert F. Cozzens
1976 David L. Venezky

Washington-Idaho Border

(Pullman, Wash.)

(*See also* Northern Intermountain)

1928 J. L. St. John
1929 J. L. Culbertson
1930-31 Ralph W. Gelbach
1932 Wm. H. Cone
1933 L. I. Gilbertson
1933 Edwin C. Jahn
1934-35 Harry L. Cole
1936 Harold P. Klug
1937 R. P. Cope
1938 Harry S. Owens
1939 Stewart E. Hazlet
1940 James B. Reed
1941 Ralph W. Gelbach
1941 E. V. White
1942 Matthew K. Veldhuis
1943-44 O. E. Stamberg
1945 Carl M. Brewster
1945 Don H. Anderson
1946-47 Harry L. Cole
1948 Alvin C. Wiese
1949 Geo. T. Austin
1950 C. C. Cowin
1951 Ralph W. Gelbach
1952 Castle O. Reiser
1953 Arnold Warren Dodgen
1954 Darwin L. Mayfield
1955 Grant G. Smith
1956 Edgar H. Grahn
1957 Carl M. Stevens
1958 Dwight S. Hoffman
1959 H. Bayard Milne
1960 Morland W. Grieb
1961 Carl J. Nyman, Jr.
1962 James H. Cooley
1963 Robert J. Foster
1964 Peter K. Freeman
1965 Bruce A. McFadden
1966 Jeanne Marie Shreeve
1967-68 Richard Allen Porter
1968-70 Kirk D. McMichael
1971 Charles O. Hower
1972 Ralph G. Yount
1973 Karl H. Pool
1974 Jeanne M. Shreeve
1975 Dennis Brown
1976 James A. Magnuson

Western Carolinas

(1946; Greenville, S.C.)

1947 Jasper W. Ivey
1948 H. P. Vannah
1949 Frank B. Schirmer
1950 Gilbert I. Thurmond
1951 Raymond A. Patterson
1952 Donald J. Godehn
1953 Albert L. Myers
1954 Charles H. Lindsley
1955 Howard L. Hunter
1956 Charles L. Henry
1957 James C. Loftin
1958 Arnold L. McPeters
1959 Joseph Gray Dinwiddie
1960 John P. Dosier
1961 John A. Southern
1962 Roger T. Guthrie
1963 William P. Cavin
1964 Thomas F. Corbin, Jr.
1965 Donald G. Kubler
1966 John Patrick Price
1967 A. Louise Agnew
1968 Joseph Y. Bassett, Jr.
1969 Michael V. Lock
1970 David C. Kirk, Jr.
1971 James C. Fanning, Jr.
1972 Royce S. Woosley
1973 Lawrence Moore
1974-75 Lloyd D. Remington
1976 Mary Jo Pribble

Local Section Chairmen (continued)

Western Connecticut

(Stamford)

1936-37 Charles R. Downs
1937 G. M. Johnstone Mackay
1938 Elliott J. Roberts
1939 Frederick H. Getman
1940 Harry L. Fisher
1941 Robert B. Barnes
1942 William H. Cope
1943 Robert J. King
1944 Wm. C. Moore
1945 Robert C. Swain
1946 Edward C. Sterling
1947 Richard O. Roblin, Jr.
1948 Henry J. Wing
1949 Jack T. Thurston
1950 Frederick R. Balcar
1951 Elmore H. Northey
1952 G. Richard Burns
1953 Robert M. Leekley
1954 Jack T. Cassaday
1955 Jackson P. English
1956 Donovan J. Salley
1957 Wm. A. Lutz
1958-59 Erwin L. Carpenter
1960 Harold K. Steele
1961 John H. Fletcher
1962 Gordon K. Wheeler
1963 Alfred L. Peiker
1964 Lloyd H. Shaffer
1965 John H Daniel
1966 John A. Barone
1967 John F. Flagg
1968 Alan P. Bentz
1969 Walter M. Thomas
1970 Robert A. Wall
1971 G. Sidney Sprague
1972 Eugene M. Scoran
1973 George A. Castellion
1974 Richard W. Zuehlke
1975 Lawrence Gallacher
1976 Harvey Steadly

Western Maryland

(Cumberland)

1940 Richard R. Sitzler
1941 Donald A. Lacoss
1942 John L. Baggett
1943 Robert W. Work
1944 Leonard J. Murphy
1945 Edgar D. Bolinger
1945 Richard R. Sitzler
1946 Julian G. Patrick
1947 John L. Baggett
1948 Herbert C. Heineman
1949 Geo. C. Ward
1950 Geo. J. Harriss
1951 Clarence Hieserman
1952 Walter R. Edwards
1953 Ralph F. Preckel
1954 Rudolp Steinberger
1955 Laurence I. Horner
1956 Chas. A. Orlick
1957 Howard W. Irwin
1958 R. Welford Phillips, Jr.
1959 Linus Max Heming
1960 William E. Kight
1961 Gordon Sutherland
1962 Ivan A. Hall
1963 James Stewart Elmslie
1964 Edward H. deButts, Jr.
1965 J. Donald Gibson
1966 D. Robert Leverling
1967 Paul P. Hunt
1968 Rocco C. Musso
1969 William A. Harding
1970 Richard J. Legare
1971 Theodore Comfort
1972 Ralph F. Preckel
1973 Anthony Pawlowski
1974 Harry Gilbert
1975 Kenneth O. Hartman
1976 Burton H. Davis

Western Michigan

(1955; Grand Rapids)

1956 Enno Wolthuis
1957 Winthrop F. Roser
1958 Bernard H. Velzen
1959 René A. Willis, Jr.
1960 Raymond L. Boxter
1961 J. Harvey Kleinheksel
1962 William G. Jackson
1963 Dennis F. Roelofs
1964 Irwin J. Brink
1965 Robert E. Jones
1966 Lem C. Curlin
1967 Eugene C. Jekel
1968 Harold A. Woltman
1969 Carl R. Meloy
1970 Larry Wilson
1971 Robert A. King, Jr.
1972 William Schroeder
1973 Frank H. Moser
1974 Darrel E. Cordy
1975 Gary D. Richmond
1976 John R. Parker

Western New York

(1905; Buffalo)

1906 John A. Miller
1907 George H. A. Clowes
1908 Francis A. J. Fitzgerald
1909 George H. A. Clowes
1910-11 W. H. Watkins
1912 Frank A. Lidbury
1913-14 A. S. Halland
1915 Lewis E. Saunders
1916-17 Jesse G. Melendy
1918 C. G. Derick
1919 Frank S. Low
1920 C. H. Childs
1921 Ralph C. Snowden
1922 Michael J. Ahern
1923 G. P. Fuller
1924 John A. Handy
1925 Frederick L. Koethen
1926 Raymond W. Hess
1927 Van L. Bohnson
1928 Ernest B. Benger
1929 Robert B. MacMullen
1930 Lester F. Hoyt
1931 Paul S. Braillier
1932 John F. Williams
1933 Andrew J. Gailey
1933 G. A. Vincent-Daviss
1934 Groves H. Cartledge
1935 Manley L. Ross
1936 John S. Fonda
1936 Albert E. Jennings
1937 Walton B. Scott
1938-39 Arthur G. Scroggie
1939 Albert E. Jennings, Jr.
1940 Howard W. Post
1941 Wilmer H. Koch
1942 Nelson Allen
1943 J. Frederick Walker, Jr.
1944 F. E. Huggins, Jr.
1945 Robert M. Fowler
1946 Wesley Minnis
1947 Maurice C. Taylor
1948 Frederick A. Gilbert
1949 Benjamin F. Clark
1950 Carl H. Rasch
1951 Geo. M. Bramann
1952 Henry M. Woodburn
1953 James S. Sconce
1954 Gerhard A. Cook
1947 Luther L. Lyon
1955 Chas. C. Clark
1956 Edward S. Shanley
1957 E. Rexford Billings
1958 Marshall W. Mead
1959 Thomas E. Londergan
1959 Richard E. McArthur
1959-60 Fred R. Whaley
1961 Roland J. Gladieux
1962 Alvin F. Shepard
1963 Robert L. Terrill
1964 Gordon M. Harris
1965 Clarence A. Weltman
1966 James L. Hecht
1967 Herman Stone
1968 Howard Tieckelmann
1969 Theodore Henry Dexter
1970 Sydney M. Spatz
1971 Carl H. Nuemberger, Jr.
1972 Edward A. Heintz
1973 Frank J. Dinan
1974 Jean Northcott
1975 Phillip D. Heffley
1976 James Economy

Western Vermont

(Burlington)

(See *also* Vermont)

1939-40 Charles E. Braun
1941 Harold C. Hamilton
1942 Francis S. Quinlan
1943 Harold B. Pierce
1944-45 Perley C. Voter
1946 Carl Lucarini
1947 Perley D. Baker
1948 Florence B. King
1949 Grant H. Harnest
1950 William V. B. Robertson
1951 Francis S. Quinlan
1952 Donald C. Gregg
1953 Edward A. Sheldon
1954 Herbert A. Bahrenburg, Jr.
1955 Walter A. Moyer, Jr.
1956 Henry P. Lemaire
1957 George C. Crooks
1958 Richard M. McNeer
1959 E. Kirk Roberts
1960 James K. Michales
1961 Wendell J. Whitcher
1962 Edwin L. Pool
1963 John O'Neill
1964 Robert C. Woodworth
1965 Robert W. Gleason
1966 William N. White
1967 John L. Gardner
1968 Michael H Gianni
1969 Elmar V. Piel
1970 Ralph D. Nelson, Jr.
1971 A. Paul Krapcho
1972 Henry Frankel
1973-74 Gilbert L. Grady
1975 John J. McCormack, Jr.
1976 Claus A. Wulff

Wichita (Kans.)

1929 John M. Michener
1930-31 Harold L. Bedell
1932 Lawrence Oncley
1933 Lloyd McKinley
1934 John M. Michener
1935-36 Roderic Grubb
1937 Robert Puckett
1937-38 J. Willard Hershey
1939-41 Worth A. Fletcher
1941 William E. Perdew
1942 Leonard C. Kreider
1943-44 George W. Berry
1945 Leonard C. Kreider
1946 Harold L. Bedell

Local Section Chairmen (continued)

Wichita (Kans.) (cont.)

1948-49 Wendell P. Lake
1950 Penrose S. Albright
1951 V. Wayne Hatchett
1952 Worth A. Fletcher
1953 James E. Myers, Jr.
1954 W. Mack Barlow
1955 Morris L. Hughes
1956 Eldon Means
1957 Robert V. Christian, Jr.
1958 Everett E. Brown
1959 Clarence G. Stuckwisch
1960 William H. Pierpont, Jr.
1961 Glenn R. Crocker
1962 John H. Rains
1963 Jack G. Steele
1964 Dale L. Noel
1965 Hugh McCranie
1966 Donald Rae Allen
1967 John G. McCarten
1968 Earl D. Flickinger
1969 Charles M. Buess
1970 John W. Wohnson
1971 Melvin E. Zandler
1972 Worth A. Fletcher
1973 Patricia H. Moyer
1973-74 Robert V. Christian, Jr.
1975 Anneke S. Allen
1976 Robert P. Hirschmann

Wichita Falls-Duncan

(1959; Wichita Falls, Tex.)

1960 Edgar B. Bloom
1961-62 Norval S. Burt
1963 Ira D. Leffler
1964 Robert E. Price
1965 Keith A. Catto, Jr.
1966 H. Charles McLaughlin
1967 Joseph E. Rose
1968 Marvin D. Misak
1969 Wayne F. Hower
1970 Ronald M. Coon
1971 Jessie W. Rogers
1972 Jerre T. Anderson
1973 Marlin D. Holtmyer
1974 Wm. G. McPherson
1975 Theodore E. Snider
1976 Eldon H. Sund

Wilson Dam (Ala.)

1937 Earl H. Brown
1937 Kelly L. Elmore
1938 Grady Tarbutton
1939 David B. Ardern
1940 L. Howard Hull
1941 Grover Leon Bridger
1942 John W. Lefforge
1943 Maurice M. Felger
1944 Theodore C. Hoppe
1945 Ralph S. Sherwin, Jr.
1946 John A. Brabson
1947 Ernest O. Huffman
1948 Herbert L. Thompson
1949 Edward P. Egan, Jr.
1950 Damon V. Royce, Jr.
1951 Edward C. Houston
1952 Archie V. Slack
1953 Walter E. Brown
1954 Franklin A. Lenfesty
1955 Julius D. Fleming
1956 J. Robert Lehr
1957 Hilland Y. Allgood
1958 James C. Barber
1959 Zachary T. Wakefield
1960 J. Alston Branscomb
1961 Weldon E. Cate
1962 Cecil Rose
1963 Nolan E. Richards
1964 Wendell D. Wilhide
1965 Oscar Wendell Edwards
1966 Raymond E. Isbell
1967 J. Clinton Brosheer
1968 Frank J. Johnson
1969 Joseph C. Thomas
1971 George A. Wieczorek
1971 William B. Stengle
1972 Richard C. Sheridan
1973 Vaughn Bollough
1974-75 Darrell A. Russel
1976 John F. Phillips

Wisconsin

(1907; Madison)

1908 Louis Kahlenberg
1909 Victor Lenher
1910 Edwin Bret Hart
1911 Richard Fischer
1912 Edward Kremers
1912 James H. Walton
1913 Francis C. Krauskopf
1914 Harold C. Bradley
1915 Joseph H. Mathews
1916 Elmer V. McCollum
1917 Paul W. Carleton
1918 Otto L. Kowalke
1919 Alfred E. Koenig
1920 James H. Walton
1921 Louis Kahlenberg
1922 Homer B. Adkins
1923 Joseph H. Mathews
1924 Francis C. Kravskopf
1925 Arlie W. Schorger
1926 Farrington Daniels
1927 Henry A. Schuette
1928 William O. Richtman
1929 William E. Tottingham
1930 Juliet H. Walton
1931 Samuel Marion McElvain
1932 William H. Peterson
1933 Karl Paul Link
1934 Norris Hall
1935 Conrad A. Elvehjem
1936 Villiers W. Meloche
1937 Alfred J. Stamm
1937-38 Olaf A. Hougen
1939 M. Leslie Holt
1940 C. Harvey Sorum
1941 John W. Williams
1942 Michael W. Klein
1943 Carl Baumann
1944 Frank M. Strong
1945 Roland A. Ragatz
1946 Alfred L. Wilds
1947 Marvin J. Johnson
1948 William S. Johnson
1949 Van R. Potter
1950 Joseph O. Hirschfelder
1951 Philip P. Cohen
1952 John E. Willard
1953 Robert H. Burris
1954 John D. Ferry
1955 David E. Green
1956 Paul Bender
1957 Harold F. Deutsch
1958 Robert A. Alberty
1959 Tekeru Higuchi
1960 Edwin M. Larsen
1961 Edward L. King
1962 John L. Margrave
1963 Robert M. Bock
1964-65 Laurens Anderson
1966 Walter J. Blaedel
1967 R. Byron Bird
1968 Robert C. West, Jr.
1969 John W. Rowe
1970 Howard W. Whitlock, Jr.
1971 Worth E. Vaughan
1972 Lawrence F. Dahl
1973 Aaron Ihde
1974 Wm. H. Orme-Johnson
1975 David Perlman
1976 Stanley H. Langer

Wooster (Ohio)

1941 Roy I. Grady
1941 Milton P. Puterbaugh
1942 Chas. H. Hunt
1943 Walter P. Arnold
1944 John W. Whittum
1945 Chas. F. Monroe
1946 Ralph H. Bescher
1947 Frederick D. Johnson
1948 J. Boyd Cook
1949 William F. Kieffer
1950 Harry E. Weidenhamer
1951 Glen J. Hartman
1952 Forest W. Dean, Jr.
1953 William T. Yzmazki
1954 John D. Reinheimer
1955 Elton Whitted
1956 Clyde H. Jones
1957 Alvin L. Moxon
1958 Andrew J. Yoder
1958 Orville G. Bentley
1959 Clyde H. Jones
1960 George Reesman
1961 Robert C. Butler
1962-63 Kenneth K. Wyckoff
1964 Burk A. Dehority
1965 Theodore R. Williams
1966 C. E. Trewiler
1967 Leory Wilbur Haynes
1968 Elliott M. Craine
1969 Harold G. Olin
1970 Ronald E. Bambury
1970-71 Richard D. Kreins
1972 Charles L. Borders, Jr.
1973 Richard H. Bromund
1974 Alvin L. Moxon
1975 John C. Weygant
1976 Thomas L. Bowman

Wyoming

(Laramie)

1946 Andrew Van Hook
1946-47 Ernest R. Schierz
1948 John S. Ball
1949 Henry G. Fisk
1950 Carl S. Gilbert
1951 David W. O'Day
1952 Gerald U. Dineen
1953 Fred C. Freytag
1954 Kenneth E. Stanfield
1955 Walter E. Duncan
1956 Clyde M. Frost
1956 Sara J. Rhoads
1957 J. Thomas Field
1958 William E. Haines
1959 Sara J. Rhoads
1960 Peter R. Tisot
1961 John E. Maurer
1962 Glenn L. Cook
1963 Harold F. Eppson
1964 Howard B. Jensen
1965 Rebecca Raulins
1966 Andrew W. Decora
1967 Vernon C. Bulgrin
1968 John W. Hamilton
1969 John Howatson
1970 R. Vernon Helm
1971-72 Owen Asplund
1973 Howard F. Silver
1973-75 J. Claine Petersen
1976 Francis P. Miknis
1975 Herbert O. House
1976 Franz Sondheimer

Local Section Officers Group

1922-23 W. Lee Lewis
1924-28 Harry N. Holmes
1928-30 Gerald L. Wendt
1930-32 Horace T. Herrick
1932-33 Hobart H. Willard
1933-35 Auburn A. Ross
1935-37 Otto M. Smith
1937-39 Ed. F. Degering
1939-40 Frank E. Brown
1940-41 Jacque C. Morrell
1941-42 Francis O. Rice
1942-43 William P. Yant
1943-44 Harry L. Fisher
1944-46 Arnon O. Snoddy
1946-47 Thomas F. Murray, Jr.
1947-48 Raymond E. Kirk
1948-49 George F. Rugar
1949-50 John B. Entriken
1950-51 Fisher Gaffin
1951-52 William A. Stanton
1952-53 Leland A. Underkofler
1953-54 Lester L. Winter
1954-55 A. Gordon Bowers
1955-56 Hugh F. Beeghly
1956-57 Desiree S. LeBeau
1957-58 Edward J. Durham
1958-59 Sherman S. Shaffer
1959-60 Carl F. Graham
1960-61 Herman S. Bloch
1961-62 Nathan Weiner
1962-63 Ellington M. Beavers
1963-64 Fred Klepetar
1964-65 Albert W. Meyer
1965-66 Howard W. Bond
1966-67 Kenneth Morgareidge
1967-68 Carleton W. Roberts
1968-70 Arthur H. Hale
1970-71 Preston H. Williams
1972-73 Maurice Clark
1973-74 Arnet L. Powell
1974-75 Marvin W. Hanson
1975 Roland Roskos
1976 Oscar R. Rodig

ACS Award Winners

Roger Adams Award in Organic Chemistry

Sponsored by Organic Syntheses, Inc. and Organic Reactions, Inc., and the Division of Organic Chemistry of the American Chemical Society.

1959 D. H. R. Barton
1961 Robert B. Woodward
1963 Paul D. Bartlett
1965 Arthur C. Cope
1967 John D. Roberts
1969 Vladimir Prelog
1971 Herbert C. Brown
1973 George Wittig
1975 Rolf Huisgen

ACS Award for Creative Invention

Sponsorship assumed by ACS Committee on Corporation Associates in 1975.

1968 William G. Pfann
1969 J. Paul Hogan
1970 Gordon K. Teal
1971 S. Donald Stookey
1972 H. Tracy Hall
1973 Carl Djerassi
1974 Charles C. Price
1975 James D. Idol, Jr.
1976 Manuel M. Baizer

ACS Award for Creative Work in Synthetic Organic Chemistry

Sponsored by Synthetic Organic Chemical Manufacturers Association

1957 Robert B. Woodward
1958 William S. Johnson
1959 John C. Sheehan
1960 Herbert C. Brown
1961 Melvin S. Newman
1962 Charles R. Hauser
1963 Nelson J. Leonard
1964 Lewis H. Sarett
1965 Donald J. Cram
1966 William von E. Doering
1967 Gilbert J. Stork
1968 Theodore L. Cairns
1969 H. Gobind Khorana
1970 Eugene E. van Tamelen
1971 Elias J. Corey
1972 Bruce Merrifield
1973 George Buchi
1974 Edward C. Taylor

ACS Award for Distinguished Service in the Advancement of Inorganic Chemistry

Sponsored by Mallinckrodt, Inc.

1965 Robert W. Parry
1966 George H. Cady
1967 Henry Taube
1968 William N. Lipscomb, Jr.
1969 Anton B. Burg
1970 Ralph G. Pearson
1971 Joseph Chatt
1972 John C. Bailar, Jr.
1973 Ronald J. Gillespie
1974 F. Albert Cotton
1975 Fred Basolo
1976 Daryle H. Busch

ACS Award for Nuclear Applications in Chemistry

Sponsored by G. D. Searle & Co.

Established in 1953 by Nuclear-Chicago Corp., a subsidiary of G. D. Searle & Co.

1955 Henry Taube
1956 Willard F. Libby
1957 Melvin Calvin
1958 Jacob Bigeleisen
1959 John E. Willard
1960 Charles D. Coryell
1961 Joseph L. Katz
1962 Truman P. Kohman
1963 Martin D. Kamen
1964 Isadore Perlman
1965 Stanley G. Thompson
1966 Arthur C. Wahl
1967 Gerhart Friedlander
1968 Richard L. Wolfgang
1969 George E. Boyd
1970 Paul R. Fields
1971 Alfred P. Wolf
1972 Anthony Turkevich
1973 Albert Ghiorso
1974 Lawrence E. Glendenin
1975 John R. Huizenga
1976 John O. Rasmussen

ACS Award for Pollution Control

sponsored by Monsanto Co.

1973 A. J. Haagen-Smit
1974 Harold S. Johnston
1975 Aubrey P. Altshuller
1976 Thurston E. Larson
1953 Nathan O. Kaplan

ACS Award in Analytical Chemistry

Sponsored by Fisher Scientific Co.

1948 N. Howell Furman
1949 G. E. F. Lundell
1950 Isaac M. Kolthoff
1951 Hobart H. Willard
1952 Melvin G. Mellon
1953 Donald D. Van Slyke
1954 G. Frederick Smith
1955 Ernest H. Swift
1956 Harvey Diehl
1957 John H. Yoe
1958 James J. Lingans
1959 James I. Hoffman
1960 Philip J. Elving
1961 Herbert A. Laitinen
1962 H. A. Liebhafsky
1963 David N. Hume
1964 John Mitchell, Jr.
1965 Charles N. Reilley
1966 Lyman C. Craig
1967 Lawrence T. Hallett
1968 Lockhart B. Rogers
1969 Roger G. Bates
1970 Charles V. Banks
1971 George H. Morrison
1972 W. Wayne Meinke
1973 James D. Winefordner
1974 Philip W. West
1975 Sidney Siggia
1976 Howard V. Malmstadt

ACS Award in Biological Chemistry

Sponsored by Eli Lilly and Co.

Administration transferred to the Division of Biological Chemistry in 1975.

1935 William Myron Allen
1937 Harold S. Olcott
1938 Abraham White
1939 George Wald
1940 Eric G. Ball
1941 David Rittenberg
1942 Earl A. Evans, Jr.
1943 Herbert E. Carter
1944 Joseph Stewart Fruton
1945 Max A. Lauffer
1946 John D. Ferry
1947 Sidney P. Colowick
1948 Dilworth Wayne Wooley
1949 Irving M. Klotz
1950 William Shive
1951 John M. Buchanan
1952 David B. Bonner

ACS Awards (continued)

Biological Chem. (cont.)

1954 **Harvey A. Itano**
1955 **William F. Neuman**
1956 **Robert A. Alberty**
1957 **Harold A. Scheraga**
1958 **Lester J. Reed**
1959 **Paul Berg**
1960 **James D. Watson**
1961 **Frederick L. Crane**
1962 **Jerald Hurwitz**
1963 **William P. Jencks**
1964 **Bruce N. Ames**
1965 **Gerald M. Edelman**
1966 **Phillips W. Robbins**
1967 **Gordon G. Hammes**
1968 **Charles C. Richardson**
1969 **Mario R. Capecchi**
1970 **Lubert Stryer**
1971 **David E. Wilson**
1972 **Bruce M. Alberts**
1973 **C. Fred Fox**
1974 **James E. Dahlberg**

ACS Award in Chemical Education

Sponsored by Scientific Apparatus Makers Association

1952 **Joel H. Hildebrand**
1953 **Howard J. Lucas**
1954 **Raymond E. Kirk**
1955 **Gerrit Van Zyl**
1956 **Otto M. Smith**
1957 **Norris W. Rakestraw**
1958 **Frank E. Brown**
1959 **Harry F. Lewis**
1960 **Arthur F. Scott**
1961 **John C. Bailar, Jr.**
1962 **William G. Young**
1963 **Edward L. Haenisch**
1964 **Alfred B. Garrett**
1965 **Theodore A. Ashford**
1966 **A. Conway Pierce**
1967 **Louis F. Fieser**
1968 **William F. Kieffer**
1969 **L. Carroll King**
1970 **Hubert N. Alyea**
1971 **Laurence E. Strong**
1972 **J. Arthur Campbell**
1973 **Robert C. Brasted**
1974 **George S. Hammond**
1975 **William T. Lippincott**
1976 **Leallyn B. Clapp**

ACS Award in Chemical Instrumentation

Sponsored by Beckman Instruments, Inc., until 1959 when sponsorship was assumed by Sargent-Welch Scientific Co.

1955 **R. Bowling Barnes**
1956 **Harold W. Washburn**
1957 **Ralph H. Müller**
1958 **Maurice F. Hasler**
1959 **Howard Cary**
1960 (No award)
1961 **Marcel J. E. Golay**
1962 **Howard K. Schachman**
1963 **Howard V. Malmstadt**
1964 **Robert Homer Cherry**
1965 **James N. Shoolery**
1966 **Leonard T. Skeggs**
1967 **Robert L. Bowman**
1968 **J. Raynor Churchill**
1969 **Dale J. Fisher**
1970 **Norman D. Coggeshall**
1971 **Fred W. McLafferty**
1972 **Edward B. Baker**
1973 **Jack W. Frazer**
1974 **Christie G. Enke**
1975 **Myron T. Kelley**

ACS Award in Chromatography

Sponsored as an award in chromatography and electrophoresis by Lab-Line Instruments, Inc., until 1970 when SUPELCO, Inc. assumed sponsorship.

1961 **Harold H. Strain**
1962 **L. Zechmeister**
1963 **Waldo E. Cohn**
1964 **Stanford Moore** and **William H. Stein**
1965 **Stephen Dal Nogare**
1966 **Kurt A. Kraus**
1967 **J. Calvin Giddings**
1968 **Lewis G. Longsworth**
1969 **Morton Beroza**
1970 **Julian F. Johnson**
1971 (No award)
1972 **J. J. Kirkland**
1973 **Albert Zlatkis**
1974 **Lockhart B. Rogers**
1975 **Egon Stahl**
1976 **James S. Fritz**

ACS Award in Colloid or Surface Chemistry

Sponsored by The Kendall Co.

1954 **Harry N. Holmes**
1955 **John W. Williams**
1956 **Victor K. La Mer**
1957 **Peter J. W. Debye**
1958 **Paul H. Emmett**
1959 **Floyd E. Bartell**
1960 **John D. Ferry**
1961 **Stephen Brunauer**
1962 **George Scatchard**
1963 **William Albert Zisman**
1964 **Karol J. Mysels**
1965 **George D. Halsey, Jr.**
1966 **Robert S. Hansen**
1967 **Stanley G. Mason**
1968 **Albert C. Zettlemoyer**
1969 **Terrell L. Hill**
1970 **Jerome Vinograd**
1972 **Egon Matijevic**
1971 **Milton Kerker**
1973 **Robert L. Burwell, Jr.**
1974 **W. Keith Hall**
1975 **Robert Gomer**
1976 **Robert J. Good**

ACS Award in Enzyme Chemistry

Established in 1945 by Paul-Lewis Laboratories, Inc., a part of Pfizer Inc.; administered since 1975 by the Division of Biological Chemistry.

1946 **David E. Green**
1947 **Van R. Potter**
1948 **Albert L. Lehninger**
1949 **Henry A. Lardy**
1950 **Britton Chance**
1951 **Arthur Kornberg**
1952 **Bernard L. Horecker**
1953 **Earl R. Stadtman**
1954 **Alton Meister**
1955 **Paul D. Boyer**
1956 **Merton F. Utter**
1957 **G. Robert Greenberg**
1958 **Eugene P. Kennedy**
1959 **Minor J. Coon**
1960 **Arthur B. Pardee**
1961 **Frank J. Huennekens**
1962 **Jack L. Strominger**
1963 **Charles Gilvarg**
1964 **Marshall Nirenberg**
1965 **Frederic M. Richards**
1966 **Samuel B. Weiss**
1967 **P. Roy Vagelos** and **Salih J. Wakil**
1968 **William J. Rutter**
1969 **Robert T. Schimke**
1970 **Herbert Weissbach**
1971 **Jack Preiss**
1972 **Ekkehard K. F. Bautz**
1973 **Howard M. Temin**
1974 **Michael J. Chamberlin**

ACS Award in Fertilizer and Soil Chemistry

Sponsored by National Plant Food Institute. (Discontinued after one presentation.)

1968 **Donald A. Rennie**

ACS Award in Inorganic Chemistry

Sponsored by Texas Instruments Inc.

1962 **F. Albert Cotton**
1963 **Daryle H. Busch**
1964 **Fred Basolo**
1965 **Earl L. Muetterties**
1966 **Geoffrey Wilkinson**
1967 **John L. Margrave**
1968 **Jack Halpern**
1969 **Russell S. Drago**
1970 **Neil Bartlett**
1971 **Jack Lewis**
1972 **Theodore L. Brown**
1973 **M. F. Hawthorne**
1974 **Lawrence F. Dahl**
1975 **James P. Collman**
1976 **Richard H. Holm**

ACS Award in Petroleum Chemistry

Sponsored by The Lubrizol Corp.

Sponsored 1948-73 by Precision Scientific Co.

1949 **Bruce H. Sage**
1950 **Kenneth S. Pitzer**
1951 **Louis Schmerling**
1952 **Vladimir Haensel**
1953 **Robert W. Schiessler**
1954 **Arthur P. Lien**
1955 **Frank Ciapetta**
1956 **Milburn J. O'Neal, Jr.**
1957 **C. Gardner Swain**
1958 **Robert P. Eischens**

ACS Awards (continued)

Petroleum Chem. (cont.)

1959 George C. Pimentel
1960 Robert W. Taft, Jr.
1961 George S. Hammond
1962 Harold Hart
1963 John P. McCullough
1964 George A. Olah
1965 Glen A. Russell
1966 James Wei
1967 Andrew Streitwieser, Jr.
1968 Keith U. Ingold
1969 Alan Schriesheim
1970 Lloyd R. Snyder
1971 Gerasimos J. Karabatsos
1972 Paul G. Gassman
1973 Joe W. Hightower
1974 (No award)
1975 (No award)
1976 John H. Sinfelt

ACS Award in Polymer Chemistry

sponsored by Witco Chemical Corp.

1964 Carl S. Marvel
1965 Herman F. Mark
1966 Walter H. Stockmayer
1967 Frank R. Mayo
1968 Charles G. Overberger
1969 Frank A. Bovey
1970 Michael M. Szwarc
1971 Georges J. Smets
1972 Arthur V. Tobolsky
1973 Turner Alfrey, Jr.
1974 John D. Ferry
1975 Leo Mandelkern
1976 Paul W. Morgan

ACS Award in Pure Chemistry

Sponsored by A. C. Langmuir through 1937. In 1938, James Kendall financed the prize. Since 1940 sponsored by Alpha Chi Sigma Fraternity.

1931 Linus Pauling
1932 Oscar K. Rice
1933 Frank H. Spedding
1934 C. Frederick Koelsch
1935 Raymond M. Fuoss
1936 John Gamble Kirkwood
1937 E. Bright Wilson, Jr.
1938 Paul D. Bartlett
1939 (No award)
1940 Lawrence O. Brockway
1941 Karl A. Folkers
1942 John Lawrence Oncley
1943 Kenneth S. Pitzer
1944 Arthur C. Cope
1945 Frederick T. Wall
1946 Charles C. Price, III
1947 Glenn T. Seaborg
1948 Saul Winstein
1949 Richard T. Arnold
1950 Verner Schomaker
1951 John C. Sheehan
1952 Harrison S. Brown
1953 William von E. Doering
1954 John D. Roberts
1955 Paul Delahay
1956 Paul M. Doty
1957 Gilbert J. Stork
1958 Carl Djerassi
1959 Ernest M. Grunwald
1960 Elias J. Corey
1961 Eugene E. van Tamelen
1962 Harden M. McConnell
1963 Stuart A. Rice
1964 Marshall Fixman
1965 Dudley Herschbach
1966 Ronald Breslow
1967 John D. Baldeschwieler
1968 Orville L. Chapman
1969 Roald Hoffman
1970 Harry B. Gray
1971 R. Bruce King
1972 Roy G. Gordon
1973 John I. Brauman
1974 Nicholas J. Turro
1975 George M. Whitesides
1976 Karl F. Freed

ACS Award in the Chemistry of Milk

Sponsored by The Borden Foundation, Inc.

1939 Leroy S. Palmer
1941 Claude S. Hudson
1942 George E. Holm
1943 Earle O. Whittier
1944 William Mansfield Clark
1945 Ben H. Nicolet
1946 Ira A. Gould
1947 George C. Supplee
1948 B. L. Herrington
1949 George R. Greenbank
1950 George A. Richardson
1951 Thomas L. McMeekin
1952 Carrell H. Whitnah
1953 Robert Jenness
1954 Donald V. Josephson
1955 Fred Hillig
1956 Samuel R. Hoover
1957 Stuart Patton
1958 William G. Gordon
1959 Charles A. Zittle
1960 E. L. Jack
1961 Vladimir N. Krukovsky
1962 David F. Waugh
1963 Serge N. Timasheff
1964 J. Robert Brunner
1965 Edgar A. Day
1966 Bruce L. Larson
1967 Dyson Rose
1968 Martin J. Kronmam
1969 Kurt E. Ebner

ACS Award in the Chemistry of Plastics and Coatings

Sponsored by Borden Foundation, Inc.

1968 Harry Burrell
1969 Sylvan O. Greenlee
1970 Raymond F. Boyer
1971 Raymond R. Myers
1972 Richard S. Stein
1973 Carl S. Marvel
1974 Vivian T. Stannett
1975 Maurice L. Huggins
1976 Herman F. Mark

Arthur C. Cope Award

For achievement in organic chemistry; administered by ACS Board of Directors.

1973 Robert B. Woodward and Roald Hoffmann
1974 Donald J. Cram
1976 Elias J. Corey

James Bryant Conant Award in High School Chemistry Teaching

Supported by E. I. duPont de-Nemours & Co., Inc., 1967-72, by the American Chemical Society, 1973-74, and by CHEM Study since then. Since 1973 the national award winner has been chosen from regional award winners of the three prior years.

1967 Dist. 1 Raymond T. Byrne
1967 Dist. 2. Elaine M. Kilbourne
1967 Dist. 3. Harry C. Taylor
1967 Dist. 4. Theodore E. Militor
1967 Dist. 5. Elaine W. Ledbetter
1967 Dist. 6. Harold E. Alexander
1968 Dist. 1. Daniel P. Corr
1968 Dist. 2. Harold W. Ferguson
1968 Dist. 3. Robert M. Sims
1968 Dist. 4. Charles F. McClary
1968 Dist. 5. Marion Nottingham
1968 Dist. 6. George T. Bazzetta
1969 Dist. 1. Eliz. V. Lamphere
1969 Dist. 2. Jos. S. Schmuckler
1969 Dist. 3. Lee R. Summerlin
1969 Dist. 4. Ben O. Propeck
1969 Dist. 5. Frank S. Quiring
1969 Dist. 6. W. Keith MacNab
1970 Dist. 1. Dorothy W. Gifford
1970 Dist. 2. James V. DeRose
1970 Dist. 3. William B. Robertson
1970 Dist. 4. Newell Smeby
1970 Dist. 5. Charles D. Mickey
1970 Dist. 6. George Birrell
1971 Dist. 1. Elizabeth W. Sawyer
1971 Dist. 2. Audrey J. Cheek
1971 Dist. 3. Bernard Toan
1971 Dist. 4. Leo J. Klosterman
1971 Dist. 5. Clara Weisser
1971 Dist. 6. Nellis G. Fletcher
1972 Dist. 1. Frank J. Tuzzolino
1972 Dist. 2. Albert J. Judge
1972 Dist. 3. Anne A. Wiseman
1972 Dist. 4. Henrietta A. Parker
1972 Dist. 5. Harold L. Pearson
1972 Dist. 6. Irma Greisel
1973 Melvin Greenstadt
1974 Wallace J. Gleekman
1975 George W. Stapleton
1976 Dorothea H. Hoffman

Peter Debye Award in Physical Chemistry

Sponsored by Humble Oil & Refining Co. until 1970 when sponsorship was assumed by Exxon Chemical Co., U.S.A.

1962 E. Bright Wilson, Jr.
1963 Robert S. Mulliken
1964 Henry Eyring
1965 Lars Onsager
1966 Joseph O. Hirschfelder
1967 Joseph E. Mayer
1968 George B. Kistiakowsky
1969 Paul J. Flory
1970 Oscar K. Rice
1971 Norman Davidson
1972 Clyde A. Hutchison, Jr.

ACS Awards (continued)

Debye Award (cont.)

1973 William N. Lipscomb, Jr.
1974 Walter H. Stockmayer
1975 H. S. Gutowsky
1976 Robert W. Zwanzig

Garvan Medal

For distinguished service to chemistry by women chemists; endowed by Francis P. Garvan.

1937 Emma P. Carr
1940 Mary E. Pennington
1942 Florence B. Seibert
1946 Icie G. Macy-Hoobler
1947 Mary Lura Sherrill
1948 Gerty T. Cori
1949 Agnes Fay Morgan
1950 Pauline Berry Mack
1951 Katherine B. Blodgett
1952 Gladys A. Emerson
1953 Leonora N. Bilger
1954 Betty Sullivan
1955 Grace Medes
1956 Allene R. Jeans
1957 Lucy W. Pickett
1958 Arda A. Green
1959 Dorothy V. Nightingale
1960 Mary L. Caldwell
1961 Sarah Ratner
1962 Helen M. Dyer
1963 Mildred Cohn
1964 Birgit Vennesland
1965 Gertrude E. Perlmann
1966 Mary L. Peterman
1967 Marjorie J. Vold
1968 Gertrude B. Elion
1969 Sofia Simmonds
1970 Ruth R. Venerito
1971 Mary Fieser
1972 Jean'ne M. Shreeve
1973 Mary L. Good
1974 Joyce J. Kaufman
1975 Marjorie C. Caserio
1976 Isabella L. Karle

James T. Grady Award for Interpreting Chemistry for the Public

Sponsored by the American Chemical Society

1957 David H. Killeffer
1958 William L. Laurence
1959 Alton L. Blakeslee
1960 Watson Davis
1961 David Dietz
1962 John F. Baxter
1963 Lawrence Lessing
1964 Nate Haseltine
1965 Isaac Asimov
1966 Frank E. Carey
1967 Irving S. Bengelsdorf
1968 Raymond A. Bruner
1969 Walter Sullivan
1970 Robert C. Cowen
1971 Victor Cohn
1972 Dan Q. Posin
1973 O. A. Battista
1974 Ronald Kotulak
1975 Jon Franklin
1976 Gene Bylinsky

Ernest Guenther Award in the Chemistry of Essential Oils and Related Products

Sponsored by Fritzsche Dodge & Olcott Inc.

1949 John L. Simonsen
1950 A. J. Haagen-Smit
1951 Edgar Lederer
1952 Yves-Rene Naves
1953 Max Stoll
1954 A. R. Penfold
1955 Hans Schinz
1956 Herman Pines
1957 D. H. R. Barton
1958 George H. Buchi
1958 Frantisek Sorm
1960 Carl Djerassi
1961 C. F. Seidel
1962 E. R. H. Jones
1963 Arthur J. Birch
1964 Oskar Jeger
1965 Konrad E. Bloch
1966 Albert J. Eschenmoser
1967 George A. Sim
1968 Elias J. Corey
1969 John W. Cornforth
1970 Duilio Arigoni
1971 Ernest Wenkert
1972 Guy Ourisson
1973 William G. Dauben
1974 Günther Ohloff
1975 S. Morris Kupchan
1976 Alastair I. Scott

Ipatieff Prize

For work in high pressure catalysis. Administered by ACS; Northwestern University is the trustee.

1947 Louis Schmerling
1950 Herman E. Ries
1953 Robert B. Anderson
1956 Harry G. Drickamer
1959 Cedomir M. Sliepcevich
1962 Charles Kemball
1965 Robert H. Wentorf, Jr.
1968 Charles R. Adams
1971 Paul B. Venuto
1974 George A. Samara

Frederick Stanley Kipping Award in Organosilicon Chemistry

Sponsored by Dow Corning Corp.

1962 Henry Gilman
1963 Leo H. Sommer
1964 Colin Eaborn
1965 Eugene G. Rochow
1966 Gerhard Fritz
1967 Makoto Kumada
1968 Ulrich Wannagat
1969 Robert A. Benkeser
1970 Robert West
1971 Alan G. MacDiarmid
1972 Dietmar Seyferth
1973 Adrian G. Brook
1974 Hubert Schmidbaur
1975 Hans Bock
1976 Michael F. Lappert

The Irving Langmuir Award in Chemical Physics

Sponsored by The General Electric Foundation

1965 John H. Van Vleck*
1966 H. S. Gutowsky
1967 John C. Slater*
1968 Henry Eyring
1969 Charles P. Slichter*
1970 John A. Pople
1971 Michael E. Fisher*
1972 Harden M. McConnell
1973 Peter M. Rentzepis*
1974 Harry G. Drickamer
1975 Robert H. Cole*
1976 John S. Waugh

*Selection and presentation made by the Division of Chemical Physics of the American Physical Society.

E. V. Murphee Award in Industrial and Engineering Chemistry

Sponsored by Exxon Research and Engineering Co.

1957 Warren K. Lewis
1958 deBois Eastman
1959 Edwin R. Gilliland
1960 Neal R. Amundson
1961 Olaf A. Hougen
1962 Eugene J. Houdry
1963 Manson Benedict
1964 Bruce H. Sage
1965 Vladimir Haensel
1966 Richard H. Wilhelm
1967 Alfred Clark
1968 Melvin A. Cook
1969 Alex G. Oblad
1970 Peter V. Danckwerts
1971 Heinz Heinemann
1972 Paul B. Weisz
1973 Thomas K. Sherwood
1974 Herman S. Bloch
1975 Donald L. Katz
1976 James F. Roth

James Flack Norris Award in Physical Organic Chemistry

Sponsored by the Northeastern Section, ACS

1965 Christopher K. Ingold
1966 Louis P. Hammett
1967 Saul Winstein
1968 George S. Hammond
1969 Paul D. Bartlett
1970 Frank H. Westheimer
1971 Cheves Walling
1972 Stanley J. Cristol
1973 Kenneth B. Wiberg
1974 Gerhard L. Closs
1975 Kurt M. Mislow
1976 Howard E. Zimmerman

ACS Awards (continued)

Charles Lathrop Parsons Award

Sponsored by the American Chemical Society for outstanding public service by a member.

1952	Charles L. Parsons
1955	James B. Conant
1958	Roger Adams
1961	George B. Kistiakowsky
1964	Glenn T. Seaborg
1967	Donald F. Hornig
1970	W. Albert Noyes, Jr.
1973	Charles C. Price
1974	Russell W. Peterson

Priestley Medal

For distinguished services to chemistry. Sponsored by the American Chemical Society.

1923	Ira Remsen
1926	Edgar F. Smith
1929	Francis P. Garvan
1932	Charles L. Parsons
1935	William A. Noyes
1938	Marston T. Bogert
1941	Thomas Midgley, Jr.
1944	James B. Conant
1945	Ian Heilbron
1946	Roger Adams
1947	Warren K. Lewis
1948	Edward R. Weidlein
1949	Arthur B. Lamb
1950	Charles A. Kraus
1951	Evan J. Crane
1952	Samuel C. Lind
1953	Robert Robinson
1954	W. Albert Noyes, Jr.
1955	Charles A. Thomas
1956	Carl S. Marvel
1957	Farrington Daniels
1958	Ernest H. Volwiler
1959	H. I. Schlesinger
1960	Wallace R. Brode
1961	Louis P. Hammett
1962	Joel H. Hildebrand
1963	Peter J. W. Debye
1964	John C. Bailar, Jr.
1965	William J. Sparks
1966	William Oliver Baker
1967	Ralph Connor
1968	William G. Young
1969	Kenneth S. Pitzer
1970	Max Tishler
1971	Frederick D. Rossini
1972	George B. Kistiakowsky
1973	Harold C. Urey
1974	Paul J. Flory
1975	Henry Eyring
1976	George S. Hammond

ACS Award for Outstanding Performance by Local Sections

Small

1968	Eastern North Carolina
1969	Mississippi
1970	Mississippi
1971	Mississippi
1972	Central Utah
1973	Western Vermont
1974	Permian Basin
1975	South Plains

Medium Small

1968	Central Arizona
1969	South Jersey
1970	South Jersey
1971	South Jersey
1972	Peoria
1973	Puerto Rico
1974	Kanawha Valley
1975	Central North Carolina

Medium Large

1968	Eastern New York
1969	Virginia
1970	Louisiana
1971	Louisiana
1972	Eastern New York
1973	Milwaukee
1974	Midland
1975	Orange County

Large

1968	Philadelphia
1969	Connecticut Valley
1970	Rochester
1971	Akron
1972	Akron
1973	Delaware
1974	Delaware
1975	Delaware

Divisional Awards

(Not including best paper, student, or travel awards)

Award for Advancement of Application of Agricultural and Food Chemistry

Sponsored by International Flavors and Fragrances

Agricultural & Food Chemistry

1973	Albert L. Elder
1974	Frank M. Strong
1975	Stuart Patton

Eli Lilly Award in Biological Chemistry

Biological Chemistry

1975	Mark Ptashne

Pfizer Award in Enzyme Chemistry

Biological Chemistry

1975	Malcolm Gefter

Claude S. Hudson Award

Division of Carbohydrate Chemistry

1946	Claude S. Hudson
1947	Frederick J. Bates
1949	Frederick W. Zerban
1950	William B. Newkirk
1951	William D. Horne
1952	Melvin L. Wolfrom
1953	George P. Meade
1954	Horace S. Isbell
1955	Kenneth R. Brown
1956	James M. D. Brown
1957	Julian K. Dale
1958	Hermann O. L. Fischer
1959	William Ward Pigman
1960	Roy L. Whistler
1961	John C. Sowden
1962	Fred Smith
1963	Nelson K. Richtmyer
1964	Dexter French
1965	C. G. Caldwell
1966	Raymond U. Lemieux
1967	W. Z. Hassid
1968	Hewitt G. Fletcher, Jr.
1969	John K. Netherton Jones
1970	Norman F. Kennedy
1971	Robert Stuart Tipson
1972	Derek Horton
1973	Roger William Jeanloz
1974	Wendell W. Binkley
1975	Hans Helmut Baer

Anselme Payen Award

Cellulose, Paper and Textile

1962	Louis E. Wise
1963	Clifford B. Purves
1964	Harold M. Spurlin
1965	Carl J. Malm
1966	Wayne A. Sisson
1967	Roy L. Whistler
1968	Alfred J. Stamm
1969	Stanley G. Mason
1970	Wilson A. Reeves
1971	Tore E. Timell
1972	Conrad Schuerch
1973	David A. I. Goring
1974	Vivian T. Stannett
1975-	John K. N. Jones

Centennial of Chemistry Award

Chemical Education

1974	Joseph Priestley
1974	Sir Derek H. R. Barton

Victor K. LaMer Award

Colloid and Surface Chemistry

1970	Charles W. Querfeld
1971	Edward McCafferty
1972	Donald E. Brooks
1973	W. Henry Weinberg
1974	Stephen Brenner
1975	Michele Flicker

Certificate of Merit

Colloid and Surface Chemistry

1952	Joseph Howard Mathews
1964	Elroy J. Miller
1964	Lloyd H. Reyerson
1965	Winfred O. Milligan
1965	Victor K. LaMer

Divisional Awards (continued)

Cert. of Merit (cont.)

1966 Donald Penrose Graham
1966 Joseph Howard Mathews
1966 Ralph A. Beebe
1966 W. Conway Pierce
1966 John W. Williams
1967 Leo Shedlovsky
1967 Thomas Fraser Young
1968 John G. Aston
1968 Jacob J. Bikerman
1968 J. Leon Shereshevsky
1968 Paul H. Emmett
1969 Curtis R. Singleterry
1969 Stephen Brunauer
1970 Walter R. Smith
1971 William A. Zisman
1972 Willard M. Bright
1975 Anthony M. Schwartz
1975 Frederick R. Eirich
1975 Herman E. Ries, Jr.

Distinguished Service Award

Environmental Chemistry

1957 Edward Bartow
Abraham S. Behrman
William D. Collins
Arthur M. Buswell
Robert C. Bardwell
1958 Floyd W. Mohlman
William D. Hatfield
1959 John R. Baylis
Dudley K. French
1960 Charles S. Howard
Otto M. Smith
1961 William M. Stericker
Fred Lindsay
1962 Hovaness Heukelekian
L. Drew Betz
1963 William Allan Moore
William L. Lamar
1965 Louise F. Warrick
Clair A. Boruff
1967 S. Kenneth Love
Richard D. Hoak
1968 John J. Maguire
H. Gladys Swope
1969 Hilding B. Gustafson
Henry C. Marks
1970 George Hatch
Abraham A. Berk
1971 John F. Wilkes
Thurston E. Larson
1972 Robert Ingols
Calvin Calmon
1973 S. Charles Caruso
James P. Lodge
1974 Henry C. Bramer
Benjamin F. Willey
1975 Frances L. Estes
Louis F. Worth, Jr.

Award for Creative Work in Fluorine Chemistry

Sponsored by PCR, Inc.
Fluorine Chemistry

1972 George Cady
1973 Joseph Simons
1974 William Miller
1975 Joseph Park

Henry H. Storch Award in Coal Research

Fuel Chemistry

1964 Irving Wender
1965 Everett Gorin
1966 R. A. Friedel
1967 Henry R. Linden
1968 J. H. Field
1969 P. L. Walker, Jr.
1971 G. R. Hill
1972 R. W. Van Dolah
1973 A. M. Squires
1974 R. Tracy Eddinger
1975 G. Alex Mills

Dexter Chemical Corp. Award

History of Chemistry

1956 Ralph E. Oesper
1957 William Haynes
1958 Eva Armstrong
1959 John Read
1960 Denis Duveen
1961 James R. Partington
1962 Henry M. Leicester
1963 Douglas McKie
1964 Edward Farber
1965 Martin Levey
1966 Earle R. Caley
1967 Mary Elvira Weeks
1968 Aaron J. Ihde
1969 Walter Pagel
1970 Ferenc Szabadvary
1971 Wyndham D. Miles
1972 Henry Guerlac
1973 Bernard Jaffee
1974 (No award)
1975 Jan W. Van Spronsen

Joseph E. Stewart Award

(Scroll of Honor Award)
Industrial and Engineering Chemistry

1953 Walter A. Schmidt
R. Norris Shreve
1954 Walter J. Murphy
1955 Francis J. Curtis
1956 William A. Pardee
1957 Charles M. Cooper
1958 Fraser H. Johnstone
1960 Melvin C. Molstad
1961 Barnett M. Dodge
1962 James M. Church
1963 Will H. Shearon
1964 Joseph E. Stewart
1966 Brage Golding
1967 Leo Friend
1968 Norbert Platzer
1969 Arthur R. Rescorla
1970 David E. Gushee
1971 Robert N. Maddox
1972 (Not given)
1973 (Not given)
1973 Vernon A. Fauver
1975 James D. Idol, Jr.

Medicinal Chemistry Award

Medicinal Chemistry

1966 Bernard R. Baker
1968 Sydney Archer
1970 James Maurice Sprague
1972 George H. Hitchings
1974 Josef Fried

Microbial Chemistry and Technology Distinguished Service Award

Microbial Chemistry and Technology

1966 Clair S. Boruff
Charles N. Frey
1968 Marvin J. Johnson
1970 James M. VanLanen
1972 Henry J. Peppler
1974 Elmer L. Gaden, Jr.

Burdick & Jackson International Award for Research in Pesticide Chemistry

Pesticide Chemistry

1969 John E. Casida
1970 Richard D. O'Brien
1971 Robert L. Metcalf
1972 Ralph L. Wain
1973 Hubert Martin
1974 T. Roy Fukuto
Michael Elliott

Charles Goodyear Medal

Rubber

1941 David Spence
1942 Lorin B. Sebrell
1944 Waldo L. Semon
1946 Ira Williams
1948 George Oenslager
1949 Harry L. Fisher
1950 Carroll C. Davis
1951 William C. Geer
1952 H. E. Simmons
1953 James T. Blake
1954 G. Stafford Whitby
1955 Raymond P. Dinsmore
1956 Sidney M. Caldwell
1957 A. W. Carpenter
1958 J. C. Patrick
1959 F. H. Banbury
1960 William B. Wiegand
1961 Herbert A. Winkelmann
1962 Melvin Mooney
1963 William J. Sparks
1964 Arthur E. Juve
1965 Benjamin S. Garvey
1966 E. A. Murphy
1967 Norman Bekkedahl
1968 Paul J. Flory
1969 Robert M. Thomas
1970 Samuel D. Gehman
1971 Harold J. Osterhof
1972 Frederick W. Stavely
1973 Arnold M. Collins
1974 Joe C. Krejci
1975 Otto Bayer

Local Section Awards

(Professional-level awards only, not including student, high-school teacher, or best paper awards)

Distinguished Service Award

Akron

1971 Glenn H. Brown
1972 Lawrence E. Forman
1973 Vernon L. Folt
1974 Walter C. Warner
1974 Henry A. Pace
1975 Otto C. Elmer

W. F. Coover Lecture

Ames

1948 Carl F. Cori
1951 Robert S. Mulliken
1952 William S. Johnson
1953 Bryce Crawford
1954 G. Frederick Smith
1955 Roger Adams
1956 Linus Pauling
1957 Henry Eyring
1958 Gilbert Stork
1959 Fritz Lipmann
1960 Henry Taube
1961 Saul Winstein
1962 Peter Debye
1963 Henry Gilman
1964 George B. Kistiakowsky
1965 Severo Ochoa
1966 Paul D. Bartlett
1967 Leo Brewer
1968 George S. Hammond
1969 William N. Lipscomb
1971 Ronald S. Nyholm
1972 Bruce Merrifield
1973 Petr Zuman
1974 Michael Kasha

Charles E. Coates Memorial Award

Baton Rouge

1958 Jesse Coates
1959 Louis L. Rusoff
1960 A. R. Choppin
1961 F. Drew Mayfield
1962 J. P. McKenzie
1963 Hulen B. Williams
1964 Alexis Voorhies, Jr.
1965 James G. Traynham
1966 Arthur G. Keller
1967 Philip W. West
1968 Henry G. Allen
1969 George F. Kirby
1970 Roger W. Richardson
1971 C. F. Gray
1972 Paul E. Koenig
1973 Joel Selbin
1974 Gordon A. Hughmark
1975 Paul Murrell

Gilbert Newton Lewis Medal

California

1951 Linus Pauling
1953 J. G. Kirwood
1956 William Giauque
1958 Joseph E. Mayer
1960 Robert S. Mulliken
1961 Lars Onsager
1963 Henry Eyring
1965 Kenneth Pitzer
1967 Louis P. Hammett
1969 E. Bright Wilson
1973 John Pople

California Section Award

1951 C. H. Li
1952 Isadore Perlman
1953 Donald Cram
1954 Norman Davidson
1955 E. F. Hammel
1956 Harold Johnston
1957 George Pimentel
1958 Howard Schachman
1959 William Dauben
1960 William T. Simpson
1961 Harden McConnell
1962 Kenneth Wiberg
1963 Jerome Berson
1964 Andrew Streitwieser
1965 Ignacio Tinoco
1966 Andreas Acrivos
1967 Daniel Kivelson
1968 Bruce Mahan
1969 David M. Grant
1970 David A. Shirley
1971 Mostafa A. El-Sayed
1972 James P. Collman
1973 David Kearns
1974 Leonard Jurd
1975 John P. McTague

Charles H. Stone Award

Carolina-Piedmont

1971 Charles N. Reilley
1972 Ruben Ellestad
1973 Oscar K. Rice
1974 John A. Dean
1975 James R. Durig
1976 George E. Boyd

Bateman Lecture

Central Arizona

1964 Linus C. Pauling
1965 Willard F. Libby
1966 Harrison S. Brown
1967 Edward Teller
1968 Arthur Kornberg
1969 Harold C. Urey
1970 Paul Ehrlich
1972 Carl Djerassi
1973 Glenn T. Seaborg
1975 Albert Szent-Gyorgyi

Achievement Award

Central Arizona

1965 George M. Bateman
1966 Arthur Lee Phelps
1968 Junia McAllister
1969 Claud McLean, Sr.
1970 Gordon D. Perrine
1971 Arnold E. Bereit
1972 J. Smith Decker

John Dustin Clark Medal

Central New Mexico

1965 Morris F. Stubbs
1967 Ralph Müller
1969 Karl Bergstresser
1971 Sherman Rabideau
1973 Kay R. Brower

Utah Award

Central Utah and Salt Lake

1958 M. D. Thomas
1959 Henry Eyring
1960 Sherwin Maeser and Joseph K. Nichles
1961 M. A. Cook
1962 John R. Lewis
1963 Emil Smith
1964 Lloyd E. Malm
1965 H. Tracy Hall
1966 Carl J. Christensen
1967 Leo T. Samuels
1968 George R. Hill
1969 Sam Kestler
1970 J. Calvin Giddings
1971 Reed M. Izatt
1972 David M. Grant
1973 Marvin Tuddenham
1974 J. Rex Goates

Distinguished Service Award

Chattanooga

1958 Emerson Poste
1959 Andrew Kelly
1960 Murray Raney
1961 William Swan
1962 (not given)
1963 Irvine Grote
1964 (not given)
1965 Benjamin Gross
1966 Thomas Street
1968 Ralph Berning
1968 John Christensen
1969 Ross Russell
1970 Raymond H. Ingwalson
1971 James Galbraith
Marion Barnes
1972 Harold Goslen
1973 William Luther
1974 Marie Black
1975 Alice Raines

Chemist of the Year

Chattanooga

1958 Harold Reeves
1959 Irvine Grote
1971 Marion Barnes

Willard Gibbs Medal

Chicago

1911 Svante Arrhenius
1912 Theodore W. Richards
1913 Leo H. Baekeland
1914 Ira Remsen
1915 Arthur A. Noyes
1916 Willis R. Whitney
1917 Edward W. Morley
1918 William M. Burton
1919 William A. Noyes
1920 F. G. Cottrell
1921 Mme. Marie Curie
1923 Julius Stieglitz
1924 Gilbert N. Lewis
1925 Moses Gomberg
1926 Sir James Colquhoun Irving
1927 John Jacob Abel
1928 William Draper Harkins
1929 Claude Silbert Hudson
1930 Irving Langmuir

Local Section Awards (continued)

W. Gibbs Medal (cont.)

1931 Phoebus A. Levene
1932 Edward Curtis Franklin
1933 Richard Willstatter
1934 Harold Clayton Urey
1935 Charles A. Kraus
1936 Roger Adams
1937 Herbert Newby McCoy
1938 Robert R. Williams
1939 Donald Dexter Van Slyke
1940 Vladimir Ipatieff
1941 Edward A. Doisy
1942 Thomas Midgley, Jr.
1943 Conrad A. Elvehjem
1944 George O. Curme, Jr.
1945 Frank C. Whitmore
1946 Linus Pauling
1947 Wendell M. Stanley
1948 Carl F. Cori
1949 Peter J. W. Debye
1950 Carl S. Marvel
1951 William F. Giauque
1952 William C. Rose
1953 Joel H. Hildebrand
1954 Elmer K. Bolton
1955 Farrington Daniels
1956 Vincent du Vigneaud
1957 W. Albert Noyes, Jr.
1958 Willard F. Libby
1959 Hermann I. Schlesinger
1960 George B. Kistiakowsky
1961 Louis Plack Hammett
1962 Lars Onsager
1963 Paul D. Bartlett
1964 Izaak M. Koithoff
1965 Robert S. Mulliken
1966 Glenn T. Seaborg
1967 Robert B. Woodward
1968 Henry Eyring
1969 Gerhard Herzberg
1970 Frank H. Westheimer
1971 Henry Taube
1972 John T. Edsall
1973 Paul John Flory
1974 Har Gobind Khorana
1975 Herman F. Mark

Julius Stieglitz Memorial Lectures

Chicago

1940 E. A. Doisy
1941 Fred C. Koch
1943 Carl S. Marvel
1944 Linus Pauling
1945 William D. Harkins
1946 Don M. Yost
1947 Dorothy Wrinch
1948 Vincent du Vigneaud
1949 B. J. Cohn
1950 C. K. Ingold
1952 Robert B. Woodward
1953 Frank R. Mayo
1954 Paul D. Bartlett
1956 F. H. Westheimer
1957 Henry B. Hass
1958 Herbert C. Brown
1959 Henry A. Lardy
1960 Louis P. Hammett
1962 Nelson J. Leonard
1963 William S. Johnson
1964 Paul M. Doty
1965 Charles C. Price
1966 H. Gobind Khorana
1967 William von E. Doering
1968 George Hammond
1969 Donald J. Cram
1970 Jerome A. Berson
1971 Carl Djerassi
1972 Jerrold Meinwald
1973 Andrew Streitwieser, Jr.
1974 Derek H. R. Barton
1975 Elias J. Corey

Cincinnati Chemist Award

Cincinnati

1950 Clarence C. Ruchhoft
1951 A. S. Richardson
1952 Martin H. Fischer
1953 Francis F. Heyroth
1954 Ralph E. Oesper
1955 Jacob Cholak
1956 Leon H. Schmidt
1957 Milton Orchin
1958 Willy Lange
1959 Harvey C. Brill
1960 Edwin S. Lutton
1961 Hans Jaffe
1962 Howard L. Ritter
1963 Milan A. Logan
1964 Elton S. Cook
1965 Oscar T. Quimby
1966 Thomas B. Cameron
1967 Aubrey P. Altschuler
1968 Eugene Sawicki
1969 Fred H. Mattson
1970 Harold Stokinger
1971 Hans Zimmer
1972 Aaron A. Rosen
1973 William Erman
1974 Jack Kwiatek
1975 Harold Petering

Morley Medal

Cleveland

1965 Ernest L. Eliel
1967 David Pressman
1968 Waldo L. Semon
1969 Melvin S. Newman
1970 George A. Olah
1971 Leo A. Paquette
1973 Richard H. F. Manske
1974 Everett C. Hughes
1975 Daryle H. Busch

Award in Chemistry

Colorado

1967 Stanley J. Cristol
1968 E .W. D. Huffman
1970 Joseph D. Park
1971 Edward L. King
1972 Perry A. Argabright
1973 William B. Cook
1974 Norman F. Witt
1975 Charles H. DePuy

Meritorious Service Award

Colorado

1962 Norman F. Witt
1963 John L. Ellingboe
1964 Dale N. Robertson
1965 William C. Stickler
1966 William T. Miller, S.J.
1967 Mrs. Alice O. Robertson
1968 John A. Beel
1969 Joe T. Kelly
1970 Larry L. Miller
1971 Richard L. Taber
1972 Charles H. DePuy
1973 Alvin L. Schalge
1974 Edward L. King

Columbus Section Award

1975 Melvin S. Newman

Eugene C. Sullivan Award

Corning

1967 Martin E. Nordberg
1968 Harrison P. Hood
1969 Robert H. Dalton
1970 David F. Fortney
1971 S. Donald Stookey
1972 Richard W. Monney
1973 John J. Rothermel
1974 A. Stuart Tulk
1975 Paul C. Saunders

Wilfred T. Doherty Recognition Award

Dallas-Fort Worth

1972 Morton F. Mason
1973 Ralph L. Shriner
1974 Gordon K. Teal
1975 Donald E. Woessner

Austin M. Patterson Award

Dayton

1949 Austin M. Patterson
1951 Arthur B. Lamb
1953 Evan J. Crane
1955 Howard S. Nutting
1957 Melvin Mellon
1959 Leonard T. Capell
1961 Malcolm Dyson
1963 W. Albert Noyes, Jr.
1965 Elmer Hockett
1967 Melville L. Wolfrom
1969 Herman Skolnik
1971 Charles D. Hurd
1973 Pieter E. Verkade
1975 William J. Wiswesser

Delaware Section Award

Delaware

1957 Rudolph Pariser
1958 Arthur B. Metzner
1959 Theodore L. Cairns, Rudolph A. Carboni, Donald D. Coffffman, Joseph R. Downing, Vaughn A. Engelhardt, Bruce S. Fisher, Richard E. Heckert, Edward G. Howard, Jr., Carl G. Krespan, Ernest L. Little, Jr., Catherine E. Looney, Edith G. McGeer, Blaine C. McKusick, Richard E. Merrifield, William J. Middleton Howard F. Mower, William D. Phillips, George N. Sausen, Richard M. Scribner, Clement W. Theobald, Hilmer E. Winberg

Local Section Awards (continued)

Del. Sec. Award (cont.)

1960 Paul W. Morgan and Stephanie L. Kwolek
1961 William R. Hasek, William C. Smith, and Vaughn A. Engelhardt
1962 John Bugosh
1963 David F. Eaton, Alden Dwayne Josep, William D. Phillips, and Richard E. Benson
1964 Edward H. Vandenberg
1965 Richard Cramer
1966 Richard Heck
1967 Richard Cramer and Richard V. Lindsey, Jr.
1968 Charles J. Pederson
1969 Ralph W. F. Hardy, Richard D. Holsten, Earl K. Jackson, and Richard C. Burns
1970 Harold Kwart, John Olsen, and William H. Garner
1971 (No award)
1972 (No award)
1973 J. Jack Kirkland
1974 Jay K. Kochi, Peter Bakuzia, and Paul J. Krusic

Thomas Midgley Award

Detroit

1965 Thomas A. Boyd
1966 Sidney M. Caldwell
1967 Summer B. Twiss
1968 Arnold M. Collins
1969 George E. F. Brewer
1970 Waldo L. Semon
1971 Henry Brown
1972 Raymond G. Rieser
1973 William M. LeSuer
1974 J. Franklin Hyde
1975 John T. Patton, Jr.

S. C. Lind Lectureship Award

East Tennessee

1948 Edward Mack, Jr.
1949 Linus C. Pauling
1950 Hobart H. Willard
1951 Izaak M. Kolthoff
1952 William A. Noyes, Jr.
1953 Samuel C. Lind
1954 Ludwig F. Audrieth
1955 George B. Kistiakowsky
1956 Harvely C. Diehl
1957 Willard F. Libby
1958 Carl S. Maryel
1959 Herbert C. Brown
1960 Michael Kasha
1961 Peter Debye
1962 F. Albert Cotton
1963 Herbert A. Laitinen
1964 Bryce L. Crawford, Jr.
1965 Konrad E. Bloch
1966 Farrington Daniels
1967 Paul von R. Schleyer
1968 John C. Bailar, Jr.
1969 Milton Burton
1970 Paul Delahay
1971 Groves H. Cartledge
1972 Roald Hoffman
1973 Harold C. Urey
1974 Joel H. Hildebrand

Florida Award

Florida

1952 Paul Gross
1953 A. E. Wood
1954 Cash Pollard
1955 Howard E. Skipper
1956 George Davis
1957 Charles R. Hauser
1958 Karl Dittmer
1959 Jack Hawkins
1960 Harry H. Sisler
1961 Michael Kasha
1962 Jack Hine
1963 George Butler
1964 Carl Bahner
1965 Werner Herz
1966 Paul Tarrant
1967 Oscar K. Rice
1968 Earl Frieden
1969 John F. Baxter
1970 Sean McGlynn
1971 Ray Lawrence
1972 James V. Quagliano
1973 Gregory R. Choppin
1974 Sidney Fox
1975 Dean F. Martin

Civic Service Award

Florida

1964 S. Cicero Ogburn, Jr.
1965 David H. Killeffer
1968 Alfred P. Mills
1969 Antonio M. Gordon
1970 Louis J. Polskin
1971 Robert B. Bennett
1974 Walter C. Funderburk, Jr.

Herty Medal

Georgia

1933 Fred Allison
1934 Charles H. Herty
1935 F. P. Dunnington
1936 W. H. McIntyre
1937 James Lewis Howe
1938 Charles E. Coates
1939 Frank K. Cameron
1940 J. Sam Guy
1941 William F. Hand
1942 Townes R. Leigh
1943 John H. Yoe
1944 James E. Mills
1945 Paul M. Gross
1946 Wilbur A. Lazier
1947 E. Emmett Reid
1948 W. R. Rudd
1949 Osborne R. Quayle
1950 Ralph W. Bost
1951 James T. MacKenzie
1952 A. E. Bailey
1953 Raymond W. McNamee
1954 John R. Sampey
1955 Frank J. Soday
1956 Mahlon P. Etheredge
1957 Stewart J. Lloyd
1958 Lucius A. Bigelow
1959 C. Harold Fisher
1960 Arthur E. Wood
1961 Howard E. Skipper
1962 Charles Hauser
1963 Jack Hine
1964 S. Young Tyree
1965 Charles T. Lester
1966 James E. Copenhaver
1967 G. H. Cartledge
1968 Charles N. Reilley

1969 George L. Drake, Jr.
1970 Robert E. Lutz
1971 S. William Pelletier
1972 Kent C. Brannock
1973 D. Stanley Tarbell
1974 John A. Montgomery
1975 Mary L. Good

Indiana Section Award

1967 Wayne W. Hilty
1969 Edward J. Hughes
1971 Perie R. Pitts
1972 Asa N. Stevens
1973 Keith M. Seymour
1974 Maurice E. Clark
1975 Robert M. Brooker

Iowa Award

1948 Frank H. Spedding
1949 Robert S. aCsey
1950 Henry A. Mattill
1951 Henry Gilman
1952 George Glockler
1953 Leo P. Sherman
1954 Harley A. Wilhelm
1955 Walter R. Fetzer
1956 Carl C. Kesler
1957 James B. Culbertson
1958 Ben H. Peterson
1959 Robert E. Rundle
1960 Stanley Wawzonek
1961 Harvey C. Diehl
1963 Clarence P. Berg
1965 Lee D. Ough
1967 W. Bernard King
1971 Glen A. Russell
1973 Henry B. Bull

Chemist of the Year

Joliet

1958 M. Joan Preising
1959 Edward H. Carus
1960 Joseph V. Karabinos
1961 Howard W. Gould
1963 Donald T. Lurvey
1966 James A. Hazdra

Scientific Award

Kanawha Valley

1970 Roy L. Pruett
1971 B. Das Sarma
1972 E. E. Marcinkowsky
1973 David J. Trecker
1974 William C. Kuryla
1975 James R. Johnson

Kenneth A. Spencer Award

Kansas City

1955 Ralph M. Hixon
1956 Conrad A. Elvehjem
1957 William C. Rose
1958 Elmer V. McCollum
1959 Karl Folkers
1960 Clyde H. Bailey
1961 Herbert L. Haller
1962 Arnold K. Balls
1963 Charles G. King
1964 Daniel Swern
1965 Aaron M. Altschul
1966 Robert L. Metcalf

Local Section Awards (continued)

Spencer Award (cont.)

1967 Melville L. Wolfrom
1968 Herbert E. Carter
1969 Edwin T. Mertz
1970 Lyle D. Goodhue
1971 William J. Darby
1972 Emil M. Mrak
1973 Esmond E. Snell
1974 Roy L. Whistler
1975 Thomas H. Jukes

Citizen Chemist Award

Kansas City

1967 Leonard V. Sorg
1968 John S. Ayres
1969 Herbert M. Steininger
1970 Ludwig C. Krchma
1971 (No award)
1972 Myron N. Jorgensen

Remsen Memorial Lecture

Maryland

1946 Roger Adams
1947 Samuel C. Lind
1948 Elmer V. McCollum
1949 Joel H. Hildebrand
1950 Edward C. Kendall
1951 Hugh S. Taylor
1952 W. Mansfield Clark
1953 Edward L. Tatum
1954 Vincent du Vigneaud
1955 Willard F. Libby
1956 Farrington Daniels
1957 Melvin Calvin
1958 Robert W. Woodward
1959 Edward Teller
1960 Henry Eyring
1961 Herbert C. Brown
1962 George Porter
1963 Harold C. Urey
1964 Paul D. Bartlett
1965 James R. Arnold
1966 Paul H. Emmett
1967 Marshall W. Nirenberg
1968 Har G. Khorana
1969 Albert L. Lehninger
1970 George S. Hammond
1971 George S. Pimentel
1972 Charles H. Townes
1973 Frank H. Westheimer
1974 Elias J. Corey
1975 Henry Taube

Maryland Chemist Award

Maryland

1962 E. Emmett Reid
1963 W. Mansfield Clark
1964 Alsoph H. Corwin
1965 John C. Krantz, Jr.
1966 Belle Otto Talbot
1967 Walter S. Koski
1968 George L. Braude
1969 Leslie Hellerman
1970 Paul H. Emmett
1971 Giles B. Cooke
1972 Arnold H. Seligman
1973 Lester P. Kuhn
1974 Joyce J. Kaufman
1975 Benjamin Witten

Southern Chemist Award

Memphis

1950 Calvin A. Buehler
1951 George E. Boyd
1952 Allan T. Gwathmey
1953 Francis Webber Sherwood
1954 Hans Boegh Jonassen
1955 James Tucker MacKenzie
1956 C. Harold Fisher
1957 Charles Newton Kimberlin, Jr.
1958 Thomas Palmer Nash, Jr.
1959 Edward S. Amis
1960 Harry W. Coover
1961 Oscar K. Rice
1962 Frederick R. Darkis
1963 Paul Tarrant
1964 Alan Bell
1965 Emmett B. Carmichael
1966 Wilson A. Reeves
1967 Hilton A. Smith
1968 Ruth Rogan Benerito
1969 Harry H. Sisler
1970 Earl W. Sutherland
1971 Gregory R. Choppin
1972 William S. Pelletier
1973 Paul K. Kuroda
1974 Michael Kasha
1975 Leo A. Goldblatt

Renaud Foundation Lectureship

Michigan State University

1949 Max A. Lauffer
1950 Milton Burton
1951 Melvin S. Newman
1952 Harvey Diehl
1953 Melvin Calvin
1954 Richard Dodson
1955 Leo Marion
1956 Joseph J. Katz
1957 Irving M. Klotz
1958 John D. Roberts
1959 Henry Eyring
1960 Herbert A. Laitinen
1961 George W. Watt
1962 Derek H. R. Barton
1963 Peter J. W. Debye
1964 Charles Tanford
1965 Elias J. Corey
1966 Manfred Eigen
1967 Ronald S. Nyholm
1968 Herbert C. Brown
1969 Harden M. McConnell
1970 F. Albert Cotton
1971 Carl Djerassi
1972 Linus Pauling
1973 Paul D. Bartlett
1974 Gerhard Herzberg
1975 William N. Lipscomb

Texaco Research Award

Mid-Hudson Section

1968 Alfred Arkell
1970 John A. Patterson
1973 Kenneth L. Kreuz
1975 Isaac D. Rubin

Milwaukee Section Award

1958 J. V. Steinle
1959 Henry B. Merrill
1960 John Biel
1961 August E. Orthmann
1962 Carroll E. Imhoff
1963 Armand J. Quick
1964 Hamilton A. Pinkalla
1965 Enos H. McMullen
1966 Henry J. Peppler
1967 Stephen E. Freeman
1968 John D. Jenkins
1969 Beatrice Kassell
1970 Merle G. Farnham
1971 Alfred Bader
1974 Allen G. Boyes
1975 William K. Miller

Minnesota Award

1958 Lee Irwin Smith
1960 Izaak M. Kolthoff
1962 Frank A. Bovey
1964 Fred Smith
1966 Courtland I. Agre
1968 Bryce L. Crawford
1970 William E. Parham
1972 A. Richard Baldwin
1974 John L. Wilson

College Award

Minnesota

1971 Olaf A. Runquist
1973 John R. Holum
1975 Emil Slowinski

Virgil B. Payne Award

Monmouth County Section

1965 Vincent Webers
1966 John N. Mrgudich
1967 Thomas J. Engelbach
1968 Arthur F. Daniel
1969 William F. Nye
1970 John P. Wadington
1971 William C. Pfefferle
1972 Raymond J. Le Strange
1973 Charles C. Worthington
1974 Philip DeCoursey Kratz
1975 Thomas S. Harvey

Nebraska Lectureship

1963 Peter Debye
1964 Vincent du Vigneaud
1967 Andrew Streitweiser
1968 James P. Collman
1969 J. Calvin Giddings
1971 William Klemperer
1972 Peter Lengyel
1973 Gilbert J. Stork
1974 Lawrence F. Dahl
1975 Lockhart B. Rogers

John Gamble Kirkwood Award

New Haven

1962 Lars Onsager
1963 Manfred Eigen
1964 Robert S. Mulliken
1965 Robert B. Woodward
1966 Henry Taube
1967 Joseph E. Mayer
1969 Neil Bartlett
1971 Paul J. Flory
1973 Sune Bergstrom

Local Section Awards (continued)

William H. Nichols Medal

New York

1903 Edward B. Voorhees
1904 (No award)
1905 Charles L. Parsons
1906 Marston T. Bogert
1907 Howard B. Bishop
1908 William H. Walker
1909 William A. Noyes
1910 Leo H. Baekeland
1911 M. A. Rosanof
1912 Charles James
1913 (No award)
1914 Moses Gomberg
1915 Irving Langmuir
1916 Claude S. Hudson
1917 (No award)
1918 Treat B. Johnson
1919 (No award)
1920 Irving Langmuir
1921 Gilbert N. Lewis
1922 (No award)
1923 Thomas Midgely, Jr.
1924 Charles A. Kraus
1925 Edward C. Franklin
1926 Samuel C. Lind
1927 Roger Adams
1928 Hugh S. Taylor
1929 William L. Evans
1930 Samuel E. Sheppard
1931 John A. Wilson
1932 James B. Conant
1933 (No award)
1934 Henry C. Sherman
1935 Julius A. Nieuwland
1936 William M. Clark
1937 Frank C. Whitmore
1938 P. A. Levene
1939 Joel H. Hildebrand
1940 John M. Nelson
1941 Linus Pauling
1942 Duncan A. MacInnes
1943 Arthur B. Lamb
1944 Carl S. Marvel
1945 Vincent du Vigneaud
1946 Wendell M. Stanley
1947 George B. Kistiakowski
1948 Glenn T. Seaborg
1949 Izaak M. Koithoff
1950 Oskar Wintersteiner
1951 Henry Eyring
1952 Frank H. Spedding
1953 Reynold C. Fuson
1954 Charles P. Smyth
1955 Wendell M. Latimer
1956 Robert B. Woodward
1957 Louis P. Hammett
1958 Melvin Calvin
1959 Herbert C. Brown
1960 Herman F. Mark
1961 Peter J. W. Debye
1962 Paul J. Flory
1963 Louis F. Fieser
1964 Arthur C. Cope
1965 Herbert E. Carter
1966 Frederick D. Rossini
1967 Karl Folkers
1968 William S. Johnson
1969 Marshall Nirenberg
1970 Britton Chance
1971 Henry Taube
1972 John D. Roberts
1973 R. Bruce Merrifield
1974 Harold A. Scheraga
1975 F. Albert Cotton

The Madison Marshall Award

North Alabama

1964 George B. Kistiakowsky
1965 Peter J. W. Debye
1966 Carl S. Marvel
1967 Joel Hildebrand
1968 Henry Eyring
1969 Linus Pauling
1971 Paul D. Bartlett
1972 Glenn T. Seaborg
1973 Carl Djerassi
1974 Gerhard Herzberg
1975 Herbert C. Brown

Leo Hendrik Baekeland Award

North Jersey

1949 Edward R. Gilliland
1947 Paul John Flory
1949 Eugene G. Rochow
1951 Lewis Hastings Sarett
1953 Leo Brewer
1955 Robert B. Woodward
1957 Bruno Hasbrough Zimm
1959 Carl Djerassi
1961 Gilbert Stork
1963 F. Albert Cotton
1965 Eugene E. van Tamelen
1967 George A. Olah
1969 Ronald Breslow
1971 Stuart Alan Rice
1973 Willis H. Flygare
1975 Christopher S. Foote

Theodore Williams Richards Medal

Northeastern

1932 Theodore W. Richards
1932 Arthur A. Noyes
1934 Gregory P. Baxter
1936 Charles A. Kraus
1938 Gilbert N. Lewis
1940 Claude S. Hudson
1942 Frederick G. Keyes
1946 Roger Adams
1947 Linus Pauling
1948 Edwin J. Cohn
1950 John G. Kirkwood
1952 Morris S. Kharasch
1954 George Scatchard
1956 Melvin Calvin
1958 Robert B. Woodward
1960 Robert S. Mulliken
1962 Saul Winstein
1964 Lars Onsager
1966 Paul D. Bartlett
1968 George B. Kistiakowsky
1970 William von E. Doering
1972 William Stein and Stanford Moore
1974 Henry Eyring

James Flack Norris Award for Outstanding Achievement in the Teaching of Chemistry

Northeastern

1951 George Shannon Forbes
1953 John Xan
1955 Harry Nicholls Holmes
1956 Norris Watson Rakestraw
1957 Emma Perry Carr
1957 Mary Lura Sherill
1957 Farrington Daniels
1959 Hermann Irving Schlesinger
1959 Louis Frederick Fieser
1960 Louis Plack Hammett
1961 Joel Henry Hildebrand
1962 Ralph Lloyd Shriner
1963 Avery Allen Ashdown
1964 James Arthur Campbell
1964 Laurence Edward Strong
1965 Walter John Moore
1966 John Arrend Timm
1966 Edgar Bright Wilson
1967 Edward Lauth Haenisch
1968 Samuel Edward Kamerling
1968 William Campbell Root
1969 Joseph E. Mayer
1970 Hubert Alyea
1971 Charles Bickel
1972 Saul Cohen
1973 Eugene Rochow
1974 Grant H. Harnest
1975 Leonard K. Nash

Chemist of the Year

Northeast Indiana

1957 Ralph E. Broyles
1958 Maurice M. Felger
1959 R. Conway Knapp
1960 Harry B. Bolson
1961 Earl L. Smith
1962 Ruth M. Wimmer
1963 Tod G. Dixon
1964 Wendell D. Mason
1965 Paul G. Fulkerson
1966 Judson West, Jr.
1967 Warren E. Hoffman
1968 William E. Donahue
1969 Stephen P. Coburn
1970 Max M. Lee
1971 Ernest C. Koerner
1972 Evert A. Mol
1973 Marvin E. Glidewell
1974 Marvin A. Peterson
1975 (not given)

Oklahoma Chemist Award

1972 Wayne E. White
1973 Otis C. Dermer
1974 Robert L. Banks
1975 Charles M. Starks

Award for Service Through Chemistry

Orange County

1972 Arie Jan Haagen-Smit
1973 James N. Pitts, Jr.
1974 Arnold O. Beckman
1975 F. Sherwood Rowland

Pauling Award

Oregon and Puget Sound

1966 Linus Pauling
1967 Manfred Eigen
1968 Herbert C. Brown
1969 Henry Eyring
1970 Harold C. Urey
1971 Gerhard Herzberg
1972 E. Bright Wilson
1973 Elias J. Corey
1974 Roald Hoffman
1975 Paul D. Bartlett

Local Section Awards (continued)

Edgar Fahs Smith Memorial Lecture

Philadelphia

1929 Charles W. Balke
1930 Charles A. Kraus
1931 N. V. Sidgwick
1932 Louis Kahlenberg
1933 James B. Conant
1934 Hugh S. Taylor
1935 Colin G. Fink
1936 Julius Stieglitz
1937 Charles H. Herty
1938 Roger Adams
1939 George B. Kistiakowsky
1940 Per K. Frolich
1941 Arthur A. Blanchard
1942 Harold C. Urey
1943 (No award)
1946 The Svedberg
1947 Wendell M. Stanley
1948 Carl S. Marvel
1949 Linus C. Pauling
1950 Edward C. Kendall
1951 Herman I. Schlesinger
1952 Paul D. Bartlett
1953 Joel H. Hildebrand
1954 Henry Eyring
1955 Melvin Calvin
1956 Paul M. Doty
1957 Peter J. W. Debye
1958 Vincent du Vigneaud
1959 Robert B. Woodward
1960 Glenn T. Seaborg
1961 Paul J. Flory
1962 Herbert C. Brown
1963 Henry Taube
1964 John T. Edsall
1965 Izaak M. Kolthoff
1966 Carl Djerassi
1967 Francis T. Bacon
1968 Cyrus Levinthal
1969 Elias J. Corey
1970 John D. Roberts
1971 Lawrence F. Dahl
1972 (No award)
1973 (No award)
1974 Gerald M. Edelman
1975 Nelson J. Leonard

Philadelphia Section Award

1962 Murray Hauptschein
1963 Charles C. Price
1964 Newman M. Bortnick
1965 Sidney Weinhouse
1966 Serge Timasheff
1967 Alan G. MacDiarmid
1968 B. Peter Block
1969 Britton Chance
1970 Edward R. Thornton
1971 Ernst Berliner
1972 Madeleine M. Joullie
1973 Lyle H. Phifer
1974 Bernard Loev
1975 Irwin A. Rose

Pittsburgh Award

1933 Ralph E. Hall
1934 Charles E. Nesbit
1935 (not given)
1936 Andrew W. Mellon
Richard B. Mellon
1937 Francis C. Frary
1938 George H. Clapp
1939 Edward R. Weidlein
1940 Alexander Silverman
1941 Webster N. Jones
1942 Charles G. King
1943 Junius D. Edwards
1944 Leonard H. Cretcher
1945 John C. Warner
1946 William P. Yant
1947 Chester G. Fisher
1948 Henry H. Storch
1949 Harry V. Churchill
1950 William A. Hamor
1951 William A. Gruse
1952 Homer H. Lowry
1953 Paul H. Emmett
1954 Paul D. Foote
1955 George D. Beal
1956 Robert F. Mehl
1957 Alfred R. Powell
1958 Max A. Lauffer
1959 Frederick D. Rossini
1960 Robert B. Anderson
1961 Earl A. Gulbransen
1962 Klaus H. Hoffmann
1963 Harold P. Klug
1964 Henry S. Frank
1965 Foil A. Miller
1966 Earl K. Wallace
1967 Robert A. Friedel
1968 Irving Wender
1969 W. Conard Fernelius
1970 Tobias H. Dunkelberger
1971 Paul C. Cross
1972 Edmund O. Rhodes
1973 W. Edward Wallace
1974 Bernard Lewis
1975 John A. Pople

Leonardo Igaravidez Award

Puerto Rico

1972 Osvaldo Ramirez Torres
1973 Justo Hernandez Mora
1974 Conrrado Asenjo
1975

Harrison E. Howe Lectureship

Rochester

1945 Linus C. Pauling
1946 Glenn T. Seaborg
1947 Carl F. and Gerti T. Cori
1948 Herman F. Mark
1949 Karl Folkers
1950 Robert B. Woodward
1951 Paul D. Bartlett
1952 Melvin Calvin
1953 Herbert C. Brown
1954 Frank H. Westheimer
1955 Paul J. Flory
1956 Vernon H. Cheldelin
1957 John D. Roberts
1958 William N. Lipscomb
1959 Paul M. Doty
1960 Henry Taube
1961 Michael J. S. Dewar
1962 Gilbert Stork
1963 Charles H. Townes
1964 Marshall W. Nirenberg
1965 Manfred Eigen
1966 Britton Chance
1967 Martin Karplus
1968 Harden M. McConnell
1969 Roald Hoffmann
1970 Elias J. Corey
1971 John A. Pople
1972 Harry B. Gray
1973 Kai Siegbahn and
David W. Turner
1974 Ronald C. D. Breslow
1975 F. Albert Cotton

Midwest Award

St. Louis

1944 Lucas P. Kyrides
1945 Carl F. Cori and
Gerty T. Cori
1946 Anderson W. Ralston
1948 Paul L. Day
1949 Robert D. Coghill
1950 William S. Haldeman
1951 Henry Gilman
1952 Edward Mallinckrodt, Jr.
1953 Roger Adams
1954 Ralph M. Hixon
1955 Cliff S. Hamilton
1956 Carroll Hochwalt
1957 Ray Q. Brewster
1958 Charles D. Hurd
1959 Melvin DeGroote
1960 Charles D. Harrington
1961 Samuel I. Weissman
1962 Oliver H. Lowry
1963 Herman Pines
1964 Harold H. Strain
1965 Richard H. Wiley
1966 Ralph G. Pearson
1967 Frank H. Spedding
1968 Byron Riegel
1969 Joseph J. Katz
1970 Irving M. Klotz
1971 John C. Bailar, Jr.
1972 Myron L. Bender
1973 Herbert S. Gutoowsky
1974 Glen A. Russell
1975 Takeru Higuchi

St. Louis Section Award

1970 David Lipkin
1971 Clayton F. Callis
1972 C. David Gutsche
1973 A. John Speziale
1974 Robert W. Murray
1975 James F. Roth

Ottenberg Service Award

Santa Clara Valley

1974 Shirley B. Radding
1975 Oliver F. Senn

Southwest Regional Award

Sections in Southwest Region

1948 Eugene P. Schoch
1949 Frederick E. Frey
1950 Roger Williams
1951 Klare S. Markley
1952 Paul L. Day
1953 Henry R. Henze
1954 Philip W. West
1955 Vladimir A. Kalichevsky
1956 Winfred O. Milligan
1957 Guy Waddington
1958 Kenneth A. Kobe
1959 Paul Delahay
1960 Edward S. Amis
1961 Harry L. Lochte
1962 Joe L. Franklin, Jr.
1963 Jacob Sacks
1964 Raymond Reiser
1965 Norman H. Hackerman
1966 Richard B. Turner
1967 Sean P. McGlynn

Local Section Awards (continued)

S.W. Reg. Award (cont.)

1968 Rowland Pettit
1969 Nugent F. Chamberlain
1970 Paul Kuroda
1971 Bruno Swolinski
1972 Ruth R. Benerito
1973 John L. Margrave
1974 George W. Watt
1975 William A. Pryor

Outstanding Chemist Award

South Carolina

1965 James Earl Copenhaver
1966 Harry Willard Davis
1967 Samuel Adam Wideman
1971 Joseph Ward Bouknight
1973 William Mellen McCord

Southeastern Texas Section Award

1969 Nugent F. Chamberlain
1970 Joe L. Franklin, Jr.
1971 Winfred O. Milligan
1972 Raymond B. Seymour
1973 Briggs B. Manuel
1974 Gilbert Chambers

Distinguished Service Award

South Jersey

1965 Henry E. Lee
1968 William E. Kirst
1969 Clyde O. Davis
1971 Lyle Hamilton
1974 (not given)

Richard C. Tolman Medal

Southern California

1960 William G. Young
1961 Anton B. Burg
1962 Ernest H. Swift
1963 Willis Conway Pierce
1964 Arie Jan Haagen-Smit
1965 Thomas F. Doumani
1966 Arthur W. Adamson
1967 Ulric B. Bray
1968 Francis E. Blacet
1969 Robert D. Vold
1970 Robert L. Pecsok
1971 Rowland C. Hansford
1972 James Bonner
1973 Howard Reiss
1974 John D. Roberts

Award for Excellence in Industrial Chemical Research

University of Michigan

1971 Paul L. Creger
1974 Horace A. Dewald

Distinguished Service Award

Virginia

1948 Wortley F. Rudd
1949 Sidney S. Negus
1950 Garnett Ryland
1951 Edwin Cox
1952 Hiram Rupert Hanmer
1953 Lloyd C. Bird
1954 Eugene D. Crittenden
1955 John Campbell Forbes
1956 John H. Yoe
1957 Robert H. Kean
1958 William G. Guy
1959 Edward S. Harlow
1960 Rodney C. Berry
1961 Alfred Burger
1962 Allan T. Gwathmey
1963 J. Stanton Pierce
1964 Robert E. Lutz
1965 Fred R. Millhiser
1966 Ira A. Updike and Winifred Wood Updike
1967 Randolph N. Gladding
1968 William E. Trout, Jr.
1969 Mary E. Kapp
1970 Lynn DeForrest Abbott
1971 Loyal H. Davis
1972 Alfred R. Armstrong
1973 James Earl York, Jr.
1974 Oscar Rudolf Rodig
1975 Lowell Vernon Heisey

Hillebrand Award

Washington

1925 R. F. Jackson
1926 George W. Morey
1927 E. P. Bartlett
1928 James H. Hibben
1929 (No award)
1930 Claude S. Hudson
1931 G. E. F. Lundell
1932 F. B. LaForge and H. L. J. Haller
1933 Edward W. Washburn
1934 Frederick D. Rossini
1935 O. R. Wulf
1936 Vincent du Vigneaud
1937 Sterling B. Hendricks
1938 Raleigh Gilchrist and Edward Wichers
1939 R. E. Gibson
1940 Ferdinand G. Brickwedde
1941 Michael X. Sullivan
1942 J. F. Schairer
1943 Ben H. Nicolet
1944 Raymond M. Hann
1945 Stephen Brunauer
1946 James I. Hoffman
1947 Nathan L. Drake
1948 Edgar R. Smith
1949 Lyndon F. Small
1950 Henry Stevens, E. J. Coulson and Joseph R. Spies
1951 Horace S. Isbell
1952 Dean Burk
1953 Bernard L. Horecker
1954 William A. Zisman
1955 Roger G. Bates
1956 Francis O. Rice
1957 Jesse P. Greenstein
1958 Bernhard Witkop
1959 Leon A. Heppel
1960 Frank T. McClure
1961 Sidney Udenfriend
1962 Philip H. Abelson
1963 Martin Jacobson and Morton Beroza
1964 Ellis R. Lippincott, Jr.
1965 Marshall W. Nirenberg
1966 Arthur A. Westerburg and Robert M. Fristrom
1967 Everette L. May and Nathan B. Eddy
1968 Earl R. Stadtman
1969 Isabella L. Karle and Jerome Karle
1970 Herbert A. Sober and Elbert A. Peterson
1971 Nicolae Filipescu
1972 Frederick A. H. Rice
1973 Daniel P. Schwartz
1974 Elizabeth F. Neufeld

Visiting Scientist Award

Western Connecticut

1967 Calvin A. Vanderwerf
1968 William F. Kieffer
1969 Leallyn B. Clapp
1970 William T. Lippincott
1971 M. H. Gardner
1972 Harold G. Cassidy
1973 Derek A. Davenport
1974 Gilbert P. Haight
1975 Clark E. Bricker

Jacob F. Schoellkopf Medal

Western New York

1931 Frank J. Tone
1932 W. Hale Church
1933 Frank A. Hartman
1934 James C. Downs
1935 F. Austin Lidbury
1936 Albert H. Hooker
1937 James G. Marshall
1938 Sterling Temple
1939 Charles F. Vaughn
1940 W. H. Bradshaw
1941 Arthur W. Burwell
1942 Lawrence H. Flett
1943 Raymond R. Ridgway
1944 Glen D. Bagley
1945 Alexander Schwarchmann
1946 Harvey N. Gilbert
1947 Leo I. Dana
1948 Marvin J. Udy
1949 Robert L. Murray
1950 Joseph H. Brennan
1951 Corneille O. Strother
1952 Henry N. Baumann, Jr.
1953 Emmette F. Izard
1954 Oliver W. Cass
1955 Hendrik de wet Erasmus
1956 Raymond W. Hess
1957 J. Frederick Walker
1958 Robert B. MacMullen
1959 Max E. Bretschger
1960 George H. Wagner
1961 Rolland J. Gladieux
1962 Clifford C. Furnas
1963 Robert M. Milton
1964 Walter H. Prahl
1965 David Pressman
1966 Leon O. Winstrom
1967 Gordon M. Harris
1968 Donald L. Bailey
1969 David Harker
1970 Calvin D. Ritchie
1971 Warren B. Blumenthal
1972 James Economy
1973 Michael Laskowski
1974 Thomas J. Bardos
1975 John E. Bristol

National Meetings

1890-1975

Number	Place and Date	Registration
1	Newport, R. I., Aug. 6 and 7, 1890	43
2	Philadelphia, Pa., Dec. 30 and 31, 1890	75
3	Washington, D. C., Aug. 17 and 18, 1891	
4	New York, N. Y., Dec. 29 and 30, 1891	
5	Rochester, N. Y., Aug. 17 and 18, 1892	
6	Pittsburgh, Pa., Dec. 28 and 29, 1892	
7	Chicago, Ill., Aug. 21 and 22, 1893	83
8	Baltimore, Md., Dec. 17 and 18, 1893	
9	Brooklyn, N. Y., Aug. 15 and 16, 1894	124
10	Boston and Cambridge, Mass., Dec. 27 and 28, 1894	84
11	Springfield, Mass., Aug. 27 and 28, 1895	65
12	Cleveland, Ohio, Dec. 30 and 31, 1895	
13	Buffalo, N. Y., Aug. 21 and 22, 1896	
14	Troy, N. Y., Dec. 29 and 30, 1896	
15	Detroit, Mich., Aug. 9 and 10, 1897	
16	Washington, D. C., Dec. 29 and 30, 1897	
17	Boston, Mass., Aug. 22 and 23, 1898	
18	New York, N. Y., Dec. 28 and 29, 1898	
19	Columbus, Ohio, Aug. 21 and 22, 1899	
20	New Haven, Conn., Dec. 27 and 28, 1899	
21	New York, N. Y., June 25 and 26, 1900	
22	Chicago, Ill., Dec. 28 and 28, 1900	
23	New York, N. Y., April 12 and 13, 1901 (25th Anniversary)	
24	Denver, Colo., Aug. 27 and 28, 1901	
25	Philadelphia, Pa., Dec. 30 and 31, 1901	
26	Pittsburgh, Pa., June 30 and July 1, 1902	
27	Washington, D. C., Dec. 29 and 30, 1902	
28	Cleveland, Ohio, June 29 and 30, 1903	
29	St. Louis, Mo., Dec. 30 and 31, 1903	
30	Providence, R. I., June 21, 22, and 23, 1904	
31	Philadelphia, Pa., Dec. 30 and 31, 1904	
32	Buffalo, N. Y., June 22, 23, and 24, 1905	
33	New Orleans, La., Dec. 29 and 30, 1905	
34	Ithaca, N. Y., June 28, 29, and 30, 1906	
35	New York, N. Y., Dec. 27, 28, 29, and 30, 1906	
36	Toronto, Canada, June 27, 28, and 29, 1907	
37	Chicago, Ill., Dec. 31, 1907, to Jan. 3, 1908	
38	New Haven, Conn., June 29 to July 2, 1908	
39	Baltimore, Md., Dec. 29, 1908, to Jan. 1, 1909	400
40	Detroit, Mich., June 29 to July 2, 1909	320
41	Boston, Mass., Dec. 27 to 31, 1909	558
42	San Francisco, Calif., July 12 to 16, 1910	290
43	Minneapolis, Minn., Dec. 28 to 31, 1910	275
44	Indianapolis, Ind., June 28 to July 1, 1911	432
45	Washington, D. C., Dec. 27 to 30, 1911	658
46	New York, N. Y., Sept., Sept. 11, 1912	
47	Milwaukee, Wis., March 25 to 28, 1913	425
48	Rochester, N. Y., Sept. 8 to 12, 1913	625
49	Cincinnati, Ohio, April 6 to 10, 1914	658
50	New Orleans, La., March 31 to April 3, 1915	300
51	Seattle, Wash., Aug. 31 to Sept. 3, 1915	225
52	Urbana-Champaign, Ill., April 18 to 21, 1916	729
53	New York, N. Y., Sept. 25 to 30, 1916	1,905
54	Kansas City, Mo., April 10 to 13, 1917	305
55	Boston, Mass., Sept. 10 to 13, 1917	700
56	Cleveland, Ohio, Sept. 10 to 13, 1918	588
57	Buffalo, N. Y., April 7 to 11, 1919	1,071
58	Philadelphia, Pa., Sept. 2 to 6, 1919	1,687
59	St. Louis, Mo., April 12 to 17, 1920	976
60	Chicago, Ill., Sept. 6 to 10, 1920	1,308
61	Rochester, N. Y., April 25 to 30, 1921	1,139
62	New York, N. Y., Sept. 6 to 10, 1921	1,557
63	Birmingham, Ala., April 3 to 7, 1922	381
64	Pittsburgh, Pa., Sept. 4 to 8, 1922	1,325
65	New Haven, Conn., April 2 to 7, 1923	1,200
66	Milwaukee, Wis., Sept. 10 to 14, 1923	1,000
67	Washington, D. C., April 21 to 26, 1924	1,934
68	Ithaca, N. Y., Sept. 10 to 13, 1924	1,087
69	Baltimore, Md., April 6 to 10, 1925	1,524
70	Los Angeles, Calif., Aug. 3 to 8, 1925	880
71	Tulsa, Okla., April 5 to 9, 1926	448
72	Philadelphia, Pa., Sept. 6 to 11, 1926 (50th Anniversary)	2,249
73	Richmond, Va., April 11 to 16, 1927	1,317
74	Detroit, Mich., Sept. 5 to 10, 1927	1,618

Number	Place and Date	Registration
75	St. Louis, Mo., April 16 to 19, 1928	1,307
76	Swampscott, Mass., Sept. 10 to 14, 1928	1,922
77	Columbus, Ohio, April 29 to May 3, 1929	1,755
78	Minneapolis, Minn., Sept. 9 to 13, 1929	1,174
79	Atlanta, Ga., April 7 to 11, 1930	1,434
80	Cincinnati, Ohio, Sept. 8 to 12, 1930	1,668
81	Indianapolis, Ind., March 30 to April 3, 1931	1,873
82	Buffalo, N. Y., Aug. 31 to Sept. 4, 1931	2,047
83	New Orleans, La., March 28 to April 1, 1932	777
84	Denver, Colo., Aug. 22 to 26, 1932	916
85	Washington, D. C., March 27 to 31, 1933	2,293
86	Chicago, Ill., Sept. 11 to 15, 1933	3,191
87	St. Petersburg, Fla., March 26 to 30, 1934	730
88	Cleveland, Ohio, Sept. 10 to 14, 1934	2,513
89	New York, N. Y., April 22 to 26, 1935	5,105
90	San Francisco, Calif., Aug. 19 to 23, 1935	1,111
91	Kansas City, Mo., April 13 to 17, 1936	1,491
92	Pittsburgh, Pa., Sept. 7 to 11, 1936	2,904
93	Chapel Hill, N. C., April 12 to 15, 1937	2,155
94	Rochester, N. Y., Sept. 6 to 10, 1937	3,483
95	Dallas, Texas, April 18 to 22, 1938	1,093
96	Milwaukee, Wis., Sept. 5 to 9, 1938	2,871
97	Baltimore, Md., April 3 to 7, 1939	4,024
98	Boston, Mass., Sept. 11 to 15, 1939	3,924
99	Cincinnati, Ohio, April 8 to 12, 1940	3,514
100	Detroit, Mich., Sept. 9 to 13, 1940	4,206
101	St. Louis, Mo., April 7 to 11, 1941	3,960
102	Atlantic City, N. J., Sept. 8 to 12, 1941	5,021
103	Memphis, Tenn., April 20 to 23, 1942	2,324
104	Buffalo, N. Y., Sept. 7 to 11, 1942	4,288
105	Detroit, Mich., April 12 to 16, 1943	3,719
106	Pittsburgh, Pa., Sept. 6 to 10, 1943	3,537
107	Cleveland, Ohio, April 3 to 6, 1944	4,033
108	New York, N. Y., Sept. 11 to 15, 1944	9,701
	NOTE: No meetings held in 1945 owing to immediate postwar conditions	
109	Atlantic City, N. J., April 8 to 12, 1946	7,587
110	Chicago, Ill., Sept. 9 to 13, 1946	8,921
111	Atlantic City, N. J., April 14 to 18, 1947	7,030
112	New York, N. Y., Sept. 15 to 19, 1947	11,649
113	Chicago, Ill., April 19 to 23, 1948	8,391
114	Washington, D.C., August 30 to Sept. 3, 1948	2,572
114	St. Louis, Mo., Sept. 6 to 10, 1948	2,322
114	Portland, Oregon, Sept. 13 to 17, 1948	847
115	San Francisco, Calif., March 27 to April 1, 1949	3,538
116	Atlantic City, N.J., Sept. 18 to 23, 1949	8,232
117	Houston, Texas, March 26 to 30, 1950	1,811
117	Philadelphia, Pa., April 9 to 13, 1950	3,881
117	Detroit, Mich., April 16 to 21, 1950	2,764
118	Chicago, Ill., Sept. 3 to 8, 1950	7,877
119	Cleveland, Ohio, April 8 to 12, 1951	6,846
120	New York, N. Y., Sept. 3 to 7, 1951 (Diamond Jubilee Meeting)	13,466
121	Buffalo, N, Y., March 30 to 27, 1952	6,021
121	Milwaukee, Wisc., March 30 to April 3, 1952	3,006
122	Atlantic City, N. J., Sept. 14 to 19, 1952	8,800
123	Los Angeles, Calif., March 15 to 19, 1953	4,284
124	Chicago, Ill., Sept. 6 to 11, 1953	10,000
125	Kansas City, Mo., March 23 to April 1, 1954	3,842
126	New York, N. Y., Sept. 12 to 17, 1954	13,514
127	Cincinnati, Ohio, March 29 to April 7, 1955	5,496
128	Minneapolis, Minn., Sept. 11 to 16, 1955	6,621
129	Dallas, Texas, April 8 to 13, 1956	4,792
130	Atlantic City, N. J., Sept. 16 to 21, 1956	11,094
131	Miami, Fla., April 7 to 12, 1957	5,601
132	New York, N. Y., Sept. 8 to 13, 1957	15,047
133	San Francisco, Calif., April 13 to 18, 1958	6,454
134	Chicago, Ill., Sept. 7 to 12, 1958	10,837
135	Boston, Mass., April 5 to 10, 1959	8,532
136	Atlantic City, N. J., Sept. 13 to 18, 1959	10,768
137	Cleveland, Ohio, April 5 to 14, 1960	7,012
138	New York, N. Y., Sept. 11 to 16, 1960	13,300
139	St. Louis, Mo., March 21 to 30, 1961	5,888
140	Chicago, Ill., Sept. 3 to 8, 1961	9,818
141	Washington, D. C., March 20 to 29, 1962	8,491
142	Atlantic City, N. J., Sept. 9 to 14, 1962	8,255
143	Cincinnati, Ohio, Jan. 13 to 18, 1963	1,174
144	Los Angeles, Calif., March 31 to April 5, 1963	5,344
145	New York, N. Y., Sept. 8 to 13, 1963	12,909
146	Denver, Colo., Jan. 19 to 24, 1964	1,304
147	Philadelphia, Pa., April 5 to 10, 1964	6,727
148	Chicago, Ill., August 30 to Sept. 4, 1964	10,306
149	Detroit, Mich., April 4 to 9, 1965	7,175

Number	Place and Date	Registration
150	Atlantic City, N. J., Sept. 12 to 17, 1965	10,304
Winter	Phoenix, Ariz., Jan. 16 to 21, 1966	1,328
151	Pittsburgh, Pa., March 22 to 31, 1966	5,517
152	New York, N. Y., Sept. 11 to 16, 1966	14,044
153	Miami Beach, Fla., April 9 to 14, 1967	7,954
154	Chicago, Ill., Sept. 10 to 15, 1967	10,658
155	San Francisco, Calif., March 31 to April 5, 1968	10,357
156	Atlantic City, N. J., Sept. 8 to 13, 1968	10,614
157	Minneapolis, Minn., April 13 to 18, 1969	6,713
158	New York, N. Y., Sept. 7 to 12, 1969	11,958
159	Houston, Texas, Feb. 22 to 27, 1970	3,728
CIC/ACS	Conf. Toronto, Canada, May 24 to 29, 1970	3,979
160	Chicago, Ill., Sept. 13 to 18, 1970	8,286
161	Los Angeles, Calif., March 28 to April 2, 1971	5,485
162	Washington, D. C., Sept. 12 to 17, 1971	8,057
163	Boston, Mass., April 9 to 14, 1972	6,536
164	New York, N. Y., August 27 to Sept. 1, 1972	8,462
165	Dallas, Texas, April 8 to 13, 1973	5,097
166	Chicago, Ill., August 26 to 31, 1973	7,978
167	Los Angeles, Calif., March 31 to April 5, 1974	6,371
168	Atlantic City, N. J., Sept. 8 to 13, 1974	7,492
169	Philadelphia, Pa., April 6 to 11, 1975	9,260
170	Chicago, Ill., August 24 to 29, 1975	7,162

Membership Statistics

Year	Members	Change	Year	Members	Change	Year	Members	Change
1876	230		1909	4,502	498	1943	36,001	4,284
1877	265	35	1910	5,081	579	1944	39,438	3,437
1878	256	—9	1911	5,603	522	1945	43,075	3,637
1879	289	33	1912	6,219	616	1946	48,755	5,680
1880	303	14	1913	6,673	454	1947	55,100	6,345
1881	314	11	1914	7,170	497	1948	58,782	3,682
1882	293	—21	1915	7,417	247	1949	62,211	3,429
1883	306	13	1916	8,355	938	1950	63,349	1,138
1884	323	17	1917	10,603	2,248	1951	66,009	2,660
1885	255	—68	1918	12,203	1,600	1952	67,730	1,721
1886	241	—14	1919	13,686	1,483	1953	70,155	2,425
1887	235	—6	1920	15,582	1,896	1954	72,287	2,132
1888	227	—8	1921	14,318	—1,264	1955	75,223	2,936
1889	204	23	1922	14,400	82	1956	79,224	4,001
1890	238	34	1923	14,346	—54	1957	81,927	2,703
1891	302	64	1924	14,515	169	1958	85,815	3,888
1892	351	49	1926	14,704	323	1959	88,806	2,991
1893	460	109	1927	15,188	484	1960	92,193	3,387
1894	722	262	1928	16,240	1,052	1961	93,637	1,444
1895	903	181	1929	17,426	1,186	1962	95,210	1,573
1896	1,011	108	1930	18,206	780	1963	96,749	1,539
1897	1,156	145	1931	18,963	757	1964	99,475	2,726
1898	1,415	259	1932	18,572	—391	1965	102,525	3,050
1899	1,569	154	1933	17,645	—927	1966	106,271	3,746
1900	1,715	146	1934	17,561	—84	1967	109,528	3,257
1901	1,933	218	1935	17,541	—20	1968	113,373	3,845
1902	2,188	255	1936	18,727	1,186	1969	116,816	3,443
1903	2,428	240	1937	20,677	1,950	1970	114,323	—2,493
1904	2,675	247	1938	22,185	1,508	1971	112,016	—2,307
1905	2,919	244	1939	23,519	1,334	1972	110,708	—1,298
1906	3,079	160	1940	25,414	1,895	1973	110,285	—423
1907	3,389	310	1941	28,738	3,324	1974	110,799	514
1908	4,004	615	1942	31,717	2,979	1975	110,820	21

INDEX

A

Abelson, P. 333
Abstractors for *CA* 132
Accounts of Chemical Research 117
ACS (*see also* American Chemical Society)
  Board of Directors 193
  "Chemical Monographs" 105
  committees of the 185
  Constitution 180
  divisions 236
  Executive Director of 234
  governance study 228
  Handbook for Authors 113
  headquarters building 33
  journals and magazines—circulation 95
  Monograph Series 119
  News Service 80, 171
  presidency of the 183
  single article announcement service 125
  studies 164
  Symposium Series 122
Actions, Council 206
Adams, Roger 33, 39, 65, 85, 199
Adhesion 284
Adsorption 280
Advances in Chemistry Series 114
Advertising 97, 119
Agricultural and Food Chemistry, Division of 238
Aid for the underprivileged 169
Alkyd resins 352
Alsberg, Carl L. 250
American Association for the Advancement of Science 3, 171
American Board of Clinical Chemistry 81
American Ceramic Society 172
American Chemical Journal 13, 340
American Chemical Society (*see also* ACS)
  first officers of 6
  governance of 179
American Chemist 3, 100, 307
American Electrochemical Society .13, 172
American Institute of Chemical Engineers 14, 172
American Journal of Historical Chemistry 308
American Journal of Science 3, 307
American Leather Chemists' Society 172
American Oil Chemists Society 172
American Society of Biological Chemists 13, 172, 250
American Water Works Association 172
Amino acid analyzer, automated ... 347
Ammonia process, Haber–Bosch ... 15
Ammonia process, synthetic 293
Ammonium phosphates 295
Analysis, environmental 290
Analysis, "wet chemical" 243
Analyst, The 244
Analytical Chemistry 98
Analytical Chemistry, Division of .. 242
Analytical chemistry, electronics in . 248
Analyzer, automated amino acid ... 347
Anderson, Stella 108
Aniline 376
Antibiotics 321
Antiinfectious agents 321
Antiinflammatory agents, nonsteroidal 324
Antihyperglycemics 324
Antihypertensive and diuretic drugs 323
Antineoplastic agents 326
Applications of nuclear energy 335
Applied journals 115, 234
Archives of Biochemistry 252
Argon 314
Armstrong, Neil 44
Arnon, Daniel 241
Arthritis 324
Arveson, M. H. 162, 183
Ashford, T. A. 265
Association of Official Agricultural Chemists 172
Atmospheric chemistry 291
Atomic bomb project 298
Atomic Energy Commission 150
Audio courses 75
Audio-tape cassettes 120
Automated amino acid analyzer ... 347
Awards (*see also* individual divisions, Chapter XI) ....441 ff

B

Bachman, W. E. 341
Beakeland, Leo 12, 160, 369
Bailar, John C., Jr. 313, 316
Bailey, G. C. 362
Bailey, William 184, 218
Bakelite 12, 369
Baker, Carl 251, 254

Baker, Dale B. 41, 133
Baker, Newton D. 160
Bancroft, Wilder D. 13, 107
Banflame process 260
Banks, R. L. 362
Bartell, F. E. 353
Bartlett, Paul D. 27
Bartow, Edward 65, 288
Bates, F. J. 255
Bates, Philip K. 116
Baumann, E. 369
Bayer, Adolf 369
Bayer, Otto 369, 380
Becquerel, A. H. 10
Beilstein's *Handbuch der Organischen Chemie* 269
Belknap, Richard H. 112
Benfey, O. T. 114
Berger, Frank 326
Beringer, Frederick M. 115
Bernier, Charles L. 133
Bertsch, Charles 113
Bevilacqua, Edward M. 107
Bigas process 305
Bigelow, W. D. 239
Billings, E. M. 65, 68
Biochemical Societies, Pan American Association of 254
Biochemistry 115, 251, 252
Biochemistry, International Congress of 251
Biochemistry, International Union of 251
Biological Chemistry, Division of .. 250
Bixler, Gordon 114, 119
Bliss, Allen D. 112
Bloch, Herman 87
Board Committee on the Chemical Abstracts Service 186
Board Committee on Chemistry and Public Affairs 162
Board–Council Committee on Chemistry and Public Affairs 185
Board of Directors, ACS 187, 193
elected and standing committees of the 186
Bock, W. 377
Bodley, Rachel L. 4
Bogert, Marston T. 16, 39, 173, 309
Bohr, A. 334
Bolton, Henry Carrington 4, 307
Books department 122
Bouchardat, G. 367
Boyer, Paul 253
Bradley, T. F. 351
Bramer, Henry C. 289
Brand, Erwin 250
Breneman, A. A. 100
British Abstracts 269
Brown, H. C. 349
Browne, Charles Albert 306
Bruni, G. 376
Buchanan, Jack 253
Bunnett, Joseph 117
Bunsen, Robert 317
Bureau of Mines 146
Burger, Alfred 111, 119
Butyl rubber 379

C

CA (See *Chemical Abstracts*)
Cady, H. P. 314
Cairns, Robert W. 52, 86, 99, 118, 166, 173, 230
Calcium fluoride 297
Callahan, M. J. 19
Calvin, Melvin 92, 200, 241
Campbell, J. Arthur 266
Cancer 326
Cancer Attack Program, National .. 254
Cancer chemotherapy 326
Capell, Leonard T. 131, 133
Captan 360
Carbamate insectocodes 358
Carbohydrate Chemistry, Division of 254
Carbon black 377
CO_2-acceptor process 304
Cards system, punched 271
Career guidance literature 77
Carnegie, Andrew 2
Carothers, Wallace H. ... 31, 341, 369, 377
Carpenter, F. B. 291
Carson, Rachel 45
Carter, H. E. 254
Carver, George Washington 239
Cassettes, audio-tape 120
Catalyst, zeolitic cracking 364
Catalysts, molecular sieve 281
Catalytic cracking 364
Catalytic reforming 363
CBA 75
Celluloid 368
Cellulose 255, 368
derivatives 258
fibers 259
technology 255
Cellulose, Paper and Textile Division 255
Certification 83
Chain reactions, radical anion 349
Chandler, Charles Frederick 4, 8, 13, 60, 307
Change, control of technological ... 156
Chemical Abstracts 14, 16, 27, 45, 96, 189, 269, 270
abstractors for 132
Condensates 142
coverage of 128
indexes to 130
production of 137
Source Index 142
Subject Index Alert 142
Chemical Abstracts Service 41, 126
modernization of 135
operations, computerization of ... 138
Chemical-Biological Activities 136
Chemical Bond Approach 75
Chemical Education, Division of .. 63, 263

Chemical education in junior colleges 76
Chemical engineering 311
Chemical and Engineering News (C&EN) 43, 51, 96, 119
Chemical fertilizers 292
Chemical Foundation, Inc., The ..17, 264
Chemical industry in the 1930s 22
Chemical Industry Notes 142
Chemical industry, U.S. 12
Chemical Information, Division of .. 267
Chemical literature 269
Chemical Marketing and Economics, Division of 272
Chemical nomenclature 140
Chemical products and public safety 153
Chemical Publishing Co. 9, 96
Chemical Registry System 140
Chemical Reviews 105, 111
Chemical societies, national 174
Chemical Society (London), the .118, 139
Chemical Society of Washington, The 7
Chemical Technology (ChemTech) 119, 120, 309
Chemical Titles 135, 139
Chemicals in food 155
Chemicshes Zentralblatt 139, 269
Chemisorption 280
Chemists Club of New York 6
founding members of the 10
Chemists, literature 270
Chemists, polymer 117
Chemists, salaries for 83
Chemistry 106, 114, 755
Chemistry, atmospheric 291
Chemistry, computers in 286
Chemistry, Conference on General . 266
"Chemistry in the Economy" 77
Chemistry, electroanalytical 246
Chemistry, electronics in analytical . 248
Chemistry for the general welfare .. 150
Chemistry, hydrocarbon 362
Chemistry, instrumentation in nuclear 337
Chemistry Leaflet, The 106
Chemistry, modern organofluorine . 298
Chemistry in the nation's defense .. 145
Chemistry and Public Affairs, Board Committee on 162
Chemistry, soil 292
Chemistry tests, cooperative 265
Chemistry, transition metal 318
Chemistry, undergraduate instruction in 266
Chemistry Week 39
Chemotherapy, cancer 326
CHEM Study 75, 266
ChemTech (See Chemical Technology)
Chlorpromazine 325
Christman, Russell F. 123
Chromatography 248, 343
preparative 280
Chymia 308
Circulation, ACS journals and magazines 95
Clarke, F. W. 4, 7
Clean Air and Occupational Safety and Health Acts of 1970 156
"Cleaning Our Environment" 164
Clinical Chemistry, American Board of 81
Clinical Laboratory Digest 121
Coal gasification 302
Coal hydrogenation 303
Coatings, nitrocellulose 352
Coatings, protective 352
Coleman, George H. 109, 112
Collat, Justin W. 74
Colleges, chemical education in junior 76
Collier, Peter 153
Collins, A. M. 31, 377
Colloids 269
Colloid and Surface Chemistry, Division of 277
lyophilic 278
lyophobic 278
Color Additive Amendment of 1960 155
Columbian Exposition 9
Commercial fermentation 329
Committee on Chemistry and Public Affairs 158, 195
Committee on Finance 187
Committee on Professional Training 65
Committees of the ACS 185
Committees of the Board of Directors, elected and standing 186
Compton, Arthur 20
Computation in Chemistry, National Resource for 287
Computer-aided typesetting 124
Computerization of CAS's operations 138
Computerized information systems .. 271
Computers in Chemistry, Division of 286
Conant, James B. 39, 40, 146
Conference on General Chemistry .. 266
Conformational studies 348
Connor, Ralph 194
Constitution, ACS 180
Consumer Product Safety and Environmental Pesticide Control Acts of 1972 156
Contact angles 284
Control, federal 156
Control of technological change ... 156
Cook, William B. 267
Cooke, Josiah Parsons, Jr. 314
Cooke, Lloyd M. 184
Cooperative chemistsry tests 265
Cope, Arthur C. 194
Cope rearrangement 349
Copyright 98
Cornucopia 239
Correspondence interaction courses . 75
Council, ACS 195, 203, 206

Council Committee on Constitution and Bylaws 185
Council Committee on Nomenclature 191
Council Policy Committee 188
Couper, Archibald 3
Coverage of *Chemical Abstracts* 128
Cracking, catalytic 364
Craig, David 107
Cramer, P. J. S. 376
Crane, E. J. 16, 39, 41, 127
Crawford, Bryce 119, 184
Crosby, D. G. 355
Crowe, James M. 108
Crum, John K. 119
Curie, I. 332
Curie, Marie 10, 332
Curie, Pierre 10, 332
Curran, Carleton E. 130

D

2,4-D 359
Davis, C. C. 106
Davis, Tenney L. 308
Davis, Watson 114
Day, G. A. 256
D-D soil fumigant 360
DDT 356
Delbridge, T. G. 360
Department of Agriculture 12, 150
Depression years 21, 81
Detectors for nuclear radiations 337
Dewey, Bradley 182
Diabetes 324
Diamond Jubilee 38
"Directory of Graduate Research" .. 70
Diuretic drugs 323
Division of
  Chemical Education 63
  Fertilizer Chemistry 291
  Fertilizer and Soil Chemistry 291
  Petroleum Chemistry 360
Divisions, ACS (*see also* under individual names) 236
Divisions, local sections and 206
Djerassi, Carl 325
DLVO theory 279
Dole, Frank 256
Domestic cooperation 175
Donaldson, Jesse 39
Driscoll, Alfred R. 39
Drug(s)
  discovery 327
  diuretic 323
  laws 154
Du Pont 12
du Vigneaud, Vincent............. 325
Dues 199
  -supported programs 52
Dynamics, nuclear 334

E

Eastman Kodak 12
Edgewood Arsenal, Md. 15
Edman sequencer 347
Edsall, John T. 250
Education in polymer science and technology 371
Egan, Thomas 27
Einstein, Albert 11, 29
Eisenhower, Dwight D. 44
Elder, Albert L. 184
Elected Committees of the Board of Directors 186
Elections 206
Electroanalytical chemistry 246
Electroanalytical techniques 245
Electrochemical Society 13
Electrode, glass 246
Electrokinetic phenomena 284
Electron microscope 277
Electron spin resonance 344
Electronics in analytical chemistry . 248
Electrophoresis 283
Elements, transuranium 316, 333
Eliot, Charles W. 61
Ellsworth, Henry L. 151
Emery, Alden H. 35, 161, 182, 232, 234
Employment Clearing House 22, 83
Emulsifiers, mixed 283
Emulsions 282
Endemann, Hermann 7, 100
Energy, nuclear 335
Energy use in the U.S. 361
Engineering, chemical 311
Englehorn, F. 369
Environmental analysis 290
Environmental Chemistry, Division of 288
Environmental Protection Agency .. 46
Environmental Science & Technology 116, 289
Enzyme action, mechanisms of 347
Epoxy resins 353
Equilibrium 285
Esselen, G. J. 255
Ethylene–propylene rubber 379
Executive Director of ACS 234

F

Fajans, K. 332
Farber, Emmanuel 254
Federal
  Charter 24
  Food, Drug, and Cosmetic Act ... 25
  incorporation 26
  program for water pollution control 290
Federation of American Societies for Experimental Biology 253
Fenn, E. F. 256
Fermentation, commercial 329
Fermentation processes in the U.S., evolution of 330
Fermi, Enrico 29

Fertilizer(s)
 chemical 292
 Chemistry, Division of 291
 nitrogen for 147
 and Soil Chemistry, Division of .. 291
Fibers, cellulosic 259
Film courses 75
First Divisional Officers Conference . 221
First officers of the American
 Chemical Society 6
Fischer, Emil 16
Fischer–Tropsch synthesis 302
Fission, nuclear 332
Fission products 335
Fitlig, R. 369
Fitzsimmons, Thomas 372
Flame resistance of textiles 260
Flory, Paul J. 370
Fluorine Chemistry, Division of 297
Fluorocarbons 299
Foams 282
Food
 Additives Amendment of 1958 ... 155
 chemicals in 155
 and Drug Administration 155
 Drug, and Cosmetic Act of 1938 .. 155
 laws 154
 technology 240
Folkers, Karl 240
Ford, Gerald 56
Founding members of the Chemists'
 Club 10
Frankland, Edward 317
Franklin, E. C. 339
Frazer, Persifor 4
Frear, J. Allen 39
Free radicals 342, 349
Frei, E. 254
French, Ethel 66
Freons 299
Fresenius' *Zeitschrift für
 analytischen Chemie* 244
Frisch, Otto 29
Fuel Chemistry, Division of 300
Fuel chemistry, research in 302, 304
Functions and organization,
 Council 203
Fundamental journals 115
Funding, public 161
Funding of science 22
Furman, N. Howell 38, 184

G

Gaessler, William G. 132
Gagarin, Yuri 44
Gardner, H. A. 350
Garvan, Francis P. 17, 106, 264
Garvan Medal 17
Gas
 and Fuel Chemistry, Division of .. 300
 for heating 301
 mustard 326
 natural 314
 warfare 15
Gasification, coal 302
Gates, Marshall 112
General Electric 12
Geneva Protocol 167
Genth, Frederick Augustus 315
Gerhart, Howard L. 118
Gibbs, Josiah Willard 3, 11, 60
Gibbs, Oliver Wolcott 314
Glass electrode 246
Glenn, John H., Jr. 44
Gluck, D. J. 140
Gmelin's *Handbuch der Anorga-
 nischen Chemie* 269
Goeppert-Mayer, Maria 333
Goldman, Albert 39
Goldschmidt, S. A. 20
Gooch, Frank Austin 60
Goodyear, Charles 367, 374
Goodyear, Nelson 375
Gordon, Gladys 108
Gordon, Neil E. 21, 106, 263, 264
Gordon Research Conference 21
Gould, Ben 265
Gould, Robert F. 108, 114
Governance
 of the American Chemical Society 179
 and operations, division 218
 and operations, local section 222
 study, ACS 228
Government, impact of 144
Grady, James T. 16, 41
Graham, Thomas 369
Green, D. E. 253
Greene, Frederick D. 112
Greigy, J. R. 356
Groves, Leslie R. 30
Gushee, David E. 115

H

Haber–Bosch ammonia process 15
Haber process 147
Hader, Rodney N. 111, 114, 119
Hahn, Otto 29, 332
Hale, Albert C. 9, 230
Hale, Arthur 90
Hallett, Lawrence T. 108
Hamor, William 105, 115
Hancock, John M. 36
Hancock report (1947) 85, 196
*Handbuch der Anorganischen
 Chemie,* Gmelin's 269
Handbuch der Organischen Chemie,
 Beilstein's 269
Handler, Philip 254
Hanson, A. B. 86
Harkins, W. D. 353
Harris, C. 367
Harris, Milton 118, 194
Harrison, Anna J. 179, 184
Hart, Edward 9, 13, 100
Harte, Robert 251, 252
Haworth, Walter 261
Hawthorne, M. Frederick 119

Headquarters, national 234
Headquarters staff and operations .. 230
Heating, gas for 301
Heidelberger, Charles 327
Helium 148, 314
Hemicelluloses 261
Heming, Art 251
Henriques, R. N. 375
Hepburn, Joseph 132
Herbicides 358
Hercules Powder Co. 12
Herreshoff, J. B. F. 20
Herty, Charles H. 15, 102, 160, 366
High school programs 75
Hill, Henry A. 179, 185
Hill, Norman C. 267
Hillebrand, William Francis 314
Hindered phenols 377
Hinrichs, Gustavus 314
History of Chemistry, Division of .. 306
Hitchings, George 327
Holst, T. G. 256
Hopkins, B. Smith 315
Hopkins, F. G. 240
Horecker, B. 254
Hormones, insect 347
Hormones, steroid 325
Hornig, Donald R. 44
Horowitz, Charles 39
Horsford, E. N. 4
Hovey, A. G. 351
Howard, John H. 68
Howe, Harrison E. 19, 34, 104, 232
Howe, James Lewis 315
Hubbard, Philip H. 98
Hulburt, Hugh 114
Hunt, Reid 18
Hydrocarbon chemistry 362
Hydrocracking 365
Hydrodesulfurization 365
Hydrogenation, coal 303
Hydrogen fluoride 298
Hygas process 304
Hypertension 323

I

I&EC (see Industtrial & Engineering Chemistry)
Incendiary bombs 30
Indexes to Chemical Abstracts 130
Indexing compounds systematically, naming and 130
Indexing, keyword-in-context 136
Industrial Chemists and Chemical Engineers, Division of 309
Industrial & Engineering Chemistry 19, 21, 95, 113, 234, 254
  Fundamentals 113
  Journal of 97
  Product Research and Development 114
  Process Design and Development . 114
Industrial and Engineering Chemistry, Division of 309
Industrial Research 122
Industry, U.S. chemical 12
Information systems, computerized . 271
Ingold, C. K. 341
Inorganic Chemistry 115
Inorganic Chemistry, Division of ... 313
Insect hormones and pheromones .. 347
Insecticide(s) 355
  carbamate 358
  organophosphorus 357
  synthetic organic 356
Insoluble monolayers 280
Instrumental techniques 342
Instrumentation in nuclear chemistry 337
International
  Congress of Biochemistry 251
  cooperation 172
  Union of Biochemistry 251
  Union of Pure and Applied Chemistry 172, 252
Intersociety
  Committee 176
  Energy Conversion Engineering Conference 177
  relations 171
Isoniazid 321
Isotopes of transuranium elements .. 333
Isotopic tracers 336
IUPAC 173

J

James Bryant Conant Award 74
James, Charles 315
Jensen, J. H. D. 333
Johnson, Samuel W. 239
Johnson, W. B. 351
Joliot, F. 332
Journal of Agricultural and Food Chemistry 109, 114
Journal of the American Chemical Society 7, 14, 105, 340
Journal of Analytical and Applied Chemistry 9, 100
Journal of Analytical Chemistry ... 9, 13
Journal of Biological Chemistry 252
Journal of Chemical Documentation 111
Journal of Chemical Education 17, 63, 105, 263, 264
Journal of Chemical & Engineering Data 111, 116
Journal of Chemical Information and Computer Sciences 122, 271
Journal of Industrial & Engineering Chemistry 14, 97, 309
Journal of Medicinal Chemistry .111, 320
Journal of Medicinal and Pharmaceutical Chemistry 111, 320
Journal of Organic Chemistry 110, 112, 339
Journal of Physical and Chemical Reference Data 120

Journal of Physical Chemistry 13, 110, 366
Journals
applied 115
fundamental 115
and magazines circulation, ACS .. 95
microfilm editions of 117
official policy for 121
Junior colleges, chemical education in 76

K

Kayser, H. 245
Kefauver–Harris Amendments 155
Kendall, James 27
Kennedy, J. 333
Kennedy, John F. 44
Kenyon, Richard L. 41, 109, 114, 166
Kerr, C. H. 132
Kesting, E. 257
Keyword-in-context indexing 136
Kharasch, Morris S. 109, 342
Kiefer, William 106
Kienle, R. H. 351
Killeffer, David 108
Killian, James R. 44
King, Edward L. 115
King, L. Carroll 114
Kirner, W. A. 242
Kistiakowksy, George 44
Koerwer, Thomas N. J. 98
Konrad, E. 377
Kraske, W. A. 256
Kuney, Joseph H. 108

L

Lacquers, nitrocellulose 19
Laitinen, Herbert 116
Lamb, Arthur B. 38, 104, 112
Langham, Cecil C. 131
Langmuir, A. C. 27
Langmuir, Irving 12, 27
Larrabee, C. B. 111
Laser 249
Latex rubber 378
Lauffer, Max 253
Lawrence, E. O. 30
Laws, food and drug 154
Lead arsenate 355
League for International Food Education 176
Lebedev, S. V. 367
Leblanc, Nicholas 11
Legal Aid Fund 92
Leicester, Henry M. 308
Leloire, L. 254
Lewis, G. N. 340
Lewis, Harry F. 266
Library of Congress 129
Lind, S. C. 65, 107
Linstead, R. P. 353
Lippincott, Thomas 106
Liquid crystals 282
List of Periodicals 142
Literature, chemical 269
Literature chemists 270
Little, Arthur D. 309, 311
Little CA, The 128
Local Section Activities, Office of ... 227
Local sections and divisions 206
Local section governance and operations 222
Luberoff, Benjamin J. 120
Lundell, G. E. F. 242
Lurgi process 302
Lyophilic colloids 278
Lyophobic colloids 278

M

McCarthy, Joseph R. 37
McCoy, Herbert Newby 315
McCurdy, Patrick P. 119
McElroy, William 254
McFarland, D. F. 314
MacIntosh, Charles 375
Mack, Harvey F. 96
Mack, Pauline Beery 106
Mack Printing Co. 96
McMillan, Edwin M. 317, 333
Macromolecules 117
Magazines circulation, ACS journals and 95
Magill, Mary A. 131
Malaria 322
Manuscript acceptance 112
Marcy, Willard 162
Mariella, Raymond P. 179
Marine Biological Laboratory 108
Mark, Herman 261, 368
Marketing research 273
sampling in 274
Marks, A. H. 375, 376
Marvel, C. S. 341
Mass spectrometers 343
Maxwell, James Clerk 3
Maynard, John T. 162
Mechanisms of enzyme action 347
Mechanism studies, reaction 340
Medicinal Chemistry, Division of .. 320
Meetings, national 210
Meetings in 1976, regional 225
Meitner, Lise 29
Member Finance 209
Members, minority 201
Membership 208
MemphIon, The 108
Mendeleev, Dmitri 3, 314
"Men and Molecules" 47, 120, 200
Metal chemistry, transition 318
Methods of structural study, new .. 316
Meyer, K. H. 261, 368
Meyer, Lothar 314
Micelles 282
Microbial Chemistry and Technology, Division of 328

Microfilm editions of journals 117
Micronutrients 296
Midgley, Thomas, Jr. 34, 65, 85, 193
Miller, John J. 127
Miller Pesticide Amendment 155
Millikan, Robert 20
Minicomputers 288
Minority members 201
Mixed emulsifiers 283
"Modern Chemical Technology" ... 76
Modern organofluorine chemistry .. 298
Modernization of CAS 135
Molecular motion 277
Molecular sieve catalysts 281
Monographs, ACS 105, 115
Monolayers, insoluble 280
Montgomery, John A. 327
Moore, Gideon 100
Morey, Sir Robert 3
Morgan, Harry L. 140
Morgan, James J. 116
Morrill Act of 1862 62
Mote, S. C. 377
Mottelson, B. 334
Muller, Paul 356
Müller, Ralph H. 108
Munitions, nitrogen for fertilizer and 147
Munroe, C. F. 8
Munroe, Charles E. 20
Murphy, E. A. 378
Murphy, Walter J. 34, 41, 107, 110
Mustard gas 326
Myatt, D. O. 108

N

Nair, John H. 38
Naming and indexing compounds
systematically 130
National Aeronautics and Space
Administration 150
National Bureau of Standards 12
National Cancer Attack Program ... 254
National chemical societies 174
National Environmental Policy Act . 157
National headquarters 234
National Institutes of Health 150
National meetings 210
National Registry in Clinical
Chemistry 176
National Resource for Computation
in Chemistry 287
National Science Foundation .44, 149, 342
Nation's defense, chemistry in the .. 145
Natta, Guilio 368
Natural gas 314
steam reforming of 293
Natural products, synthesis of 347
Nernst, Walter 16
Neurath, Hans 111, 253
Newlands, John 314
Newman, Pauline 162
Newton, Helen 108
New York College of Pharmacy 7
New York Lyceum of Natural
History 4
New York Section, Society of
Chemical Industry 172
News Service, ACS 80, 141, 171
Nichols, William H. 5
Nielsen, Arthur C. 275
Niese, H. E. 20
Nieuwland, J. A. 377
Nightingale, A. F. 61
Nitric acid 294
Nitrocellulose 369
coatings 352
lacquers 19
Nitrogen 147, 292
program, wartime 147
Nixon, Alan C. 177, 184
Nixon, Richard 44
Nomenclature 271
chemical 140
Nominations and elections 206
Nonsteroidal antiinflammatory
agents 324
Noyes, Arthur A. 13, 102
Noyes, W. A., Sr. 126
Noyes, W. A., Jr. 105, 107, 112,
115, 173, 231
Nuclear chemistry, instrumentation
in 337
Nuclear Chemistry and Technology,
Division of 331
Nuclear dynamics 334
Nuclear energy, applications of 335
Nuclear fission 332
Nuclear magnetic resonance
(NMR) 343
Nuclear radiations, detectors for ... 337
Nucleosides 348
Nucleotides 348
Nutrition 240
Nyholm, Ronald S. 316
Nylon 369

O

Oenslager, George 376
Office of Local Section Activities ... 227
Office of Scientific Research and
Development 342
Official policy for journals 121
Olah, G. A. 362
Operations
division governance and 218
headquarters staff and 230
local section governance and 222
Oppenheimer, J. Robert 30
Optics 244
Orbital symmetry 349
Organic
chemicals 148
chemistry 3, 341
insecticides, synthetic 356
synthesis 341
Organic Chemistry, Division of 339

Organic Coatings and Plastics Chemistry, Division of 350
Organization, council functions and 203
Organofluorine chemistry, modern . 298
Organometallic chemistry 317
Organophosphorus insecticides 357
Osborn, S. J. 255
Ostwald, W. 16, 376
Overberger, Charles G. 174
OXO process 302

P

Page charges 112, 118, 123
Paint industry 351
Paint, Plastics and Printing Ink Chemistry, Division of 350
Paint, Varnish and Plastics Chemistry, Division of 350
Paints, water-base 354
PAMCO process 305
Pan American Association of Biochemical Societies 254
Paneth, F. A. 332
Papermaking 257, 258
Parathion 357
"Paring" reaction 365
Parkes, Alexander 368, 375
Parkinson, Nellie 108
Parr, S. W. 300
Parry, Rober W. 111, 115
Parsons, Charles L. 14, 35, 36, 107, 146, 160, 172, 231, 232
Patrick, J. C. 377
Patterson, Austin M. 126, 130
Pauling, Linus 27
Payen, Anselme 255
Pearson, Osborn 39
Pecsok, Robert L. 76
Penicillin 321, 342
Pension benefits 90
Periodic Law 314
Persson, S. H. 256
Pesticide Chemistry, Division of 355
Petroleum Chemistry, Division of .. 360
Petroleum refining 2
Petroleum Research Fund 34, 73, 193
Pharmaceutical Chemistry, Division of 320
Pharmaceutical industry 31
Phenolic resins 369
Phenols, hindered 377
"Phenomenology" 287
Pheromones, insect hormones and .. 347
Phosphate rock 294
Phosphates, ammonium 295
Phosphoric acid 295
Phosphorus 294
Photocomposition 124
Photometers 244
Physical Chemistry, Division of 366
Physical and Inorganic Chemistry, Division of 366
Physicochemical treatment processes 291
Pigford, Robert L. 113
Pigments, research on 353
Pinkus, Gregory 325
Planck, Max 11
Plant, Albert F. 122
Plastics Industry Notes 142
Poison gas 298
Polarograph 246
Polymer Chemistry, Division of 367
Polymer chemists 117
Polymer industry, synthetic 370
Polymer Preprints 367
Polymer Science & Technology 142
Polymer science and technology, education in 371
Polymer structure 368
Polynucleotides 348
Polyurethane 369
Potash 147, 151
Potassium 295
ores 296
Preparative chromatography 280
Presidency of the ACS 183
Price, Charles C. 159, 163, 183, 194
Priestley Centennial 3
Priestley Museum 20
Printers Ink 111
Printing 96
Proceedings of the American Chemical Society 6, 94
Production of *CA* 137
Products and public safety, chemical 153
Professional concerns 212
"Professional," definition of a 79
"Professional Employment Guidelines" 92
Professional Enhancement Program (PEP) 184
Professional programs in Chemistry 66
Professional Relations, Division of 236, 372
Professionalism 185
Progesterone 325
Prohibition 19
Proteins, ultracentrifugation of 281
Psychopharmacological agents 325
Public affairs 159, 167, 216
Public funding 161
Public safety, chemical products and 153
Publications 94, 212
research and development 124
Pulping 256
Punched cards system 271
Pure Food and Drugs Acts of 1906 .. 12
Putnam, F. 253, 254
Pyrethrum 356

Q

Quigley, Stephen T. 166
Quill 177

R

Radiations, detectors for nuclear ... 337
Radical anion chain reactions 349
Radioactivation 336
Rakestraw, Norris 106
Raman, C. V. 246
Randall, Lowell O. 326
Rane, Leo 322
Rare earth elements 315
Rayon 258
Reaction mechanism studies 340
Recording spectrophotometer 247
Refining, petroleum 2
Regional Meetings in 1976 225
Reinhold, Ralph 98
Reinmuth, Otto 106
Relations, intersociety 171
Remsen, Ira 2, 13, 60
Reprint collections 123
Research
- and Development Publications ... 124
- in fuel chemistry 302, 304
- marketing 273
- on pigments 353

Review of American Chemical Research 102, 126
Reynolds, Bertha 108
Rheniforming 364
Richards, T. W. 12, 15
Richardson, W. D. 102
Robins, Roland K. 327
Roentgen, Wilhelm 10
Romani, E. 376
Roosevelt, Franklin 25
Ross, W. H. 315
Rouiller, Charles A. 132
Rowlett, Russell J., Jr. 136
Rowley, T. 376
Royalties 123
Rubber 367
Rubber Age 374
Rubber, butyl 379
Rubber Chemistry, Division of 373
Rubber Chemistry & Technology 105, 373
Rubber, ethylene–propylene 379
Rubber industry, synthetic 31
Rubber, latex 378
Rubber, synthetic 377
Rush, Benjamin 59, 60
Rutherford, Ernest 246, 332

S

Safety, chemical products and public 153
Sage, Bruce 116
Salaries 83, 84
Sampling in marketing research 274
Saybolt, George 2
Sayre, Charlotte 108
Schidowitz, Paul 378
Schoellkopf Aniline & Chemical Co. 12
Schrader, Gerhard 357
Schuber, J. 256
Schultz, Julius 251
Scientific research, federal interest in 149
Seaborg, Glenn T. ... 30, 184, 266, 317, 333
Sections, local 206, 222, 227
Sequencer, Edman 347
Shearon, Jr., Will H. 111
Sherley Amendment of 1912 155
Short Courses 75
Shull, Gilbert M. 329
Silliman, Benjamin, Sr. 151
Silliman, Benjamin, Jr. 2, 4, 8, 307
Simon, E. 369
Single Article Announcement service, ACS 125
Skolnik, Herman 111
Smith, Edgar Fahs 263, 306
Smith, J. Lawrence 4, 315
Smith, Otto M. 265
Smutz, Morton 331
Society of Chemical Industry 176
Soddy, Frederick 315, 332
Soil chemistry 292
Soil fumigant, D–D 360
Space race 43
Sparks, William J. 31, 378
Spectrophotometer, recording 247
Spectroscopy 244, 278
Stack, James H. 41
Stadman, E. 252
Staff and operations, headquarters .. 230
Stanerson, B. R. 50
Staudinger, Hermann 261, 368, 377
Steam reforming of natural gas 293
Stemen, W. Russel 128, 131
Sternbach, Leo 326
Steroid hormones 325
Stevens, A. B. 320
Stewart, Joseph 90
Stiles, Martin 118
Strassmann, Fritz 29, 332
Strong, L. E. 266
Structural organic chemistry 341
Structural study, new methods of .. 316
Structure, polymer 368
Student affiliate program 73
Sulfanilamide 321
Superphosphate process 294
Surface science 277
Surface states 281
Sutherland, Earl 324
Svedberg technique 281
Swain, R. E. 65
Synthane process 305
Synthesis of natural products 347
Synthetic
- ammonia process 293
- Liquid Fuels Act 303
- organic insecticides 356
- *cis*-polyisoprene 379
- polymer industry 370
- rubber 377
  - industry 1, 31

Szent-Gyorgyi, Albert 10
Szilard, Leo 29

T

Tariff Act of 1922 148
Tarrant, Paul 297
Tate, Fred A. 118, 136
Technical affairs 211
Technological change, control of ... 156
Technology, education in polymer
science and 371
Technology Quarterly 102
Teflon 299
Terrant, Seldon 125
Tests, cooperative chemistry 265
Textile development 258
Textiles, flame resistance of 260
Thermal cracking 360
Thermobalance 248
Thermodynamics 3
Thiazides 323
Thiokol 377
Thomas, Charles A. 182
Thomas, R. M. 31, 378
Thomson, J. J. 10
Thomson, Robert 375
Tires, urethane 380
Tishler, Max 92, 184
Titrimetry 243
Toluene 31
Tracer techniques 247
Tracers, isotopic 336
Tranquilization 326
Transition metal chemistry 318
Transuranium elements 316, 333
Trevor, J. E. 13
Truman, Harry S. 33
Tschunker, E. 377
Tuberculosis 321
Tuttle, J. B. 373
Typesetting, computer-aided 124

U

Ultracentrifugation of proteins 281
Ultramicroscope 277
Undergraduate instruction in
chemistry 266
Underprivileged, aid for the 169
Unionization 82, 87
U.S. Bureau of Mines 12, 301
U.S. chemical industry 12
USDA 152
U.S. Patent Office 150
U.S. rubber 12
Uranium hexafluoride 298
Urbain, Georges 315
Urea 2, 294
Urethane tires 380
Urey, Harold C. 27
Utermark, W. L. 378

V

Van Antwerpen, F. J. 108
Vanderbilt, Byron 114
Vane, John R. 324
Vinyl resins 352
Volwiler, Ernest H. ..133, 161, 182, 184, 194
von Hevesy, G. 332
Vulcanization 374

W

Wahl, A. 333
Wall, Frederick T. 50, 115, 119, 234
Walling, Cheves 123
Walter Reed Army Medical
Research Center 322
Walz, Isidor 5
Warfare, gas 15
Warfarin 360
Warren, R. M. 177
Wartime nitrogen program 147
Wastewater renovation 290
Water-base paints 354
Water, Air, and Waste Chemistry,
Division of 288
Water Pollution Control Act 156
Water pollution control, federal
program for 290
Weber, C. O. 375
Weeks, Mary Elvira 265
Weidlein, E. R. 65
Weiser, H. B. 65, 277
Weiss, M. L. 376
Wendt, Gerald K. 105
Werner, Alfred 315
Westheimer, Frank H. 48, 162
Westheimer report 150
"Wet chemical" analysis 243
Wheeler, D. H. 351
Whitmore, F. C. 362
Whitney, Willis R. 12, 160
Whittaker, M. C. 102
Wiley, Harvey 7, 8, 153, 172
Willard, F. W. 65, 105
Wilson, Charles 20
Wilson, Woodrow 17, 264
Winslow, Field H. 117
Winthrop, John 3
Wise, Louis E. 132
Wittenberg, Robert B. 275
Wöhler, Friedrich 2
Woods Hole, Mass., Marine Bio-
logical Laboratory 108
World Chemical Conclave 38
World's Congress of Chemists
(1893) 9, 172
World War I 146, 351
World War II 28, 129, 149, 160, 302,
312, 315, 321, 342, 356, 378

X

Xylans 262

Y

Younger Chemists Task Force 189

Z

Zeitschrift für analytischen Chemie, Fresenius' 244
Zeolites 286
Zeolitic cracking catalyst 364
Ziegler, Karl 368
Zisman, W. A. 353

Taming the Time Stealers

Tricks of the Trade From Organized Teachers

M. C. Gore ▪ John F. Dowd

For information address:

Corwin Press, Inc.
A Sage Publications Company
2455 Teller Road
Thousand Oaks, California 91320
E-mail: order@corwinpress.com

SAGE Publications Ltd.
6 Bonhill Street
London EC2A 4PU
United Kingdom

SAGE Publications India Pvt. Ltd.
M-32 Market
Greater Kailash I
New Delhi 110 048 India

Printed in the United States of America

Library of Congress Cataloging-in-Publication Data

Gore, M. C.
Taming the time stealers: Tricks of the trade from organized teachers / by M.C. Gore, John F. Dowd.
p. cm.
Includes bibliographical references.
ISBN 0-8039-6843-4 (cloth)
ISBN 0-8039-6844-2 (paper)
1. Teachers—Time management. 2. Effective teaching. 3. Teaching—Decision making. 4. Classroom environment.
I. Dowd, John F. II. Title.
LB2838.8 .G67 1999
371.102—dc21 98-58137

This book is printed on acid-free paper.

99 00 01 02 03 04 9 8 7 6 5 4 3 2 1

Corwin Editorial Assistant: Julia Parnell
Production Editor: Wendy Westgate
Production Assistant: Nevair Kabakian
Typesetter/Designer: Janelle LeMaster
Cover Designer: Tracy E. Miller

CONTENTS

Preface viii

Acknowledgments x

About the Authors xi

1. Organizing Your Time and Memory **1**

First-Generation Organizational Aids 1

Second-Generation Organizational Aids 17

2. Buried Under Administrative Paperwork? **21**

Streamlining Daily Tasks 21

Accommodating Occasional Administrative Tasks 23

Organizing Student Data 26

Setting Up the Classroom Library 29

Creating the Paperless Classroom 31

3. Managing Instructional Print Material **33**

Streamlining Print Resources and Planning 33

Organizing Student Work Print 37

Transporting Materials 41

Organizing for Substitute Teachers 43

4. Corralling Nonprint Materials **45**
Containers 45
Containers for Small Items 46
Storing Small But Relatively Flat Items 50
Storing Bulletin Board Materials 52
Organizing Tools 54
Organizing Physical Education Equipment 57
Moving Nonprint Materials 58

5. "Dejunking": Taking the Cure **60**
Basic Dejunking 60
Minimizing Memento Junk 65
Weaning Yourself Away From Junk 66

6. Decisions: From Simple to Complex **69**
Quick and Easy Decision-Making Techniques 70
More Complete Decision-Making Techniques 74
The Best and Most Complete Model of Decision Making 77

7. Organizing Your Teaching **88**
The Content 89
The Lesson-Planning Process 92
The Teaching Act 98
Improving Standardized Testing 106

8. Helping Students Learn to Organize Themselves **108**
Materials 109
Time and Memory 110
Learning Behavior 112

9. Teaching Students With Special Needs **115**
Establishing the Behavioral Environment 115
Organizing the Learning Environment 123
Developing the Support System Communication 131

10. Working With Parents **133**
Writing to Parents 133
The Parent Conference 136
Parent Involvement 143

11. Requesting Assistance **147**
Deciding What Kind of Help You Need 147
Organizing to Get the Help You Need 151

12. Organizational Strategies for Your Soul **154**

References 161

PREFACE

Disorganization is a teacher's great enemy. We call it The Time Stealer (Figure I.1). The Time Stealer can take control of your life as a teacher. The Time Stealer causes you to take work home at night. It causes you to lose kids' papers, notes from parents, and sometimes even your gradebook!

The Time Stealer embarrasses you. It causes you to forget to attend IEP meetings, parent conferences, and team meetings. It causes you to have to apologize to your principal for failing to attend a meeting. It causes you to have to ask your school secretary for second copies of bulletins because you have misplaced your first ones. It causes you to have to send home second copies of permission slips because you lost the signed ones. It causes you to do a sloppy job of writing up narrative progress reports because you kept putting them off until you had more time. The Time Stealer can take control of your life unless you tame it. This book will teach you how.

We became interested in asking organized teachers about their best Time Stealer Tamers after a survey that Millie conducted with one of her education classes. Millie asked her teacher interns: What is the greatest problem that you have to overcome in order to become a more effective teacher? Fully a third of the class said that disorganization and procrastination were their biggest barriers to becoming more effective as teachers. As the students talked about their difficulty with disorganization and procrastination, they decided that their procrastination was more often than not a result of their disorganization.

In order to help the interns become more effective teachers, Millie went to see John, one of her fellow education professors. She told John what their students had said.

Figure I.1. The Time Stealer

John said, "I'm not surprised." Then he pointed out another important reason for teachers to increase their organizational repertoire. He teaches educational administration in addition to techniques of teaching courses, and he said, "With the increasing emphasis on accountability, teachers are going to have to become more organized in order to become more effective. Principals whose teachers don't score well on achievement tests are themselves going to be evaluated poorly. Therefore, principals are going to have to increase the pressure on their teachers for accountability. Disorganized teachers are going to find themselves dismissed. Organized teachers are going to do well."

Millie and John then decided to survey organized teachers about their Time Stealer Tamers and share what they found with their students. As Millie and John began to collect Time Stealer Tamers, they decided that in addition to their students, other teachers would find the information useful. This book is the result.

In the book, we have used the names of teachers who gave us Time Stealer Tamers. Some teachers, however, preferred not to have their real names used, and for those we created fictitious names. All students' names are fictitious. In stories that teachers gave us about their Time Stealers, we have created fictitious names in order to avoid embarrassment to those teachers who were willing to share their experiences with us. In the examples we created to illustrate Time Stealers and Time Stealer Tamers, we used fictitious names. Any resemblance to real persons in our fabricated examples is purely coincidental.

We continue to seek Time Stealer Tamers. If you have a great Time Stealer Tamer, send it to us. We will include it in future editions of this book.

Figure I.2. The Tamed Time Stealer

You may either e-mail it to us or send it by direct mail to either of the following addresses.

Dr. Millie Gore or Dr. John F. Dowd
Midwestern State University
Wichita Falls, Texas 76308
e-mail: millie.gore@nexus.mwsu.edu or e-mail: fdowdj@nexus.mwsu.edu

Happy Time Stealer Taming! (Figure I.2)

ACKNOWLEDGMENTS

In addition to the teachers whose individual names are listed with their contributions, we also would like to thank the following teachers whose organizational tricks of the trade are included in this book: Marvin Peevey, Marjean Cox, Pat Collins, Susan Mayo, Anita Chaney, Margaret Jones, Van Green, Paula Tilker, Jan Carpenter, Shelly Cunningham, Rae Gillan, Kari O'Brien, and April Moreland.

We are also proud to acknowledge our illustrator, Adam M. Chavez.

We gratefully acknowledge Louise MacKay and JoAnn Hohenbrink for their thoughtful reviews.

ABOUT THE AUTHORS

M. C. Gore is Associate Professor at the Gordon T. and Ellen West Division of Education at Midwestern State University in Wichita Falls, Texas, where she has taught a variety of methods courses since 1993. She received the undergraduate and master's degrees from Eastern New Mexico University and the doctorate from the University of Arkansas. She taught for 16 years in the public schools from first grade through high school special education. Her research interests include adolescent relationships. She is the author of *A Parent's Guide to Adolescent Friendship: The Three Musketeer Phenomenon,* and the coauthor of *Raising Other People's Kids: A Guide for Houseparents, Foster Parents, and Direct Care Staff.* She and her husband Don enjoy volunteering and taking their dogs for drives.

John F. Dowd is Associate Professor at Midwestern State University where he has taught courses in instructional techniques and educational leadership since 1989. He earned his advanced degrees at Syracuse University. He taught in the public schools for 9 years, was a middle school principal for 3 years, and a superintendent for 11 years. His research interests include decision making, qualitative techniques, and leadership behavior. He conducts many studies for local public schools. He has been married to Sandy Dowd for 35 years. They enjoy skiing, golfing, and visiting their two daughters, Colleen and Erin.

To *Don*, who helps me organize my life around the things that are important. With love, MG

And to *Sandy*, for all her patience, help, and understanding. With love, JFD

1

ORGANIZING YOUR TIME AND MEMORY

Time. What a precious commodity the Time Stealers steal from us! In this chapter, we will tell you about some ways of taming the Time Stealers by organizing your time so that you eliminate wasted minutes and therefore have more time for the things in life that are important to you as a teacher: good teaching, mentoring your students, listening to them, and encouraging them.

While it's obvious that organizing time will help you tame the Time Stealers, what's less obvious is that organizing your memory is also crucial in taming the Time Stealers. Forgetting things you're supposed to do and appointments that you're supposed to keep are terrible Time Stealers, too, so the techniques in this chapter are designed to help you organize your memory also. These kinds of techniques have been called First-Generation and Second-Generation organizational aids by Stephen R. Covey in his book, *The 7 Habits of Highly Effective People.* We like Covey's terms!

FIRST-GENERATION ORGANIZATIONAL AIDS

In this section of Chapter 1, we will discuss what are called first-generation organizational aids. They include:

- ☐ To-do lists
- ☐ Checklists
- ☐ Rubrics
- ☐ Telephone memo books

To-Do Lists

To-do lists come in three varieties: (a) basic, (b) ordered, and (c) prioritized. All three types ensure that a task will be remembered, and remembering is a central skill to good organization.

Basic To-Do Lists

Much of what teachers do when they plan is to make to-do lists. Understanding the nature of to-do lists will help teachers receive greater benefit from their to-do lists. The basic to-do list is one that is designed only to make certain that tasks are remembered. When John makes a list of all of the materials that he wants to photocopy before he goes home on Monday night so that they are ready before school on Tuesday morning, he is making a to-do list. He does not care in what order he photocopies the materials, nor does he need to prioritize the photocopying. He simply needs an aid to ensure that he remembers to photocopy everything. If John doesn't write out his photocopying to-do list when he writes out his lesson plan, chances are good that he'll fail to remember something that he's going to need on Tuesday. At best, he will probably have to make an extra trip or two from the workroom back to his classroom to retrieve something that he left behind. Taking a few moments to make a to-do list will save him valuable minutes and effort.

Ordered To-Do Lists

A second type of to-do list orders the tasks without reference to their importance. Millie often had trouble remembering to do all the tasks required at her school before going home in the afternoon. Several times she forgot to lock the windows, and on occasion, she even forgot to close them. She also often forgot to turn down the thermostat and to lock the file cabinet.

One night in the early fall it all came to a head. The room was cold in the morning, but heated up during the day and then cooled off quickly at night. Millie had turned on the heater first thing in the morning. Then, after the room heated up from the southern exposure during the day, she'd

opened the windows. She left at the end of the day with the thermostat on, the windows open, and the file cabinet unlocked. At 10:30 p.m., her phone rang. Her principal was on the other end. "The police called me about open windows in your classroom. I'm at the school now. Your file cabinet is open. You better come on down and look through it to see whether anything important is missing. Also inventory your equipment when you get here so we can tell if anything has been stolen. By the way," he added, "you left the heater on, and all the hot air blew right out the windows. You know we're watching the utility bill!"

Millie dressed frantically and broke the speed limit racing to the school. The policeman and her principal (his hair uncombed, wearing a jumpsuit and house slippers) were waiting for her. She checked out the room and assured them that everything was in order. Driving home, white-knuckled, red-faced with embarrassment, and exhausted, she vowed to write a to-do list of end-of-day activities and post it by the door. Although the list was a great help, Millie discovered that ordering the list was even more useful, because several times after she locked the file cabinet, she discovered that sensitive materials were still on her desk. By listing the tasks in the order that she needed to accomplish them, she saved time and effort.

A second way to use to-do lists is to use them to help students organize their work. When Millie taught first grade, students who were not in the reading group with which she was working had several tasks to accomplish independently or in groups. By numbering the tasks and using rebuses, Millie could make a to-do list that even a first grader who could not read could follow (Figure 1.1).

Prioritized To-Do Lists

Much of what teachers write in their lesson plan book is an ordered to-do list. In actual practice, however, we need prioritized to-do lists because activities often take more or less time than a teacher has allotted for them. When asked what she had learned during her internship semester, one of our interns responded, "I learned that nothing takes the kids as long to do as I think it will take them. Activities I think the kids will finish in a few minutes take half an hour, while they fly through activities that I thought they'd need a half hour to finish!" Prioritizing to-do lists helps us see that the crucial tasks are tackled before the less crucial tasks.

We asked Marilyn, a first grade teacher, how she used prioritized to-do lists. Marilyn said, "I like to have at least ten activities for teaching a new consonant blend or digraph. When I prioritized my list on the day I taught

Figure 1.1. Ordered To-Do List

the *sh* digraph, I found I had three activities that were essential: whole-group introduction to the digraph by an analytic lesson, students generating words that have the new sound while I write them on the board, and whole-group practice reading the list of words that the kids nominated. I realized that I had three activities that were important but not essential: singing a song I made up called 'Shh! Don't Shout!,' dancing a dance I call 'The Ship Dance,' and cutting pictures of things that start with the *sh* sound out of magazines. I decided that I had three activities that were useful but not important: doing the three different workbook pages in the basal reading series. I also realized that I had two activities that were fun if time permitted them for enrichment or reinforcement: making three-dimensional ships out of construction paper and making a mural of ships at sea." Marilyn told us that by prioritizing her list, she could ensure that the essential and most important activities were accomplished.

Checklists

Six kinds of checklists are useful for teachers. The first one is the class skills checklist. The second type of checklist is the student accountability checklist that teachers use when they give credit for work without assigning a grade. Third is one that is used to record information about the frequency of a student's discrete behavior. The fourth kind of checklist is the continuous behavior checklist that is used to document more subtle student behaviors. The fifth kind is the health and welfare checklist. The sixth is a checklist of the essential elements of a lesson checklist, a checklist that a teacher makes for herself to ensure that she has met the criteria for a lesson.

Class Skills Checklists

Class skills checklists allow you to identify which skills in a subject area have been mastered by which students. You use the checklist to determine who needs reteaching. You can either group the students together to reteach them the skill, or you can assign a student who has mastered the skill to reteach it to a student who has not mastered the skill.

Kayla, a veteran second grade teacher, likes to indicate which students she will group together and reteach herself, and which students will peer tutor other students individually (Figure 1.2). The student number of the student who will be the tutor is inserted in the empty check box of the student who needs to be retaught the skill. The capital T means that the teacher herself will reteach the student, usually in a small group of students having similar skills.

Student Accountability Checklists

A student accountability checklist allows the teacher to hold students accountable for completing a task that does not require an evaluation of the quality of the assignment. For example, Randa, a third grade teacher, assigned all of her students to make and decorate a portfolio box at home and bring it to school. Any way that a student chose to decorate the box was acceptable. "There was no wrong way to do it," said Randa. "Some kids used all the bells and whistles. One girl glued pictures of dalmatians all over hers and then added a rhinestone dog collar. A boy covered his with seashells on a blue background, and in the center he had a big starfish. A very quiet girl chose to cover hers in plain white paper and leave it that way. However a kid wanted to do it, that was okay." In Randa's case, an accountability checklist was an appropriate tool. She simply checked off that a student had turned in a portfolio box.

#	*Name*	*cap sent*	*cap noun*	*period*	*quest mrk*	*exclam*	*sub/vrb*
01	Marti	✓	✓	✓	✓	✓	✓
02	Tomas	T	16	16	✓	✓	16
03	Anita	✓	✓	✓	✓	✓	✓
04	Ebony	✓	✓	✓	✓	✓	✓
05	Skye	T	✓	✓	✓	✓	✓
06	Tonisha	✓	✓	✓	✓	4	✓
07	Andree	✓	✓	✓	✓	✓	✓
08	Mikki	✓	✓	✓	✓	✓	✓
09	Paul	T	10	✓	✓	✓	✓
10	David	✓	✓	✓	✓	✓	✓
11	Phillip	✓	✓	✓	✓	✓	✓
12	Miguel	✓	✓	✓	✓	✓	✓
13	Gabe	T	15	T	T	15	T
14	Toni	✓	✓	✓	✓	✓	✓
15	Summer	✓	✓	✓	✓	✓	✓
16	Lesley	✓	✓	✓	✓	✓	✓

Figure 1.2. Class Grammar Skills Checklist

Discrete Student Behavior Checklists

The discrete and continuous student behavior checklists are tools of the special education teacher who specializes in working with students who have severe behavior problems, but they are very helpful to the general education teacher as well. First we'll discuss the discrete behavior checklist.

Rosemary, an award winning social studies teacher, said, "Focusing on good behaviors and ignoring bad ones is a good rule for teachers, and I pride myself on doing just that. I don't sweat the small stuff and I like to catch the kids being good and focus on the things they do well. But," she cautioned, "when a problem behavior is too important to be ignored because it interferes with the student's ability to be successful, or with other students' ability to be successful, the teacher's job is to intervene. That's when I use a student behavior checklist." She pulled a document out of her file cabinet (Figure 1.3).

"My classroom is highly interactive," she said, "but sometimes the students must work individually and without talking to other students. Todd would turn around and talk to the girl behind him. Not only could he not do his own work, but he was keeping her from doing her work, and he was distracting the students around them. I used a discrete behavior checklist to

Monday
𝍸 ||
Tuesday
||||
Wednesday
𝍸 ||||
Thursday
𝍸 ||
Friday
𝍸 𝍸

Figure 1.3. Todd's Checklist of Turning Around and Talking

document how many times he engaged in the behavior over a week's time. Then I sat down with him and showed him the checklist. We went from there."

Continuous Student Behavior Checklists

Millie used a related checklist to document Amanda's problems with staying on task. A continuous behavior checklist differs from a discrete behavior checklist in that it documents behaviors that are not easily counted. For example, frequency of getting out of a seat can be documented by a discrete behavior checklist. In contrast, being off task is a more difficult behavior to document. A student might be off task only once in an hour—all hour long! In order to deal with continuous behaviors, a continuous behavior checklist is the appropriate tool.

Millie gave her second graders an assignment that she anticipated would take 20 minutes. Then she used a checklist to compare Amanda's off-task behavior to that of an average student. Using a two-column checklist, Millie checked whether or not Amanda was on task at 15 and 45 seconds after each minute. She checked whether or not the comparison student was on or off task at 00 and 30 seconds after each minute (Figure 1.4). After

November 15
10:15 English Class

Time	*Amanda*	*Comparison Student*
15	+	
30		+
45	+	
0		+
15	+	
30		+
45	0	
0		+
15	0	
30		+
45	0	
0		0
15	0	
30		0
45	0	
0		+
15	+	
30		+
45	0	
0		+
15	0	
30		+
45	0	
0		+
15	+	
30		+
45	+	
0		+
	6+	12+

Figure 1.4. On-Task Behavior Checklist for Amanda Mitchelet

keeping the checklists for a week, Millie was ready to try some interventions. (Note that it is important to have a comparison subject in order to make your checklists more valid.)

Another way that a discrete or a continuous behavior checklist helps is by helping a teacher decide whether she has "Worst Kid Syndrome." There will always be one child in a class whom, in your heart of hearts, you like a little less than you like the other children. You may find yourself focusing on that child's shortcomings and ignoring his strong points. When you find yourself criticizing every move he makes, a behavior checklist will help you decide whether the problem lies with his behavior or with the way you are looking at him. When you use a behavior checklist in this way, you must do two checklists: one documenting the behaviors of the target student and one documenting the behaviors of an average, solid student. If you find that the frequency of behaviors of the target student is not more than 40% greater than that of the average student, then the problem may lie with the way you feel about the student, rather than with the student himself.

Health and Welfare Checklists

Another type of student behavior checklist, the health and welfare checklist, documents a case of child neglect. While the two previous checklists recorded frequency of discrete behaviors or duration of continuous behaviors, the checklist for documenting neglect records one or more problems a day for an extended period of time.

Because neglect cases are difficult to make, keep a daily checklist of health and welfare problems of a child whom you suspect is being neglected, problems such being sent to school with a runny nose, wearing dirty clothes, smelling of urine, and so forth (Figure 1.5). If you will keep the list for 3 or 4 weeks, then when you turn your checklist over to the school nurse, she can contact Child Protective Services with enough evidence to get them to open a case.

Essential Elements Checklists

The essential elements checklist helps ensure that you are writing a test that covers everything it needs to cover. When Carl Amandez wrote a test for his social studies students, he wanted to be certain that he tested all five concepts that the unit covered. He listed those five concepts, and every time he wrote a question that assessed one of the concepts, he wrote the number of the question under the concept. When the teacher next door to Carl tried the checklist strategy, she discovered that her test covered three of the concepts thoroughly (four or five questions), one concept nominally (two questions), and omitted one concept altogether. Checklists are an essential tool for writing valid tests (Figure 1.6).

Date	*Runny Nose*	*Dirty Clothes*	*Urine Smell*
February 01	+		
February 02	+	+	+
February 03	+	+	+
February 04	+		
February 05	+		
February 08	+	+	+
February 09	+	+	+
February 10	+	+	+
February 11	+		
February 12	+		
February 15	+	+	+
February 16	+	+	+
February 17	+		
February 18	+		
February 19	+		

Figure 1.5. Health and Welfare Checklist for Anthony Alabi

Location	*Place*	*.Regions*	*Movement*	*Relationships*
1	3	9	2	10
5	4		8	
6	7			

Figure 1.6. Essential Elements Checklist for India Exam

Rubrics

Rubrics are an important part of organizing your instruction, but rubrics are not merely helpful for the teacher. Students appreciate them, too. They like knowing what is expected. They appreciate being well informed about how their work will be graded.

The term *rubric* originally referred to the notes in the margins of hymnals and books of prayer. The notes were often printed with red ink. The notes were reminders of what to do or how to do it. (Examples: "bow your head as you say this," or "sotto voce here.")

A rubric for the classroom is a checklist of what the teacher expects. For example, if students are assigned book reports, the teacher should produce

Name ______________________________ Date ______________

Objective: The student will: (1) choose two books in the same genre, and
(2) compare and contrast them.

Content:

The introduction explains your book selection	________
Similarities of the two books are explained	________
Differences between the two books are explained	________
The conclusion summarizes your comparison	________

Mechanics:

Punctuation is correct	________
Capitalization is correct	________
Spelling is correct	________
Paragraphing is appropriate	________
Pages are numbered	________
Margins are all 1″	________
Handed in on time	________
At least five pages	________
Grade for this book report	________

Figure 1.7. Book Report Rubric

a book report rubric and distribute it to the students well ahead of the due date.

Book Report Rubrics

To create your first book report rubric, quickly review some of the book reports that you have received from students in the past. List the aspects that you liked, for example, Serena's large margins that provided plenty of room for you to write encouraging notes. Don't forget Sam's page numbers that made it easier to determine when he had written a large enough report. Then create a second list; this time note the things that bothered you. Start with the terrible spelling that made Morey's report tough to read. Include the smudged and messy report that Vengung handed in to you. Now you are well on your way to developing a rubric that will efficiently communicate your preferences to your students. Your secondary grade level book report rubric might look like the example in Figure 1.7.

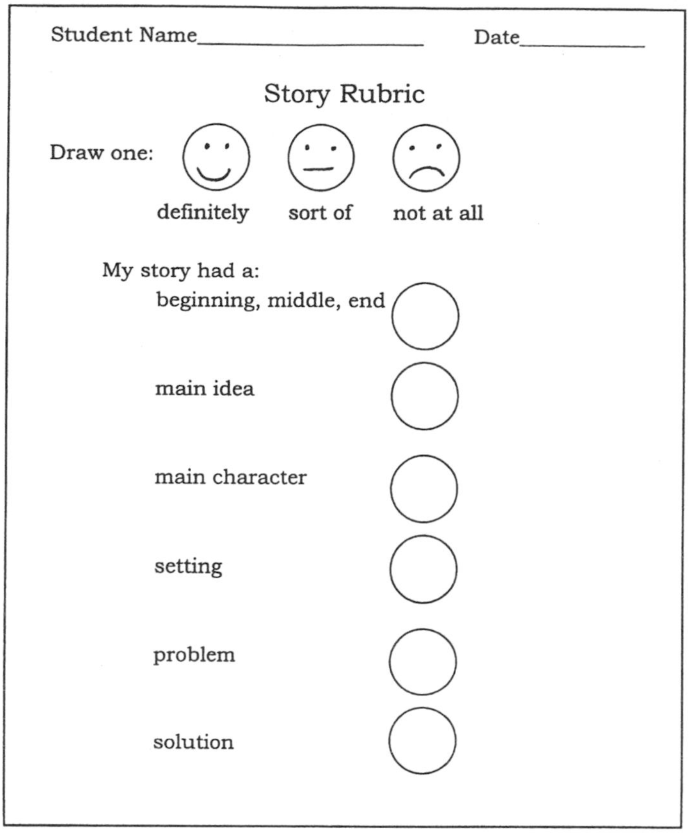

Student Name____________ Date________

Story Rubric

Draw one: ☺ ☹

definitely | sort of | not at all

My story had a:

beginning, middle, end ◯

main idea ◯

main character ◯

setting ◯

problem ◯

solution ◯

Figure 1.8. Happy Face Writing Rubric

Writing Assignment Rubrics

The book report sample provided a space next to each factor to place an evaluation code. Other teachers prefer to circle a yes or a no.

The next rubric is a writing assignment rubric simplified for elementary school use. We like to use happy faces for this level (Figure 1.8).

The next sample writing rubric (Figure 1.9) uses the Y and N abbreviations. It was contributed by a teacher who teaches English and writing by having her class produce a newspaper.

Rather than a yes or no, other teachers prefer a rating scale. The sample in Figure 1.10 is a rubric a teacher created for her students' group project called In Other Words. Here, a 5 is the highest rating and a 1 is the lowest rating. Notice that this teacher included subtotals to help in counseling with each student group.

Name of Student Author ______________ Date ________

Content		
Facts checked	Y	N
Interesting "gotcha" first sentence	Y	N
5 W's covered	Y	N
Adequate supporting details	Y	N
Title		
Catchy	Y	N
Mechanics		
Complete, grammatically correct sentences	Y	N
Capitals & punctuation correct	Y	N
Spell checked	Y	N
Written in past tense	Y	N
At least two paragraphs	Y	N
Creativity		
Interesting to read	Y	N
Strong verbs used	Y	N
Strong descriptors	Y	N

Figure 1.9. Newspaper Rubric

Some teachers like to have the students rate themselves before the teacher's evaluation. Then they conference and discuss their perceptions. This sample (Figure 1.11) is from a teacher who taught an aeronautics unit.

An even more sophisticated approach is to involve other students in doing peer evaluations. The student author shows his report to another student, who rates it on each of the criteria. Then the same student is the rater for someone else. In this way, students learn writing skills more quickly and thoroughly. Figure 1.12 is an example of a zoo careers report rubric that incorporates a peer review.

Once you begin developing rubrics, the process becomes much easier. You can make your rubrics simpler for slower learning students and more demanding for faster learners. Rubrics are seldom "done." Using word processing, teachers find rubrics convenient and quick to revise as new ideas arise when students ask questions and hand in assignments.

No matter what the topic of your rubric, other teachers will want a copy. This provides you with opportunities for sharing, which is another important dimension of being organized. For example, if four sections of the sixth

Names of Students in the Group: ______________________________

1.	Content						
	Meaningful	5	4	3	2	1	
	Effective words	5	4	3	2	1	
							C. Subtotal ________
2.	Mechanics						
	Correct punctuation	5	4	3	2	1	
	Correct capitalization	5	4	3	2	1	
	Correct spelling	5	4	3	2	1	
							M. Subtotal ________
3.	Visual Aids						
	Informative	5	4	3	2	1	
	Attractive	5	4	3	2	1	
	Colorful	5	4	3	2	1	
	Error free	5	4	3	2	1	
							V. Subtotal ________
4.	Task Commitment						
	Used time wisely	5	4	3	2	1	
	Contributed to project	5	4	3	2	1	
							T. Subtotal ________
5.	Oral Presentation						
	Adequate projection	5	4	3	2	1	
	Adequate pacing	5	4	3	2	1	
	Innovative	5	4	3	2	1	
							O. Subtotal ________
							Total Score ________

Figure 1.10. "In Other Words" Rubric

grade are all studying zoo careers, the most efficient way to teach effectively is to share the work. One teacher might develop a rubric, another teacher might reference all the activities to the school's curriculum guide, another might make the arrangements for a field trip, and the last teacher might make all the parent contacts. When teachers cooperate they all save time, and their students benefit from better instruction. This is a dynamite way to get organized.

Name ____________________ Date ____________

Topic ______________________________

Circle the number of points earned in each category.

	Self-Rating	*Teacher Rating*
Content:		
accurate	1 3 5	1 3 5
interesting	1 3 5	1 3 5
enough information	1 3 5	1 3 5
feelings shown	1 3 5	1 3 5
use of own words	1 3 5	1 3 5
Task Commitment:		
stayed on task	1 3 5	1 3 5
completed on time	1 3 5	1 3 5
Product:		
creative	1 3 5	1 3 5
neat	1 3 5	1 3 5
Presentation:		
well prepared	1 3 5	1 3 5
creative	1 3 5	1 3 5
projected voice	1 3 5	1 3 5
Total score	______	______

Figure 1.11. Aeronautic Rubric

Telephone Memo Books

In the age of accountability, it is important to be able to document each parent contact, each time a parent leaves a message for you, and each time you attempt to contact a parent. A telephone memo book is inexpensive and allows you to keep all of your parent phone calls neatly documented. It also helps remind you to return calls to parents who have left you a message. The memo books can be bought with carbonless copy paper that makes duplicates for you. Then you can put the original in the student's file and keep the copies of all your calls together in your own files.

Name of Student Author ____________________

Career ____________________ Date __________

Name of student rater ____________________

	Self-Rating	*Peer Rating*
Content:		
accurate	1 3 5	1 3 5
sufficient information	1 3 5	1 3 5
organized	1 3 5	1 3 5
interesting	1 3 5	1 3 5
feelings demonstrated	1 3 5	1 3 5
Task commitment:		
stayed on task	1 3 5	1 3 5
completed on time	1 3 5	1 3 5
Product:		
used own words	1 3 5	1 3 5
creative	1 3 5	1 3 5
neat	1 3 5	1 3 5
Presentation:		
well prepared	1 3 5	1 3 5
loud enough for everyone to hear	1 3 5	1 3 5
used expression well	1 3 5	1 3 5
creative	1 3 5	1 3 5

Figure 1.12. Zoo Careers Rubric

Summary

In this section of Chapter 1, we have discussed the purposes of and educational uses for first generation organizational aides. Specifically, we have discussed three kinds of to-do lists (basic, ordered, and prioritized), six kinds of checklists (class skills, student accountability, discrete behavior, continuous behavior, health and welfare, and essential elements), rubrics, and telephone memo books. In the next section, we will discuss second-generation organizational aids.

Check those elements that you now know how to use. Reread the information for any elements about which you feel uncertain.

- ☐ To-do lists
- ☐ Checklists
- ☐ Rubrics
- ☐ Telephone memo books

SECOND-GENERATION ORGANIZATIONAL AIDS

In the last section, we identified first-generation organizational tools. In this section, we will introduce second-generation organizational tools:

- ☐ Calendars
- ☐ Day planners
- ☐ Databases

Whereas first-generation tools organize only the immediate future, second-generation tools organize for the distant future as well as the immediate future. Hyon Lee uses her to-do list to plan what she is going to do in the next hour; she uses her second-generation tool to plan what she is going to do tomorrow or next year.

Calendars

Having a lesson plan book doesn't preclude the need for a portable calendar. The lesson plan book is reserved for school activities and displays only one week at a time. Seldom does a teacher flip forward in her lesson plan book; her upcoming commitments are therefore not constantly being seen to remind her to prepare for them. In addition, the lesson plan book usually stays on the desk; like a faithful old dog, the portable calendar goes wherever the teacher goes.

Day-at-a-Glance

When deciding on a calendar, the first decision is whether you want to buy a one-day-at-a-glance appointment calendar, a week-at-a-glance calendar, or a month-at-a-glance calendar. While day-at-a-glance calendars are useful for lawyers and physicians, who have each hour partialed out for

appointments, they don't work as well for teachers. Teachers find week-at-a-glance or month-at-a-glance calendars more useful.

Week-at-a-Glance

American life revolves around the week as the primary unit of time, so using a week-at-a-glance calendar is appealing. In addition, given that both the weekly and monthly calendars have the same size page, the weekly calendar affords four times more writing space. When Consuela used a weekly calendar, she swore she'd never switch to a monthly one. However, she decided to make herself try a monthly calendar one year, and fell in love with being able to view the entire month at a time. She uses a monthly calendar to this day.

Month-at-a-Glance

The month-at-a-glance calendar has special appeal for teachers because so much of what we do is planned far in advance. For example, Harry scheduled all his guest speakers for the year as soon as school started. When Harry knew he would be teaching a unit on career education in March and April, he would call his resources in September, schedule them to come talk to his kids, and be able to get a feel for the whole month whenever he turned to the appropriate page. Because Harry used his calendar only for appointments (such as parent conferences) or important dates (such as report card time), the smaller space afforded by the monthly calendar, as compared to the weekly, was sufficient.

Day Planners

Day planners, also called day runners or personal planners, differ from calendars in that they include other tools in addition to a calendar. You'll be surprised how much better you'll like a day planner than a stand-alone calendar. The contents of your day planner can vary with your specific needs. You can buy one already assembled or you can buy the empty cover and custom make your own. If you make you own, you'll find that in addition to the calendar, you will want to include a telephone-address book and some blank paper for notes and to-do lists.

A useful way to organize the telephone-address book is to categorize and cluster, rather than alphabetizing, the people you contact. For example, you may want to list all of the people who do presentations for your class in one cluster. Then when you're scheduling your speakers for the year, you

won't forget one. You may want to categorize and cluster the parents who will supervise field trips, those who will send food for parties, and so on. In addition to parents clustered in those groups, Hyon Lee also keeps the addresses and phone numbers of all students in one list. Then when she makes a quarterly call to each child's home, she has all of the names at hand and won't overlook someone.

Teachers who keep a spiral-bound telephone message log in their planners find it an important organizational tool. When the teacher returns a call, if it is necessary to set a follow-up meeting or teleconference date, he has both the message log and the calendar handy.

Student Databases

In addition to their addresses and phone numbers, you'll want a copy of a student database in your day planner. The student database should include parents' work phones, emergency phone numbers, the child's birth date, and any other important data such as medical information. Some teachers include such information as names of brothers and sisters so when they call a parent, they can show interest in the family by inquiring about the other children.

You'll want to create the database on your computer so that you can update it easily and have the computer version handy as well as the hard copy in your planner. Simply use the database program that comes with your basic office program. Figure 1.13 shows an example of a student database.

Summary

In this section, we have discussed calendars, day planners, and databases. We have suggested the tools that you want to include in your day planner.

Check off those elements that you now know how to use. Reread about those that you are not certain that you now know how to use.

- ☐ Calendars
- ☐ Day planners
- ☐ Databases

	Last Name	*First Name*	*Parents*	*Street Address*	*Mailing Address*	*Phone*
1	Barber	Billy Jack	Pat & Jo Kincade	123 Alabama	same	555-1234
2	Coe	Ann	Mary O'Riley	234 Pennsylvania	same	555-1237
3	Dart	C.J.	Ed & Wanda	345 Florida	same	555-1238
4	Emerson	Oscar	Bandy & Rose	123 Oklahoma	same	none
5	Estrada	Andrea	Mike & Anita	456 Alaska	PO 678	555-2345
6	Fry	Freddie	William	Rt 4 Box 444	same	555-5678
7	Griffiths	Gay Su	Rocky & May	789 Ohio	same	555-2346
8	Hawthorne	V.J.	Vera Bell	234 Texas	PO 456	555-1235
9	Hudson	Jonnie	Jeff & KC	123 Kansas	same	555-2347
10	Kitchen	Melissa	M.C.	345 Washington	PO 567	555-2348
11	Land	Mac Jr.	Mac	Rt. 3 Box 305	same	555-4567
12	Newton	Galen	Newt & Liz	567 Trenton	PO 234	555-6789
13	Tettleton	Dawn	Mollye Scotland	456 Windthorst	same	555-4789
14	Williams	Phillip	Phyllis LeBlanc	567 Buffet	same	none

Figure 1.13. Student Database

RECOMMENDED READING

Covey, S. R. (1989). *The 7 habits of highly effective people.* New York: Simon & Schuster.

2

BURIED UNDER ADMINISTRATIVE PAPERWORK?

Many teachers who burn out say, "It's not the kids that made me give up on teaching. It's the paperwork! I went to college to be able to teach kids. Not to push papers around." Because the stress of administrative paperwork can be so debilitating, it's imperative that teachers learn good ways to streamline these nonteaching duties. This chapter will help you with that task.

STREAMLINING DAILY TASKS

In the last chapter, we discussed how to organize your time and memory. In this section, we will discuss how to streamline your daily administrative tasks. We will discuss:

- ☐ Delegating authority
- ☐ Student numbers
- ☐ Clipboards
- ☐ Photocopy box

Delegating Authority

Having to take roll is a chore for teachers, but students love to get to take roll. Delegate that authority!

Teachers tend to want to be all things to all people. No one can be all things to all people! We all have limits and have to take care of ourselves. One way that we can take care of ourselves is by delegating responsibilities. Even primary children can learn to help with administrative paperwork. You can put children in charge of roll call, checking in and out class library books, reminding pull-out children to go to their special classes, and a host of other administrative chores. Not only does it lessen the load you carry, but it gives the children opportunities to use the skills learned in school in real-world situations. Children receive a great sense of class ownership when they help with the daily routine tasks of running a class.

Student Numbers

Student numbers are a huge time-saver! Assign each student the number that his name falls on in your gradebook. For example: Abelson, Aaron might be number 1 in your grade book, and Zywicki, Zoe might be number 26. Once you've assigned the student numbers, have each student write his number on his paper in addition to his name. Then, when students turn in their papers, you can quickly place them in numerical order and they are ready to record in your gradebook. By placing them in numerical order, you save time over having to alphabetize them.

Clipboards

First grade teacher Sylvia Broussard told us, "Clipboards stack up in a small space, keep items securely separated, and are highly mobile. On my desk, one clipboard holds my current lesson plans, one holds a roster of students' names on a checklist, and one is for future lesson plans. The checklist roster of students helps keep track of items students are supposed to turn in (picture money, permission slips, homework). The current lesson plan clipboard gives me easy access to my lesson plans throughout the day. I can carry it with me to the reading area and don't have to run back to my desk to check the lesson plan book. The future lesson plan clipboard is right at my fingertips to jot down ideas and reminders, or to clip on information about programs or schedule changes that may affect next week's planning. When I begin to plan the next week's lessons, all my ideas and reminders

are in one place." We recommend a clipboard with a clip that won't dump all your pages when you bump into something.

Photocopy Box

The photocopy box is a tool you'll use every day. When you are planning and come across a page that you want to photocopy for the students, place it in the photocopy box. Then when you get ready to go to the photocopy machine, you have all the papers you need together. If you don't use a photocopy box, you'll spend time looking for all the papers that you'd intended to photocopy that day, because you'll have laid them on your desk and covered them up with other papers.

We find ourselves highlighting the parts of articles we most want to use. If you do that too, be sure you use highlighters that are invisible to photocopy machines. Beware of highlighters that will copy as dark gray smudges.

Summary

In this section, we have discussed ways to deal with daily administrative tasks. Specifically, we've covered the items following. Read over them and make sure that you understand each.

- ☐ Delegating authority
- ☐ Student numbers
- ☐ Clipboards
- ☐ Photocopy box

ACCOMMODATING OCCASIONAL ADMINISTRATIVE TASKS

In the last section, we discussed some ways to streamline administrative tasks that face you every day. In this section, we discuss administrative tasks that arise on a more occasional basis. Specifically, we will cover the following:

- ☐ Teacher's Bulletin Board
- ☐ Color coding classes
- ☐ Form letters
- ☐ To-be-sent-home box
- ☐ Carbonless copy paper

- ☐ Self-duplicating memo forms
- ☐ Documentation

The Teacher's Bulletin Board

We capitalized the term The Teacher's Bulletin Board to discriminate between it and the regular classroom bulletin boards. The Teacher's Bulletin Board is hers alone. It holds reminders, notices, and memos that she needs to consider, and manila envelopes into which she can slip related information. A useful size is 2 × 3 feet. Such bulletin boards can be purchased for under 10 dollars at a discount store.

Color Coding Classes

Pam Alexander teaches in a departmentalized elementary school. Pam tells us that she color codes her different classes. She uses neon paper, and each class has its own color. She uses the neon papers for roll sheets, grade sheets, and so on. That way she doesn't have to go through a whole stack of roll sheets, for example, to find the correct one. She just makes a beeline for the appropriate color in the stack.

Form Letters

Title a computer folder Form Letters. Keep all of your various and assorted form letter masters in the folder. Keep it on your hard drive and on a backup disk. If you don't have form letter masters, make some. Writing an original letter each time you need to communicate with a parent is a terrible waste of time. By using the form letter, you can simply insert the information that applies to the particular situation and save yourself lots of time.

Susan Babbitt, a Tulsa, Oklahoma, school administrator, told us, "Unless there's some reason that it's essential for a form letter to look formal, I like to use clip art to make them seem warmer and more friendly and personal. I also like to use a font with more flair than typical business letters use, and I use 18 point size if feasible. Who could possibly be threatened by a form letter with a fun font and a merry piece of clip art?"

To-Be-Sent-Home Box

Whenever you have things to be sent home with children, such as form letters, place them in a to-be-sent-home box. Keep the box by the door, and

assign a student each week to help you remember to check the box before you dismiss class.

Carbonless Copy Paper

John is a great aficionado of carbonless copy paper. He uses the smaller size to hand-write notes to students when he wants to keep a copy. Of course, he could photocopy the notes, but that would require a trip out of his room, down the hall, up the stairs . . . to the copier. He could also wordprocess the note. He does wordprocess notes when they should be more formal; however, when he specifically wants the note to appear informal and nonthreatening, he prefers to hand-write it, so he uses carbonless copy paper.

With its built-in carbon, the 8.5 × 11 inch size carbonless copy paper is also an excellent way to provide a copy of a good notetaker's notes to a special education student whose Individualized Education Plan requires that he be provided with a copy of class notes. It's also a good way to provide a copy of the notes for a student who has been ill and missed class.

Self-Duplicating Memo Forms

John says that the most handy tool he uses is a book of self-duplicating memo forms. They are made of carbonless copy paper. John prefers the memos that are 5½ × 4 inches and come two memos to a page. As with the carbonless copy paper, he uses these when he doesn't want the memo to be threatening to the recipient, yet he wants to keep a copy of the memo.

They make a great documenting tool for working with the student who is in special education, particularly the student who has emotional or behavior problems. When an incident arises in class that you think the parent needs to know about, write a short memo, send it home to the parent, and keep your copy in the memo book. In the same way, if you want to send the principal a short memo documenting an incident, you'll have a copy of your own memo if you use the book. Having a book for each of the students in your class for whom documenting behaviors is going to be essential is a great organizational idea.

Documentation

That leads us into our discussion of documenting. You may find that for most documentation, you don't need to send a note to anyone. For this type of documentation, save space in the lower righthand corner of your

lesson plan pages each week. Note a day, time, and behavior in that area, and then at the end of each grading period, transfer that information to individual student folders.

You can also transfer that information to your student database on the computer.

Wendy Hazlett is a second grade teacher and reading specialist who has a nifty way to document. Wendy keeps a page of blank address labels on a clipboard. As she walks about her classroom, when she notices something that she wants to document, she writes it on an address label. At the end of the day, Wendy sits down with her student folders and the address labels. She sticks each label onto the inside of the appropriate folder.

Summary

In this section, we have discussed ways to streamline tasks that you need to do only occasionally. Make certain that you understand all of these organizational techniques before proceeding:

- ☐ Teacher's Bulletin Board
- ☐ Color coding classes
- ☐ Form letters
- ☐ To-be-sent-home box
- ☐ Carbonless copy paper
- ☐ Self-duplicating memo forms
- ☐ Documentation

ORGANIZING STUDENT DATA

In the last section, we discussed tasks that you need to do only occasionally. In this section, we will discuss a task that you need to do only at the beginning of the year and when you get new students. We didn't include it in the last section because we think it deserves its own.

- ☐ Student data Smart Books

Student Data Smart Books

We both find our student information data banks invaluable. On the first day of class, we have each student complete a personal data sheet. The sheet

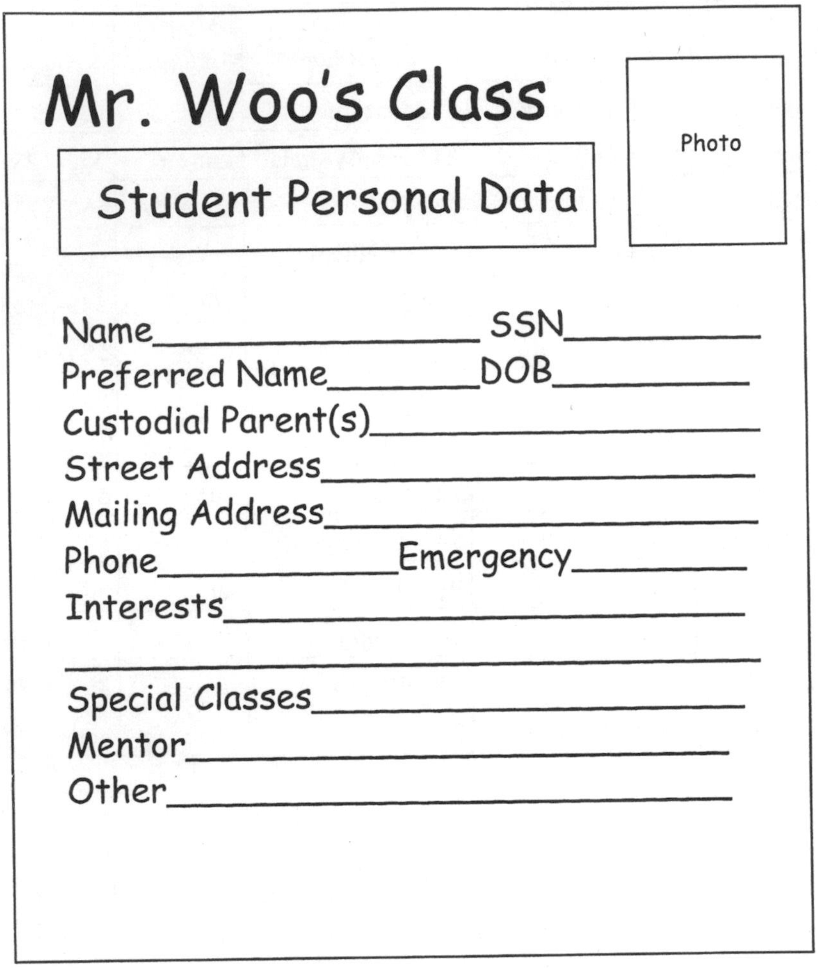

Mr. Woo's Class

Photo

Student Personal Data

Name________________ SSN__________
Preferred Name________ DOB__________
Custodial Parent(s)__________________
Street Address_____________________
Mailing Address____________________
Phone____________ Emergency________
Interests_________________________
____
Special Classes____________________
Mentor___________________________
Other____________________________

Figure 2.1. Mr. Woo's Class

includes phone number, address, birthday, all the usual stuff. We also include a previous GPA so we will be able to start the year with heterogeneous groupings. Then we take a photo of each student. We enter all the written data into a database, and then scan in the students' pictures. This allows us either to keep the database paperless, or else to print out a paper copy of each individual student's information, a list of all the students' information, or both, and place it/them in a three-ring binder (Figures 2.1 and 2.2).

An advantage of having a paper copy of each student's datasheet in a three-ring binder is that you can carry it with you and have instant access to the information whenever you might need it. You can also log such

	Last Name	*First Name*	*Parents*	*Street Address*	*Mailing Address*	*Phone*
1	Barber	Billy Jack	Pat & Jo Kincade	123 Alabama	same	555-1234
2	Coe	Ann	Mary O'Riley	234 Pennsylvania	same	555-1237
3	Dart	C.J.	Ed & Wanda	345 Florida	same	555-1238
4	Emerson	Oscar	Bandy & Rose	123 Oklahoma	same	none
5	Estrada	Andrea	Mike & Anita	456 Alaska	PO 678	555-2345
6	Fry	Freddie	William	Rt 4 Box 444	same	555-5678
7	Griffiths	Gay Su	Rocky & May	789 Ohio	same	555-2346
8	Hawthorne	V.J.	Vera Bell	234 Texas	PO 456	555-1235
9	Hudson	Jonnie	Jeff & KC	123 Kansas	same	555-2347
10	Kitchen	Melissa	M.C.	345 Washington	PO 567	555-2348
11	Land	Mac Jr.	Mac	Rt. 3 Box 305	same	555-4567
12	Newton	Galen	Newt & Liz	567 Trenton	PO 234	555-6789
13	Tettleton	Dawn	Mollye Scotland	456 Windthorst	same	555-4789
14	Williams	Phillip	Phyllis LeBlanc	567 Buffet	same	none

Figure 2.2. Student Database

information as when a student has been referred to the principal for discipline, when telephone calls have been made home, and more. For each student who has an IEP, you can keep a copy of the objectives and modifications in the folder.

One of the teachers we surveyed told us that he calls his student data notebook his Smart Book from a term used in the military that tells a newcomer everything he needs to know about his new assignment. We like the name!

Summary

In this section, we have discussed how to organize your student data.

☐ Student data Smart Book

SETTING UP THE CLASSROOM LIBRARY

In the last section, we discussed how to organize your student data. In this section, we will share some ideas on how to organize your classroom library.

- ☐ Classroom library card system
- ☐ Color coding classroom library books
- ☐ Use inventory of classroom library books

Classroom Library Card System

You probably already use library cards and library card envelopes to keep track of your class library books, but here's a wonderful addition to that idea.

On a bulletin board or a poster, place a library card envelope for each child. When a child checks out a book, have her slip the library card into her envelope. That way you can see at a glance who is reading what (Figure 2.3). You'll be able to note if someone seldom checks out a book, and you'll be prompted to encourage him to check one out.

Color Coding Classroom Library Books

While you probably already organize your nonfiction classroom library books by topic, try color coding your fiction titles by topic using colored dots on the spines of books. For example, put a blue dot on the spine of all books that are animal stories, a purple dot on all the sports stories, and so on. A student who is interested only in animal fiction could tell at a glance which books she'd like to read most. Some of the books may need two or three different dots. For example, an animal sports story would have both a blue and a purple dot.

Use Inventory of Classroom Library Books

A use inventory of classroom library books is entirely different from an inventory that identifies what books are missing. In the use inventory, you chart how many times each book is checked out in a semester. You can then use that data to make decisions about categories of books and individual titles. For example, if you find that nonfiction books on space are checked out the most frequently of any category, then you know to buy more

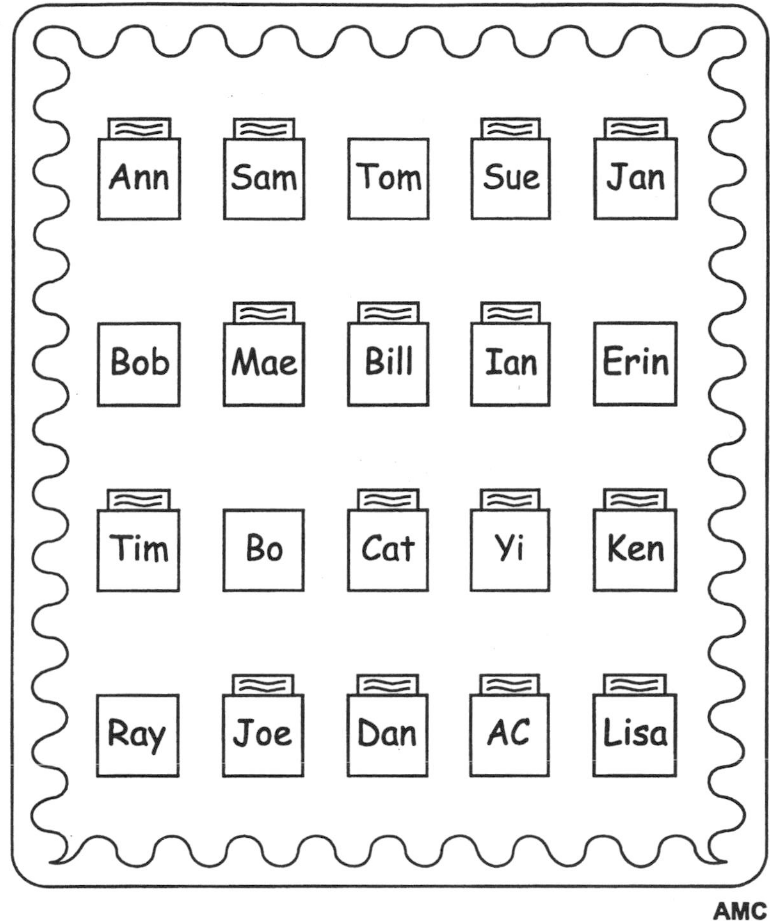

Figure 2.3. Classroom Library Card System

nonfiction books on space. If you find that *Island of the Blue Dolphins* is checked out more often than your other fiction titles, then you know to buy duplicate copies of that particular book.

In contrast, if you found that books on gardening are never checked out, you would know: (a) not to buy more gardening books, and/or (b) to launch a campaign to encourage the kids to learn about gardening.

Kids love to do the use inventory for you, so delegate that authority!

Summary

In this section, we have discussed some ways to streamline your classroom library. Specifically, we discussed:

- ☐ Classroom library card system
- ☐ Color coding classroom library books
- ☐ Use inventory of classroom library books

CREATING THE PAPERLESS CLASSROOM

In the last section, we discussed how to streamline your classroom library. In this section, we will discuss the paperless classroom. Specifically, we will address:

- ☐ The paperless classroom
- ☐ Computerized gradebooks
- ☐ Using computers to make students' work available to parents

The Paperless Classroom

We are excited about the potential that the paperless classroom has. By moving most of your administrative work to computer, you will cut down on an amazing amount of clutter. Attendance, for example, can be kept on computer and sent by e-mail to the office. Lunch count can be handled the same way. Notes to the principal, counselor, or other staff can, and probably should be, sent e-mail. If you want to save a copy of the note and document that you sent it, you can send yourself a copy or a blind copy when you mail it to the other person.

Inventory is such a pain, but computer databases make taking inventory more tolerable. Being able to search for a book number by computer is far preferable to the old way when the list is alphabetical by name of child. Likewise, searching for the name of a child on a book inventory is far preferable to the old way of organizing by book number.

The classroom library checkout system should also be computerized, and students can handle the "administrivia" involved with the class library. By using a database program, it is not only easy to know who has read what, but you can quickly identify the titles that justify a second copy or that should be discarded.

Computerizing your classroom library, however, does not mean that the classroom library bulletin board we discussed earlier isn't a valuable tool. It is! However, by also having the records on the computer, you will be able to analyze data in a way that the bulletin board alone would not allow.

Computerized Gradebooks

Computerized gradebooks are nothing more than spreadsheets. Anyone who has had a basic computing course can develop a computer gradebook; however, they can be purchased for a very reasonable price. Some of the features of computerized gradebooks are: (a) individual printouts of a student's grades and the subject of the assignment for which she received that grade; (b) which assignments have not been turned in; (c) a grade average to date for each student; (d) a class grade average on each assignment; and (e) a dozen other things you never knew you needed until you found out about them. Some commercial computer gradebooks have ready-made letters to parents that allow you to insert the grade information you want to share.

Think of the benefit of being able to print out a copy of a child's grades for his parents on the spot! You could show parents how the student has improved this 6 weeks over the past 6 weeks. You could show which assignments he had missed. On low grades, you could show the parents the class's grade average as compared to their child's grade on an assignment. . . . The list goes on and on.

Using Computers to Make Students' Work Available to Parents

As schools become more sophisticated in their use of computers, more possibilities for communicating with parents will be possible for teachers. One way is scanning student papers into your classroom's web page file and inviting parents to review them via their home computers. Many programs for doing this will likely be on the market soon. Currently we know of one that could be used for this purpose, SERF®, from the University of Delaware.

Summary

In this section, we have discussed the potential of and some specifics of the paperless classroom. Reread any item you are unsure about.

- ☐ The paperless classroom
- ☐ Computerized gradebooks
- ☐ Using computers to make students' work available to parents

3

MANAGING INSTRUCTIONAL PRINT MATERIAL

As teachers, we can find ourselves drowning in a veritable pool of paper! We have piles of paper resources that help us with our planning. We have baskets of kids' papers to examine, grade, and edit. Learning how to organize the papers that make up our teacher-lives is a crucial task. After all, the terrific diagram you have on the circulation system does you absolutely no good if you don't know where you stashed it. In addition, if you don't have good ways to handle kids' papers, you come across as disorganized and inefficient. This chapter will help you with those tasks.

STREAMLINING PRINT RESOURCES AND PLANNING

In the last chapter, we identified ways to organize administrative print material. In this section, we will show you how to organize resources and planning print materials:

- ☐ Three-ring notebooks
- ☐ Index dividers

- ☐ Plastic page protector sleeves
- ☐ Computer storage disk sleeves
- ☐ Color coding
- ☐ Workshop folders
- ☐ Unit library materials list
- ☐ Folders within folders
- ☐ Organizing the lesson plan book
- ☐ Hanging files for daily planning

Three-Ring Notebooks

Many organized teachers like using three-ring notebooks to organize teaching material. There are several advantages of three-ring notebooks over file folders. First, all of your material on a given topic can be kept together instead of in several folders. Second, the three-ring notebook is convenient to carry with you when you go home, to the teachers' lounge, or to a grade level meeting. Third, if you drop it, you don't scatter your papers all over the parking lot and lose them.

Keep your three-ring notebooks on your bookshelves. Write the title of each notebook on its spine in large letters so it's easily identifiable.

Index Dividers

Index dividers are those tagboard sheets with tabs that section off your three-ring notebook. Good index dividers cost quite a bit more than the cheap ones, but they're worth the extra investment. The cheap ones have clear plastic tabs. You write the title of the section on the small pieces of paper that come with the dividers and slip them into the clear plastic tabs. The more expensive index dividers are either numbered or lettered. They come with a blank table of contents page. Being able to write the title of the section on the cheap dividers is nice, but the more expensive numbered or lettered dividers last longer and look more professional.

Plastic Page Protector Sleeves

When you have materials that are fragile (such as pages from magazines), place them in plastic page protector sleeves and pop them into the notebook. A page protector sleeve is simply an envelope of clear plastic with holes punched in one side to allow it to be inserted into a three-ring notebook. We've tried the cheap page protectors and have found that we much prefer the more expensive ones. They're much more durable than their

cheaper cousins. "Never buy cheap canned tomatoes, cheap paint, or cheap page protectors!" said Virginia Klapperich, a retired art and special education teacher.

Computer Disk Storage Sleeves

Computer disk storage sleeves that fit into a three-ring notebook are a good investment. Keep the disks for each subject in your notebook for that particular subject; however, you may prefer to have a three-ring binder strictly for your disks.

Color Coding

One way to organize your three-ring notebooks, your file folders, or other paperwork systems is through the use of color coding. For example, Margaret uses different color notebooks for different subjects; red denotes math units, blue denotes language arts units, and so on. Within large notebooks, say for example, her notebook on teaching social studies, she color codes different sections by using colored dividers and by attaching colored sticker dots to various handouts. Yellow dots denote material to teach geography, blue denote citizenship, green denote economics, and so on.

Unit Library Materials List

When you create a thematic unit, have the librarian make you a list of all library materials that are related to your unit. Keep the list in your unit folder or unit three-ring notebook. As you review each of the materials, highlight the ones that you've decided to use. That way, you don't have to ask her every year what materials are available on your topic. You'll not only have a list of all the materials you used, you'll also have a list of all available materials on your topic in case your needs change.

Be sure to ask your librarian to let you know when new materials related to your topics are purchased by the library.

Folders Within Folders

Folders within folders are an excellent way to organize lesson material on your computer. For example, within a computer folder on literature, you might have one folder on books set in the antebellum South, another set in the Progressive Era, another set in the Great Depression, and so forth. There

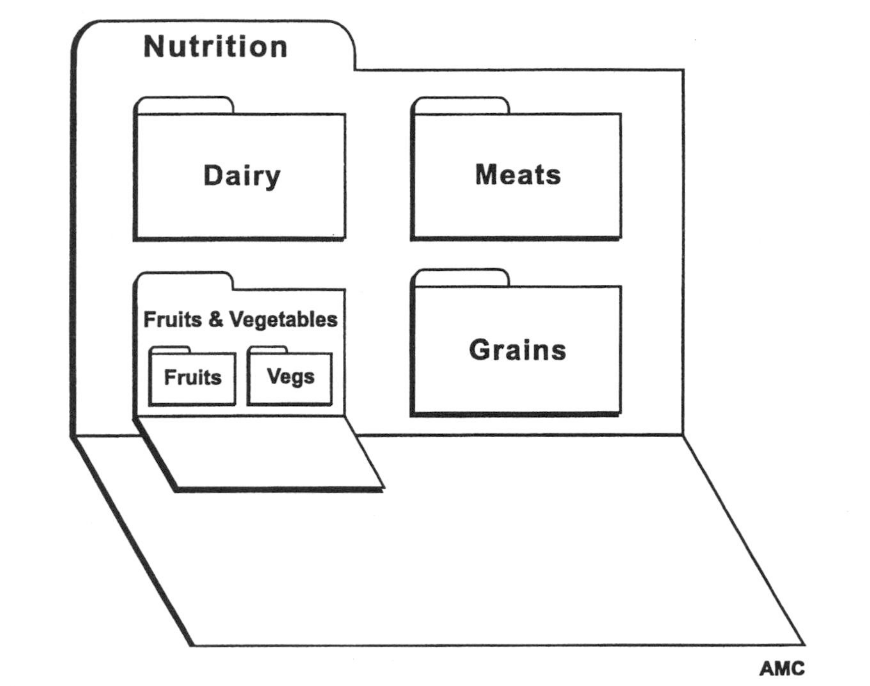

Figure 3.1. Folders Within Folders

is no limit to the number of folders that you can put inside of virtual folders (Figure 3.1).

Organizing the Lesson Plan Book

J.D., a curriculum director, gave us this idea. Before school starts, sit down with your district calendar and your lesson plan book. Record the dates of each week. Then record all the important dates: every holiday, the dates that report cards are due, the dates of special school programs, and so on.

If your district has identified specific time frames for major units (as J.D.'s did), schedule your units at this time. For example, if you're supposed to teach state history for 12 weeks and then move on to American history, record the dates for beginning and ending each unit. Also schedule any special speakers and record the dates that they'll be coming. You may also want to note to call them several days ahead of time in order to make certain that they haven't forgotten you!

When she taught first grade, Millie used to change thematic units on the first of each month. At the beginning of the school year, she gave her

librarian the monthly topics so she could reserve all related materials. J.D.'s idea would help you in planning for thematic units.

Hanging Files for Planning

Several teachers told us that they use a hanging file to help organize their planning. They advised us to tell you to have a folder in the hanging file for each day. Place all that day's papers and plans in the file instead of in stacks all over your desk.

Summary

In this section, we have discussed ways to streamline your print resources for teaching and for planning. Check the following list to determine which ideas you know how to use and those about which you are unsure. Go back and reread any that you are unsure about before you proceed further.

- ☐ Three-ring notebooks
- ☐ Index dividers
- ☐ Plastic page protector sleeves
- ☐ Computer storage disk sleeves
- ☐ Color coding
- ☐ Unit library materials list
- ☐ Folders within folders
- ☐ Organizing the lesson plan book
- ☐ Hanging files for daily planning

ORGANIZING STUDENT WORK PRINT

In the last section, we discussed ways of streamlining your resource and planning print. In this section, we'll take a look at ways that you can deal with student classwork and homework more efficiently. Specifically, we will cover:

- ☐ Individual student folders
- ☐ Locker magnets for homework
- ☐ Homework mailboxes
- ☐ Hanging files for returning homework
- ☐ Makeup work for absent students

- ☐ Color coding edited writing

Individual Student Folders

Michelle Cassetty, an elementary teacher, and Traci Cook, a seventh grade reading teacher, each use individual student folders. Michelle files her pre-assessments in the folder, as well as any notes she receives from parents.

On the front of her folders, Traci writes the student's name, class period, and seating assignment (as in Row 3, Seat 2). The folders are color coded by class period. They are stored in a milk crate by the door. The first person who comes to class each period picks up the class's folders and passes them out. It doesn't matter whether or not the student knows the names of all the students in the class; the seating assignment takes care of showing him what folder goes to what seat.

Traci explains that she doesn't allow students to have anything on their desktops except the folders. Inside the folder, the student has a journal record, reading record, extra paper, discipline card, and any uncompleted work. At the end of class, students pass their folders to the front and one student places them back in the milk crate.

Locker Magnets for Homework

Here's a neat idea. Michelle Cassetty uses her hot glue gun to glue a small magnet on the back of a clothespin. Each student places a magnet clothespin on her locker. Then when she comes in each morning, she clips her homework to the clothespin. Michelle can tell at a glance who has completed their homework.

Homework Mailboxes

Homework mailboxes can either be cereal boxes or stacking desk trays. If you opt for cereal boxes, staple them together five wide by as tall as necessary. If you use stacking trays (and they are a good investment), line them up five abreast and as tall as necessary (Figure 3.2). Each child has her own homework mailbox. As each student enters the classroom in the morning, she will place her homework in her mailbox. By looking at the mailboxes, you can tell in an instant which kids haven't turned in their homework.

Use student numbers to assign mailboxes. Then when a student collects the homework for you, the papers are in order for your gradebook.

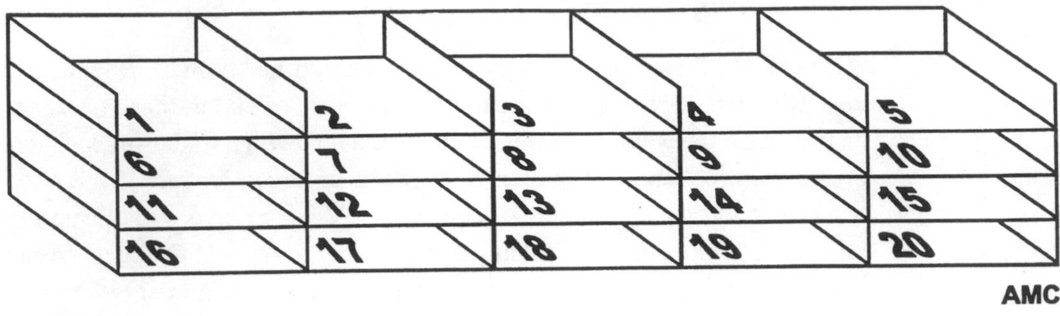

Figure 3.2. Homework Mailboxes

When your student collects the homework from homework mailboxes, have her ask students whose homework is missing where their assignments are.

Kimberly Smith uses her student mailboxes (she uses stacking file trays) to send home notes as well as to pass papers back to students.

We think that stacking trays are a good investment not only because they are sturdy, but because they look very professional.

Hanging Files for Returning Homework

A hanging file for each student, with the student's name on it, is an efficient way to take care of returning homework. Instead of taking class time to return dozens of homework papers, slip each student's homework into his hanging file. You can either let students come up before and after class, or you can return several assignments together, rather than one assignment at a time. If you held the papers until there were 10 in a student's stack, you would hand back papers only one tenth of the time that you do when you hand back each paper! Of course the problem with that is the delay in time for corrective feedback. You would want to pull any papers that showed the student needed reteaching and act on that immediately. All students could be told that if they don't receive their homework back immediately, they can rest assured that they are doing satisfactory work.

Makeup Work for Absent Students

Dealing with makeup work is such a hassle! Here are several ideas to help you manage makeup assignments.

Homework Calendar

Lara Matthews, a secondary teacher, shared her homework calendar idea with us. Lara has a laminated homework calendar that she keeps displayed on the wall. Below the calendar is a box with a folder for each day of the month. Lara writes each day's work on the calendar. If there are worksheets or papers, they are placed in the folder for that day of the month. Lara said, "For example, if Johnny is absent on the third of October, he can look on the calendar for the third and write down his assignments. Then he can look in the folder with the 3 on it and find the worksheets he needs."

When you are passing out worksheets, have a student assigned to the job of placing the extras in the folder for that day.

Individual Absentee Folders

First grade teacher Lisa Palmer uses a red laminated file folder for each absent student. She places the folder on the student's desk and has a seatmate place each worksheet in the folder. Lisa has written, "Absent today! We missed you!" on the folder, so if a sibling or parent comes to pick up the work, the child knows that his teacher is thinking of him.

Lisa explains that if the work missed is to be made up at school instead of at home, the student keeps his red folder on his desk until all of the work has been completed. That way, Linda knows at a glance who should be working on makeup work during free time.

Color Coding Edited Writing

When you are editing a student's writing, using colored felt tip pens allows you to color code your remarks. For example, purple notes things you thought were great. Grammar and spelling errors can be noted in red. Green edits note a problem in organization. Blue notes mean you are sharing a thought that came into your head as you were reading. Brown comments indicate that you question whether a fact is correct.

Summary

In this section, we've discussed ways to organize paper teaching materials. Check off the methods that you know how to use and reread those about which you are uncertain.

- ☐ Individual student folders
- ☐ Locker magnets for homework

- ☐ Homework mailboxes
- ☐ Hanging files for returning homework
- ☐ Makeup work for absent students
- ☐ Color coding edited writing

TRANSPORTING MATERIALS

In the last section, we discussed ways of streamlining student classwork and homework. In this section, we will give you a couple of ways to deal with printed materials when you are mobile. Specifically, we will present three ideas:

- ☐ Workshop folders
- ☐ Small rolling suitcases with handles
- ☐ Audiovisual carts

Workshop Folders

We all find ourselves with piles of papers when we come home from workshops or conventions. Here are two good ways to deal with those papers when you first get them, and later when you have time to go through them.

Go to the workshop with a folder. Before you go, label the folder on the outside. Then draw lines down the inside of the folder so you have three columns on the inside of the front cover and three columns on the inside of the back cover. Take your notes at the workshop in these columns rather than on edges of handouts, old napkins in your purse, or any of the other inappropriate places we make notes at workshops and conventions.

Then place any handouts you get inside the folder.

As soon as you get home, go through the folder and discard any handouts that aren't useful to you. Either keep those that you find useful in the folder or, better yet, pull them out and place them in the appropriate three-ring binder. If the workshop is on something for which you don't have a three-ring binder (say, special education modifications), then start a new three-ring binder for those and related handouts.

Small Rolling Suitcases With Handles

For the materials that you have to take home, small rolling suitcases with handles are wonderful (see Figure 3.3). Ann Estrada introduced us to them.

Figure 3.3. Small Rolling Suitcase With Handles

Once you try one of these great inventions, you'll abandon your briefcase in favor of it. It'll save your back and allow you to take home much more material than you'd otherwise be able to carry. You can attach your laptop to it by using the strap at the top of the suitcase, or carry it inside of the suitcase for more protection. We got ours for 30 dollars.

Audiovisual Carts

If you are one of those teachers who must move from room to room, such as an Inclusion Teacher, an audiovisual (AV) cart may provide the answer you need. Because AV carts are usually two-tiered, you can place a hanging file on the top tier and your other materials on the bottom tier. Then, instead of trying to carry your materials from room to room, you can wheel them down the hall.

If you don't know where a spare AV cart in your building can be found, check with your custodian. She will almost certainly know where one is stored.

Summary

In this section, we have discussed how do deal with paperwork when you must be mobile. Check the following list to be certain that you understand each idea:

- ☐ Workshop folders
- ☐ Small rolling suitcases with handles
- ☐ Audiovisual carts

ORGANIZING FOR SUBSTITUTE TEACHERS

In the last section, we discussed organizational ways of increasing your mobility from room to room, room to home, and room to workshop. When you are going to be gone to a workshop with that workshop folder and/or roll-aboard, you'll need to plan for your substitute teacher. Here are a couple of organizational ideas to help you:

- ☐ Substitute teacher folders
- ☐ Photo seating chart

Substitute Teacher Folders

Kimberly Smith tells us that she keeps a substitute teacher folder ready at all times. Kim's sub folder includes student IDs, student name tags, classroom procedures, where to go for help, class schedule, names of students who may need special help, and names of students who can be counted on for help.

Kim then inserts her daily lesson plans in the sub folder, and she's ready to take off for that professional workshop feeling confident that things will run smoothly while she's gone.

Photo Seating Chart

Sometimes kids find it fun to switch seats with each other when a sub is in charge. One way to help a substitute teacher keep on top of that problem

is a photo seating chart. If you scanned in students' pictures when you created your student data Smart Book, you can take those pictures and copy them onto a page reflecting your seating chart. That way, the kids can't tell the teacher that Henry is Barry and Tammy is Zoe.

Summary

In this section, we have discussed two ways to help you organize for your substitute teacher. Check off those ideas that you know how to use:

- ☐ Substitute teacher folders
- ☐ Photo seating chart

4

CORRALLING NONPRINT MATERIALS

A teacher (who wishes to remain anonymous!) told us recently that she had a colleague who had so many nonprint teaching materials in her room that her principal got fed up with her. He told her on several occasions that her room looked unprofessional, but she was unable to get her nonprint materials organized. Finally, he marched into her room after school with boxes, and he began to crate up her materials himself!

Teachers need oodles and oodles of nonprint materials. The problem is finding ways to organize them so that the classroom looks professional. The ideas in this chapter will help you do that.

CONTAINERS

In the previous chapter, we discussed how to organize print teaching materials. In this chapter, we will discuss how to organize nonprint materials. In this section, we will address only two important ideas:

- ☐ First-class containers
- ☐ Marking containers

First-Class Containers

The first thing we will say about containers is that you need to invest in first-class containers. You're going to be teaching for a lot of years. Don't skimp on your materials. It costs only a little more money to buy good containers than to purchase flimsy ones. Millie was reminded of this several days ago when a colleague picked up a flimsy milk-crate type container out of her car trunk, the container's handle broke, and the contents scattered in the breeze all over the parking lot.

Not only are first-class containers sturdier than cheap ones, but they look professional. A cereal box is not only a flimsy container, but it looks unprofessional. Teachers can't afford to look unprofessional to students, parents, or administrators.

Note that a container can be first class without being used for its original purpose. What we mean by this is that you can recycle sturdy containers, such as two-pound processed cheese boxes and metal adhesive bandage boxes, which we'll discuss later. But if you do use such containers, cover them so that they don't look like two-pound processed cheese boxes and metal adhesive bandage boxes.

Marking Your Containers

Mark all containers on all sides so that you can identify them regardless of how they are stored. For large, hard-to-get-to containers, you'll want to write large with a felt marker. For smaller and easier-to-get-to containers, type or write on adhesive backed labels.

Summary

In this section, we have discussed containers. Specifically, we have discussed the following:

- ☐ First-class containers
- ☐ Marking containers

CONTAINERS FOR SMALL ITEMS

In the last section, we discussed some general ideas about containers. In this section, we will specifically discuss containers for small things.

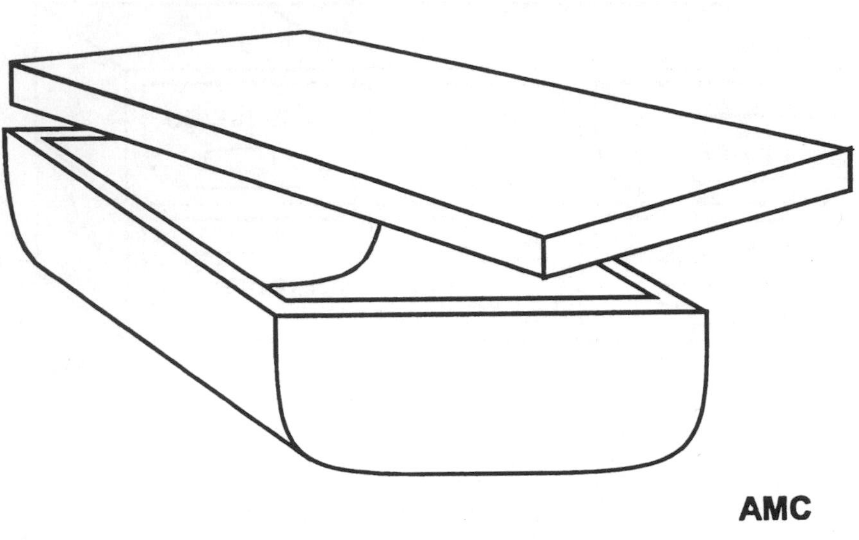

Figure 4.1. Plastic Shoe Boxes

- ☐ Plastic storage boxes and drawers
- ☐ Clear plastic organizer boxes
- ☐ Two-pound processed cheese boxes
- ☐ Metal adhesive bandage boxes

Plastic Storage Boxes and Drawers

Buy a variety of high-quality plastic storage boxes and storage drawers. Some of the brands we like are Tupperware,® Rubbermaid,® Callaway,® and Trend Basics.® Rubbermaid and Trend Basics are available at your local discount store. Callaway products are available though Callaway House, Inc. (1-800-233-0290 or www.callawayhouse.com). The individual clear plastic boxes resemble shoe boxes; the plastic storage drawers are transparent and come four to six encased in a plastic shell. Some of them have opaque shells; others have transparent shells. Rubbermaid even makes a column of storage boxes with transparent sides and mounted on wheels. What a great tool, especially for a teacher who moves from classroom to classroom, such as an Inclusion Teacher or an art teacher (Figures 4.1, 4.2, and 4.3).

The shoe boxes are heavier and sturdier than the drawers, which can be an advantage. Another advantage of the shoe boxes is that they can be carried around with the lids on; if you want to go down to the lounge to work on

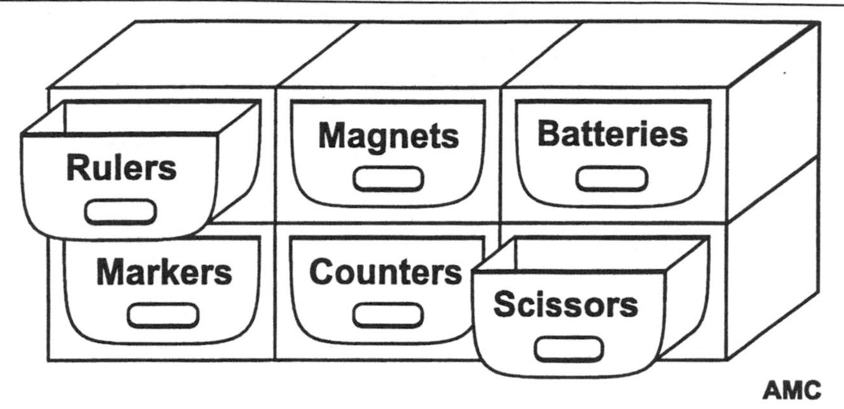

Figure 4.2. Plastic Storage Boxes

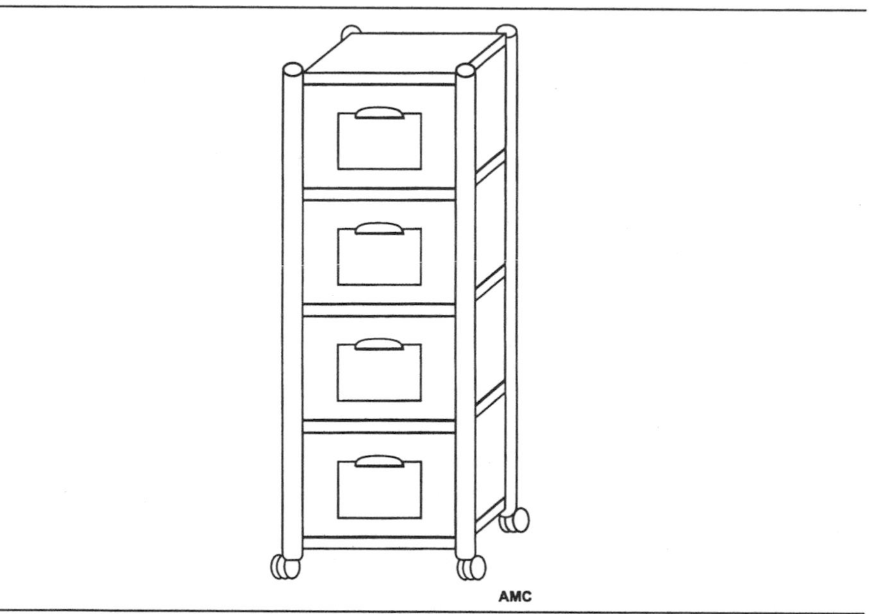

Figure 4.3. Rubbermaid's® Column Storage Drawers With Rollers

something, you can carry a shoe box with you. Joan, a fifth grade teacher, keeps her blank 3 × 5 index cards in a shoe box; when she wants to take them home to create flash cards or game cards, she doesn't worry about them flying away in the wind. She also uses the shoe boxes for materials that she needs only a few days a year. These she stores in the back of a shelf or cabinet until she needs them.

The plastic storage drawers are ideal for items that are used on a regular basis. Charlie, who teaches first grade, uses them to store supplies that he

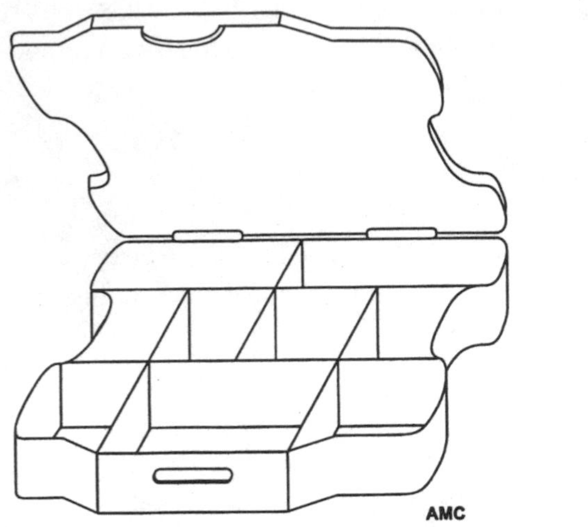

Figure 4.4. Rubbermaid's® Organizer

uses frequently in modeling science lessons. For example, he uses metal washers in a number of experiments, so he stores them in the plastic storage drawers where he has convenient access to them. Joan and Charlie both store their classroom sets of scissors, glue sticks, rulers, plastic counters, and so on, in the storage drawers.

Clear Plastic Organizer Boxes

For your little treasures, such as paper clips, staples, gold stars, and so on, a clear plastic organizer box is a treasure box. Organizer boxes have lots and lots of small compartments of varying sizes. They are an organizer's dream come true! A second grade teacher told us that she requires each of her students to have one of them in their desks. She is committed to helping her students learn to organize themselves.

Two-Pound Processed Cheese Boxes

Two-pound processed cheese boxes are the perfect size and shape to fit on your chalk tray. Our friend Sylvia Broussard covers hers with contact paper and glues a magnetic strip on the side for extra stability. She stores her magnetic backed counters and flash cards that she uses on the dry erase board in the boxes.

Sylvia also makes her regular flash cards the right size to store in the box in her desk. Make divider cards a different color and a little taller than the flash cards to keep them organized.

Metal Adhesive Bandage Boxes

Sylvia also uses metal adhesive bandage boxes on her chalk tray. She removes the lid and stores her dry erase markers inside.

Summary

In this section, we have discussed containers for small items. Specifically, we have presented information on the following:

- ☐ Plastic storage boxes and drawers
- ☐ Clear plastic organizer boxes
- ☐ Two-pound processed cheese boxes
- ☐ Metal adhesive bandage boxes

STORING SMALL BUT RELATIVELY FLAT ITEMS

In the last section, we discussed how to store small things that are of various shapes. In this section, we will discuss how to store small, relatively flat items, such as books. Specifically, we will discuss:

- ☐ Math unit material envelopes
- ☐ Plastic gallon zipper freezer bags
- ☐ Plastic sandwich zipper bags
- ☐ Magnetic clips
- ☐ Copier paper box tops

Math Unit Material Envelopes

Several teachers gave us this idea: When you have small math unit materials such as flash cards or card-like game pieces, keep them together in large manila envelopes. You can even store small books related to the unit in the envelopes.

Label the unit at the top and bottom. Then tape a list of the contents on the front of the envelope before you file it.

Plastic Gallon Zipper Freezer Bags

Instead of the manila envelopes, you may wish to store items in gallon zipper freezer bags. Punch holes in the bottoms of the bags so they will fit in a 4-inch deep three-ring notebook. Keep them in the notebook where you can find them easily. Because the bags are transparent, you can quickly find the cards you are seeking.

Our friend Sylvia requires her first graders to keep a magazine and a book in their desks at all times. These are kept in gallon zipper freezer bags to keep them in good condition.

Sylvia also recommends having children carry gallon zipper bags to collect specimens on nature walks.

Plastic Sandwich Zipper Bags

Plastic sandwich zipper bags are perfect for storing individual students' flash cards or crayons. They don't get spilled inside the child's desk, on the floor, or elsewhere. Primary teachers tell us that sandwich bags are a treasure. The bags, do, however, wear out, so you will have to replace them occasionally.

We also use them to store sets of things. For example, we have game boards for review games. We have enough of them for each group of four to have one. We keep the dice, markers, and game cards that go with each game together in a sandwich bag.

In addition, we keep our sets of flash cards in sandwich bags, and color code the bags with colored dots. The cards for geography are color coded red, those for generic higher-order thinking skills blue, and so on.

Magnetic Clips

Magnetic clips are used to hold student work for exhibit on magnetic boards, as well as for storing schedules and important announcements on the side of your file cabinet.

Copier Paper Box Tops

Photocopier paper box tops provide a good way to organize materials for interest centers. For example, our friend Sylvia keeps a list of spelling words and a plastic bag of rubber letters in one box top so that a student can take this mobile interest center to her desk to work on it.

Summary

In this section, we have discussed ways of storing small, relatively flat items. Specifically, we have presented information on the following:

- ☐ Math unit material envelopes
- ☐ Plastic gallon zipper freezer bags
- ☐ Plastic sandwich zipper bags
- ☐ Magnetic clips
- ☐ Copier paper box tops

STORING BULLETIN BOARD MATERIALS

In the last section, we presented ways to store small, relatively flat things. In this section, we will talk about ways to organize your bulletin board materials. We will present information on streamlining your storage by using:

- ☐ Monthly bulletin board boxes
- ☐ Poster rolls
- ☐ Large trash cans
- ☐ Moving mirror and painting boxes

Monthly Bulletin Board Boxes

We asked teachers to give us their best organizational ideas, and one of our favorites was for monthly bulletin board boxes. Monthly bulletin board boxes should be the same size and shape for neat and attractive storage. You can use cardboard boxes and have each one decorated to represent its month, or you can buy plastic storage tubs with lids and simply write the month on the front of the tub. The teacher who told us about them uses cardboard boxes that she has decorated.

At the beginning of each month, she brings down the box during class. The children are always excited and eager to see what is in the box, and she allows them to decorate the bulletin boards using the materials she has provided. We especially like this idea, because getting to decorate an area helps give students ownership of the area, and ownership is a key to motivation.

Figure 4.5. Monthly Bulletin Board Boxes

The teacher who gave us the idea lined her boxes up along the top of the lockers in her classroom (Figure 4.5). What a great way to use space that's usually wasted!

Poster Rolls

Maps and posters made of paper can be a pain to store. We have tried laying them flat in the bottom of a large open shelf; but when we put other things in the shelf, we wrinkled or tore the posters. We have tried rolling them up with a rubber band around them and storing them upright in a closet, but they got ruined. We even tried rolling them up and placing them inside of a mailing tube, but we inadvertently tore them when we tried to get them out. We found that the perfect solution is to roll them around a fat mailing tube and secure them with a rubber band. In this way they can be placed almost anywhere. Be sure to color code a corner on the outside of the rolled up poster so you will know to which subject it relates. Below the colored dot, write the name or a brief description of the poster in very small letters so that you will know exactly which poster you want when you look through them.

In your monthly bulletin board boxes, bulletin board border trim can be stored by wrapping it around a mailing tube, a paper towel roll, or a toilet paper roll.

Large Trash Cans

One of the best ideas we know of for storing posters came to us by way of an architect. Architects store blueprints in large trash cans with lids. After you roll your posters around poster rollers, place them in large trash cans and secure the lids.

Word of warning! Label the trash can and show it to your custodian! Explain to him that it is not a trash receptacle, but a poster storage container. Otherwise, he may dutifully throw away five hundred dollars worth of your favorite posters!

Moving Mirror or Painting Boxes

Large, flat materials such as cardboard posters or game boards that you have made out of poster board can be slid into the mirror or painting boxes that can be bought from moving companies. If you decide to use the mirror boxes, you will want to color code them, also. The only problem about the mirror or painting boxes is that they may cost a little more than you want to pay for a cardboard box. If you remember, however, that you will be able to use them for years, the investment is negligible. Of course, you can always watch the classified ads for someone who is selling their moving boxes or ask someone at your local art store to save some for you.

Summary

In this section, we discussed how to organize bulletin board material. Specifically, we discussed the following strategies:

- ☐ Monthly bulletin board boxes
- ☐ Poster rolls
- ☐ Large trash cans
- ☐ Moving mirror and painting boxes

ORGANIZING TOOLS

In the last section, we discussed how to store bulletin board materials to which you need access only once each year. In contrast, in this section, we will discuss how to store tools that you need frequently.

- ☐ Pegboards

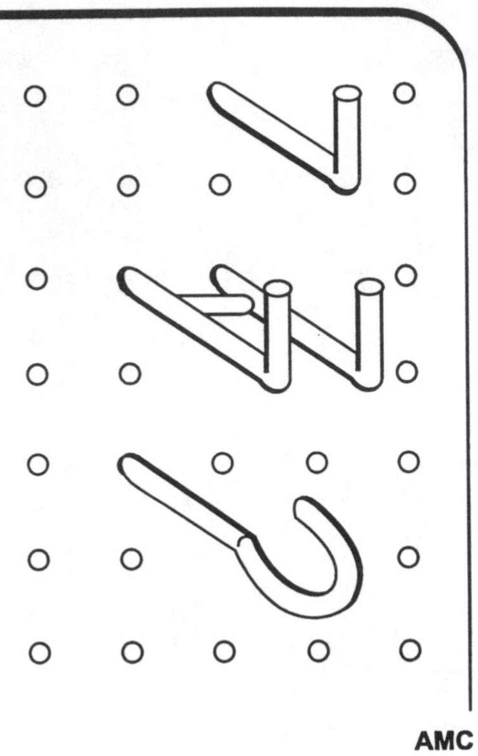

Figure 4.6. Pegboard Hooks

Pegboards

While we keep small tools that are great in number (such as class sets of children's scissors) in plastic storage boxes or drawers, we use pegboards to organize large tools that are few in number (such as the teacher's large scissors). We fell in love with pegboards years ago. They end digging through a deep box or drawer to find things. The variety of peg hooks available at hardware stores will accommodate almost any type of tool that you need to hang: a yardstick, a T-square, a desk stapler, a tape dispenser, and so on. For example, a simple hook will store a yardstick with a hole at one end. If the yardstick doesn't have a hole in one end from which to hang it on a hook, then two hooks placed horizontally 2 feet apart make a bed in which the yardstick can lie. To store a desk stapler, place two long, flat hooks 6 inches apart. The base of the desk stapler makes its own shelf when placed on the hooks. Figure 4.6 shows some types of hooks.

In order to be able to tell at a glance whether everything has been returned to the pegboard, carefully draw around each tool with a black felt

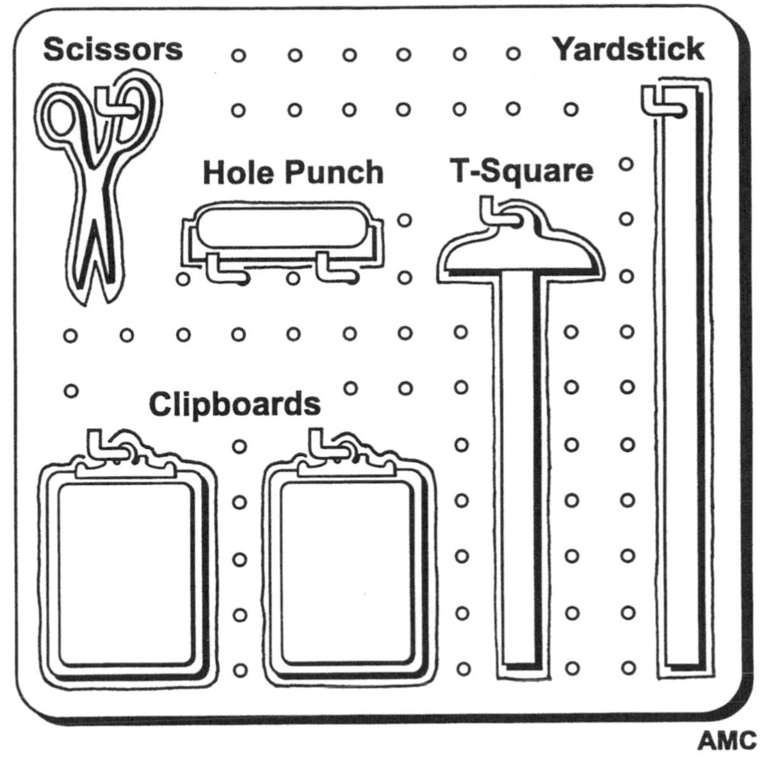

Figure 4.7. Pegboard With Tools Outlined

tip marker (see Figure 4.7). The shape tells you what belongs in that spot. You can see at a glance before the students leave at the end of class if something is missing. Students love to help with finding the missing items.

You might even want to mark off a section of the board with "Teacher Supervision Only!" written on it, in order to prevent children from accidentally taking tools that they shouldn't be using independently, such as the teacher's scissors.

Summary

In this section, we have discussed how to store large tools that are few in number. Specifically, we have discussed:

- ☐ Pegboards

ORGANIZING PHYSICAL EDUCATION EQUIPMENT

In the last section, we discussed pegboards as a way to organize large tools that are few in number. In this section, we will discuss how to organize physical education (PE) equipment. Specifically, we will discuss the following:

- ☐ PE buckets, washtubs, and laundry baskets
- ☐ Color coding equipment
- ☐ Marking equipment
- ☐ Disseminating equipment

PE Buckets, Washtubs, and Laundry Baskets

A PE teacher gave us this idea: PE buckets. Recess or physical education supplies can be stored in plastic buckets that children can carry out to the playground. Smaller pieces of equipment can be stored in kitchen-sized buckets, but larger pieces (such as soccer balls) can be carried either in plastic washtubs with rope handles, or in laundry baskets.

Tape a laminated list of supplies stored in the bucket/washtub/basket to the side so that the child in charge of the bucket can be sure that she has collected all the supplies when recess is over.

Color Coding Equipment

If possible, color code the equipment that your class uses on the playground. The teachers who share the playground with you can agree to a color coding system, such as green for fourth grade, blue for fifth grade, and red for sixth grade. Then you can tell at a glance if the fifth graders have taken away the fourth graders' ball. Of course, not all equipment is available in an array of colors, but take advantage of the equipment that does come in different colors.

Marking Equipment

Each piece of your class's equipment should be marked with your name or initials if you purchased it out of your own pocket. If you purchased the equipment out of your school's funds, then it should be marked with your

room number. Then Mr. Garcia's fourth graders can't say that your green ball belongs to them because your kids (and you) can show them the marking.

Disseminating Equipment

We mentioned earlier that the children can carry the equipment to the playground. In order to make certain that all children get the opportunity to play this important role, tape a laminated list of weeks and student numbers to the side of each bucket. If you have four buckets and 24 students, start one list with the number 1, the second list with the number 7, the third list with the number 13, and so on, having 30 to 36 numbers on each list. You'll want the extra numbers to take care of the transfer students you will receive during the year. The reason we use student numbers instead of names is that with names, we'd have to make a new list each year. A laminated number list can be reused year after year.

In order to make certain who is in charge each week, at the end of the last recess of each week, have that week's equipment people pass the buckets on to the next week's equipment people. That will prevent the "He did it last week!" confusion from arising.

Summary

In this section, we have discussed methods for organizing PE equipment. Specifically, we have discussed the following:

- ☐ PE buckets, washtubs, and laundry baskets
- ☐ Color coding equipment
- ☐ Marking equipment
- ☐ Disseminating equipment

MOVING NONPRINT MATERIALS

In the last section, we discussed how to organize PE equipment. In this section, we once again address the issue of organizing for mobility of materials.

- ☐ Small rolling suitcases with handles

Small Rolling Suitcases With Handles

Finally, once again we want to champion small rolling suitcases with handles as a way to transport nonprint materials. If you are a speech therapist, diagnostician, or other school professional who goes from school to school, you know well the frustrations of trying to move books and equipment back and forth. Small rolling suitcases with handles are for you! Not only are they amazingly convenient, but they look much more professional than cardboard boxes overflowing with materials. And they'll save wear and tear on your back!

Summary

In this section, we have discussed ways of organizing for the mobility of nonprint materials, specifically:

- ☐ Small rolling suitcases with handles

5

"DEJUNKING": TAKING THE CURE

Classrooms seem like they have one-way valves; objects flow in, but they don't flow out. Organized teachers construct psychological valves to let objects flow out of their classrooms. One way they construct these valves is by dejunking. Dejunking is hard to do, but makes you feel wonderful once you get started.

BASIC DEJUNKING

In the last chapter, we discussed ways of organizing nonprint materials. In this section, we will discuss organizing by basic dejunking, because many of those materials you are trying to organize are junk. We will cover the following:

- ☐ Identifying junk
- ☐ Getting rid of nonsentimental junk

Identifying Junk

Millie discovered a wonderful book about 15 years ago: *Clutter's Last Stand* by Don Aslett. She was listening to a radio program in the car when the emcee began interviewing Aslett. Millie was feeling overwhelmed by the

clutter in both her home and her classroom; she went straight to the bookstore and bought a copy. Since that time, she has bought a copy of every one of the books Aslett has published. She has given copies of *Clutter's Last Stand* for birthdays, anniversaries, and wedding presents. She also used *Clutter's Last Stand* as a textbook for her high school special education students. Buy Aslett's book. It might be the best thing you do for yourself this year.

The important ideas presented in this section are gleaned from Aslett's books.

Aslett defines junk as: (a) things that we don't use, or (b) things that don't give us warm fuzzies every time we see them.

Things-That-We-Don't-Use Junk

Connie Deutchland's video of the life of Stalin was junk because she didn't use it. In contrast, her video of desert life wasn't junk, because she used it every year. She dejunked and gave the Stalin video to the library.

Second-Best-Anything Junk

Likewise, a second-best anything is a piece of junk. Aslett uses the example of china. He explains that if you have two sets of china, you should start using the better one and give away the second best. Why have something lovely if you aren't going to use it? If it is unused, it is junk, so use the good china and give the other one away.

Vern, a junior high social studies teacher, used to have two pairs of adult scissors in his desk. He always used the second-best one. He was saving the better pair. He didn't know what he was saving it for, but saving it he was, and he was using the poorer pair. Finally he gave away the poorer pair and began using the better pair. What a pleasure to use a nice sharp pair of scissors! He wondered why he had waited so long.

Memento Junk

As for the second kind of junk, if something nonutilitarian gives you a warm glow inside whenever you see it, it's not junk. If you don't get the warm glow, it is.

Connie's poster of the beach at Galveston was junk because she didn't feel a little glow inside whenever she saw it. Conversely, her poster of a Norwegian troll is not junk, because she does glow inside every time she sees it. Aslett explains that even your fine crystal vase is junk if you don't get a warm glow whenever you see it. In contrast, your refrigerator magnet

of a hummingbird is not junk if you do get the warm fuzzies every time you open the fridge.

Identifying Teacher Junk

For a teacher, classroom junk is (a) anything that we can conduct our class without or (b) anything that we can easily access in the building when we do need it. For example, at one point when Millie was teaching first grade, she found an enormous wooden box outside by the Dumpster. The box was 6 feet long, 4 feet wide, and 3 feet deep. It had a hinged wooden lid. Originally, the box had been made to store playground balls and checkout equipment for recess. Millie dragged the box into her classroom and thought she had a prize.

The Box sat empty for a couple of weeks. Then Millie had a brilliant idea. Or so she thought. She would send a note home with her first graders asking their parents to send recyclable items such as cardboard paper towel cores, tin pie plates, empty milk cartons, and so on. Then she and the children would always have plenty of objects for various types of art activities and for indoor recesses during inclement weather. The Box was full within 2 weeks.

The only problem was that after a couple of weeks, the kids lost interest in the objects in the box. Millie couldn't find any craft ideas to use the materials, but she had collected them, by golly, and she didn't want to let them go. So they took up an enormous amount of space for the rest of the year.

Millie kept The Box and its contents over the summer and into the next school year. During the spring of the second year of having The Box full of unused recyclable trash taking up space, she threw it out. She felt liberated, light, and free. Looking back, she can't understand why she burdened herself with The Box and its contents for so long. In classifying her junk, The Box itself was something that she could teach without, and the recyclables were something that if she did need for a project, the children could easily supply within a few days.

It's initially hard to accept the idea that something is junk in our classrooms if we don't use it, but junk is a relative concept. For example, Millie kept an overhead projector in her classroom for years. She never did use it; she's highly auditorily distractible and can't think when an overhead is on because of the loud fan. Somehow it didn't seem American for a classroom not to have an overhead. But after she read *Clutter's Last Stand,* Millie realized that the overhead was junk to her, so she put it in the equipment storage room and freed up some counter space. Not only did she

not use the overhead, but one would be easy to get if she did need it. The whole thing was relative; to a teacher who did use an overhead, it wasn't junk, but to Millie it was.

That brings us to a touchy subject: trade books. While we do need textbooks for every child, and while it is nice to have a few trade books in the classroom, we don't need many. The school library can take care of our trade book needs. Connie used to spend a ton of money buying trade books, and she used a ton of space storing them. She bought them on every subject for which she had developed a unit: pioneers, wild animals, the ocean, and so on. She used a great deal of space to store the books, and she used each book only one period of one day a year. How inefficient! She finally donated all of her trade books to the school library. Then when she was ready to teach her pioneer unit, she simply gave the school librarian her topic, and the librarian gathered not only the pioneer books that Connie had donated, but all the other pioneer books in the library. Connie had dejunked and freed up space, the library had more books, more children had access to the books, and Connie still had the books when she needed them.

Ditto Connie's copies of teaching magazines, professional books, and the like. She gave those to the school library, too. Other teachers could share her materials, yet she still had access to them when needed, and she dejunked her classroom.

Teachers are particularly susceptible to certain categories of junk. Coffee mugs are a big item. One year, John received nine My Favorite Teacher coffee mugs. John doesn't even drink coffee. Give extra coffee mugs to your church garage sale.

Connie started keeping extra clothes at school. She never, ever used them, but she thought that it was such a good idea to keep a change of clothes at school that she couldn't bear the thought of taking them home. Take them home or give them away.

Ends of border trim rolls are tough things with which to part. You can't do much with a 9-inch piece of border trim, but it seems like such a sin to throw it away. Toss it.

We bet you have at least a dozen pens that don't work or felt tip markers that are nearly dead. Junk 'em.

Do you have a cord to some unknown piece of equipment? Or some odd computer parts from antiquated computers? Out with them.

Do you have old computer hardware or software manuals? Or disks that have errors? Can 'em.

Anything that doesn't work must be thrown away. If you don't have the authority to throw it away, take it down to your principal and let him store it.

Containers are junk if you are not using them. Even first-quality containers are junk if they are not being used. Pass 'em on.

Do you have a stack of magazine articles, convention handouts, or workshop handouts that you are going to read someday? Read them now and then either throw them out or transfer them to the appropriate notebook. If you don't have time to read them right now, then throw them out without reading them. You've done a good job teaching so far without reading them, you'll probably do just fine if you never read them.

Getting Rid of Nonsentimental Junk

Now that you've identified your junk, it's time to get rid of it.

Aslett says to start with things that belong to other people. If you have something borrowed from someone else, a book, video, or whatever, return it to that person today. If that person is another teacher in your building, send a child to him with the whatzit right now!

If you're dealing with your own junk that falls into one of the nonsentimental categories, throw it into the trash immediately. If it's junk that another teacher might need, put it in a "Take What You Want" box in the lounge or workroom. If it's something that the library/media center could use, take it to the librarian/media center specialist. Otherwise put it in the trash.

If it's appropriate to do so, you might even give some of your junk to students. But get rid of it.

You might need to do lots of self-talk in order to start getting rid of junk. Aslett explains that it helps some people to say, "My life is going to be much simpler and more organized. I'm going to have more time for important things when I am not trying to cope with all this STUFF!" When you start seeing the trash can fill with junk, when you start seeing your desk look neater, when you start seeing your room look more professional, you'll have added motivation to continue tossing your clutter.

Summary

In this section we discussed ways of recognizing junk when you see it and of trashing nonsentimental junk.

- ☐ Identifying junk
- ☐ Getting rid of nonsentimental junk

MINIMIZING MEMENTO JUNK

In the last section, we discussed identifying junk and getting rid of nonsentimental junk. Sentimental junk is more difficult, so in this section we address the special problems of mementos.

- ☐ Minimizing mementos with scrapbooks
- ☐ Trashing mementos

Minimizing Mementos With Scrapbooks

Mementos can either be junk because they don't give you a warm glow when you see them, or not be junk because they do give you a glow.

When you're a teacher, it is hard to let go of mementos; however, if you don't cull them, you'll drown in pictures, notes, wildflowers, and other gifts that children give you. Here's a way to minimize the mementos.

Don Aslett points out that we keep mementos to remind us of a special time. He teaches us to keep scrapbooks. In our nonprofessional lives, play programs, pictures of our vacation to Maui, and a ribbon from our daughter's hair when she was baptized are all kept in order to take us back to that special moment. However, Aslett explains that we don't need the entire play program, all the pictures, or the whole ribbon. Cut out just the title of the play, or maybe the title and the date and slip it in the scrapbook. Throw away the rest of the program. Find the best 2 or 3 pictures out of the 40 you took in Maui, trim those into attractive shapes, and put them in the scrapbook. Throw away the rest. No one else wants to be burdened with them. As for the ribbon, 1 inch will go as far in reminding you of the baptism as 2 feet will. Snip off one end of the ribbon, put it in the scrapbook, and throw away the rest. Thank you, Don Aslett!

As for using scrapbooks in our professional life, you can easily see how the program for your third grade play can be minimized. However, you are probably wondering how to get rid of children's gifts to you of pictures and notes. Connie did it this way: She bought a scrapbook with self-stick pages and plastic overlays. She put each child's name on one page. When that child brought her a picture or a love note, Connie first taped it to her filing cabinet for a few days. Then she either placed the whole picture/note in the scrapbook until the child brought another present, or she cut out the most important piece of the picture/note and placed it in the scrapbook. Of course, if she placed the whole picture/note in the scrapbook, she trashed the old picture/note (out of sight of the child) when she replaced it with the

new picture/note. If she simply cut out and saved part of the picture/note, she continued to add to the page until she had a delightful collage that represented her year with that child. She could retrieve a pleasant memory without taking up much space.

Trashing Mementos

If the memento gives you the warm fuzzies, keep it, or keep part of it, in a scrapbook. If it doesn't give you warm fuzzies, put it in the trash or in a garage sale. You have kept the darn thing because you think you will feel guilty if you throw it away or sell it. Just do it!

Paige Smith, a 30-year veteran sixth grade teacher, told us to think about it this way. "I used to collect ceramic pumpkins. After several years of kids knowing I collected pumpkins, I had dozens of them. I felt guilty about getting rid of any of them, but they were driving me crazy. I finally said to myself, 'The kids who gave me pumpkins gave them to me because they love me. They would never want to cause me stress. The pumpkins are stressing me, so the kids who gave them would really want me to get rid of them.' That kind of self-talk helped me put them in a garage sale."

Take your memento junk, that is, the mementos that don't give you the warm fuzzies every time you see them, and throw them in the trash, sell them in a garage sale, or give them away. But whatever you do, do it now.

Summary

In this section we discussed ways of helping you get rid of sentimental junk.

- ☐ Minimizing mementos with scrapbooks
- ☐ Trashing mementos

WEANING YOURSELF AWAY FROM JUNK

In the last section, we discussed getting rid of memento junk. However, you may find it impossible to go cold turkey with your memento junk and even with your nonsentimental junk. This section will help you wean yourself away from junk. Specifically, we will address the following:

- ☐ Sorting to wean
- ☐ The weaning box

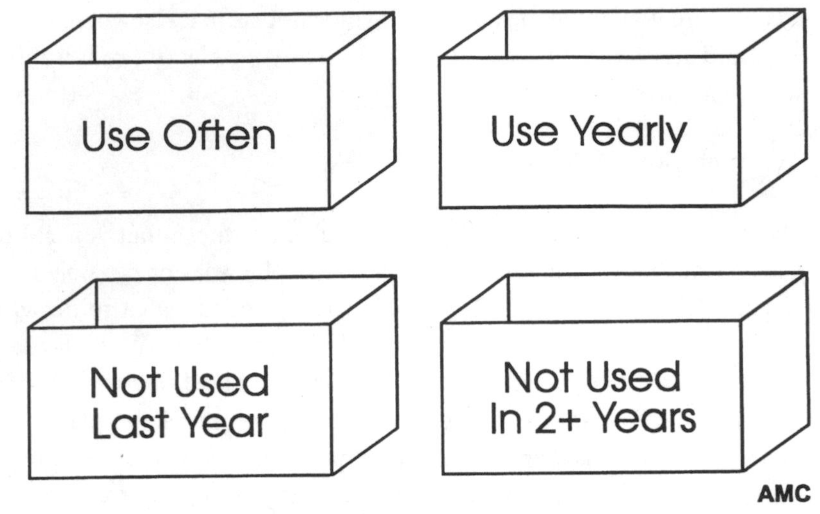

Figure 5.1. Sorting Boxes for Dejunking and Weaning

Sorting to Wean

At first you will find it difficult to get rid of classroom clutter. Don Aslett suggests a way to help you wean yourself. As you go through your classroom shelves/closets, make four stacks (Figure 5.1): (a) things you use frequently (such as scissors, stapler, ruler, etc.); (b) things you use once a year every year (such as the Christmas tree); (c) things you haven't used in the past school year (such as the plastic Eiffel Tower or the poster of Benji); and (d) things you haven't used in 2 or more years (such as the aquarium filter, plastic seaweed, and gravel).

Of course, you will keep the things that you use frequently and you will throw away the things that you haven't used in the past several years. But what of the things used yearly and the things you didn't use last year but might use again this year?

If you use something once a year every year, such as the Christmas tree, decide whether your classroom is actually the best place to store it. Is there a storage room in your building where teachers may store such things (as Millie did the overhead projector)? John says to ask your custodian where you can store these things. He explains that custodians know the locations of little nooks and crannies in the building that you can use, such as under the stairs or in the building's attic.

Could you take the Christmas tree home and store it in the attic? Or is there a hard-to-reach place in your classroom where it would be out of the way? As a rule of thumb, store used-yearly-stuff elsewhere in the building

if you can, store it at home in your attic if you can't store it elsewhere in the building, and store it in a hard-to-get-to place in your classroom if it must stay in the classroom.

The Weaning Box

The items that will give you the most trouble are those that you didn't use last year but that you might want to use later this year or possibly next year, such as the plastic Eiffel Tower. Aslett recommends what to do with those items. Box them up. Secure the box with strapping tape. Write the date on the box in large letters with an indelible marker. Label the box "Stuff I haven't used in the past year." Put the box away in one of the places we mentioned when we talked about the Christmas tree. If you decide you need something from that box at some time in the next year, go get the box, get the item you want, and reseal the box. Then 1 year, or at the most 2 years after you initially sealed the box, give or throw it away. Don't even look inside of it. If you didn't use the items in the year before you initially sealed it, and you haven't used them in the year after you sealed it, they're junk and need to be discarded. The ensuing year should have been long enough to wean you away from your junk.

Make another weaning box for items that are mementos that you simply can't quite part with. By the time they've sat in a sealed box for a year, you'll be weaned from them and can toss the box in the trash.

Summary

In this section, we have discussed ways of getting rid of those items that you can't quite let go. Specifically, we have discussed the following. Check off those about which you feel certain. If you don't feel certain of an item, be sure to reread it.

- ☐ Sorting to wean
- ☐ The weaning box

RECOMMENDED READING

Aslett, D. (1984). *Clutter's last stand: It's time to dejunk your life!* Cincinnati, OH: Writer's Digest Books.

Aslett, D. (1994). *The office clutter cure: How to get out from under it all!* Pocatello, ID: Marsh Creek.

6

DECISIONS: FROM SIMPLE TO COMPLEX

One of our colleagues has this sign on her wall: "If you don't have time to do the job right the first time, when are you going to find more time to do it over again?" One of the motivations for teachers to get organized is to escape once and for all from that terrible need to do things more than once. This is especially applicable to making decisions.

A teacher's life is full of decisions. We once heard that an elementary teacher makes over 800 decisions a day! Not being organized at making decisions is a terrible Time Stealer for anyone, but for a teacher, it's disastrous.

First, it steals time that you could be using implementing whatever decision you eventually make. If you know that you want to implement a peace education program, but you can't decide which program you want to use, then your inability to make a decision is keeping you from implementing any peace education program at all.

Second, not being organized at making decisions keeps you from doing other things you want to be doing (such as mentoring kids) because you're all wrapped up in the unmade decision. You can't truly listen to a child if your mind is occupied with an unmade decision.

Change gears now!

Figure 6.1.

Third, not being organized at making decisions results in your often making poor decisions. That steals time, because if you make a poor decision about what peace education program to use, then your quality teaching time is stolen because you're investing it in trying to teach a poor quality program.

Therefore, organizing your decision making is a crucial Time Stealer Tamer!

As you begin reading this chapter, you're going to need to change gears.

The previous chapters have dealt with very simple, concrete techniques: Make a list. Put supplies in plastic storage drawers. Throw away objects you don't use. In contrast, this chapter, and much of the rest of the book, deals with more abstract organizational strategies. Because the ideas are more abstract, our way of talking about them seems rather different than the way we talked about the more concrete strategies. You might prefer to implement the ideas in the previous chapters for several months before you begin reading and implementing the ideas in this and the following chapters. By implementing lots of concrete organizational strategies, you'll begin to think of yourself as an organized teacher. That change in self-concept may then make it easier for you to begin tackling more abstract organizational ideas.

In this chapter, we'll move from relatively simple abstract ideas of decision making to more complex ideas, but even the most complex decision-making strategies we teach can, and should be, mastered by kids as well as by teachers.

QUICK AND EASY DECISION-MAKING TECHNIQUES

☐ Evaluating by external criteria (elimination by aspects)

Evaluating by External Criteria

Evaluating by specified external criteria is a quick, easy, efficient way to make decisions. In 1972, Amos Tversky, a psychologist, began calling this

technique elimination by aspects. The term we use, evaluating by external criteria, is the one used by Benjamin Bloom in his taxonomy of educational objectives.

Here's what you do to evaluate by external criteria:

1. Decide on the criteria you will use in making the decision.
2. Rank your criteria in order of importance to you.
3. Compare two or more alternative choices on each of the criteria in order of importance.
4. Eliminate an alternative when it fails to meet a criterion.

Your choice should be the alternative that meets the most criteria before it is eliminated.

When Fr. Francis Xavier, a parochial school elementary teacher, lived in upstate New York, he was going to develop a new thematic unit. His decision had to do with whether he would create the unit around lighthouses or around caves. Fr. Francis listed his criteria:

> I will be able to find sufficient information to create a unit.
> My theme hasn't been used by teachers at a grade level lower than mine.
> My theme isn't used by teachers at a grade level higher than mine.
> The children will be interested in the theme.
> I will be able to arrange a field trip related to the theme at an affordable price.

Next, Fr. Francis ranked his criteria (Figure 6.2).

1. The children will be interested in the theme.
2. I will be able to find sufficient information to create a unit.
3. My theme isn't used by teachers at a grade level higher than mine.
4. I will be able to arrange a field trip related to the theme at an affordable price.
5. My theme hasn't been used by teachers at a grade level lower than mine.

Fr. Francis then compared each of the themes to each of his criteria. He started with the lighthouse theme. He evaluated it on criterion #1: The children will be interested in the theme. "Yes," he decided, "The kids will like a unit on lighthouses."

Option	*Option*	*Criteria*
Lighthouses	*Caves*	
_______	_______	1. Kids will be interested
_______	_______	2. Sufficient info available
_______	_______	3. Theme not used by later teacher
_______	_______	4. Field trip affordable
_______	_______	5. Theme not used by previous teacher

Figure 6.2. Decision Making by External Criteria, Step 1

Then he evaluated lighthouses on criterion #2: I will be able to find sufficient information to create a unit. "No problem," he thought. "We have some books on lighthouses in the library and I can get more through interlibrary loan. . . . I can teach one science unit on light waves and another on buoyancy. . . . Aquatic birds would be perfect!

"In social studies, we can study the role of ocean transportation in the settlement of America, and we can study the coastal states on both sides of the continent. . . . We can do the Seven Wonders of the Ancient World, because Pharos of Alexandria was the first lighthouse, and it was one of the Seven Wonders . . .

"There's plenty of fodder for math instruction, and for language arts, each group can develop a report on a lighthouse. . . . Art will be a blast because we'll use our math to create scale models of our lighthouses!

"For music, we might do old sailors' songs, or maybe I can find a lighthouse song . . ."

Next, Fr. Francis compared lighthouses to his third criterion. "No, no one in fifth or sixth grade does a lighthouse unit. Ms. Fincher does one on crabbing, but that won't be a problem."

On the fourth criterion, the affordable field trip, Fr. Francis had to make a couple of phone calls. He found that he could take the children to a lighthouse for only the cost of the bus. His principal agreed to provide the transportation to a lighthouse that was 50 miles away. The school cooks would make sack lunches. "Okay," Fr. Francis said to himself, "We have an affordable field trip."

On the fifth criterion, Fr. Francis checked with the kindergarten through fourth grade teachers. He found that one of the first grade teachers did a thematic unit on lighthouses. "Lighthouses don't meet my fifth criterion," he said to himself (Figure 6.3).

Next, Fr. Francis began to compare caves as a theme to each of his criteria.

Option	*Option*	*Criteria*
Lighthouses	*Caves*	
+	______	1. Kids will be interested
+	______	2. Sufficient info available
+	______	3. Theme not used by later teacher
+	______	4. Field trip affordable
0	______	5. Theme not used by previous teacher

Figure 6.3. Decision Making by External Criteria, Step 2

"Yes," he said to himself, "The kids will like a unit on caves."

Then he compared caves to his second criterion: "We have scads of information on caves. My geology unit would fit perfectly here . . .

"In math, we could pretend like we were colonizing a cave, and that would have endless possibilities . . It would also be good in social studies for learning about various types of governments and the ways that governments meet the needs of their people . . .

"For language arts, we could develop a newspaper to inform the rest of the world about the goings-on in our cave colony . . .

"For art, we could study cave paintings, and make a mural of our own cave paintings out in the hall . . .

"Music . . We could study some state songs and write our own colony song!

"Caves meet this criterion!" he announced to himself, who was all ears.

Caves were then compared to the next criterion, whether or not a fifth or sixth grade teacher used caves as a thematic unit. He asked around and found out that Mrs. Rodriguez in sixth grade had just finished developing a unit on caves.

"That's it," Fr. Francis said to himself. "Caves didn't meet my third criterion." His decision was made. Lighthouses would be the theme (Figure 6.4).

Summary

In this section, we discussed a fast and simple way of organizing your decision making. Before you read on, make a simple decision (such as what you want to eat for dinner) using this strategy to make sure that you understand it. Then check off the strategy and read on.

☐ Evaluating by external criteria (elimination by aspects)

Option Lighthouses	Option Caves	Criteria
+	+	1. Kids will be interested
+	+	2. Sufficient info available
+	0	3. Theme not used by later teacher
+		4. Field trip affordable
0		5. Theme not used by previous teacher

Figure 6.4. Decision Making by External Criteria, Step 3

MORE COMPLETE DECISION-MAKING TECHNIQUES

In the last section, we presented a quick and easy model for organizing your decision making. In this section, we examine two more complete models of decision making:

- ☐ The additive model of decision making
- ☐ The PMI

The Additive Model of Decision Making

In the additive model of decision making (see Reed, 1996), you rate each of your alternatives from 1 to 5 on each of your criteria. The alternative with the higher score is the one you select. Here's an example.

Brannin wanted to implement a peace education program developed by his school district. The district had developed two programs over the years: Peace Full and Kids for Peace. Brannin had to decide which program he was going to implement.

First he had to decide on his criteria. He decided that the criteria were: that the program (a) include role-playing, (b) community involvement, (c) school involvement outside of the classroom, and (d) conflict resolution training. He listed his criteria under each alternative program.

Brannin then examined each program and rated it from 1 to 5 on each of the criteria. For example, Peace Full included extensive role-playing, while Kids for Peace included only moderate role-playing. After Brannin rated each program on each criterion, he totaled the scores and selected Kids for Peace, the program with the higher overall score (Figure 6.5).

An advantage of the additive model over the elimination model is that it indicates whether one alternative is significantly better than another, or only

Option	*Option*	*Criteria*
Peace Full	Kids for Peace	
5	3	Role-playing
4	5	Community involvement
4	5	School (out of class) involvement
0	5	Conflict resolution training
13	18	Total

Figure 6.5. Additive Model of Decision Making

slightly better than another: Does one alternative score 22 and the other 21, or does one alternative score 22 and the other 11? However, a disadvantage to the model is that it does not prioritize criteria, assuming instead that all criteria are equally important.

The PMI

A teacher of decision making, Dr. Edward de Bono, wrote about the PMI, another way to evaluate something. PMI stands for plus, minus, and interesting.

Let's go back to Fr. Francis and his unit on lighthouses. At the end of the unit, Fr. Francis wanted to evaluate each child's performance in the project. As her part of her group's project, Ebonee had mapped the East Coast and located all of the well-known lighthouses. Fr. Francis did a PMI to decide what grade Ebonee should receive. He started with the positive aspects (pluses) of her work.

"Positive things . . . let's see . . .

"Well, first, her map is very accurately drawn . . The coastline looks like a coastline, it's not smooth and artificial looking. . . . She drew the states along the coastline completely, instead of drawing only the coastline. . . . She labeled the states. . . . She numbered each lighthouse and along the borders of the map she drew a good accurate reproduction of each house keyed so that it can be located on the map.

"Those are very positive things." Fr. Francis recorded the positive things under the *plus* category (Figure 6.6).

Next, Fr. Francis evaluated the negative things about the map.

"Negatives . . . let's see . . .

"Well, first, she misspelled Massachusetts. . . . Also, she switched the location of two of the lighthouses. . . . She omitted the directional key and a scale . . .

Plus	Interesting	Minus
accurately drawn		
good coast		
complete states		
states labeled		
lighthouses numbered		
lighthouses well drawn		

Figure 6.6. Positive Attributes in PMI Evaluation

Plus	Interesting	Minus
accurately drawn		misspelled MA
good coast		2 locations wrong
complete states		omitted key & scale
states labeled		
lighthouses numbered		
lighthouses well drawn		

Figure 6.7. Positive and Negative Attributes in PMI Evaluation

"That's all that I can see that's negative," he said to himself. He then recorded the negative elements under the minus category (Figure 6.7).

Turning his thoughts to the interesting things, Fr. Francis said to himself, "The fact that she colored the water green instead of a more traditional blue is certainly interesting . . .

"Drawing crabs, whales, and octopi in the ocean was an interesting touch . . .

"Putting a Japanese whaler on the east coast together with the USS Teddy Roosevelt was certainly an interesting touch . . ." (Figure 6.8).

Based upon his observation of the positive, negative, and interesting aspects of her project, Fr. Francis made the decision to give her a grade of A.

Summary

In this section, we have discussed two models of decision making that are more thorough than the elimination model. Before you proceed, make a simple decision using each of the models. You might again want to use the subject of your dinner in order to get a feel for how the different models

Plus	*Interesting*	*Minus*
accurately drawn	colored H_2O green	misspelled MA
good coast	crabs, whale, etc.	2 locations wrong
complete states	Japanese whaler & USS T. R.	omitted key & scale
states labeled		
lighthouses numbered		
lighthouses well drawn		

Figure 6.8. Positive, Negative, and Interesting Attributes in PMI Evaluation

would work. Once you have successfully used each model, check it off and read the next section.

- ☐ The additive model of decision making
- ☐ The PMI

THE BEST AND MOST COMPLETE MODEL OF DECISION MAKING

In the last section, we discussed two moderately thorough models of organizing your decision making, one of which was a basic approach by Dr. Edward de Bono, who has conducted a great deal of research in decision making. In this section, we will thoroughly discuss de Bono's more sophisticated "hats" approach, a very thorough but elegantly simple method of organizing your decision making.

- ☐ De Bono's Six Thinking Hats

De Bono's Six Thinking Hats

Whereas competent and motivated teachers habitually use a complete decision-making process, some teachers make inadequate decisions on important matters by using only one or two dimensions of thinking. For example, some teachers make their decisions based primarily on information; they have heard that knowledge is power, and once they have what seems to be enough information, they decide in favor of the direction that the information seems to be heading. Sometimes this reliance on facts is given a sparkle of credibility by turning it into a computer printout.

In contrast, some teachers make decisions by just following their hearts. Their decisions take the direction of their emotions and intuitions at the time. They might spend their whole year's supply budget on darling bulletin board materials, and then not have money to replace worn-out math materials. The emotional appeal of the bulletin board materials loses much of its luster when there are more kids to be taught math than there are sets of math materials with which to teach them.

Some teachers are attracted to novelty. They delight in finding a different approach to a problem. In making decisions, they favor the most creative option, regardless of whether or not it is workable.

Making the right decision the first time saves money and (most important) time.

It's possible that a few people have been born with the gift of organized thinking that leads to good decisions, but it's an acquired skill for all the rest of us. As you've seen throughout this chapter, we learn to organize our thinking, just as we learn how to construct a complete sentence, solve an equation, or make chocolate chip cookies. So, if we all can learn to organize our thinking, which of the many approaches is the very best one to learn?

The best approach to organized thinking is extensively used by successful corporations whose decisions have millions of dollars riding on them. In addition, the organized thinking used by successful businesses is uncomplicated enough for us to use individually as we make decisions for our classroom and for our personal lives. We discovered such an approach. It has an excellent track record of success and works as well in the classroom as it does in the boardroom. We believe the best approach to organized thinking is Dr. Edward de Bono's *Six Thinking Hats.* Dr. de Bono's decision-making skills are good not only for teachers, they also enable students to use organized thinking in social studies, English, science, and all other subjects.

De Bono's approach is generally called "the hats" because he urges us to "put on your thinking hat." The hats are easy to learn and have many uses for teachers and for students, too. Because we are so thoroughly sold on the hats model of organizing decision making, we believe that teachers have an obligation to teach their students how to use this model to organize their own thinking. Instead of a moderate discussion of the hats in this chapter, and then another moderate discussion in the chapter on helping students to organize themselves, we will cover the hats very thoroughly in this chapter, and then simply remind you about them in the chapter on helping students to organize themselves.

De Bono uses six "thinking hats" to organize thinking. Employing the full set of six hats guarantees that all important dimensions of a decision

have been considered. This confidence of thoroughness calms that terrible nagging feeling that your decision has overlooked something that will surely come back to haunt you. When students learn to organize their thinking by using all six hats, their thinking is thorough, and also gender free.

Gender stereotypes pop up all over, even among school people, who should know better. We would like to believe that girls and boys would be spared typecasting in school. But, until all students are carefully taught to organize their thinking with the full set of six thinking skills, boys will be assumed to be engineering types and girls will be steered toward English and the arts. de Bono's desire is for all of us to organize our thinking by mastering all six thinking skills, regardless of gender, age, race, or ethnic background. Perhaps the worst aspect of gender stereotyping is that some thinking skills are considered more noble than others.

We teachers are all familiar with the hierarchy of Benjamin Bloom: knowledge, comprehension, application, analysis, synthesis, and evaluation. His vertical paradigm implies that some skills are more important than others. Indeed, he clearly says that the top four (application, analysis, synthesis, and evaluation) are the more desirable higher-order thinking skills, and that the bottom two (knowledge and comprehension) are less desirable skills.

In contrast to Bloom's vertical structure, if we were to illustrate de Bono's six hats thinking skills the arrangement would be horizontal. This implies that all of the six hats thinking skills are equally important for everyone. The red hat, the blue hat, the yellow hat, all the hats are equally important. Let's think about each hat in turn.

The blue hat indicates process control. When wearing our blue hats, we look not at the subject of our thinking, but at the thinking itself. This is similar to the conductor of an orchestra, responsible for maintaining an overview of what all the instruments are doing. We wear this hat when planning and organizing.

The white hat is neutral. It is for facts. "Let's do some white hat thinking" is a request for information, not opinions. While wearing this hat you could also list the information you need but do not have. (Note that the white hat may be colored beige in some schools where racial or ethnic misunderstandings would be likely.)

The red hat invites us to share our feelings, emotions, hunches, and intuitions. These are valuable for making good decisions. Ordinarily, we believe we have to create logic to support our feelings. We often disguise our feelings as facts. Red hat thinking allows you to express your personal opinion without having to defend it at all. There also is an understanding that your gut feeling may alter over time, and you will not have to justify

that change, either. For some people it is difficult to imagine a CEO and a corporate Board of Directors tuning into their feelings. But de Bono's research shows that successful businesses do it regularly.

The gray hat represents cautious thinking and judgment. It is the one we have to use a lot in making decisions, especially in schools. It prevents us from making mistakes. When wearing a gray hat you think of all the reasons why a thing may not succeed. It must be logical, otherwise it would be emotional (red hat). (Note that de Bono's book originally called cautious thinking "black hat thinking." We prefer the gray hat to avoid unintended racial or ethnic overtones.)

The yellow hat is optimistic, hopeful, and positive. When wearing your yellow hat, you give all the reasons why a thing will surely succeed. It must be logical, otherwise it would be emotional (red hat).

The green hat represents energy and growth. This is the creative hat. This hat calls for many varied and unusual ideas. Even the wildest proposals are listed without prejudice. Sometimes the best idea eventually emerges from the suggestion that was initially considered very weird. Some people (mainly adult people) are not used to thinking creatively and find it quite uncomfortable at first. Courses or workshops in creativity are helpful. With practice, everyone can learn new and better ways to think.

Learning new and better ways to think is a challenge for some people. For adults, thinking habits may be deeply ingrained and using real hats (or visors or headbands) at first can be essential. Old habits are often tough to overcome, even when you understand why they should be changed. For kids in Grades 2 through 12, using real hats (or visors or headbands) is fun and the learning is fast. Some adults, however, are too uptight to use learning aides such as real hats. For any age group, role playing is an effective way to learn something new as quickly as possible. The idea is to play-act yourself as a creative thinker (or a factual thinker, etc.). Once the new habits have been mastered, wearing real hats (or visors or headbands) is replaced with a single set of six hats (one of each color) saved as symbols to remind everyone to think all the same way at the same time.

Everyone thinking the same way at the same time is especially important for business decisions, since many are made in groups. These days plenty of school decisions are made or influenced by school improvement committees, vertical teams, and so on. When working in a group, it is critically important that all are on the same page at the same time. Ineffective decision-making groups take this literally, and ask members to use only the same factual information at the same time. The maxim means much more than that. Effective decision-making groups make sure that everyone is using the same type of thinking at the same time. For example, everyone is doing

red hat thinking (feelings) at the same time. Then the group may switch to another type of thinking (maybe white hat) and again all use the same thinking skill at the same time. It's amazing how much more quickly a group works with this one improvement.

Improvements can be made in our classroom instruction, too. Good decision-making skills can be easily taught to your students. Organized thinking enhances your instruction, no matter what subject or grade level. The research on using hats in the classroom is just beginning, but the comments so far indicate that elementary and secondary students learn the skills readily. Here's an example of how the organized thinking of the hats approach can help structure a discussion of a K-2 trip to a fire station. de Bono mentioned a trip to a fire station and we developed it like this:

white	Tell us a fact about the firefighters' boots.
red	How did you feel about their rain coats?
red	How did their trucks make you feel?
yellow	What are some good things about fire stations?
gray	What might go wrong at a fire station?
green	Let's list all the different ways we can say thanks to the firefighters.
blue	Let's plan our thank-you notes. What should we say first?

In addition to structuring a class discussion, a teacher can easily teach the hats skills by verbalizing how her thinking is organized. This is a great technique; students often don't realize the thinking processes an adult uses. Here's an example of a classroom decision verbalized by a teacher in Grades 3-5.

"With my white hat on I saw the fact of dark clouds moving in. Using my green hat, I thought of a creative way we could avoid losing our playground time this afternoon without missing math. Employing yellow hat thinking, changing our time for math would be good because we could get outside to play. Wearing a gray hat, changing our math time might not work if we are not serious about math after playing outside. With my red hat on, I feel good about giving it a try. How about you?" The children responded that they'd like to give it a try, too, so the teacher continued by asking, "Using our blue hats, what should we do first before going outside? What should we plan to do as soon as we come inside?"

The hats approach to organized thinking is appropriate for nearly any subject. Below is an example of its use in a Grade 3-5 science lesson concerning amphibians.

________	I hate that idea.	(red)
________	It won't work because of parents.	(gray)
________	We would have lots of money to do other things.	(yellow)
________	Which sports? Gym class, too?	(white)
________	Maybe we could rent the gym advertising to businesses, or sell candy.	(green)
________	We need more info. How can we get other people's opinions about this?	(blue)

Figure 6.9. Example of de Bono's Thinking Hats With Students in Grades 6 Through 8

"Using our white hats, one fact we found was that some turtles can live to be a hundred years old. With your yellow hats on, what do you think are some advantages to living that long?

"Putting on your gray hats, what do you think are some disadvantages of living that long?

"Using your green hats, what are some creative things you could do if you lived for a hundred years or more?

"Do some red hat thinking and tell me how you feel about living that long yourself.

"Now put on your blue hats and tell us how you plan to find out more about longevity. Will you be ready to discuss it more tomorrow?

One of the most important reasons for teaching your students the hats structure is to equip them to organize their thinking so they will make good decisions and judgments. To do this well, they need to be able to recognize the different types of thinking. Figure 6.9 gives a sample that teachers in Grades 6-8 might use. The students are asked to imagine that a controversial proposal has been made that would ban interscholastic sports from the school budget and during the school day. An answer for each item is included on the right in parentheses. The students' task would be to identify the type of thinking represented by each of the items in Figure 6.9.

As you get into teaching your students to organize their thinking, you may encounter negative peer influences, especially among teens. Sometimes organized thinking is not cool. The hats approach gives teens a way of avoiding peer blame for thinking. Here are some examples of things kids in Grades 6-8 might say to keep their thinking organized while minimizing criticisms from other teens:

blue	What exactly is the problem?
red	What are your initial feelings about it?
white	What are the facts?
green	What are some possible things you could do?
yellow	What are some advantages of the green hat ideas?
gray	What are some disadvantages of the green hat ideas?
blue	Based on the advantages and disadvantages, summarize the green ideas in order of "most likely to work."
red	Which green hat idea do you like best now?
blue	If necessary, revise your summary chart.
gray	Review the risks of the top-ranking ideas.
red	What are your final feelings?
blue	What is your plan of action?

Figure 6.10. Example of de Bono's Hats With High School Students

"That isn't me, it's just some wild and crazy creativity."

"I don't know what I'll end up doing, but my feeling right now is . . ."

"Hey, aren't we supposed to think of some possible problems with that?"

"You *do* want me to say, 'on the other hand,' don't you?"

"Hey, aren't we supposed to think of some possible advantages to that?"

As the early teens of junior high mature into late teens of senior high, real-life problem situations can be a teacher's main vehicle to show students how to organize their thinking for good decision making. Figure 6.10 gives a sample of a typical full hats sequence to deal with the problem of adults intimidated by groups of teens hanging around in the neighborhood.

As Figure 6.10 illustrates, an individual needs to be skilled in the use of every hat. That is an important idea for group decision making, too. Each member of the group must be able to organize her thinking from all six perspectives. This is crucial to avoiding stereotypes such as, "Jose's our creative one," or "Sally is the only person on our committee who is a good planner." To help your students master the full set of thinking skills, we suggest doing a needs assessment. This is similar to the classic diagnose-and-prescribe technique promoted by John Dewey. Figure 6.11 shows a sample profile of a student's thinking skills. Notice that it is in the form of a bar graph. To have well-organized thinking, what individual skills does this student need to work on?

Gray	Blue	Green	Red	Yellow	White
X					
X			X		
X	X		X		
X	X	X	X		
X	X	X	X	X	X

Figure 6.11. Bar Graph of Student's Thinking Skills #1

Gray	Blue	Green	Red	Yellow	White
		X	X		
		X	X	X	
	X	X	X	X	
X	X	X	X	X	X

Figure 6.12. Bar Graph of Student's Thinking Skills #2

All teachers are familiar with student diversity and individualized needs. In the area of organized thinking skills, student differences may be the greatest. Fortunately, remediation in this area is typically very easy. Let's consider another student with a unique profile (Figure 6.12). What skills should this student practice to become a well-organized thinker?

A good way to help secondary students begin to practice organizing their thinking is to focus them on a problem and conduct a class meeting that develops a solution to the problem. We recommend giving the students an agenda for the class meeting. This agenda should contain all six types of thinking. The agenda organizes their thinking and provides a model for them to use in other meetings. Your goal as a teacher is to encourage your students to develop the habit of thinking in an organized way. Figure 6.13 shows a sample agenda that includes all six types of thinking.

As they become more skillful, you may help your students learn to organize their thinking by challenging them to use all six thinking skills to study an issue in a small group and create a written report to be handed in to you. See Figure 6.14 for a rubric for the report to the teacher from a group of students. Notice that it requires the whole group to exercise all six of the thinking skills. If they make (and use) real hats (or visors or headbands), you can tell at a glance which skills they are using and whether all members of the group are thinking from the same perspective at the same time.

1. Organize the meeting.
   A. Select recorder(s) for this topic. (Use easel or overhead projector.)
   B. Select a chair for this topic.
   C. Clarify the topic (specific problem).
2. What are the facts?
   A. List what we're sure of.
   B. List additional facts we'll need.
3. Analyze the proposal.
   A. List favorable aspects.
   B. List unfavorable aspects.
4. Revise list of facts we'll need.
5. Look for alternatives that might overcome the more serious disadvantages without losing the major advantages. List all the creative options we can, even the crazy ideas.
6. List our current gut-level feelings about each of the alternatives. We can revise these later. Accept all of our feelings without criticism.
7. Organize for the next meeting.
   A. When and where?
   B. Homework assignments.
8. Adjourn.

Figure 6.13. Sample Agenda Using de Bono's Hats

The topic our committee studied was:
The members of our committee were:
The date of this report is:
Our recommendation is:
The process we used was:
The facts we found were:
The feelings we uncovered were:
The creative alternatives we considered were:
Optimistically, the reasons our proposal should work are:
Cautiously, the reasons our proposal might not work are:

Figure 6.14. Sample Rubric for a Write-Up of a Committee Meeting

After your students practice using the hats to organize their thinking a few times, you may want to quiz them to diagnose their understanding. Figure 6.15 gives a sample quiz that will quickly reveal misunderstandings and allow you to reteach effectively where necessary.

In addition to organizing your thinking as a teacher and to teaching your students to organize their thinking, don't overlook organizing your own personal decisions. This would especially benefit you on major decisions. Once you become comfortable with this organized approach, you will

DIRECTIONS: For each situation below:
A. Decide if true or false.
B. Explain your decision.

1. We can't use the six thinking hats in our group, because there are only four of us in the group.
2. We can't make any decisions today, because Booker is absent, and he's our creative one and we need all the new ideas we can get.
3. Our new chairperson will likely be a good leader because she knows the facts.
4. Vasha will never be a good decision maker because he's so emotional.
5. When organizing your thinking for making good decisions, green hat thinking is at the top of the hierarchy.
6. Consuela is an important member of our decision-making team because sometimes we need some girl-type thinking.
7. Using a method of organizing our thinking that has been proven successful in businesses is a good idea for the school classroom.
8. Adults may be able to learn to organize their thinking, but it's extra hard for us kids to learn it.
9. When you start learning the hats approach, you start out wearing real hats, visors, or headbands, but you don't have to wear them forever.
10. Some people are born with well-organized thinking, but the rest of us are out of luck.
11. The hats approach may work for a corporation, but it's too complex for us to use for school subjects.
12. The hats approach helps me identify the types of thinking other people are using, and also the types they are leaving out.

Figure 6.15. Sample Quiz for de Bono's Hats

tend to use it even for minor decisions, and your whole personal life will start to be more organized and have lower stress.

Summary

In this section, we have provided a thorough discussion of de Bono's technique of organizing decision making. We have included a detailed description of how you can use the technique to help your students organize their own decision making. In addition, we have encouraged you to try the technique not only in your profession as a teacher, but in your personal life as well.

Before you continue reading, try using de Bono's method on a simple decision, again perhaps what you're having for dinner, to be certain that you understand the concepts. Then check off the technique and read the next

chapter. But please, please, at your first opportunity, teach your students to use the strategy themselves.

☐ De Bono's Six Thinking Hats

RECOMMENDED READING

de Bono, E. (1985). *Six thinking hats.* Boston: Little, Brown.

de Bono, E. (1991). *Six thinking hats for schools: K-2 resource book.* Logan, IA: Perfection Learning.

de Bono, E. (1991). *Six thinking hats for schools: 3-5 resource book.* Logan, IA: Perfection Learning.

de Bono, E. (1991). *Six thinking hats for schools: 6-8 resource book.* Logan, IA: Perfection Learning.

de Bono, E. (1991). *Six thinking hats for schools: 9-12 resource book.* Logan, IA: Perfection Learning.

de Bono, E. (1991). *Six thinking hats for schools: Adult educators resource book.* Logan, IA: Perfection Learning.

Reed, S. K. (1996). *Cognition: Theory and applications* (4th ed.). Pacific Grove, CA: Brooks/Cole.

Tversky, A. (1972). Elimination by aspects: A theory of choice. *Psychological Review, 80,* 281-299.

7

ORGANIZING YOUR TEACHING

Of all the things teachers need to have well organized, lesson planning is one of the most crucial. The teacher with world-class teaching skills is helpless without good lesson plans, and having an organized approach to lesson planning can help ensure good plans.

Once you have planned a well-organized lesson, using good organization to teach the lesson is a key to being an effective teacher. We'll cover both planning and the teaching act in this chapter.

Last, we'll discuss an organizational strategy for increasing your students' performance on standardized assessments.

Like the previous chapter and those following, this chapter deals with more abstract concepts than the earlier chapters dealt with, so you'll need to continue reading in the gear you shifted into when you began reading the decision-making chapter. Stay in that low gear while you work your way through this and the following chapters.

One of the strategies that we will use to help you think about the abstract concepts in this chapter is that of offering you both positive and negative examples of good organization of planning. Bruce Joyce and Marsha Weil, leading researchers in learning, tell us that research on concept learning shows us that in order to learn a new concept, we need not only clear positive examples, but clear negative ones as well. Only by seeing the positive examples in contrast to the negative ones will you be fully able to grasp the concept.

THE CONTENT

In the last chapter, we discussed ways to organize your thinking. In this section, we will discuss ways to organize your content to prepare for your lesson planning. We will present the following:

- ☐ Starting with the curriculum
- ☐ Concept analyses and hierarchies

Starting With the Curriculum

We interviewed an elementary principal and asked: What is the biggest problem you see with your teachers?

The principal answered, "Teachers don't start their planning with the curriculum. They plan activities without giving a thought to the curriculum required by the state for their grade level. Teachers want to teach fun things, not things that are dry. Suppose a couple of teachers go off to a workshop. They learn some fun activities, so they come back and want to do those activities. No matter that the activities don't match the curriculum for their grade, the teachers will use those activities anyway. What you end up with is that every teacher at every grade level in your building teaches a unit on dinosaurs, while no one teaches oil production."

So, point number one: Start your planning with the curriculum. Don't start it with an idea for a fun activity or with what's written in a textbook. Start your planning with the curriculum.

Concept Analyses and Hierarchies

Without a doubt, you learned how to do a task analysis when you were a teacher education student. However, you may not have been taught how to construct a *concept* analysis and a concept hierarchy.

The most useful thing we do in planning is start with a concept analysis and a concept hierarchy. A concept analysis describes a concept's characteristics, related concepts, and examples. Nonexamples added to a concept analysis help define the concept. Best examples of a concept are called prototypes.

Related concepts are of three types: superordinate, coordinate, and subordinate. Superordinate concepts are larger categories into which a concept fits; coordinate concepts are parallel concepts; and subordinate concepts are subsets of the concept we're examining.

Hair

Characteristics: body covering of mammals; various textures and colors
Related concepts:

Superordinate:	animal body coverings
Coordinate:	scales, feathers, exoskeletons
Subordinate:	function, structure, types

Examples:	prototype–bear's fur
	human head hair
	human beard
	sheep's wool
	dog's fur
Non-examples:	chicken's feathers
	lizard's scales
	grasshopper's exoskeleton

Figure 7.1. Concept Analysis of Hair

A concept hierarchy is a graphic representation of a concept's superordinate, coordinate, and subordinate concepts.

Suppose that the concept that we're going to teach is *hair.* The first thing we do is go to a source to learn enough about hair to conduct a concept analysis and construct a concept hierarchy. Don't assume that you already know enough about a concept to conduct a concept analysis and construct a concept hierarchy. This is an error that some interns make. They create very poor concept analyses and hierarchies because they think, "Oh, I know enough about hair [or whatever the concept is] to make a concept analysis and hierarchy." They don't. Unless you're regarded as an authority in a field, always go to a book about the topic, an encyclopedia, or an Internet resource prepared by an authority in the field.

Only after you've sought out a source of authority on the topic will you be prepared to make a concept analysis and hierarchy. Since we started with the concept *hair,* let's use it as an example. Figures 7.1 and 7.2 provide an example of a concept analysis and a concept hierarchy, respectively, that we constructed after we researched the topic in our multimedia encyclopedia. Only when you can see the big picture of your concept will you be able to plan how to teach it.

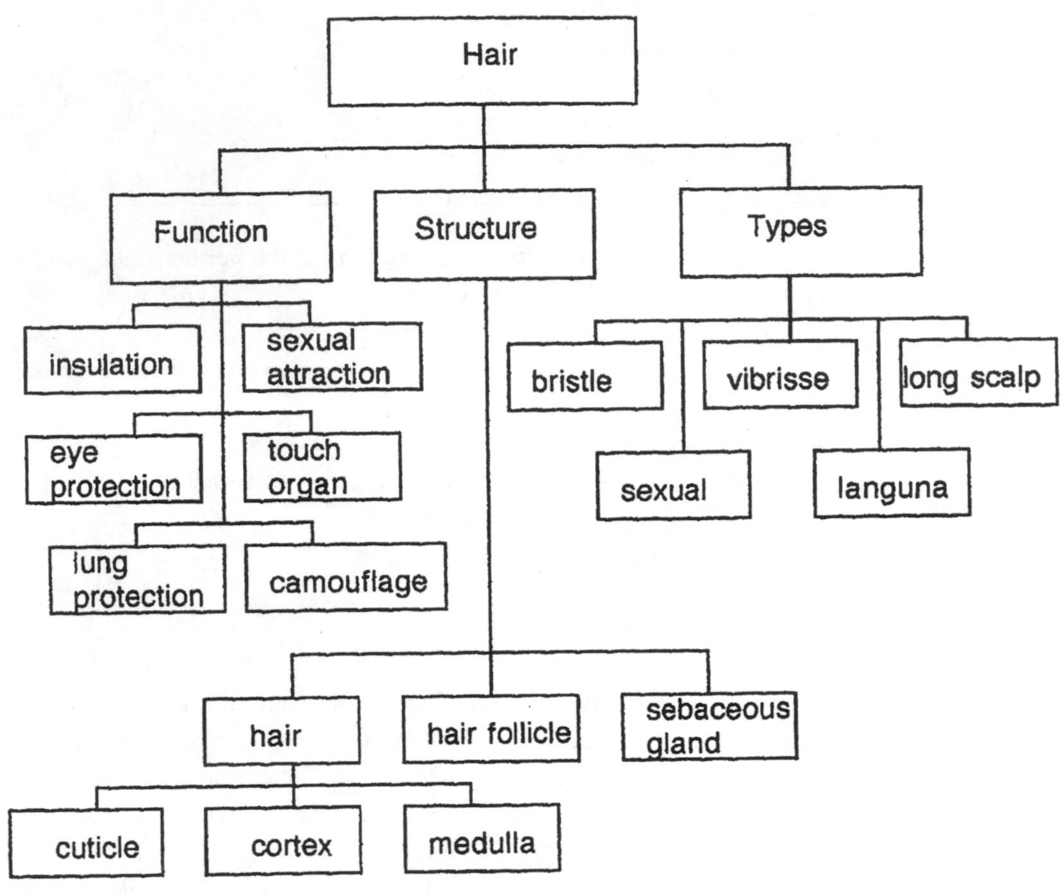

Figure 7.2. Partial Concept Hierarchy of Hair

Summary

In this section, we have presented information concerning organizing your content to prepare for lesson planning. Specifically, we have presented information on the following topics. Stopping now to review your grade level curriculum and to conduct a concept analysis and hierarchy on a topic in the curriculum is a good strategy before proceeding. When you have reviewed the curriculum and conducted a concept analysis and hierarchy, check off the following items and continue reading.

- ☐ Start with the curriculum
- ☐ Concept analyses and hierarchies

THE LESSON-PLANNING PROCESS

In the previous section, we discussed organizing the content that you are going to teach. In this section, we discuss ways of organizing the planning of how you are going to teach the content.

Specifically, we will discuss the following organizational approaches:

- ☐ Interdisciplinary lesson plans
- ☐ Matching goals to objectives to objectives to activities
- ☐ Matching assessments to objectives
- ☐ Personalizing your lesson plan forms

Interdisciplinary Plans

If you are not already using interdisciplinary planning, you have missed an excellent way to organize yourself for teaching. You probably already know what a unit plan is: one in which you develop lessons across the curriculum around a theme, such as Fr. Francis Xavier did with lighthouses. Making a graphic display of your interdisciplinary plan will help you to organize it more effectively. Figure 7.3 shows Fr. Francis's graphic display of his interdisciplinary unit.

Matching Goals to Objectives to Objectives to Activities

We use the strange term "matching goals to objectives to objectives to activities" to emphasize the fact that disorganization can cause us to fail in teaching what we're supposed to be teaching anywhere along the track from goal to long-term objective to the day's objective to the lesson activities of the day. But don't despair! Good organization can help us ensure that everything matches correctly!

Matching Long-Term Objectives to Goals

You may wonder why we included this section at all. We included it because we are constantly surprised by how many teachers don't match their long-term objectives to their goals.

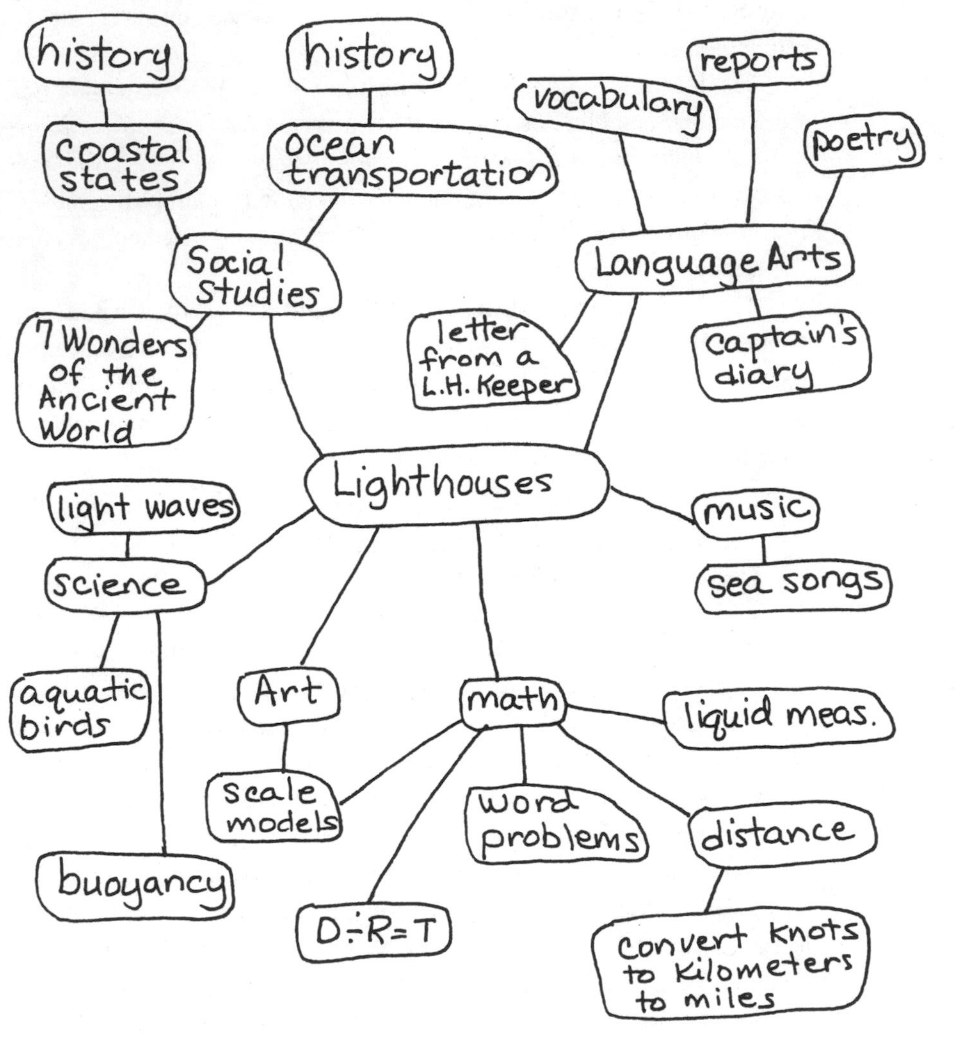

Figure 7.3. Web Organizer of Interdisciplinary Curricula

For example, one science goal in many states' curricula is: *to use scientific processes to conduct investigations.* Suppose that is the goal Ms. Herrera thinks she is addressing. But Ms. Herrera takes a look at her textbook and writes a long-term objective to match what is in the textbook, stating: *The students will identify the various ecosystems in and around bodies of water.* Absolutely no connection exists between the goal and the objective.

We see this again and again when we visit in teachers' classrooms. Teachers tend to fall into this trap when they teach the textbook; the textbook

may not be a good tool at all to use in addressing your state's or district's curricular goals.

Ms. Herrera could write this long-term objective to match her goal: *The students will conduct observational research on the kinds of and numbers of animals living around the pond behind the school.* While Ms. Herrera's textbook doesn't help her organize her teaching to match the state curriculum, the environment in which she and her students live can help her make the match.

This is not to say that the textbook might not be an excellent approach to meeting the state curriculum. To the extent that it helps a teacher meet her curricular goals, it should be used. We simply offer the caveat that the textbook may disrupt your organizational plan of matching goals to objectives to objectives to activities.

Matching Instructional (Short-Term) Objectives to Long-Term Objectives

Just as we have found that many teachers do not make their long-term objectives match their goals, we have also found that even more frequently, teachers do not match their instructional objectives to their long-term objectives. (Other terms that are used interchangeably with the term *instructional objectives* are *short-term objectives* and *daily objectives.*)

An example from our practice: A teacher's long-term objective was: *Given a world map and a gazetteer, students will use map interpretation skills (with 80% accuracy) to: find locations, calculate distance between two locations, and determine locations on the map where the land areas are distorted in size by more than 30% due to the projection.*

The day we observed her teach, the teacher's instructional objective was that students would *pick out a country that they would like to visit, read about it in the encyclopedia, pretend they were visiting there, and write a postcard to a friend.* That would be a fine instructional objective for some other long-term objective, but it was not a good instructional objective for learning to use map interpretation skills to find locations, calculate distance between two locations, and determine where on the map land areas are distorted due to the projection.

You may be thinking that the students were told to use map interpretation skills to find locations and calculate the distance from their town to the place they were visiting, but the teacher did not require that they do so. Even if she had required that they do so, having a child only find one place and calculate the distance to that one place is not sufficient practice in a lesson. We know that Academic Learning Time (ALT) is the best predictor of

student achievement. In order to have had good ALT for this objective, the students should have spent the entire instructional time practicing one or more of the skills in the long-term objective.

Here, then, are some examples of instructional objectives used by another teacher we observed whose instructional objectives matched the long-term objective of using the map skills identified when given a world map and a gazetteer.

1. Given the names of the capitals of 12 African countries, students will state the location (latitude and longitude) of each.
2. Given the names of capitals of eight European countries, students will calculate the distance between Baghdad and each capital.
3. Given a Mercator map and a blank map worksheet, students will outline in red the areas in which the projection distorts the size of the land by more than 30%.

Matching Lesson Activities to Instructional Objectives

We also see problems when we examine the activities that some teachers use in a lesson. Some teachers use activities that are totally unrelated to their instructional objectives. Case in point: One of our interns, we'll call her Maxine, conducted the following demonstration lesson:

Goal:	To develop social studies concepts and skills
Long-Term Objective:	The students will employ time lines to hypothesize cause and effect.
Instructional Objective:	After listening to the story *One Cold Windy Day,* the students will be able to identify events in time relative to other events, using the terms *first, second, third,* and *before, next,* and *after* by constructing a time line.

First Maxine read the story to her students. Then Maxine and the students discussed the events and their relation to each other. So far so good! But then she had the students draw and color a picture of a windy day! The students never made a time line; instead they spent the time that they should have spent constructing the time line on drawing and coloring a picture!

Sherilyn was another intern who used *One Cold Windy Day* to address the same objective, but she did plan activities that matched her objective.

First Sherilyn read the story aloud. Then she had the class as a group name each event, which she then represented by a picture on tagboard. Next, two students went to the board and each chose a tagboard event. Each student held up the tagboard on which their event was depicted. The students then asked the class which event came first, and they then arranged their bodies (holding the tagboard) in the correct order.

Other students joined them one at a time and the class told the new students where to stand on the time line. The teacher kept asking the class such questions as, "In which place does this event go?" "What happened next?" "What happened first?" "Then what happened?" "What happened before . . . ?" "What happened after . . . ?" "What happened last?" "When did this happen compared to that?" And so on.

By the time the class was finished with this activity, they had constructed a human time line, and before very many days, they would be ready to try using time lines to hypothesize cause and effect. Sherilyn was on a roll!

Please note that if Maxine had had an interdisciplinary plan and had identified as an art objective that the students were to draw and color a windy day, that would have been fine after they had constructed their time line. Our objection to Maxine's lesson is that she did not match her instructional activities to her instructional objective.

Don't be a Maxine! Do be a Sherilyn!

Matching Assessments to Objectives

Assessment! In special education circles, a well-known maxim is: If you ain't tested it, you ain't taught it! Assessment is the only way we know whether or not our teaching has been effective.

The key to organizing for assessment is curriculum based assessment, or CBA. To use CBA, take the short-term objective: If students can perform the short-term objective, they've passed the assessment. If students can't perform the short-term objective, they've failed the assessment. It's as simple as that.

Eddie and Chadd were interns at the sixth grade level. They both had the following objective: The student identifies the purposes of different types of texts (to express, to entertain, to inform, to influence). Eddie's assessment consisted of eight short readings. At the end of each reading, the student was to answer a multiple choice question: Which of the following was the purpose of this reading? (a) to express; (b) to entertain; (c) to inform; (d) to influence. He had a great assessment! His assessment matched his objective.

In contrast to Eddie's, Chadd's assessment didn't match the objective. First, he had a series of questions about the plot and characters of a short story the students had read. Then he asked what must be included in a newspaper-type report. He wrote a wonderful assessment of something, but it didn't assess his objective at all!

Ron Finley, a Civil Service employee who works for the Department of Defense in the training area, told us that the training in the United States military begins with an assessment. The trainers ask, "What does this student need to be able to do?" The answer may be, "Break down this engine and put it back together again." The instructors make the assessment first, before they identify objectives. From the assessment, the instructors identify clusters of behaviors, and those become the behavioral objectives. This is sometimes known as backwards planning; it ensures that the assessment matches the objectives, something that public school teachers sometimes fail to do.

For school teachers, the key to organizing for assessment is CBA: Write your assessment to measure whether or not the students can perform the objective.

Personalizing Your Lesson Plan Forms

First grade teacher Sylvia Broussard told us, "Develop your own form, which can be photocopied each week. Enter as much of your repeated routine as possible on your lesson plans so you are not writing the same thing over and over. I also leave a space for enrichment and reteach. As the lesson is taught or papers are corrected, I note students' needs in these two areas. I jot down what is needed to meet their individual needs for the next day right on the lesson plan."

Bully idea!

Summary

In this section, we have discussed how to organize for good planning. Examine the following checklist and check off the items about which you feel confident. Reread any about which you need to continue thinking.

- ☐ Interdisciplinary lesson plans
- ☐ Matching goals to objectives to objectives to activities
- ☐ Matching assessments to objectives
- ☐ Personalizing your lesson plan forms

THE TEACHING ACT

In the last section, we discussed ways of organizing your lesson planning. In this section, we will discuss some ways of organizing the teaching act itself.

Specifically, we will cover:

- ☐ Using graphic organizers
- ☐ Using checklists
- ☐ Emphasizing important points
- ☐ Selecting students systematically
- ☐ Using talking beanbags
- ☐ The learning stage/personnel/activity chart
- ☐ Ensuring that all students know the answer
- ☐ "Each one teach one"

Using Graphic Organizers

Graphic organizers are an excellent way to open and close a lesson. A number of graphic organizers may meet your needs for any specific lesson.

Web Graphic Organizers

The majority of Millie's interns (about 8:1) prefer her to use a web organizer when she teaches. She starts the lesson by asking the students what they know about a topic. The students tell her, and she webs that material. After the lesson, she adds to the web the new material that the students have learned. If she is teaching a unit, she makes the graphic organizer on butcher paper and keeps it to start and end each day's lesson. If the lesson won't go beyond one day, she uses the chalkboard.

Figure 7.4 gives an example of a web from one of Fr. Francis Xavier's classes that was starting the lighthouse unit.

The advantage of the web type of organizer is that the teacher can include every student's input. The disadvantage is that it is difficult to conceptualize the knowledge tree in this form. Contrasting the web to the concept hierarchy organizer shows that it is generally more simple to conceptualize benefits and drawbacks of each method.

Figure 7.4. Web Graphic Organizer of Students' Knowledge of Lighthouses

Concept Hierarchies as Graphic Organizers

In contrast to the majority of her students, Millie and the minority of her students prefer a concept hierarchy type of graphic organizer. The advantage of this type of organizer is that it better allows students to see how information fits in the tree of knowledge (Figure 7.5).

The disadvantage of this type of organizer is that you may get into trouble if you start the lesson with the organizer by asking students what they already know about the topic. The things that they tell you may not fit very well into your hierarchy.

If you use this type of organizer, you will probably want to put it up on the wall on butcher paper when you first begin a unit. Then each time you teach a piece of the concept over the course of the unit, color in that box at the end of the lesson to show how what the students learned fits into the big picture, as Figure 7.6 shows. You probably will not want to try to incorporate what the children say they know about the topic into the hierarchy. Instead, you can ask the children what they know, make a web on the chalkboard, and keep the hierarchy separate from the web.

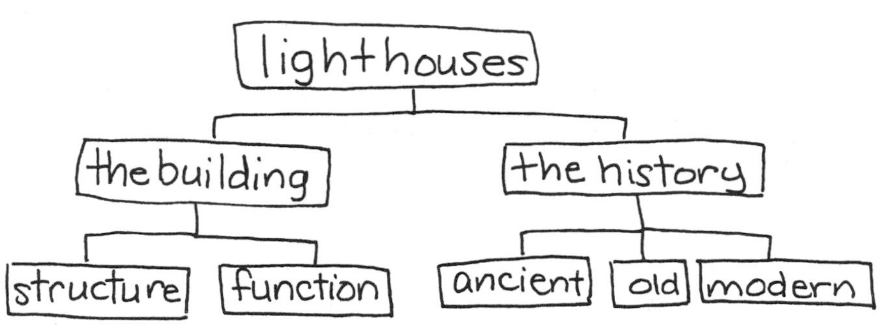

Figure 7.5. Concept Hierarchy Graphic Organizer of Lighthouses

KWL Charts as Graphic Organizers

KWL, or what do you Know, what do you Want to know, and what did you Learn, charts are a type of graphic organizer. One of the strengths of this type of organizer is that the students identify what they want to learn about the topic (which may or may not be what you intend to teach them!). A problem, however, is that KWL charts seem to be weaker on helping students construct a graphic picture of the concept. Each type of graphic organizer has its assets and liabilities. Here's a KWL chart for the lighthouse unit (Figure 7.6).

Using Checklists

Ms. Herrera likes to use checklists to open and close a class or a unit. One benefit of this strategy is that it reminds you to include all the steps in the lesson that you planned to include. Without a checklist, Ms. Herrera finds that she often forgets to include lesson activities! By using a checklist, she can make certain to include all the parts of the lesson.

You can use checklists in two ways: publicly or privately. Using the checklist publicly means to write it on the chalkboard, overhead, butcher paper, or class computer monitor. The students will then know exactly what they will be doing in the lesson. As you complete each part of the lesson, check it off. Then at the end of the lesson, quickly review each item on the checklist. One advantage of the public checklist is that the students will perceive you as very organized if you post the list and check off each item.

Using the checklist privately simply means that your checklist is strictly for your use, so it isn't displayed where the students can see it. The private

K	*W*	*L*
for safety on shore all different some are old show the right way	How do they work What happens in a hurricane? Why are they all different? How old are they?	

Figure 7.6. KWL Chart of Lighthouses

checklist simply serves to help you make certain not to forget part of the lesson.

Emphasizing Important Points

Many children have difficulty discriminating important points from unimportant. Students who have disabilities often find determining what is important from what is not important to be very challenging. Teach your students that the following behaviors on your part mean that something is crucial in a lesson: (a) you repeat it verbatim three times; (b) you say, "This is very important. Remember it"; (c) you have them repeat it back to you verbatim multiple times as a chorus; and (d) you have them repeat it to their seatmates, the ceiling, the floor, Ms. Pencil, Mr. Desk, and any other available person or object.

Even as college professors, John and Millie find that their students often have trouble discriminating between what is of crucial importance and what is of lesser import. Explicitly teaching students these strategies for knowing what is important will increase their achievement.

Selecting Students Systematically

Research shows that we tend to call on the same students over and over. After all, it's hard to ask someone who often doesn't know the correct answer when you know that Consuela Griego does know the right answer. We have to work hard to make ourselves call on all students equitably.

A wonderful idea that we received from a teacher helps us to select students systematically without going down the rows or using our gradebook to call them in alphabetical order.

Poker Chips

Write each student's name on a poker chip. Drop all the chips in a five-pound coffee can. Draw a name and call on that person. Drop the chip back in the can so that person won't know whether or not he needs to stay alert; his name might be called again!

Clothespins

When it is more important to be certain that every person gets called on rather than that every person has an equal chance to be called on each time, use clothespins. Write each student's name on a pin. Drop them into the coffee can. Draw out a pin and call the name. Then clip the used pin to the side of the can. Everyone gets a turn.

Using Talking Beanbags

Teacher Pat Page gave us great advice. She said that when you have students who have a hard time remembering to raise their hands, you should use talking beanbags. The rule is that no one may talk unless holding a beanbag. The teacher keeps one beanbag in order to be able to talk at will. The second beanbag is tossed to the student who is to speak next. Pat says that whenever someone starts to blurt something out, she holds up her beanbag as a nonverbal cue for the student to hold his tongue. She tells us, "You would not believe how many people get interested in class when the beanbags are flying out and flying around the room." By using beanbags, Pat doesn't have to keep telling the kids to raise their hands.

The Learning Stage/ Personnel/Activity Chart

With more and more individual and small group instruction required in today's schools, knowing how to teach multiple students multiple things at one time is a necessity. By taking the learning stages and thinking about what activities are appropriate for each and what personnel can be involved in each, the chart in Figure 7.7 emerges to help you organize your classroom.

Note that we argue that interacting with the teacher is not necessary during fluency/proficiency instruction or during maintenance activities. We do recommend the teacher direct adaptation level activities, because if a skill/knowledge is not to be used or is to be altered in some important way, we think you should be the purveyor of that information, or, if you are using Cooperative Learning activities, that you monitor very closely.

	Stage of Learning	*Activity*	*Personnel*
5	Adaptation	Direct Instruction	Teacher
		Cooperative Learning Activities	Cooperative Learning Groups
4	Generalization/Transfer	Cooperative Learning Activities	Teacher Aide Capable Volunteer Cooperative Learning Groups
3	Maintenance	Games	Peers
		Group Projects	Cooperative Learning Groups
		Individual Projects	
		Computer Aided Instruction	
2	Fluency/Proficiency	Cooperative Learning Activities	Cooperative Learning Groups
		Computer Aided Instruction	
1	Acquisition	Cooperative Learning Activities	Cooperative Learning Groups
		Inductive Technique Models	Teacher
		Direct Instruction Models	Teacher

Figure 7.7. Activities and Personnel Optimal for Stage of Learning

Your direction is also required at the acquisition level, although again, close monitoring of Cooperative Learning activities is advised. However, you shouldn't have to involve yourself much with students who are engaged in fluency/proficiency or maintenance activities. You don't even have to monitor closely. Those groups can take care of themselves.

You will notice that we use Cooperative Learning extensively. Cooperative Learning is one of the best organizational tools in existence. Because the students are teaching and giving feedback to each other, often in a ratio of 1:1, but never in a ratio more than 3:1, behavior problems are minimized. That brings us to the next section: ensuring that every student knows the answer.

Ensuring That Every Student Knows the Answer

Cooperative Learning is the key to ensuring that every student knows the answer. You're probably either enthusiastically nodding your head, or you're rolling your eyes and saying, "Yeah. Right." If you are among the head-nodder group, hooray for you! You have probably received instruction in Cooperative Learning. If you're one of the eye rollers, you simply have not been taught how to conduct Cooperative Learning Lessons.

The eye rollers often say this: "I've tried Cooperative Learning. One kid ends up doing all the work, and the others are slackers who get a good grade because of the kid who did the work." That statement alone tells us that you weren't doing Cooperative Learning. You were doing group work. It isn't the same thing.

The reason that we knew from one sentence that you weren't doing Cooperative Learning is that group grades are NEVER given in Cooperative Learning. Group rewards are given, such as 5 minutes free time, but grades are never given to a group. One of the hallmarks of Cooperative Learning is Individual Accountability. If individual accountability isn't present, you've got group work, not Cooperative Learning.

We are going to teach you one Cooperative Learning structure from Dr. Spencer Kagan's book, *Cooperative Learning.* Try it. You'll find you want to know all the dozens of structures in his book. You can purchase the book by calling 1-800-Wee Co-op. We can't think of a more powerful investment.

The structure we are going to teach you is called Numbered Heads Together. In this structure you do the following: First, assign students to heterogeneous groups of four. Assign a number to each group, starting with Group #1. Next, have the students number off within their groups, so that each group has a Person #1, a #2, a #3, and a #4.

Next, using material to which the students have been exposed (say, a chapter in their social studies book), get ready to ask the students a question, which you will pose in problem form by saying, "Make sure that everyone in your group can tell me . . ."

Pose the problem, such as "Make sure that everyone in your group can tell me what the Executive Branch of government does." Then say, "When everyone in your group can tell me the answer, raise your hand." Have the students work together to find the answer and make certain that all of them know it. (Some students call this structure "Butts Up" because that's the position they tend to assume when they're all talking and working to decide on the answer.)

When all hands in the class are raised, roll a die or spin a spinner to determine a person number, for example, #3. All others lower hands. Then roll or spin again to decide a group, for example, #5. That person, #3 in group #5, answers the question.

No grades are given for Numbered Heads Together. It's simply a learning and study strategy. The grades would come from individual quizzes over the material at the end of the period, or at a later time. Individual accountability is key.

If you choose to, you could give group or class rewards of something such as one minute free time for each five questions answered correctly by the class. But don't give group grades under any circumstances.

Kids love it! Our college students love it! We love it!

You'll love it, too.

You will also want to explore David Johnson and Robert Johnson's books on Cooperative Learning, such as *Cooperation and Competition: Theory and Research* (1989) and Robert Slavin's books, such as *Cooperative Learning: Theory, Research, and Practice* (1995). These are excellent resources that every professional teacher should have.

"Each One Teach One"

A good way to teach complex skills, such as using the computer, is by using an "each one teach one" approach. Sylvia Broussard gave us an example of how she uses this strategy with her first graders: "I introduce new computer skills by focusing on teaching two students until they become proficient. Then they each teach another student, and so on, until most of the students learn the skill. These students can help solve many problems and the students are not dependent solely on the teacher to help them."

Summary

In this section, we have discussed ways to organize your teaching more efficiently. We stressed the importance of using Cooperative Learning to assist you in organizing your teaching. We also covered the following items. Check off those you understand. Reread any about which you are uncertain.

- ☐ Using graphic organizers
- ☐ Using checklists
- ☐ Emphasizing important points
- ☐ Selecting students systematically
- ☐ Using talking beanbags
- ☐ The learning stage/personnel/activity chart
- ☐ Ensuring that all students know the answer
- ☐ "Each one teach one"

IMPROVING STANDARDIZED TESTING

In the last section, we discussed ways that you can organize the teaching act more effectively. In this section, we will discuss one way that you can organize your teaching to improve your students' standardized test scores.

- ☐ Weekly practice tests

Weekly Practice Tests

Teachers tell us that their state-mandated standardized testing disrupts the life of their classrooms for the month that precedes the testing and that their students know more than the tests reveal. We talked to organized teachers about how they handle this problem. We consistently heard the same answer.

Organized teachers don't wait until the month before the state-mandated assessments to prepare for them. They begin the first month of school. Here's what they do.

Starting the first month, they set aside half an hour to an hour one afternoon a week. They give students timed practice assessments of the material that they've covered that week that is of the type that will appear on the state tests. They include all of the types of formats of questions that

are used on their state tests. The students become accustomed to the format of the questions, filling in bubble sheets, and taking timed tests.

The assessments don't count for a grade. The students understand that they are taking the practice assessments in order to prepare for the state test. After the assessments, the students correct their own papers.

These teachers are organizing their teaching so that their students will excel on the state tests by beginning early in the year and helping the students overcome the anxiety that is usually associated with such tests.

Summary

In this section, we have discussed one way to organize your teaching in order to help your students improve their standardized test scores. Specifically, we discussed the following:

- ☐ Weekly practice tests

RECOMMENDED READING

Johnson, D., & Johnson, R. (1989). *Cooperation and competition: Theory and research.* Edina, MN: Interaction Books.

Joyce, B., & Weil, M. (1986). *Models of teaching* (3rd ed.). Englewood Cliffs, NJ: Prentice Hall.

Kagan, S. (1994). Cooperative learning. USA: Resources for Teachers. (Available by calling 1-800-Wee Co-op)

Slavin, R. (1995). *Cooperative learning: Theory, research, and practice.* Boston: Allyn & Bacon.

8

HELPING STUDENTS LEARN TO ORGANIZE THEMSELVES

A teacher who complains of her disorganization told us recently, "I know we're all responsible for our own behavior, but I can't help but wish that someone had taught me to organize myself when I was a child. Mom, my teachers . . . *someone* should have helped me. I remember my second grade teacher criticizing me in front of the class about having the messiest desk in the room. And my desk was awful, I'll admit it. I remember my face feeling hot, and wanting to hide as the other kids stared at me with this haughty look. I played sick the next day so I wouldn't have to face everyone.

"In retrospect, I blame my second grade teacher for not teaching me how to organize myself, instead of simply criticizing me for not being organized. Since she saw the problem, she should have helped me with it.

"Same with my mom. Don't get me wrong, I adore her, but she only criticized me for my messiness and disorganization. Why didn't she teach me HOW to organize my things and myself?"

Because our emphasis as teachers is less and less on rote memorization and more and more on teaching skills that will help students be successful in the real world, we must teach kids to organize themselves. Being organized is a crucial skill for success. Regardless of how smart and creative students are as adults, if they miss crucial appointments, fail to meet deadlines, and can't produce quality products because of their disorganization, they'll never be successful. Therefore, in this chapter, we will address a few ideas that will help you help your students organize themselves.

MATERIALS

In the last chapter, we discussed organizing your teaching. In this section, we will discuss helping students learn to organize their materials and papers. We will discuss:

- ☐ Pencil exchange
- ☐ Organizer boxes
- ☐ Three-ring notebooks
- ☐ Dejunking

Pencil Exchange

Sylvia Broussard told us that the inordinate amount of time that first graders spend on pencil sharpening led her to develop the following organizational strategy: All students are to have two sharpened pencils ready at the beginning of school. If they break a pencil or it becomes too dull, it's exchanged for a sharpened pencil that the teacher keeps in a mug on her desk. The helpers sharpen the pencils in the mug at noon and at the end of the day. A little organizational problem solved!

Organizer Boxes

We talked about the joys of organizer boxes in our chapter on organizing nonprint teaching materials. We think that all children in self-contained classrooms should have organizer boxes in their desks. Students who are in departmentalized classes should have organizer boxes in their backpacks. The old cigar boxes were works of art, but everything in them got mixed up with everything else. For the same reason, the new plastic school boxes are less useful than the organizer boxes. Organizer boxes are a crucial tool for kids to begin to organize themselves more effectively.

Three-Ring Notebooks

In order for your students to be able to take advantage of three-ring notebooks, you have to help them by holepunching all papers that you give them! That's important enough to say again! HOLEPUNCH ALL PAPERS YOU GIVE STUDENTS!!!

For those papers that for some reason you haven't holepunched, have three or four good quality holepunchers distributed throughout the room

for student use. Insist that students have all papers in their desks filed in a three-ring notebook, and that none be floating about the desk.

Because little kids hate to throw away their papers, insist that they take home the papers that you return each day. Explicitly state that by taking the papers home daily, they are keeping their desks well organized.

You can choose not to hand the papers out until the end of the day. That solves the problem of disorganization from papers. But by handing the papers out earlier in the day, you can instruct the students to put their papers in their three-ring notebook, and begin teaching them the habit of organizing themselves in this way. We like the idea of taking this proactive role in teaching students to organize themselves.

Dejunking

In addition to insisting that students take papers home every day, teach them to dejunk their desks regularly. Help kids learn to identify junk using Don Aslett's definitions. Then teach them to wean themselves from desk junk using the techniques we discussed in Chapter 5.

You may wish to have a dejunking day periodically in which you distribute small paper bags to substitute for weaning boxes. Kids can put ambivalent items in the bags, write their names on the outside, and staple the bags shut. Then you can take up the bags and announce that any unclaimed bags will be trashed at the end of the 2-week period. Helping kids learn to dejunk is a favor that will last them a lifetime!

Summary

In this section, we have discussed ways of helping students learn to organize objects. Specifically, we have discussed the following:

- ☐ Pencil exchange
- ☐ Organizer boxes
- ☐ Three-ring notebooks
- ☐ Dejunking

TIME AND MEMORY

In the last section, we discussed how to help kids learn to organize their materials. Here we will discuss two ideas that will help kids learn to organize time and memory.

- ☐ Student checklists
- ☐ Student day planners

Student Checklists

A second grade teacher told us, "It made me crazy when my own child would come home at the end of the day and I asked him what he learned that day. He always shrugged his shoulders and said, 'Nothing.'

"I knew he'd learned lots of things, but try as he might, he didn't seem to remember what any of those things were. When I went back to college and became a teacher, I didn't want my students to go home and tell their parents that they hadn't learned anything in my class each day, so I devised a plan.

"At the end of each day, I discussed with the kids specific things we'd learned. Then we put those things in order on the board and made a checklist. The kids took the checklist home and checked off each item as they told their folks about it.

"By this strategy, I was able to accomplish several things. First, I was able to incorporate practice on sequencing such as 'What did we learn first? Second? Third?' I was also able to use the list as an authentic literacy activity. Then, I was able to make sure that when their parents asked them what they'd learned that day, nobody would say, 'Nothing!' "

We would argue that this world-class teacher was also teaching her kids how to organize themselves by using checklists!

Several teachers told us that they list all the subjects in the upper lefthand corner of the board each day. Then as the class finishes each subject, the teacher checks it off. This would be a good example of modeling the use of a checklist for the students. Wouldn't it be a great idea to have kids attach small checklists to the upper lefthand corner of their desks and check off each subject as they finished it? What a great way to teach the skill of using checklists! The potential for teaching kids to use checklists is endless.

Student Day Planners

The idea of having students carry day planners was a new one for us. We first learned about this strategy when Greg Gibson, one of our graduate students in educational leadership, told us about instituting mandatory student day planners in his high school. He wanted to conduct his master's degree research on the planners, and we were intrigued. He surveyed all the students in his high school and had great results.

Greg found that some of his students already used day planners on their own. However, after requiring day planners (provided by the school), a significant percentage of his students said that they have better grades now than they did before they used the day planners. Students also reported that the day planners help them keep up with extracurricular activities and commitments.

As far as we know, no one is researching the use of day planners with younger children, but we think it is a grand idea. They could be made inexpensively on the computer and duplicated for children who could carry them in three-ring notebooks. Teachers at different grade levels could tailor make the day planners at increasing levels of sophistication as their students increased their organizational skills.

Summary

In this section, we have discussed ways to help students learn to organize their time and memory. Specifically, we have discussed the following:

- ☐ Student checklists
- ☐ Student day planners

LEARNING BEHAVIOR

In the last section, we discussed how to help kids organize their materials, memory, and time. A more abstract notion is to teach them to organize their learning behavior. This section will briefly discuss a way to approach helping students to organize their own learning behavior.

- ☐ Choice Theory

Choice Theory

As teachers become more organized, they tell us that it feels good finally to start having some control over their lives. Adults have no difficulty admitting they need to have some control over their lives, usually the more the better. In addition to feeling good, they find themselves accomplishing more, seeming more professional, and teaching more effectively. But those same teachers may overlook similar needs in their students.

Students who have no control over their classroom environment are likely to react in undesirable ways. Even capable students may withdraw and

do as little as possible. Some bright students become rebellious disciplinary problems. The major work in this field was conducted by William Glasser, M.D. After years of studying low achievers, Dr. Glasser found that when problem students are given some control over their learning, they often begin to see school as a rewarding place, start to accomplish more, find that they are learning, and begin to feel good about themselves. These improvements result in better academic achievement and fewer discipline problems, two outcomes dear to a teacher's heart.

But if those dear teachers can accomplish all that, why isn't every teacher giving more control to students? The answer is that extending more control to students is not easy. It takes more organization and requires more creativity.

Creativity is necessary to find things over which students can exert control without becoming monsters. This is a realistic concern, since some students are immature and can mishandle even the best idea, turning it into a problem instead of a help. Since most teachers' two main goals are to increase achievement and decrease discipline problems, let's address each.

Achievement increases when more than one way of learning is available, and the student can select the way that best fits her learning style or preference. The challenge for the teacher is to create alternate ways to learn. For example, some visually oriented students might learn $3 \times 4 = 12$ by just reading it in an arithmetic book. For kinesthetically oriented students, this concept will not be understood until they arrange three sets of four (and four sets of three) paper clips on their desks and repeatedly count the total. A musically oriented student might learn it best by working it into a song or a rap. Howard Gardner and David Lazear are two authors who help us understand at least seven of the different learning styles.

Don't throw this book! We're not saying that every good teacher should have seven different learning activities ready for every concept. We are recommending that a couple of ways to do some things will have a noticeable impact when your students can elect which they will do. Teachers who develop alternates for a few concepts quickly find themselves generating options for other lessons, too. Your students are likely to suggest even more alternatives. The second time you teach a course you easily add to the variety of learning options you used the first time. Once your creativity is tickled, watch out!

Watching for opportunities for multiple learning options, a disorganized teacher quickly panics and gives up. This is why it is so important to become as organized as possible: Effective teachers will increasingly employ choices for their students. And offering students these choices will teach them to begin organizing their own learning behavior.

Summary

In this section, we have discussed one way to teach students to organize their own learning behavior. Try incorporating choice theory as a device for helping students learn to organize their own learning behavior.

☐ Choice Theory

RECOMMENDED READING

Gardner, H. (1983). *Frames of mind: The theory of multiple intelligences.* New York: Harper & Row.

Gardner, H. (1991). *The unschooled mind: How children think, and how schools should teach.* New York: Basic Books.

Gardner, H. (1993). *Multiple intelligences: The theory in practice.* New York: Basic Books.

Glasser, W. (1986). *Control theory in the classroom.* New York: Perennial Library.

Glasser, W. (1998). *Choice theory in the classroom.* New York: Harper Perennial.

Lazear, D. (1991). *Seven ways of knowing: Teaching for multiple intelligences.* Arlington Heights, IL: IRI/Skylight Training and Publishing.

9

TEACHING STUDENTS WITH SPECIAL NEEDS

First-year teachers (and many experienced teachers) tell us that the biggest problem that they have is working with students who have behavioral and learning problems. Many teachers feel poorly equipped to work with students who have challenges. This chapter will tell you how to organize your teaching to help remediate children who are challenged. You'll be surprised to find how effective these organizational strategies are, even for students without disabilities!

ESTABLISHING THE BEHAVIORAL ENVIRONMENT

In the last chapter, we discussed how to help students organize themselves. In this section, we will outline ways to organize the behavioral environment for students who have behavior problems. Specifically, we will examine:

- ☐ The keys to working with students who have problems
- ☐ Creating structure for students who have learning problems
- ☐ Systematically teaching social skills
- ☐ Ensuring transfer of skills

The Keys to Working With Students Who Have Problems

There are two keys to working with students who have problems: (a) explicitness and (b) repetition. That's a pretty simple answer for a big problem, but you will find that explicitness and repetition are powerful tools.

Explicitness

Special needs students don't learn inductively. For example, Antonio, who is not a special needs student, came to class 5 minutes late from recess. When Antonio entered the classroom, his teacher said, "You are to report to detention hall after school for being late to class." Antonio never came in late from recess again. He inductively learned that if he were ever late to class again, he would have to go to detention.

In contrast, Bobby, a special needs student, came in 5 minutes late from recess. When Bobby entered the classroom, his teacher said, "You are to report to detention hall after school for being late to class." Bobby was late to class again later in the week. Bobby had not learned inductively that he would have to go to detention if he were late for class again.

Does that sound confusing? To those of us who learn inductively it is confusing, but to a person who doesn't learn inductively, it makes perfect sense. Bobby only learned that he had to go to detention on that particular day because he was late to class. He did not infer that he would have to go to detention every time he was late for class. He did not even infer that he would have to go to detention *ever again* for being late for class. Bobby does not have the ability to connect cause and effect. That's an almost universal problem for students who have problems. Bobby will be late for class again and again; his teacher will send him to detention again and again; he will never connect the two events without explicit teaching.

By explicit teaching, we mean that you must tell Bobby explicitly, "Bobby, you were late to class. Because you were late to class, you have to go to detention. Every time you are late to class, you will have to go to detention. The next time you are late, you will have to go to detention. The time after next that you are late, you will have to go to detention. Being late causes you to have to go to detention. Going to detention is the effect, or result, of your being late to class. The two events are connected to each other as surely as if there were a rope tying them together."

The only way that Bobby will ever "get it" is by being told explicitly over and over what the consequences of his behaviors are.

Repetition

Closely aligned to explicit teaching is repetition. Researchers tell us that regular children must be exposed to something many times (some say 40 times) before they learn it. Children who have special needs may require repetition four thousand times before they learn it! Unless we expose students who have special needs to important information over and over and over and over and over, they aren't going to learn it. If they could learn in a typical number of repetitions, they wouldn't be students who had problems; they would be typical students. They require endless repetition.

One of the ways that we can ensure adequate repetition is by using a HyperCard™ program to create drill and practice programs for our special needs students. The computer never tires or gets cranky when it has to give corrective feedback for the five hundredth time. Computer aided instruction is invaluable in working with students who have special needs.

Creating Structure for Students Who Have Problems

When Millie was an undergraduate in special education, her professors told her repeatedly, "Structure, structure, structure." She mastered the idea. But when she took her first teaching job (junior high school behavior disorders), she quickly realized, "My professors told me to create structure, but they didn't tell me how! What is structure? I can't create it if I don't know what it involves!"

After a trek through the research and many years of trial and error, Millie developed what she calls the 4Rs of Structure: rules, responsibilities, routines, and rituals (Figure 9.1). We'll address each in turn.

Rules

Millie defines *rules* as *guidelines that govern our behavior toward others. Don't hurt other people's property* is a rule because it governs one's behavior toward others. *Don't hurt other people's bodies or their feelings* is a rule, because it governs one's behavior toward others.

You probably noticed that these rules were not stated positively in the way that we were all taught in our elementary education classes. That is because special needs students don't learn inductively. If we were to instead use the rule: *Respect others,* our special needs students wouldn't get it. Case in point:

Monty was a high school student who had a behavior disorder. He was mainstreamed for several classes, including sophomore English. His English

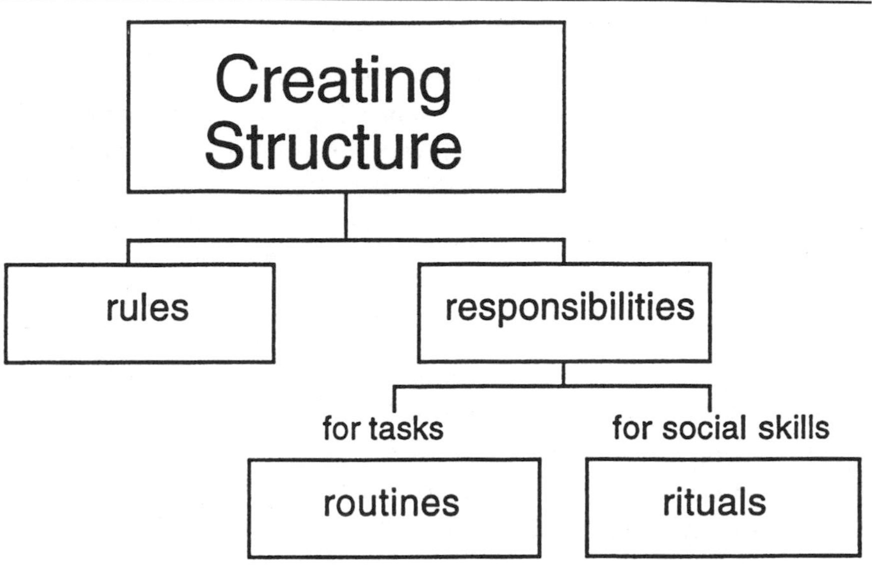

Figure 9.1. The 4Rs of Structure

teacher, Coach, contacted Millie about his disruptive behavior in class. Millie arranged to come the next day and conduct an antecedent-behavior-consequence observation.

Millie entered the class during break the following day. Monty entered and interacted with the other students in an appropriate, friendly, teenage manner. When Coach asked the students to quiet down and listen to their instructions, Monty did so. When Coach was finished giving the instructions and was ready for the students to begin working independently on their assignments, he said, "Get to work and don't disturb others."

In about one minute, Monty began talking to the two girls by whom he was sitting. Then they all started laughing. Coach sent Monty out of the room.

Millie interviewed Monty back in the special ed room. "Why did you disturb the girls after Coach had just told you not to?" she asked.

Monty looked offended. "I didn't disturb anyone," he objected. "I was only telling them a joke!"

If even a negatively phrased direction such as "Don't disturb" wasn't effective in letting Monty know what the rules were, you can be certain that he wouldn't have understood something as vague to him as "Respect others."

When we're teaching rules to students with problems, we have to start with the most basic list of "Don'ts" we can. "Don't touch anyone." "Don't touch anyone else's belongings." "Don't say rude or mean things to anyone." "Don't tease anyone." Those are the kinds of very basic rules with which we must start. After the child has mastered those four rules, we can then teach him explicitly that we are subsuming those rules under the rule: "Respect others." Then when a rule infraction occurs, we say, "You didn't respect others. You know that not touching anyone else's belongings is part of the 'respect others' rule. You touched someone else's belongings, so you broke the 'respect others' rule."

When we want to teach a rule to our student who has problems, we write down the rule, talk with him about the rule, demonstrate the rule, and then have him practice it over and over (see the section on "Systematically Teaching Social Skills"). For example, in the case of "Don't touch anyone," demonstrate the rule by role-playing and saying out loud what your thoughts would be, such as: "It sure is boring sitting here. Monty's sitting next to me. It would be fun to reach over and poke him on the arm. But that's against one of the rules. One of the rules is 'Don't touch anyone,' so if I touch Monty, I'll be breaking the rule. I better not touch him. I guess I'll draw a picture instead."

Don't work on too many rules at once with the student who has special needs. Three rules at a time would be the maximum, and you may have students who can work on only one rule at a time. And remember, the rules must be posted where the student with special needs can see them at all times.

Responsibilities

In contrast to rules, Millie defines responsibilities as "Tasks I gotta do." She doesn't define responsibilities as being related to behavior toward others, since that is the definition of a rule, although sometimes the two may overlap. For example, putting the ice cream back in the freezer is a task, but it could also be a rule, because the next person will be affected if we leave it on the counter. She has never had a student with special needs who had problems understanding that occasionally a rule and a responsibility overlap.

Routines

Routines and responsibilities are intimately entwined. Routines are the little steps required in carrying out a responsibility. The routine for doing a

responsibility in Ms. Grant's social studies class may be quite different from doing the same responsibility in Mr. Noland's math class. For example, getting ready for class in Ms. Grant's class may require sitting down with a pencil and journal and writing an entry about the morning news. Getting ready for class in Mr. Noland's class may involve sharpening your pencil, getting out your book, paper, and calculator, and visiting quietly with your neighbor until Mr. Noland steps up to the chalkboard.

Like rules, responsibilities should be posted, discussed, modeled, and practiced. Millie posted the responsibility that she was teaching along with those that had been taught, then had her students practice it; for example, coming into the building, entering the classroom, going to the lockers, getting out a pencil and paper, going to the desk, sitting down, and beginning to work on the sponge activity on the board. Her students practiced and practiced it until everyone had it down pat. When all the students except a couple had mastered the responsibility, then only those who still needed help continued practicing for as many days as necessary.

Rituals

Rituals are to social behaviors what routines are to tasks. A big problem for teachers of students who have behavior problems is those students' rituals, or lack thereof. Here are some examples of social behaviors for which students may need to be taught rituals: taking correction from a teacher, apologizing, asking a question, making a request, using appropriate eye contact, disagreeing, and so on. Don't assume that because a student does something that you consider rude, he intended to be rude.

Millie's best example had to do with a student in first grade who passed gas in class in a loud and distracting manner. She thought he was being rude. What she discovered was that he didn't know it was considered rude behavior. In his family, passing gas was considered amusing and was met with comments of manly approval. He simply did not know that a different ritual existed for passing gas in school. He wanted to please his teacher, and responded more quickly than she had expected to learning the new ritual.

Even very young children can understand that certain rituals are required at school that are different from those required at home. If the child is having trouble understanding this concept, ask him, "Do you play football?" "Do you play basketball?" "Are the rules the same?" "Well, the 'rules' for burping at school are different from your mom's 'rules' for burping at home. You have to play by the school 'rules' when you are at school."

Teaching rituals is exactly like teaching routines: Write them down, talk about them, model them, and have the child practice them.

Systematically Teaching Social Skills

When you need to teach a social skill, you will generally detect the child's deficiency based upon her behavior in your classroom; however, if you become impatient with just putting out fires, you need a good resource for a social skills curriculum. We suggest Arnold P. Goldstein's *Prepare Curriculum,* a hierarchy of 60 skills that every student should master. This curriculum would be ideal for guiding your school to implement a comprehensive social skills training program from kindergarten through junior high. You will find it listed in the recommended reading list at the end of this chapter.

The information that we will now discuss applies to teaching both routines and rituals. It is imperative that you follow this strategy; the research on teaching social skills has dismal results. We think that has occurred because of the ways in which social skills instruction has been approached. Because the special education research says that explicitness and repetition are the keys to teaching learners with special needs, and because direct instruction and cognitive behavior modification, two strategies that use explicitness and repetition, have research support, we suggest that teaching social skills using those strategies should be effective.

As we said earlier, it is essential to write out the steps to the ritual/routine, talk about them, model them, and then have the student practice, practice, practice.

Here are the steps in detail:

1. Write down the steps and post them.
2. Talk to the student about the need for learning this new way of doing things.
3. Get the student to commit to learning the strategy.
4. Model the strategy, talking yourself through each step aloud as you go.
5. Have the student practice the strategy while you talk him though each step. Do this until he has demonstrated mastery at least 10 times.
6. Have the student practice the strategy as he talks himself through each step aloud. Do this until he has demonstrated mastery at least 10 times.
7. Have the student practice the strategy as he whispers himself through each step. Do this until he has demonstrated mastery at least 10 times.
8. Have the student practice the strategy as he subvocalizes himself through each step. Continue until the strategy becomes automatic.

In the case of both routines and rituals, the practices will probably have to be conducted through role play. For example, the student role-plays walking into the classroom and getting ready for class or role-plays making a request of the teacher. When the practice is of a ritual (social skill), then after the student becomes proficient at the ritual within your classroom (*analog practice* is the term for practice in such a lab situation), then you will want him to practice the skill in a real-life situation outside the classroom (*in vivo practice* is the term for real-life practice).

Be sure to reinforce the student each time his practice is correct. When he uses the skill in real-life situations, he should receive natural reinforcement from the person with whom he uses the skill. By natural reinforcement, we mean that if his skill is making a request, compliance with the request is his natural reinforcement.

The system used for systematic social skill instruction is very nearly the same as that used for academic strategy instruction, which we'll cover in the next chapter in the section titled "Cognitive Behavior Modification."

Ensuring Transfer of Social Skills

You can't assume transfer of social skills from your lab situation to the real world. That's why in vivo practice is so important. However, you can't even assume that a social skill learned correctly in one in vivo situation will transfer to another in vivo situation. That's because the student does not learn inductively, so he can't infer, "Because I learned that I should say please and thank you when making a request of the music teacher, I should use that same skill when making a request of the recess aide, the ladies in the cafeteria, and the principal."

The way we deal with this transference problem is by MEMC, or using *multiple exemplars in multiple contexts.* It takes a long time to teach a student with disabilities to transfer a skill (*transferring* is to a *skill* what *generalization* is to a *concept*), because he must practice the skill in all of the contexts in which you can imagine he will need to use it, but it's worth the investment in terms of an organizational strategy; it will save you time and effort in the long run.

Summary

In this section, we have discussed organizing the behavioral environment for the student who has learning problems. Specifically, we have examined the following items. Check off each item about which you are clear. Reread any about which you are uncertain.

- ☐ The keys to working with students who have problems
- ☐ Creating structure for students who have learning problems
- ☐ Systematically teaching social skills
- ☐ Ensuring transfer of skills

ORGANIZING THE LEARNING ENVIRONMENT

In the last section, we discussed organizing the behavioral environment for the student who has learning problems. In this section, we will discuss organizing the learning environment. Specifically, we will discuss:

- ☐ The keys to teaching students who have problems
- ☐ Direct Instruction
- ☐ Ensuring transfer/generalization of a skill/concept
- ☐ Cognitive Behavior Modification
- ☐ First Line modifications

The Keys to Teaching Students Who Have Problems

These keys are exactly the same keys that we discussed in the section on organizing the behavioral environment. In this section, we have used examples keyed to learning content rather than to learning behavior.

Explicitness

Explicitness is the first key to teaching students with disabilities, because they do not learn inductively. Making certain to tell the content/concept explicitly to the student with disabilities is crucial in organizing your teaching for him.

Suppose that the content/concept you want to teach is that family is the most important value in the Hispanic culture. You might have asked your students to read three short stories about Hispanic people, construct a Venn diagram, and then write an essay on the primary value of Hispanic people. Your regular ed kids would flower under this type of assignment. Your student with special needs would flounder because it requires inductive thinking.

In contrast to this lesson for regular students, if you want the student with problems to learn that Hispanic people value family, tell him, "The most important value to Hispanic people is family." Then you might have him read the stories and look for examples illustrating the concept, or you might

show him the examples that illustrate the concept. He won't "get it" if you expect him to discover it for himself inductively.

Repetition

The second key to teaching students who have special needs is repetition. Researchers report that children without disabilities must be exposed to something many times before they learn it. Children who have special needs may need 10 times as many or a hundred times as many repetitions of new content before they learn it. Unless we expose students who have special needs to important information over and over and over and over and over, they aren't going to learn it. If they could learn in a typical number of repetitions, they wouldn't be students who had problems; they would be typical students. Students with learning problems require endless repetition.

Millie was delighted when she learned how to make HyperCard stacks on computer. HyperCard programs provide us with ways to ensure adequate repetition for our special needs students. The computer never gets cranky, tired, or frustrated when it has to give corrective feedback for the five hundredth time. Computer aided instruction is invaluable in working with students who have special needs.

Direct Instruction

There are five research supported techniques for teaching students with learning problems. A concept hierarchy of those techniques is in Figure 9.2.

Direct Instruction is one of the five research-supported ways of teaching students who have problems. We have all learned how to teach Direct Instruction lessons. For the purposes of this discussion, we'll use the definition of Direct Instruction that we like best: *Tell 'em what you're gonna teach 'em, teach 'em, and then tell 'em what you taught 'em.* That wraps it up nicely. There is nothing implicit or inductive in Direct Instruction.

Remember that part of the Direct Instruction model involves guided practice, or repetition of use of the concept/skill until it is mastered. Direct Instruction is explicit and repetitive.

Ensuring Transfer/Generalization of a Skill/Concept

An important part of Direct Instruction (or any instruction) for kids with special needs is ensuring transfer of a skill or generalization of a concept. Kids with special needs don't transfer skills or generalize concepts. When new teacher Lily Wong taught Brett that dogs were mammals because they had a four-chambered heart, were warm-blooded, had live babies, were

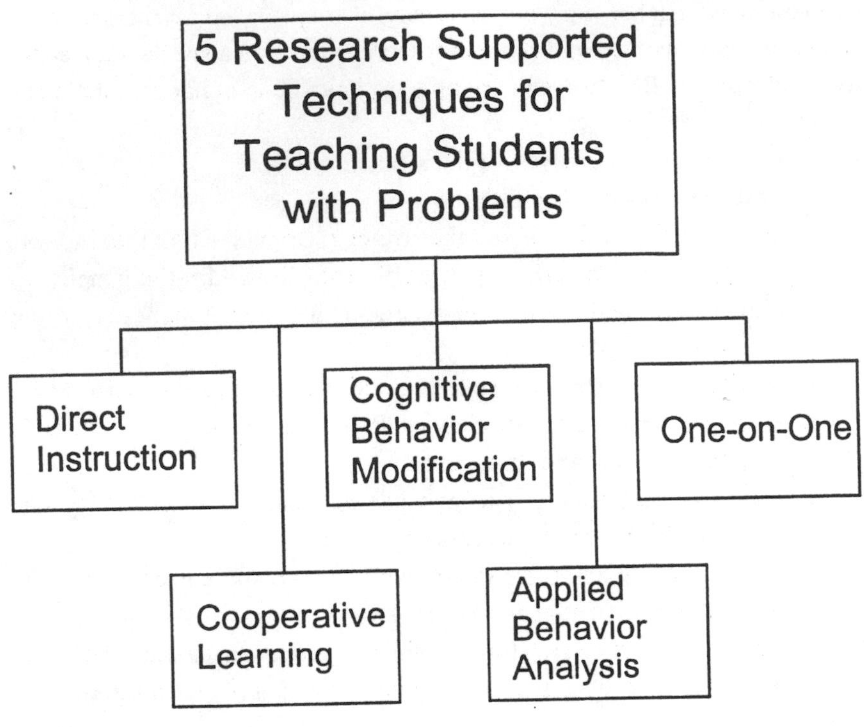

Figure 9.2. Five Research-Supported Techniques for Teaching Students With Problems

fur-bearing, and gave milk, she thought he would generalize that concept to cats. He didn't.

In special education, we use the MEMC Principle when we want students to generalize/transfer information. MEMC stands for *multiple exemplars in multiple contexts.* MEMC marries explicitness with repetition. Using the MEMC Principle, Lily explicitly taught Brett all of the animals that she wanted him to know were mammals.

Similarly, when Lily wanted Brett to learn to use an indention to start a paragraph, she explicitly had to teach the skill to him in each subject and in every situation: multiple exemplars in multiple contexts.

Cognitive Behavior Modification

Cognitive Behavior Modification (CBM) is the special ed teacher's term for what general ed teachers call Strategy Instruction. CBM is the teaching model we use to teach a skill or strategy. The process is the same: Conduct a task analysis and then explicitly teach the task. The definition of CBM that we like is: *I do it, we do it, you do it.*

General ed teachers may not be as careful as special ed teachers to make their instruction explicit and repetitive. Here is the way a special ed teacher would approach CBM. You can see how explicit she is in her instruction.

1. Conduct a task analysis of the strategy.
2. Post the steps of the strategy.
3. Talk with the student about the strategy. Convince him that the way he approaches the strategy is inefficient. Show him the benefits to him of learning the new strategy. Secure his commitment to learning the new strategy.
4. Model the new strategy, stating aloud each step as you perform it.
5. Have him practice the strategy while you read aloud each step as he performs it. Repeat as necessary.
6. Have him practice the strategy while he reads aloud each step as he performs it. Repeat as necessary.
7. Have him practice the strategy while he whispers each step as he performs it. Repeat as necessary.
8. Have him practice the strategy while he subvocalizes each step as he performs it. Repeat as necessary until the skill is internalized.

You will note that CBM is both explicit and repetitive, the keys in organizing your teaching for students who have problems.

It is important to remember MEMC when you use CBM. Because a student has mastered a skill or a strategy in one context does not mean that he will know to use it or be able to use it in another context. For example, if you teach him the strategy of outlining a chapter in his social studies book, that does not mean that he will know to or be able to outline a chapter in his science book. You must use the MEMC Principle and teach him the skills/strategies you want him to use in each context in which you want him to use it.

Modifications

Millie classifies modifications into two families. She calls the first group First Line Modifications. First Line Modifications are those that are easily incorporated into all lessons for all students. She uses that term because she likens them to the first line of defense that is called out in time of war. Millie calls the second family Specific Modifications. Specific Modifications are those that are restricted to the student for whom they were intended. First Line Modifications consist of: The Big Ideas, emphasizing important points, graphic organizers, guides, teaching vocabulary, and mnemonics.

The Big Ideas

The Big Ideas are those huge, overlying ideas that pervade a particular discipline. For example, The Big Ideas in science are: *structures, models, cycles of change, stability, evolution,* and *scale*. Those six concepts pervade all of science education. The Big Ideas in history are: *problem, solution, outcome.* There are two kinds of problems: human rights and economics. There are five kinds of solutions: dominate, accommodate, move, innovate, and tolerate. There are three kinds of outcomes: the problem gets better, the problem gets worse, or a new problem arises. These Big Ideas drive all of history.

If you don't know The Big Ideas in a discipline, you can go to the standards created for that discipline by the flagship organizations in the discipline to find out what they are. In science, the document *Science for All Americans* explicitly states The Big Ideas. The National Council of Teachers of Mathematics and the National Council for the Social Studies have clearly stated their Big Ideas. Other disciplines may have standards that require you to tease out the half dozen or so Big Ideas.

The student with special needs requires explicit instruction in The Big Ideas and constant reference to them throughout his lessons. If all he knew when he left your class was The Big Ideas in each of the disciplines you teach, he would have gained something important.

Emphasizing Important Points

Emphasizing important points is an essential organizational tool in working with students who have difficulties. Students who have problems learning new information cannot discriminate between what information is important and what is not. If you told the story of the Boston Tea Party to your students, the general ed students might figure out that the important thing was that the colonists rebelled against the stamp tax being imposed upon them without their representation in the government that imposed it. The special student would think the brand of tea dumped in the bay was important and might altogether miss the idea of rebellion against taxation without representation.

Explicitly teach your students that you will emphasize important points by one of the following four strategies: (a) your repeating it three times verbatim; (b) your saying, "This is very important. Remember it"; (c) your having them repeat it back to you multiple times as a chorus; and (d) your having them repeat it to their seatmates, the ceiling, the floor, Ms. Pencil, Mr. Desk, and any other available person or object.

Graphic Organizers

Graphic organizers are very powerful tools in helping students with disabilities learn. We have thoroughly discussed graphic organizers in Chapter 7, in the section titled "The Teaching Act," and refer you to that discussion.

Guides

Guides are important tools in helping students who have problems get the most out of reading a chapter, watching a video, or listening to a lecture. Millie tested this notion with her college students. She showed a video and tested her students over it. The class average was 30% (the students knew that the test was not recorded in the grade book). The next week, she showed a video of like difficulty, but this time, she gave students a guide to help them recognize important points. This time, the test average (which they knew was not recorded in the grade book) was 80%. Made a believer out of her.

A textbook guide can be drawn directly from the questions at the end of the chapter. Have the student copy the questions and reread them before she starts reading the chapter so that she can identify the important ideas as she encounters them.

Highlighting

Set aside a copy of each textbook for your students who have disabilities. Go through the texts and highlight the most important material in each chapter. That way the students can spend their time learning what is important rather than trying to read and understand the entire chapter.

Explicitly Teaching Content Vocabulary

Explicitly teaching vocabulary in the content areas is a powerful strategy. General education students may pick up the new vocabulary words when they encounter them, but students with special needs require your explicit and repetitive instruction if they are to learn. We know that teaching vocabulary in the content areas is one of the best ways to improve reading ability and especially comprehension.

Mnemonics

Finally, mnemonic devices are the last of what Millie calls First Line Modifications. She first started teaching her students with mnemonic devices when she was teaching high school special ed. She read in Dear Abby

that Canadians are offended that we know so little about their country when they know so much about ours and we are such close neighbors both geographically and ideationally. The letter to Dear Abby recounted a survey in which almost none of the Americans surveyed could name the provinces and territories in Canada. Millie decided that her students' self-esteem would grow if she could teach them this information, and after they had mastered it, have them survey their parents, neighbors, and other students on how many of the provinces and territories they could name. The project was a wonderful success.

The particular mnemonic device that she used for that lesson was the Keyword strategy. In this strategy, the student is taught the target word and then pairs that with a keyword that sounds like the target word. A strong visual picture is then presented to accompany the keyword, and a story weaves all of the keywords together.

Here are the keywords, pictures (Figure 9.3), and story Millie used to teach the Canadian provinces and territories.

Keywords	*Target Words*
Yucky stuff	Yukon Territory
New vest	Northwest Territory
B. C. tablet	British Columbia
Albert Gore	Alberta
Sash to catch a swan	Saskatchewan
Manischewitz	Manitoba
On a terrace	Ontario
Ice cubes	Quebec
Newfoundland dog	Newfoundland
Prince Edward	Prince Edward Island
New Scottie	Nova Scotia
Brunswick stew	New Brunswick

Here is the story that accompanies the keyword pictures: I got yucky stuff on my new vest! I was so upset, that I had to take a B.C.™ Tablet. Here's how it happened: Albert Gore invited me to his condo for a wine cooler. What started the disaster was that he had to use his sash to catch a swan. He had to catch the swan because it stole the bottle of Manischewitz™ wine that he left on a terrace when he went to get some ice cubes for our glasses. He was running so fast that he tripped over his Newfoundland dog, Prince Edward. This caused him to fall forward over his new Scottie, who was eating a bowl of Brunswick stew. The bowl of stew flipped up into the air and landed all over my new vest, making it yucky and giving me a headache!

Figure 9.3. Keyword Mnemonic Device

3CROD and Graphing

The term 3CROD refers to Close Continuous Contact with Relevant Outcome Data. That simply means to monitor constantly the student's progress by a testing, prescribing, teaching cycle. In special ed, we say, "If you ain't tested it, you ain't taught it." This cycle of testing, prescribing, and teaching is known by a variety of names—Diagnostic Teaching, Precision Teaching, Prescriptive Teaching—but it all boils down to the same thing: Test the child to find out what she doesn't know, teach it to her by Direct

Instruction or CBM, and then test her to see if she learned it. If she didn't, reteach the part she didn't learn.

An integral part of 3CROD is graphing. Graphing a child's progress is an amazingly powerful way to motivate her. When they start graphing their children's performance, interns constantly tell us it improves. "A boy in my class didn't care about anything, but when I started graphing his performance, he came up to me and asked when we could work together again so he could improve his graph." We're not making this up. We hear it all the time! It is hard to understand why all teachers don't use this strong tool for organizing their teaching.

Summary

In this section of the chapter, we have covered ways to modify the teaching environment for students who have learning and behavior problems. Specifically, we have covered the following items. Check off those you feel prepared to use. Reread any which you don't feel competent to try.

- ☐ The keys to teaching students who have problems
- ☐ Direct Instruction
- ☐ Ensuring transfer/generalization of a skill/concept
- ☐ Cognitive Behavior Modification
- ☐ First Line Modifications

DEVELOPING THE SUPPORT SYSTEM COMMUNICATION

In the last section, we discussed ways to organize the learning environment for students who have learning and behavior problems. In this section, we will discuss how to organize the support system for students who have learning and behavior problems. Specifically, we will discuss the use of:

- ☐ Memo books
- ☐ Checklists

Memo Books

Confusion can reign when a student who has learning or behavior problems moves about among special ed classes, general ed classes, and home. Poor communication becomes a problem that allows the student to

manipulate his general ed teachers, special ed teachers, and home. How many of us have faced a student in a pull-out program who has told us, "My other teacher told me to do this," and we haven't been able to confirm or deny his assertion?

The answer to the problem is simple: a memo book. Lily Wong required each of her special students to carry a memo book in his pocket at all times. When her student Julio entered the general ed class, the first thing his teacher did was ask for his memo book. The general education teacher checked all entries since her last entry. At the end of the period, she noted anything that she wanted either the special ed teacher or the parents to know. The special ed teacher did the same thing, and the parents also participated. In this way, everyone knew what was going on everywhere else, and manipulation of adults became much more difficult.

Checklists

When you are organizing for children with special needs, it is essential to organize for communicating with parents about how to help their children. A checklist helps you do this.

Send a checklist home with the child with special needs with each assignment. Don't assume that the parent already knows how you want her to help the child. Do you want her to dictate spelling words to the child? Write that on your checklist. Do you want her to have the child spell the words aloud to her three times? Write that on your checklist. Most parents want to help their kids; they just don't know what you expect. A checklist will ensure that you have communicated to the parent how to help the child.

Summary

In this section, we discussed how to organize the support system for the student who has special needs. If you feel ready to try using these techniques, check them off and move on to the next chapter.

- ☐ Memo books
- ☐ Checklists

RECOMMENDED READING

Goldstein, A. P. (1988). *The Prepare Curriculum: Teaching prosocial competencies.* Champaign, IL: Research Press.

10

WORKING WITH PARENTS

Many teachers are uncomfortable working with parents. Once you organize yourself for working with parents, you will find that much of your anxiety will evaporate. This chapter will prepare you for your parent interactions.

WRITING TO PARENTS

In the last chapter, we discussed organizing for working with the student who has learning and behavior problems. The parents of the student who has learning and behavior problems are often the parents to whom a teacher must communicate unpleasant information. Therefore, organizing for writing to parents is a crucial task. In this section, we will cover:

- ☐ Starting with accolades
- ☐ Making assertions
- ☐ The blueprint paragraph
- ☐ Citing evidence and using reasoning
- ☐ Summarizing

Starting With Accolades

When you need to write to parents about a problem, start with accolades to the child. Starting with favorable comments softens the impact of the unpleasant news that may follow. Starting with accolades is an important part of the organizational formula for writing to parents.

Making Assertions

After the initial accolades, state your assertion. An assertion is your conclusion. Here's an example: Kat taught seventh grade English. When Kat was concerned about Cheryl's failure to bring homework to school, she began the letter to Cheryl's parents with accolades, and then stated, "I need for you to read in Cheryl's memo book every afternoon to find out what homework she has. I then need you to help her with her homework and make certain that she brings it to school the next day."

The Blueprint Paragraph

After making the assertion, write a blueprint paragraph. The blueprint paragraph tells the parents exactly what evidence/reasoning you are going to present. Here is the blueprint paragraph that Kat wrote to Cheryl's parents: "First, I will report the number of times that Cheryl has had homework to do this quarter. Second, I will report the number of times that she has turned in her homework. Third, I will then report the number of times that she has not turned in her homework. Fourth, I will report the percentage of the grade that homework represents. Finally, I will summarize the information."

Note that in the blueprint paragraph, we use the terms first, second, third, and so on. That helps the readers organize their minds to read the information.

Citing Evidence and Using Reasoning

Evidence refers to the facts that you present. The facts must be empirical. Think of evidence as the bricks in a wall. Each brick is a piece of evidence. In contrast to evidence, reasoning refers to the way you organize and treat the evidence in order to come to a conclusion, or, as we said earlier, an assertion. You can think of reasoning as the mortar that holds the bricks together to form the wall. The wall is the assertion you make from the bricks and the mortar.

When you are citing evidence, stick to the facts for which you can provide documented records. Be exact. Don't say, "Cheryl doesn't turn in her homework half the time." State only what you can document with exact figures. In the case of Cheryl, Kat could document evidence that resulted in the following:

"Of the 32 days this quarter, I have assigned homework 25 days. The homework assignments have ranged from answering the questions in the grammar book to writing a story to writing a newspaper article. I did not give two homework assignments due on any one day.

"Cheryl turned in her homework on 7 of the 25 days in which she had homework to do. She did not turn in homework on 18 of the 25 days. I asked her each morning for her homework. For the past week, I have written down her explanations for not having her homework. She responded 'I forgot,' on two occasions; 'We went to my cousin's house,' on one occasion; and 'My mom wouldn't help me and I couldn't do it by myself,' on two occasions. When Cheryl has not done her homework, she is unable to participate in class activities because she is not prepared.

"Of the seven times that Cheryl did turn in homework, the homework was late two times. Of the two times, one homework assignment was turned in 1 day late, and the other was turned in 6 days late. By being late, she was unable to participate in the class activities for which the homework prepared the students. The opportunity for learning the information with the group is gone when Cheryl is as much as 1 day late with her homework.

"The homework assignments to date represent 25% of Cheryl's grade. With 25 of 32 assignments not handed in, and with two of the five assignments receiving only half credit for being late, Cheryl's homework brings her overall grade average down to a 57 out of 100 possible points. She could be passing the class if she were bringing her homework assignments to school completed and on time."

Summarizing

Kat then summarized by writing: "Therefore, as I stated at the beginning of the letter, I need for you to read in Cheryl's memo book every afternoon to find out what homework she has. I then need you to help her with her homework and make certain that she brings it to school the next day."

As you can see, Kat laid out the evidence for the parents clearly and concisely. She left nothing to conjecture. She used reasoning to construct an argument leading to her assertion. She began with the assertion, and ended with a summary paragraph.

Cheryl's parents may not be happy that Kat wrote to them, and they may not help to ensure that their daughter does her homework, but Kat can provide solid evidence to her principal that she tried to work with the parents in a very professional way.

Summary

In this section, we have discussed organizing your writing to communicate more effectively with parents. Specifically, we discussed the following items. Check off each item that you feel confident using, and reread those about which you feel uncertain.

- ☐ Starting with accolades
- ☐ Making assertions
- ☐ The blueprint paragraph
- ☐ Citing evidence and using reasoning
- ☐ Summarizing

THE PARENT CONFERENCE

In the last section, we discussed writing a letter to parents. In this section, we will discuss organizing for the parent conference. Specifically, we will discuss:

- ☐ Using an agenda for the routine conference
- ☐ Praise and problem cards
- ☐ Using an agenda for the problem conference

Using an Agenda for the Routine Conference

Research tells us that most teachers dread parent conferences and working with parents. The teacher's dread is born of a feeling of uncertainty of what is expected of her in a parent conference. A couple of simple organizational strategies will reduce the anxiety associated with the parent conference.

The purpose of the parent conference will dictate what you will need to do to organize for it. We will start by addressing how to organize for a routine parent conference that is required by the school for all parents.

For the routine parent conference, first write up an agenda (Figure 10.1). This is essential, and something that teachers are not used to doing.

Purpose: Required conference for receiving student's report card:

1. To inform parent of student's strengths and areas of concern.
2. To tell parent how to support child's education.
   a. How student is performing academically
   b. How parent can help student's academic performance
   c. How student is performing behaviorally
   d. How parent can help student improve behavior
   e. Scheduling follow-up conference if desired

Figure 10.1. Fall Parent Conference Agenda

The agenda should include the purpose of the meeting and the steps that you will follow in conducting the meeting. You will post the agenda on the table during the meeting and refer to it as you move from step to step through it during the meeting.

If the parents stray away from the agenda, refocus them by pointing to the agenda and saying, "Mr. and Mrs. Lake, I'd be happy to discuss this issue with you later, but in order to stay on schedule today, we have to stay with the agenda. If you'd like to discuss this issue further, we'll make an appointment before you leave this afternoon."

Second, write up an elaborated agenda (Figure 10.2). Send a copy of the elaborated agenda home to the parents so that they will know what to expect. This will also allow them to prepare themselves with information that you might require.

On the table where you conduct your conference (and do sit at a table with the parent, not behind your desk) display sample copies of excellent, average, and poor work in a variety of subject areas (Figure 10.3). Of course, the papers must have the names removed. You may wish to use papers from a previous year or from another teacher's class so that no parent can identify their child's (or another child's) work.

For each student, have a folder of work samples ready. You should have marked these papers with a colored pen so that the strengths and problem areas are immediately observable. Point out and discuss the strengths first. Try to find at least three positive things to share. Then point out one or two things that the child most needs to improve.

Finally, give the parent a list of the child's strengths and problem areas in the academic arena and a list of three things that she can do to help her child improve academic performance (Figure 10.4).

Dear Parent,

As you know, it's time for our fall parent conference! You must come to this conference to get your child's report card. The conferences are limited to fifteen minutes. If we need more time to talk, we'll need to set up another meeting for next week.

Here's the agenda for this conference:

1. First, we'll talk about your child's academic performance. I'll show you samples of excellent, average, and poor work from other third graders. Then I'll show you a sample of your child's work so we can compare your child's progress with that of other students.
2. Next, I'll tell you from one to three specific ways you can help your child increase academic performance.
3. Next, we'll talk about your child's behavior. I will tell you your child's strengths and problem areas.
4. Then I'll tell you from one to three specific ways you can help your child improve in the area of behavior.
5. Finally, I'll ask you if you want to schedule a follow-up meeting to discuss any concerns you may have.

Thanks for helping me keep our parent conferences on schedule!

Sincerely,

Figure 10.2. Fall Parent Conference

Praise and Problem Cards

Next, discuss the child's behavior. In advance, prepare index cards demonstrating positive aspects of the child's behavior and cards demonstrating areas that are of concern (Figure 10.5). Make the positive cards one color, and the concern area cards another color. Present the positive cards one at a time, discussing each. Then present the concern area cards one at a time, discussing each. Finally, review the positive cards again, gather up all the cards, and give them to the parents (de Bono's hats would suggest that you use yellow for praise cards and gray for concern cards).

Suggest that the parent display the cards on the refrigerator so both the parents and the child are constantly reminded of the child's strengths and areas targeted for growth.

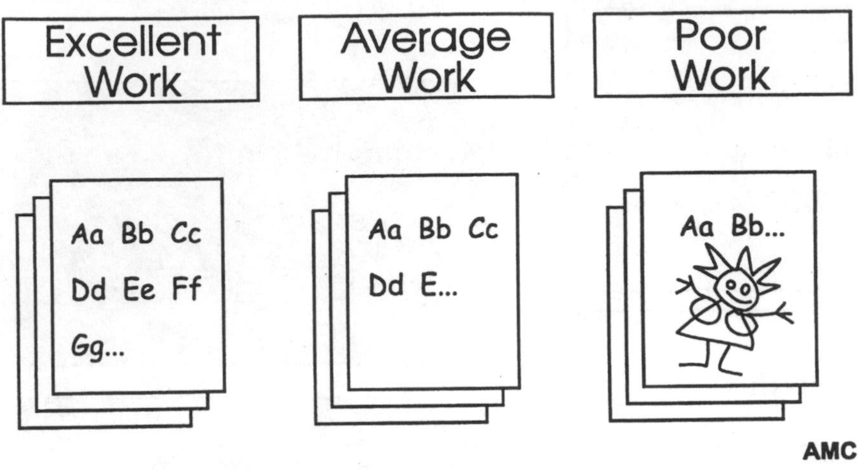

Figure 10.3. Samples of Student Achievement

Tina's Academic Strengths

1. Tina has many good ideas in social studies.
2. Tina is a strong speller.
3. Tina can multiply and divide accurately.

Tina's Problem Areas

1. Tina has trouble understanding word problems in math.

How Parents can help Tina improve her understanding of word problems in math:

1. Each night, ask to see the math that Tina did at school that day. Go over every word problem with her, both those she missed and those she did correctly. Make certain that she understands why she missed the ones she missed, and how to do them correctly.
2. When it's time for Tina to do her math homework, sit down at the table with her. Have Tina tell you how she is working each problem and why she's doing the things she's doing to work it.
3. Have Tina draw pictures to represent word problems. If the pictures show that she doesn't understand how to do the problem, show her how to do it correctly.

Figure 10.4. Fall Parent Conference: Academic Strengths, Problems, Recommendations

Strength	Strength Area for Growth
turns in all work on time	*needs to improve organizational skills*
Strength *is eager to help others*	**Strength Area for Growth** *needs to be more patient*
Strength *demonstrates strong leadership skills*	**Strength Area for Growth** *needs to accept correction more positively*
Strength *demonstrates strong creativity*	
Strength *shares with others*	

Figure 10.5. Praise and Problem Cards

Problem:	Tomas doesn't turn in homework
Conference purpose:	Increase frequency with which Tomas turns in homework

Teacher establishes rapport by serving coffee and discussing Tomas's strengths.

1. Teacher defines problem.
2. Parent provides information about problem.
3. Parent and teacher generate possible solutions.
4. Parent and teacher select solution and write it down in contract form.
5. Parent, teacher, and Tomas commit to implementing solution and keeping data. Parent, teacher, and Tomas sign contract.
6. Parent and teacher schedule a follow-up meeting within 2 weeks.

Figure 10.6. Parent Problem Conference Agenda

Using an Agenda for the Problem Conference

If the conference is not a routine one, but is necessary to deal with a problem, you'll need to follow a different format for the agenda. The agenda still needs to have a goal and the steps you intend to follow at the meeting; however, the steps are different for a problem conference than for a routine one (Figure 10.6).

The first step in the problem agenda is stating the problem and explaining why it is a problem. For example, John told a parent, "The problem that Tomas is having is that he is not doing his homework in math. Students need lots of practice with math problems before they master how to do them. The students who are doing their math homework are mastering the process. Because he is not doing his math homework, Tomas is falling farther and farther behind."

The second step in the problem agenda is asking the parents if they can provide any information about the problem. Sometimes parents can shed a whole new light on what is happening with the child.

The third step is exploring possible solutions to the problem. You propose some possible solutions, and ask the parents to propose some.

The fourth step is selecting one of the solution alternatives and writing it down in a contract form. The fifth step is for the teacher, the parents, and the student to make a commitment to implementing the solution by signing the contract.

Dear Mrs. Fuentes,

Thank you for agreeing to meet with me on Tuesday at 3:00 to discuss a problem Tomas is having with his schoolwork. Here is the agenda I have prepared for the conference. If you would like to add anything to the agenda, please let me know.

Agenda for Conference

1. I will explain the problem of Tomas not doing his homework.
2. You will give me information about the problem. (Please talk to Tomas and be thinking what might be causing him to fail to do his homework.)
3. Together, we'll think up possible solutions to the problem. (Please be thinking about what you or I might do to help Tomas with this problem.)
4. Together, we'll select one solution.
5. Together, we'll commit ourselves to implementing the solution.
6. We'll schedule a follow-up meeting in two weeks.

Thanks for agreeing to meet with me! I'll see you Tuesday.

Sincerely,

Figure 10.7. Elaborated Problem Conference Agenda Sent to Parents

The last step is to agree on a follow-up contact within 2 weeks. The follow-up contact agreement is essential. It helps keep everyone honest. As teachers, it helps keep us focused on the behavior that we are trying to change. For the parent, it ensures vigilance in working on the problem at home. For the student, it motivates the child to improve, because Mom and Teacher are going to be checking with each other to see if the target behavior has improved.

After you have drawn up the agenda, you should send home an elaborated agenda (Figure 10.7) in the same manner that you did for the routine conference.

Summary

In this section, we have discussed organizing for the parent conference. Specifically, we have presented information related to the following items.

Check off each item about which you feel confident. If you are unsure about any item, reread it.

- ☐ Using an agenda for the routine conference
- ☐ Praise and problem cards
- ☐ Using an agenda for the problem conference

PARENT INVOLVEMENT

In the last section, we discussed organizing for the parent conference. In this section, we will discuss how to organize for involving parents in their children's education. Specifically, we will discuss the following items:

- ☐ Checklists for working with the child at home
- ☐ Creating the master list of parent volunteer tasks
- ☐ Having parents choose tasks
- ☐ Ensuring quality control
- ☐ The Volunteer Notebook

Checklists for Working With the Child at Home

Research tells us that many parents are afraid to work with their children on homework because they don't know what is expected of them. As we noted in Chapter 9, sending home a checklist with each assignment assures the parents that they are doing exactly the right things with their child. What an easy solution to a problem!

You can either have the students hand write the checklist or you can wordprocess it on the homework assignment if it's in worksheet form.

You can also send home a checklist for certain subjects at the beginning of the year. For example, in spelling, you may want the parents to follow the same procedure each week. In that case, one checklist sent home and placed on the refrigerator the first week of school should do the trick.

Creating the Master List of Parent Volunteer Tasks

When you have parents who are willing to be involved with their children's education in ways other than helping with homework, dance a jig in celebration. Unfortunately, many teachers eschew having parents who want to help with the class. The reason that such teachers ignore this gold

We need parents who will volunteer to do the following:

- ☐ Listen to individual students read
- ☐ Listen to small groups of students read
- ☐ Read to individual students
- ☐ Read to small groups of students
- ☐ Dictate sentences for individual students to write
- ☐ Dictate sentences for small groups of students to write
- ☐ Take dictation from individual students
- ☐ Give students practice tests
- ☐ Drill individual students on flash cards
- ☐ Drill small groups of students on flash cards
- ☐ Help individual students select library books
- ☐ Help individual students with art projects
- ☐ Record books for students with learning problems
- ☐ Proofread and edit students' reports

Figure 10.8. Parent Volunteer Wish List

mine resource is the anxiety of not knowing what to have the parents do to help.

The first step in organizing for parent involvement is to create a master list of all of the tasks for which you could possibly use an extra pair of hands. The sky's the limit. Any wild idea you have can be included on the list. You may not have any parent volunteer to do some of the tasks, but you never know until you try! Millie once had a parent make 32 decorative pillows for the antique bathtub in her classroom. She could have lived without them, but she put them on her Parent Volunteer Wish List, and a parent decided to volunteer to make them. Figure 10.8 shows a master list that John used in his classroom.

Having Parents Choose Tasks

After you have constructed your master list of volunteer tasks, have the interested parents identify all of the tasks that they would like to do. Let them examine the book alone at first. Tell them to note any tasks about which they are not certain so that you can tell them more about what's involved with that particular task. Don't rush them.

Ensuring Quality Control

After the parents select their tasks, draw up a rubric for each one so that you can be sure that they will know how to do the task exactly as you want it done. Don't use any teacher jargon on the rubric. Write it so that someone with absolutely no knowledge of the teaching profession will understand it. If you write it in teacherese, you'll scare off tentative volunteers, who think, "I'm not smart enough to be a volunteer."

After you've developed the rubric, take the volunteers and the rubric and explicitly demonstrate to the volunteers exactly what you mean by each instruction on the rubric. Don't assume that the volunteers know how to operate the photocopying machine at all, much less that they know how to copy two one-sided documents to make them into one two-sided document. Don't assume that the volunteers know what you want when they are supposed to "work with Marleena on her sight words." Yes, it takes time to train volunteers, but it's time well invested. If you spend one hour teaching a volunteer how to do something that she can do for you one hour every week for 30 weeks, you have a 30:1 return on your time.

Note that you can also teach volunteers a task that they can then teach other volunteers in an "each one teach one" routine. If you teach one volunteer who teaches one other volunteer, and each performs the task for you one hour a week for 30 weeks, you have a 60:1 return on your time! Our stockbroker should do so well!

The Volunteer Notebook

Develop a Volunteer Notebook using our old friend the three-ring notebook. Your Volunteer Notebook should contain the master list of volunteer tasks. That way you don't have to go hunting it every time a volunteer inquires about working for you.

As you develop each rubric, place it in your Volunteer Notebook. You only have to develop a rubric as you need it. You will find that most of your volunteers will volunteer for the same half-dozen tasks, so that the first half-dozen rubrics you develop will serve you over and over again.

Your notebook should also serve as storage for your volunteers' name tags. We require all of our interns to wear name tags in their field placements. By making an official name tag for each parent and letting other teachers know that you have volunteers who will be wearing the name tags, you can

avoid an embarrassing situation when a colleague challenges a volunteer who is a strange face in the building.

In addition, having their own name tags gives volunteers an additional feeling of belonging to the school and your class.

We recommend that you also have parents log in the notebook the time they spend volunteering so that you can recognize their efforts at the end of the year.

In this section, we have discussed how to organize for using parent volunteers in the classroom. Specifically, we have discussed the following items. Check off any about which you now feel confident; reread any about which you remain uncertain.

- ☐ Checklists for working with the child at home
- ☐ Creating the master list of parent volunteer tasks
- ☐ Having parents choose tasks
- ☐ Ensuring quality control
- ☐ The Volunteer Notebook

11

REQUESTING ASSISTANCE

Hillary Rodham Clinton borrowed a wonderful old African proverb. She has made it a battle cry: It takes a village to raise a child. No teacher, no matter how organized, can possibly serve the many needs of students without plenty of assistance from others. By asking for assistance in an organized way, you will be more likely to get the help that you need.

Who receives the best service from the various offices in a school? Generally, it's the teacher who is organized and asks early for specific assistance. Teachers who appear scatterbrained will be at the bottom of everyone's priorities; when you act like a professional, other professionals will treat you like one, and being organized is part of acting like a professional.

DECIDING WHAT KIND OF HELP YOU NEED

- ☐ Things you need to ask for from the principal
- ☐ Things you need to ask for from the counselor
- ☐ Things you need to ask for from other teachers

Things You Need to Ask for From the Principal

Some schools keep a database of student information. The database can include everything from addresses to medical information to standardized test scores. In some schools, however, data such as individual students'

longitudinal standardized test scores are not routinely maintained. Organized teachers to the rescue! Many principals admit to being pushed by organized teachers to provide more and better services. But they admit this with smiles on their faces, because the teachers who are continually assertive for assistance for the harder-to-teach students are typically the best teachers in the building. So if your school doesn't keep such a database, begin lobbying your principal to do so!

In the past, a teacher had to take each student's folder and study each piece of information. That's a big task for an elementary teacher in a self-contained classroom with 24 children; it's a monumental task, an undoable task, for a secondary teacher with over 100 students. In fact, in this technological age, having to invest such time in examining student records is undefensible. By using a computer database, a school can record each student's records so that you can be given a printout of requested information on every student in your class. For example, you could ask for all of your students' standardized achievement test scores for the past 3 years, and have all that information on one piece of paper. You could even ask that that information be sorted from highest to lowest achievement score. Having that information before the first week of school would be of enormous help in setting up your heterogeneous cooperative groups in which you want a high-achieving student, a low-achieving student, and two average-achieving students in each group.

You could ask for a printout of your five students who scored the lowest on their achievement tests in the previous year, and have that information in one minute, rather than in the hours and hours it would take you to ferret out that information for yourself by studying every student's record. Having that information in advance of the school year would allow you to begin marshaling your resources and preparing for early intervention with those students.

Things You Need to Ask for From the Counselor

Ask your counselor to provide you with a list of your community's referral sources. For example, if you have a child whose parents are going through a divorce and the parents tell you that they know their child is having problems because of it, a list of referral sources would allow you to give them the name of an agency and a phone number that could help them by providing counseling or a support group for their child.

If you have a student who needs clothes but can't afford them, you could give the parents the name of a resource that could help. If you have parents who want to learn to speak English, you could provide them with the name

Dear Colleagues:

Here are some materials I have that you might want to borrow. I'll be glad to loan them to you to preview and/or use!

Lily Wong

- ☐ 8 U.S. geography game boards and materials
- ☐ 3 Cooperative Learning books on social studies
- ☐ 6 generic game boards and generic pieces for any subject

Figure 11.1. Checklist of Teaching Resources for Colleagues

of a community resource. Your counselor knows the community resources, but may not think to distribute a list of those resources to you unless you request them.

In addition, ask your counselor to give you a list of what types of counseling are provided at school and what services are available. Does the counselor offer group counseling to kids who are experiencing divorce? Individual counseling? What about social skills training for shy children? Mediation training to teach kids how to solve playground conflicts? Training to implement a no tolerance program toward bullying? Assistance in working out a problem between you and a parent?

Suggest to the counselor that other teachers would appreciate having such a list, too.

Things You Need to Ask for From Other Teachers

If you are a new teacher, ask your grade level colleagues to supply you with a list of field trips and guest speakers that are commonly employed for your grade. Ask your colleagues to share with you outstanding materials that they may have, and be sure to share yours with them, too, by providing them with a list of things you have in which you think they might be interested. A sample of a list you might provide is in Figure 11.1.

Your special education teacher and 504 coordinator will supply you with a list of IEP modifications that you must make for each child; however, ask them to make a one-page master list of the modifications required for all of the students in your class requiring modifications. It's a dynamite tool that will allow you to tell at a glance how many copies of modified assignments to photocopy and what needs to be done for whom. Figure 11.2 shows a sample of one.

Modification	*Jasmine*	*Eric*	*Samuel*	*Cheryl*	*Brad*
Teacher Instruction					
preteach vocabulary	✓	✓	✓	✓	✓
teach Big Ideas	✓	✓	✓	✓	✓
graphic organizers	✓	✓	✓	✓	✓
lecture guides	✓	✓	✓	✓	✓
emphasize important points	✓	✓	✓	✓	✓
mnemonics	✓	✓	✓	✓	✓
take to special class					✓
Assignments					
reading guides	✓		✓		
reduced		✓			
extra time	✓		✓		✓
alternative				✓	
oral				✓	
peer helper	✓		✓		
teacher's aide				✓	
take to special class				✓	✓
Textbooks					
highlighted	✓	✓	✓		✓
read by peer helper					
read by teacher's aide				✓	
tape recorded					
alternate					
Tests					
multiple choice only	✓	✓	✓	✓	
reduce 4 choices to 2	✓	✓	✓		
reduce # problems	✓		✓		
simplify wording				✓	
test only Big Ideas				✓	
Graph improvement	✓	✓	✓	✓	✓

Figure 11.2. IEP Modifications Checklist

Summary

In this section we identified some things you need to request from other professionals. Specifically, we suggested:

- ☐ Things you need to ask for from the principal
- ☐ Things you need to ask for from the counselor
- ☐ Things you need to ask for from other teachers

ORGANIZING TO GET THE HELP YOU NEED

In the last section, we discussed the types of assistance you need from other staff in order to help you be a more organized teacher. In this section, we will present two ways to see that you get the help you need. Specifically, we will discuss:

- ☐ Checklists
- ☐ Memos

Checklists

Checklists are a powerful tool to help you get the assistance you need from others. By giving a requestee a checklist, you accomplish several things. First, you make absolutely certain that the person knows what you need done. Second, you provide a concrete reminder to them that remains after you have walked out of the office. In other words, you may be out of sight, but your request is not out of mind. Third, you show that your request is serious and that you have given much consideration to it. Fourth, you communicate to others that you are a well-organized professional.

Give the person a copy of the checklist and let them see that you have a copy, too. This is important, and you'll see why in a minute. Then ask, "When may I expect to receive this material?" When they give you a date, jot it down on your copy of the checklist so they see you do it. This will carry the powerful message that you take their word seriously, and it will set the stage for what we will tell you about in the next section.

Figure 11.3 gives examples of checklists you might give your principal and counselor.

Memos

Several days before the date that the principal or counselor told you that you could expect the information you requested, follow up with a memo. Using e-mail if your school has it, or a note pad if you don't have e-mail, send a memo reminding them that you are looking forward to receiving the information. Don't be pushy or arrogant; be warm and friendly, but insis-

Aug. 13

Dear Ms. Griego,

Help! In order to be the most effective teacher I can be, I need the following info from your office in database form. Please help me secure it ASAP so I can do a bang-up job for our kids!

Thanks!

Lily Wong

- ☐ Students' standardized test data by subject for past three years, sorted by highest reading score to lowest
- ☐ Parents' names
- ☐ Addresses & telephone numbers
- ☐ Special class placements

Aug. 13

Dear Ms. Weiderstein,

Help! In order to be the most effective teacher I can be, I need the following info from your office in database form. Please help me secure it ASAP so I can do a bang-up job for our kids!

Thanks!

Lily Wong

- ☐ List of free/inexpensive community resources
- ☐ List of kinds of counseling you/others provide at our school
- ☐ List of services you provide for teachers
- ☐ Information on my students that you think I should know

Figure 11.3. Checklists for Requests for Assistance

tent. By writing something like, "I'll be a more organized and effective teacher, and the kids will ultimately benefit from your helping me get this information!" your message will come across without sounding like you are hostile.

The day before, drop by the office and casually mention that you'll see them tomorrow when you come to pick up the information they've prepared for you.

Summary

In this section, we discussed ways to make certain that your requests for assistance in becoming a more organized teacher are honored. Specifically, we have discussed:

- ☐ Checklists
- ☐ Memos

12

ORGANIZATIONAL STRATEGIES FOR YOUR SOUL

In the preceding chapters, we offered you (a) very simple concrete strategies of organizing your time, money, and objects; and (b) more complex strategies of organizing your decision making, planning, and teaching. In this final chapter, we want you to take a quantum leap from those types of strategies and begin thinking of what might be called meta-organizational strategies, or organizational strategies by which to organize organizational strategies. Steven Covey calls these topics third and fourth generational strategies. As our chapter title reflects, we think of them as organizational strategies for your soul.

We have chosen to break from the pattern of the previous chapters and not use a checklist, because the soul doesn't seem to us very amenable to breaking down into checklists. Therefore, we will take a more holistic approach to talking about the soul.

You begin the organizational strategy for your soul by assessing your value system (a) as a teacher and (b) as a human being. What is your core value as a teacher? As a human being? If your core teacher value is to become a more caring teacher, then start with that one teaching value. If your core teacher value is to become a more effective teacher, then start with that one teaching value. Whatever your core teaching value is, write it in the center

of a piece of paper and draw a bull's eye around it. Then on a second paper, write your core value as a human being outside of your teaching life. Is your core nonteaching life value being emotionally available for your significant other? Is your core value being trustworthy? Nurturing to your children? Whatever that one value is, write it in the middle of a piece of paper and draw a bull's eye around it (of course, you may have the same value in both your professional and out-of-school domains, and that's fine).

Once you've decided on your core teacher value and your core value in your life outside of school, draw three rings around the bull's eyes. In the first ring, write down at least five things that you can do TODAY to enact your core values. Keep them simple and doable. If your core value is to be a more caring teacher, you might write down such things as "hold a child's face in my hands, look intently into her eyes, and tell her that she means the world to me," "write a note to a child who is having a problem, and encourage him to persevere," and "ask a child whom I have embarrassed by my sharpness to forgive me." Then do those things today. In the second circle around the bull's eye, write three things that can't be accomplished today to implement that value, but that can be accomplished within one week. For example you might write, "take cookies to the home of a child whose family is in distress," "make a home visit to the homes of the three children who are at the bottom of the class in popularity," and "hand-write and mail postcards to each of my students telling them one thing about them I really appreciate." Then start implementing those activities.

In the third circle, write three things that you can't accomplish in one week, but you can accomplish in one month. For example, you might write, "teach my kids conflict resolution skills," "develop and implement a strategy to get students to embrace the three least popular children in the class," and "make and bring a lunch to share each day with one individual student." A drawing of what we have in mind is in Figure 12.1.

Third-circle tasks sound like a lot of work. They are. Third-circle tasks should require a high degree of commitment. So where are you going to find the time? Kayla Wade, a second grade teacher, said, "You make time to do the important things. Period. If it's important to you, you do it. Everyone has the same number of hours in the day. Make them count."

Steven Covey explains that there are two dimensions to how we spend our time: urgency and importance. Tasks can be urgent and important, urgent but not important, not urgent but important, and not urgent and not important. Stop and create a diagram that will show these four combinations of the two dimensions. It will help you understand the rest of this chapter.

An example of urgent and important is taking care of a child who is bleeding profusely. An example of urgent but not important is getting

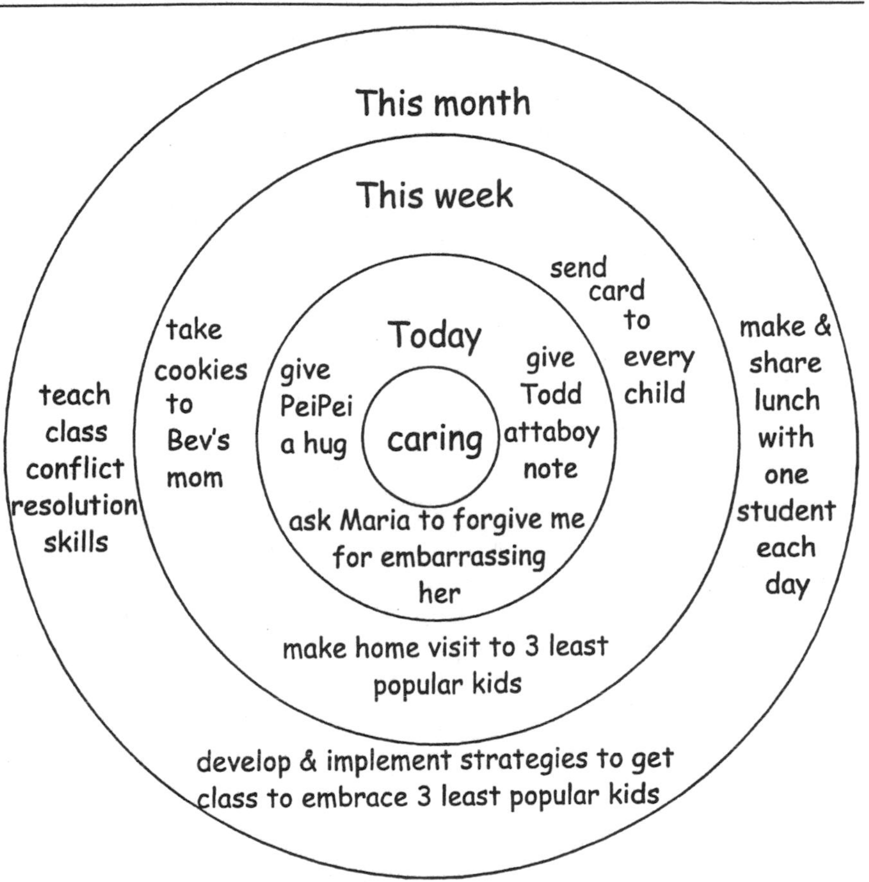

Figure 12.1. Target of Organizing for Your Soul

attendance taken when the attendance aide is standing at the door waiting for it. An example of not urgent and not important is flipping through old stacks of educational supply catalogs. An example of not urgent but important is baking cookies and taking them to the home of a student whose family is in distress.

Covey explains that we spend far too much of our time dealing with urgent matters, be they important or not important. If we're disorganized, we find ourselves overwhelmed by all the things we must do immediately (urgent, whether important, like stopping a hemorrhage, or not important, like taking attendance), so we don't have time to do the *not* urgent, but important things.

Covey explains that the first step in becoming a person who focuses on important things is to assess our values. Then we must commit ourselves to spending the lion's share of our time doing those things that are important

and not urgent. If we're taking care of the things that are important but not urgent, then far fewer things will become urgent. In other words, if you spend time doing the important but not urgent task of creating a caring community in your classroom, then you'll have to spend far less time doing the important and urgent task of stopping one of your students from beating another one of your students into a bloody pulp on the playground. We've created pie charts (Figure 12.2) to help you visualize moving from where you are to where you want to be in the urgency and importance typology.

A crucial move that will help you make time to do important but not urgent tasks is to take Don Aslett's advice. In *Clutter's Last Stand,* Aslett explains that negative people clutter up our lives and bad habits clutter up our lives. One bad habit that is a life clutterer is gossip. Another is smoking. Aslett says that even junk food is a bad habit that clutters up our lives.

If we're going to be authentic Time Stealer Tamers, we must tame ourselves, our very souls. If you smoke, stop. If you eat junk food, give it up. And go cold turkey on the gossip. Don't let those things waste your valuable time.

As for negative people, are there any wasting the precious minutes of your life? Junk them. Sounds outrageous? It's necessary. Divest your life of negative people.

Some school faculties have a negative teacher, or two, or three. These folks seem to seek out each other. A synergy takes place that makes the group more negative than any one of the individuals is alone. This kind of negative behavior is contagious! It catches! Don't risk cluttering up your life with a Time Stealer like negativism.

Once you eliminate the negative people from your life and your own negative habits from your life, you can begin to create a school environment that will nourish your soul. Research tells us that the best way to prevent burnout is to create a supporting, nurturing network of fellow teachers. A very simple way to start creating a nurturing network is by acts of kindness. Kindness to colleagues is a virtue that is lacking in many schools.

Dr. Rosemary Grant conducted research on acts of kindness that teachers performed for each other. As she interviewed teachers, they said, "Something kind that I've observed another teacher do? Forget it! I can't think of any! But boy can I tell you of mean acts that I've seen teachers do to each other!" The major finding of Dr. Grant's work was that about the only time teachers do something kind for a colleague is in time of tragedy: a life-threatening illness of a daughter or son, the death of a spouse, and so on. Those are important and urgent acts. We think teachers should spend more time doing important but not urgent acts, that is, doing kindnesses for each other every day, for no special reason.

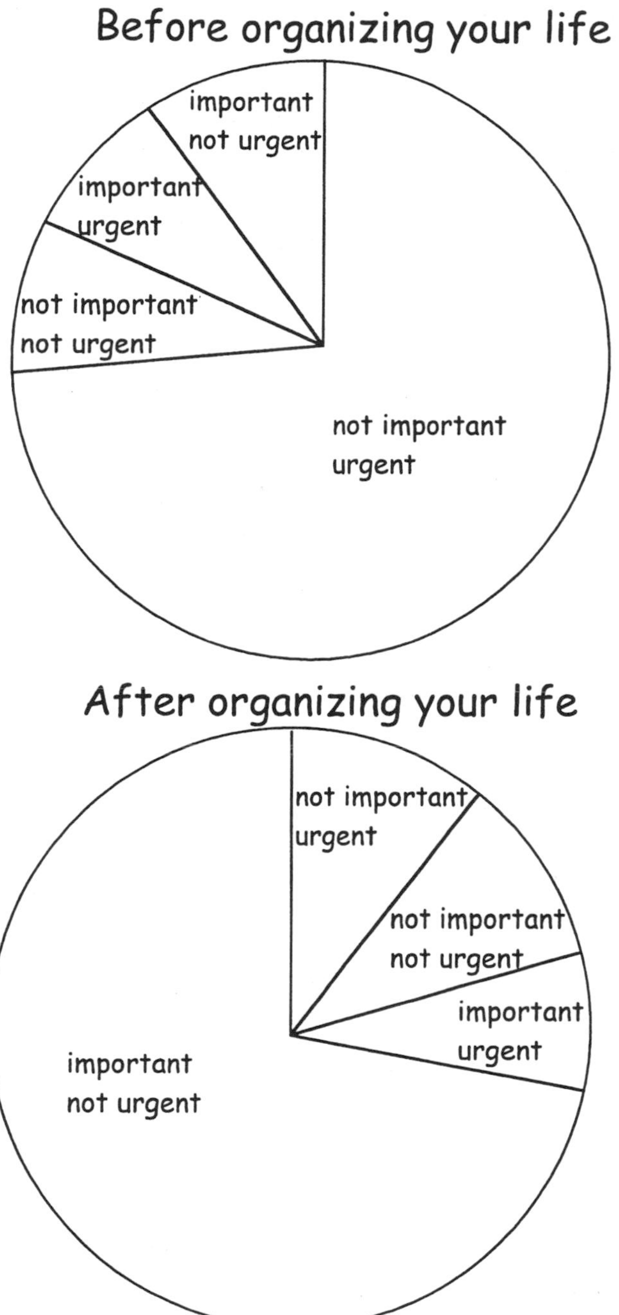

Figure 12.2. Pie Charts of Time Spent Now and After Organizing for Your Soul

Today, in order to start creating your more nurturing network, brew a cup of tea before school and take it to a teacher who is working down in her room. Drop a note in the box of a colleague telling him something that you appreciate about him. Make a copy of a great resource you've found and take it to a teacher who would like it. Move from doing urgent and important things to doing more not urgent but important things.

As you begin your new life as a Time Stealer Tamer, taming your time and memory, your materials, your decision making, teaching, and your very soul, know that we support your efforts, and we ask you to support our efforts at those things. One way you can do that is to send us your ideas and experiences as a Time Stealer Tamer. Please make a point to do that. Our e-mail addresses are in the Preface to the book.

Even more crucial than sending us your ideas and experiences, know that we support your efforts at becoming a Time Stealer Tamer in our hearts, and we ask you to support our efforts in your hearts. Because only in that way, only together, can we accomplish the not urgent but truly important things in the world.

REFERENCES

Aslett, D. (1984). *Clutter's last stand: It's time to dejunk your life!* Cincinnati, OH: Writer's Digest Books.

Aslett, D. (1994). *The office clutter cure: How to get out from under it all!* Pocatello, ID: Marsh Creek.

Covey, S. R. (1989). *The 7 habits of highly effective people.* New York: Simon & Schuster.

de Bono, E. (1985). *Six thinking hats.* Boston: Little, Brown.

de Bono, E. (1991). *Six thinking hats for schools: K-2 resource book.* Logan, IA: Perfection Learning.

de Bono, E. (1991). *Six thinking hats for schools: 3-5 resource book.* Logan, IA: Perfection Learning.

de Bono, E. (1991). *Six thinking hats for schools: 6-8 resource book.* Logan, IA: Perfection Learning.

de Bono, E. (1991). *Six thinking hats for schools: 9-12 resource book.* Logan, IA: Perfection Learning.

de Bono, E. (1991). *Six thinking hats for schools: Adult educators resource book.* Logan, IA: Perfection Learning.

Gardner, H. (1983). *Frames of mind: The theory of multiple intelligences.* New York: Harper & Row.

Gardner, H. (1991). *The unschooled mind: How children think, and how schools should teach.* New York: Basic Books.

Gardner, H. (1993). *Multiple intelligences: The theory in practice.* New York: Basic Books.

Glasser, W. (1986). *Control theory in the classroom.* New York: Perennial Library.

Glasser, W. (1998). *Choice theory in the classroom.* New York: Harper Perennial.

Goldstein, A. P. (1988). *The Prepare Curriculum: Teaching prosocial competencies.* Champaign, IL: Research Press.

Johnson, D., & Johnson, R. (1989). *Cooperation and competition: Theory and research.* Edina, MN: Interaction Book Co.

Joyce, B., & Weil, M. (1986). *Models of teaching* (3rd ed.). Englewood Cliffs, NJ: Prentice Hall.

Kagan, S. (1994). *Cooperative Learning.* San Juan Capistrano, CA: Resources for Teachers.

Lazear, D. (1991). *Seven ways of knowing: Teaching for multiple intelligences.* Arlington Heights, IL: IRI/Skylight Training and Publishing.

Reed, S. K. (1996). *Cognition: Theory and applications* (4th ed.). Pacific Grove, CA: Brooks/Cole.

Slavin, R. (1995). *Cooperative learning: Theory, research, and practice.* Boston: Allyn & Bacon.

Tversky, A. (1972). Elimination by aspects: A theory of choice. *Psychological Review, 80,* 281-299.

The Corwin Press logo—a raven striding across an open book—represents the happy union of courage and learning. We are a professional-level publisher of books and journals for K–12 educators, and we are committed to creating and providing resources that embody these qualities. Corwin's motto is "Success for All Learners."

2092

DISCARDED

The text of this book for pages 1–381 is set in 10 point Baskerville with two points leading. Pages 383–456 are set in 6 point News Gothic with one point leading. Pages 395–456 are reduced 15 percent. The chapter numerals are set in 12 point Baskerville; the chapter titles are set in 18 point Baskerville italic.

The book is printed offset on Danforth Natural White text, 50-pound. The cover is Skivertex.

Jacket design by Norman Favin.
Production by Joan Comstock.

The book was composed by The Service Composition Co., Inc., Baltimore, Md., printed and bound by The Maple Press Co., York, Pa.